Science of Hard Materials

Sponsors

The Conference was sponsored by the Substitution and Materials Technologies sub-programme of the Commission of the European Communities R and D programme in the Raw Materials Sector (1982–1985), De Beers' Industrial Diamond Division, G E Carboloy, Iscar Ltd Nahariya, Kennametal Inc., Mitsubishi Metal Corporation, Sumitomo Electric Industries Ltd, Toshiba Tungaloy Co Ltd, USAF European Office of Aerospace Research and Development and US Army Research and Development, and Standardization Group—UK.

Organising Committee

E A Almond (NPL, UK), P A Bex (De Beers, UK), C A Brookes (University of Exeter), U Dworak (Feldmuhle A G, FRG), H E Exner (Max Planck Institute, FRG), H F Fischmeister (Max Planck Institute, FRG), H Holleck (Nuclear Research Centre, FRG), B Lux (Technical University of Vienna), A Neckel (University of Vienna), R Warren (Chalmers University of Technology, Sweden)

International Liaison Committee

H Doi (Mitsubishi Metal Research Institute, Japan), J Gurland (Brown University, USA), P S Kisly (Institute for Superhard Materials, USSR), M Lee (GEC, USA), S B Luyckx (University of the Witwatersrand, RSA), I Minkoff (Israel Institute of Technology, Israel), C M Perrot (CSIRO, Australia), V K Sarin (GTE, USA), B Uhrenius (Sandvik, Sweden), R K Viswanadham (Reed Rock Bit Co, USA)

Conference Helpers

D Bonnell, P J Burnett, A Christofouldis, M G Gee, R M Hooper, B Johannesson, M Laugier, T F Page, A Parry, H G Schmid, I Smid, G Vekenis, Mrs E Almond, Mrs J Brookes, Miss M Diacogis, Miss R Selischkar

Honorary Editors

E A Almond, C A Brookes, R Warren

Science of Hard Materials

Proceedings of the International Conference on the Science of Hard Materials held in Rhodes, 23–28 September 1984

Edited by E A Almond, C A Brookes and R Warren

Institute of Physics Conference Series Number 75

Adam Hilger Ltd, Bristol and Boston

CODEN IPHSAC 75 1–1091 (1986)

British Library Cataloguing in Publication Data

International Conference on the Science of
  Hard Materials (*2nd: 1984: Rhodes*)
  Science of hard materials 1984.—
  (Conference series, ISSN 0305–2346; no. 75)
  1. Hard materials
  I. Title   II. Almond, E. A.   III. Brookes, C. A.
  IV. Warren, R.   V. Series
  620.1'1   TA418.45

  ISBN 0-85498-166-7
  ISSN 0304-2346

Published on behalf of The Institute of Physics by Adam Hilger Ltd
Techno House, Redcliffe Way, Bristol BS1 6NX, England
PO Box 230, Accord, MA 02018, USA

Printed in Great Britain by J W Arrowsmith Ltd, Bristol

Preface

The Rhodes Conference on Science of Hard Materials was the second of a series which started at Jackson Lake, Wyoming in 1981. It differed from its predecessor by defining the term hard as greater than 1000 HV and extending the scope accordingly to diamond, cubic boron nitride and ceramics, rather than concentrating simply on WC-Co hardmetals. Likewise, 'the science' was emphasised by considering the fundamentals of atomic bonding and wear mechanisms and paying less attention to processing and specific tool applications.

The conference was opened by the Governor of the Dodecanese, Mrs Jenny Karavelli. A rewarding feature of the attendance of 153 delegates, was the presence of representatives from 23 countries, including China, Czechoslovakia, Hungary, Israel, Yugoslavia, Mexico, South Africa, USSR and Venezuela. A major attraction was the inclusion of numerous internationally renowned speakers from a very wide range of specialities covering atomic bonding, fundamental properties, superhard materials, ceramics, cermets, manufacture and wear. A high attendance of 70% or more was maintained at all the four-hour sessions which were held in the morning and evening, leaving the afternoons for recuperation or sight-seeing in temperatures in the high 20°C's.

Delegates appeared to benefit from the cross-fertilisation of ideas in the discussion periods. For example, researchers on fundamental properties were surprised to learn that it was meaningless to relate atomic bond strength and crystal structure to hardness since the latter is not an intrinsic materials property. Delegates were presented with the latest developments in theories for atomic bonding, doping and tough ceramics. They were made aware of advances in titanium based cermets, in diamonds and cubic boron nitride, and in coatings, and of the primitive state of understanding of wear processes in hard materials.

In compiling the conference proceedings, the editorial committee has attempted to impose high standards on the requirements for the scientific content and presentation of papers. The objective was to produce a volume with the high quality and long 'shelf life' of the classic conference proceedings of the early sixties. We express our gratitude to all the authors for their patience, understanding and co-operation in helping to achieve this aim.

We acknowledge generous funding from the many sponsors, the hard work and assistance of the other members of the conference organising committee, the international liason committee and the numerous enlisted helpers at the conference. Without their support the conference and proceedings would not have been possible.

E A Almond
National Physical Laboratory, TW11 0LW
C A Brookes
University of Hull, HU6 7RX
R Warren
Chalmers University of Technology, Sweden

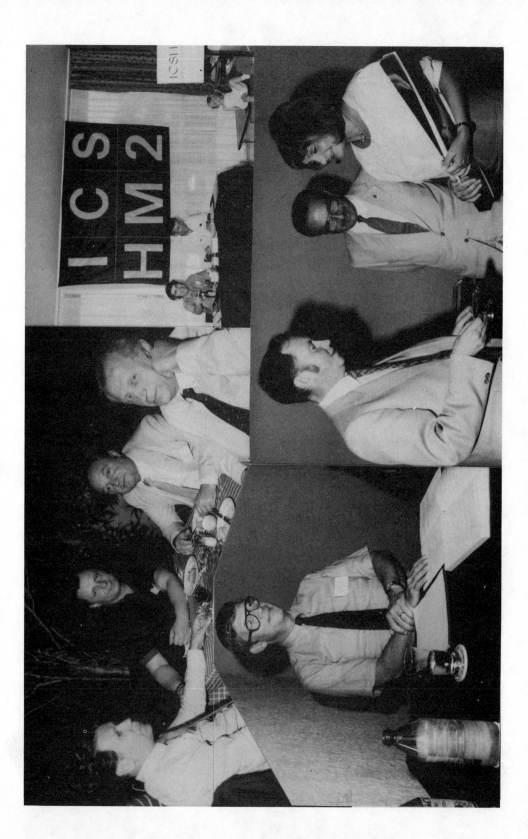

Contents

Chapter 4: Fundamentals of microstructure and deformation

Chapter 7: Cermets and hardmetals — Part two

Chapter 10: Wear of metals and hardened surfaces

Chapter 11: Manufacture

Contents

ICS
2nd International
Conference – Science
of Hard Materials
HM2

Inst. Phys. Conf. Ser. No. 75: Chapter 1
Paper presented at 2nd Int. Conf. Science Hard Mater., Rhodes

Ab initio calculations of structural properties of solids

MARVIN L COHEN

Department of Physics, University of California, and Materials and
Molecular Research Division, Lawrence Berkeley Laboratory, Berkeley, CA
94720

ABSTRACT

A review is presented of the total energy-pseudopotential method and its
application. The emphasis is on studies of structural properties of
semiconductors and insulators.

INTRODUCTION

Early applications of quantum mechanics focused on atoms and model
systems. In particular, optical spectra of atoms, especially hydrogen and
helium, provided the testing systems for the theory. Applications to
solids have been more difficult. Except for a few cases, only in the last
25 years have theorists using quantum mechanics been able to deal with
properties of real materials (Cohen, Heine, and Phillips, 1982). Again,
optical properties were of central importance to test the theoretical
results, and in turn, calculational approaches like the pseudopotential
method were used to decipher the observed spectra. This area was a very
active one in the 1960's and 1970's.

Applications to optical spectra were not straightforward as in the atomic
cases because solid state spectra are broad. However, computer techniques
allowed an interpretation of the optical structure from the visible to the
ultraviolet in terms of transitions between electronic energy bands.
These studies were applied mostly to semiconductors and insulators. For
metals, a great deal of information was obtained from static electronic
properties like Fermi surface measurements. As the data were analyzed, it
became clear that the band structure picture of solids had a firm basis
and that predictions and explanations of optical and electronic structure
using quantum mechanics could be trusted.

The pseudopotential method was used extensively in the studies described.
This method was invented by Fermi (1934) (see also Cohen, 1984a) to
examine high-lying states of atoms, and many variations of the approach
have been developed in the last 50 years (Hellmann, 1935; Phillips and
Kleinman, 1959; Cohen and Heine, 1970). A pseudopotential describes the
interaction between the valence electrons and atomic cores. For optical
studies, the approach relied on experimental input to fix the
pseudopotential. These potentials were then used in turn to interpret
data coming from optical and photoemission studies.

At first, structural information was not obtained from these theoretical studies since the electronic properties were determined for the experimentally observed structural arrangement of atoms. However, in the early 1970's, the electronic wavefunctions obtained from the optical studies were used to produce electronic charge density maps (Walter and Cohen, 1970; Cohen, 1973) which gave information about bonding and hence structure.

Direct attempts to obtain structural information using pseudopotentials began in the late 1970's, and the last five years have seen considerable growth of research in this area (Cohen, 1982). Most of the calculations are based on accurate computations of the total energy of a system for a specific structure. By comparing the energies of different structures, the lowest energy structure is expected to correspond to the observed structure. Deformations of the structures give information about static structural, vibrational, and solid-solid phase transition properties. The method has been applied successfully to semiconductors, insulators, and metals.

This review will focus on applications of the total energy-pseudopotential approach. After a description of the theoretical methods, the results of calculations for some prototype systems will be described. The emphasis will be on results for semiconductors and insulators with some reference to research on metals.

THEORETICAL METHODS

The pseudopotential method relies on a model of a solid in which the system is composed of a periodic array of cores (nucleus plus core electrons) and valence electrons (hereafter referred to as electrons) which are free to move throughout the fixed lattice of cores. For example, in the case of Si, a core contains the nucleus plus the ten $1s^2 2s^2 2p^6$ core electrons; the free (valence) electrons are $3s^2 3p^2$. The core electrons are assumed to be inert, and their properties are taken to be identical to those associated with isolated atoms. Hence, we have a system composed of positively charged cores (+4 in the case of Si) and negative electrons. To simplify the many-body problem, the Born-Oppenheimer approximation is made. The interactions of the cores will be dealt with later when discussing structural and vibrational properties. At this point, we focus on the electrons.

In a one-electron or Hartree model, an electron moves in the average potential created by the cores and the other valence electrons. The Empirical Pseudopotential Method (EPM) (Cohen and Heine, 1970) is based on this model, and it relies on experimental input to obtain the pseudopotential, $V(\vec{r})$. Hence, the pseudopotential Hamiltonian is

$$\left(\frac{p^2}{2m} + V(r)\right)\psi = E\psi \qquad\qquad 1$$

where p, m, ψ, and E are the momentum, mass, wavefunction, and eigenenergy for the electron. Because of the symmetry of the lattice, the pseudopotential can be expressed as a Fourier sum (over reciprocal lattice vectors $\vec{G}$) of the pseudopotential form factors, $V(\vec{G})$ and the structure factor $S(\vec{G})$,

$$V(\vec{r}) = \sum_{\vec{G}} V(\vec{G})S(\vec{G})e^{i\vec{G}\cdot\vec{r}} \qquad\qquad 2$$

Hence, the crystal structure is put in through the structure factor, and all that is needed to solve Eq. 1 for the wavefunctions and eigenvalues are the form factors, $V(\vec{G})$, in Eq. 2.

A schematic drawing of a typical pseudopotential appears in Fig. 1. A Coulomb or ion potential is shown for comparison. Because of the Pauli exclusion principle, the valence electrons experience a repulsive interaction when they approach the core electrons. This repulsive potential cancels (Phillips and Kleinman, 1959) much of the attractive Coulomb potential in the core region leaving a net weak pseudopotential. This cancellation in the core region results in small form factors in Eq. 2 at large $\vec{G}$, and these can be neglected. For example, the sum in Eq. 2 can be cut off after three form factors in the case of Si, and using only these three numbers, an excellent description of the electronic structure can be obtained. Similar results are found for a wide variety of systems, and the EPM has produced a large number of accurate band structures (Cohen and Bergstresser, 1966; Cohen and Heine, 1970; Cohen and Chelikowsky, 1982).

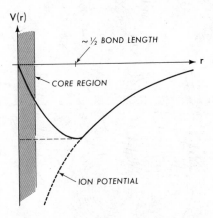

Fig. 1   Schematic drawing of a typical pseudopotential as a function of position.

As mentioned earlier, the contribution of the EPM to structural properties came mostly through the investigations of the electronic charge density. By examining the wavefunctions obtained from Eq. 1, the position dependent charge density, $\rho_n(\vec{r})$, could be calculated for a specific band n by summing over the k-states in that band

$$\rho_n(\vec{r}) = \sum_{\vec{k}} |\psi_{n,\vec{k}}(\vec{r})|^2 \qquad\qquad 3$$

The total charge density for the valence electrons, $\rho(\vec{r})$, can be obtained by summing over occupied bands

$$\rho(\vec{r}) = \sum_{n} \rho_n(\vec{r}) \quad .$$

occupied
bands

Charge density maps indicated the locations and magnitudes of bonds and provided information about the nature of the chemical bonding.

To illustrate the use of charge density plots, Fig. 2 contains maps of the charge density of Ge, GaAs, and ZnSe. These systems are isocoric and contain two cores and eight valence electrons in a primitive cell. The transition from Ge to GaAs to ZnSe can therefore be viewed as a process of transferring first one and then two positive charges (protons) from the cation to the anion. For Ge, where both atoms are equivalent, the charge piles up midway between the atoms forming the covalent bond. In GaAs, the more positive As anion attracts more electronic charge forming a partially ionic bond. The ZnSe case is more extreme, and there is little left of the covalent bond concentrated between the atoms.

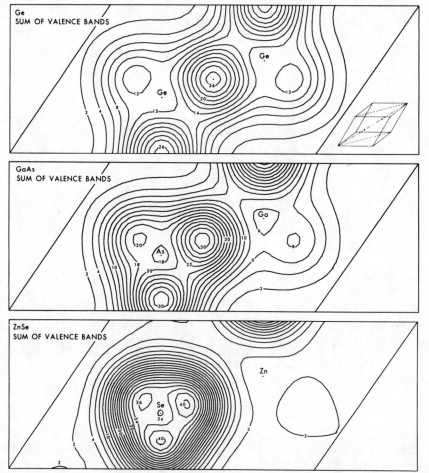

Fig. 2   Total valence charge density maps for Ge, GaAs, and ZnSe plotted in the (110) plane. The contours are in units of e/Ω where Ω is the cell volume.

By integrating the charge near the peak of the bond, it is possible to
estimate the strength of the bond charge. As is evident from Fig. 2, the
covalent bond charge decreases as the system becomes more ionic. Phillips
(1973) developed a scale for ionicity based on the spectral properties of
semiconductors and insulators. A plot of the bond charge versus Phillips'
ionicity appears in Fig. 3 for the Ge, GaAs, and ZnSe results given in
Fig. 2 and another isoelectronic series, Sn, InSb, and CdTe. At an
ionicity of about 0.8, the curves extrapolate to vanishing bond charge.
This result was supportive of the idea of a critical ionicity -0.8
(Phillips, 1973) which separates four-fold and six-fold coordinated
structures. It is believed that the covalent bond stabilizes the
four-fold coordinated structures like diamond, zincblende, and wurtzite
whereas the six-fold coordination is more consistent with electrostatic or
ionic bonding as found in the NaCl or **rocksalt** structure. The vanishing
bond charge near an ionicity of 0.8 indicates that systems with higher
ionicity will not form four-fold coordinated structures.

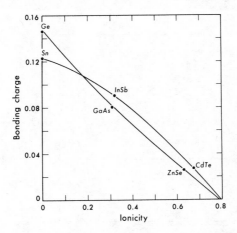

Fig. 3 Bond charge for two isoelectronic series plotted as a function of
        ionicity using the Phillips scale of ionicity.

Although the EPM did give structural information of the kind described
above and charge density plots did bring theorists back to position space
from momentum space, direct calculations of structural properties based on
the EPM were not satisfying. Structural energies are small, and total
energy calculations for the EPM had not reached a state of high
precision. The method which evolved was based on a self-consistent
approach to the calculation of electronic interactions. In the EPM, the
electron-core and electron-electron interactions were combined when
forming the pseudopotential used in Eq. 1. One physical situation which
required a separation of these two contributions was the study of
surfaces. At a surface, the electronic charge readjusts, and hence a new
electronic potential results. It is necessary to compute the electronic
wavefunction for states near a surface and use the charge density to form
the potential for electron-electron interactions. Using this potential
plus the electron-core potential, electronic wavefunctions can be obtained
by solving the Schrödinger equation, and the procedure is repeated until
input and output charge densities or potentials agree. At this point, the
solution is self-consistent (Appelbaum and Hamann, 1976; Schlüter and
co-workers, 1975).

Two other developments, which are integral parts of the current approach
used to calculate structural properties, are the procedures for obtaining
ab initio pseudopotentials (Starkloff and Joannopoulos, 1977; Zunger and
Cohen, 1978; Hamann, Schlüter, and Chiang, 1979; Kerker, 1980; Louie,
Froyen, and Cohen, 1982; Yin and Cohen, 1982a) and a method for
calculating the total energy of a solid system in momentum space (Ihm,
Zunger, and Cohen, 1979). The construction of the ab initio
pseudopotentials requires only the atomic number of the atom considered.
A potential is formed which reproduces an all-electron atomic wavefunction
for distances far from the core. This is shown in Fig. 4 for the 3s
radial wavefunction of Si. The pseudowavefunction is identical to the
all-electron wavefunction down to distances between the outermost maximum
and the outermost node. At this point, the pseudowavefunction is
extrapolated to zero keeping the normalization fixed. Because the two
wavefunctions differ only in the core region, either can be used to
compute most solid state properties since the latter depend primarily on
the outer regions of the atom.

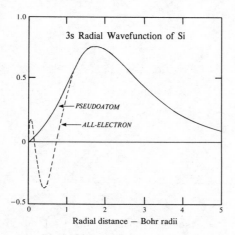

Fig. 4   A plot of the normalized Si 3s radial all-electron and
         pseudopotential wavefunction versus radial distance from
         the nucleus.

The ab initio pseudopotentials are often taken to be angular momentum or
ℓ-dependent. These ℓ-dependent potentials are usually called "non-local
pseudopotentials." For the Si example, pseudopotentials with ℓ values
0,1,2 corresponding to s, p, and d wavefunctions are constructed; these
are shown in Fig. 5. The s, p, and d potentials are generated by
augmenting a computer program which calculates atomic wavefunctions. Once
the pseudopotential is fixed, it is possible to compute the electron-core
contributions to the total energy. The total energy has additional terms
coming from the electron-electron Coulomb interaction which depend on the
electronic charge density and terms arising from the exchange and
correlation energies of the electrons. These latter terms can be
approximated using a local density approach (Hohenberg and Kohn, 1964;
Kohn and Sham, 1965) to compute functionals which depend only on the
position-dependent density of the electrons. The electronic wavefunctions
are used to calculate the kinetic energy of the electrons, and Madelung
sums can account for the core-core interactions. Hence, all the main

energy contributions can be obtained for the pseudopotential model of
cores and electrons. The calculation is done self-consistently, and the
total energy is computed for a specific periodic array of cores. More
details on the procedures along with results for specific systems are
given in the following sections.

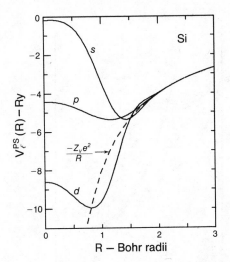

Fig. 5   Nonlocal pseudopotentials and the Coulomb potential (dashed line)
         for Si.

SILICON AND GERMANIUM

The first detailed applications of the theory were made to Si and Ge. Ab
initio pseudopotentials were generated and the total energy computed. The
total energy versus volume curves are given in Fig. 6 for Si and Fig. 7
for Ge. Seven crystal phases are considered: diamond, hexagonal diamond,
white tin (β-Sn), simple cubic (sc), body-centered cubic (bcc),
face-centered cubic (fcc), and hexagonal close packed (hcp). The volume
is normalized to the observed volume of the diamond phase at atmospheric
pressure. Since the Si and Ge results are similar and more studies have
been done on Si, we focus on the calculations for Si and point out
important differences in the conclusions for the two semiconductors when
they occur.

As shown in Fig. 6, the minimum energy structure is diamond. Hexagonal
diamond lies higher in energy than diamond over the entire range of
volumes shown. Hence, a transition to the hexagonal phase is not
expected. At smaller volumes (higher pressures), the β-Sn structure has a
lower energy than the diamond structure, hence a pressure-induced
solid-solid phase transition is expected to occur. We first concentrate
on the diamond structure. By computing the volume at the minimum energy,
the lattice constant for this structure can be obtained. The curvature of
the $E(V)$ curves in Figs. 6 and 7 near the minimum give the dependence of
the total energy on small changes in volume and therefore determine the
compressibility or bulk modulus. By computing the total energy for a very
large lattice constant, it is possible to compute the cohesive energy of
the solid structures. At large lattice constants, the system resembles an

array of isolated atoms, hence the difference in the total energy between
this configuration and the energy for a lattice constant corresponding to
a solid structure gives the cohesive energy. The results are shown in
Table 1 (references to the experimental data are given in Yin and Cohen
(1982b)).

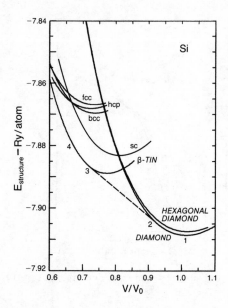

Fig. 6   Total energy versus volume for seven phases of Si. The volume is
normalized to the experimental value for the diamond structure at
atmospheric pressure. A dashed line is used to indicate the
common tangent between the diamond and white tin phases.

The solid-solid phase transition between the diamond and β-tin structures
can also be studied using the results of Figs. 6 and 7. In both figures,
a dashed line is drawn as the common tangent to the diamond and β-Sn E(V)
curves. Under pressure, the system (e.g. Si) will follow the path 1→2→3→4
indicated by the numbers in Fig. 6. At a volume corresponding to 1, the
system is in the diamond structure. At point 2, the transition to β-Sn
begins; the volume at this point is the diamond to β-Sn transition
volume, V_t^d. At point 3, the phase transformation ends for the transition
volume, V_t^β, and at point 4, the system is in the β-Sn phase. The slope of
the dashed line gives an estimate of the transition pressure. Comparisons
with observed data are given in **Table 2** (references to the experimental
data are given in Yin and Cohen (1982b)).

The results presented in **Tables 1 and 2** are impressive considering that
only the atomic numbers for Si and Ge were used as input. No information
about the solid state was used beyond the choice of a subset of crystal
structures to compare. The agreement between the calculations and
experimental data found here is typical of results for most semiconductors
and insulators studies. Usually lattice constants are computed to within
about 1% of their measured values. (For the β-Sn structure, the axial
ratio was also determined through the minimum energy scheme.) Bulk moduli

TABLE 1 <u>Comparison of Calculated and Measured Static Properties of</u>
<u>Si and Ge</u>

	Lattice constant nm	Cohesive energy eV/atom	Bulk modulus Mbar
Si			
calculation	0.5451	4.84	0.98
experiment	0.5429	4.63	0.99
Ge			
calculation	0.5655	4.26	0.73
experiment	0.5652	3.85	0.77

TABLE 2 <u>Comparison of Calculated and Measured Transition Volumes ($V_t^{d,\beta}$)</u>
<u>of the Diamond and β-Phases, their Ratios (V_t/V_{t_d}), and the Transition</u>
<u>Pressures (P_t) for Si and Ge. The Volumes are normalized to the Measured</u>
<u>Zero-Pressure Volumes.</u>

	V_t^d	V_t^β	V_t^β/V_t^d	P_t kbar
Si				
calculation	0.928	0.718	0.774	99
experiment	0.918	0.710	0.773	125
deviation	1.1%	1.1%	0.1%	-20%
Ge				
calculation	0.895	0.728	0.813	96
experiment	0.875	0.694	0.793	100
deviation	2.3%	4.9%	2.5%	-4%

can be obtained to within about 6%, and 6% is also a reasonable estimate
of the accuracy of calculations of the cohesive energy. The limitation on
the latter calculation appears to be the estimate of the energy of the
isolated atomic system. Transition volumes appear to be in better
agreement with observed values than transition pressures. Typical
deviations for transition volumes are of the order of a percent or two.
Estimates of the theoretical errors for the transition pressures in Si and
Ge for the diamond to β-Sn transition are of the order of 10%. The larger
discrepancy given in **Table 2** for Si may not be correct. More recent
experimental observations of this transition by Olijnyk and co-workers
(1984) and by Hu and Spain (1984) report lower values.

The bcc and fcc phases shown in Figs. 6 and 7 have not been observed as yet. However, for Si two new phases have been observed and one is hcp. The hcp phase was found (Olijnyk and co-workers, 1984) at 400 kbar, and in addition, these researchers find a simple hexagonal (sh) phase at 160 kbar. Hu and Spain (1984) also report a sh phase but at a pressure of 130 kbar. Pseudopotential calculations have been done for these phases by Needs and Martin (1984), Yin (1984), and Chang and Cohen (1984a). Some results of the latter authors are given here.

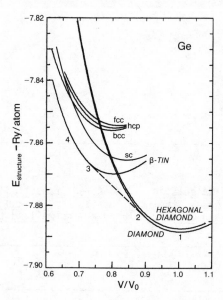

Fig. 7   Total energy versus volume for seven phases of Ge. The volume is normalized to the experimental value for the diamond structure at atmospheric pressure. A dashed line is used to indicate the common tangent between the diamond and white tin phases.

Fig. 8 displays the E(V) curve for Si in the diamond, β-Sn, sh, and hcp structures. The transition pressure from diamond to β-Sn is calculated to be 93 kbar which is within the 10% theoretical estimate when comparing with the results of Yin and Cohen (1982b) given in **Table 2**. Because estimates of the transition pressure from β-Sn to sh require the construction of a common tangent between two E(V) curves which are almost coincident (Fig. 8), this value is not determined as accurately as for the diamond to β-Sn case. The calculated result is 120 kbar; this should be compared with the measured values of 130 and 160 kbar discussed above. For the sh to hcp transition, a value of 410 kbar is calculated for the transition pressure. The measured result is 400 kbar. Calculated transition volumes and the c/a ratios for both solid-solid phase transitions are in good agreement with the observations.

Both the sh and hcp Si phases are metals. The c/a ratio for the sh phase is computed to be 0.955, and the bonding between hexagonal planes appears to be somewhat stronger than the intraplane bonding. This is in contrast to the situation for graphite. A plot of the electronic charge density for the sh phase is given in Fig. 9. The pile up of covalent charge between layers is evident in the plot. A density of states calculation

(Chang and Cohen, 1984a) reveals a fairly large value at the Fermi
energy. The sizable density of states and the evidence for a covalent
charge distribution suggest that the electron-phonon interaction may be
strong enough in this system to cause a transition to the superconducting
state. The β-Sn phase of Si is known to be superconducting (Wittig, 1966).

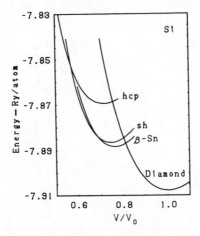

Fig. 8 Total energy versus volume for Si normalized to the diamond volume
at atmospheric pressure.

Fig. 9 Electronic charge density contour map in the [10$\bar{1}$0] plane for Si
in the sh structure. The volume chosen is 12.16 Å^3/atom in units
of electrons/cell volume.

A metastable phase of Si has recently been studied using pseudopotentials
by Yin (1984) and by Biswas and co-workers (1984). This phase is called
Si III and/or BC-8; it has a bcc structure with eight atoms in a unit
cell. Si BC-8 is formed when the high pressure β-Sn phase is unloaded to
ambient pressure (Wentorf and Kasper, 1963). The results of Yin (1984),
shown in Fig. 10, indicate that the BC-8 phase is unstable even at high
pressures. If a common tangent were drawn between the diamond and BC-8

E(V) curves, its slope would be larger than that of the diamond to β-Sn common tangent (see Fig. 10), and a calculated pressure of 130 kbar would be needed. Hence, diamond Si transforms first to β-Sn. Since the BC-8 phase is observed when the pressure is reduced from the β-Sn phase, the barrier between these two phases must be smaller than between diamond and BC-8.

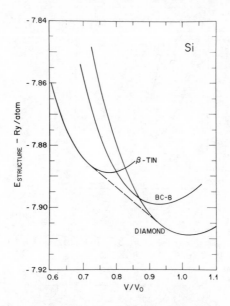

Fig. 10 Total energy versus normalized volume for the diamond, BC-8, and β-tin phases of Si. The dashed line is the common tangent between the diamond and β-Sn curves.

The question of the stability of the graphite phase of Si has been examined (Yin and Cohen, 1984) using similar methods. This calculation gives a negative result indicating that graphitic Si is weakly bound compared to the diamond phase with an increase in structural energy of 0.71 eV/atom. It is estimated that a large negative pressure of -69 kbar is required to form this phase.

The total energy-pseudopotential method can also be applied to the calculation of lattice vibrational frequencies. By distorting the structure corresponding to a phonon mode which has been "frozen in," the total energy increase over the undistorted system can be evaluated. Using this energy difference and the atomic mass, the restoring forces and phonon frequencies can be calculated. The first applications were to Si and Ge (Yin and Cohen, 1982), and the phonons corresponding to a zone center (Γ) and zone edge (X) symmetries were studied. Differences of only a few percent (see Table 3) were found between the calculated and observed frequencies. The calculated constants (Table 4) were also found to be in excellent agreement with the measured values. (For Tables 3 and 4, references to the experimental data are given in Yin and Cohen (1982c).) The required inputs for the lattice vibrational calculations are the atomic number, atomic mass, and the crystal structure.

TABLE 3 Comparison of Calculated Phonon Frequencies (in TH$_z$) of Si and Ge at Γ and X with Experiment (f$_{expt}$). The Values for f$_E$ (f$_F$) are obtained from Energy (Force) Calculations. The Deviations from Experimental Values are given in Parentheses

	LTO(Γ)	LOA(X)	TO(X)	TA(X)
Si				
f$_E$	15.16 (-2%)	12.16 (-1%)	13.48 (-3%)	4.45 (-1%)
f$_F$	15.14 (-3%)	11.98 (-3%)	13.51 (-3%)	4.37 (-3%)
f$_{expt}$	15.53	12.32	13.90	4.49
Ge				
f$_E$	8.90 (-3%)	7.01 (-3%)	7.75 (-6%)	2.44 (2%)
f$_F$	8.89 (-3%)	6.96 (-3%)	7.78 (-6%)	2.45 (2%)
f$_{expt}$	9.12	7.21	8.26	2.40

The phonon dispersion curve, $\omega(q)$, can also be obtained using this ab initio approach. By calculating the interlayer force constants, the phonon frequencies in a symmetry direction can be calculated. For Si (Yin and Cohen, 1982d), the calculation was applied to $\omega(q)$ along the [001] direction from Γ to X. The results are shown in Fig. 11 (references to the experimental data are given in Yin and Cohen (1982d)).

Because a local density functional scheme is used to determine electron-electron exchange and correlation, the results discussed here should be considered accurate only for ground-state properties of the kind discussed above. The valence charge densities (Fig. 12) are expected to be reliable, and the successes of the calculated static structural properties serve as tests of the local density approach for calculations of ground-state properties. The failure of the method for excited state properties is easily shown by calculating the minimum band gaps for Si and Ge. These calculations yield results which are 30 to 50% smaller than the observed values.

CARBON

The application of the total energy-pseudopotential method to C has yielded some interesting results related to this material. In the previous section, many of the calculational results were similar for Si and Ge. Figs. 6 and 7 showed many of the same features except at small volumes. Low pressure phase solid-solid transitions like the diamond to β-Sn phase transition are expected to be similar in Si and Ge. However, for C the results are drastically different.

One major difference between C and other group IV elements is its simple core. Only s electrons ($1s^2$) are contained in the core, and this has important consequences for the C pseudopotential. Using the arguments of Phillips and Kleinman (1959), the repulsive Pauli force in the core is angular momentum or ℓ-dependent. Hence, the C core will repel electrons

TABLE 4 <u>Comparison of Calculated Mode-Gruneisen Parameters for Phonons at</u>
<u>Γ and X of Si and Ge with Experiment</u>

	LTO(Γ)	LOA(X)	TO(X)	TA(X)
<u>Si</u>				
calculation	0.9	1.3	0.9	-1.5
experiment	0.98	1.5	0.9	-1.4
<u>Ge</u>				
calculation	0.9	1.4	1.0	-1.5
experiment	1.12 ± 0.02	-	-	-
	0.88 ± 0.08			

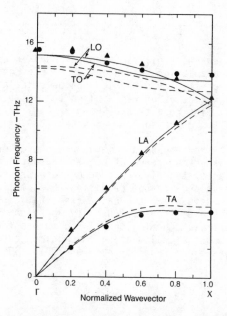

Fig. 11 Calculated Si phonon dispersion curve along the [001] direction.
The solid (dashed) line corresponds to the theoretical results
with large (small) plane wave basis sets. The experimental
points are given by dots (transverse) and triangles
(longitudinal).

in s-states but not those in p- or higher states. This allows the 2p
electrons in C to come closer (relatively) to the core than in the cases
of Si or Ge. Another feature of the electronic structure is the absence
of ℓ = 2 (d-states) in the n = 2 valence electron quantum level. Unlike
Si and Ge, the C $2s^2 2p^2$ valence electrons are far removed in energy from
the next d-state which is 3d. These features result in a different bond

charge configuration for C. A comparison of the bond charges of C and Si
(Yin and Cohen, 1981) is given in Fig. 13. As discussed above, the
absence of p-core states results in the p-valence electrons moving closer
to the core. This movement "splits" the bond charge into two peaks. For
Si and Ge, a single peak is found as shown in Fig. 12 (note that the
charge density scale changed by a factor of 2). The double-hump feature
of the C-C bond which is also expected for graphite may be related to the
ability of C to form multiple bonds.

Valence charge density (110 plane)

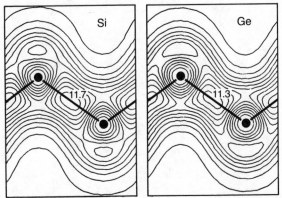

Fig. 12 Electronic charge density contour maps for diamond structure Si
         and Ge plotted in the (110) plane. The units for the charge
         density is electrons/atomic volume with a contour step of 1. The
         black dots represent atomic positions and straight lines denote
         bonds.

Valence charge density (110 plane)

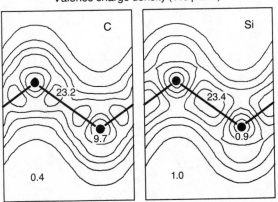

Fig. 13 Valence electron charge density for diamond and Si plotted in the
         (110) plane. The units are electrons/cell with a contour step of
         4. Bond distances are scaled for ease of comparison.

Based on the discussion above, we expect differences in the structural
properties of diamond compared to Si and Ge. By applying the total

energy-pseudopotential approach, the E(V) curve can be calculated. The
results are given in Fig. 14 for six structural phases of C. For the
diamond phase of C, the energy minimum and curvature near the energy
minimum give the lattice constant and bulk modulus. The results are
0.3602 nm (1%) and 4.33 Mbar (-2%) where the percentages in parenthesis
give the deviations from the observed values (references are given in Yin

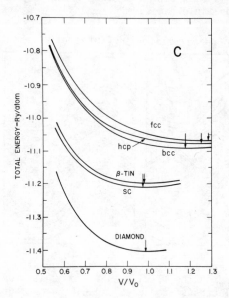

Fig. 14 Total energy versus normalized volume for six phases of C.

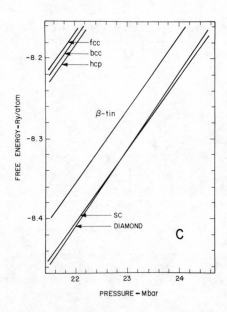

Fig. 15 Gibbs free energy versus pressure at zero temperature for six
structural phases of C.

and Cohen (1981)). Using the same computational methods as in the Si and Ge calculations, the cohesive energy is foudn to be 7.58 eV/atom (3%). Hence, the calculated static structural parameters are in good agreement with experiment.

Unlike Si and Ge (Figs. 6 and 7), the E(V) curves for C suggest that a solid-solid phase transition from diamond to the -Sn structures is unlikely. In addition, the minima for several of the structures lie at larger volume than the diamond minimum. To determine the possible structural transitions, the Gibbs free energy is computed as a function of pressure for the six phases (Fig. 15). The lowest pressure intersection is between the diamond and simple cubic (sc) curves. This point corresponds to 23 Mbar, and within the subset of phases studied, this is the lowest transition pressure.

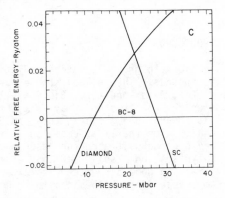

Fig. 16   Relative free energy versus pressure for C in three structural phases.

More recently, two studies of C in the BC-8 structure (Yin, 1984; Biswas and co-workers, 1984) have shown that diamond will transform to the BC-8 structure at about 12 Mbar. The results (Yin, 1984) for the pressure dependence of the relative free energy for the diamond, BC-8, and sc structures is given in Fig. 16. These curves illustrate the transitions among this subset of structures. Hence, at this time, the lower limit on the pressure at which diamond will transform is approximately 12 Mbar. However, other structures may give a lower limit.

The most stable form of C is graphite, and the above methods were applied to study this phase and its relation to diamond (Yin and Cohen, 1984). This calculation is more difficult than those done for the other structures discussed because the energy difference between diamond and graphite is very small. For graphite, the calculated lattice constants are a = 0.247 nm and c = 0.673 nm which are in good agreement with the measured values a = 0.2461 nm and c = 0.6709 nm (experimental data is referenced in Yin and Cohen, (1984)). The calculated isotropic bulk modulus for graphite is also in good agreement with experiment. It is difficult to get an accurate estimate for the diamond graphite energy difference since the theoretical uncertainty in each of the cohesive energies is larger than the difference. The calculation of Yin and Cohen (1984) gives a value of 0.014 eV/atom; the experimental value is 0.025 eV/atom. As mentioned previously, the calculated difference for graphite

Si is higher than diamond (i.e., it is negative, -0.71 eV/atom).

Recently, a detailed study of the anharmonic phonon-phonon coupling constants for diamond was done using a total energy pseudopotential approach (Vanderbilt, Louie, and Cohen, 1984). For all of the previous calculations described in this review, the electronic wavefunction was expanded in a basis set of plane waves. The diamond phonon-phonon calculation used a localized orbital approach developed by Chelikowsky and Louie (1984). This calculation is based on a frozen phonon approach, and it yielded the phonon frequencies at specific points in the Brillouin zone and higher order anharmonic terms; e.g., the third- and fourth-order anharmonic coupling constants for optical phonons. Because the calculated fourth-order coupling constants are negative, the earlier suggestion that a two-phonon bound state can exist is ruled out. This suggestion had been made as an attempt to interpret an anomalous peak in the Raman spectrum of diamond.

PARTIALLY IONIC AND IONIC COMPOUNDS

The successes of the applications to group IV materials motivated a study of III-V, II-V, and I-VII compounds. Theoretical studies were done for four III-V compounds, GaAs, GaP, AlAs, and AlP. Since the III-V semiconductors are considered the "first step" away from the group IV homopolar covalent materials toward ionic solids, they provide a test of the theory for ionic materials. For the II-VI case, one wurtzite structure compound, BeO, and one rocksalt structure compound, MgO, were tested. Finally, the prototypical ionic solid NaCl was examined. For all cases studied, the results were in good agreement with experiment; in fact, the agreement was comparable to that obtained for the group IV solids.

For the III-V compounds (Froyen and Cohen, 1983), five structural phases were considered: zincblende, β-Sn, rocksalt (NaCl), CsCl, and NiAs. All four compounds, GaAs, GaP, AlAs, and AlP, were tested in the zincblende, β-Sn, and NaCl structures, but AlAs and GaAs were also considered to be in the CsCl and NiAs structures. Several initial studies demonstrated the similarity of GaAs and GaP, and the same was true for AlAs and AlP. Hence, GaAs and AlAs were considered prototypes. It was found that the anions, As and P, determined the equilibrium volumes but not the minimum energy structure for the structures studies.

In all four cases, the zincblende structure was found to have the lowest energy among the structures tested (Fig. 17). **Table 4** contains the lattice constants and bulk moduli determined from the zincblende E(V) curves. Since measurements of the bulk moduli of AlP and AlAs were not available, the entries for these in **Table 5** were obtained by extrapolation from other compounds. Hence, the calculated values serve as predictions. (References to experimental data for **Tables 5 and 6** are given in Froyen and Cohen, 1983)

Experimentally, the high pressure phases of these compounds are not as well-determined as for Si and Ge. It is often assumed that because of the similarity of the II-V compounds to Si and Ge, the high pressure phase should be β-Sn. High pressure studies by Baublitz and Ruoff (1982) suggest that GaAs transforms to an orthorhombic structure. GaP is known to transform to β-Sn, but the high pressure structures of these compounds are not yet established with certainty.

According to the theory, all the high pressure phases studied are close
packed metals. As shown in Fig. 17, for the two prototype compounds, GaAs
transforms to either the rocksalt, ß-Sn, or NiAs structure, whereas
for AlAs the rocksalt and NiAs phases have shifted down in energy and
should therefore occur at lower transition pressures. At present, it is
not possible to distinguish between these competing structures. In
addition, the relative stability should also depend on temperature and the
experimental conditions. The values for the transition pressure, volumes,
and total energy differences for the zincblende to rocksalt transition are
given in **Table 6**.

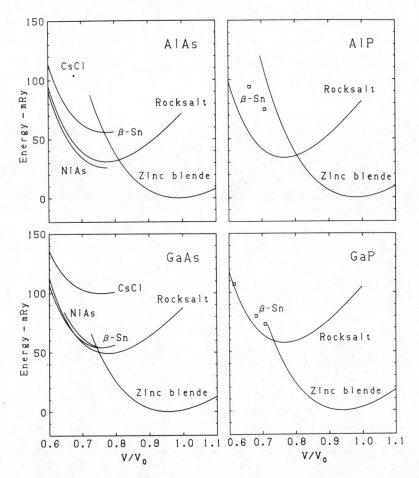

Fig. 17 Total energy per molecule versus normalized volume for several
         structural phases of III-V compounds.

Although the results of this theoretical study of high pressure structures
are not conclusive, i.e. a single structure is not suggested at high
pressure phases. In particular, it was found that the cation is important
for determining the crystalline phases. For Ga compounds, it is likely
that the ß-Sn, NaCl, and NiAs structures will be prime candidates. Since
several structures lie in a narrow range of energy, perhaps combinations

TABLE 5 Lattice Constants and Bulk Moduli. The Experimental Lattice Constants are all measured at Room Temperature, and the Bulk Moduli for AlP and AlAs have been obtained by Extrapolation from Other Components

	AlP	AlAs	GaP	GaAs	
$\underline{a}$					
calculation	0.5420	0.5641	0.5340	0.5570	
experiment	0.5451	0.5662	0.5451	0.5653	nm
	-0.6%	-0.4%	-2.0%	-1.5%	
B_0					
calculation	0.865	0.741	0.897	0.725	
experiment	0.86	0.77	0.887	0.748	Mbar

TABLE 6 Transition Pressures, Volumes and Total Energy Differences for the Zincblende to Rocksalt Transition. $V_t(zb,rs)$ are the Transition Volumes for the Two Phases, $V_o(rs)$ is the Equilibrium Volume for the (Metastable) Rocksalt Phase, and ΔE_o is the Energy Difference per Molecule between the Minima of the Two Phases. The Volumes are all given as Fractions of the Calculated Equilibrium Volume for the Zincblende Phase

		AlP	AlAs	GaP	GaAs	
$\underline{P}$						
	calculation	0.093	0.076	0.217	0.160	
	experiment	0.14-0.17		0.20-0.24	0.16-0.19	Mbar
$V_t(zb)$	calculation	0.90	0.92	0.84	0.86	
$V_t(rs)$	calculation	0.73	0.73	0.71	0.71	
$V_o(rs)$	calculation	0.78	0.78	0.81	0.81	
ΔE_o	calculation	0.46	0.43	0.78	0.67	eV

of these or phases mixtures are possible. One should also make the caveat that a lower energy structure which has not yet been tested may exist.

For Si and Ge, it was found (Figs. 6 and 7) that the hexagonal diamond phase was above the cubic diamond phase for all volumes studied. In the case of III-V semiconductors, the energy differences between the wurtzite and zincblende structures were found to be extremely small and difficult to compute because of the uncertainties in the total energies for each phase. Because it is known that zincblende has a lower energy than wurtzite for the III-V's, this difference was not studied in detail for each compound. However, some II-VI materials can be formed in either the zincblende or wurtzite structure while others prefer one or the other. Wurtzite BeO was chosen as a prototype of the hexagonal structure because it was expected that the zincblende-wurtzite energy difference should be relatively large. (It is not surprising that the zincblende-wurtzite energy separation is usually small since these two structures have the same nearest neighbor and next nearest neighbor environments.)

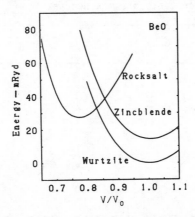

Fig. 18 Total energy per molecule versus volume normalized to the equilibrium volume of wurtzite BeO.

The BeO study (Chang and Cohen, 1984b) considered three structural phases: zincblende, wurtzite, and rocksalt. The results for the E(V) curves are given in Fig. 18 which shows that the wurtzite structural phase is more stable than zincblende over the range of volumes studied. The calculated lattice constants for wurtzite, given in **Table 7** , are found to be in excellent agreement with the measured values (references in Chang and Cohen (1984b)). The bulk modulus is larger than the observed values, but the computation in this case had some limitations. To reduce the computer time used, the evaluation of the charge density was limited to a few points in the Brillouin zone, and the equilibrium c/a ratio was used for every volume tested. Based on the results shown in Fig. 18, a transition from the wurtzite to the rocksalt structure is predicted at a pressure of 217 kbar. The transition volumes are 0.934 and 0.745 for wurtzite to rocksalt and rocksalt to wurtzite respectively (normalized to the equilibrium wurtzite volume). At the minimum of each phase, the total energy difference per molecule between the wurtzite and rocksalt phases is 0.37 eV.

MgO is a II-VI material which exists in the rocksalt structure. Following the proposals of Phillips (1973), which were discussed earlier, the difference in structure between BeO and MgO arises from the higher ionicity of the latter. MgO is more ionic than BeO, and less covalent charge is available to stabilize a four-fold coordinated structure. Table 8 , (experimental references in Chang and Cohen (1984c)) gives the calculated and measured results for the lattice constant, bulk modulus, cohesive energy, and frequency of the transverse optical phonon mode at the zone center for MgO in the rocksalt structure. Once again the agreement with the measurements is good.

There has been considerable speculation concerning the high pressure phase

of MgO. A transition from the rocksalt (B1) to CsCl (B2) phase is
expected for this material, but estimates of the transition pressure have
varied. In particular, the question of whether MgO could be stable in the
B2 phase inside the earth has been raised.

TABLE 7 Lattice Constants, Bulk Moduli and Cohesive Energies. The c/a
Ratios are compared at the Experimental Equilibrium Volume V_o. The
Experimental Lattice Constants are measured at Room Temperature

	a nm	c nm	c/a	B_o Mbar	E_c Ryd
wurtzite					
calculation	0.2664	0.4324	1.623	2.83	1.06
experiment	0.2698	0.4380	1.623	2.49	0.9
zincblende					
calculation	0.3766	-	-	2.97	1.05
hypothetical	0.3793	-	-	-	-

TABLE 8 Lattice Constant, Bulk Modulus, Cohesive Energy and Frequency of
the TO(Γ) Mode for Rocksalt MgO. The Experimental Lattice Constant and
Frequency are measured at Room Temperature and 8 K, respectively

	a nm	B_o Mbar	E_c eV	f TH_z
calculation	0.4191	1.46	9.96	12.69
experiment	0.4211	1.62	10.33	12.23
		1.55		

The pseudopotential E(V) curve is given in Fig. 19, and a plot of the
Gibbs free energy versus pressure appears in Fig. 20. As is evident from
Fig. 19, the B1 and B2 phases are well-separated down to normalized
volumes of 0.7. To compute the transition pressure between the two
phases, the Gibbs free energy is calculated (Fig. 20). A crossing occurs
at 10 Mbar, and this is the predicted pressure for a transition from the
rocksalt to CsCl structure. Hence, it is unlikely that MgO exists in the
CsCl structure even in the lower mantle of the earth.

The standard prototype for ionic crystals is NaCl. This I-VII compound
has been studied extensively as a model illustrating electrostatic bonding
in ionic compounds. It is not evident at first that a plane wave
pseudopotential approach would be applicable to this system since the
common view is that NaCl is composed of positive (Na^+) and negative (Cl^-)
ions on a periodic lattice with localized electrons. Hence, simple models
have evolved to explain structural, electronic, and vibrational

properties. In the total energy-pseudopotential study (Froyen and Cohen, 1984), the same approach was applied in this case as in earlier calculations on Si, Ge, etc. A plane wave basis set was used, and a local density approximation was assumed despite the localized nature of the charge. As expected, a large number of plane waves are necessary to reproduce the Cl orbitals.

Total energies for the B1 and B2 phases versus lattice constant are given in Fig. 21. The solid lines are fits to the calculated points. Based on the E(V) curve for B1, the calculated lattice constant and bulk modulus are 0.556 nm and 28.4 GPa. These compare well with the measured values of 0.560 nm and 26.6 GPa. These results were corrected for zero-point motion. References to the experimental data discussed are given in Froyen and Cohen (1984).

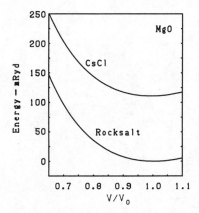

Fig. 19   Total energy per molecule versus volume normalized to the equilibrium volume for MgO in the rocksalt phase.

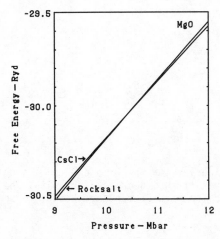

Fig. 20   Zero-temperature Gibbs free energy (per molecule) for MgO as a function of pressure.

At a pressure of 27 GPa, the E(V) curves for the B1 and B2 phases cross in Fig. 21. The experimental value for this transition is around 30 GPa. Transition pressures and volumes are given in **Table 9**. To study the B1 to B2 transition, a calculation of the volume versus pressure was done. The results along with the experimental data of Sato-Sorensen (1983) are given in Fig. 22. There appears to be a systematic deviation between experiment and theory at high pressures. It was suggested (Froyen and Cohen, 1984) that this could arise form nonhydrostatic effects in the experiment. More recently, data from Heinz and Jeanloz (1984) are in much closer agreement with the theoretical curve.

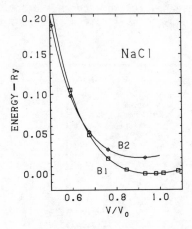

Fig. 21   Total energy per molecule versus volume normalized to the equilibrium B1 volume for NaCl.

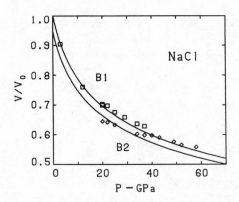

Fig. 22   Pressure-volume curves for the B1 and B2 phases of NaCl. The solid lines represent the theoretical results, and the squares and diamonds are from measurements of Sato-Sorensen (1983).

CONCLUSIONS

This review emphasized semiconductors and insulators, however it should be added that comparable results have been obtained for metals (e.g. see Cohen (1982)). Hence, it is reasonable to conclude that the total energy-pseudopotential approach can be applied to calculated structural properties of solids in cases where the bonding is metallic, covalent, or ionic. In addition, we conclude that a local density approximation for calculating the effects of exchange and correlation for the electron-electron interactions is sufficiently accurate to give structural information. Improvements are clearly needed for excited states, but as can be seen from the examples presented here, the local density approximation works well for ground-state properties.

TABLE 9 Transition Pressures and Transition Volumes for the B1 - B2 Phase Transformation in NaCl. The Volumes are given as Fractions of the Room Temperature Zero Pressure Volume (4.486 nm^3/molecule)

	calculation	experiment	
p	0.27	0.30 0.29	Mbar
v_{B1}	0.646	0.643 0.650 0.63-0.70	
v_{B2}	0.617	0.606 0.618 0.60-0.64	
$\dfrac{\Delta v}{v_{B1}}$	-4.5	-5.8 -5.0	%

It should be noted that considerable research has been done with this method in the area of surfaces (see, for example, Cohen and Louie, 1984). This approach has been used to determine surface reconstruction. Predicting the atomic structure of surfaces is an ambitious undertaking, but it is probably less difficult than predictions of bulk solids because of the reduced dimensionality. This brings up one of the major limitations of the method. It is difficult to determine whether a lower energy structure is the lowest energy structure for a given volume. By calculating the Hellmann-Feynman forces on each atom in a unit cell, it is possible to determine the direction that the atoms "want" to move in. In the next iteration of the calculation, the atoms can be moved in the directions indicated and the forces recalculated. After several cycles, a zero force geometry can be achieved, and this geometry corresponds to a low energy structure, but it is still not certain that this is the lowest energy structure. Monte-Carlo sampling of configuration space may provide more insight, but not much has gone into this approach as yet.

We are therefore at a stage where a subset of structures can be tested and many detailed properties of these structures can be obtained. Hopefully, the theory will continue to develop and wider applications made possible in the future. However, it should be emphasized that even at this point it is possible to predict new crystalline phases of materials knowing only the atomic numbers of the constituent atoms.

Acknowledgements

This work was supported by National Science Foundation Grant No. DMR8319024 and by the Director, Office of Energy Research, Office of Basic Energy Sciences, Materials Sciences Division of the U.S. Deparment of Energy under Contract No. DE-AC03-76SF00098.

REFERENCES

Appelbaum, J. A. and Hamann, D. R. 1976, Rev. Mod. Phys. 48 479-496
Baublitz, M. A. and Ruoff, L. 1982, J. Appl. Phys. 53 6179-6185
Biswas, R., Martin, R. M., Needs, R. J., and Nielsen, O. H. 1984, Phys. Rev. B. 30 3210-3213
Chang, K. J. and Cohen, M. L. 1984a, Phys. Rev. B 30 5376-5378
Chang, K. J. and Cohen, M. L. 1984b, Solid State Comm. 50 487-491
Chang, K. J. and Cohen, M. L. 1984c, Phys. Rev. B 30 4774-4781
Cohen, M. L. 1973, Science 179 1189-1195
Cohen, M. L. 1982, Physica Scripta T1 5
Cohen, M. L. 1984a, Am. J. Phys. 52 695-703
Cohen, M. L. and Bergstresser, T. K. 1966, Phys. Rev. 141 789
Cohen, M. L. and Chelikowsky, J. R. 1982, Handbook on Semiconductors. Ed. W. Paul. North-Holand, Amsterdam, p. 219.
Cohen, M. L. and Heine, V. 1970, Solid State Physics. Eds. H. Ehrenreich, F. Seitz and D. Turnbull. Academic Press, New York, pp. 37-248.
Cohen, M. L., Heine, V., and Phillps, J. C. 1982, Sci. Am. 246 82.
Cohen, M. L. and Louie, S. G. 1984, Ann. Rev. Phys. Chem. 35 537-562
Fermi, E. 1934, Nuovo Cimento 11 157-166
Froyen, S. and Cohen, M. L. 1983, Phys. Rev. B 28 3258-3265
Froyen, S. and Cohen, M. L. 1984, Phys. Rev. B 29 3770-3772
Hamann, D. R., Schlüter, M., and Chiang, C. 1979, Phys. Rev. Lett. 43 1494
Heinz, D. L. and Jeanloz, R. 1984, Phys. Rev. B 30 6045-6050
Hellmann, H. J. 1935, J. Chem. Phys. 3 61.
Hohenberg, P. and Kohn, W. 1964, Phys. Rev. 136 B864-B871
Hu, J. Z and Spain, I. L. 1984, Solid State Comm. 51 263-266
Ihm, J., Zunger, A., and Cohen, M. L. 1979, J. Phys. C 12 4409-4422
Kerker, G. 1980, J. Phys. C 13 L189-L194
Kohn, W. and Sham, L. J. 1965, Phys. Rev. 140 A1333-A1338
Louie, S. G., Froyen, S., and Cohen, M. L. 1982, Phys. Rev. B 26 1738-1742
Needs, R. J. and Martin, R. M. 1985, Proc. 17th Int. Conf. Phys. of Semiconductors, 1984, Springer-Verlag, New York, p. 965.
Olijnyk, H. J., Sikka, S. K., and Holzapfel, W. B. 1984, Phys. Lett. 103A 137-140
Phillips, J. C. 1973, Bonds and Bands in Semiconductors. Academic Press, New York.
Phillips, J. C. and Kleinman, L. 1959, Phys. Rev. 116 287-294
Sato-Sorensen, Y. 1983, J. Geophys. Res. 88 3543-3548
Schlüter, M., Chelikowsky, J. R., Louie, S. G., and Cohen, M. L. 1975, Phys. Rev. B 12 4200-4210
Starkloff, T. and Joannopoulos, J. D. 1977, Phys. Rev. B 16 5212-5215
Vanderbilt, D., Louie, S. G., and Cohen, M. L. 1984, Phys. Rev. Lett. 53 1477-1480
Walter, J. P. and Cohen, M. L. 1970, Phys. Rev. B 4 1877
Wentorf, R. H. and Kasper, J. S. 1963, Science 139 338-339
Wittig, J. 1966, Z. Phys. 195 215-238
Yin, M. T. 1985, Proc. 17th Int. Conf. Phys. of Semiconductors, 1984. Springer-Verlag, New York, p. 927.

Yin, M. T. and Cohen, M. L. 1981, Phys. Rev. B 24 6121-6124
Yin, M. T. and Cohen, M. L. 192a, Phys. Rev. B 25 7403-7412
Yin, M. T. and Cohen, M. L. 1982b, Phys. Rev. B 26 5668-5687
Yin, M. T. and Cohen, M. L. 1982c, Phys. Rev. B 26 3259-3272
Yin, M. T. and Cohen, M. L. 1982d, Phys. Rev. B 25 4317-4320
Yin, M. T. and Cohen, M. L. 1984, Phys. Rev. B 29 6996-6998
Zunger, A. and Cohen, M. L. 1978, Phys. Rev. B 18 5449-5472

Inst. Phys. Conf. Ser. No. 75: Chapter 1
Paper presented at 2nd Int. Conf. Science Hard Mater., Rhodes

29

Fermi surface properties and bonding nature of TiB$_2$ and WC

Y ISHIZAWA AND T TANAKA

National Institute for Research in Inorganic Materials,
1-1 Namiki, Sakura-mura, Niihari-gun, Ibaraki 305, Japan

ABSTRACT

The de Haas-van Alphen (dHvA) effect has been studied in the refractory
compounds TiB$_2$ and WC using a field modulation technique. Observed dHvA
frequencies of both compounds range from the order of 10^2 to 10^3
tesla. The angular dependences of the dHvA frequencies of TiB$_2$ closely
resemble those of the Fermi surface of semimetallic ZrB$_2$. The
semimetallic nature of TiB$_2$ and WC is discussed on the basis of the
properties of experimental Fermi surfaces.

INTRODUCTION

The compounds TiB$_2$ and WC are noted as highly refractory compounds
because of their high melting temperatures, great hardness and chemical
stability. Our research purpose is to clarify the Fermi surface
properties experimentally and get information on the electronic structures
and bonding nature of these compounds.

We selected TiB$_2$ and WC in order to investigate their Fermi surfaces for
the following reasons.

(1) The compounds TiB$_2$ and WC have similar crystal structures. The space
group of both compounds is simple hexagonal.
(2) The similarities in electronic properties of TiB$_2$ and WC are
remarkable.

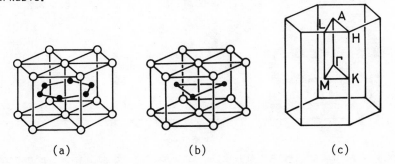

(a) (b) (c)

Fig. 1 Crystal structures and Brillouin zone of TiB$_2$
and WC. (a) TiB$_2$, (b) WC, (c) Brillouin zone.

(3) These compounds have the same 10 valence electrons.
(4) There are no direct measurements of the Fermi surfaces of these compounds.

We will look into the similarities of first two points in more detail. As shown in Fig. 1, the crystal structure of TiB_2 is very similar to that of WC except that carbon vacancy sites order in WC. Metals and metalloids take alternatively layer arrangements. A unit cell of TiB_2 is a little larger than that of WC. The space group of both compounds is simple hexagonal (P6/mmm for TiB_2, a = 0.3030 nm, c = 0.3227 nm (Higashi and co-workers, 1976), and $P\bar{6}m2$ for WC, a = 0.2906 nm, c = 0.2837 nm (Toth, 1971)). These structures contain a single formula unit per primitive cell. The Brillouin zone of these two is hexagonal as shown in Fig. 1. The Γ or A point is in the center of the hexagon, and the K point is on the edge of the hexagon. These A and K points are important to the location of the Fermi surface of TiB_2 as shown later.

Next, we point out the similarities in electronic properties . However, there are only a few data on the electronic properties of TiB_2 and WC. Especially, there are less data on WC. We show the electrical resistivities, the Hall coefficients (Juretschke and Steinitz, 1958 ; Bachmann and Williams, 1971), the γ –values of electronic heat capacity (Tyan and co-workers, 1969 ; Toth, 1971) of TiB_2 and WC in Table 1, including those of ZrB_2 whose electronic structure and Fermi surface were clarified recently by both experimentally and theoretically. (Tanaka and co-workers, 1978 ; Ihara and co-workers, 1977 ; Johnson and co-workers, 1980). It is noticed that carrier mobilities of TiB_2 and WC at room temperature calculated by the electrical resistivity and the Hall coefficient using a single band model, are quite high. This high mobility and small γ –value indicate that carrier densities of the free carriers of TiB_2 and WC are small in contrast to usual metals. Therefore, it is expected that the electronic structure of TiB_2 is very close to that of WC.

TABLE 1   Some Electronic Properties of TiB_2, ZrB_2 and WC

	TiB_2	ZrB_2	WC
Space group	P6/mmm	P6/mmm	$P\bar{6}m2$
Lattice constant [nm]	a = 0.3030 c = 0.3227	a = 0.3170 c = 0.3531	a = 0.2906 c = 0.2837
Electrical Resistivity [10 nΩm]	5.7	3	17
Hall coefficient [μm^3/C]	-2.4×10^{-3}	-2.0×10^{-3}	-4.0×10^{-3}
Carrier mobility [10^{-4} m^2/V sec]	369	666	235
γ –value [mJ/K^2-mol]	1.40	0.93	0.79

The Fermi surface investigation is most useful to study the electronic structure at the Fermi level. The de Haas–van Alphen (dHvA) effect is very powerful method to investigate the Fermi surface, that is, the Fermi

surface shape, dimension, location in the Brillouin zone, and carrier
density and cyclotron mass (Springford,1980). Hitherto, there are no
dHvA data for TiB_2 and WC, except our preliminary dHvA data for TiB_2
(Tanaka and Ishizawa, 1980).

The de Haas-van Alphen Effect

The observation of the dHvA effect in inorganic compounds which gives
direct measurements of the electronic structure at the Fermi level,
becomes possible if they satisfy the strong magnetic field condition,
$\omega_c \tau \gg 1$, where ω_c and τ are the cyclotron frequency and the relaxation
time, respectively. Recent progress of preparing high quality single
crystals makes observation of the dHvA effect possible in these
refractory compounds.

Fig. 2 is an example of dHvA oscillations of TiB_2. In this case,
magnetic fields were applied along the [11$\bar{2}$0] direction. Diamagnetism of
conduction electrons oscillates with magnetic fields at low temperatures
which we call the dHvA effect.

dHvA frequencies and amplitude depend on the magnetic field directions.
From a study of oscillation frequencies F, we are able to determine
extremal cross-sectional areas S of the Fermi surface using an Onsager
relation,

$$F = (\hbar c/2 \pi e)S .$$

Here, $\hbar$, c, and e are Planck's constant over 2π, the velocity of light,
and the electronic charge, respectively.

Meanwhile, we are able to determine the cyclotron mass m^* of carriers from
the temperature dependence of the amplitude A of dHvA oscillations,

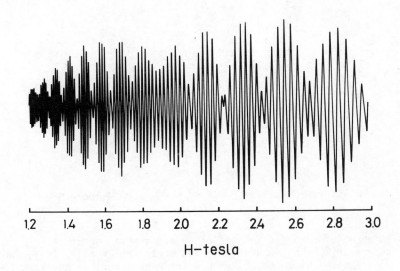

H-tesla

Fig. 2 dHvA oscillations of TiB_2. Magnetic fields
were applied along the [11$\bar{2}$0] direction.

$$A/T \propto [\sinh(2\pi^2 ckm^*T/ehH)]^{-1}.$$

Here, we point out that among borides, carbides and nitrides, dHvA effect
was detected only in LaB_6 (Ishizawa and co-workers, 1977 ; Arko and
co-workers, 1976), CeB_6, PrB_6, NdB_6 (van Devrsen and co-workers, 1982),
YB_4 (Tanaka and Ishizawa, 1984), ZrB_2 (Tanaka and Ishizawa, 1978) and TiN
(Winzer and Reichelt, 1984) in addition to the present presentation of
TiB_2 and WC. This is because of difficulty of growing high quality single
crystals.

EXPERIMENTAL

Single Crystal Preparation

Single crystals of TiB_2 were prepared by a floating zone technique. A
conceptual diagram of the furnaces for sintering and crystal growth is
shown in Fig. 3. Firstly, we prepared polycrystalline rods. Boron powder
was added 2 wt.% in excess to TiB_2 powder in order to compensate the
preferential vaporization of boron during the zone pass. After adding a
ethanol- camphor solution and mixing, they were compressed into a
rectangular rod with dimensions of about 10 mm x 10 mm x 200 mm. The
compressed rod was then re-pressed hydrostatically in order to achieve a
higher and more uniform density. After the rod had been shaped into a
cylinder of diameter of 10 mm, it was sintered at 2300 K in a vacuum for
30 min. using a BN crucible inserted in a graphite susceptor. The

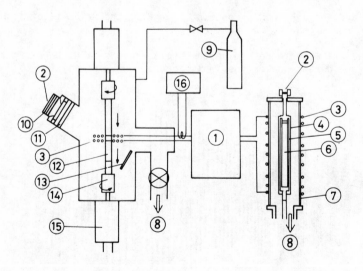

Growth Furnace   Sintering Furnace

Fig. 3 Diagram of furnaces for sintering and crystal growth.
(1) RF generator(200 kHz, 40 kW), (2) view port, (3) RF coil, (4) graphite
wool, (5) graphite susceptor, (6) sample for sintering, (7) quartz tube,
(8) vacuum system, (9) gas cylinder, (10) shading filter, (11)
IR-absorption filter, (12) crystal rod, (13) mirror, (14) BN holder,
(15) part for drive and rotation of shaft, and (16) recorder.

obtained polycrystalline rod had a density of   about 60 % of theoretical one.

The crystals were grown in a high pressure type furnace (10 MPa, 200 KHz, 40 Kw). The floating zone procedure was carried out with the feed rod driving downwards through a work coil of inner diameter of 16 mm (three turns and two steps) which was used for r.f. induction heating. The vaporization of TiB_2 from the molten zone was very severe. Thus, the furnace was filled with helium or argon at a pressure of 1.5-2.0 MPa in order to prevent vaporization. In the case of argon, discharge occurred sometimes between turns of coil or the work coil and the molten zone, destroying the work coil. Helium is superior to supress the discharge. However, more adhesion of vaporized TiB_2 on the grown crystal occurred. Such adhesion deteriorated the crystal quality at periphery of the crystal rod. The growth speed of the crystal was 10 mm/h and the feed speed was 15-20 mm/h. This difference in speeds was necessary to compensate the density difference between the feed polycrystalline rod and the crystal rod, taking the vaporization loss from the molten zone into consideration.

The TiB_2 crystals had dimensions of about 9 mm in diameter and 30 mm long. The crystal rods consisted of a large single crystal at the center of the rod and a polycrystalline rim about 1mm thick at the peripheral region. Metal impurities were cheked by fluorescent X-ray analysis. No impurities were detected from the crystal. However, Fe, Cr, Ni, Mn and Co were detected from the vaporized TiB_2 powder. This indicates the purification effect for these volatile impurities by the zone pass. The highest RRR, residual resistance ratio ($\rho(300\ K)/\rho(4.2\ K)$) was 25.

For the dHvA measurements, two cubic specimen with a volume of about 8 mm^3 were spark-cut. Their faces were parallel to the $(10\bar{1}0)$, $(11\bar{2}0)$ and (0001) crystallographic planes and were chemically polished to remove the surface damage. The RRR of the crystals for dHvA measurements were 15 and 25.

Single crystals of WC were grown by the flux growth method. The method of crystal growth is the same as that of TiB_2, where a carbon susceptor was inductively heated. Cobalt-tungsten carbide mixtures containing 14 at.% WC were melted in an alumina crucible inserted in the carbon susceptor at 1920 K in a helium atmosphere. This temperature was held constant for 2 hours to assure complete homogeneity. The temperature was rapidly lowered from 1920 K to 1720 K at a rate of 100 K/h. Then, the crucible was slowly cooled from 1720 K to 1520 K at a rate of 4 K/h. The temperature at the bottom of the crucible was set to 10 K lower than the top of the crucible. The mixture solidified completely at 1520 K.

The obtained crystals were embedded in a cobalt flux. After dissolution of the cobalt matrix in boiling (H_2O + HNO_3 + 3HF) solution, we found the crystal form mainly equilateral triangular plate-like. The largest dimension of the crystals was approximately 5 mm in edge length. The reflection Laue photograph showed that the [0001] axis was perpendicular to the plane of the triangular-shaped crystals. The RRR of the crystals were scattered from 10 to 70 even in the same batch. The crystal having the RRR of 70 was used for the dHvA measurements.

dHvA Measurements

The dHvA measurements were carried out using a field modulation technique

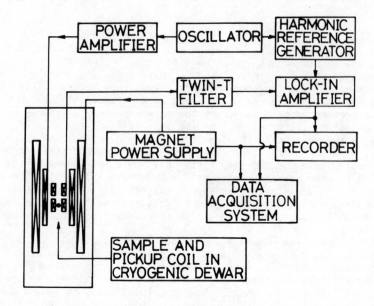

Fig. 4 A block diagram of dHvA spectrometer

at liquid helium temperatures and magnetic fields up to 6 tesla(T). The
second harmonic frequency was used for detecting signals with a modulation
frequency of 400 Hz. A sample and a pick-up coil assembly are in a
superconducting magnet. dHvA signals of the pick-up coil was amplified
by lock-in amplifier through twin tee filter. A block diagram of dHvA
spectrometer is shown in Fig. 4. The dHvA frequencies and amplitudes were
obtained by fast-Fourier analysis.

EXPERIMENTAL RESULTS AND DISCUSSION

1. dHvA Effect of TiB$_2$

dHvA Results

The dHvA oscillations have been observed in the two frequency ranges
$(0.7-4.6) \times 10^2$ T and $(1.5-5.7) \times 10^3$ T. Observed frequencies of the
order of 10^2 T in the $(10\bar{1}0),(11\bar{2}0)$ and (0001) planes are shown in
Fig. 5. The oscillations labelled α and β were most easy to detect
because of their large amplitude.

The α frequencies have four branches in the vicinity of the [0001]
direction in the $(10\bar{1}0)$ plane. However, only three branches were observed
for the nearly same direction in the $(11\bar{2}0)$ plane. Near the [11$\bar{2}$0]
direction in the $(10\bar{1}0)$ plane, three branches were observed and for most
of the remaining region, two branches were observed. A minimun value of
the α frequencies, 1.41×10^2 T, was observed at 38° from the [11$\bar{2}$0]
direction in the $(10\bar{1}0)$ plane. The angular dependence of the α
frequencies shows that the α Fermi surface is nearly ellipsoidal as far
as the observed frequencies are concerned.

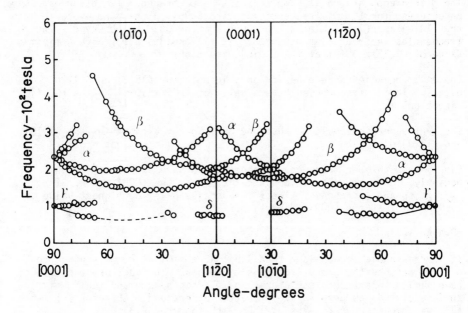

Fig. 5   dHvA frequencies with the order of 10^2 T of TiB_2 in the $(10\bar{1}0)$, $(11\bar{2}0)$ and (0001) planes.

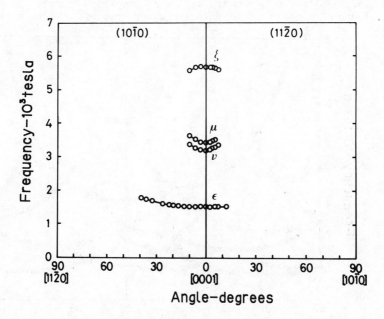

Fig. 6   dHvA frequencies with the order of 10^3 T of TiB_2 in the $(10\bar{1}0)$ and $(11\bar{2}0)$ planes.

The β frequency branch is simpler than the α branch. Two branches could be observed for the β frequencies in the (0001) plane. However, in both the (10$\bar{1}$0) and (11$\bar{2}$0) planes, only one branch could be observed. The β frequencies whose angular dependence is ellipsoidal also, have a minimum value of 1.75 x 10^2 T at the [10$\bar{1}$0] direction.

The γ frequencies were observed only in both the (10$\bar{1}$0) and (11$\bar{2}$0) planes. The γ frequencies have a value of 1.00 x 10^2 T at the [0001] direction.

The δ frequencies could be observed only in the vicinity of the [10$\bar{1}$0] and [11$\bar{2}$0] directions.

Fig. 6 shows higher frequency branches which have been observed quite recently. However, obsevation is limited near the [0001] axis. They are ξ, μ, ν and ε branches. The ξ frequencies have the highest value of 5.69 x 10^3 T at the [0001] direction.

dHvA Analysis

It is reasonable to consider that TiB$_2$ has a similar Fermi surface to that of ZrB$_2$ because the group IVa transition-metal diborides are expected to have similar electronic structures and bonding nature and the Fermi surface of ZrB$_2$ is comparatively better understood. Therefore, first of all, we compare dHvA results of TiB$_2$ with those of ZrB$_2$.

We performed the dHvA experiments of ZrB$_2$ several years ago (Tanaka and Ishizawa, 1978) and obtained the following frequency branches as shown in Figs. 7 and 8. Comparison of dHvA frequency branches of TiB$_2$ with those

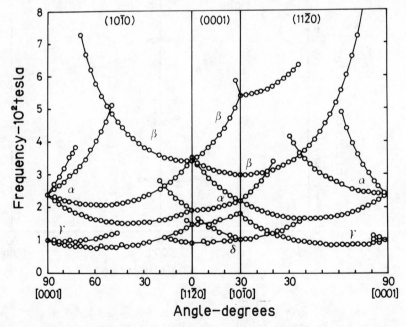

Fig. 7 dHvA frequencies with the order of 10^2 T of ZrB$_2$ in the (10$\bar{1}$0), (11$\bar{2}$0) and (0001) planes.

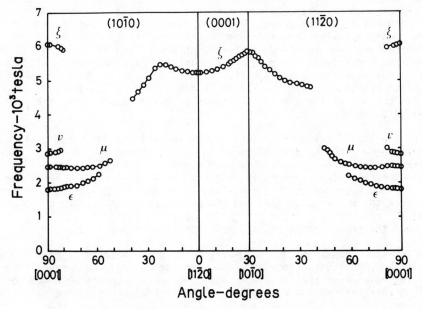

Fig. 8 dHvA frequencies with the order of 10^3 T of ZrB_2 in the $(10\bar{1}0)$, $(11\bar{2}0)$ and (0001) planes.

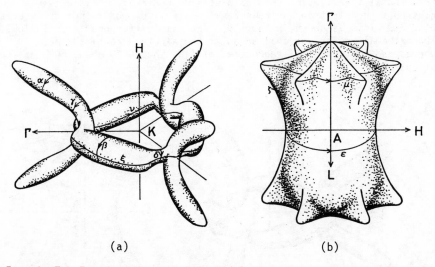

(a)   (b)

Fig. 9 The Fermi surface of ZrB_2. (a) The electron Fermi surface around the K point. (b) The hole Fermi surface at the A point.

of ZrB_2 shows that α, β, γ and δ branches corresponds to those of ZrB_2. The angular dependence of each branch is almost the same. The ζ, μ, ν and ε branches of TiB_2 are also considered to be one part of the corresponding branches of ZrB_2.

The experimental Fermi surface of ZrB_2 is shown in Fig. 9. The electronic
structure and theoretical Fermi surfaces calculated by Ihara(1977) using
the APW method and by Johnson and co-workers (1980) using the KKR method
are consistent with the experimental Fermi surface. As shown in the
figure, there are two kinds of Fermi surfaces. One is a ring-like
electron Fermi surface around the K point to which the nearly ellipsoidal
Fermi surfaces are joined together at the point where the ring-like Fermi
surface crosses the ΓK lines. The other is a wrinkled dumbbell-like hole
Fermi surface at the A point which has an anisotropy between the [10$\bar{1}$0] and
[11$\bar{2}$0] directions.

The ring-like Fermi surface consists of the ellipsoidal Fermi surfaces
corresponding to the β frequencies which cut across the KM line. These β
Fermi surfaces are connected to each other by δ Fermi surfaces. These
make the ring-like Fermi surface around the K point. The α Fermi surfaces
are joined together to this ring-like Fermi surface. Then, the connection
between the α Fermi surface and the ring-like Fermi surface corresponds to
the γ frequencies. The ν and ζ frequencies come from the orbits running
around the inner and outer parts of this ring, repectively.

On the other hand, the ε and μ frequencies for magnetic fields (H) along
near the [0001] direction, correspond to the central and noncentral orbits
of the hole Fermi surface, repectively. The frequencies observed for H
along near the c-plane correspond to the central orbit of the hole Fermi
surface.

Fermi Surface of TiB_2

From the similarities of dHvA data of TiB_2 and ZrB_2, it is possible to
construct the electron and hole Fermi surfaces by modifying Fermi surfaces
of ZrB_2. This is shown in Fig. 10. The main difference is as follows.
The extremal cross sectional areas of the electron Fermi surface
corresponding to the β frequencies of TiB_2 shrank to about 60 % of those
of ZrB_2.

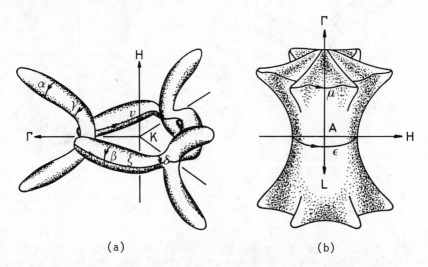

(a)   (b)

Fig. 10 The Fermi surface of TiB_2. (a) The electron Fermi surface
around the K point. (b) The hole Fermi surface at the A point.

TABLE 2   dHvA Frequencies of TiB_2 and ZrB_2

Field direction	Branch	Carrier	Frequency–10^2 tesla TiB_2	ZrB_2	$F(TiB_2)/F(ZrB_2)$
[0001]	ζ	electron	56.9	60.9	0.94
[0001]	μ	hole	34.4	24.6	1.40
[0001]	ν	electron	31.9	28.4	1.12
[0001]	ε	hole	15.0	18.1	0.83
[0001]	α	electron	2.30	2.38	0.97
[0001]	γ	electron	1.00	0.99	1.01
[1010]	ζ	hole	––	48.6	––
[1010]	β	electron	3.4	5.42	0.63
[1010]	α	electron	2.01	2.20	0.91
[1010]	β	electron	1.75	3.00	0.58
[1010]	γ	electron	––	1.79	––
[1010]	δ	electron	0.83	1.02	0.81
[1120]	ζ	hole	––	42.3	––
[1120]	α	electron	3.2	3.50	0.91
[1120]	β	electron	2.02	3.43	0.59
[1120]	α	electron	1.76	1.92	0.92
[1120]	γ	electron	––	1.49	––
[1120]	δ	electron	0.74	0.94	0.79
F_{min} in the (1010) plane					
	α	electron	1.41(38)	1.54(38)	0.92
	γ	electron	––	0.76(55)	––

In the case of the hole Fermi surface, cross sectional areas corresponding to the ε frequencies of TiB_2 shrank to 83 % of those of ZrB_2 while those of the μ frequencies expanded to 140 % of those of ZrB_2. Comparison of dHvA frequencies of TiB_2 with those of ZrB_2 are summarized in Table 2.

The carrier concentration of TiB_2 was estimated from the observed Fermi surface to be about 0.02 per unit cell for electrons and holes, respectively. This low concentration and close compensation of carriers are characteristic of a semimetal, as the case of ZrB_2.

Electronic Structure and Bonding Nature of TiB_2

Electronic band structure of TiB_2 was calculated by Ihara(1977b) using the APW method, and by Perkins and Sweeney(1976) using the tight binding approximation. The present experimental Fermi surfaces are consistent with theoretical Fermi surfaces (Ihara and Gonda, 1981).

Agreement of dHvA data of TiB_2 with band structure calculations by Ihara(1977b) is quite good. Therefore, in order to understand the bonding nature of TiB_2, we use his bandstructure calculation which explains well X-ray photoelectron spectra (Ihara,1977b) and our dHvA results. On the basis of his calculation, the density of states (DOS) of TiB_2 is schematically shown in Fig. 11. The Fermi level is located at the DOS minimum between the large peaks of P_3 and P_4. The two peaks P_1 and P_2

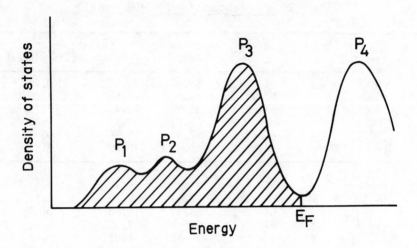

Fig. 11 Schematic density of states of TiB$_2$

come from boron 2s and 2p bands. The two peaks P$_3$ and P$_4$ are bonding and anti-bonding bands constructed by hybidization of the transition metal d orbitals and boron 2p orbitals. These bands have a p$_z$ character in the vicinity of the Fermi level and a d character near the peak. The valence bands consisting of peaks of P$_1$, P$_2$ and P$_3$, can accommodate 10 electrons per unit cell. The p–d hybrid bonding band is just filled in TiB$_2$ so that the Me–B covalent bond is very strong. On the other hand, the s–p band, corresponding to the peaks P$_1$ and P$_2$, has insufficient states to construct the full sp^2 band which constructs the graphite structure. Considerable parts of the p orbitals of the boron atom are divided into the p–d hybrid bands. Therfore, the B–B bond in the network structure is not so strong as expected. Moreover, the conduction band and the valence bands of TiB$_2$ overlap slightly resulting to the semimetallic band structure.

2. dHvA Effect of WC

dHvA Results

The dHvA oscillations have been observed in the frequency range of 10^2 to 10^3 T (Ishizawa and Tanaka, 1984). The dHvA frequencies in the (0001), (11$\bar{2}$0) and (10$\bar{1}$0) planes are shown in Fig. 12. The results consist of the low-frequency branches α, β, γ, δ and the high-frequency branches ε, μ, ν, λ.   The amplitudes of the former branches are strong and those of the latter branches are rather weak. It is easily recognized that the angular dependences of observed frequencies are almost the same in the (11$\bar{2}$0) and (10$\bar{1}$0) planes.

The α and β branches disappear about 78° from the [0001] direction in both the (10$\bar{1}$0) and (11$\bar{2}$0) planes. The γ and δ branches have constant frequencies within ± 0.5 % in the (0001) plane and disappear at about 76° from the [11$\bar{2}$0] direction in the (10$\bar{1}$0) plane and at about 70° from the [10$\bar{1}$0] direction in the (11$\bar{2}$0) plane. The α and β , and γ and δ branches seem to be pair-branches, respectively.   The ε branch has a minimum cross

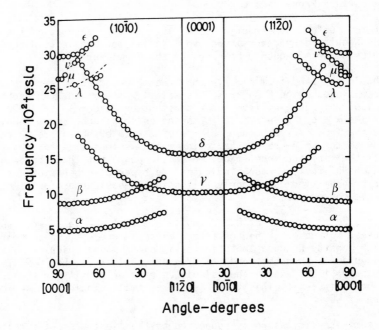

Fig. 12   dHvA frequencies of WC in the (10$\bar{1}$0),(11$\bar{2}$0) and (0001) planes.

TABLE 3   dHvA Frequencies and Cyclotron Mass Ratios in WC

Field direction	Branch	Frequency–10^2 T	m^*/m_0
[0001]	α	4.78	0.18
[0001]	β	8.66	0.34
[10$\bar{1}$0]	γ	10.08	0.35
[10$\bar{1}$0]	δ	15.59	0.43
[11$\bar{2}$0]	γ	10.08	0.35
[11$\bar{2}$0]	δ	15.59	0.43
[0001]	ε	29.77	
[0001]	μ	26.56	

section for H along the [0001] direction which can be observed up to about 30° from the [0001] direction.   The λ and ν branches were observed from 6° to 40° for the λ branch and to 23° for the ν branch.   These λ and ν signals were much weaker in the (10$\bar{1}$0) plane than in the (11$\bar{2}$0) plane. The μ branch was observed only in the limited range around the [0001] direction.   The amplitude of the μ branch is relatively large just along the [0001] direction.

Table 3 shows the values of the dHvA frequencies at high symmetry directions.   Cyclotron masses for the low-frequency branches are also shown in Table 3.

Fermi Surface Properties of WC

The Fermi surface construction of WC is difficult because of their simple angular dependences. However, it is worthwhile to describing features of the frequency branches or the Fermi surface.

It should be noted from Fig. 12 that the observed Fermi surface is fairly isotropic around the Γ A axis. Especially, the γ and δ branches are most isotropic judging from their constant behaviour in the (0001) plane. Pair-like α and β, and γ and δ branches show conspicuous behaviour among the frequency branches. These pair-like branches are interpreted to be due to the spin-orbit interaction which is important for heavy metals. As for the Fermi surface shape, we should point out that the angular variation of the ν branch indicates the hyperbolic behaviour and the other branches indicate nearly ellipsoidal behaviour as far as observed data are concerned.

That the Fermi surface of WC is small is one of the very important features. If no other higher frequencies are present, we can estimate the carrier density using a tentative ellipsoidal model which consists of combined $\delta - \varepsilon$ branches, and the frequency values at the [0001], [10$\bar{1}$0] and [11$\bar{2}$0] axes. The calculated carrier density is the order of 0.01 per unit cell. This value is fairly small, indicating semimetallic nature of WC.

This semimetallic nature is supported by the light mass of carriers, the small γ-value of the electronic heat capacity, and the large Hall coefficient. The present dHvA data show that light mass carriers exist. The γ-value is reported to be 0.79 mJ/K^2-mole (Toth, 1971) which is comparable to the γ-value of 0.5-1.0 mJ/K^2-mole of the group IVa transition-metal carbides which are semimetals. The Hall coefficient is negative with a value of 4.0 x 10^{-9} m^3/C at 300 K (Bachmann and Williams, 1971) which is also comparable to those of the group IVa transition-metal carbides.

Electronic Band Structure of WC

Recently, band structure calculations of WC were reported by a few groups (Mattheiss and Hamann,1984 ; Stefan and co-workers, 1984 ; Alekseev and co-workers, 1982). Mattheiss and Hamann have reported the band structure calculations of WC with the use of a self consistent, scalar relativistic version of the linear APW method. Their results show that WC is a semimetal with Fermi surfaces of several small electron and hole pockets. According to their calculations, strong W5d-C2p hybridization is important with the Fermi energy located at the minimum between the bonding and anti-bonding bands. These calculations are qualitatively consistent with present dHvA data.

CONCLUSIONS

Fermi surface properties of TiB$_2$ and WC have been studied by means of the dHvA effect. Observed dHvA frequencies of both compounds are the order of 10^2 to 10^3 tesla. The Fermi surfaces corresponding to the observed frequencies are fairly small, indicating that TiB$_2$ and WC are semimetals. It was shown that the p-d hybridization is very important in these refractory compounds through the combination of the present dHvA data and recent band structure calculations.

REFERENCES

Alekseev, E. S., Arkhipov, R. G. and Popova, S. V. 1982, Phys. Stat. Sol. (b) 110 K151–154.

Arko, A. J., Crabtree, G., Karim, D., Muller, F. M., Windmiller, L. R., Ketterson, J. B. and Fisk, Z. 1976, Phys. Rev. B13 5240–5247.

Bachmann, K. and Williams, W. S. 1971, J. Appl. Phys. 42 4406–4407.

Higashi, I., Takahashi, Y. and Atoda, T. 1976, J. Cryst. Growth 33 207–211.

Ihara, H., Hirabayashi, M. and Nakagawa, H. 1977a, Phys. Rev. 16 726–730.

Ihara, H. 1977b, Researches of the Electotechnical Laboratory, No. 725.

Ihara, H. and Gonda, S. 1981, Solid State Phys. (Japan) 16 583–588.

Ishizawa, Y., Tanaka, T., Bannai, E. and Kawai, S. 1977, J. Phys. Soc. Jpn. 42 112–118.

Ishizawa, Y. and Tanaka, T. 1984, Solid State Commun. 51 743–745.

Johnson, D. L., Harmon, B. N. and Liu, S. H. 1980, J. Chem. Phys. 73 1898–1906.

Juretschke, H. J. and Steinitz, R. 1958, J. Phys. Chem. Solids 4 118–127.

Mattheiss, L. F. and Hamann, D. R. 1984, Phys. Rev. B30 1731–1738.

Perkins, P. G. and Sweeney, A. V. 1976, J. Less–Common Metals 47 165–173.

Springford, M. 1980, Electrons at the Fermi Surface, Cambridge Univ. Prees, Cambridge, pp 319–508.

Stefan, P. M., Shek, M. L., Lindau, I., Spicer, W. E., Johansson, L. I., Herman, F., Kasowski, R. V. and Brogen, G. 1984, Phys. Rev. B29 5423–5444.

Tanaka, T., Ishizawa, Y., Bannai, E. and Kawai, S. 1978, Solid State Commun. 26 879–882.

Tanaka, T. and Ishizawa, Y. 1980, J. Phys. C 13 6671–6676.

Tanaka, T. and Ishizawa, Y. 1984, to be published in J. Less–Common Metals.

Toth, L. E. 1971, Transition Metal Carbides and Nitrides, Academic Press, New York.

Tyan, Y. S., Toth, L. E. and Chang, Y. A. 1969, J. Phys. Chem. Solids 30 785–792.

van Deursen, A. P. J., Fisk, Z. and de Vroomen, A. R. 1982, Solid State Commun. 44 609–612.

Winzer, K. and Reichelt, J. 1984, Solid State Commun. 49 527–529.

Inst. Phys. Conf. Ser. No. 75: Chapter 1
Paper presented at 2nd Int. Conf. Science Hard Mater., Rhodes

45

Chemical bonding in refractory transition metal compounds

K SCHWARZ(1) AND A NECKEL(2)

(1) Institut für Technische Elektrochemie
Technische Universität Wien, A-1060 Vienna, Austria
(2) Institut für Physikalische Chemie
Universität Wien, A-1090 Vienna, Austria

ABSTRACT

Energy band calculations by the (linearized) augmented plane wave method are used to describe the electronic structure of transition metal compounds. First, energy bands, density of states and total energy results are illustrated for ZrN, then the binding mechanism in terms of metallic, ionic, and covalent contributions is discussed in the series TiC, TiN, and TiO. Comparison of calculated quantities with experimental data is made for photoelectron and x-ray emission spectra as well as with optical measurements.

INTRODUCTION

Refractory transition metal carbides and nitrides have been thoroughly investigated because of their interesting physical and chemical properties such as high melting points, ultrahardness, and metallic conductivity. Some of them are superconductors with transition temperatures as high as 18 K (niobium carbonitride). For a recent review of these compounds see Neckel (1983).

The unusual combination of properties has challenged theorists to study the chemical bonding in these systems. Calculations on the electronic structure have been performed for several refractories crystallizing in the NaCl structure, so for example by means of the augmented plane wave (APW) method using the local density approximation for treating exchange and correlation: the compounds with 3d-transition metals, ScN, ScO, TiC, TiN, TiO, VC, VN and VO, have been investigated by Neckel and co-workers (1975,1976). Recently additional calculations on the titanium compounds have been performed by Blaha and Schwarz (1983) and Schwarz and Blaha (1984) using the linearized version of the APW scheme, the LAPW method. Compounds with 4d-transition metals have also been studied, for example ZrC and ZrN (Schwarz, Ripplinger and Neckel 1982) and NbC and NbN (Schwarz 1977).

These calculations yield energy bands, the corresponding densitites of states (DOS), and electron densities $\rho(r)$, quantities from which valuable information on the bonding can be obtained. It is found that all three main types of bonding contribute in the refractories:

i) metallic contributions with a finite density of states at the Fermi-energy, E_F;

ii) ionic bonding caused by a transfer of electrons from the metal to the non-metal atom;

iii) covalent bonds consisting mainly of strong interactions between non-metal p and metal d states, but also of metal-metal interactions. These types of chemical bonding are illustrated below.

ZrN AS A FIRST EXAMPLE

Energy Bands and Density of States

The energy bands and the corresponding density of states of ZrN are shown in Fig.1. The low lying energy band at about -15 eV below the Fermi-energy E_F originates mainly from the atomic nitrogen-2s states. The next three bands, up to about -3 eV, come primarily from the nitrogen-2p states, but one can already see from the band structure that there is a mixing with the higher lying Zr-4d bands, which dominate at E_F. As a short-hand notation we shall call these three types of bands 's-', 'p-', and 'd-band', but this notation indicates only the main character of these bands or of the corresponding DOS.

There is an appreciable DOS at E_F, so that ZrN is a good conductor. It turns out that ZrN is even a superconductor with a transition temperature $T_C = 10$ K. There are 9 valence electrons in ZrN, 4 from Zr and 5 from N. The low lying 's-band' accommodates 2 electrons, and the 'p-band' is occupied by 6 electrons. If the 'p-' and 'd-band' overlap as in ZrN, then it is found that, in this class of refractories, there is always a pronounced minimum in the DOS at 8 valence electrons.

Total Energy, Lattice Constant and Bulk Modulus

In Fig.2 we illustrate a total energy calculation carried out by the augmented spherical wave (ASW) method (Williams, Kübler and Gelatt 1979) treating all electrons. This ASW as well as the APW calculations are true ab initio methods, i.e. the only inputs required for such calculations are the atomic numbers of the constituent atoms and the assumed crystal structure. By minimizing the total energy for ZrN an equilibrium lattice constant of 0.4598 nm is obtained (Schwarz and Moruzzi 1984) in excellent agreement with the experimental value of 0.4585 nm. The bulk modulus $B = 0.24$ MN/mm^2 is derived from the curvature of the total energy as a function of volume.

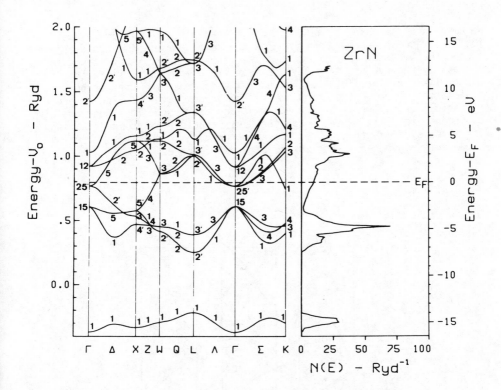

Fig.1 Band structure and density of states N(E) of ZrN (Schwarz, Ripplinger and Neckel 1982). N(E) is given in states of both spins per Ryd and unit cell (1 Ryd = 13.6eV); the energy scale is shown twice, on the left in Ryd with respect to the muffin-tin potential zero V_0, on the right in eV with respect to the Fermi-energy E_F.

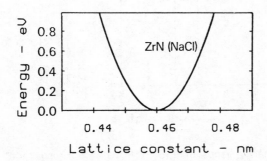

Fig.2 Total energy of ZrN (with respect to its minimum) as a function of lattice constant obtained by ASW calculations (Schwarz and Moruzzi 1984).

CHEMICAL BONDING OF TiC, TiN, TiO

We have chosen the series TiC, TiN and TiO as a representative example
to illustrate the chemical bonding in the refractories. It should be mentioned that none of these compounds have the ideal rock-salt structure
assumed by theory, but they deviate from stoichiometry. This fact is
especially true for TiO, where about 15% vacancies are found on both
sublattices, i.e. on the titanium and the oxygen sites. In the case of
TiC and TiN, smaller vacancy concentrations are found. They occur mainly
on the non-metal sites. Although the NaCl structure is an idealization
for these systems, these three compounds provide an interesting series
to determine general trends for related refractories.

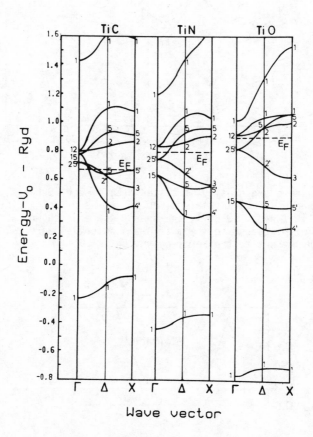

Fig.3 Section of energy bands of TiC, TiN and TiO along the Γ-Δ-X
direction in the first Brillouin zone according to Blaha and Schwarz
(1983). Energies (in Ryd) are with respect to the constant muffin-tin
potential V_o.

Band Structure

The energy bands of these three compounds are shown in Fig.3 for one symmetry direction in the first Brillouin zone. Note the similarity between TiN and ZrN for which a more complete band structure is shown in Fig.1. Before discussing trends in the series we characterize the energy bands for TiO: there is the low lying 'O-2s band', then there are 3 bands derived from the O-2p states which originate at Γ_{15} (the Δ_5 bands are doubly degenerate); since Ti is coordinated octahedrally by oxygen, the Ti-3d states are split by the ligand field into the t_{2g} ($\Gamma_{25'}$) and the e_g (Γ_{12}) symmetry at Γ. This manifold of 5 bands constitutes the Ti-3d band. The highest band shown in Fig.3 originates from the Ti-4s state which is occupied in the atom and also in bulk Ti. In the compounds, however, the repulsive interaction with the non-metal-s band shifts this band to higher energies. As a result, it lies above the Fermi level. It has also strong, free-electron-like character.

In the series TiC, TiN, TiO, the non-metal s and p bands are lowered in energy and simultaneously the Ti-4s band is also lowered because of its reduced repulsive interaction with the non-metal-s band. At the same time the Fermi-energy moves up in energy, since the number of valence electrons increases from 8 to 10.

Density-of-States

The total DOS, such as the one shown for ZrN (Fig.1), can be decomposed using additional information derived from the corresponding wave functions. Fig.4 shows such a decomposition of the DOS generated from an LCAO fit to the APW energy bands (Neckel and co-workers 1976). In the series carbide, nitride and oxide, the non-metal (X) valence states (2s and 2p) decrease in energy because of the increased nuclear charge. This atomic effect can be clearly seen in the low-lying band which originates almost entirely from the X-2s states (Fig.3). The X-2p states interact with the Ti-3d states. In the case of TiO, only a small admixture of Ti-d DOS is found in the 'p-band' which is separated by a gap from the 'Ti-d band'. For TiN, and even more for TiC, the p-d interaction increases, i.e. the 'p-' and 'd-band' overlap (Fig.3) and there is a higher Ti-d DOS in the 'p-band' (Fig.4). These aspects will be discussed below in connection with covalency.

As mentioned above there is always a sharp minimum in the DOS when the bands are filled with 8 valence electrons as e.g. for TiC; this minimum separates the bonding from the antibonding bands having a p-d interaction. In the case of TiN or TiO, the 1 or 2 additional valence electrons cause the Fermi energy to rise above this minimum. In all three cases a finite DOS exists at E_F, leading to metallic conductivity.

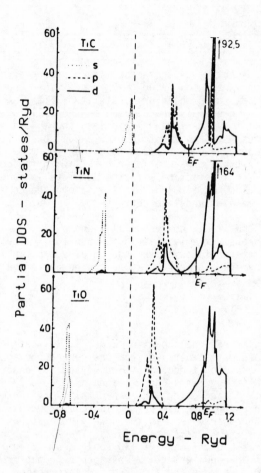

Fig.4 LCAO partial DOS for the non-metal 2s and 2p, and the Ti-3d states (Neckel and co-workers 1976).

Ionicity

One aspect of ionicity is illustrated in Fig.5, which shows the charge transfer defined by the difference in the (spherically averaged) radial electron densities $\sigma(r)$ taken between the crystal (as calculated by APW) and a superposition of atomic densities of the corresponding neutral atoms. Inside the non-metal atomic sphere $\Delta\sigma$ is positive, but inside the titanium sphere it is negative. This observation holds throughout this series, and shows that in the crystal there is more electronic charge around the non-metal atom (and less around Ti) than there would be in a hypothetical crystal of superposed non-interacting neutral atomic densities; three observations should be made:

   i) there is definitely an electronic charge transfer (CT) from Ti
to the non-metal atom;

   ii) the total CT depends on the atomic-sphere radius R_i up to which
the integral over $\Delta\sigma$ is carried out, but these atomic sphere sizes are
not uniquely defined; however, even for a different choice of the atomic
sphere radii the same qualitative result, namely a transfer of electronic
charge from the Ti atom to the non-metal atom would be obtained.

   iii) in the series TiC, TiN, TiO the CT remains about constant
(about half an electron), indicating that in terms of CT the oxide is
not more ionic than the carbide.

The last result is at first surprising, since the difference in electronegativity between titanium and the non-metal atom increases drastically
from the carbide to the oxide. Therefore, one would expect that the oxide
has to have a much larger charge transfer than the carbide. Ionicity,
however, has various aspects and CT is only one of them. Another aspect
is the localization of the orbitals involved and the absence of directed
covalent bonds. In terms of localization and decreased covalency, TiO is
certainly more ionic than TiC. The 'anomaly' of the constant CT can be
understood on the basis of band structure calculations and is discussed
below after the covalent bonds in these compounds have been described.

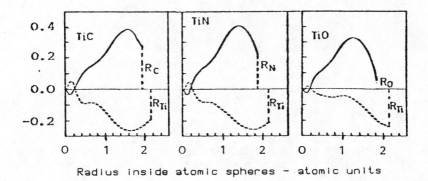

Radius inside atomic spheres – atomic units

Fig.5 Charge transfer $\Delta\sigma$ where $\sigma(r) = 4\pi r^2 \rho(r)$ is the (spherically
averaged) radial electron density (at.u.); the difference $\Delta\sigma$ is taken
between the crystal and the superposed atomic densities (Neckel and co-
workers 1975). R_X and R_{Ti} are the radii of the respective atomic spheres
around the non-metal (X) and Ti sites.

Valence Electron Density

The valence electron density is another quantity which allows further insight into the binding mechanism. Recently these densities have been calculated for all three titanium compounds by Blaha and Schwarz (1983) using the LAPW method. The results for TiC are shown in Fig.6. It can be seen that the density around carbon is spherically symmetric and it is comparable in magnitude to the density near the Ti atom; the latter, however, deviates from spherical symmetry indicating covalent interactions which will be discussed in the next paragraph.

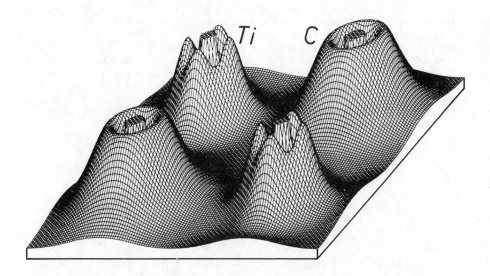

Fig.6   Valence electron density of TiC in the (100) plane according to Blaha and Schwarz (1983)

The fact that the electron density remains relatively high between the atoms is consistent with the metallic character of TiC, in contrast to an insulating highly ionic compound in which the density between the ions would almost vanish.

The computed electron densities (or structure factors) have been confirmed for TiC and TiN by high-precision x-ray diffraction measurements carried out by Dunand, Flack and Yvon (1985). In this comparison between a measurement on a real crystal and a calculation for TiC assuming the ideal NaCl structure (Blaha, Redinger and Schwarz 1985), it turned out to be very important to consider non-stoichiometry effects especially in the case of TiC. It should be stressed that vacancies or impurities on either sublattice are often observed in these compounds and they strongly affect many properties of these materials.

Covalent Bonding

Covalent interactions in these compounds mainly occur between X-2p and Ti-3d atomic states. Recently Schwarz and Blaha (1984) were able to illustrate these interactions by calculating electron densities of selected energy states. Fig.7 is taken from their work and shows the three types of covalent bonds (pd_σ, dd_σ, and pd_π) which are found in TiC for the three occupied states (just below E_F) at $\vec{K}= \Delta$(half-way between Γ and X). The d-orbitals centered on a titanium site can be classified according to e_g and t_{2g} symmetry; the e_g orbitals point towards the corners, the t_{2g} orbitals towards the faces of an octahedron formed by the non-metal ligands.

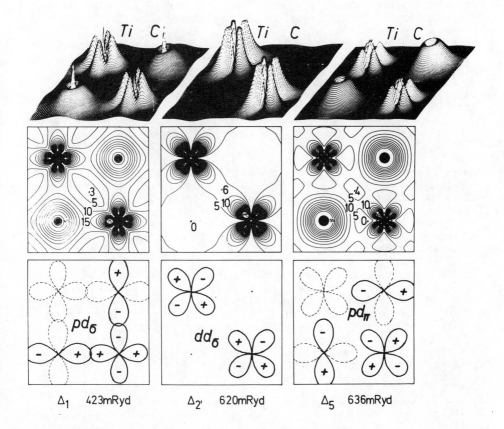

Fig.7 Charge densities in the (100) plane corresponding to 3 occupied band states for $\vec{K}=\Delta=\frac{\pi}{\alpha}(1,0,0)$ taken from Schwarz and Blaha (1984).

From Fig.7 it can be seen that the Ti-d orbitals of e_g symmetry can form pd_σ bonds with the neighbouring C-p orbitals (Δ_1 state). If the Ti-d orbitals have t_{2g} symmetry, either metal-metal dd_σ bonds can form ($\Delta_{2'}$ state which has hardly any admixture from carbon wave functions) or pd_π bonds (Δ_5) are possible.

The fact that the density around carbon is almost spherically symmetric is caused by the cubic symmetry, i.e. the p_x, p_y, and p_z orbitals are equally occupied and the sum has a spherically symmetric electron density.

For the two types of p-d interactions, pd_σ and pd_π the weights (charge densities) of carbon and titanium are of comparable magnitude and the wave functions overlap significantly. Furthermore, the C-p and Ti-d partial DOS (Fig.4) are both large in the energy range of the 'p-band'. Therefore all essential criteria for a strong covalent interaction (comparable orbital energies, strong overlap and similar size of the relevant orbitals) are fulfilled, so that we can interpret the interaction to be an efficient covalent bonding between C and Ti.

In simple molecular orbital (MO) theory, two orbitals of appropriate symmetry can form bonding and antibonding MO's. This picture can also be used in connection with energy band calculations as is shown in Fig.8. In TiC the two Δ_5 states correspond to such a bonding and antibonding pair, for which the respective wave functions can be written (in simplified form) as

$$c_a \, \chi_a \pm c_b \, \chi_b,$$

where χ_a represents a Ti-3d orbital and χ_b a C-2p orbital; c_a and c_b are linear combination coefficients; the plus sign corresponds to the bonding the minus to the antibonding MO. Since c_i^2 is related to the charge on center i, we can estimate the coefficients in the representative MO from the charges around Ti and C. This is done in the lower part of Fig.8 and we see that the coefficients c_{Ti} and c_C are balanced indicating a relatively strong covalency.

In TiO there are also two Δ_5 states, but the situation is different. Since the O-2p orbital energy is lower than that of C-2p, the energy separation between the X-p and the Ti-d state is greater in TiO than it is in TiC. Consequently, in TiO the bonding MO, which is closer to O-p orbital energy, has predominantly O-p character, whereas the antibonding state near the Ti-d energy has a large coefficient for the Ti-d wave functions as is apparent from Fig.8. Such a change in weights between bonding and antibonding states is also present in the case of TiC, but to a much lesser extent.

Since the wave functions change sign between Ti and C for the antibonding states, there is a node which causes the density to vanish, but otherwise the antibonding character is not apparent from a density plot.

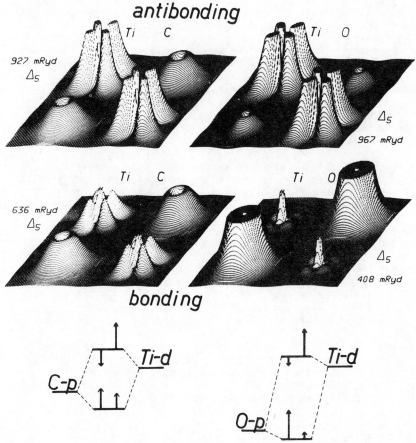

Fig.8 Charge densities corresponding to two states of Δ_5 symmetry of TiC (left) and TiO (right); at the bottom the schematic MO diagram is shown in which the spectral weight of the wave functions is indicated by arrows pointing up or down corresponding to the sign of the coefficients; taken from Schwarz and Blaha (1984).

Binding Mechanism

In light of all the discussions given above, we can summarize the binding mechanism of the titanium compounds in the following way:
i) there is always a non-vanishing density of states at the Fermi-energy and an appreciable electron density throughout the unit cell, so that metallic conductivity is guaranteed;
ii) the charge transfer from Ti to the non-metal leads to electrostatic interactions and adds an ionic component to the binding;
iii) there are several covalent interactions which can be classified into bonding and antibonding parts. The most important covalent inter-actions are between Ti-d and nonmetal-p orbitals, but metal-metal interactions also occur.

The covalent component is strongest in the carbide, weaker in the nitride, but much reduced in the related oxide. Although in this series the reduction in covalency is accompanied by an increase in ionicity, the charge transfer remains about constant or even slightly decreases. This observation can be understood from the partial DOS (Fig.4). Going from TiC to TiO the Ti-d component decreases in the 'p-band'. However, this reduction is compensated, because in TiO the two extra valence electrons (provided by oxygen) occupy additional states of predominantly Ti-d character and the corresponding wave functions are spatially localized near the Ti site. This mechanism of compensation leads to an almost constant charge transfer in this series.

If we take TiC as an example, the hardness or the high melting point can be qualitatively understood by observing that the Fermi-energy falls in a pronounced minimum of the density of states, and all bonding states are occupied, whereas the antibonding states remain unoccupied. We have found metal-metal and Ti-C interactions, where in particular the latter have a strong covalent component, since the orbitals involved are close in energy and their radial wavefunctions are comparable in size and thus overlap, so that the usual criteria for strong covalent bonding are fulfilled. The observed charge transfer could add an ionic stabilization through electrostatic interactions. In pure metals it is found that the stability is highest for about half-filled d shells. We can argue analogically in TiC the complex of the 'p-' and 'd-bands' is about half-filled too, so that also the metallic bonds favour the stability.

As we move towards TiN or TiO, however, the covalent interactions decrease and more antibonding states must be occupied. Consequently the stability is reduced and such systems progressively avoid these antibonding states by vacancy formation. TiO is a typical example, because about 15% vacancies are found on both sublattices.

COMPARISON WITH RELATED EXPERIMENTAL DATA

Band structure calculations are often the basis for an interpretation of experimental data. Three types of such experiments, which are closely related to band structure results, are illustrated below and provide a good test for the quality of electronic structure calculations described above.

Photoelectron Spectra (XPS or UPS)

Photoelectron spectra induced either by x-rays (XPS), by UV (UPS) or by synchrotron radiation reflect the main features of the DOS of the occupied states. The comparison of the photoelectron spectra with the calculated DOS reveals the origin of different peaks in the experimental spectrum and yields information on the relative position of the valence bands.

As an example we show in Fig.9 the photoelectron spectra of TiN using synchrotron radiation as measured by Johansson, Stefan. Shek and Christensen (1980). At the bottom of Fig.9, the partial DOS of TiN (see Fig.4) are included which reproduce the experimental spectra well. The intensities are influenced by the transition probabilities and depend on the photon energy, an aspect not discussed here.

The peak between the Fermi level and the minimum at about 2.5 eV is due to transitions from the 'Ti-d band'; the 'p-band' gives rise to the structure showing a peak at 5.3 eV and a shoulder at 7 eV. The peak at 16.4 eV in the experimental spectrum originates from the 'N-s band'. For further discussions see Neckel (1983).

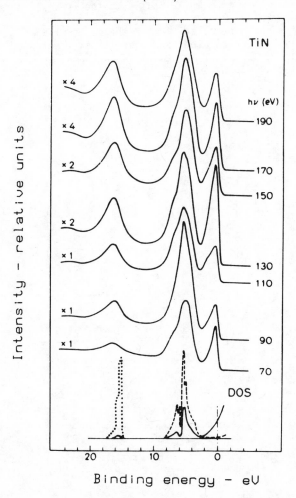

Fig.9 Experimental photoelectron spectra of TiN (Johansson, Stefan, Shek, Christensen 1980); photon energies are indicated on the right side; at the bottom the calculated partial DOS are included (see Fig.4).

Valence Band X-ray Emission Spectra

Valence band x-ray emission spectra (XES) are caused by transitions of electrons from occupied valence states into a core hole, leading to the emission of x-ray radiation. The intensities of such spectra are related to the local partial DOS and are governed by a $\Delta\ell = +1$ selection rule; thus, e.g. a C-K XES (transition to C-1s core hole) is determined by the p-like DOS corresponding to the carbon sphere; similarly, the $Zr-L_{III}$ or $Zr-N_{III}$ spectrum reflectes the Zr-d (and Zr-s) DOS weighted by appropriate transition probabilities to the core states $Zr-2p_{3/2}$ and $Zr-3p_{3/2}$, respectively. The $Zr-M_V$ spectrum originates from the small components in the valence states of Zr-p and Zr-f character and corresponds to transitions from these states to the $Zr-3d_{5/2}$ core hole.

The intensities of XES can be calculated on the basis of band structure results. A comparison of calculated and experimental spectra of ZrC and ZrN is presented in Fig.10. Good agreement in peak position and relative intensities is found.

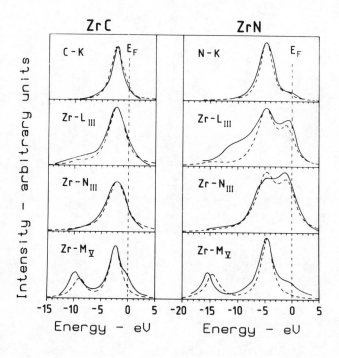

Fig.10 X-ray emission spectra of ZrC and ZrN; theoretical spectra (dashed lines); experimental spectra (full lines) taken from Schwarz, Ripplinger and Neckel (1982)

Optical Properties

The optical properties of metallic compounds can be described by the complex dielectric function, whose imaginary part ε_2 is determined by two contributions:
i) the intraband transitions correspond to excitations of free electrons and can be described by the classical Drude picture; they occur at low photon energies;
ii) the interband transitions from occupied to unoccupied energy bands are directly related to the band structure.

Results of energy band calculations can be used to calculate the contributions of the interband transitions to ε_2. In Fig.11 the results of such a calculation for TiN are compared with experimental data derived from reflectivity measurements. The onset of the ε_2 curve at about 2.5 eV is due to transitions from the top of the 'p-band' to the Ti-d states just above the Fermi-level. The peak in ε_2 at about 6 eV corresponds to the peak in the DOS of the 'p-band' (Fig.4) which is about 6 eV below E_F.

Since the interband transitions start to be important above 3 eV (an energy almost outside the range of visible light), the color of TiN is determined by the intraband transitions which sharply decrease beyond the plasma frequency.

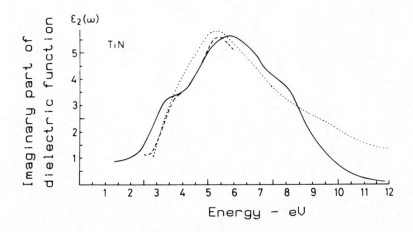

Fig.11 Interband transitions contributing to the imaginary part of the dielectric function $\varepsilon_2(\omega)$; comparison between the experimental (dotted and dashed curves) and theoretical (full curve) data taken from Eibler, Dorrer and Neckel (1983).

REFERENCES

Blaha P. and Schwarz K. 1983, Int. J. Quantum Chem. 23 1535-1552
Blaha P., Redinger J. and Schwarz K. 1985, Phys. Rev. B 31 2316-2325
Dunand A., Flack H.D. and Yvon K. 1985, Phys.Rev. B 31 2299-2315
Eibler R., Dorrer M., and Neckel A. 1983, J. Phys. C: Solid
    State Phys. 16 3137-3148
Johansson L.I., Stefan P.M., Shek M.L., and Christensen A.N.
    1980, Phys. Rev. B 22 1032
Neckel A., Schwarz K., Eibler R., Weinberger P. und Rastl P.
    1975, Ber. Bunsenges. Phys. Chem. 79 1053-63
Neckel A., Rastl P., Eibler R., Weinberger P. and Schwarz K.
    1976, J. Phys. C: Solid State Phys. 9 579-592
Neckel A. 1983, Int. J. Quantum Chem. 23 1317-1353
Schwarz K., Ripplinger H. and Neckel A. 1982,
    Z.Physik B 48 79-87
Schwarz K. 1977, J. Phys. C: Solid State Phys. 10 195-210
Schwarz K. and Blaha P. 1984, Local Density Approximations
    in Quantum Chemistry and Solid State Physics.
    Eds. J.P.Dahl and J.Avery. Plenum Press, New York,
    pp 605-616
Schwarz K. and Moruzzi V.L. 1984, (unpublished)
Williams A.R., Kübler J. and Gelatt C.D.Jr. 1979,
    Phys. Rev. B 19 6094-6118

Inst. Phys. Conf. Ser. No. 75: Chapter 1
Paper presented at 2nd Int. Conf. Science Hard Mater., Rhodes

61

The strength of interatomic forces

G GRIMVALL AND M THIESSEN

Department of Theoretical Physics, The Royal Institute
of Technology, S-100 44 Stockholm, Sweden

ABSTRACT

The frequency spectrum of atomic vibrations in a solid depends on the
atomic masses and the interatomic forces. In the logarithmic average,
$<\ln\omega>$, of the vibrational frequencies ω, the masses may be separated
from the forces. Since $<\ln\omega>$ is also directly related to the vibrational
entropy, one can obtain an average effective force constant k_e of a
compound from thermodynamic data. We introduce this scheme and apply it to
elements, transition metal carbides, nitrides and borides and alkali
halides. A comparison is made between k_e and the microhardness of some
metal carbides and nitrides.

INTRODUCTION

The hardness of materials depends on deformation mechanisms but also on the
strength of the interatomic forces. It is of interest to have a simple
one-parameter measure of the intrinsic magnitude of these forces, referring
to a material without defects. The parameter should be uniquely related to
that fundamental interaction between the atoms which results from a quantum
mechanical treatment of the system. A quantity which meets some of these
requirements is the cohesive energy, U_{coh}. For, say, a diatomic ionic
compound A^+B^-, U_{coh} is the energy of separate and neutral atoms A and B
minus the energy of the compound AB. The heat of formation, ΔH_f, measures
the energy of (a mole of) A and B in their standard states minus the energy
of the compound AB. Another similar quantity is the binding energy U_b. For
the ionic compound AB, U_b measures the energy of separated ions A^+ and B^-
minus the energy of AB.

In spite of ambiguities in the identification of an energy which is suitable
to relate to hardness, it is certainly true that hard materials tend to have
a large ΔH_f as well as a large U_{coh} and U_b. Plendle and Gielisse (1962)
noted that in technical hardness testing, hardness can be defined as a
pressure, i.e. with the physical dimension (force)/(area) or (energy)/
(volume). This lead them to consider U_b/V where V is the molar volume. The
quantity U_b/V was empirically shown to correlate well with Mohs' hardness
scale for a large number of non-metallic compounds. A similar approach,
using ΔH_f, has been used by Gilman (1970) to account for the hardness of
carbides.

The introduction of the molar volume to form U_b/V was done <u>ad hoc</u>, in order to get a quantity with the desired physical dimension. However, the atomic size is only indirectly related to hardness. It would be better first to define a parameter with the dimension of a force constant, i.e. (force)/(length) and then to rescale it according to the spacing between the interatomic bonds in the material. That is the line we shall follow.

A SIMPLE EXAMPLE

The key idea of this paper is to derive an average effective force constant from the measured vibrational entropy at intermediate or high temperatures. As an illustration, we first consider a monatomic solid in which the lattice vibrations are described by independent isotropic harmonic oscillators (i.e. an Einstein model). Let k be the force constant and M be the atomic mass. The vibrational frequency is given by

$$\omega_E = \sqrt{k/M} \tag{1}$$

The vibrational entropy of a solid with N atoms is

$$S(T) = 3Nk_B \left\{ \frac{xe^x}{e^x - 1} - \ln(e^x - 1) \right\} \tag{2}$$

where $x = \hbar\omega/(k_B T)$. The high temperature expansion of (2) yields

$$S(T) \simeq 3Nk_B \left[1 + \ln(k_B T/\hbar\omega_E) \right] \tag{3}$$

If this simple model were adequate, we could use heat capacity data $C_p(T)$ to obtain the entropy

$$S(T) = \int_0^T C_p(T') \, (dT'/T') \tag{4}$$

From S and (2) or (3) we get ω_E and then, by (1), the force constant k. We shall now generalise this idea to allow for an arbitrary vibrational spectrum in a solid with several different atomic masses.

A GENERAL EXPRESSION IN LATTICE DYNAMICS

In a general case of a vibrating lattice, the high temperature expansion of the entropy, (3), becomes

$$S(T) \simeq 3Nk_B \left[1 + \ln(k_B T/\hbar) - <\ln\omega> \right] \tag{5}$$

where $<\ln\omega>$ is an average over all vibrational modes. Usually, the force constants and the atomic masses are not separable in the frequency ω of a particular mode. However, such a separation occurs in the average $<\ln\omega>$. One can write (cf. equation (1))

$$<\ln\omega> = (1/2) \left[\ln(\det\underline{k}) - \ln M_e \right] \tag{6}$$

Here $\det\underline{k}$ is the determinant of the force constant matrix $\underline{k}$, which contains elements measuring all the interatomic forces in the solid. M_e is an effective mass;

$$M_e = (M_1)^{c_1} (M_2)^{c_2} \ldots (M_j)^{c_j} \qquad 7$$

where M_1, M_2, ..., M_j are the masses of the j different atoms and c_j are the concentrations ($^j\Sigma\ c_j = 1$). We now define an 'entropy Debye temperature' Θ_S by

$$\ln(k_B\Theta_S) = <\ln\hbar\omega> \qquad 8$$

and an effective force constant k_e by (cf. eq. (1))

$$k_B\Theta_S = \hbar \sqrt{k_e/M_e} \qquad 9$$

A theoretical justification for the steps leading to eq. (9) is given by Grimvall and Rosén (1983).

A few comments are in order. The concept of an entropy Debye temperature Θ_S does not mean that we have resorted to the approximation of a Debye model. It is just a convenient way to express $<\ln\omega>$ through eq. (8). In the case of a strict Debye model for the lattice vibrations, Θ_S equals the usual parameter Θ_D that characterises the Debye model. The interatomic forces which enter the matrix $\underline{k}$ may be of any range. They are not restricted to be of a central pair-wise form, but may include, e.g., torsional forces.

EMPIRICAL FORCE CONSTANTS FROM ENTROPY DATA

We have used the recommended entropy data $S_{exp}(T)$ from Hultgren and co-workers (1973a,b) and from JANAF Thermochemical Tables (1971) to obtain Θ_S and then, by eq. (9), the effective force constant k_e. More specifically, we put $S_{exp}(T)$ equal to the entropy $S_D(\Theta_S/T)$ of a Debye model. Since the true vibrational spectrum is not of the Debye form, the solution to

$$S_{exp}(T) = S_D(\Theta_S/T) \qquad 10$$

yields a Θ_S which varies with the temperature T. However, $\Theta_S(T)$ approaches our desired Θ_S when $T \gtrsim \Theta_S/2$. To get a well defined value of Θ_S to be used in eq. (9) we have taken $T \simeq \Theta_S(T)$ in eq. (10). This means an iterative process, but in practice $\Theta_S(T)$ varies so slowly with T at these temperatures that no iteration is needed. For typical curves of Θ_S versus T, see Rosén and Grimvall (1983). At $T > \Theta_S$, anharmonic effects cause Θ_S to decrease with T. We note that the effective force constant usually is evaluated from entropy data referring to ambient or somewhat higher temperatures.

EFFECTIVE FORCE CONSTANTS IN SOME ELEMENTS

We expect k_e to be qualitatively correlated to the cohesive energy U_{coh}. Fig. 1 gives a plot of k_e versus U_{coh} for several elements. For the so called simple metals, i.e. non-transition metals, the cohesive properties are largely determined by the electron density. An analogous correlation is expected also for the effective force constant k_e. Fig. 2 shows k_e versus r_s. The parameter r_s is a measure of the electron density;

$$Z(4\pi/3)r_s^3 = V/N = \Omega_a \qquad 11$$

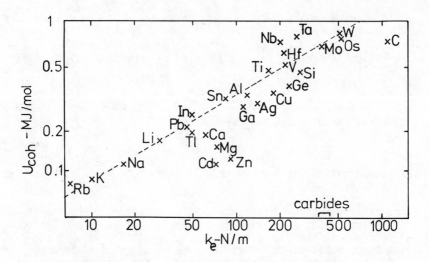

Fig. 1 The effective force constant k_e introduced in this paper, plotted versus the cohesive energy U_{coh}. The straight dashed line is a guide to the eye. U_{coh} is from Gschneidner (1965). The typical range of k_e for group IVB and VB transition metal carbides is also marked.

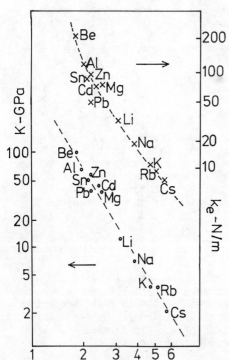

Fig. 2 The effective force constant k_e (right scale) and the bulk modulus K (left scale) of non-transition metals, versus the electron density expressed through the parameter r_s/a_o. r_s is defined in eq. (11) and a_o is the Bohr radius.

Here Ω_a is the atomic volume and Z is the number of conduction electrons per atom (e.g. Z = 1 for alkali metals, Z = 2 for Zn and Mg, Z = 3 for Al and Z = 4 for Pb). Thus r_s can be considered to be the radius of a sphere which contains on the average one conduction electron. For a comparison we also show in Fig. 2 the bulk modulus K versus r_s.

FORCE CONSTANTS IN COMPOUNDS

In the case of a pure element, the separation of the atomic mass M from the vibrational frequencies ω is trivial since all $\omega \sim 1/\sqrt{M}$. However, our method of analysis is of particular interest in polyatomic solids, especially when there are large mass differences between the constituents of the compound, such as is the case for carbides and borides. Then, $<\ln\omega>$ is the only combination of (all) the phonon frequencies which allows a unique separation of the interatomic forces and the masses. (But cf. the elastic limit discussed later.) Table 1 gives some examples of k_e for compounds with large mass differences. For comparison we also include k_e for alkali halides with NaCl structure. We have denoted the compounds by their stoichiometric formula even when there are deviations from that composition. We can estimate the effect of non-stoichiometry on k_e as follows. Entropy data by Storms (1967) give a decrease in Θ_S by about 10% when going from $NbC_{0.98}$ to $NbC_{0.87}$. The change in M_e is small. Hence the effective force constant k_e per atom has been reduced by about 20% when the number of carbon atoms is reduced by 11%, i.e. a reasonable result. See also Grimvall and Rosén (1983) for some comments on the calculation of Θ_S.

TABLE 1 The Effective Force Constant k_e

Material	k_e Nm^{-1}	Material	k_e Nm^{-1}
TiC	445	TiB_2	479
ZrC	398	ZrB_2	471
HfC	445	HfB_2	427
VC	391	LiF	113
NbC	382	NaF	104
TaC	403	NaCl	60
		KCl	52
TiN	372	KI	39
ZrN	345		
VN	303		

ELASTIC LIMIT FORCE CONSTANTS

Above, it was remarked that $<\ln\omega>$ is the only average over ω for which the atomic masses in polyatomic compounds separate from the force constants. However, in the limit of small ω the material behaves as an elastic continuum, with $\omega = c(q,\lambda)|q|$ where q is the wave vector, λ is a mode index (longitudinal or transverse modes) and $c(q,\lambda)$ is a sound velocity. One has $c = f(c_{ij})/\sqrt{\bar{M}}$ where $\bar{M}$ is the average atomic mass in a crystallographic unit cell and $f(c_{ij})$ is a function of the elastic constants c_{ij}. (For instance, in a cubic crystal structure only c_{11}, c_{12} and c_{44} enter.) The elastic limit Debye temperature Θ_E, which gives the low temperature limit of the vibrational heat capacity, averages over $[c(\underset{\sim}{q},\lambda)]^{-3}$. We can write

$$k_B\Theta_E = \hbar\sqrt{k_E/\overline{M}} \qquad\qquad 12$$

where k_E is an effective force constant. In a strict Debye model for an elemental solid with atomic mass M, $\Theta_S = \Theta_E = \Theta_D$ and $M_e = \overline{M} = M$. Then, $k_e = k_E$.

EFFECTIVE FORCE CONSTANTS AND THE HARDNESS OF CARBIDES AND NITRIDES

Group IVB and VB transition metal carbides and nitrides form a class of materials with unusual physical properties. They have the hardness and high melting temperature characteristic of covalent bonds, but some of them also show the metallic feature of superconductivity, with a high critical temperature. The acoustic phonon branches of the superconductors have pronounced anomalies. Roedhammer and co-workers (1976) measured the heat capacity of TiC, ZrC, NbC and TaC at low temperatures (2 < T < 100 K). From the corresponding Debye temperature Θ_E, and eq. (12), we get k_E = 436 N/m (TiC), 362 N/m (ZrC), 370 N/m (NbC) and 590 N/m (TaC) respectively.

It is difficult to assign reproducible measured hardness values to real materials, and hardness is certainly not in a unique relation to our effective force constants, so we can only expect a qualitative correlation between hardness data and k_e. Before a comparison is made, we should also make a rescaling of k_e to account for the difference in the spacing between the bonds in different materials. In analogy to the approach by Plendle and Gielisse (1962) and Gilman (1970) we divide k_e by some characteristic length in the lattice and get a quantity with the same physical dimension as microhardness.

TiC, ZrC, HfC, VC, NbC, TaC, TiN, ZrN and VN all have the NaCl structure. In Fig. 3 we plot k_e/a for these compounds (a is the lattice parameter) versus the microhardness. The uncertainty in k_e is a few times larger than the uncertainty in the entropy from which k_e is derived, implying an uncertainty in k_e of about 5%. Different sources of hardness data often give quite different values, even when the method of measurement is the same. To be unbiased in our selection of references, we have chosen to rely on a single source and take the hardness values recommended by Toth (1971). Fig. 3 shows a clear correlation between hardness and k_e/a. An analogous plot of hardness versus k_E/a shows no correlation. (Note the high value of k_E for TaC).

In this context we note that Nørlund Christensen and co-workers (1983) have fitted models involving a large number of force constants to phonon data for several carbides and nitrides. Those force constants define the force constant matrix in eq. (6). It is a certain average of these force constants that enters our parameter k_e.

CONCLUSIONS

From the vibrational entropy of a solid at intermediate temperatures (T = Θ_D), we have derived an average effective measure, k_e, of the strength of the interatomic forces. It has been shown that k_e may provide a useful parameter to compare with the intrinsic hardness of materials.

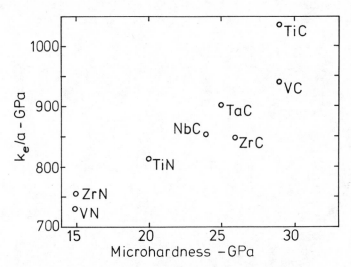

Fig. 3 The effective force constants k_e (related to the entropy), rescaled by the lattice parameter a to form k_e/a and plotted versus the microhardness.

ACKNOWLEDGMENT

This work has been supported in part by the Swedish Natural Science Research Council. One of us (G.G.) gratefully acknowledges discussions with Torsten Lundström.

REFERENCES

Gilman, J.J. 1970, J. Appl. Phys. 41, 1664-6.
Grimvall, G. and Rosén, J. 1983, Int. J. Thermophys. 4, 139-47.
Gschneidner, K.A., Jr. 1965, in Solid State Physics vol. 16, eds. F. Seitz and D. Turnbull, Academic Press, New York. p. 275.
Nørlund Christensen, A., Kress, W., Miura, M. and Lehner, N. 1983, Phys. Rev. B 28, 977-81.
Hultgren, R., Desai, P.D., Hawkins, D.T., Gleiser, M., Kelley, K.K. and Wagman, D.D. 1973a, Selected Values of the Thermodynamic Properties of the Elements. American Society for Metals, Cleveland.
Hultgren, R., Desai, P.D., Hawkins, D.T., Gleiser, M. and Kelley, K.K. 1973b, Selected Values of the Thermodynamic Properties of Binary Alloys. American Society for Metals, Cleveland.
JANAF Thermochemical Tables, 2nd ed. 1971. National Bureau of Standards, Washington D.C. Supplement 1975, J. Phys. Chem. Ref. Data 4, 118.
Plendle, J.N. and Gielisse, P.J. 1962, Phys. Rev. 125, 828-32.
Roedhammer, P., Weber, W., Gmelin, E. and Rieder, K.H. 1976, J. Chem. Phys. 64, 581-5.
Rosén, J. and Grimvall, G. 1983, Phys. Rev. B 27, 7199-208.
Storms, E.K. 1967, The Refractory Carbides (Refractory Materials, vol. 2), Academic Press, New York.
Toth, L.E. 1971, Transition Metal Carbides and Nitrides (Refractory Materials, vol. 7). Academic Press, New York.

Inst. Phys. Conf. Ser. No. 75: Chapter 1
Paper presented at 2nd Int. Conf. Science Hard Mater., Rhodes

69

Molecular orbital studies of refractory metal carbides

J E LOWTHER

Physics Dept, University Witwatersrand, Johannesburg, South Africa

ABSTRACT

Molecular Orbital Theory has been employed to compute local atomic charge configurations, total energies for a given lattice geometry and thus lattice constants and elastic moduli. The method has also been applied to surface bonding of WC with some transition metals and investigate energetics of bond breaking taking place during the process of crack propagation in the sintered material.

INTRODUCTION

Despite their considerable technological importance a detailed understanding of the refractory metal carbides is still lacking. Disagreement as to whether these materials are ionic or covalent in character was discussed by Ramqvist (1969) but it was Toth (1971) who debated that the strong cohesive nature of the carbides probably arises because of a covalent metal-carbon bonding. This suggestion led to a quantum chemical approach being applied to the problem (Ivanovsk, Gubanov and Shveikin (1981),Gabanov, Zurmaev and Ellis (1981) with the Iterative Extended Huckel (IEH) method being used to examine the role of vacancies in materials like TiC. The IEH method has also been applied to study surface adsorption on tungsten (Anders, Hansen and Bartell, 1973) where the deficiency of the method to calculate total energies was partly overcome by including an empirical replusive term into the expression for the total energy. More extensive molecular orbital schemes are available and, in particular, the Complete Neglect of Differential Overlap (CNDO) method which given a set of parameters derived from atomic or molecular data, computes one electron energies and a total energy for a given lattice geometry. The approach has been widely used in treating molecular configurations (Baetzhold (1971), Clack Hush and Yandle (1971)) as well as solid state problems (Mainwood and Stoneham (1976)). Recently the method has been applied to study surface adsorption of carbon monoxide on tungsten (Kobayashi, Yamaguchi and Yoshida (1982)).

With a view to obtaining a microscopic understanding of the various electronic and cohesive processes in the hard metal carbides the CNDO method has been applied to a representative cluster of atoms. All interactions; - metal-metal, metal-carbon and carbon-carbon have been included and the model also applied to study the surface bonding between WC and some transition metals. In light of some results of the calculations we briefly discuss bonding aspects of the cracking process in the sintered WC - Co material.

TiC AND WC

These materials have different structures in that TiC is cubic whereas WC
is a hexagonal layered structure. To specify our choice of cluster we view
these structures as indicated in Fig.1

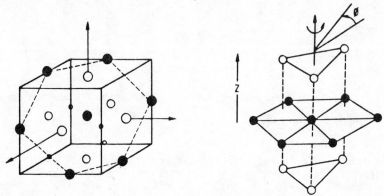

Fig.1 Cubic structure of TiC and hexagonal structure of WC. Viewed
along the (111) cubic axis the structure of TiC corresponds to using
ϕ = 60 in the figure to the right which is referred to as the
staggered configuration. ϕ = o is the eclipsed configuration.

The CNDO parameters were atomic values taken from Pople and Beveridge
(1970), Clack Huch and Yandle (1971) and Kobayashi, Yamaguchi and Yoshida
(1982).

To study the cohesive aspects of the TiC and WC clusters we have computed
CNDO total energies as the metal atoms are displaced away from the plane
of C atoms and also for relative rotations of the metal planes, the
results being as follows:-

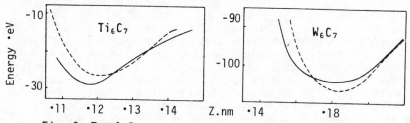

Fig. 2 Total Energy as separation of metal atoms is increased:———
staggered; --- eclipsed configuration.

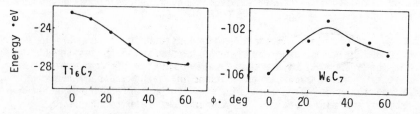

Fig. 3 Total energy as angle ϕ, shown in Fig. 1 is varied.

As is evident from the Figures above, the cluster and CNDO
parameterization, are capable of correctly predicting the correct relative
orientation of the metal atoms. Furthermore the slope of the total energy
curve at the minimum value affords an estimate of the elastic modulus.
These values together with the computed charge configurations are given in
Table 1.

TABLE 1   Calculated Elastic Moduli and Charge Configurations

	Calculated	Experiment	Metal	Carbon
TiC	$9.7 \times 10^5 Nmm^{-2}$	$2.4 \times 10^5 Nmm^{-2}$	$d^{2 \cdot 4} s^{\cdot 24} p^{\cdot 80}$	$s^{1 \cdot 7} p^{2 \cdot 8}$
WC	$19.1 \times 10^5 Nmm^{-2}$	$5.3 \times 10^5 Nmm^{-2}$	$d^{4 \cdot 8} s^{\cdot 27} p^{\cdot 76}$	$s^{1 \cdot 6} p^{2 \cdot 5}$

neutral atom configurations are Ti; $3d^2 4s^2$:W:$5d^4 6s^2$:C;$2s^2 2p^2$

BONDING OF WC WITH TRANSITION METALS

There are several positions at which an additional atom could be adsorbed
on the surface of WC. The experimental results reported by Ramqvist
indicated the carbon surface to be the most likely and for the present
case we have considered various positions on this surface as shown in
Fig. 4

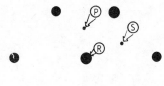

Fig.4 Different positions
considered for adsorbtion of a
single Co atom; it was found that
Site R was much less likely than
Sites P or S.

We therefore consider the electronic structure of a cluster $W_3 C_7 M_3$ where M
is the transition metal. Following the approach in the previous section we
thus calculate the total energy of the cluster as the transition metals
are moved away from the carbon plane and also as the angle ϕ, which was
depicted in Fig. 1   is varied. These results together with the charge
configuration of the metal, at the lowest energy, are shown in Fig.5 .

The net charges on the W and C atoms also are changed slightly from the
values appropriate to the cluster $W_6 C_7$ and these values are shown in
Fig.6 .

The results of Fig.6   are rather striking when we reflect on the
importance of the sintered material WC-Co. We see that in the Fe-Co region
of the transition metal series that the matching of Co and W potentials is
very close, with a small ionic charge developing across the interface
being indicative of a strong covalent bonding. But accompanying this
rather good match the CNDO results indicate an abrupt change in the
orientation of Co atoms which seem to prefer an almost staggered
configuration rather than the eclipsed configuration as would be the case
for a plane of W atoms.

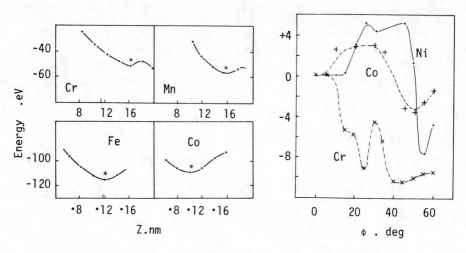

Fig.5 Total energy as the transition metal (M) is moved toward the carbon plane and the metal atoms are rotated.

TABLE 2   Electronic Charge Configurations

Metal (M)	Assumed atomic configuration	Configuration in $W_3 C_7 M_3$
W	$d^4 s^2$	$d^{4.83} s^{0.27} p^{0.76}$
Cr	$d^4 s^2$	$d^{4.42} s^{0.32} p^{0.92}$
Mn	$d^5 s^2$	$d^{5.05} s^{0.51} p^{1.16}$
Fe	$d^6 s^2$	$d^{6.19} s^{0.38} p^{1.40}$
Co	$d^7 s^2$	$d^{7.17} s^{0.42} p^{1.49}$
Ni	$d^8 s^2$	$d^{8.00} s^{0.45} p^{1.70}$
Cu	$d^{10} s^1$	$d^{9.00} s^{0.50} p^{1.90}$

Bearing in mind that cobalt exists in both hexagonal and cubic forms - the interchange of forms being an important contributing factor to the usefulness of WC-Co as a hard material (Lui and coworkers, 1983) - we have extended the size of the cluster to investigate the relative orientation of another plane of Co atoms. Results for the cluster $W_3 C_7 Co_6$ are indicated in Fig.7 .

In this case the calculation has implied that the second plane of Co atoms eclipses the lower plane of W atoms but in turn is at a staggered configuration to the lower plane of Co atoms. This latter orientation would be consistent with that found in the f.c.c. form of cobalt.

BOUNDARY STRUCTURE AND CRACK PROPAGATION

The mechanism of crack propagation in strained WC-Co carbides is not really well understood, although generally viewed as involving some form of process along matching slip systems in the WC and Co.

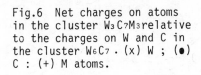

Fig.6 Net charges on atoms in the cluster $W_3C_7M_3$ relative to the charges on W and C in the cluster W_6C_7. (x) W ; (●) C : (+) M atoms.

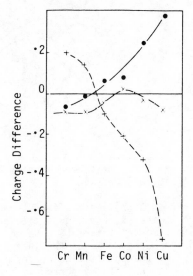

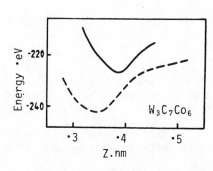

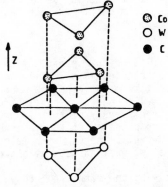

Fig.7 Total energy as second plane of atoms is displaced from carbon surface and the configuration of atoms in cluster $W_3C_7Co_6$ staggered; eclipsed configuration

The slip systems are defined relative to the translational symmetry of either WC or Co and really correspond to directions in which the strength of bonding between atoms is low. It is on this latter interpretation that we have employed our CNDO model to consider the energy as we break different bonds in the representative cluster. This is, of course, a very idealistic way of visualising the process of cracking - but essentially leads to some insight of the relative contributions from different atoms in the bonding process. The present results are summarised in Fig.8 .

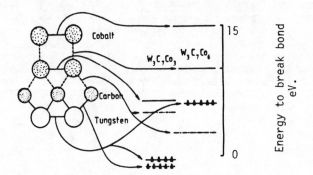

Fig.8 Relative energies as different bonds are broken in the two clusters we have used.

These results reflect different variations in bonding strength taking place at the surface and particularly point to two conclusions in relation to the interface bonding. Firstly the energy to break the various bonds decreases towards the interface and, secondly, that this energy is larger on the cobalt side of the interface. The implication of this result would be that, for a perfectly sintered surface in which no broken bonds present, that cracking occurs near the interface possibly involving the breaking of a metal-carbon bond.

CONCLUSION

The results of our investigation have led us to conclude in the case of WC, that bonding is primarily covalent with the constituent atoms having an almost zero net electronic charge. However even though almost electrically neutral, the electronic atomic configuration of the metal strongly differs from that in the carbide with more electrons occupying the d and p shells at the expense of a reduction in occupancy of the metal s shell. The CNDO parameterization we have adopted also reasonably computes several structural and cohesive properties of these materials.

In applying the method to study bonding with transition metal atoms we found that hardly any charge polarisation developed across the interface with WC for those transition metals coming from near the middle of the series. This we again interpret as being indicative of strong covalence effects. There does not seem to be any simple reason for this though since the covalent bond formed involves occupancy of the d, s and p shells of the transition metal atom and is thus a relatively complex hybrid. Furthermore we found that carbon plays a more important role in this process than does tungsten.

One particular interesting feature which came from our model relates to the surface topology at the interface between WC : Co. Despite the close similarity in geometry across the boundary an angular shift in the Co atoms relative to the W atoms in the carbide was found. Further results then revealed that the next layer of Co atoms reverts back to a similar configuration as that in the W plane, with the orientation of Co atoms then being as expected for f.c.c. cobalt when viewed along the (111) cubic direction.

Finally we have also indicated how the model affords some insight into the relative strength of bonding between carbide and metal binder. The results, which indicated a lowering of bonding strength to take place at the interface and away from the binder, are consistent with some recent deformation measurements on the WC : Co system by Liu and coworkers (1983). However the fact that fracture proceeds away from the interface points to the fact that fracture paths are primarily associated with the **presence** of defects which have not, of course, been considered here.

In concluding we have seen that even the small cluster of atoms employed in this study gives some insight into electronic and cohesive aspects of carbide and metal binder systems. The success of this, we believe, is simply because the bonding mechanism in these materials is primarily covalent thus involving short range interactions.

ACKNOWLEDGEMENTS

Helpful discussion with S B Luyckx. H E Le Roux and F R L Schöning

REFERENCES

Anders, L.W. 1973, J. Chem. Phys. 59 5277-5284
Baetzhold, R.C. 1971, J. Chem. Phys. 55 4355-4362
Gubanov, V.A., Kurmaev, E.Z., Ellis, D.E. 1981, J. Phys. C. 14 5567-5576
Ivanovsky, A.L., Gubanov, A.V., Shveikin, G.P. 1981, J. Less-Common
    Metals., 1 78-91
Kobayashi, H., Yamaguchi, M., Yoshida, S. 1982, Theor. Chim. Acta, 61
    473-481
Lui Manlang., Huang Xiaoying., Duan Shitian, Shao Daqin., Cui Yumei.,
    Yao Zhenmei. 1983, J. Ref. Hard Metals 3 129-133
Mainwood A., Stoneham A.M. 1976, J. Less-Common Metals 49 271-281
Pople, J.A., Beveridge, D.L. 1970. "Approximate Molecular Orbital Theory"
    McGraw-Hill New York
Ramqvist, L. 1969, Jernkontorest Ann. 152 1-28
Toth, L.E. 1971, "Transition Metal Carbides and Hydrides". Academic Press,
    New York.

Discussion on Atomic bonding

Rapporteur: W S WILLIAMS (1)
Session Chairman: E D HONDROS (2)

(1) University of Illinois, Urbana-Champaign, USA
(2) National Physical Laboratory, Teddington, Middlesex

The session on atomic bonding, which opened the conference, was organized to explore the theoretical basis for a deeper understanding of the fundamentals of hardness. The five papers comprising the session evoked enthusiastic comment and lively discussion.

The lead paper by Professor Cohen on "Ab Initio Calculation of Structural Properties" described a refined pseudopotential approach to calculating the total energy of a solid starting with only the atomic number of the constituent atoms. Of particular interest to the conference was the ability of the method to predict the correct crystal structure and lattice parameter of simple solids by comparing the energies of alternative structures - an achievement of theory unthinkable only a few years ago. Electron charge densities in momentum space and real space can also be calculated accurately. In principle, the method can treat ionic, metallic and covalent solids and hence should be applicable eventually to hard materials exhibiting mixed bonding types.

In answer to a question by Professor Heuer concerning the use of this theory to tell us about the sizes of atoms and ions in compounds, Professor Cohen mentioned two ways of addressing this venerable problem: (1) the pseudopotential itself gives a measure of ion sizes, and tables of these exist; and (2) the charge density contour plots suggest a definition of ionic size from the last closed contour around an ion.

Responding to a challenge by Professor Heuer to explain the large energy difference between the zincblende and wurtzite forms of BeO, Professor Cohen noted that the difference is not very large in energy or structure, being defined by third nearest neighbours. The wurtzite form is generally favoured when the row difference in the periodic table is large, he explained.

Dr Exner asked whether the pseudopotential method could predict the properties of interfaces (e.g. WC/Co; TiC/Ni). Professor Cohen said that so far he has only treated interfaces between semiconductors and metals, vacuum and other semiconductors, but believes that the systems mentioned could, in principle, be considered.

Dr Buckley enquired as to whether the theory has considered the apparent difference in bond strength between the surface and the bulk of a solid. In reply, Professor Cohen said that he has done so and finds a difference theoretically, but that the details depend on the amount of surface reconstruction and hence are specific to specific materials.

Dr Fischmeister noted that we are looking for a modern successor to Pauling's rules to allow us to predict the influence on bonding of non-stoichiometry and of solid solution formation between carbides and nitrides and between different transition metals. We also want to predict the Peierls potential and cohesion across an interface. Professor Cohen accepted the challenge and will try to work such problems into his programme.

Dr Page asked whether the pseudopotential theory could rationalize Russian reports of a pressure-induced "metallic" phase transformation in silicon at tens of GPa. Professor Cohen suggested that the β-Sn transformation might be the one observed, as it happens at about 100 kbar.

Professor Kisly said that there are data indicating that nitrogen and oxygen assume the diamond structure at high pressures and asked whether appropriate calculations had been done. Professor Cohen answered no, but said it would be interesting to see the data.

A paper on the electronic properties of two refractory compounds, WC and TiB_2, was presented by Y Ishizawa and T Tanaka. They prepared single crystals for experimentation by the Co flux method and the floating zone method, respectively. From the de Haas-van Alphen oscillations in these crystals at low temperatures, the authors computed Fermi surfaces for both compounds. They also estimated the number of conduction electrons at the low value of approximately 0.01 electron/unit cell, making these materials semi-metals, like TiC. The relevance of this study to the present conference on hard materials lies in the experimental verification of recent energy band calculations which lead to conclusions about the source of the strong bonding in WC and TiB_2. As for the cubic IVB transition metal carbides, p-d mixing occurs for TiB_2 and produces bonding and anti-bonding states with the Fermi energy near the minimum in the density of states.

Professor Cohen asked whether the authors had measured the temperature dependence of the amplitudes of the de Haas-van Alphen oscillations to evaluate the ratio of the effective masses. The answer was no.

The paper by Professor K Schwarz and A Neckel reported on a theoretical approach to chemical bonding in transition metal compounds specifically. They use the augmented plane wave (APW) method, which employs a muffin tin potential. Results are discussed in terms of s, p, d, f atomic states, and electron energy bands and densities of states are generated. Total energies versus lattice constant are calculated and lead to agreement better than 1% for the equilibrium lattice constant for ZrN - again, a notable success in the field of band calculations. The calculations are mostly on ideal structures though some work has been done on vacancies. Substantial mixing near the Fermi energy occurs between metal d-states and non-metal p-states as found by other workers. Transition metal compounds generally have minima in the densities of states between maxima created by this p-d mixing to form both bonding and anti-bonding states. The authors suggested that vacancies occur in these non-stoichiometric compounds

because anti-bonding p-d states would need to be occupied to fill all
atomic positions. The covalent component for TiC is stronger than for TiO.
The charge density around the Ti site deviates from spherical symmetry for
the oxide, and changes its character from e_g to t_{2g} in the series TiC to
TiO. The charge densities are beautifully displayed in computer-generated
contour plots.

Dr Fischmeister asked whether it is possible to quantify the relative
contributions of ionic, covalent and metallic bonding to the cohesive
energy of transition metal compounds. Dr Schwarz answered that this can
be done in simple systems, but these concepts are not well-defined for
more complex systems, and the results are model-dependent.

Dr Hondros asked whether metal vacancies and impurities affect the ionic
or the covalent contribution more. Dr Schwarz replied with the example of
NbO; their calculation predicts that NbO does not have the NaCl structure
- rather, it reduces the number of covalent bonds by having fewer nearest
neighbours (4 instead of 6) and thereby does not have to fill anti-bonding
p-d states, but replaces them by metal-metal interactions.

Dr Lundstrom enquired about the use of the rigid band approximation for
the series TiC-TiN-TiO. Dr Schwarz explained that the approximation is
dangerous here and compared TiC and ZrC: the bands are similar but their
ordering is reversed at the Fermi energy. Professor Cohen joined in
cautioning against using the rigid band approximation and enthused over
the example given.

The paper by Dr Grimvall and Dr Thiessen defined a high-temperature
effective Debye temperature for real solids derived from experimental
values of the entropy. This quantity sums over all vibrational modes and
includes anharmonic contributions. An effective force constant within the
harmonic approximation is then introduced and related to the internal
cohesive energy, bulk modulus and microhardness.

Dr Schwarz noted that the agreement reported by the authors between the
conventional elastic-limit force constant and the new effective force
constant proposed by the authors is close for NbC but not for TaC. As
both of these solids are B1 superconductors with soft phonon modes, why
should the results be different? Dr Grimvall replied that the larger mass
of Ta pushes the phonon anomaly to lower frequencies, which may affect the
elastic-limit force constant more. He felt that the effective force constant would be the better choice.

Dr Hondros asked what bulk data are required for evaluating the effective
force constant. Dr Grimvall replied that for many materials, the standard
entropy at 298 K is sufficient, but for hard materials (with high Debye
temperatures) high-temperature values are required.

Dr Hondros also remarked that years ago, the Russians had proposed relating hardness to surface free energy. Is there a correlation? Or with
cohesive energy? Dr Grimvall answered that all of these quantities are
correlated but it is difficult to track down the details.

Dr Page observed that since microhardness is not a simple material
property, depending on test conditions, surface, etc., the choice of
hardness data to correlate with the effective force constant would be

important. Dr Grimvall replied that he took all the microhardness values from the book by Louis Toth, <u>Transition Metal Carbides and Nitrides</u>.

Dr Chatfield asked whether the approach described by Dr Grimvall could be used to predict, even roughly, the adhesion or bonding between different materials. Dr Grimvall replied in the negative but noted that as the approach illuminates systematic trends in bonding forces, it may be of some value in analyzing the intrinsic strength (as opposed to defect-controlled values) of bonding at interfaces.

The fifth and last paper of this session on atomic bonding was given by Dr Lowther on molecular orbital calculations on refractory metal carbides. He employed the CNDO (Complete Neglect of Differential Overlap) version of molecular orbital theory, which includes empirical information on mole-cules and atoms. Calculating on the clusters Ti_6C_7 and W_6C_7, he found charge transfer from titanium to carbon but relatively little transfer in the tungsten carbide case; i.e. TiC is more ionic. (The sense of the charge transfer for TiC agrees with other cluster calculations and with all but one band calculation - Rapporteur.) Dr Lowther also used a cluster approach to simulate the interface of WC and a transition metal. With the cluster $W_3C_7M_3$, where M is a transition metal, the CNDO method predicted that optimum bonding would be with elements from the upper part of the transition metal series, because of increasing occupancy of the p shell of the metal ion, the location of the Fermi level and the degree of ionic polarization across the boundary. The polarization goes to zero at about the position of Fe and Co.

Dr Hondros asked whether this result would not imply a tendency for WC/Co to fracture at the interface. Dr Lowther agreed and cited experimental evidence. However, Dr Almond asserted that Auger electron spectroscopy shows that the fracture occurs a few lattice parameters away from the interface, indicating that the WC/Co interface is strong. (Evidently the depopulation of Co 4s states and the filling of Co 4p states in the $W_3C_7Co_3$ cluster compared with the Co atom allows for strong interactions with the W atoms, which, according to Dr Lowther, are in a staggered con-figuration with respect to the Co atoms; a strong bond forms then even in the absence of an ionic interaction across the interface - Rapporteur.)

Dr Fischmeister admired Dr Lowther's attempt to model interfaces but was concerned about the influence of cluster size on the results. Dr Lowther replied that the results for WC should not be very sensitive to cluster size because there is little ionic character to the bond; however, for TiC, there is considerable charge transfer so there might be a dependence on cluster size as the Coulomb interaction is long range. (However, cluster calculations done at the University of Illinois show the same general features for a TiC_5 cluster as for the infinite solid - Rapporteur.)

Dr Fischmeister also worried about overstretching cluster calculations in trying to treat fracture at interfaces in view of the deformation which would occur in both phases. Dr Lowther felt that such deformations could be introduced into the cluster geometry and their impact on bond strengths calculated.

In the general discussion following the five presentations, Dr Hondros called for some application of the atomic bonding ideas presented to

practical consideration of hardness. Dr Gurland reminded the conference
that hardness testing involves plastic deformation, whereas the calculations treat only elastic deformation. There is a relation, based on
models of dislocation behaviour, but it is not so direct and fundamental.
Dr Brookes also noted a lack to this point in the conference of treatment
of extended defects - e.g. dislocations. (In later papers and discussions
at the conference, considerable attention was paid to plastic deformation
and the participation of dislocations - Rapporteur.)

Inst. Phys. Conf. Ser. No. 75: Chapter 2
Paper presented at 2nd Int. Conf. Science Hard Mater., Rhodes

Effects of doping on mechanical properties of semiconductors

P.B. HIRSCH, P. PIROUZ, S.G. ROBERTS AND P.D. WARREN

Department of Metallurgy and Science of Materials, University of Oxford, Parks Road, Oxford, OX1 3PH, U.K.

ABSTRACT

Dislocation velocities in semiconductors depend strongly on the concentration of electrically active impurities, at temperatures in which the intrinsic lattice resistance is rate controlling. The effect is explained in terms of deep acceptor and donor levels in the band gap associated with kinks. Doping affects yield stress, indentation hardness and the pattern of plastic flow and cracking around indentations. New insight has been gained about the nature of the plastic zone under an indenter, and the difference in hardness of (111) and (ĪĪĪ) faces of GaAs has been explained.

INTRODUCTION

In 1966 Patel and Chaudhuri showed that there is a pronounced effect of doping on the dislocation velocity in Ge. Since then similar effects have been observed in Si and III-V compounds (Patel, Testardi and Freeland 1976, Erofeev and Nikitenko 1971, George and Champier 1979, Erofeeva and Ossipyan 1973, Choi, Mihara and Ninomiya 1977). Fig. 1 shows some recent results by Pirouz and Freeland on Ge, using a wide range of doping levels and the single stress pulse dislocation etching technique. These results extend the original work of Patel and Chaudhuri (1966) and show that the dislocation velocity for n-type material is greater and for p-type material smaller than that for intrinsic material. It should be noted from Fig. 1 that around 340°C the velocity of dislocations in n-type is nearly fifty times greater than for p-type Ge. For Si n- and p-type doping tends to increase dislocation velocity, although n-type doping is more effective than p-type (George and Champier 1979). There is strong evidence from the earliest experiments that the effect is due to an electronic mechanism, i.e. dependent only on electron concentration or the position of the Fermi level, and not on the particular doping element used. The activation energy controlling the temperature dependence of the dislocation velocity is found to depend on doping, as is clear from the changes in slope of the various lines in Fig. 1, and for Ge the activation energy for strongly p-type Ge is greater than that of strongly n-type material by about 0.55 ev (Patel and Chaudhuri 1966). This is of the same order, albeit somewhat smaller, as the width of the bandgap in Ge in the temperature range of the experiments (~500°C).

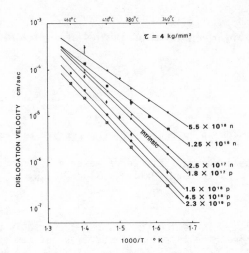

Fig. 1   Log dislocation velocity as a function of reciprocal temperature in Ge specimens doped with As or Ga (Pirouz and Freeland, unpublished).

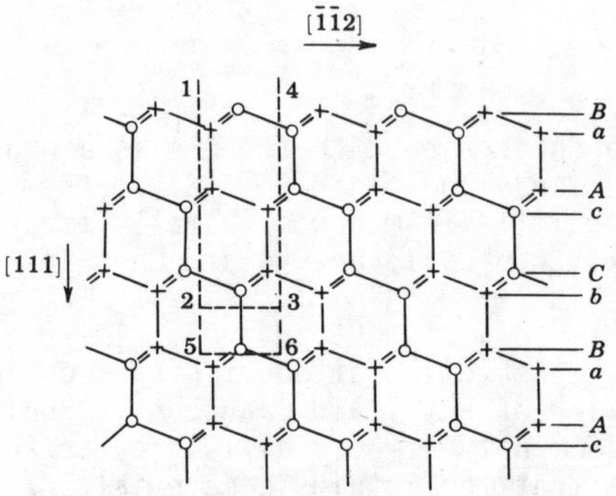

Fig. 2   Diamond cubic lattice projected normal to (110). o represents atoms in the plane of the paper, and + atoms in the plane below (From Hirth and Lothe 1968).

Strong materials such as SiC, diamond, Si_3N_4, and BN have pronounced covalent character, and in principle their mechanical properties should be dependent on small amounts of electrically active impurities. The objectives of the research programme are (1) to assess the potential of the doping effect for such materials, particularly with a view to improving toughness, and (2) to understand the mechanism and its effect on mechanical properties. The work described in this paper is concerned with the second of these objectives, with semiconductors being used as model substances.

DISSOCIATION OF DISLOCATIONS IN SEMICONDUCTORS

In the diamond cubic or zincblende structures the glide dislocations lie on {111} planes and have the ½ 110 Burgers vector. They could lie either between the narrowly spaced (e.g. between planes b and C) or the widely spaced (e.g. between planes B and b) planes making up the {111} plane sandwich (see Fig. 2); these two configurations have been called glide and shuffle respectively(Hirth and Lothe 1982 pages 373-383). If they lie between the narrowly spaced planes, dissociation can occur into two Shockley partials, leading to a lowering in energy. Weak beam electron microscopy has shown that glide dislocations in Si, Ge and diamond (Ray and Cockayne 1971, Ray and Cockayne 1973, Haussermann and Schaumburg 1973, Pirouz and coworkers 1983), and in a number of III-V and II-VI compounds, GaAs, GaP, GaSb, InAs, InP, InSb, CdTe (Gai and Howie 1974, Gomez and Hirsch 1979, Gottschalk, Patzer and Alexander 1978, Hall and Vander Sande 1978) are generally dissociated into partials. A 60° dislocation dissociates into a 90° and a 30° partial; Fig. 3 shows the atomic configuration for such partials, assuming that their cores are not reconstructed. Hornstra (1958) suggested an alternative configuration for the 30° partial, in which it climbs into the neighbouring plane by absorbing a line of either vacancies or interstitials. High resolution lattice images of well characterised 30° partial dislocations in Si and Ge have ruled out the first of the Hornstra configurations, and are consistent with the simple glide configuration (Anstis and coworkers 1981, Bourret, Desseaux and D'Anterroches 1981). Although the possibility of the shuffle configuration derived from interstitial climb has not yet been thoroughly tested, the weight of the evidence suggests that the dislocations are generally in the glide configuration.

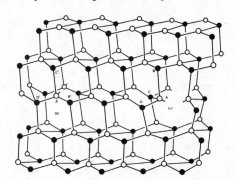

Fig. 3   60° dislocation with Burgers vector ½[1Ī0] on a (111) plane in a III-V compound dissociated into a 90° partial at (a) and a 30° partial at (b).

CORE RECONSTRUCTION

Shockley's (1953) original structure of the core of an undissociated 60° dislocation on the "shuffle set" planes consists of a row of dangling bonds along the core, corresponding to a one dimensional half-filled band (Read 1954) (Fig. 4). But with the glide partials reconstruction by bonding along the core for 30° partials and across the core for 90° partials is possible (Hirsch 1979, Jones 1979). Calculations due to Marklund (1983), Jones (1979), Altmann, Lapiccirella and Lodge (1983) suggest that for Si reconstruction for 30° partials is likely, although the conclusion for 90° partials is more uncertain. Fig. 5a shows the

reconstructed core of a 30° partial running along the length of the
diagram, but transferring from one [110] row to the next at a kink; the
core is viewed normal to the (111) slip plane, but only one set of atoms
above and below the slip plane is shown. Reconstruction of this type
doubles the periodicity along the core, and any particular atom has the
choice of bonding with one of two neighbours in equivalent positions,
leading to different relative displacements. Where the bonding changes
from one set of neighbours to the other along the core, an antiphase
defect (APD) occurs, associated with a dangling bond (Hirsch 1980). This
defect has been called a soliton by Heggie and Jones (1982, 1983). Such

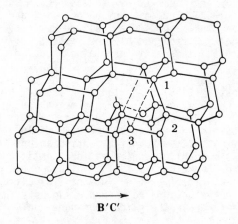

B′C′

Fig. 4   Shockley's (1953) model of 60° dislocation on shuffle plane,
          showing row of dangling bonds.

an APD/soliton is shown in Fig. 5b, where it is situated at a kink. Kinks
can be either reconstructed, as in Fig. 5a, or associated with dangling
bonds, as in Fig. 5b (Hirsch 1980, 1981, Jones 1980). An APD can react

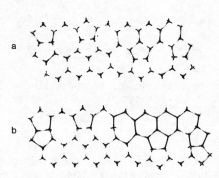

Fig. 5   Projection on (111), with only one layer of atoms above and
          below glide plane, showing reconstructed core of 30° partial with
          reconstructed kink in (a), and dangling bond kink in (b). The
          latter consists of a dangling bond antiphase defect and a
          reconstructed kink.

with a reconstructed kink to form a dangling bond kink (see Figs. 5a, b)
and conversely APD's can be generated by decomposition of a dangling bond
kink.

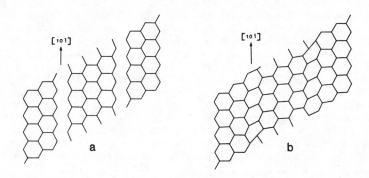

Fig. 6   Projection on to (111) glide plane of cores of two 90° partials
           a) in unreconstructed configuration, b) in reconstructed
           configuration. Only single planes of atoms immediately above and
           below the glide plane are shown (From Lodge and coworkers 1984).

For a 90° partial, the unreconstructed and reconstructed configurations
are shown, in projection on the (111) plane, in Figs 6a and 6b. The
calculations of Marklund (1983), Altmann, Lapiccirella and Lodge (1983)
and Lodge and coworkers (1984) suggest that for Si the reconstructed
configuration has the lower energy. In this case also APD's can occur,
and kinks can be reconstructed or have dangling bonds (Hirsch 1980, 1983,
Jones 1980).

The effect of reconstruction on electronic energy levels is that the half
filled band for the undissociated shuffle dislocation model is replaced by
split acceptor and donor levels, the latter filled and the former empty in
the neutral state. Fig. 7 shows schematically a set of energy levels
which might be expected from reconstructed dislocations and kinks.
Acceptor and donor levels for reconstructed dislocations (E_{da}, E_{dd}) and
kinks ($_1E_{Ka}$, $_1E_{Kd}$) are likely to be relatively shallow levels; dangling
bond levels associated with dangling bond kinks ($_2E_{Ka}$, $_2E_{Kd}$) are expected
to be associated with deeper levels. When a kink migrates along the
dislocation line, bonds are broken in the saddle point configuration
between neighbouring sites for the kink. The energy levels for the saddle
point configurations ($_{1,2}E_{Ksa}$, $_{1,2}E_{Ksd}$) are again expected to be deep
levels, and are shown in Fig. 7.

There are many results in the literature on energy levels deduced from
deep level transient spectroscopy, Hall conductivity data, optical
spectroscopy, and other data, but at the present time there is still some
uncertainty regarding the identification of various levels. However, for
dislocations in Si there is good evidence for the occurrence of both
acceptor and donor levels although the correlation with particular sites
on the dislocation is not clearly established.

For III-V compounds dislocations with edge components of opposite sign are not equivalent; for example the extra half planes of the 30° and 90° partials of the 60° dislocation in Fig. 3 end on one particular type of atom (say group III); for a 60° dislocation of opposite sign the extra half planes end on the other type of atom (say group V). Since the bonding characteristics of the two types of element are different, the energy levels associated with dislocations of opposite sign are likely to be different, and the dislocation velocities are therefore expected to be different from each other, and from that of the screws, which consist of two 30° partials, one of each type. The two types of edge dislocations in the glide configuration in a compound AB are called A(g) and B(g) respectively depending on whether the extra half planes terminate on a row of A atoms (Group III) or B atoms (Group V). (Many workers have interpreted their results on the assumption that slip takes place on the shuffle planes, and the dislocations with their extra half planes ending on A and B atoms are called α (or A(s)) or β (or B(s)) respectively. Assuming slip actually occurs in the glide configuration, dislocations previously assumed to be α and β are actually B(g) and A(g) respectively.) (For explanation of the notation see foreword to Conference on Dislocations in Tetrahedrally Coordinated Semiconductors, Hunfeld, 1978; Journ. de Physique, 40, Colloque C-6, 1979.)

The differences in dislocation velocities for A(g), B(g) and screw dislocations are well established experimentally for a number of III-V compounds (Choi, Mihara and Ninomiya 1977, Erofeeva and Ossipyan 1973, Mihara and Ninomiya 1975, Ninomiya 1979, Choi, Mihara and Ninomiya 1978, Maeda and coworkers 1977).

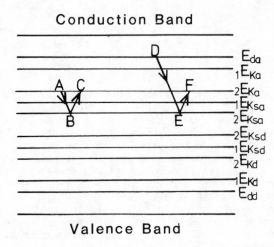

Conduction Band

Valence Band

Fig. 7   Schematic set of acceptor and donor energy levels for reconstructed dislocations (E_{da}, E_{dd}), reconstructed kinks ($_1E_{Ka}$, $_1E_{Kd}$), dangling bond kinks ($_2E_{Ka}$, $_2E_{Kd}$), saddle point configurations ($_{1,2}E_{Ksa}$, $_{1,2}E_{Ksd}$). An electron on a charged dangling bond kink moves from A to B to C as the kink moves one atomic distance. For a double kink nucleated at a negative charge on a dislocation, an electron moves from D to the saddle point energy level E, to the kink level F.

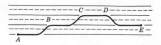

Fig. 8   Movement of dislocations by generation and motion of kinks.

EFFECT OF DOPING ON DISLOCATION VELOCITY

At relatively low temperatures the dislocation velocity in semiconductors
is controlled by the generation and motion of double kinks (Fig. 8).
Assuming that the dislocation cores are reconstructed, but that deep
levels occur at kinks in their equilibrium and/or saddle point
configurations, Hirsch (1979, 1983) has developed a theory of the doping
effect applicable both within the framework of the drift model for kink
controlled dislocation velocity (Hirth and Lothe 1982, pages 531-545),
valid at relatively high temperatures and low stresses, and at low
temperatures and high stresses when the velocity is controlled by the
nucleation of double kinks. On the drift model the velocity is
proportional to the concentration of kinks (exp $-F_k/KT$, F_k is the kink
formation energy), the velocity of the kinks (α exp $-W_m/kT$, W_m is the
activation energy for kink migration along the dislocation line assumed to
be controlled by the secondary Peierls potential), and the driving force
τ^n (τ is the resolved shear stress and n ~ 1→2). In the model doping
affects the concentration of charged kinks, and the activation energy of
migration of charged kinks. The theory follows closely that for the
doping effect on diffusion, kinks replacing point defects. If we neglect
the concentration of positively charged kinks, $+c_k$, the concentration of
negatively charged kinks, $-c_k$, relative to that of neutral kinks, $_0c_k$, is
given by Fermi-Dirac statistics as

$$-c_k/_0c_k = \exp\ (E_F - {}_iE_{Ka} - eV)/kT \qquad\qquad 1$$

where E_F is the Fermi energy and eV is the electrostatic energy of the
charged dislocation, i = 1, 2 depending on the type of kink. The
concentration of charged kinks therefore depends on doping through E_F.
The concentration of neutral kinks $_0c_k$ is constant, at a given
temperature, independent of the doping level, provided the interaction
between charged and neutral kinks is neglected, and the concentrations of
all kinks are small. The velocity of the dislocations controlled by
negatively charged kinks (assuming $+c_k$ to be negligible) is given by

$$v_-/v_0 = \exp\ (E_F - {}_iE_{Ka} - eV - \Delta W_m)/kT \qquad\qquad 2$$

where $\Delta W_m = W_m' - W_m$ is the difference in migration energy between charged
(W_m') and uncharged (W_m) defects. During the motion of the kinks over the
saddle point, assuming the Born-Oppenheimer approximation holds, the
electron will follow the motion of the ion and move from energy level
$_iE_{Ka}$ to $_iE_{Ksa}$ and back to $_iE_{Ka}$ during the activation process (e.g. A to B
to C in Fig. 7). Thus

$$\Delta W_m = {}_iE_{Ksa} - {}_iE_{Ka} + \Delta W_m' \qquad\qquad 3$$

where $\Delta W_m'$ is the contribution to ΔW_m from changes in position of the ions
due to charge. Thus, the activation energy term ΔU in (2) becomes

$$\Delta U = (E_F - {}_iE_{Ksa} - eV - \Delta W_m') \qquad\qquad 4$$

This expression is similar to that obtained by Jones (1980) who assumed that equilibrium with the electron gas is established during the activation at the saddle point. The present treatment avoids this difficulty (Hirsch 1981) since equilibrium is established already in the ground state of the kink. However, as pointed out by Jones (1980), the electron energy level important in the doping effect is that corresponding to the saddle point configuration. This energy level is not necessarily the same as that found from DLTS, Hall conductivity or optical data from specimens containing static dislocation distributions.

At higher stresses and lower temperatures, the dislocation velocity is controlled by the nucleation of double kinks (Hirsch and Lothe 1982, pages 531-545). For the experimental conditions under which the doping effect on dislocation velocity has been measured, the dislocation velocity v is independent of the length of the dislocation segment which is advancing; most of the kinks formed by nucleation at different points along the segment annihilate each other before reaching the ends of the segment; i.e. the mean free path for the kinks is less than the length of dislocation segment. Then

$$v = 2h\sqrt{(Jv_K)} \qquad\qquad 5$$

where h is the distance between neighbouring Peierls valleys, J is the rate of nucleation of double kinks per unit length of dislocation line, and v_K is the kink velocity (Hirth and Lothe 1982, pages 531-545). Hirsch (1981, 1983) suggested that preferential nucleation occurs at a charged point on the dislocation line; Heggie and Jones (1982, 1983) proposed that preferential nucleation occurs at an APD. Several alternative reaction paths seem possible, but in the simplest case for the former mechanism an electron moves from a dislocation level E_{da}, to a saddle point level ($_2E_{Ksa}$, see Fig. 7), corresponding now to the saddle point for nucleation of the double kink, and beyond the saddle point to a charged kink level $_2E_{Ka}$, i.e. corresponding to D to E to F in Fig. 7. The result is the nucleation of a double kink, with one charged and the other uncharged. The charged kink is likely to be more mobile, but after equilibrium is reestablished with the electron gas, both kinks may be charged, depending on the position of E_F. The resulting expression for the difference in activation energy of the velocity of dislocations controlled by charged and neutral double kinks is identical to that derived for the drift model, assuming that the velocity is independent of the segment length. Heggie and Jones (1982, 1983) have obtained the same result for their model.

For an extensive set of data for 60° dislocations in Si (George and Champier 1979) it has been possible to neglect the eV term in (4) and derive a consistent pair of acceptor and donor levels (Hirsch 1981, 1983); an acceptor level $E_a = 0.67\pm0.04$ eV, with $\partial E_a/\partial T = (-1.6\pm0.5)\times10^{-4}$ eV K^{-1} and a donor level $E_d = 0.28\pm0.17$ eV with $\partial E/\partial T = (-2.0\pm2.1)\times10^{-4}$ eV K^{-1}, all measured relative to the valence band. For the Ge data in Fig. 1 it has not been possible to obtain a consistent fit by neglecting eV and this term will probably have to be taken into account.

EFFECT OF DOPING ON YIELD STRESS

Rabier, Veyssière and Demenet (1983) have measured the lower yield stress σ_{ly} as a function of temperature for heavily doped n-type Si and for intrinsic Si. Fig. 9 shows their results in the temperature range in which velocity measurements are available (George and Champier 1979). Good agreement is obtained between the experimentally observed ratios of yield stress for the two materials at a given temperature and those calculated using the microscopic theory of the yield point (Alexander and Haasen 1968), and the known dislocation velocity data (Rabier, Veyssière and Demenet 1983). At about 500°C $\dfrac{\sigma_{ly} \ \text{(doped material)}}{\sigma_{ly} \ \text{(intrinsic material)}} \sim 0.3$, i.e. under these conditions the softening due to the electronic effect is considerable.

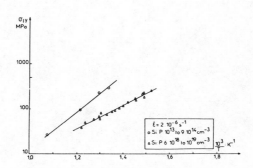

Fig. 9   Log lower yield stress σ_{ly} against reciprocal temperature for intrinsic and n-type Si (Rabier, Veyssière and Demenet 1983).

EFFECT OF DOPING ON INDENTATION PLASTICITY

The effect of doping on dislocation velocity is reflected in hardness, plasticity as observed for example by the extent of dislocation rosettes, and cracking behaviour in microindentation tests. Results for Si, Ge and GaAs have been published elsewhere (Roberts, Pirouz and Hirsch 1983, Roberts, Pirouz and Hirsch 1985, Hirsch and coworkers 1985).

Microindentation of (001) Ge surfaces

Roberts, Pirouz and Hirsch (1985) report results of indentation
experiments on Ge in the temperature range 20-420°C. Dislocation rosettes
are formed along the ⟨110⟩ directions on the (001) surface, produced by
slip on the {111} planes shown in Fig. 10. The extent of the rosettes
varies with temperature and doping, as shown in Fig. 11. It should be
noted that the rosettes for n-type are longer and for p-type shorter than
for intrinsic material, exactly as expected from the dislocation velocity
measurements (Fig. 1). Table 1 shows hardness values for the three types
of Ge at different temperatures (there is a substantial difference at
400°C between the results obtained with a cold and hot indenter,
corresponding for the former approximately to a temperature near the
indenter lower by about 60°C relative to that of the temperature of the
bulk of the specimen). Again, as expected, n-type is found to be softer
and p-type harder than intrinsic material. The effect is a maximum at
~400°C using the cold indenter, and amounts to a difference of about 36%
between the hardness of n- and p-type material.

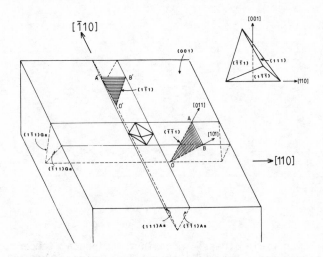

Fig. 10   Geometry of dislocation rosettes formed by indentation of (001)
          plane.

Radial cracks are observed along the ⟨110⟩ rosette directions,
corresponding to cracking along {110} cleavage planes ; the geometry of
the indentation leads to cleavage along these planes rather than along the
{111} planes usually preferred in elemental semiconductors. Fig. 12 shows
the variation of radial crack span with temperature. There is a "brittle-
ductile" transition for radial cracking, and the "transition temperature"
is about 100°C higher for p-type material than for n-type material.

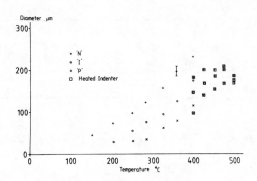

Fig. 11   Variation of rosette diameter for (001) indentations in Ge, normalised to a 20μm indentation size, with doping and temperature. Note the strong dependence of rosette size on doping, with data for all doping converging at ~500°C (Roberts, Pirouz and Hirsch 1985).

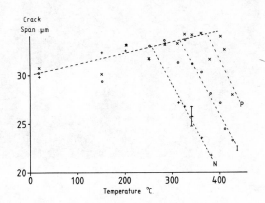

Fig. 12   Variation in radial crack size (total span as measured on the surface) at (001) indentations in Ge with temperature and doping for 50g load. The data from each doping diverge from a common line at a distinctive "ductile/brittle transition temperature" varying by ~±50°C from intrinsic to p- and n-type material (Roberts, Pirouz and Hirsch 1985).

TABLE 1   Hardness of Germanium at Various Temperatures

| Type | Temp. °C | | | |
| | 250 | 400 | 400 | 500 |
	(cold ind.)		(hot ind.)	
N	684	404	265	112
I	697	520	275	125
P	732	581	334	131
Typical Error (2σ)	±60	±50	±40	±20

Hardness Units are HV(kgmm)$^{-2}$; all measurements made at 50g load.

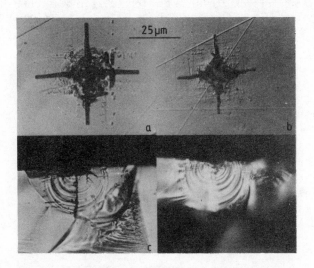

Fig. 13   Cracks around 50g indentations on (001) faces of p-type (a,c) and
n-type (b,d) Ge at 360°C, seen in plan (a,b) and cross-section
(c,d) (Roberts, Pirouz and Hirsch 1985).

Fig. 13 shows cracks around 50g indentation in p-type (a,c) and n-type
(b,d) Ge at 360°C, seen in plan (a,b) and cross-section (c,d). The median
cracks below the surface are not significantly affected by doping or
temperature in this temperature range, but experiments using a hot
indenter at higher temperatures suggest that the median cracks eventually
disappear at a higher temperature.

Various possible models exist for the elastic stress fields beneath and around an indenter, for example those due to Boussinesq (ideally sharp contact) and Hertz (contact between spheres) (see e.g. Lawn and Swain 1975). Resolution of such stress fields onto available slip systems would be expected to give a reasonable indication of the active slip systems around an indenter (in the first stages of slip, at least). However, the Boussinesq field is not appropriate for an indenter of finite extent, especially in the region of interest close to the indenter, suggesting patterns of slip very different from those actually observed. The form of the Hertzian field, while more appropriate, is analytically quite complex. In this study, we have used a field due to Nadai (1931), that for a long flat punch, to estimate slip patterns close to the edges of an indenter.

Details of the procedure are given by Hirsch and coworkers (1985) for {111} surfaces, and by Roberts, Warren and Hirsch (1986) for {001} surfaces. Fig. 14a shows the type of slip predicted; slip under and close to the indenter would be generated on {111} planes with Burgers vectors inclined to the suface. The other type of slip with Burgers vectors parallel to the surface (AB and CD in Fig. 14a) is expected to be the predominant component of slip in the rosettes far from the indenter. Fig. 14b shows a cross-section through the rosette arm under and close to the indenter; under the indenter slip takes place on converging slip planes, leading to the formation of Lomer-Cottrell locks along the lines of intersection of slip planes a, and resulting in high workhardening rates on these planes. Outside the indenter slip takes place on diverging slip planes b; the geometry of the slip is such that locks are not formed along their lines of intersection, and workhardening will be low.

The Lomer-Cottrell locks act as nuclei for the nucleation of median cracks parallel to {110} planes, beneath the indenter. On unloading, these cracks extend towards the surface to form the 'median/radial' crack system. The driving force for this extension is the residual stress field after indenter removal, arising from the elastic mismatch of the plastically-deformed region near to the indenter and the elastically-deformed 'hinterland'. Theories exist for the derivation of K_{IC} from the lengths of such cracks and their variation with load (e.g. Lawn and Marshall 1979): however, these are based on stress fields derived from an 'expanding spherical cavity' model (e.g. Chiang, Marshall and Evans 1982). The applicability of such models to cases where cracks nucleate and grow completely within a highly structured, non-centrosymmetric 'plastic zone' is probably limited. As can be seen from Fig. 13, the crack profile does not change, but the extensions to the surface show substantial changes in size as the dislocation structures around the indentation vary in extent and density.

To account for the variation of radial crack length with temperature and doping, two possible mechanisms of interaction between dislocation structures and the cracks should be considered:-

1. Dislocations near the crack tips (either nucleated there or already present, e.g. from rosettes) move so as to absorb energy from the crack-opening mechanism, thus effectively raising K_{IC};

2. The extent and form of the plastic zone change, so that the stress-intensity factors controlling crack growth are modified, by stresses from dislocation arrays.

The first of these mechanisms is unlikely to be responsible in this case, since indentation tests on pre-deformed specimens (Roberts and Pirouz 1984) showed no variation of crack length with bulk dislocation density. On the other hand the second mechanism provides an explanation for the results in Fig. 12. The driving force for radial crack extension at the surface will be the circumferential hoop stress outside the plastic zone, which itself is a region of compression. In this case at and near the indentation the plastic zone is in two parts, an inner region due to converging slip a, and an outer region due to diverging slip b, in Fig. 14b. For a given doping, as the temperature increases, the amount of plastic strain on the diverging slip planes will increase relative to that on the converging slip planes, since at a given applied stress the slip line lengths for the former will increase with increasing temperature as the lattice friction stress decreases, while for the latter the slip line length is determined by the position of the Lomer-Cottrell locks, which are governed by the size of the indentation. With increasing temperature the size of the indentation increases for a given load; this will extend both the regions of converging and diverging slip, but over the temperature range in which the crack lengths vary rapidly, these changes will be relatively small. Thus, with increasing temperature, the stresses from the outer plastic zone should become increasingly important. Consideration of the movements, e.g. on plane ABC, of dislocations with Burgers vectors $\overline{BC}$ and $\overline{AC}$, shows that interaction leads to the formation of dislocations with Burgers vector $\overline{AB}$, with the dislocations lying in the plane normal to AB, i.e. the plane containing the other rosette direction parallel to $\overline{CD}$. These dislocations produce tensile stresses on the inside of the outer plastic zone acting on the cracks nucleated at the Lomer-Cottrell locks at the edges of the inner plastic zone, helping these cracks to propagate at least to the edges of the outer plastic zone at and near the surface. Hence it would not be expected that slip on planes of type b would affect the extent of the 'median' cracks - as is observed. Outside the outer plastic zone the stress from these dislocations is compressive, and further propagation of the cracks depends on the relative strengths of the net tensile hoop type stresses originating from the inner region of compression and of the compressive stresses outside the outer plastic zone along the rosette arms. With increasing temperature the latter should become increasingly important and crack propagation will be inhibited, for the portions of the median/radial crack system above the diverging (type b) slip planes. More extensive slip along the rosettes by dislocations with Burgers vectors $\overline{AB}$ and $\overline{CD}$, carrying material away from the centre, should also lead to a reduction in the tensile hoop stress with increasing temperature. For n-type material crack propagation will be inhibited at a lower temperature than in p-type or intrinsic material, because the dislocation velocities are faster and thus the dislocation configuration inhibiting cracking forms at lower temperatures, in agreement with the observations.

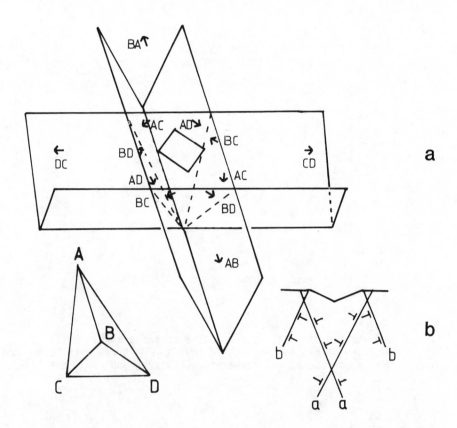

Fig. 14 a) Directions of Burgers vectors inclined to the surface around a
(001) indentation of Ge; the arrow indicates the directions in
which dislocations move from the indentation centre. b) Cross-
section through indentation showing geometry of slip.

Microindentation of {111} GaAs surfaces

Following studies of microindentation of {100} faces (Warren, Pirouz and
Roberts 1984), microindentation experiments have been carried out on {111}
surfaces of n- and p-type GaAs in the temperature range 20-400°C (Warren
and coworkers 1985).   The following results have been obtained:-

1. For n-type GaAs the Ga face is harder than the As face at all
temperatures tested (see Fig. 15); for p-type GaAs the As face is
harder than the Ga face at high temperatures, but the difference is
negligible below about 150°C. Experiments on other III-V compounds
have generally shown the group III element face to be harder than the
group V element (Shimizu and Sumino 1970, Gatos and Lavine 1965, Maeda
and coworkers 1977)

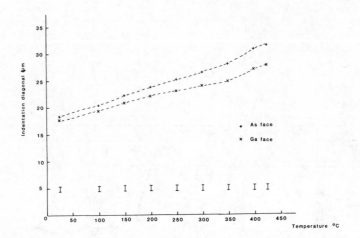

Fig. 15   Indentation diagonal as a function of temperature on {111}
surfaces of n-type GaAs. Note that the Ga surface is harder than
the As surface at all temperatures (Warren and coworkers 1984).

2. At 400°C n-type GaAs is about 40% harder than p-type GaAs, comparing
the mean hardness values for the Ga and As faces.

3. The crack, rosette and slip patterns have approximate 3-fold symmetry.
Fig. 16a shows the cracks (C), rosettes (R) and slip line triangles,
surrounding the indentation on the As face in n-type GaAs indented at
350°C, using a load of 100g. The slip lines are found experimentally
to form on recovery after the load is taken off. The crack traces
along ⟨112⟩ imply that they are parallel to {110} cleavage planes.
Fig. 16b shows the pattern at 25°C, and it should be noted that
compared with the configuration at 350°C the cracks are now extended
in the opposite directions. Fig. 17 illustrates the pattern
schematically for the As and Ga faces. For p-type GaAs, the behaviour
on the Ga face is similar to that on the As face for n-type material,
and for the As face is similar to that on the Ga face for n-type
material at 25°C, but the crack directions are reversed at 350°C.

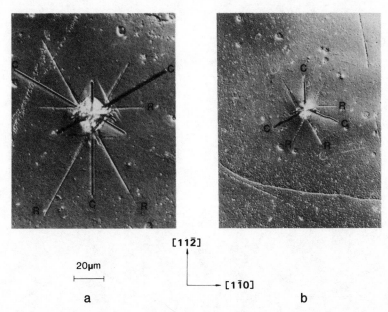

[11$\bar{2}$]

20μm

[1$\bar{1}$0]

a   b

Fig. 16   Cracks, rosettes and slip lines on the As(111) face in n-type
          GaAs indented a) at 350°C, b) at 25°C, using a load of 100g.

To understand this complex behaviour we note that for n-type GaAs, As(g)
dislocations are about 100 times faster than Ga(g) dislocations at about
350°C, while for p-type GaAs, Ga(g) dislocations are about 10 times faster
than As(g) dislocations and screws (Choi, Mihara and Ninomiya 1977). The
{111} slip planes define two tetrahedra in the crystal as shown in Fig.
18, one ABCD with its apex inside when viewed from the indentation
direction, the other DEFG with its apex outside when viewed from the same
direction. Stress field analysis shows that under and around the indenter
slip will occur by dislocations whose extra half planes lie on the side of
the slip plane facing the indenter moving into the crystal from the
surface on these tetrahedra (and also away from the indenter along the
rosettes) (Hirsch and coworkers 1985). Dislocations whose extra half
planes are on the inside of one of the tetrahedra, and on the outside of
the other, are all of one type, which is defined by the nature of the
indented surface. Thus, for the As face, defined by the surface plane of
atoms, e.g. the open circles for the top face in Fig. 3, Ga(g)
dislocations glide on the internal apex tetrahedron with their extra half
planes inside, As(g) dislocations on the external apex tetrahedron with
their extra half planes inside. For the Ga face the opposite geometry
applies.

Consider now the slip pattern expected for As face indentations in n-type
GaAs. It follows from the above discussion that the faster As(g)
dislocations glide on the external apex tetrahedron. Slip on the inclined
planes with Burgers vectors EF, FG, GE produce the rosettes, and a
dislocation arrangement shown in Fig. 19a, the signs of the dislocations

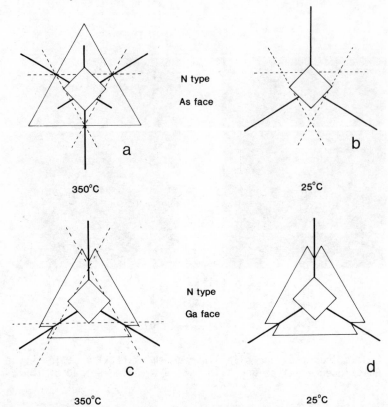

Fig. 17   Geometry of pattern of cracks (thick full lines), rosettes
(dotted lines) and slip lines (thin full lines) around
indentations on {111} As face and Ga face.

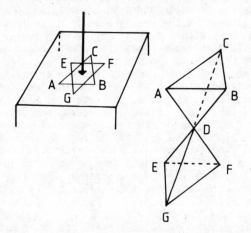

Fig. 18   Tetrahedra of {111} slip planes with apex inside (ABCD) and
outside (DEFG) crystal, when viewed from the indentation
direction.

being determined by the elastic stress field (Hirsch and coworkers 1985). If interaction takes place at their intersection, Lomer-Cottrell dislocations are produced.   With regard to slip with inclined Burgers vectors, the internal network shown in Fig. 19b is produced, assuming no interaction with the rosette dislocations.   If the inclined dislocations interact with the rosette dislocations, dislocations with inclined Burgers vectors are formed (Fig. 19c).

The slower Ga(g) dislocations glide on the planes of the internal apex tetrahedron, with their extra half planes inside. For the rosette slip the dislocation arrangement is similar to that in Fig. 19a, except that the orientation of the triangle on the surface is that shown in Fig. 19d. The rosettes observed at 25°C and 350°C are those produced by slip by the fast As(g) dislocations, as expected (see Figs. 17a,b). For slip with inclined Burgers vectors, the dislocation arrangement is as shown in Fig. 19d; Lomer-Cottrell dislocations are now formed where the slip planes intersect, causing high workhardening. The Lomer-Cottrell dislocations form the edges of a tetrahedral network of dislocations, with the apex inside the crystal.

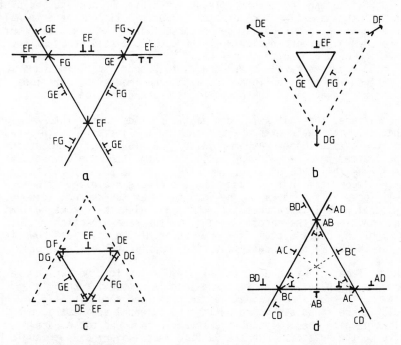

Fig. 19 Dislocation arrangements around {111} indentations in n-type
         GaAs. a) Rosettes on As and Ga face consisting of fast
         dislocations moving on planes of appropriate inclination. b)
         Internal networks below the indentation formed by slip with
         inclined Burgers vector on planes EFGD (Fig. 18) for indentation
         on As face. c) Reactions between rosette dislocations and those
         with inclined Burgers vectors, and surface traces of slip planes
         of type ABCD (Fig. 18) (shown dashed) active on recovery when
         load removed (As face). d) Arrangement of dislocations due to
         slip with inclined Burgers vectors on planes ABCD, showing Lomer-
         Cottrell dislocations at slip plane intersections (Ga face).

At low temperatures the rosettes, even those involving slip by the fast
As(g) dislocations, will be very shallow, since their depth is determined
by the motion of the screws (parallel to the surface) which are thought to
be slow (Choi, Mihara and Ninomiya, 1977), although there is some
uncertainty about this for n-type GaAs. The important cracks will then be
nucleated at the Lomer-Cottrell dislocations formed along the edges of the
network generated by slip on the internal apex tetrahedron (Fig. 19d).
The tensile stresses outside the Lomer-Cottrells will tend to propagate
the cracks, but the network of dislocations formed by glide on the
external apex tetrahedron (DEFG) (see dislocations EF, FG, GE in Fig. 19b)
generates compressive stresses across these cracks outside this network,
inhibiting crack propagation. At low temperatures, when slip on the
external apex tetrahedron is not well developed, the tensile stresses from
the dislocations on the internal apex tetrahedron predominate, and the
cracks formed at the Lomer-Cottrells in Fig. 19d propagate, giving rise to
radial cracks passing through the mid points of the rosette triangle (Fig.
19a), exactly as observed experimentally (see Fig. 17b for the As face in
n-type GaAs). At higher temperatures (350°C) there is much more slip by
the fast As(g) dislocations on the planes of the external apex
tetrahedron, and the compressive stresses inhibit the propagation of these
cracks. At the same time the rosettes formed by slip on the external apex
tetrahedron are much more developed because the screws penetrate further
into the crystal, cracks nucleate at the Lomer-Cottrell locks at the
intersection of the rosette slip lines (Fig. 19a), and the tensile stress
outside the rosettes causes these cracks to propagate. This leads to the
crack pattern in Fig. 17a, in accord with experimental observations,
including the reversal of the predominant crack direction.

For the Ga face, the fast As(g) dislocations glide on the internal apex
tetrahedron, with their extra half planes on the inside. In this case
slip on the external apex tetrahedron, by the slow Ga(g) dislocations,
will be limited both at 25°C and 350°C. Consequently the compressive
stresses from this type of slip, important for the case of the As face,
are of no importance in this case, and the cracks nucleated at Lomer-
Cottrell locks at the edges of the internal apex tetrahedron network will
propagate at both temperatures. In this case also the cracks nucleated at
the intersection of the rosettes are along the same directions. The
geometry of the crack system is therefore predicted to be as shown in Fig.
17c,d for the Ga face, in agreement with experimental observations.

For the As face in n-type GaAs the slip steps in Fig. 16a (which are found
experimentally to be formed after the load is removed) form a triangle
opposite to that defined by the rosette slip (see also Fig. 17a). The
slip planes are found always to define the internal apex tetrahedron in
Fig. 18, in which the inside of the triangle is found to be raised
relative to the outside. For the Ga face the recovery slip triangle has
again the same orientation as that of the internal apex tetrahedron, but
in this case has the characteristic appearance of a triangle with part of
the apices missing (see Fig. 17c,d). All these features have been shown
to arise from a recovery mechanism involving the cross-slip of screws
(Hirsch and coworkers 1985).

The difference in hardness of the As and Ga faces in n-type GaAs (Fig. 15)
is due to the fact that the faster As(g) dislocations glide on the
diverging external apex tetrahedron planes for the former, and on the
converging internal apex tetrahedron planes for the latter. The
dislocation network formed by the diverging slip does not involve the

formation of locks (Fig. 19b), and the penetration of the slip lines into the crystal is determined by the lattice friction stress; workhardening is low. The amount of diverging slip will therefore depend sensitively on the lattice friction stress, i.e. on the mobility of the dislocations. Thus the amount of diverging slip will be considerably greater if it is generated by the faster As(g) dislocation, which is the case for the As face. By contrast the penetration of converging slip is limited by the formation of Lomer-Cottrell locks and is relatively independent of the dislocation mobility, i.e. the amount of converging slip will be similar for the As and Ga faces. We conclude therefore that the difference in hardness of the As and Ga faces is due to the difference in plastic strain on the diverging slip system, the hardness being lower the greater the plastic strain, i.e. the higher the dislocation mobility, on this system. It follows from the dislocation velocity data that for n-type GaAs the As face is softer, while for p-type GaAs the Ga face is softer, as observed experimentally in the temperature range in which the velocity data are likely to be applicable (see e.g. Fig. 15).

In general for p-type GaAs the deformation and cracking behaviour is in line with that expected, bearing in mind that the Ga(g) dislocations are now faster than As(g) dislocations. However, the cracking reversal and the appearance of a slip step triangle opposite to that of the rosettes for the As face at 200°C requires further consideration.

The lower hardness of p-type GaAs compared to that of n-type GaAs is presumably due to the higher dislocation velocities for the former than for the latter (Choi, Mihara and Ninomiya 1977).

CONCLUSIONS

The results on the deformation of semiconductors show that doping with small concentrations of electrically active solutes causes significant changes in dislocation velocity, yield stress, hardness and fracture behaviour in the temperature range in which plastic flow becomes important and is controlled by the Peierls mechanism. Similar effects are expected for SiC, diamond, BN, Si_3N_4 and other strongly covalent materials. Larger band gaps (e.g. for diamond) should lead to larger changes in activation energy for glide on doping, although the changes in activation energies relative to that for intrinsic material may not be greater than for semiconductors. Since doping affects basically the dislocation velocity, it should result in changes in brittle-ductile transition temperatures, with the possibility of toughening the strong materials by softening them in the brittle-ductile transition regime. It is recognised of course that in practice, it will be difficult to obtain materials of sufficient purity to make doping possible, and the usefulness of the effect may in any case be limited for polycrystalline materials whose properties are governed by grain boundary mechanisms.

ACNOWLEDGMENTS

The support of the B.P. Venture Research Unit and of the SERC is gratefully acknowledged.

REFERENCES

Alexander, H. and Haasen, P. 1968, Rev. Solid State Physics 22, 28-158
Altmann, S.L., Lapiccirella, A. and Lodge, K.W. 1983, Inst. J. Quant.
    Chem. 23, 1057-1063
Anstis, G.R., Hirsch, P.B., Humphreys, C.J., Hutchison, J.L. and
    Ourmazd, A. 1981, Microscopy of Semiconductor Materials, Oxford 1981,
    Inst. of Phys. Confer. Series 10, p. 15 (Bristol, London: Institute of
    Physics) 15-22
Bourret, A., Desseaux, J. and D'Anterroches, C. 1981, Microscopy of
    Semiconductor Materials, Oxford 1981, Inst. of Phys. Confer. Series 10,
    p. 9 (Bristol, London: Institute of Physics) 9-14
Chiang, S.S., Marshall, D.B. and Evans, A.G. 1982, J. Appl. Phys. 53,
    298-311
Choi, S.K., Mihara, M. and Ninomiya, T. 1977, Jap. Journ. Appl. Phys. 16,
    737-745
Choi, S.K., Mihara, M. and Ninomiya, T. 1978, J. Appl. Phys. 17, 329
Erofeev, V.N. and Nikitenko, V.J. 1971, Sov. Phys. Solid State 13,
    116-120
Erofeeva, S.A. and Osipyan, Yu. A. 1973, Sov. Phys. Solid State 15,
    538-540
Gai, P.L. and Howie, A. 1974, Phil. Mag. A30, 939-943
Gatos, H.C. and Lavine, M.C. 1965, Progr. in Semiconductors (Editors
    A.F. Gibson and R.E. Burgers) 9, p. 1 (Heywood, New York) 3-45
George, A. and Champier, G. 1979, Phys. Stat. Solidi (a) 53, 529-540
Gomez, A.M. and Hirsch, P.B. 1978, Phil. Mag. A38, 733-737
Gottschalk, H., Patzer, G. and Alexander, H. 1978, Phys. Stat. Solidi
    (a)45, 207-217
Hall, E.L. and Vander Sande, J.B. 1978, Phil. Mag. A37, 137-145
Haussermann, F. and Schaumburg, H. 1973, Phil. Mag. 27, 745-751
Heggie, M. and Jones, R. 1982, J. de Physique, Suppl. C1, 43, C1-45-50
Heggie, M. and Jones, R. 1983, Microscopy of Semiconductor Materials,
    Oxford 1983, Inst. of Phys. Confer. Series 67, p. 45, Bristol, London:
    Institute of Physics 45-53
Hirsch, P.B. 1979, J. de Physique, Suppl. C6, 40, C6-117-121
Hirsch, P.B. 1980, J. Microscopy 118, 3-12
Hirsch, P.B. 1981, Proc. Mater. Res. Soc. Symp. on Defects in
    Semiconductors (ed. J. Narayan and T.Y. Tan) p. 257-271 (New York:
    North Holland)
Hirsch, P.B. 1983, Microscopy of Semiconductor Materials, Oxford 1983,
    Inst. of Phys. Confer. Series 67, p. 1-10 (Bristol, London: Institute
    of Physics)
Hirsch, P.B., Pirouz, P, Roberts, S.G. and Warren, P.D., 1985, Phil. Mag.
    B, Vol. 52, 761-786
Hirth, J.P. and Lothe, J. 1982, Theory of Dislocations 2nd edn. New
    York: McGraw Hill
Hornstra, J. 1958, J. Phys. Chem. Solids 5, 129-141
Jones, R. 1979, J. de Physique, Suppl. C6, 40, C6-33-38
Jones, R. 1980, Phil. Mag. B42, 213-219
Lawn, B.R. and Marshall, D.B. 1979, J. Amer. Ceram. Soc. 62, 347-350
Lawn, B.R. and Swain, M.V. 1975, J. Mat. Sci. 10, 113-122
Lodge, K.W., Altmann, S.L., Lapiccirella, A. and Tomassini, N. 1984,
    Phil. Mag. 49, 41-61
Maeda, K., Ueda, O., Murayama, Y. and Sakamoto, K. 1977, J. Phys. Chem.
    Solids 38, 1173-1179
Marklund, S. 1983, J. de Physique, Suppl. C4, 44, C4-25-35
Mihara, M. and Ninomiya, T. 1975, Phys. Stat. Solidi (a)32, 43-52

Nadai, A., 1931, Plasticity (New York: McGraw-Hill)
Ninomiya, T. 1979, J. de Physique, Colloque C6, Suppl. No. 40, C6-143-145
Patel, J.R. and Chaudhuri, A.R. 1966, Phys. Rev. 143, 601-608
Patel, J.R., Testardi, L.R. and Freeland, P.E. 1976, Phys. Rev. B 13, 3548-3557
Pirouz, P., Cockayne, D.J.H., Sumida, N., Hirsch, P.B. and Lang, A.R. 1983, Proc. Roy. Soc. A. 386, 241-249
Rabier, J., Veyssière, P. and Demenet, J.L. 1983, J. de Physique, Colloque C4, Suppl. 9, 44, C4-243-253
Ray, I.L.F. and Cockayne, D.J.H. 1971, Proc. Roy. Soc. A. 325, 543-554
Ray, I.L.F. and Cockayne, D.J.H. 1973, J. Micros, 98, 170-173
Read, W.T. 1954, Phil. Mag. 45, 1119-1128
Roberts, S.G. and Pirouz, P. 1985, Proc. Int. Metall. Soc. 17th Annual Meeting, Philadelphia, U.S.A., in press
Roberts, S.G., Pirouz, P. and Hirsch, P.B. 1983, J. de Physique, Colloque C4, Suppl. 9, 44, C4-75-83
Roberts, S.G., Pirouz, P. and Hirsch, P.B. 1985, J. Mat. Sci., 20, 1739-1747
Roberts, S.G., Warren, P.D. and Hirsch, P.B. 1986, J. Mat. Res. Soc. 1, in press
Shimizu, H. and Sumino, K. 1970, J. Phys. Soc. Japan 29, 1096
Shockley, W. 1953, Phys. Rev. 91, 228
Warren, P.D., Pirouz, P. and Roberts, S.G. 1984, Phil. Mag., 50, L23-L28
Warren, P.D., Roberts, S.G., Pirouz, P. and Hirsch, P.B. 1985, Report on "Fracture at Stress Concentrators" Conference, 17-19 Sept. 1984, Churchill College, Cambridge.

Inst. Phys. Conf. Ser. No. 75: Chapter 2
Paper presented at 2nd Int. Conf. Science Hard Mater., Rhodes

107

The chemical bond strength and the hardness of high melting point compounds

P S KISLY

Institute for Superhard Materials of the
Ukrainian Academy of Sciences, Kiev, USSR

ABSTRACT

The paper deals with calculations of the chemical bond strength of nonmetallic high melting point compounds, using Harrison bonding orbitals
technique (a simplified version of LCAO-method), and of metal-like high
melting point compounds of NaCl-type structure taking into account the
electrostatic interaction of s- and d-electrons and Van der Waals interaction for atom skeletons. The correlation between the binding energy,
its covalency and material properties has been established and the
classification of substances by their hardness is proposed.

INTRODUCTION

High melting point compounds comprise a wide class of inorganic substances
used in current technology for the production of machines and apparatus
working at temperatures above 1000°C. These compounds have strong
chemical bonds and consequently possess extreme physical and mechanical
properties. Some of the compounds formed by boron, carbon, nitrogen,
oxygen, aluminium, silicon and phosphorus are nonmetallics and are mostly
covalent substances with wurtzite or zinc blende structures.

High melting point compounds formed by boron, carbon, nitrogen, oxygen,
silicon and phosphorus with transition metals are metal-like substances
which are covalent-ionic ones largely of NaCl-type (carbides, nitrides,
oxides, phosphides) and of complex (borides, silicides) structures. The
nature of chemical bonds in high melting point compounds is the subject of
constant investigation. The correlation between the binding energy and
the properties of substances has not been adequately established and
requires additional in-depth studies.

CHEMICAL BOND IN NONMETALLIC COMPOUNDS

Nonmetallic high melting point compounds are predominantly covalent substances but chemical bonds in these compounds show a definite ionicity,
readily detectable by a simple correspondence of atomic spacings in crystals to the sum of covalent (ΔC) and ionic (Δj) radii of atoms (Table 1).

TABLE 1   Atomic Spacings and their Deviation from the Sum of
Covalent (ΔC) and Ionic (Δj) Radii in Nonmetallic
High Melting Point Compounds

Compound	Atomic spacing, d Å	ΔC,%	Δj,%
C	1.54	-0.0	-
Si	2.35	-0.4	-
SiC	1.89	-2.5	-36.8
B_4C	1.64	-0.6	-42.8
AlB_{12}	1.98	-7.4	-
BN	1.58	0.0	-6.0
AlN	1.89	-3.5	-7.8
Si_3N_4	1.75	-6.4	0.0
BP	1.97	2.5	-4.9
AlP	2.36	2.6	-2.9
SiO	1.61	-12.0	-8.0
Al_2O_3	1.85	-3.6	-5.5

Not all of the above compounds are tetrahedral crystals. Only some of the atoms in boron carbide, aluminium dodecaboride, silicon and aluminium oxides have nearest neighbours with tetrahedral coordination. In spite of this, the LCAO-method can be used to describe chemical bonds in these compounds. With this end in view we have used a simplified method, the bonding orbitals technique, developed by Harrison (1980), that readily permits the establishment of the relation between the electronic structure and the properties of crystals. The binding energy in nonmetallic high melting point compounds is defined in terms of the energies of covalent E_c and ionic E_p bonds as the energy of bonding orbitals involving sp^3-hybridized orbitals:

$$E_c = 4.37 \hbar^2/m \, d^2 \qquad\qquad 1$$

$$E_p = (\varepsilon_h^A - \varepsilon_h^B)/2 \qquad\qquad 2$$

where $\hbar^2/m = 7.62$ eV A^2, d is the atomic spacing, ε_h^A and ε_h^B are the energies of the hybridized states of A and B atoms respectively. The calculation of the interaction energy concerned with the overlap of the first nearest neighbour wave functions is a very complicated procedure and we, therefore, have proposed to approximate the total chemical bond energy by the geometrical sum:

$$E_v = \sqrt{E_c^2 + E_p^2} \qquad\qquad 3$$

This energy is somewhat less than the experimental one defined by:

$$E_o = \Delta H + E_a^A + E_a^B \qquad\qquad 4$$

where ΔH is the compound formation heat, E_a^A, E_a^B are the atomization heats of the elements. The theoretical and experimental binding energies for nonmetallic high melting point compounds taken from tables in Samsonov (1976), and Samsonov and Vinitzky (1976) are summarized in Table 2.

TABLE 2   Chemical Bond Energies for Nonmetallic
High Melting Point Compounds

Compound	E_c,eV	E_p,eV	E_v,eV	E_0,eV
C	14.04	-	14.04	16.14
Si	6.03	-	6.03	10.45
SiC	9.42	1.42	9.52	12.69
B_4C	12.38	1.49	12.69	13.52
AlB_{12}	8.49	1.94	8.65	-
BN	13.33	3.12	13.69	13.20
AlN	9.32	4.09	10.16	11.50
Si_3N_4	10.87	3.04	11.28	17.11
BP	8.58	1.20	8.68	10.0
AlP	5.97	2.17	6.38	8.50
SiO_2	12.84	4.80	13.72	16.3
Al_2O_3	9.72	5.85	11.34	22.75

For the nontetrahedral compounds (B_4C, AlB_{12}, SiO_2, Al_2O_3) the calculation of bond energy is merely an estimate because it is based on sp^3 hybridization. Experimental values of E_0 might be expected to give better correlations with the mechanical properties of these particular compounds.

CHEMICAL BOND IN METAL-LIKE COMPOUNDS

The metal-like high melting point compounds are ionic substances by their structure (NaCl-type), metals and semiconductors by their electronic properties and covalent crystals by their mechanical characteristics. The electron density distribution in the structure of these substances indicates that the Me-Me bonds are metallic and Me-X bonds are covalent-ionic ones. The X-X bonds are probably absent because the spacings between the nonmetal atoms are too large for the atomic interaction due to the electron shells overlap. The contribution of each chemical bond component in these compounds can be also readily evaluated by utilizing the geometrical constants of the structures and atoms (Table 3).

As can be seen from Table 3 the Me-Me spacing in carbides and nitrides is 3-8% greater than that in the structure of pure metals and, approximately by the same value, the Me-X spacing in the compounds structures is greater than the sum of covalent radii in metal and nonmetal. The departure of the bond length Me-X from the sum of ion radii of metal and nonmetal is a substantial one, but for oxides it differs by less than 2%.

TABLE 3 Atomic Spacings and their Deviation from the Sum of
Covalent (ΔC) and Ionic (Δj) Radii in Metal-like
High Melting Point Compounds

Compound	Me-X d, Å	ΔC,%	Δj,%	Me-Me d, Å	ΔMe,%
TiC	2.158	3.20	-33.4	3.065	4.9
ZrC	2.341	5.4	-32.5	3.324	3.9
HfC	2.319	4.9	-32.2	3.294	3.6
VC	2.059	3.5	-35.8	2.923	7.5
NbC	2.216	5.0	-34.2	3.147	7.0
TaC	2.205	4.5	-34.5	3.131	7.2
TiN	2.124	9.5	-11.5	3.016	4.0
ZrN	2.268	9.6	-	3.221	0.6
HfN	2.250	9.2	-	3.195	0.5
VN	2.196	12.0	-	3.118	6.0
NbN	2.068	12.4	-12.4	2.936	7.9
TaN	2.172	10.8	-	3.084	5.6
TiO	2.081	5.1	-1.8	2.955	1.2
VO	2.046	8.8	0.3	2.806	6.8
FeO	2.155	17.7	0.2	3.060	20.5
MnO	2.222	21.4	-0.3	3.155	21.3
MgO	2.100	2.9	0	2.982	-6.8

The correspondence of atomic spacings to the sums of covalent and ionic
radii in compounds indicates that the Me-X bond is a mostly covalent one
in carbides, ionic one in oxides and covalent-ionic one in nitrides. From
the geometrical data it follows that the s-electrons provide a metal bond
formation both in the compounds and in the metals, the d-electrons of the
metals may be considered to be electrons of ionic skeletons as they are
located much nearer to the nucleus of atoms. Bearing in mind that the
energies of d-electrons are rather near to the energies of valence electrons it is possible to calculate the binding energy for these compounds
by taking into account the electrostatic interaction of s- and d-electrons
following the principles of the bonds calculation in pure ionic compounds.
We propose, therefore, to define the total potential energy of NaCl-
structure high melting point compounds in the manner of calculating the
energy for ionic crystals (Gray, 1965):

$$E_p^o = -\frac{q_1 \cdot q_2 \cdot e^2}{d} + be^{-ad} - \frac{c}{d^6} \qquad 5$$

where q_1, q_2 are the charges of metal and nonmetal ions respectively, d is
the atomic spacing, a,b,c are the constants for each group of components.
For carbides, nitrides and oxides these values are unknown but they could
be obtained as follows. The atoms of nonmetals, as a result of the electrons liberated by atoms of metals, form ions with the electron pattern of
corresponding inert gases. Carbon, nitrogen and oxygen take up four, three
and two electrons respectively, forming the electron pattern of neon.

Transition metals Ti, V and Cr liberate electrons and form the electron pattern of argon, Zr, Nb and Mo form that of krypton and Hf, Ta and W form the xenon-type pattern of electrons.   Thus, to calculate the energy of compounds as for the ionic crystals using Eq. 5 it is possible to utilize the coefficients of the Van der Waals interaction of neon with argon, krypton and xenon (Table 4).   The a, b and c constants for the interaction of these gases were experimentally derived by Mason (1955).

TABLE 4   Coefficients of Van der Waals Interaction of
Metals with Carbon, Nitrogen and Oxygen

Metals	a	b	c	Inert gases
Ti, V, Cr	2.8	242	30.6	Ne-Ar
Zr, Nb, Mo	2.02	132	42.5	Ne-Kr
Hf, Ta, W	2.00	214	66.1	Ne-Xe

The potential energy defined using Eq. 5 is the energy required for divorcing Me^+ and X^- ions. The energy of dissociation of compounds into E_D atoms

$$MeX \longrightarrow Me + X \qquad\qquad 6$$

differs from the potential energy E_p^o by an energy value required for taking electrons off the $X^-(E_e^X)$ ion and their attachment to the $Me^+(E_U^{Me})$ ion

$$E_D = -E_p^o - E_U^{Me} + E_e^X \qquad\qquad 7$$

where E_{UX}^{Me} is the potential of the electrons ionization in the transition metal, E_e^X is the energy of the electron affinity in carbon (1.25 eV), in nitrogen (-0.1 eV) and in oxygen (1.47 eV).   The results of the binding energy calculation using Eqs. 5 and 7 are given in Table 5.

The covalency level of the compounds was defined in terms of ionicity as follows:

$$a_c = 1 - a_p \qquad\qquad 8$$

where a_p is the ionicity of compounds according to Pauling (1960):

$$a_p \cdot 100 = 16\,[X_A - X_B] + 3.5\,[X_A - X_B]^2 \qquad\qquad 9$$

where X_A, X_B are the values of electronegativity of the elements.

TABLE 5 The Binding Energy and the Covalency of
Monocarbides, Nitrides and Oxides

Compounds	E_p^o,eV	E_D,eV	E_o,eV	a_c,%
TiC	107.61	17.05	14.66	81
ZrC	98.05	20.77	15.75	78
HfC	98.61	20.96	17.01	76
VC	110.74	13.34	13.75	73
NbC	103.50	17.57	16.32	73
TaC	103.39	20.75	16.98	71
TiN	60.13	11.49	13.24	68
ZrN	56.67	12.02	14.96	66
HfN	56.70	13.70	15.98	63
VN	61.56	10.83	12.79	60
NbN	60.39	11.42	14.81	60
Ta	58.41	12.41	15.33	57
TiO	26.59	7.66	6.79	54
VO	26.94	7.47	6.43	44
FeO	25.98	3.37	4.28	63
MnO	25.54	3.94	4.19	54
MgO	26.44	5.24	4.41	45

As can be seen from Table 5, the agreement between the calculated and experimental binding energy values is quite reasonable. This is the reason that the transition metal compounds comprising carbon, nitrogen and oxygen and having NaCl-type structure can be considered to be ionic crystals. In any case it is possible to assume that these compounds are formed due to electron transfer from metal to nonmetal atoms resulting in M^+ and X^- ions formation. It should be noted here that the assumptions that these compounds are interstitial phases (the atoms of nonmetals embedded within the octahedral spacings of f.c.c. lattice of a metal) is an illogical one. The transition metals forming the NaCl-type compounds do not form f.c.c. phases and the metals forming f.c.c. structures do not produce compounds of NaCl-type.

From the structural point of view the compounds of boron with transition metal are complex substances. Me-B bonds are also covalent-ionic bonds and Me-Me bonds are metallic ones. However, in contrast to carbides, nitrides and oxides where nonmetal atoms do not form chemical bonds, the boron atoms in borides form strong covalent bonds. In view of the complexity of making covalent-mechanical calculations of binding energy values for these compounds, even rough ones, only the experimental binding energy values E_o are used for the further analysis.

BINDING ENERGY AND PROPERTIES OF SUBSTANCES

Nonmetallic and metal-like high melting point compounds, as can be seen from the above Tables, have chemical bonds of high energies. These define their mechanical and thermal properties. However, unlike the binding energy and the melting point, which are related directly, the mechanical properties are interrelated in a complicated way. Thus the binding energy of carbides and nitrides of metals of one group in the periodic system increases with the order number of the metal, and the hardness even goes down despite the fact that in substances of these types a dependence of a hardness enhancement on the binding energy rise is generally manifested (Table 6). The date in Table 6 shows that though the binding energy does define the properties of substances this interdependence requires, to define the direct correlation, a special consideration.

TABLE 6   Melting Points and Mechanical Properties of High Melting Point Compounds (Samsonov and Vinitzky, 1976)

Compounds	E_o,eV	T_{melt},K	E,GPa	G,GPa	H,GPa
TiC	14.20	3530	510	196	31.7
ZrC	15.58	3803	355	165	29.5
HfC	16.45	4163	514	225	28.3
VC	14.63	2921	430	159	24.8
NbC	16.62	3886	345	-	21.7
TaC	16.92	4258	-	-	17.2
TiN	13.20	3223	-	-	20.5
ZrN	14.50	3253	397	159	16.7
HfN	16.06	3273	-	-	16.0
VN	12.68	3223	357	142	13.1
NbN	15.02	2573	493	-	15.2
TaN	12.41	3360	-	-	22.4
TiO	6.79	2010	-	-	19.6
VO	6.43	2350	-	-	-
FeO	4.28	1641	-	-	5.4
MnO	4.19	2058	-	-	1.9
MgO	4.41	3073	-	-	7.4
SiO_2	16.3	1983	-	-	11.4
Al_2O_3	22.74	3333	380	-	20.6

Ions in solids are maintained in the dynamic equilibrium and vibrate with a definite frequency. According to Debye the phonon vibration frequencies range from 0 to a certian value ν_m which is characteristic for each substance, i.e. is defined by the bond strength (Ashcroft and Mermin, 1979):

$$E_o \propto 4\pi \nu_m^2 M \qquad\qquad 10$$

where M is the atomic mass. The limiting frequency is related to the force constants and the melting temperature T_{melt} through the Lindeman equations

$$\nu_m \propto \sqrt{\frac{2\beta}{M}} \qquad\qquad 11$$

$$\nu_m = 2.8 \cdot 10^{12} \sqrt{\frac{T_{melt}}{MV^{2/3}}} \qquad\qquad 12$$

where $\beta = E/d$, E is the modulus of elasticity, d is the atomic spacing and V is the atomic volume. Eqs. 10, 11 and 12 enable the correlation of the binding energy and the elastic and thermal properties of solids. It can be readily seen that the elastic modulus and the melting temperature of solids are proportional to the binding energy.

Harrison (1980) has established the following correlation between the binding energy, the elastic constants E and G and free surface energy γ :

$$E \propto \frac{E_o a_c}{d^2} \qquad\qquad 13$$

$$G \propto \frac{E_o a_c}{d^3} \qquad\qquad 14$$

$$\gamma \propto \frac{E_o a_c}{d^2} \qquad\qquad 15$$

HARDNESS OF SOLIDS

The hardness of solids is a complicated and complex property. Components of the work done during indentation will include - to a greater or lesser degree - elastic deformation, plastic deformation, crack initiation and the production of new surfaces. Nevertheless, if the relevant stresses are expressed through the known relationships using proportionality coefficients, then the material hardness may be expressed as the sum of these four components:

$$H = K_1 \sqrt{\frac{E\gamma}{d}} + K_2 \frac{Gb}{d_o} + K_3 \sqrt{\frac{2E\gamma}{L}} + K_4 \gamma \qquad\qquad 16$$

where E is the elastic modulus, G is the shear modulus, b is the Burgers vector, d, d_o are the atomic spacing and the distance between atoms of the adjacent shear planes respectively, L is the Griffith crack critical length and K_i are the proportionality constants.

Having taken into account Eqs. 13, 14 and 15 it is possible to define the hardness of substances in terms of the binding energy:

$$H = M_1 \frac{E_o a_c}{d^2\sqrt{d}} + M_2 \frac{E_o a_c}{d^3} \cdot \frac{b}{d_o} + M_3 \frac{E_o a_c}{d^2\sqrt{L}} + M_4 \frac{E_o a_c}{d^2} \qquad 17$$

Thus, the hardness of substances is defined by the binding energy, the covalency level, the atomic spacing and by the parameters of deformation characteristics and fracture, i.e. the Burgers vector and the Griffith crack critical length.

The dependence of the hardness on the atomic spacing is more conspicuous in the case of various binary compounds of the same atoms (Table 7). In all the cases, with no exception, the hardness of both monoborides and diborides is higher when the atom spacing is smaller.

TABLE 7   Atomic Spacings and Hardness of Borides

Compounds	Atomic spacing d, Å	Binding energy E_o, eV	Hardness H, GPa	a_c, %
TiB	3.81	11.95	27.0	
TiB_2	2.37	13.58	33.7	92
ZrB	2.32	13.64	35.0	
ZrB_2	2.54	15.28	22.5	90
HfB	2.31	14.98	-	
HfB_2	2.51	16.65	29.0	87
VB	2.96	12.32	-	
VB_2	2.31	13.67	28.0	80
NbB	3.16	13.58	21.9	
NbB_2	2.43	15.70	26.0	80
TaB	3.15	13.80	31.8	
TaB_2	3.41	15.75	25.0	82
CrB	3.00	-	12.0	
CrB_2	2.30	-	21.0	80

The dependence of the covalent substance hardness on the binding energy is non-linear (Fig. 1). The covalency dependence of the hardness can be rather well approximated by a straight line (Fig. 2); the higher the covalency level in the compounds the higher is the hardness. As follows from Eq. 17, all the components of the sum include $E_o a_c/d^2$ - i.e. the hardness of a substance should be linearly dependent on this value and this is well illustrated in Fig. 3. The correlation coefficient in the latter case is equal to 0.80, whereas the correlation between covalency and hardness is 0.72. It should also be noted that the correlation coefficient is underestimated because the material hardness value is defined with accuracy of ±10.%.

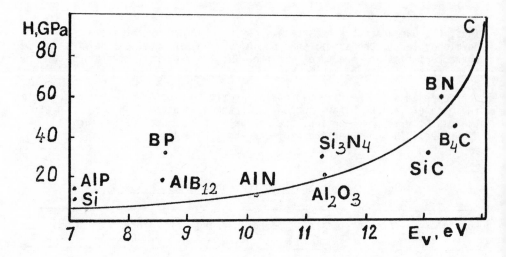

Fig. 1   The correlation of the hardness and the binding
energy in covalent substances.

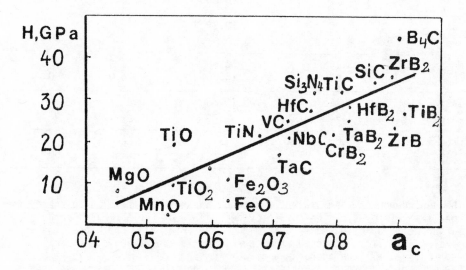

Fig. 2   The covalency dependence of hardness
for high melting point compounds.

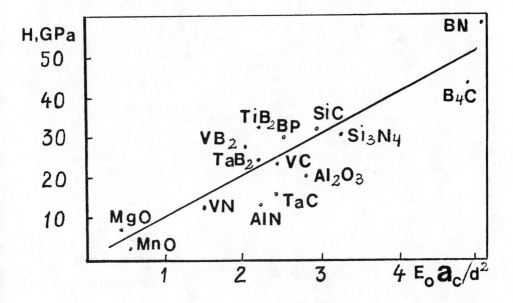

Fig. 3 The dependence of hardness on $E_0 a_c/d^2$ parameter for high melting point compounds.

CLASSIFICATION OF SUBSTANCES BY THEIR HARDNESS

High-melting point compounds include substances of very high hardness (e.g. BN - 60 GPa) and those of comparatively low hardness (e.g. MnO - 1.9 GPa) which is lower than that of many metals. But due to the entirely different nature of their deformation we must consider all of them to be very hard and all the metals to be hard substances.

In very hard substances and in hard ones the shear strain is accompanied, respectively, by the rupture of bonds and by sliding without forming new surfaces. The substances are generally assumed to be very hard if the yield stress/elastic modulus ratio exceeds 0.01 ($\sigma_y/E > 0.01$) and hard if $\sigma_y/E < 0.01$.

It is useful to single out superhard substances from the class of very hard ones because the term 'superhard' is widely accepted and usually refers to diamond and cubic boron nitride (CBN). At the same time, the alloyed boron carbide has a hardness of up to 75 GPa, which is higher than that of cubic BN, but it is not termed a 'superhard' material.

Unfortunately, it is not possible to find any physical principle instrumental for the delineation of the superhard materials class. This could be explained by the fact that the strain resistance both in metals and in ionic and covalent crystals decreases monotonically with increasing temperature. The abrupt jump in strength at the brittle-ductile fracture transition temperature T_x cannot be taken as a basis for their differentiation by hardness as well, because this is typical for all the classes of substances.   Therefore, the restriction in terms of 'superhard' materials can be introduced only by the level of a substance's hardness.

The delineation of 'superhard' substances could be done on the basis of the Mohs hardness scale introduced as far back as 1820. The scale comprises ten substances which are bearers of a standard hardness level. The scale is arbitrary in absolute hardness units. This becomes quite obvious from Fig. 4 where the ratios of Mohs hardness and microhardness values are shown. The first nine of the Mohs hardness scale units cover the hardness of all the known natural compounds. The microhardness for all of them is lower than 20.0 GPa. It is logical, therefore, that all the materials with hardness higher than that of corundum (i.e. > 20.0 GPa) should be referred to as 'superhard' materials, the materials with hardness higher than that of metals (i.e. from 5.0 to 20.0 GPA) should be considered 'very hard' materials and those with hardness lower than 5.0 GPa should be referred to 'hard' substances.

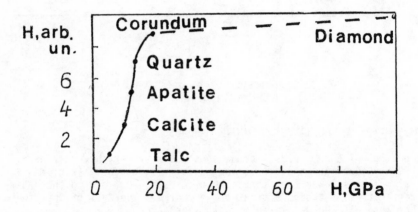

Fig. 4   The correlation of the Mohs hardness and microhardness in substances.

Among the nonmetallic high melting compounds there are about ten superhard substances which form a pyramidal diagram of hardness (Fig. 5). The metal-like high melting point compounds include a rather large number of superhard substances (Fig. 6). They also form a pyramid of hardness with boron compounds being at the vertex.   Seventy-four metal-like superhard compounds in total are known nowadays: fifty-seven borides, fifteen carbides and two nitrides. The 'pyramids' of hardness place a sharp restriction on the elements which can be utilized in the design of new superhard materials using the principles of isovalent substitution and the formation of ternary and quaternary compounds with high pressure methods and technology.

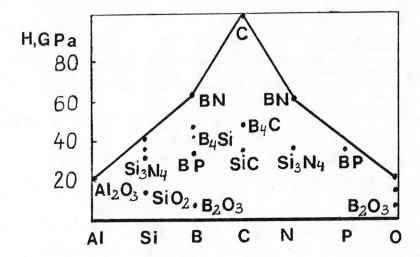

Fig. 5 The pyramidal diagram of hardness for covalent substances.

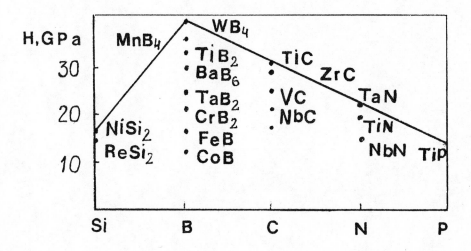

Fig. 6 The pyramidal diagram of hardness for metal-like substances.

REFERENCES

Ashcroft, N. and Mermin, N. 1979, Physics of Solids (in Russian), Mir Moscow
Gray, H.B. 1965, Electrons and Chemical Bonding. W.A. Benjamin Inc. New York - Amsterdam

Harrison, W.A. 1980, Electronic Structure and the Properties of Solids
    W.N. Freeman and Co. San Francisco
Mason, E.A. 1955, J. Chem. Phys. 23 49-55
Pauling, L. 1960, The Nature of the Chemical Bond. Cornell University
    Press, Ithaca, New York
Samsonov, G.V. 1976, The Properties of Elements (in Russian),
    Metallurgiya, Moscow
Samsonov, G.V. and Vinitzky, I.M. 1976, High Melting Point Compounds
    (in Russian), Metallurgiya, Moscow

Inst. Phys. Conf. Ser. No. 75: Chapter 2
Paper presented at 2nd Int. Conf. Science Hard Mater., Rhodes

121

Bonding of metal/ceramic interfaces

E D HONDROS

Commission of the European Communities
Joint Research Centre Petten Establishment
P.O. Box 2 1755 ZG Petten, the Netherlands

ABSTRACT

An attempt has been made to provide the basis for predictive guidelines for designing metal/ceramic couples of a given interfacial bond strength. The approach is based on interfacial thermodynamics. A reasonable inverse correlation is shown to exist between the work of adhesion and the metal/ceramic interfacial free energy. From this, it is predicted that segregates at equilibrium to such interfaces will reduce the interfacial free energy and thus improve adhesion. We describe experiments to test this model using Fe/Al_2O_3 and Fe/ThO_2 couples with specific, surface-active additives.

INTRODUCTION

There is wide scientific and technological interest in the bonding of metals to ceramics, which is reflected in the needs of a variety of materials-based industries. In hardmetals, for example, reliable adhesion is expected between tungsten carbide particles and the cobalt matrix. In diverse fields of metallurgy, this adhesion is a basic factor - in the stability of precipitates or second phase particles, in the materials science of composites and in thin film metallisation layers in microelectronics devices of complex architecture.

In general, adhesion in metal/inorganic compound systems is a central controlling parameter, both for mechanical properties as well as for kinetic properties, such as growth or ripening processes or phase morphology instabilities, where adhesion is related to the interfacial free energy between the metal and the ceramic. The interest in adhesion stems primarily from a need for high strength - the general case - as in seals and cemented carbides. However, ironically, there are also requirements for designs with low adhesion: for example, cutting tools such as cemented carbides or alumina operating on metals require low adhesive properties to obviate wear; and in another industrial example, the glass-to-metal adhesion should be low in order to avoid the sticking of hot glass to steel rolls during production. Indeed there are situations where bonds of intermediate strength are required, as in the synthesis of certain composite materials designed to avoid catastrophic failure in which the interface acts as an energy absorber by decohering before the onset of sudden fracture.

Ideally, we require a theory of design of materials pairs with the means of predicting a defined interfacial bond strength, given the performance variables such as the loading conditions, as well as the materials chemistry conditions, such as the interfacial composition and expressed in terms of a practical processing schedule. This paper aims to explore one direction which might lead to the development of such a design methodology.

The Metal/Compound Interface

Because of the atomistic complexity of the structure of the metal/ceramic interface and because the bonding derives from fundamental physical and chemical interactions as well electrical polarisation, we approach this phenomenon from the point of view of interfacial thermodynamics. Conceptually this should encompass all the diverse physical and chemical sources of bonding. This thermodynamic "Work of Adhesion" as defined in the classical Yound-Dupré formulation is simply the algebraic difference between the surface free energies of the two separated surfaces and the pre-existing interface. Traditionally, in the surface chemistry of solid/liquid systems, this expression has been accepted as a measure of the real strength of the joint.

Here, we extend this concept to adhesion at a solid/solid interface. This notion has been suspect in the past because of the observation that in many adhering metal/compound systems, chemical reactions take place at the interface and hence there does not exist a well defined, two dimensional surface of contact but a chemically diffuse interlayer. Clearly, where such reactions occur, adhesion as defined here cannot be meaningful and the strength of the bonded layer is determined by mechanical events, such as internal stress concentrations, fracture nucleation and propagation. Here we shall be concerned with the general case, that is, forms of adhesion at interfaces between dissimilar solids in which contact is established at a two-dimensional plane of separation, with atom-to-atom bonds: this interface is in every sense the analogue of the grain boundary in crystalline single phase systems.

There is now copious evidence, both indirect and direct observations, that in many systems the interfaces are of the type described above. The sharpness of the interface has been revealed by a variety of high resolution electron/optical techniques. The example shown in Fig. 1(a) is of a field ion microscopy view of a tip on iron with individual atoms resolved, the dark bands representing lamellae of the carbide phase (Smith, 1982). The corresponding chemical analysis in manganese is indicated by the traverse at right angles to one of the lamellae across the field of view Fig. 1(b). This was measured with an atom-probe time-of-flight mass spectroscope attachment. The analysis shows the manganese concentration profile across the interfaces separating the carbide from the metal. The point of interest here is that the thickness of the interface zone over which high localised concentration exists is compatible with the dimensions of the interface zones over which segregation has been measured in grain boundaries of polycrystalline materials.

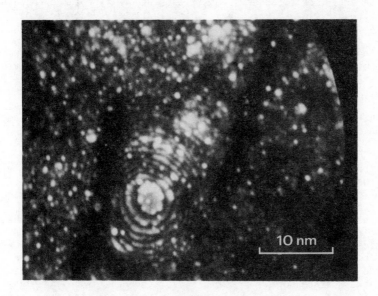

Fig. 1(a) Field ion microscope image of iron (ferrite)
   tip with carbide lamellae (black bands) (Smith).

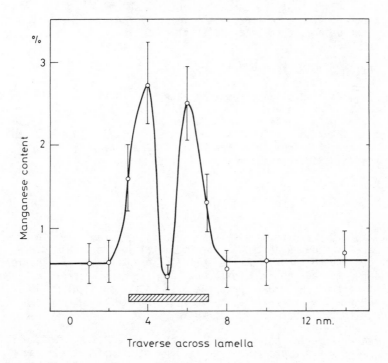

Fig. 1(b) Composition in Mn across a lamella of Fig. 1(a) (Smith).

From diverse experimental observations, it is generally assumed that the practical strength of a metal/ceramic joint is largely determined by extrinsic factors such as the manner of loading, stress concentrations, plastic relaxation work terms and other elastic/plastic terms. It is pertinent then to ask why this study of metal/ceramic bonding is approached from the thermodynamic standpoint. First, this is an amenable way of modelling this type of complex interface, second, and of more practical value, empirical observations lead to the view that, in general, systems with a high work of adhesion as measured by the classical approach are also strongly bonded, even though the experimentally determined strength of adhesion does not correspond absolutely to the thermodynamic work of adhesion.

One way of broaching this apparent inconsistency is through the paralellism with the brittle fracture of crystalline solids. In a perfectly brittle solid, the work of fracture corresponds to the free energy requirements for creating the exposed surfaces of rupture, γ_s, whereas in practice, the surface work term γ_p is orders of magnitude higher, due to the plastic relaxation accompanying the opening of the crack. In a number of theoretical treatments, it has been established that the value γ_p is controlled by that of γ_s. Thus, the higher the value of γ_s, which reflects the intrinsic thermodynamic cohesion, the higher will be the localised stress state during a fracture event and therefore the higher the probability of stress relief through a suitably oriented slip system. In the case of polycrystalline metals, the precise relation between γ_p and γ_s will depend on orientation and stress state terms: in their analysis, McMahon and Vitek (1979) show that $\gamma_p \, \alpha \, \gamma_s^n$ where n lies in the range 2-5. Hence applying these concepts to the metal/ceramic interface, it is not difficult to see how the value of the thermodynamic work of adhesion can determine the real joint strength.

WORK OF ADHESION OF METAL/CERAMIC INTERFACES

One reason for the lack of an acceptable scientific basis for understanding adhesional phenomena in the present systems arises from the confusion caused by projecting data obtained in metal wetting situations to solid/solid interfaces. This is due to the paucity of values of solid-state surface energies from which the Work of Adhesion, W_{ad}, is determined according to the Young/Dupré equation:

$$W_{ad} = \gamma_M + \gamma_C - \gamma_{M/C}$$ 1

Where M, C and M/C refer respectively to the metal, ceramic and interface. Thermodynamically, the quantity W_{ad} for the metal in the liquid state should not correspond to W_{ad} for the solid/solid interface. Furthermore, Nicholas (1968) tested the hypothesis that W_{ad} for temperatures just above the melting point of the metal approximate to solid state values of W_{ad}. His measurements show clearly that a correlation between wetting behaviour values and solid state bonding is not valid.

Here we consider experimentally determined evaluations of W_{ad} in the solid state. This requires measurement of the solid state surface free energies. The valid techniques which exist for these measurements have been extensively reviewed elsewhere (Hondros, 1970). In summary, measurement of the surface free energy of metals requires the use of the "zero-creep" approach for the most consistent, absolute measurements, which can be obtained with an accuracy of within ±5%. For the energies of internal interfaces, no simple absolute technique exists and in general, polyphase equilibria have been applied for relative interfacial energy measurements. By selecting a suitable reference surface, the absolute energy of which can be measured, the relative interfacial quantities can be converted to absolute values.

This is illustrated for one type of equilibrium configuration in Fig. 2. This shows schematically a metal bonded to a ceramic, each member of which has a grain boundary intersecting the interface normally. Where such interfaces intersect, there will be an adjustment to an equilibrium configuration of a minimum free energy, where the local surface tensions are in vectorial balance. Some typical microgeometries are shown in Fig. 2, from which, McLean and Hondros (1971) measured the grain boundary groove angles θ_1 to θ_4 by interference microscopy, noting that the sapphire used in these experiments is transparent in these conditions. In the above configurations, the equilibria hold:

$$\gamma_{M/M} = 2\gamma_{M/V} \cos \frac{\theta_1}{2} = 2\gamma_{M/C} \cos \frac{\theta_2}{2} \qquad 2$$

$$\gamma_{C/C} = 2\gamma_{M/C} \cos \frac{\theta_3}{2} = 2\gamma_{M/C} \cos \frac{\theta_4}{2} \qquad 3$$

Where the subscripts relate to the interface and are defined in Fig. 2. Thus, from a knowledge of $\gamma_{M/V}$ obtained by the absolute method mentioned above, all other quantities can be measured.

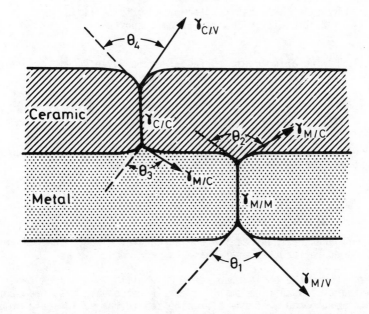

Fig. 2 Equilibrium configurations in a metal/ceramic couple where grain boundaries intersect the surfaces.

A technique which has yielded much useful data is the solid state equivalent of the classical sessile drop resting on a ceramic substrate. This is illustrated in Fig. 3, which is drawn schematically to indicate the polyhedral form assumed by a solid particle at equilibrium.

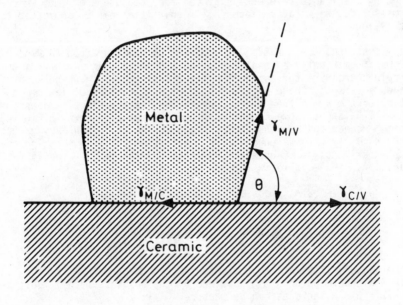

Fig. 3 Schematic diagram of a solid particle
at equilibrium on a ceramic substrate.

The crystal, having anisotropic surface free energy exposes low energy facets. At equilibrium, the relation holds:

$$\gamma_{M/C} = \gamma_{C/V} + \gamma_{M/V} \cos \theta \qquad\qquad 4$$

where the energy terms are as defined in the diagram. From the Young-Dupré equation, the thermodynamic work of adhesion is:

$$W_{ad} = \gamma_{M/V} (1 + \cos \theta) \qquad\qquad 5$$

In this configuration, besides the contact angle, clearly the absolute surface free energy of the metal phase is required.

In order to ensure the achievement of true equilibrium in such experiments, the particles must be small to permit reasonable mass transport and shape adjustment because of the slow kinetics in the solid state compared with the liquid state. The method has been used successfully by a number of workers, including Pilliar and Nutting (1967) who measured the drop dimensions by electron microscopy shadowgraphs and by Murr (1976) who applied Scanning Electron Microscopy. An example from the work of Murr is shown in Fig. 4 which demonstrates nickel particles of sub-micrometre size resting on a thoria substrate and equilibrated for 200 hours at 1473K.

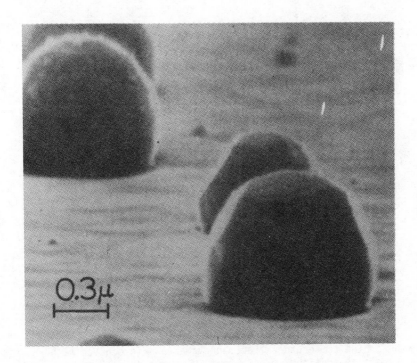

Fig. 4 Ni particles at equilibrium on a ThO$_2$ substrate
(from Murr, 1976).

This indicated clearly the faceted particles from which the contact angle can be measured with a fair accuracy. In general, much care must be taken with these techniques to obtain the equilibrium configuration and to prevent inaccuracies arising from contamination in the ambient atmosphere. The problem here is to ensure that the reference absolute surface energy of the metal, which is measured separately, is strictly that pertaining to the atmospheric conditions of equilibrium of the fine particles. These are produced either by the break-up of a thin deposited film of metal on the substrate or by fine individual liquid droplets which are allowed to cool through the melting point.

A variant of the equilibrium drop configuration introduced by Hancock, Dillamore and Smallman (1968) is shown in Fig. 5. This represents a particle entrained in a solid metallic matrix which had previously been subjected to deformation, so that the particle decoheres from the matrix.

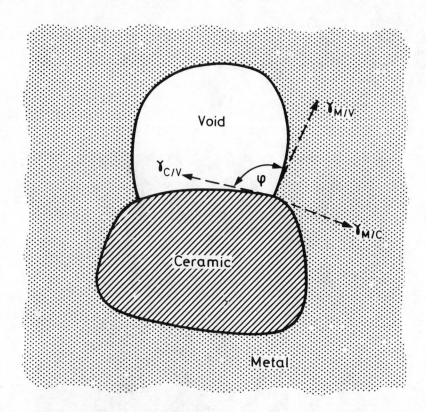

Fig. 5 Equilibrium configuration of a void at a particle,
entrained in a metal matrix.

Following a high temperature treatment, the void in contact with the particle assumes an equilibrium form from which the contact angle φ can be
measured by transmission electron microscopy. The equations above can be
employed to yield, W_{ad}, the value of M/V being obtained independently as
before. This approach has much to commend it, principally because the
void should be free of the contamination that besets the exposed droplet
configuration of Fig. 3.

ESTIMATED WORK OF ADHESION IN METAL/CERAMIC SYSTEMS

Data arising from various sources and involving the application of
measurement techniques of the type described above have been assessed
and presented in Table 1.   With the exception of Fe/Fe_3C, these refer to
metal/oxide systems, all the data being obtained for couples in the solid
state, and the sources are Nicholas, 1968 (ref 1 in the Table);   Palmer
and Smith, 1968 (2); Feingold, Blokely and Che-Yu Li, 1964 (3);   Martin
and Sellars, 1970 (4); McLean and Hondros, 1971 (5); Hancock and Dillamore,
1968 (6); Navara and Easterling, 1971 (7); Murr, 1976 (8). The thermodynamic work of adhesion was calculated in each case either directly from
the algebraic sum of the interfacial energies (Eq. 1) or from contact

TABLE 1 Interfacial Energetics in Metal/Compound Systems

System	Technique	Temperature K	Interfacial Energy mJm^{-2}	Work of Adhesion mJm^{-2}	(Ref)
Au/Al_2O_3	Equilibrium profile of metal particle	1273	1725	530	(1)
Ag/Al_2O_3	"	973	1630	435	(1)
Cu/Al_2O_3	"	1123	1925	475	(1)
Ni/Al_2O_3	"	1273	2140	645	(1)
$\gamma Fe/Al_2O_3$	"	1273	2065	800	(1)
Cu/SiO_2	Contact angle measurement	1273	740	1440	(2)
Ni/Al_2O_3	Equilibrium grain boundary groove at M/O interface	1623-1698	2500	226	(3)
Fe/Fe_3C	Equilibrium particle shape		1040	1928	(4)
Pt/Al_2O_3	Grain boundary groove at M/O interface		1050	1817	(5)
Ni/Al_2O_3	Contact angles on decohered particles	748	2026	700	(6)
$Co-40Ni/Al_2O_3$	"	1273		2815	(6)
$Fe-40Ni/Al_2O_3$	Contact angle on decohered particle	1273	2505	400	(7)
$Fe-40Ni-10Cr/Al_2O_3$	"	1373	1219	1030	(7)
$Fe-5Cr/Al_2O_3$	Critical strain to fracture	1373	269	1980	(7)
$Fe-15Cr/Al_2O_3$	"	1373		2850	(7)
$Fe-5Mo/Al_2O_3$	"	1373	1580	1120	(7)
Fe/Al_2O_3	"	1373	2440	760	(7)
Ni/ThO_2	Equilibrium profile of metal particle	1473		1100	(8)
$80Ni-20Cr/ThO_2$	"	1473		700	(8)
$304SS/Al_2O_3$	"	1473		1440	(8)

angle measurements (Eq. 3). Clearly in the derivation of W_{ad} by these methods, we presuppose adhesion across an ideal, two dimensional interface.

The data for W_{ad} are plotted as a function of the interfacial free energies for each metal/compound pair and presented in Fig. 6. Within the scatter of the data points, there exists a reasonable inverse correlation between the work of adhesion and the interfacial free energy. This is consistent with the notion of adhesion across a planar interface, with atom to atom matching across it: the more ordered the interface, the lower will be the associated free energy and the stronger the bonding.

The practical, or observed strength of the interfacial bond need not be simply reflected in the computed thermodynamic work of adhesion, for the reasons noted earlier. While at this stage it has not been found possible to measure in a quantitative manner the real strength of the interfaces represented in Fig. 6, there is a fair qualitative evidence based on observations reported in the literature that in the present systems, the most strongly adherent pairs correspond to those with the highest values of the thermodynamic work of adhesion. Thus, consistent with observed

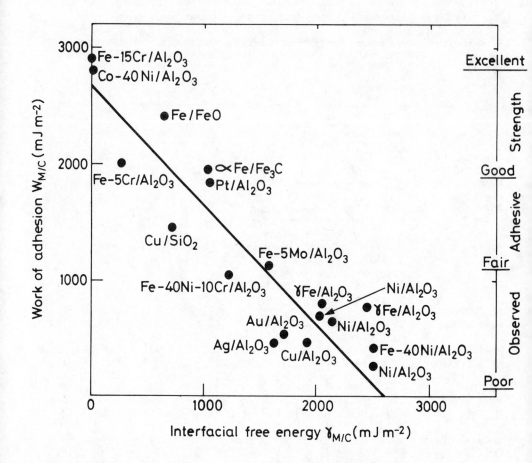

Fig. 6 Empirical relationship between the thermodynamic work of
adhesion and the interfacial free energy for metal/compound pairs.
Details of data are in Table 1.

practice, pure Fe and Ni do not adhere well to alumina and as shown in the
diagram these tend to have a low work of adhesion. The addition of Cr to
Fe lowers the interfacial energy and increases the work of adhesion.

The important deductions from the correlation of Fig. 6 are; first, the
fact that experimental data, obtained from various sources and systems and
based on different measurement techniques, can be plotted to yield such a
correlation confirms that in the general case, adhesion at a metal/ceramic
interface occurs across a narrow zone for which the concepts of interfacial thermodynamics are valid - this would not be the case if the bonding occurred across a diffuse reaction interlayer. Second, a reasonable
guideline for the strength of adhesion in such systems is the thermodynamic work of adhesion W_{ad} and the quantity that reflects this is the
interfacial free energy $\gamma_{M/C}$. Thus by manipulating $\gamma_{M/C}$, such as by
introducing segregating dopants, it is possible in principle to effect
changes in the adhesive strength. Below we shall test this hypothesis.

INTERFACIAL SEGREGATION AND ADHESION

As is evident from Fig. 6, alloying additions may influence profoundly the adhesion of a metal to a ceramic. Thus, whereas the work of adhesion of Fe to Al_2O_3 is, according to published data, of the order of 700 mJ/m^2, the addition of Cr to the Fe will progressively improve the adhesion to values above 2,000 mJ/m^2. Indeed, in the case of Fe/15% Cr, Navara and Easterling (1971) observed that the adhesion to Al_2O_3 was so strong that particles with associated voids could not be seen after 75% deformation. The presence of 40% Ni in the Fe/10% Cr alloy reduced the adhesion to well below that of Fe/5% Cr.

These observations can be rationalised in terms of the microchemistry of the metal/Al_2O_3 interface. To test these ideas, we prepared samples of Fe containing levels of Cr, Ni, Mn and Mo corresponding to the alloys examined by Navara and Easterling whose data are shown in Fig. 6. Their samples contained Al_2O_3 particles, the small size of which make it very difficult to examine the Al_2O_3/metal interface by standard techniques of surface analysis (although in recent years instruments are becoming available which permit this level of spatial resolution). Here, simulated samples were prepared suitable for Auger electron Spectroscopy analysis by melting beads of the alloy on flat sapphire substrates and following a heat treatemnt at 1273K, they were debonded on a specially designed fracture stage, **in situ** in ultra high vacuum in the chamber of the spectrometer. The metal component of each pair was analysed sequentially with distance from the debonded interface, using argon ion sputtering to remove the surface layers gradually. The concentration profile for Cr with distance from the interface is shown in Fig. 7. This confirms that Cr is highly enriched at the interface over a very narrow zone of a few nm, for the Fe/15% Cr alloy. The addition of 1% Mn reduces this Cr segregation significantly, while in the alloys containing 40% Ni, the interfacial enrichment is barely above the level of Cr in the bulk.

The segregation at equilibrium of Cr to the metal/Al_2O_3 interface has reduced the interfacial energy - this is the behaviour expected in all interfacial systems in which Gibbsian adsorption occurs at the interface. Structurally, the interface has re-ordered into a more relaxed structure with a lower free energy. This favours a high interfacial adhesion.

The corresponding concentration profiles in terms of Ni show that from a bulk concentration of 40% Ni, there are approximately equal numbers of Fe and Ni atoms in the interfacial layer adjacent to the Al_2O_3. This is again consistent with the work of adhesion measurements of Fig. 6 which show that the adhesive strength of this alloy lies between that of Fe/Al_2O_3 and Ni/Al_2O_3.

Experiments to Adjust Bonding Strength by Interfacial Doping

The correlation in Fig. 6 implies that the work of adhesion can be increased by any process that reduces the interfacial energy. As was shown in the previous section, the equilibrium adsorption of bulk species at the interface should reduce the free energy. This was demonstrated for alloys that contain a substantial fraction of the active species, here Cr in Fe.

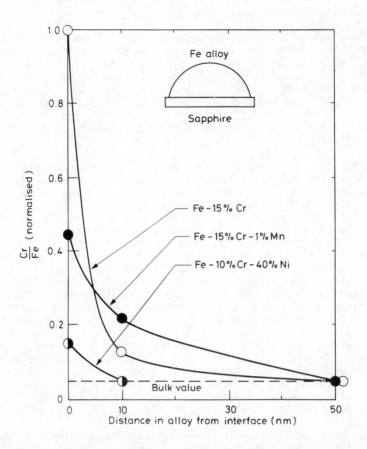

Fig. 7 Concentration profile for Cr with distance from the
interface in Fe alloys. Zero point on the abscissa corresponds
to the plane of the interface.

There are advantages in the engineering of metal/compound couples in
introducing highly surface active species at the level of dopants rather
than alloying elements. At very low levels the dopants should not affect
the mechanical properties of the matrix and should serve as selective
agents for improving the adhesion by enriching the interface. There are
reliable observations in the literature which affirm that surface active
species adsorb at precipitate/matrix interfaces, thereby reducing the
interfacial energy. In principle, this should be possible if the reduction in interfacial energy is much greater than the reduction in the
surface free energies of the separated metal and ceramic components.
Below we have examined this possibility, using Fe as the metal component,
on Al_2O_3 and ThO_2 substrates, and P and Sn as the surface active species.
The main reasons for the choice of these dopants is that they are known to
be surface active in metallurgical systems i.e. they adsorb at interfaces
and, in addition, the absolute values of the surface free energies of Fe
containing these species have been measured.

The sample of pure Fe containing fine dispersions of Al_2O_3 and ThO_2 were first hot extruded and annealed at 1273K for 5 hr, followed by the doping procedure. This consisted of heating the samples at 1373K for 10 h in a high vacuum and in close proximity to master alloys of Fe containing Sn and P respectively. The latter were in the form of thin sheet wrapped around the samples which permitted the transfer of the dopant species in closely controlled conditions. The doped samples were subsequently deformed at room temperature with reductions of up to 70%, then annealed at 1373K for 2 hr in order to equilibrate the voids about the particles. Thin foil specimens for transmission electron microscopy were prepared and examined by high voltage electron microscopy. For each sample, a large number of representative events showing particles with equilibrated voids were recorded and analysed.   This statistical approach was necessary because of variations in the extent of decohesion and the attendant measurement errors. The series of micrographs in Figs. 8-9 are representative of the encountered behaviour. Fig. 8(a), for undoped Fe/Al_2O_3, is typical for this system, giving a rather average capability for measuring the contact angle with accuracy. Up to fifteen measurements at this level of resolution or better were recorded and the value of W_{ad} was calculated using Eq. 5 and the known surface free energy of the metal.

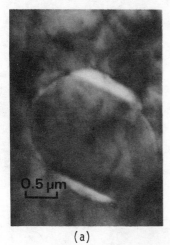

(a)   (b)

Fig. 8   Equilibrated voids on Al_2O_3 particles in:
(a) pure Fe (X 50K);   (b) pure Fe doped with Sn (X 43K).

The Fe/Al_2O_3 series with P additions showed on the average an increase in the measured contact angle from 42° for the pure Fe sample to 52° for the doped sample. However, the value of the surface energy of Fe in the presence of P is significantly lower than for the pure metal so that the calculated net effect on W_{ad} is rather modest. The effect on Sn dopant is however measurably higher. An example of the microstructures measured is shown in Fig. 8(b). From these data and the known value of γ_M in the presence of Sn, the work of adhesion appears to have been improved significantly.

The Fe/ThO_2 series without dopants indicated such prolific decohesion that only some rare measurable events could be found. A typical microstructure, Fig. 9(a), indicated complete decohesion of the ThO_2 particle whereas the

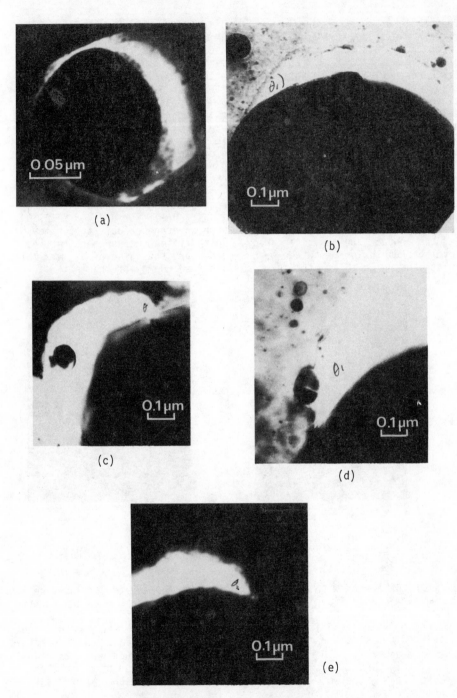

Fig. 9 Equilibrated voids on ThO$_2$ particles in (a);
(b) pure Fe (X 50K); (c) (d) pure Fe doped with Sn (X 70K);
(e) pure Fe doped with P (X 30K).

rarer event of Fig. 9(b) allows a reasonable measure of the contact angle and thus the work of adhesion. The dopant additions had a marked visual improvement in the microstructures, with fewer disbonded structures and many samples showing larger contact angles than in the case of the pure metal, (Fig. 9(c) to 9(e)).

The results of these experiments are summarised in Table 2. We note that the value of W_{ad} for Fe/Al_2O_3 measured here is somewhat lower than reported elsewhere. Nevertheless, additions of Sn and P procure an increase in W_{ad} in accordance with the earlier predictions. The best measured effect, that of Sn in Al_2O_3 is significant but not remarkable compared with the large effect of several percent of Cr. In the Fe/ThO_2 system, the low measured adhesion can be doubled by dopant additions. This however is not a very high adhesive strength when viewed on the scale of Fig. 6.

TABLE 2 Work of Adhesion in the Systems Fe/Al_2O_3 and Fe/ThO_2 in the presence of Sn and P dopants

Metal/Ceramic Couple	Number of Measurements	Work of Adhesion (mJm^{-2}): mean value
Fe/Al_2O_3	11	513
Fe/Al_2O_3 + Sn	9	837
Fe/Al_2O_3 + P	7	615
Fe/ThO_2	6	207
Fe/ThO_2 + Sn	8	460
Fe/ThO_2 + P	7	470

In conclusion, these experiments confirm that the basic model of thermo-dynamic adhesion in metal/ceramic systems holds in terms of the predictions of the model, although the dopant species selected here do not effect a remarkable improvement in adhesion.

ACKNOWLEDGEMENT

The author is grateful for the contribution of Mr C Hunt, especially in the elegant microscopy measurements.

REFERENCES

Feingold, A.H., Blakely, J.M. and Che-Yu Li, 1969, Grain Boundary Grooving at a Solid-Solid Interface in 'Kinetics of Reactions in Ionic Systems', Vol. 4, Materials Science Research, Ed. T.J. Gray and V.D. Frechette, Plenum Press, N.Y. pp 304-314
Hancock, J.W., Dillamore, I.L. and Smallman, R.E. 1968, 6th Plansee Seminar, pp 467-493
Hondros, E.D. 1970, Physicochemical Measurements in Metals Research, Ed. R.A. Rapp, Techniques of Metals Research Series, Interscience Publishers, N.Y. Vol. IV, Part 2, pp 293-348

Hondros, E.D. and Seah, M.P. 1983, Physical Metallurgy, Ed. R.W. Cahn
    and P. Haasen, Elsevier Science Publishers B.V., Ch. 13
Martin A.G. and Sellars, C.M. 1970, Metallography, 3 259-273
McLean M. and Hondros, E.D. 1971, J. Mat. Sci, 6 19-24
McMahon, C.J. and Vitek, V. 1979, Acta Metall, 27 507
Murr, L.E. 1976, Adhesion Measurements of Thin Films, ASTM, STP 640, p 82
Navara, E. and Easterling, K.E. 1971, Jernkont Ann, 155 437-441
Nicholas, M. 1968, J. Mater. Sci, 3 571
Palmer, I.G. and Smith, G.C. 1968, Proc. Second Bolton Landing Conference
    on Oxide Dispersion Strengthening, Gordon and Breach, N.Y.
Pilliar, R.M. and Nutting, J. 1967, Phil. Mag, 16 181-188
Smith, G. University of Oxford, Metallurgy Dept (private communication)

Inst. Phys. Conf. Ser. No. 75: Chapter 2
Paper presented at 2nd Int. Conf. Science Hard Mater., Rhodes

Development of accurate estimation methods for calculating thermal expansivities of hard materials

O. H. KRIKORIAN

University of California, Lawrence Livermore National Laboratory,
P. O. Box 808, L-369, Livermore, California 94550

ABSTRACT

On the basis of qualitative theoretical arguments and correlation studies, we find the thermal expansivities, β, of borides, carbides, nitrides, and oxides can be predicted to within about ±10% based on two parameters: ΔE_a, the atomization energy, which is a measure of bond strength; and h, the microhardness, which is a measure of the steepness of the extension side of the potential energy well. The correlation is described by the relation $\beta = 21.1 \ \Delta E_a^{-1} (T/n)^{1/3}$, where ΔE_a is in kJ/g-atom and h is in HK units.

INTRODUCTION

As new fabrication technologies and applications areas continue to develop for hard materials, we find that we have an increasing need for better physical and mechanical property data. These data are needed not only for hard materials in monolithic forms, but also for their use as coatings, cermets, composites, and as cementing materials for joining of parts.

Accurate data on thermal expansivities are especially useful for the selection and development of hard materials for coatings, where both the coating adherence to the substrate and the resistance of the coating to thermal stress fracture depend strongly on obtaining a good thermal expansion match between coating and substrate. Thus, we see that accurate values of thermal expansivities fulfill an important need in the proper selection and design of hard material coatings. Similarly, accurate thermal expansivities of hard materials can be shown to fulfill a need for proper materials selection and design of cermets, composites, and cements for joining of parts. In reviewing the literature, we find that accurate experimental thermal expansivity data for hard materials are limited to a very small fraction of the hard materials that exist (Touloukian and co-workers, 1977). Thus, the calculation of accurate thermal expansivities by theoretical or empirical methods would allow us to narrow the selection of candidate materials to a relatively small number that can be more readily tested and evaluated for a given potential application.

Unfortunately, there are no generally applicable prediction methods that fulfill our needs for thermal expansivities of hard materials at the pre-

*Work performed under the auspices of the U.S. Department of Energy by the Lawrence Livermore National Laboratory under contract number W-7405-ENG-48.

sent time. Theoretical approaches usually use a quasiharmonic approximation such as the Debye equation of state that requires input values of
Grüneisen parameters, heat capacities, and isothermal bulk moduli, or equivalent input data in terms of other parameters (Touloukian and co-workers,
1977; Zharkov and Kalinin, 1971; Kittel, 1966). These parameters are not
generally available for hard materials, so that the theoretical approaches
are currently of very limited value as a general predictive tool.

An empirical method that has been useful for predicting the thermal expansivities of metals and ionic compounds that have simple structures, is
based on a correlation between thermal expansivity and melting point. This
type of correlation shows that for a pure metal with a simple bcc, hcp, or
fcc structure there is a volume expansion of about 8% between absolute zero
and the melting point (Krikorian, 1971). For alkali halides, a similar
correlation shows a volume expansion of about 14%. (Touloukian and co-
workers, 1977; Krikorian, 1971). However, for hard materials the correlation is poor. For example, the volume expansions to the melting point for
MgO and Al_2O_3 are 14% and 5%, respectively (Touloukian and co-workers,
1977). Hence, we need a better predictive method for estimating thermal
expansivities for hard materials.

EQUATION OF STATE OF SOLIDS BY THE METHOD OF POTENTIAL ENERGY FUNCTIONS

According to theory (Touloukian and co-workers, 1977; Zharkov and Kalinin,
1971; Kittel, 1966), the thermal expansion of a solid is a consequence of
the anharmonic behavior of lattice vibrations, which with increasing temperature and a concurrent increase in the amplitude of the lattice vibrations lead to a time-averaged increase in lattice bonding distance. In a
very qualitative way, the potential energy of bonding between pairs of
atoms can be expressed as a power series in terms of the vibrational displacements. There will not be any linear displacement terms in this series
since there is no net force acting on the atoms in their equilibrium positions, and the second order terms give us a harmonic oscillator. This harmonic oscillator approximation is very useful for representing certain
material properties, such as heat capacities, but is not useful for thermal
expansivity calculations since it lacks the anharmonicities that are necessary to account for thermal expansion (see curve A of Fig. 1).

As higher order terms are added to the potential function power series, the
curve becomes assymetric and shows a more rapid rise of potential energy
on the contraction side of the vibrations and a more gradual rise on the
extension side as compared to a harmonic oscillator. The assymetry finally
leads to a dissociation of the atomic bond at an energy D_e above the potential minimum. This behavior is illustrated in curves B and C in Fig. 1,
where both curves have the same shape but curve B has a substantially
higher dissociation energy than curve C. We therefore have a potential
function of the proper form to account for the anharmonicities that lead to
thermal expansivity. In examining curves B and C, we note that for a given
degree of thermal excitation in the vibrational levels, curve C shows a
greater increase in the average bond distance (r) above the potential minimum (r_e) than does curve B, i.e., $r - r_e$ is greater for curve C than for
curve B at a given temperature. Further, as a consequence of the assumption of equivalent shapes for curves B and C, the values of $r - r_e$ at a
given temperature are roughly in inverse proportion to the bond dissociation energies (D_e) for the two cases. Extending these observations to real
solids, we would expect that if two solids have similar structures and

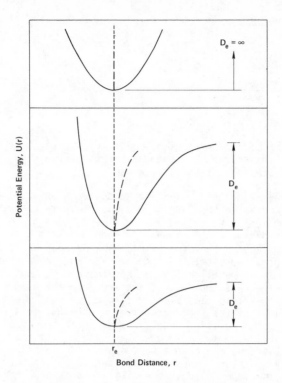

Fig. 1 The variation of average
internuclear distance r_e with thermal
excitation is illustrated for 3 types of
potential energy functions: A is a
harmonic oscillator, and B and C are
anharmonic oscillators having equivalent
shapes but different dissociation
energies, D_e.

similar types of bonding, but differ in bond strength, we would expect
their thermal expansivities to be in inverse proportion to their atomization energies.

The actual shape of the potential energy function will depend on the type
of bonding in the solid. This is usually complicated by the fact that more
than one type of bonding is usually present in a given solid element or
compound. In hard materials we can expect to have bonding contributions
from ionic-, metallic-, and covalent-type bonds. Potential energy functions, $U(r)$, as a function of the internuclear distance (r) can be expressed approximately for the three bond types by the following functional
forms (Zharkov and Kalinin, 1971):

Ionic Bonding:

$$U(r) = a_i \exp[b_i(1 - r)] - c_i r^{-1},$$ 1

Metallic Bonding:

$$U(r) = a_m \exp[b_m(1 - r)] + d_m r^{-2} - c_m r^{-1}, \qquad\qquad 2$$

Covalent Bonding:

$$U(r) \underset{\sim}{} (a_c r^{-1} - c_c) \exp[b_c(1 - r)]. \qquad\qquad 3$$

The letters a, b, c, and d represent parametric constants that need to be theoretically or experimentally established for the various bonding types as indicated by the subscripts i, m, and c.

Ionic and metallic bonding are similar in several aspects. In both cases the attractive forces in the lattice are Coulombic in nature and give a term of the type $c_i r^{-1}$, which gives a relatively slow variation of potential energy with distance. In ionically bonded crystals the net Coulombic attractive force results from the attractions between positive and negative ions, less the repulsions between ions of like charge, all properly summed over the lattice. In metallically bonded crystals the net Coulombic attractive force results from attractions between the conduction electrons and the positive metallic ions plus an attractive force because of exchange energy interactions between the conduction electrons, less the Coulomb repulsion between the positive metallic ions. The principal repulsive forces in both ionic and metallic bonds are due to the overlap of the electron shells of the ions and give a term of the form $a_i \exp[b_i(1 - r)]$ in the potential energy function. The repulsive term $d_m r^{-2}$ in metallic bonding is only of significant importance in the alkali metals and derives from the repulsive force due to the Fermi kinetic energy of the conduction electrons. Thus, the general features of the potential energy curves for both ionic and metallic bonding are a relatively steeply rising curve on the repulsive force side and a relatively slowly rising curve on the attractive force side. Ionic bonds are generally considerably stronger than metallic bonds and hence have shorter bond distances and larger dissociation energies. We need to note that pure ionic bonding does not exist in real solids, but is accompanied by some degree of covalent bonding. In the case of metallic bonding, we also find for transition elements and for intermetallic compounds that we have a combination of metallic and covalent bonding as evidenced by a substantial increase in bonding strength as compared to pure metallic bonding.

In covalent bonding, the bonds are highly directional because of the overlap requirements for the orbitals. Both sides of the potential well are steep and are best represented by exponential functions. Thus, the attractive force contribution to the potential energy is given by $-c_c \exp[b_c(1 - r)]$ and the repulsive force contribution by $a_c r^{-1} \exp[b_c(1 - r)]$. The factor $a_c r^{-1}$ is a screening factor that compensates for the negative space charge of electrons in the covalent bond between the ion cores. Although the potential energy function described here shows the main features expected in such a bond, theoretical calculations are quite complex for covalently bonded solids and therefore we are not aware of any quantitative calculational results in this area. Experimental data indicate that covalent bonds are generally very strong. Also, as is the case with ionic and most metallic solids, pure covalent bonding does not exist in a solid compound but is accompanied by various degrees of ionic and/or metallic bonding.

Thus, we see that potential energy functions are useful in giving us insight as to the bonding characteristics of hard materials, but the theory has not reached the point where we can calculate accurate thermal expansivities as a general matter. In particular, we infer from the foregoing description of bonding characteristics that it is the covalent bonds with their associated high bond strengths and high directionality that give hard materials properties such as high hardness and high rigidity. Also, since pure covalent bonding does not occur in a real solid, we expect that the amount of ionic or metallic bonding that is present will reduce the hardness or brittleness of the material by some degree which is related to the proportionate amount of ionic or metallic bonding.

EQUATION OF STATE OF SOLIDS BASED ON THE QUASIHARMONIC APPROXIMATION

Instead of approaching the equation of state of solids by the method of potential energy functions, an alternative approach is to make the quasiharmonic approximation (Touloukian and co-workers, 1977; Zharkov and Kalinin, 1971; Kittel, 1966). Here we assume that a change in volume of a solid alters only the spectrum of the lattice vibrations but that the vibrations themselves remain harmonic. Thus, as an example of the quasiharmonic approach, using the assumptions for the spectral density of states and normalization requirements as set forth by Debye, we have the Debye equation of state,

$$P = P_0 + \gamma R[9\theta_D/8 + 3T\,D(\theta_D T)]/V = P_0 + \gamma E_D/V, \qquad 4$$

where P_0 derives from a potential energy function for the crystal and is dependent only upon volume. The functions θ_D (the Debye temperature), $D(\theta_D/T)$ (the Debye function), and E_D (the Debye internal energy), are all dependent on lattice vibration frequencies; and γ the Grüneisen parameter, depends on both lattice vibration frequencies and volume. To obtain an explicit expression for the volume thermal expansivity, β, we note that

$$\beta = \frac{1}{V}\,(\partial V/\partial T)_P = -\frac{1}{V}\,\frac{(\partial P/\partial T)_V}{(\partial P/\partial V)_T} = \frac{(\partial P/\partial T)_V}{B_T}, \qquad 5$$

where B_T is the isothermal bulk modulus. We obtain $(\partial P/\partial T)_V$ and B_T by appropriately differentiating equation 4, thus:

$$(\partial P/\partial T)_V = \gamma 3R\big(4\,D(\theta_D/T) - (3\theta_D/T)/[\exp(\theta_D/T) - 1]\big)/V, \qquad 6$$

$$B_T = -V(\partial P/\partial V)_T = B_{T,0} + \gamma R\big([9\theta_D/8 + 3T\,D(\theta_D/T)]\,(1 + \gamma - \partial \ln\gamma/\partial \ln V)$$
$$- 12\gamma T\,D(\theta_D/T) + 9\gamma\theta_D/[\exp(\theta_D/T) - 1]\big)/V, \qquad 7$$

where $B_{T,0}$ is dependent only on volume. We see here however that an explicit value for B_T cannot be obtained from the Debye approximation alone, since the potential energy function needs to be known before $B_{T,0}$ can be evaluated. This seriously limits the application of the Debye method for calculating thermal expansivities from equation 5.

Thus, we conclude that although methods such as the Debye approach based on the quasiharmonic approximation give us considerable insight as to the contributions of thermally excited vibrations to the physical behavior of solids, they do not permit us to directly calculate thermal expansivities from theory without an explicit expression for the potential energy function along with experimentally measured Debye temperatures.

AN EMPIRICAL APPROACH FOR PREDICTING THERMAL EXPANSIVITIES

We believe that a useful empirical approach can be developed for predicting thermal expansivities of hard materials by selecting correlating parameters that are based on materials properties that reflect upon (1) the shape of the potential energy function, and (2) the strength of bonding between atoms in the solid.

Two material properties that show promise as correlating parameters for the shape of the potential well are hardness and elastic modulus. We assume here that high hardness or a high elastic modulus are properties associated with the high rigidity and directionality that is characteristic of covalent bonds and in direct contrast with the less directional ionic and metallic bonds. Since hardness data are more readily obtained and more hardness data are generally available for hard materials than are elastic modulus data, we select hardness as the parameter to study here for representing the potential energy shape factor. Thus, we assume that the higher the material hardness is, the greater will be the proportion of covalent bonding in the material and the steeper the vibrational extension side of the potential energy curve, and hence the lower the thermal expansivity. We use Knoop microhardness or a variant of it as the most quantitative measure of hardness for this study.

Next, we need to select a material property that will reflect upon the bonding strength of the atoms in the crystal. As we found earlier, for a fixed shape of the potential energy curve (i.e., for a fixed degree of covalent bonding), we can expect the atomization energy of the crystal to be inversely related to the degree of thermal expansion of the crystal for a given temperature of excitation of the vibrational levels (see Fig. 1). We actually find in preliminary correlations of thermal expansivities for hard materials that percentage expansion up to the melting point is not a good indicator, but that average atomization energy is a fairly good indicator. Considering that hard materials contain a mixture of bonding types, perhaps some combination of average atomization energy and melting point as parameters with proportionate contributions from the different bonding types would give an improved fit over the average atomization energy alone. But, the present degree of accuracy of thermal expansivity data is not sufficient to allow us to make this refinement based on a statistical analysis of the data. Therefore, we select average atomization energy as the parameter for the bonding strength of the atoms in the crystal.

DERIVATION OF A CONSISTENT SET OF MATERIAL PROPERTY VALUES

In order to proceed properly with the correlation study on thermal expansivities of hard materials, we need to first establish a consistent approach for deriving the material properties to be used in the correlation. The properties of concern are the thermal expansivity and its variation with temperature, microhardness, and atomization energy. These properties will now be discussed in the above order.

In order to obtain a universal form for the temperature dependence of β, we first refer back to the examples of metals with simple structures and ionic compounds such as alkali halides where for each class of materials we can expect a fixed percent expansion from absolute zero to the melting point. Assuming that this behavior can be described in a corresponding states form with a power dependence of volume on temperature gives (Krikorian, 1971)

$$(V - V_0)/(V_m - V_0) = (T/T_m)^n,$$ 8

where the subscript 0 refers to absolute zero, m refers to the melting point and n is an empirical parameter. Taking $1/V$ times the temperature derivitive of V in equation 8 and rearranging terms gives

$$\beta = (1/V)(dV/dT) = [n(V_m - V_0)/VT_m^n] \ T^{n-1} \approx \text{cons't} \times T_m^{-n} T^{n-1}.$$ 9

Examining the data on metals (Touloukian and co-workers, 1975) and alkali halides (Touloukian and co-workers, 1977) we find that n has an average value of about 1.4 and generally falls within the range of 1.3 to 1.5. Within the temperature range of 500-2000 K, n remains reasonably constant for a given material (excluding phase changes). Hence, taking an overall expansion of 7.5% for metals and 13.5% for alkali halides, we obtain as universal expressions

$$\beta \approx 0.105 \ T_m^{-1.4} \ T^{0.4} \text{ for metals,}$$ 10

$$\beta \approx 0.19 \ T_m^{-1.4} \ T^{0.4} \text{ for alkali halides,}$$ 11

where the percentage difference between observed and calculated thermal expansivities is usually within ±10% of that given by expressions 10 and 11.

For hard materials, we find that even though the correlation of thermal expansivity with melting point is poor, nonetheless the temperature dependence of thermal expansivity is generally within the range of $T^{0.3-0.4}$ above room temperature. Therefore for purposes of extrapolating thermal expansivity data beyond the measured temperatures, and for developing parametric correlations of thermal expansivity, we will assume

$$\beta = \text{cons't} \times T^{1/3} \text{ for hard materials.}$$ 12

We will next discuss the property of microhardness. We find that although microhardness is a relatively simple measurement to make, many times there are large variations in values reported for hard materials by different investigators. Research into the factors that affect microhardness has clarified many of the reasons for these variations, and methods have been developed for obtaining more accurate and reproducible microhardness values (Brookes, 1983; Ivan'ko, 1974). The important factors affecting microhardness break down into three areas: material purity, method of preparation, and conditions of measurement. We will mention some of the problems that are encountered in each of these.

Purity can have different effects. If the material is single phase, the hardness can vary according to the level and type of dissolved impurities. Impurities can also influence the types and amounts of defects in the structure and consequently the elastic versus plastic deformation behavior of the material. If the impurities occur as a second phase, a macrodispersion can lead to considerable scatter in the observed microhardness values because of the difficulty of isolating a large enough pure grain to obtain a reliable measurement. A microdispersion of a second phase could lead to dispersion hardening and inordinately high microhardness values; or if the second phase occurs in a preferred orientation in the primary structure, it could lead to weakening and cracking of the material during the indentation when the indentor is aligned along that orientation.

The method of preparation can lead to materials of differing grain size, porosity, grain orientation, and retention of strains in the material. Hot compaction is a convenient method for reducing porosity, and annealing to relieve strains is usually necessary before making measurements on hard materials.

Measurement techniques include selecting the optimum conditions of loading mass and loading time for the particular type of material being tested. Too high a load can lead to microfractures in the material in the vicinity of the indentation and give too low a microhardness value. Too low a load can give erroneously high microhardness values. The dwell time needs to be minimized to avoid the possibility of creep during the test which would lead to too deep an indentation. Optimum loading time is usually about 10-20 s. The Knoop indentor, because of its elongated-pyramid shape is the preferred indentor to use to minimize stresses and fracturing of the material. The Vickers indentor is a symmetrical pyramid so that lower loads are usually required than with the Knoop indentor to avoid fracturing, but also to minimize the errors introduced by elastic recovery of the Vickers impression. It is always a good procedure to survey a range of loading masses and times on a new material in order to determine the optimum conditions for microhardness measurements.

In evaluating the available microhardness data in the literature, we use the following criteria as a guide: for Knoop 2000 HK or higher, we select the data from studies that have used loads within the range of 20-80 g, for Knoop 1000 to 2000 HK from loads of 30-100 g, and for Knoop 500-1000 HK from loads of 50-200 g. For Vickers microhardness we generally prefer somewhat lower loads to avoid fracturing effects. Whenever possible we also try to establish whether the material is free of impurities and look for densities of greater than 85% of theoretical. We tend to favor the highest reported microhardness values if loading conditions have been met.

The atomization energy is the other parameter that needs to be determined, and we do this calculationally. For convenience, we define the atomization energy, ΔE_a, as the enthalpy of vaporization of a given substance to form 1 gram-atom of gaseous atoms at room temperature. Thus, as an example, we calculate the atomization energy of Al_2O_3 as follows:

$$1/5\ Al_2O_3(s) = 2/5\ Al(s) + 3/10\ O_2(g) \qquad \Delta H^\circ_{298} = 335.14\ kJ/mole$$

$$2/5\ Al(s) \quad = 2/5\ Al(g) \qquad\qquad\qquad \Delta H^\circ_{298} = 131.72\ kJ/mole$$

$$3/10\ O_2(g) \quad = 3/5\ O(g) \qquad\qquad\qquad \Delta H^\circ_{298} = 149.52\ kJ/mole$$

$$\overline{1/5\ Al_2O_3(s) = 2/5\ Al(g) + 3/5\ O(g) \qquad \Delta E_a \quad = 616.4\ kJ/mole}$$

The first line above represents the negative of the enthalpy of formation of the compound, the next two lines are the atomization energies of the elements, and the fourth line, which is the sum of the reactions, represents the atomization energy of the compound per gram-atom. Data for enthalpies of formation are taken from various sources, or estimated. Data on atomization energies are from Hultgren (Hultgren and co-workers, 1973).

RESULTS OF CORRELATION STUDIES OF THERMAL EXPANSIVITIES OF HARD MATERIALS

Based on the methods outlined above, we summarize data in Table 1 on the parameters of atomization energy (ΔE_a), microhardness (h), and a parameter based on thermal expansivity $(\beta T^{-1/3})$ for borides, carbides, nitrides and oxides. Data are given only for those cases where experimental data are available for deriving all three parameters. Solid symbols are used for substances with cubic structures to distinguish them from the nonisotropic structures, for which we use open symbols.

First, following our assumption that thermal expansivity is in inverse proportion to atomization energy and using the temperature dependence of thermal expansivity given by equation 12, we test the expression,

$$\beta T^{-1/3} = \text{cons't} \times \Delta E_a^{-1}. \qquad 13$$

Thus, plotting $\beta T^{-1/3}$ versus ΔE_a^{-1} (see Fig. 2), we find that the data define a very broad scatter band. Closer examination shows that certain orderly features are present. For example, the transition and actinide metal monocarbides and mononitrides with the fcc structure can all be described by the expression

$$\beta T^{-1/3} = 1.87 \times 10^{-3}\ \Delta E_a^{-1}, \qquad 14$$

as indicated by the solid lines in Figs. 2 and 3. We also find, as shown in Fig. 4, that all of the borides, ranging from the monoborides through the hexaborides, can be described by

$$\beta T^{-1/3} = 1.53 \times 10^{-3}\ \Delta E_a^{-1}. \qquad 15$$

Thus, from the available data, we can conclude that β for all of the borides, except for metal-rich borides such as Co_2B, is predicated fairly well by equation 15.

More importantly, we note that the hard materials with large degrees of metallic bonding (e.g., V_2C and Co_2B) or with significant amounts of ionic bonding (e.g., $UC_{1.9}$, TiO, MgO, and Li_2O), tend to give generally high values of $\beta T^{-1/3}$. With highly covalent bonding, such as for diamond, cubic BN, SiC, and B_4C, we generally find low values of $\beta T^{-1/3}$. Some compounds, such as Si_3N_4 and SnO_2, give unusually low values of $\beta T^{-1/3}$, which may be (at least in part) a consequence of their anisotropic structures. We thus anticipate that use of a parameter, such as microhardness, to differentiate the degree of covalent bonding as contrasted to metallic or ionic bonding, should improve the correlation over the use of ΔE_a alone.

From a preliminary examination of an expression of the type

$$\beta\ \Delta E_a^{-1} T^{1/3} = \text{cons't} \times h^{-n}, \qquad 16$$

we find that a best fit is obtained with an n value of about 1/3. We can therefore write the universal expression

TABLE 1 Listing of Input Data for Correlation Studies of the Thermal Expansivity β versus Atomization Energy ΔE_a and Microhardness h

Compound (structure)	$-\Delta H^\circ_{f,298}$, kJ/mole	$\Delta H^\circ_{a,298}$ of metal, kJ/g-atom	ΔE_a, kJ/g-atom	h, HK	$\beta T^{-1/3}$, $K^{-4/3} \times 10^6$
β-B(rhd)	0.0[a]	571.1	571	3400[b,c,d]	1.98[e]
BeB_6 (tetr)	193 (est)	324.3	563	2600[f]	2.36 (est)
CaB_6 (cub)	301 (est)	178.2	558	2700[b,g]	2.36[h]
SrB_6 (cub)	301 (est)	164.0[i]	556	2900[b,g]	2.42[h]
BaB_6 (cub)	301 (est)	182.0	559	3000[b,g]	2.47[h]
ScB_2 (hex)	126 (est)	377.9	548	2600[g]	2.57[e,b]
YB_6 (cub)	301 (est)	424.7	593	3300[b]	2.28[h]
LaB_6 (cub)	301 (est)	431.0	594	2800[b]	2.33[h]
CeB_6 (cub)	301 (est)	422.6	593	3100[b]	2.65[h]
NdB_4 (tetr)	209 (est)	327.6	564	1950[j]	2.12[j]
NdB_6 (cub)	301 (est)	327.6	579	2600[b]	2.63[h]
GdB_4 (tetr)	209 (est)	397.5	578	1830[j]	2.53[j]
GdB_6 (cub)	301 (est)	397.5	589	2350[g]	3.16[h]
HoB_4 (tetr)	209 (est)	300.8	559	1680[j]	2.76[j]
YbB_6 (cub)	301 (est)	152.1	554	3800[g]	2.12[h]
SiB_4 (hex)	63 (est)	455.6	561	2400[k]	2.00[e]
SiB_6 (o-rh)	84 (est)	455.6	567	3300[k]	1.78[e]
TiB_2 (hex)	279.5[l]	469.9	630	3500[m,n]	2.53[e]
ZrB_2 (hex)	305.4[l]	608.8	686	2300[m,n]	2.25[e]
HfB_2 (hex)	334.7[l]	619.2	699	3000[m,n]	2.20[e]
VB_2 (hex)	203.8[o]	514.2	620	2800[b]	2.55[e]
NbB_2 (hex)	175.3[l]	721.3	680	2600[b]	2.34[e]
TaB (o-rh)	138 (est)	781.6	745	3100[b]	1.89[e]
TaB_2 (hex)	209.2[l]	781.6	711	2600[p,b]	2.10[e]
CrB (o-rh)	84 (est)	397.5	526	2100[p]	3.13[e,b]
CrB_2 (hex)	138 (est)	397.5	559	2100[b,c]	3.58[e,b]
MoB_2 (hex)	96.2[q]	658.1	632	2350[c]	2.84[e]
WB (tetr)	71.1[r,s]	849.4	746	3700[b]	2.02[e,c]
Co_2B (tetr)	84 (est)	428.4	504	1150[b]	4.49[e]
ThB_4 (tetr)	217.6[t,s]	575.3	616	2700[c]	2.29[e,c]
ThB_6 (cub)	228.0[l]	575.3	604	2600[c]	2.87[e,h]
UB_4 (tetr)	263.6[s]	523.0	614	2500[c]	2.42[c]
C (cub)	-1.9[a]	716.7	715	8500[u]	1.38[e]

TABLE 1 (continued)

Compound (Structure)	$-\Delta H^\circ_{f,298}$, kJ/mole	$\Delta H^\circ_{a,298}$ of metal, kJ/g-atom	ΔE_a, kJ/g-atom	h, HK	$\beta T^{-1/3}$, $K^{-4/3} \times 10^6$
Be_2C (cub)	117.2[v]	324.3	494	2700[b]	3.54[b,c]
B_4C (rhd)	71.5[v]	571.1	614	4500[b,w]	1.85[e]
ScC (cub)	100.0 (est)	377.9	597	2700[b]	3.80[b]
Y_2C_3 (cub)	206 (est)	424.7	641	900[b]	3.59[e]
YC_2 (tetr)	113.0[x]	424.7	657	700[b]	3.35[e]
SiC (hex)	66.9[v]	455.6	620	3300[p,b]	1.65[e]
TiC (cub)	183.7[v]	469.9	685	3200[p,y]	2.50[e]
ZrC (cub)	202.0[v]	608.8	764	3000[p,b]	2.29[e]
HfC (cub)	218.8[v]	619.2	777	2900[b]	2.18[e]
V_2C (hex)	136 (est)	514.2	627	2100[z]	3.74[e]
VC (cub)	100.8[v]	514.2	666	2100[p,b]	2.23[e]
Nb_2C (hex)	186.2[v]	721.3	782	2100[b]	2.60[e]
NbC (cub)	138.1[v]	721.3	788	2300[p,z]	2.25[e]
Ta_2C (hex)	202.9[v]	781.6	828	1700[b]	2.66[e]
TaC (cub)	143.1[v]	781.6	821	1600[b,z]	2.13[e]
$Cr_{23}C_6$ (cub)	580.0[v]	397.5	484	1650[b,c]	3.66[b]
Cr_7C_3 (trig)	228.0[v]	397.5	516	2100[c]	3.83[b]
Cr_3C_2 (o-rh)	109.6[v]	397.5	547	2300[aa]	3.53[e,c]
$\beta\text{-}Mo_2C$ (hex)	46.0[v]	658.1	693	1700[aa]	1.96[e]
W_2C (hex)	26.4[v]	849.4	814	3200[p,b]	1.81[p]
WC (hex)	38.1[v]	849.4	802	2200[c,p]	1.56[e]
ThC (cub)	125.5[v]	575.3	709	1000[y]	2.38[e,c]
ThC_2 (mond)	117.2[v]	575.3	709	700[b]	3.35[e,c]
UC (cub)	90.8[v]	523.0	665	950[b]	3.58[e,c]
U_2C_3 (cub)	205.0[v]	523.0	680	800[c]	3.55[e,p]
$UC_{1.9}$ (tetr)	96.2[v]	523.0	683	600[c]	4.46[e,c]
$PuC_{0.8}$ (cub)	45.5[v]	351.9	539	900[z]	3.56[e,c]
Pu_2C_3 (cub)	110.5[v]	351.9	593	700[z]	5.10[e]
β-BN (cub)	250 (est)	571.1	647[bb]	8000[p,cc]	1.81[e]
AlN (hex)	318.4[v]	329.3	560	1300[b]	1.76[e]
Si_3N_4 (hex)	744.8[v]	455.6	572	3100[b,c]	1.03[e]
TiN (cub)	336.4[v]	469.9	640	2100[p,b]	2.90[e]
ZrN (cub)	368.2[v]	608.8	725	1800[p,b]	2.46[e]
HfN (cub)	369.0[v]	619.2	730	1700[b]	2.31[e]

TABLE 1   (continued)

Compound (Structure)	$-\Delta H^\circ_{f,298}$, kJ/mole	$\Delta H^\circ_{a,298}$ of metal, kJ/g-atom	ΔE_a, kJ/g-atom	h, HK	$\beta T^{-1/3}$, $K^{-4/3} \times 10^6$
$VN_{0.4}$ (hex)	127.0[v]	514.2	593	1900[b]	2.70[b]
VN (cub)	217.1[v]	514.2	602	1500[b]	2.70[b]
Ta_2N (rhm)	272.8[v]	781.6	770	1200[b]	1.99[c]
TaN (hex)	252.3[v]	781.6	753	1000[b]	1.95[e]
Cr_2N (hex)	114.2[v]	397.5	461	1600[b]	3.61[b]
UN (cub)	294.6[v]	523.0	645	500[c,p]	3.07[e]
Li_2O (cub)	598.3[dd]	160.7[i]	390[ee]	180[ff]	9.90[gg]
BeO (hex)	598.7[dd]	324.3	586	1300[c,p]	3.14[e]
MgO (cub)	601.7[dd]	146.4	499	1000[hh]	4.41[e]
CaO (cub)	635.1[dd]	178.2	531	600[p,hh]	4.09[e]
Al_2O_3 (rhd)	1675.7[dd]	329.3	616	2400[hh]	2.73[e]
Y_2O_3 (cub)	1905.8[dd]	424.7	701	690[hh]	2.55[e]
Sm_2O_3 (cub)	1814.4[dd]	206.7	595	440[c]	2.57[e]
Eu_2O_3 (moncl)	1648.5[dd]	175.3	549	440[c]	3.26[e]
Dy_2O_3 (cub)	1865.2[dd]	290.4	639	700[c]	2.65[e]
SnO_2 (tetr)	580.7[dd]	301.2	460	1400[hh]	1.93[e]
TiO (cub)	519.7[dd]	469.9	619	1300[ii]	4.40[e]
TiO_2 (tetr)	955.7[dd]	469.9	638	1075[hh]	2.90[e]
ZrO_2 (moncl)	1100.4[dd]	608.8	736	1000[p,hh]	2.65[e]
HfO_2 (moncl)	1144.7[dd]	619.2	754	900[hh]	2.60[e]
Cr_2O_3 (rhd)	1141.0[dd]	397.5	537	2900[hh]	2.15[e]
MnO (cub)	385.1[dd]	283.3	459	570[hh]	4.70[e]
$Fe_{0.95}O$ (cub)	266.9[dd]	415.5	467	540[hh]	5.18[e]
Fe_3O_4 (cub)	1120.5[dd]	415.5	480	700[hh]	5.36[e]
Fe_2O_3 (rhd)	823.6[dd]	415.5	480	1000[hh]	4.04[e]
ThO_2 (cub)	1226.7[dd]	575.3	767	1000[p,hh]	2.95[e]
UO_2 (cub)	1083.7[dd]	523.0	702	700[c,hh]	3.32[e]
PuO_2 (cub)	1058.1[dd]	351.9	636	450[c]	3.40[e]

[a]Hultgren and co-workers, 1973; [b]Samsonov, 1964; [c]Lynch, Ruderer, and Duckworth, 1966; [d]Samsonov, 1968; [e]Touloukian and co-workers, 1977; [f]Wilkins, 1977; [g]Samsonov and Paderno, 1961; [h]Zhuravlev and co-workers,

1962: [i]JANAF Thermodynamic Tables, 1971; [j]Samsonov and Kovenskaya, 1977; [k]Samsonov and Sleptsov, 1964; [l]Schick, 1966; [m]Clougherty and Pober, 1964; [n]Gurin and Sinelnikova, 1977; [o]Spear, Shäfer, and Gilles, 1969; [p]Shaffer, 1964; [q]Touloukian, 1967; [r]Brewer and Haraldsen, 1955; [s]Krikorian, 1971; [t]Aronson and Auskern, 1966; [u]Jahns, 1960; [v]Kubaschewski and Alcok, 1979; [w]Neshpor, Nikitin, and Rabotnov, 1974; [x]DeMaria and co-workers, 1965; [y]Toth, 1971; [z]Storms, 1967; [aa]Ivan'ko, 1974; [bb]For nitride ΔE calculations, we take $\Delta H^\circ_{a,298}$ of $N(g)$ as 472.7 kJ/g-atom (Kubaschewski and Alcock, 1979); [cc]Holleck, 1983; [dd]Brewer and Rosenblatt, 1969; [ee]For oxide ΔE_a calculations, we take $\Delta H^\circ_{a,298}$ of $O(g)$ as 249.2 kJ/g-atom (Kubaschewski and Alcock, 1979); [ff]Nasu, Fukai, and Tanifuji, 1978; [gg]Kurasawa and co-workers, 1982; [hh]Samsonov, 1973; [ii]Denker, 1964.

$$\beta = 21.1 \ \Delta E_a^{-1} \ (T/h)^{1/3} \qquad\qquad 17$$

which applies to the hard material borides, carbides, nitrides and oxides, as illustrated in Fig. 5. We find that the average uncertainty in the predicted thermal expansivity is less than ±10%, and the maximum uncertainty (with a few exceptions) is within ±30%. Substances with anisotropic structures give low values of β in some cases (e.g., Si_3N_4 and SnO_2), which is within our expectations.

A portion of the scatter in Fig. 5 can be attributed to uncertainties in the input data. The uncertainties in experimental data on β range from about ±3-15% over the temperature region of measurement, with most of the β values in Table 1 being uncertain by about ±5%. Use of the $T^{1/3}$ relation to extrapolate the data on β beyond the measured region increases this uncertainty somewhat in some cases. Errors in ΔE_a are generally less than ±1% when experimental data are available on $\Delta H^\circ_{f,298}$, and are about ±5-10% when $\Delta H_{f,298}$ is estimated. We estimate the microhardness data to be uncertain by ±10% on the average, and in a few instances by as much as ±50%. Since h is raised to the 1/3 power, the corresponding errors that are translated to the calculated β are reduced from a range of ±10% to ±50% in h to a range of ±3.5% to ±17% in β. These estimated input errors suggest about a ±7% average uncertainty in the data plotted in Fig. 5. This is surprisingly close to the observed average uncertainties of less than ±10% to the fit to equation 17, which include as well the systematic errors of the correlation.

CONCLUSIONS

We conclude the following from this study on estimating the thermal expansivities of hard materials:

o   The temperature variation of β above room temperature can be described reasonably well by a $T^{1/3}$ dependence. Use of this $T^{1/3}$ dependence provides us with a good basis of extrapolating β data and intercomparing β values for different materials.

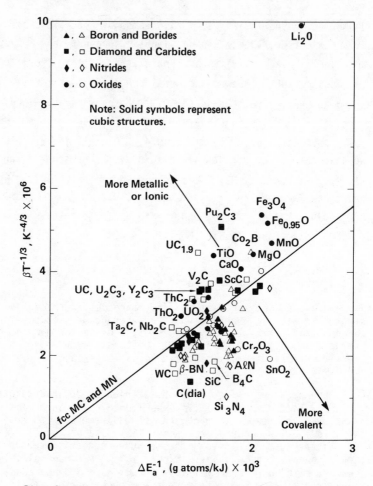

Fig. 2 Illustrated here for hard materials is the
dependence of the thermal expansivity parameter,
$\beta T^{-1/3}$, on the inverse of the energy of atomization, ΔE_a. The fcc monocarbides and mononitrides define a linear curve, with significantly
metallic or ionic materials lying above the curve,
and highly covalent materials lying below the curve.

o   Both ΔE_a and h are statistically significant parameters that influence
    β values for hard materials. The dependence of β on these parameters
    can be expressed approximately by equation 17, namely,

$$\beta = 21.1 \; \Delta E_a^{-1}(T/h)^{1/3}.$$

o   There is a qualitative theoretical basis for showing that β should have
    an inverse power dependence on both ΔE_a and h for hard materials.

o   Equation 17 should prove useful for predicting thermal expansivities to
    within an average uncertainty of ±10% for isotropic hard materials, and
    to set an upper bound for thermal expansivities of nonisotropic hard
    materials.

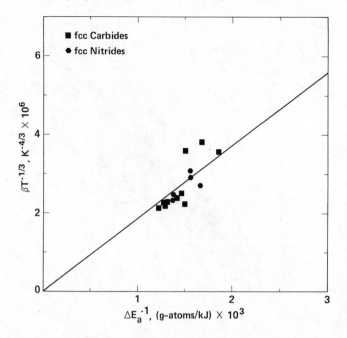

Fig. 3 Thermal expansivity parameter versus ΔE_a for the fcc monocarbides and mononitrides.

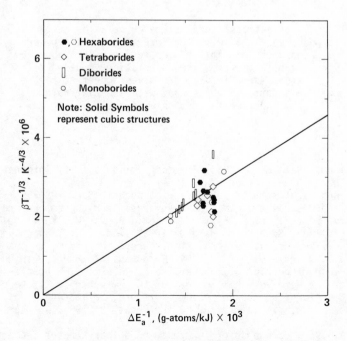

Fig. 4 Thermal expansivity parameter versus ΔE_a for the borides.

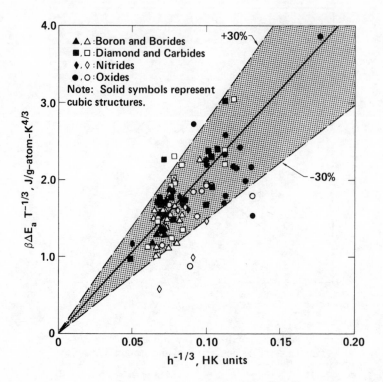

Fig. 5   Illustrated here is the correlation of the
product $\beta\Delta E_a T^{-1/3}$ with the inverse one-third power
of microhardness h. Taking microhardness into account
significantly improves the correlation compared to
ΔE_a alone (see Fig. 2), and gives agreement to
within a range of about ±30% with the available
experimental expansivity data.

REFERENCES

Aronson, S. and Auskern, A. 1966, Thermodynamics, vol. 1, International
    Atomic Energy Agency, Vienna, pp 165-170
Brewer, L. and Haraldsen, H. 1955, J. Electrochem. Soc. 102 399
Brewer, L. and Rosenblatt, G.M. 1969, Advances in High Temperature
    Chemistry, Vol. 2, Ed. L. Eyring, Academic Press, New York, pp 1-83
Brookes, C.A. 1983, Science of Hard Materials, Ed. R.K. Viswanadham,
    D.J. Rowcliffe, and J. Gurland, Plenum Press, New York, pp 181-199
Clougherty, E.V. and Pober, R.L. 1964, Compounds of Interest in Nuclear
    Reactor Technology, Ed. J.T. Waber and P. Chiotti, IMD Special Report
    No. 13 of the Metallurgical Society of the American Institute of
    Mining Metallurgical and Petroleum Engineers, pp 423-443
DeMaria, G., Guido, M., Malaspina, L. and Pesce, B. 1965, J. Chem. Phys.
    43 4449
Denker, S.P. 1964, Compounds of Interest in Nuclear Reactor Technology,
    Ed. J.T. Waber and P. Chiotti, IMD Special Report No. 13 of the American
    Institute of Mining Metallurgical and Petroleum Engineers, pp 51-62

Gurin, V.N. and Sinelnikova, V.S. 1977, Boron and Refractory Borides,
  Ed. V.I. Matkovich, Springer-Verlag, Berlin, pp 377-389
Holleck, H. 1983, Science of Hard Materials, Ed. R.K. Viswanadham,
  D.J. Rowcliffe and J. Gurland, Plenum Press, New York, pp 849-861
Hultgren, R., Desai, P.D., Hawkins, D.T., Gleiser, M., Kelley, K.K. and
  Wagman, D.D. 1973, American Society for Metals, Metals Park, Ohio
Ivan'ko, A.A. 1974, Refractory Carbides, Ed. G.V. Samsonov, Consultants
  Bureau, New York, pp 367-370
Jahns, R.H. 1960, Industrial Minerals and Rocks, third edition, American
  Institute of Mining, Metallurgical, and Petroleum Engineers, New York
  pp 383-441
Kittel, C. 1966, Introduction to Solid State Physics, third edition,
  Wiley, New York
Krikorian, O.H. 1971, Lawrence Livermore National Laboratory Report,
  UCRL-51043, April 27
Kubaschewski, O. and Alcock, C.B. 1979, Metallurgical Thermochemistry,
  fifth edition, Pergamon Press, Oxford
Kurasawa, T., Takahashi, T., Noda, K., Takeshita, H., Nasu, S. and
  Watanabe, H. 1982, Thermal Expansion of Lithium Oxide, J. Nucl. Mater.
  107 334
Lynch, J.F., Ruderer, C.G. and Duckworth, W.H. 1966, Engineering Properties
  of Selected Ceramic Materials, Battelle Memorial Institute, Columbus,
  Ohio, published by the American Ceramic Soc. Columbus, Ohio
Nasu, S., Fukai, K. and Tanifuji, T. 1978, J. Nucl. Mater. 78 254
Neshpor, V.S., Nikitin, V.P. and Rabotnov, V.V. 1974, Refractory Carbides,
  Ed. G.V. Samsonov, Consultants Bureau, New York, pp 221-229
Samsonov, G.V. and Paderno Y.B. 1961, English transl. AEC-tr-5264, U.S.
  Atomic Energy Commission Division of Technical Information, Oak Ridge,
  Tennessee Oct. 1962; a transl. of Boridy Redkozemel'nykh Metallov,
  Izdatel'stvo Akademii Nauk Ukrainskoi S.S.R., Kiev
Samsonov, G.V. 1964, Plenum Press Handbooks of High-Temperature Materials,
  No. 2 - Properties Index, Plenum Press, New York
Samsonov, G.V. and Sleptsov, V.M. 1964, Poroshkovaia Metallurgia 6, 58
Samsonov, G.V. 1968, Handbook of the Physicochemical Properties of the
  Elements, Ed. G.V. Samsonov, IFI/Plenum, New York
Samsonov, G.V. 1973, The Oxide Handbook, IFI/Plenum, New York
Samsonov, G.V. and Kovenskaya, B.A. 1977, Boron and Refractory Borides,
  Ed. V.I. Matkovich, Springer-Verlag, Berlin, pp 19-30
Schick, H.L. 1966, Thermodynamics of Certain Refractory Compounds,
  vols. 1 and 2, Academic Press, New York
Shaffer, P.T.B. 1964, Plenum Press Handbooks of High-Temperature
  Materials, No. 1 - Materials Index, Plenum Press, New York
Spear, K.E., Schafer, H. and Gilles, P.W. 1969, High Temperature
  Technology, Butterworth and Company, London
Storms, E.K. 1967, The Refractory Carbides, Academic Press, New York, p 234
Stull, D.R. and Prophet, H. (Project Directors) 1971, JANAF Thermodynamic
  Tables, second edition, U.S. National Bureau of Standards, Washington DC
Thermophysical properties of High Temperature Solid Materials - vols. 1-6
  Ed. Y.S. Touloukian, McMillan, New York, 1967
Touloukian, Y.S. Kirby, R.K., Taylor, R.E. and Desai P.D. 1975, Thermo-
  physical Properties of Matter, vol. 12, Thermal Expansion Metallic
  Elements and Alloys, IFI/Plenum, New York

Touloukian, Y.S., Kirby, R.K., Taylor, R.E. and Lee, T. 1977,
  Thermophysical Properties of Matter, vol. 13, Thermal Expansion Non-
  metallic Solids, IFI/Plenum, New York
Toth, E. 1971, Transition Metal Carbides and Nitrides, Academic Press,
  New York, pp 6-7 and 176-181

Wilkins, M.L. 1977, Boron and Refractory Borides, Ed. V.I. Matkovich, Springer-Verlag, Berlin, pp 633-648

Zharkov, V.N. and Kalinin, V.A. 1971, Equations of State for Solids at High Pressures and Temperatures, English transl. by A. Tybulewicz, Consultants Bureau, New York

Zhuravlev, N.N., Stepanova, A.A., Paderno, Y.B. and Samsonov, G.V. 1962, Soviet Physics-Cryst. 4 636

Inst. Phys. Conf. Ser. No. 75: Chapter 2
Paper presented at 2nd Int. Conf. Science Hard Mater., Rhodes

155

Mechanical testing of hard materials

E A ALMOND, B ROEBUCK AND M G GEE

National Physical Laboratory, Teddington, UK

ABSTRACT

Difficulties in obtaining reliable information on mechanical properties of hard materials originate in testing and in procedures for converting measurements into data for ranking and design purposes. Headway made in measuring strength and fracture toughness by conventional and indentation methods has been partly eroded by the realisation that failure conditions may not be identical in different tests. R-curve behaviour and specimen geometry affect fracture toughness results, and data used in probabilistic fracture mechanics analyses may be inapplicable. Systems analyses, used for wear-studies, are often appropriate for designing components.

INTRODUCTION

Mechanical tests are performed to obtain information either for using directly as test data or for incorporating into a design or an analysis of a component. Once the requirements for information have been defined, a sequence of decisions is made on the properties to be measured and on an appropriate test method. Although this procedure is of general application the detailed aspects for hard materials are considerably different to those for metals. Thus there are the interrelated problems: firstly, that the most utilised characteristic of hard materials is their high wear resistance, for which there are no truly relevant laboratory tests; whilst secondly, even the simplest properties such as strength and toughness present complications in measurement because of the difficulties in testing low ductility materials and uncertainties in the analyses for converting test results into meaningful data.

The test itself involves superimposing a definable stress field or energy input onto an existing internal stress field in a solid with the object of disturbing the state of equilibrium sufficiently to produce a measurable permanent or temporary distortion in the shape or volume of the solid. The internal stress field comprises the macroscopic residual stress gradients between the outer surface and interior of the solid, and periodically varying microstresses

generated, for example, as a result of anisotropy in thermal shrinkage
rates in neighbouring grains or general differences in elastic and thermal
properties of constituents.

The difficulties in testing hard materials can be traced back to the atomic
bonding. This may be predominantly covalent or ionic or a combination of
both, together with a metallic component as in the transition metal
carbides. As a result a high resistance to dislocation motion and a low
energy for fracture are usually observed. These characteristics are
restrictive in applications requiring toughness and as a result it is
necessary to introduce microstuctural features such as grain boundaries or
second phase particles to restrict crack propagation. Residual internal
stresses can be very high since they are not easily relaxed by plastic
deformation. In some ceramics with anisotropic coefficients of expansion it
is found that the microstresses are sufficiently large to produce
intergranular cracking during cooling if the misorientation between
neighbouring grains is sufficiently high and the grain size is large
(Buresch, 1984).

The intrinsic brittleness, the difficulty in relieving residual stresses
and stress concentrations by plastic deformation, and the presence of
inherited defects from the sintering process are all contributory factors
to the characteristic high variability in mechanical properties of hard
materials. The effects of these limitations and the costs of testing are
normally reduced by selecting the simplest possible specimen shape within
the following categories:
   i)   Smooth specimens, which are used for measuring strength,
   ductility, creep-rate, and notch-strength (for smoothly profiled
   notches).
   ii) Cracked specimens, which are used for measuring fracture
   characteristics in static-loading and in fatigue, and at high
   temperatures and in corrosive environments.
   iii) Surfaces, which when indented or scratched provide
   characteristically shaped impressions and crack phenomena which provide
   information on the properties covered by i) and ii).
   iv) Various shapes employed in wear tests.

Also, when choosing a test for hard materials it is very important to give
careful consideration, not only to the ease of testing and sensitivity of
measurement but also to the accuracy and reliability of the available
analyses for converting the measurements into property data.

STRENGTH TESTS

Embrittlement by internal and surface defects is mainly responsible for the
dependence of strength on the shape, volume and surface finish of specimens
and components manufactured from hard materials. Where practical, this
weakness is partly suppressed by designing components to operate in
compression, and consequently compressive testing is more relevant to
service conditions than tensile testing. Compressive specimens are normally
parallel sided cylinders and can be provided with profiled waists (Fig. 1a)
or soft end-shims to reduce stress concentrations (Johansson, Persson and
Hiltscher, 1970; Lueth and Hale, 1970).

True uniaxial tensile stressing occurs very rarely in tools and components. When they do exist, tensile stresses are normally components of a predominantly compressive multiaxial stress field. The closest practical approximation to uniaxial testing is obtained in bend tests. The simplest configuration is in 3-point bend but 4-point bend gives a uniform stress field on the surface within the inner rollers (Fig. 1b and c). The specimen can be chamfered (Fig. 1d) if it is required to subject only a small volume of the material to a high tensile stress (Roebuck, 1979).

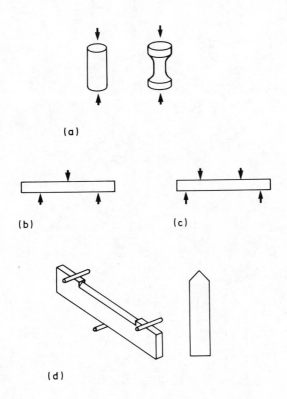

Fig. 1 Common geometries used in uniaxial strength tests: a) cylindrical and waisted compressive specimens; b) 3-point bend specimens; c) 4-point bend specimens; d) chamfered bend-test specimen.

Biaxial stress fields are produced by loading one side of a disc with a punch (Fig. 2a), with a pressurised medium, or by a system of three spherical supports (Wachtman, Capps and Mandel, 1972). Triaxial stress fields can be obtained by internally pressurising a cylinder and applying compression or tension axially (Fig. 2b).

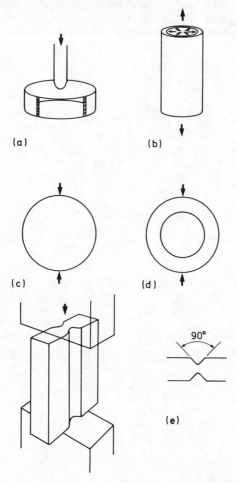

Fig. 2 Multiaxial stress tests: a) biaxial tension of a disc; b) internally pressurized cylinder with axial loading; c) diametral compression of a disc, and d) of a ring; e) indentation of a notched-square.

Multiaxial stress fields can also be generated by designing an appropriate specimen geometry. Thus in diametral compression tests on discs or rings (Figs.2c and d) a lateral tensile stress is generated when the specimen is compressed but a limiting feature of the test is the fixed ratio of tensile to compressive stress (Shaw, Braiden and De Salvo, 1975). This limitation is not applicable in the notched-square indentation-test (Fig. 2e) where by varying the ratio of the width of the anvil to specimen width, ratios of 2:1 to 1:4 can be obtained for the tensile to compressive stress ratio (Almond, Irani and Roebuck, 1980).

The results of multiaxial stress tests are normally presented in the form of a failure-stress diagram which often assists in identifying the failure mechanism by its dependence on a critical value of shear stress, energy, or other quantity. More recently it has been used to investigate various maximum stress criteria used in statistical analyses (Fig. 3a; Stout and Petrovic, 1984).

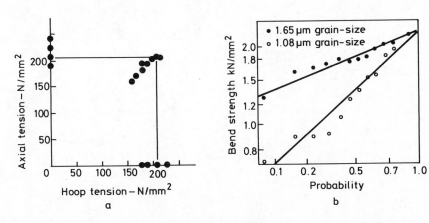

Fig. 3 a) Failure strength diagram for internally pressurized alumina tube subjected to uniaxial tension and compression (after Petrovic and Stout, 1984).
b) Bend strength of WC/Co cemented carbides versus cumulative failure probability, plotted on a logarithmic scale to determine Weibull parameters.

Because of the high variability in results from nominally identical specimens, the information obtained from average values of results is of very limited value. The variability can be expressed phenomenologically using probability theory by fitting the data to an equation of the form:

$$S = 1 - \exp -(Vk\sigma^m) \qquad\qquad 1$$

where S is the fractional cumulative probability of failure, V is the stressed volume of the specimen, k is a constant and σ is the maximum principal tensile stress. There are many variants of equation 1 but they all contain an exponent m, known as the Weibull parameter, which expresses the variability of the material's strength, and is 1.04-1.44 times the reciprocal of the coefficient of variation for tensile test results (Weibull, 1939). The Weibull parameter and k in equation 1 can be obtained by plotting data for nominal strengths as in the example shown in Fig. 3b for results of bend tests on cemented carbides.

However to use probabilistic analyses confidently it is necessary to know the mechanistic source in order to differentiate between materials variability and variability arising from a weakness in the test procedure. As an example, examination of fracture surfaces in bend specimens of WC/Co cemented carbides show defects that have initiated the failure. By plotting the size of these versus stress, a linear relation is obtained (Fig. 4) and much of the scatter in the results can then be explained (Almond and Roebuck, 1977). This approach enables effects of volume, shape and surface finish to be explained. It also recognizes the possibility that several competing failure mechanisms may be operating simultaneously and eliminates ambiguities inherent in the empirical approach such as having a volume dependent m (Shaw, Braiden and De Salvo, 1975).

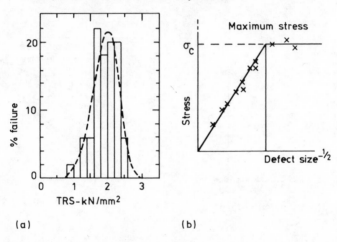

(a)   (b)

Fig. 4 a) Typical transverse rupture strength (TRS) distribution for bend tests on WC/Co cemented carbides; b) results for bend strength of WC/Co cemented carbides plotted against size of defect at fracture intiation site.

This improved explanation of a material's deformation-behaviour in a test, enables the interpretative stages to provide data of increased scientific value. Nevertheless, the traditional simplistic analyses are difficult to displace. To distinguish them from intrinsic properties the data they provide are given the status of "nominal" strength properties or are described in terms of the test, eg transverse rupture strength, modulus of rupture, bend strength, etc.

FRACTURE TOUGHNESS

The fracture mechanics concept has been developed through various energy and stress considerations for predicting crack equilibrium conditions based on a stress intensity factor K, which in its simplest form for remote loading of a crack of length a by a stress σ is given by:

$$K = Y \sigma a^{1/2}$$

2

for linear elastic fracture mechanics and Y embodies the dependence of
the function on the geometry of the specimen and can often be expressed as
a polynomial of the parameter a/w, the depth of the crack in the section
thickness w.

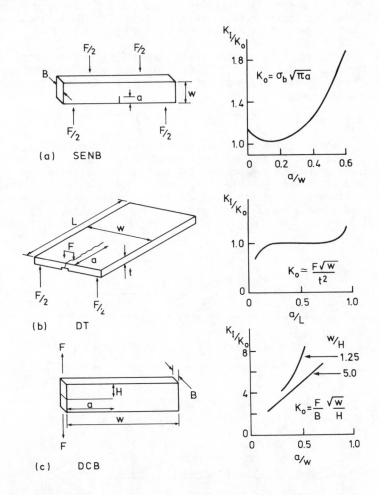

Fig. 5 Specimen geometries used for K_{IC} measurements, and
corresponding dependence of K_{IC} on crack length (Y functions) for:
a) single edge notched beam; b) double torsion; c) double cantilever
beam specimens (after Rooke and Cartwright, 1976; Pletka, Fuller and
Koepke, 1979)

Also,

$$G = K^2/E \qquad\qquad \text{in plane stress} \quad 3$$

and

$$G = K^2 (1-\nu^2)/E \qquad\qquad \text{in plane strain} \quad 4$$

where G is the rate at which strain energy is released into the crack
path with increase in crack length, E and ν are Young's modulus and
Poisson's ratio respectively. Examples of some of the popular geometries
used in testing are shown in Figs. 5a-c together with the Y-functions
(Rooke and Cartwright, 1976; Pletka, Fuller and Koepke, 1979).

Normally it is adequate to use the basic assumptions that the mode-I
fracture tendency predominates (direction of crack opening normal to the
crack faces) and to use an appropriate subscript for K. Thus the specimen
or component fails when:

a) $\qquad K_{IC} = K_R \qquad\qquad$ b) $\qquad G_{IC} = R_C \qquad\qquad\qquad$ 5

where K_R is the fracture toughness of the material and, with some
exceptions (described below), can be treated as an intrinsic property which
is independent of specimen geometry. The analogous expression for G_{IC} and
R_C can be derived from equations 3 ,4 and 5. Confusingly, the parameters
for fracture toughness and the critical stress intensity factor are often
used in a sense which makes them indistinguishable such that K_{IC}
represents both K_{IC} and K_R, and G_{IC} represents both G_{IC} and R_C.

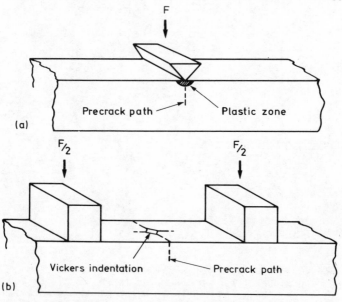

Fig. 6 Techniques for precracking fracture-toughness specimens:
a) wedge indentation; b) bridge indentation.

To utilise fracture mechanics analyses effectively, it is necessary to adopt a reliable procedure for introducing sharp precracks into specimens (Jones and Rowcliffe, 1979). Two suitable methods are the wedge indentation and the bridge indentation techniques. In the former, a wedge shaped indenter is driven into the specimen surface until a crack appears. The crack grows stably under increasing load until it reaches the required length (Fig. 6a); then the specimen is unloaded and the surface is carefully ground to remove the indentation and its associated stress field (Almond and Roebuck, 1978). In the bridge indentation method, a preliminary precrack is introduced into the surface by indentation with a Vickers pyramid indenter. When the specimen is loaded near to its extremities, the radial indentation cracks propagate to the edges of the specimen (Fig. 6b) and then propagate downwards under increasing load. The indentation is removed by lapping prior to testing (Warren and Johannesson, 1984).

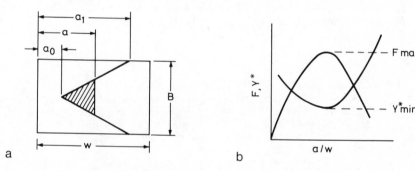

Fig. 7 a) Dimensional parameters for chevron notch specimens.
       b) Dependence of applied load and of geometric function, Y^*, on crack length in chevron notch specimens.

A specimen geometry of current interest uses a chevron notch in a DCB (double cantilever beam) or SENB (single edge notch beam) specimen to introduce a sharp crack (Barker, 1977; Shih, 1981; Munz, 1983). The crack initiates at the tip of the chevron and propagates subject to the condition

$$K_I = (AF/Bw)\ Y^*\qquad\qquad 6$$

where B is the specimen breadth and A is constant for a given geometry. The basic approach for chevron-notch testing can be understood by considering an approximate form for Y :

$$Y^* = Y\left\{\frac{a_1 - a_0}{a - a_0}\right\}^{1/2}\qquad\qquad 7$$

where a_0 and a_1 are shown in Fig. 7; Y is the geometric factor for a straight through-crack of length a, and Y has a minimum at a fixed crack position for a given specimen geometry. Consequently according to expression 6, F will reach a maximum when Y^* is a minimum and this makes it possible to determine K_R from measuring F_{max} and determining A separately.

Since choice of specimen shape is often dictated by the geometry of the components, the short-rod version of chevron-notch tests is becoming popular for cylindrical drill-bits.

Subcritical Crack Growth

One of the requirements in testing a hard material is to determine whether it exhibits a susceptibility to sub-critical crack growth. The phenomenon is normally an effect of an interaction between the environment and material at the crack tip. Under such conditions the instability criterion complies with equation 5 under normal loading rates. However, stable crack extension occurs at $K_I < K_{IC}$ with a dependence of velocity on the stress intensity factor which is usually approximated as:

$$da/dt = C \, K_I^n \qquad\qquad 8$$

where C and n are materials constants and vary with environment. It is possible to use equation 8 to derive explicit expressions for estimating the time for a crack to grow to a length when it propagates unstably to failure, by substituting for a from expression 2, and assuming a constant stress and a Y function that does not vary significantly with crack length (Davidge,1979).

A family of tests has become popular for determining the parameters C and n. They rely on geometries that require no gripping or support-pin-holes, and favour the establishment of constant-K versus crack-length. This is found in the double torsion specimens (Fig. 5b) but can also be obtained in DCB specimens (Fig. 5c) by using a tapered configuration.

Sources of Variability in Fracture Toughness Measurements

There are two major dangers in adopting the simplified approach of equations 2-5. Thus the conditions for unstable crack-growth are determined by the stiffness of the test system which is dependent on specimen geometry and machine-stiffness. Secondly, K_R may be a function of crack length and of a crack-tip process-zone that alter in properties during a test, so that the quantity measured cannot be regarded as an intrinsic materials property.

Clausing (1969) has shown that a crack can satisfy equations 2-5 but can still be stable if

$$\frac{1}{R}\frac{dR}{da} > \frac{1}{a}\left\{1 + \frac{2aY'}{Y}\right\} - \frac{2Y^2a}{\int_0^a Y^2a \, da + M(C)} \qquad 9$$

where Y' is the differential of Y with respect to a, the integration term is the compliance of an uncracked specimen, $M(C)$ is the dimensionally-corrected compliance of the loading system, and R is the fracture toughness (energy-based definition). Under fixed or controlled loading conditions, the second term in equation 9 is zero, and for monotonically increasing curves of Y versus a (as in Figs. 5a and c) a crack will propagate unstably if it satisfies equations 2-5 and R and K_R are single-valued. For constant Y or decreasing Y versus a, stable propagation can be obtained. For fixed or controlled displacements the second term is non-zero and increases with increase in machine stiffness. As a result, for some specimen geometries such as SENB tests, it is possible to obtain stable crack growth by using long crack-lengths and stiff systems (Mai and Atkins, 1980).

It is clear from equation 9 that crack stability will also be increased if R increases with increase in crack-length. This type of response to crack growth is sometimes referred to as 'R-curve behaviour', and there is evidence that it occurs in alumina and calcia/partly-stabilized-zirconia (Hubner and Jillek, 1977; Steinbrech, Khehans and Schaarwachter, 1983; Green, Nicholson and Embury, 1973).

The consequences of these limitations on using the simple equation 2-5 for describing crack susceptibility of hard materials are especially important since they form the basis of uncertainties in determining fracture parameters for design purposes from conventional fracture mechanics tests using specimens containing macrocracks. The following examples illustrate some of the difficulties.

As shown by equation 9, for some test geometries where machine stiffness is important, there will be a very high sensitivity to perturbations in the load versus displacement curve (arising from sticking, misalignment, etc.). Also, unless a sharp notch is used in an SENB test, too high a value is obtained for K_R because the stress at initiation and consequently the strain energy release rate is higher than that required to propagate a crack at any stage of the conditions in equation 5b governing the rate of change of R with a.

Secondly, if 'R-curve behaviour' is displayed, specimen-types which are relatively insensitive to machine stiffness will give higher values than those that are sensitive. Of special interest is the chevron notch geometry, since the test's attraction of a geometry-independent critical-crack-size is lost; the crack-size at maximum load is greater than assumed by theory and equation 6 gives too high a value for K_R (Munz, 1983).

INDENTATION TESTS

The advantage of simplicity of indentation tests is balanced by the absence of agreement on the procedures for converting measurements into meaningful scientific data. Also, there is a need for rigorous attention to testing procedures on account of the small size of the volume deformed when a hard material is tested with standard loads and indenter geometries used in hardness testing. As a result, specimen surface-finish and test-rig rigidity assume an increased importance, whilst the techniques' accuracy of measurement of 5-10% becomes a major limitation when investigating differences in hardnesses above 1000 HV.

Nevertheless, a hardness value is often the only indication that can be obtained on the likely resistance to plastic deformation of a material that is too brittle to test in conventional strength tests. Hardness is not an intrinsic property since the results depend on the procedure adopted, and comparative data can be reversed in ranking materials by changing the test or procedure. Nor is it yet possible to convert hardness data into an intrinsic strength property by an analytical procedure, although empirical correlations indicate that the hardness of metals is generally 2-3.5 times the uniaxial flow stress at 8% strain (Geach 1974). Unfortunately this relationship does not hold for hard materials where there is no systematic correlation and a factor of 5-10 is more common between hardness and strength (Almond, 1982).

Also associated with the measurement of hardness is a dependence on indentation size. Thus for many hard materials, it is found that with indentation loads below 20N, the pyramid indentation diagonal, L, varies with load, F, as:

$$F = k_2 L^q \hspace{4cm} 10$$

where k_2 is a constant and $q \leqslant 2$.

Thus, it follows that:

$$\frac{dH}{dF} = \left\{\frac{q-2}{q}\right\} k_1 \left\{\frac{k_2}{F}\right\}^{2/q} \hspace{3cm} 11$$

where H is the hardness and k_1 is 1.854 for a Vickers indenter and 14.22 for a Knoop indenter. As an example of the deviation from the normal value for q of 2 in most metals, for polycrystalline SiC, WC/Co cemented carbides and bearing steels, it was found that q was 1.70, 1.88 and 1.92 respectively and was as low as 1.54 in some single crystals (Brookes, 1983; Upit and Varchenya, 1973). Unawareness of this effect is often the reason for claims for the achievement of outstanding hardness values in thin coatings (Almond, 1984).

When an indication of a hard material's retention of rigidity at high temperatures is required, hot-hardness is the simplest parameter to measure. The results can be used for rating the potential high temperature performance of materials provided that the obvious precaution is taken of performing ancillary experiments to ensure that alterations in the time of application of the indentation load do not reveal significant creep effects. It is more informative to use the test to obtain data on creep-rate and on creep parameters such as the stress-exponent r and the activation energy Q in expressions for the secondary creep-rate of the form:

$$\frac{de}{dt} = A'\sigma^r \exp -Q/R'T \hspace{3cm} 12$$

where R' is the gas constant, A' is a constant and e is the

strain.

There are various analyses for converting indentation-creep measurements into the stress and creep-rate values for insertion into the appropriate expressions. These differ mainly in the manner they define strain-rate in relation to the diameter of the indentation or more accurately, the size of plastically-deformed volume associated with the indentation (Atkins, Silverio and Tabor, 1966). As an example, in a recent analysis by Roebuck and Almond (1982), the creep-rate was defined as:

$$\frac{de}{dt} = \frac{1}{d}.\frac{dd}{dt} = \frac{d(\ln\ d)}{dt} \qquad\qquad 13$$

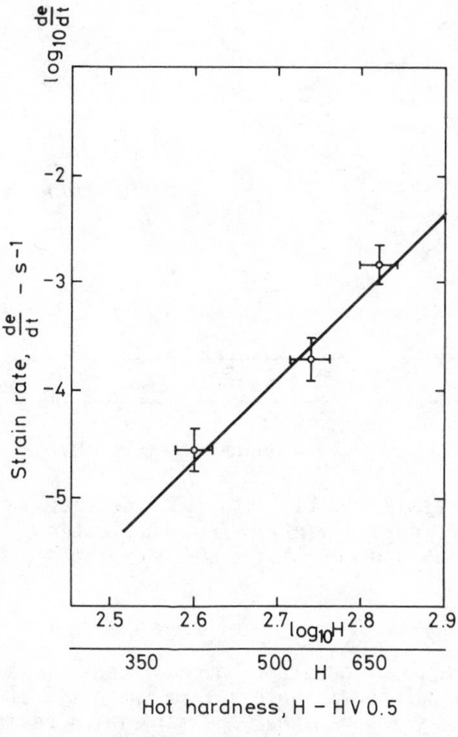

Fig. 8 Variation in strain rate with hot hardness of WC/Co cemented carbide indented for different times at 873 K.

Thus the creep-rate at various diameters 2d(t) can be determined from slopes of plots of either 2d against t, or ln d against t, and plotted as ln(de/dt) against hardness as in Fig. 8. When this technique was used to compare data from indentation and compressive tests on a WC-10%Co cemented carbide, the analysis gave values of 7 for the stress-exponent and enabled data to be extended into the high-stress/high creep-rate range (Fig. 9). Similar results have been obtained by Santhaman, Mizgalski and McCoy (1985).

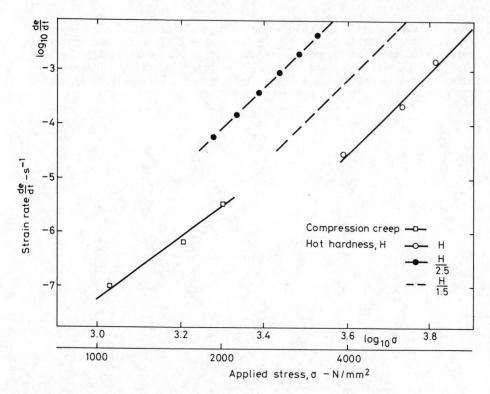

Fig. 9 Results of compressive tests, data for indentation-creep (from hot hardness tests), and curves derived from the latter by using constraint factors of 1.5 and 2.5, for a WC/Co cemented carbide at 873 K.

Measurements of the length of cracks produced in specimens by indenting with hemispherical and pointed indenters form the basis of indentation fracture mechanics. It is assumed that the material's resistance to cracking can be defined by a K_r parameter which is identical to K_R, the conventional fracture toughness parameter. Initially attention was focussed on the Hertzian circumferential cracks produced near the periphery of the zone of contact of the ball and specimen (Fig. 10a). The constraining criterion for analyses of this phenomenon is that they must be able to predict Auerbach's law (1891) that the critical load for crack formation is proportional to the radius of the indenter. To meet this requirement it has been necessary to invoke the prior existence of a distribution of very small surface flaws to initiate the circumferential crack which subsequently undergoes a stage of unstable propagation to form a cone-crack.

The analyses provide solutions of the form:

$$F_c = DG_r R''$$ 14

where F_c, R'', G_r are respectively the critical load for crack formation, the ball radius and the indentation fracture toughness. D can be regarded as a constant. It incorporates a stress-function, the crack-length, and elastic constants of the material and indenter. Nominally the analytical procedures provide a method for obtaining an explicit determination of fracture toughness. However, the G_r values are generally much lower than those obtained in conventional fracture mechanics tests (Warren, 1978) and this is regarded as a weakness in the method. Also the measurements required for calculating G_r are tedious, and recent analyses have concentrated on developing a method for calculating the K_r parameter from measurements of the radial cracks produced during indentation with Vickers pyramid indenters (Figs. 10b and c).

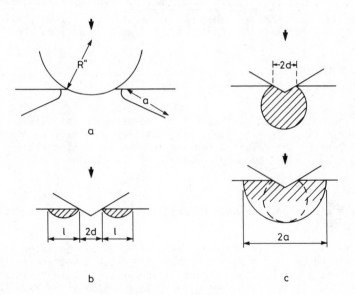

Fig. 10 Geometry of indenter and associated cracks for:
a) ball-indenter and Hertzian cracks; b) pyramid indenter with radial Palmqvist cracks; c) pyramid indenter with median cracks and radial cracks.

In current approaches it is acknowledged that significant extension of the radial cracks occurs during unloading. This is associated with the residual stress field which remains around the plastically deformed enclave after removal of the indenter load and enables penny-shaped median cracks to extend to the surface as semicircular radial cracks (Fig. 10c). The indentation cracks in cemented carbides are normally referred to as Palmqvist-cracks (Fig. 10b) and although their geometry differs in many respects from that of cracks in ceramics, this is ignored in the majority of models. The general theory can be illustrated by the model of Evans (1979). This assumes that if the maximum crack extension occurs when the specimen is fully unloaded, it can be shown that K_r can be expressed in terms of two independent dimensionless functions, g_1 and g_2 as follows:

$$\frac{K_r}{Hd^{1/2}} = g_1(a/d) \cdot g_2(E/H) \qquad 15$$

where g_2 is a plasticity term which reflects the magnitude of the residual stress. The same relation can be derived from a dimensional argument. The analysis is compromised if stresses within the plastic zone exert an important effect on crack extension.

Since the analysis is not intrinsically explicit, to develop it further it is necessary to assume that K_r equals K_R. Then, when $K_R/Hd^{1/2}$ data for various hard materials are plotted versus g_1 to see if a satisfactory functional relationship can be based on g_1 alone, it is found that the best-fit functions for data for ceramics and cemented carbides are respectively:

$$g_1 = (a/d)^{-3/2} \qquad\qquad 16a$$

$$g_1 = (a/d)^{-1/2} \qquad\qquad 16b$$

This is tantamount to assuming that the g_2 dependence on (H/E), the plasticity correction, is very small as in earlier models. Thus equation 16a is equivalent to the relation derived by Lawn and Fuller (1975) for a penny-shaped crack wedged open by a force F at the centre:

$$K_R = XF/a^{3/2} \qquad\qquad 17$$

where X is a dimensionless function and incorporates geometric, friction and surface effects.

To incorporate a plasticity correction various forms can be tried for g_2 which when plotted as $K_R/g_2Hd^{1/2}$ versus a/d give a better fit than equations 16a and b and provide a generalized curve for both ceramics and cemented carbides. It is found that a suitable function would have the form:

$$g_2 = (E/H)^{2/5} \qquad\qquad 18$$

and that the g_1 function can be described by a polynomial of a/d or approximately by $0.55 f^{-3/5} \log (8.4 d/a)$ where f is the constraint factor (Niihara, Morena and Hasselman, 1982).

Another approach employs the g_1 function from equation 16a and the g_2 function from equation 18 to provide an expression which is applied to both ceramics and cemented carbides (Anstis and co-workers, 1981).

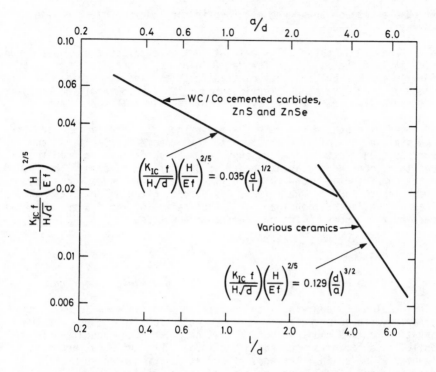

Fig. 11 Relation between K_{IC} (multiplied by plasticity correction factor) and radial crack length for ceramics, cemented carbides, ZnSe and ZnS (after Niiharaya, Morena and Hasselman, 1982)

Alternatively it might be postulated that the differences in crack geometries and plasticities of cemented carbides and ceramics require different models. As an example, Niihara, Morena and Hasselman (1982) propose that equations 16a and 16b are true reflections of this difference and suggest that this is related to the low values of a/d in the tougher materials. Thus:

$$\frac{K_R f^{3/5}}{g_2\, H\, d^{1/2}} \quad = \quad g_1(1/d) \qquad\qquad 19$$

as above, with $1 = a - d$. As can be seen in Fig. 11 this scheme works well in distinguishing between the two crack geometries except it requires that ZnS and ZnSe are categorised with cemented carbides although the former do not exhibit Palmqvist-cracking. An additional weakness according to Lankford (1982) is that if 'hard' ceramics are indented with low loads they give low a/d values and the data fit the generalized curve but not equation 19. However the significance of the low-load data may be questionable because they were obtained in the load-dependent regime of hardness.

From the above discussion it is evident that indentation techniques do not provide an independent method for determining K_R since equations 16-19 are nominally for K_r but they all contain functions and geometrical parameters which have been determined empirically by determining the best fit of indentation data to K_R values obtained in full-scale tests.

In a procedure proposed recently for measuring K_R, the radial indentation cracks are used to initiate fracture in SENB specimens (Chantikul and co-workers, 1981). It is claimed that K_R can be calculated from the indentation load and the breaking stress of the SENB specimen. The method uses the combined functions from equations 16a and 18 and assumes that the precrack length and residual stress field for a given material are uniquely defined by the indentation load, F. Application of the fracture mechanics expression for a semi-elliptical surface crack gives:

$$K_R = q \left\{ \frac{E}{H} \right\}^{1/8} . (\sigma F^{1/3})^{3/4} \qquad \qquad 20$$

where σ is the breaking stress and q is a geometrical constant.

Similar problems in deriving intrinsic materials data from indentation tests arise with scratch tests, but for coated specimens there are few alternative test methods. In a test that has attracted much attention recently, a specimen is mounted on a sliding table which can be moved in a horizontal plane against the apex of a vertically mounted Rockwell-C diamond indenter of 0.2mm radius at increasing indenter loads (Fig. 12b). The principle is to determine by observation or acoustic emission, a critical load at which the coating becomes detached from the substrate, and to use the result to calculate the shear strength, τ, of the interface from:

$$\tau = \frac{k}{R''} \left\{ \frac{FH}{\pi} \right\}^{1/2} \qquad \qquad 21$$

where R'', H, F, and k are respectively the indenter radius, the mean indentation pressure, the applied load, and a constant. The k-parameter has been altered from 1 to 0.2 to 5.5 in successive modifications of the models, to accommodate effects from friction and plastic deformation and the expression gives values of $1-10kN/mm^2$ which are within the expected orders of magnitude for the strength of bonded interfaces (Laugier, 1981). Unfortunately, the agreement is partly fortuitous since the mechanisms of deformation in hard materials are irreconcilable with those used in the model for which the expression was derived, and it is now generally believed that the test measures an empirical property that can be used for quality control purposes but not for calculating intrinsic properties.

Another indentation method that has been used for hard coatings on low-ductility substrates relies on the phenomenon of lateral cracking (Fig. 12a), which occurs on planes that are approximately parallel to the surface (Chiang, Marshall and Evans, 1981).

The procedure is to plot the relation between a, the length of the lateral crack and the indentation load, F, and to find values for the parameters K_s and F_0 from the best fit of the plotted curve to:

$$K_s = \frac{bt^{3/2} H^{1/2} (1-F_0/F) F^{1/2}}{a^2} \qquad\qquad 22$$

where b is a constant, t is the coating thickness, F_0 is the threshold load for cracking and K_s is the fracture toughness of the weakest plane for a crack path near the interface, be it along the interface, in the coating or in the substrate.

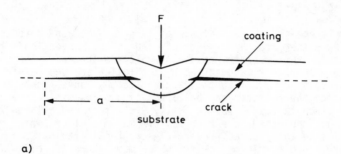

a)

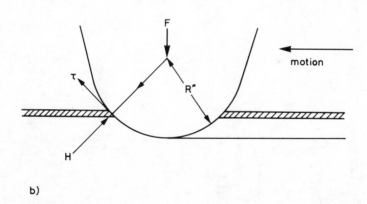

b)

Fig. 12 Interface strength tests for coatings: a) pyramid indentation test with lateral cracking; b) indentation scratch test.

APPLICATION

The choice of test is normally geared to the final destination and required accuracy of the data. However the user of hard materials is faced with deciding not only on a test but also on the appropriateness of the results and their interpretation with respect to his own problem. For example much of the data generated on fracture toughness are intended to be used in calculating the strength of components in terms of an assumed or characterized population of surface and volume defects. This normally involves the use of probabilistic fracture mechanics, which provide mechanistically based expressions of the same form as those employed empirically in Weibull statistics. The implicit assumption in the approach is that defects such as pores, inclusions and surface scratches can be treated as non-interacting Griffiths-cracks which propagate under the combined applied and residual stress fields according to the same stability conditions as prevail for macrocracks. Unfortunately this area is still beset with uncertainties.

Consider first the assumed equivalence of large cracks, and defects. This involves using either fracture toughness values from conventional tests or from indentation crack measurements. And while the former may be geometry dependent and encounter R-curve effects, the latter are semi-empirical in nature and need not be equal to the values given by conventional tests. The dangers of using the approach can be illustrated by examining the growth of a crack in a strength test. Thus initially the defect and a large volume of the specimen are subjected to a high uniform tensile stress which may be sufficiently large to produce irreversible changes in the bulk microstructure, such as microcracking, which may change the effective toughness of the material into which the defect will ultimately propagate. In contrast, in a conventional fracture toughness test up to and during crack growth the average stresses are relatively low and unlikely to affect the microstructure.

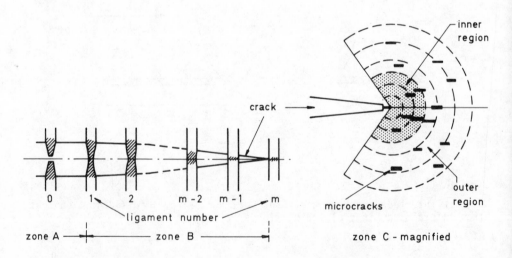

Fig. 13 Simplified model of 3-zone nature of a diffuse crack tip in a cermet or toughened ceramic: A) end of continuous crack; B) discontinuous region; C) microcracks in preparatory zone.

A second complication can be illustrated by examining the growth of a defect in a cermet or ceramic. Thus for WC/Co cemented carbides the crack front is diffuse and extends over a region containing part and wholly severed Co ligaments (Fig. 13). The establishment of an equilibrium crack-front configuration involves several stages of development if it originated from an internal or surface defect (Almond, 1983). Thus initially the defect would need to overcome a periodically varying stress field of comparable wave length to its own diameter and subsequently build up its own crack-tip damage characteristics and its own discontinuous 'diffuse-zone'. This will involve extra energy during the early stages of growth. There is also the possibility that the crack will not be able to overcome the increased resistance to growth and will be arrested. In cemented carbides and in an analogous process in alumina, it may be easier for an equilibrium crack tip configuration to be generated by a group of neighbouring microcracks linking together (Okada and Sines, 1983; Hoshida and co-workers, 1984).

These considerations are equally pertinent when considering indentation cracks and the macrocracks used in conventional fracture toughness testing. Since the crack size, the average stress field and the microstructural environment need not be the same in the two situations there is no reason to expect that the two approaches are measuring the same parameter. Nor is it possible to say which of the methods gives information of more relevance to the practical problem of fracture initiation from a manufacturing defect.

In view of the problems of testing and interpretation it is still worth considering whether empirical strength testing of components using empirical statistical analysis may produce faster results for life prediction than the fracture mechanics approach. Clearly the problem can also be tackled by combining the two types of data in an analysis. Thus in all types of test programme it is necessary to be aware that the goals may often be achieved by other routes than the mechanistic approach favoured by materials scientists. As an extreme, a systems analysis often proves effective where the purpose of the test is an assessment of a material's suitability for use in an engineering component which forms part of a machine or engine. This option is most commonly adopted in wear testing whereby a wear-problem or a wear-test is treated as a system within which the material is only one of a number of interacting elements (Czichos, 1978). The tests are normally performed on components under conditions which closely simulate the service conditions.

REFERENCES

Almond, E. A. and Roebuck, B. 1977, Met. Sci. <u>11</u> 458-461
Almond, E. A. 1978, Met. Sci. <u>12</u> 587-592
Almond, E. A. and Roebuck, B. 1978, Met. Technol. <u>5</u> 92-99
Almond, E. A., Irani, R. S. and Roebuck, B. 1980, Mater. Sci. Eng. <u>44</u>
    173-183
Almond, E. A. 1982, Powder Metall. <u>25</u> 146-152
Almond, E. A. 1983, Int. Conf. on Speciality Steels and Hard Materials. Ed.
    N. R. Comins and J. B. Clark.
Almond, E. A. 1984, Vacuum <u>34</u> 835-842

Anstis, G. R., Chantikul, P., Lawn, B. R., and Marshall, D. B. 1981, J. Amer. Ceram. Soc. 64 533-538

Atkins, A. G., Silverio, A. and Tabor, D. 1966, J. Inst. Met. 94 369-378

Auerbach, F. 1891 Ann. Phys. Chem. 43 61-75

Barker, L. M. 1977, Eng. Fracture Mech. 9 361-369

Brookes, C. A. 1983, Science of Hard Materials, Ed. R. K. Viswanadham, D. Rowcliffe and J. Gurland. Plenum Press, New York, pp 181-198

Buresch, F. E. 1984, Science of Ceramics, Vol. 12, Ed. P. Vincenzi. Ceramurgica srl Italy, pp 513-522

Chantikul, P.,Anstis, G. R., Lawn, B. R., and Marshall, D. B. 1981, J. Amer. Ceram. Soc. 64 539-543

Chiang, S. S., Marshall, D. B. and Evans, A. G. 1981, Materials Science Research, Vol. 14, Plenum Press, New York, pp 603-617

Clausing, D. P. 1969, Int. J. Fracture Mech. 5 211-227

Czichos, H. 1978, Tribology Series, Vol. 1, A Systems Approach, Elsevier, Amsterdam, The Netherlands

Davidge, R. W. 1979, Mechanical Behaviour of Ceramics. Cambridge University Press, Cambridge UK, pp 139-150

Evans, A. G. 1979, Fracture Mechanics Applied to Brittle Materials, ASTM STP 678; Ed. S. W. Freiman. ASTM Philadelpia USA, pp 112-135

Geach, G. A. 1974, Int. Metall. Rev. 19 255-267

Green, D. J., Nicholson, P. S. and Embury, J. D. 1973, J. Amer. Ceram. Soc. 56 619-623

Hoshida, T., Furuya, H., Nagase, Y. and Yamada, T. 1984, Int. J. Fracture 26 229-239

Hubner, H. and Jillek, W. 1977, J. Mater. Sci. 12 117-125

Johansson, I., Persson, G. and Hiltscher, R. 1970, Powder Metall. Int. 2 119-123

Jones, R. L. and Rowcliffe, D. J. 1979, J. Amer. Ceram. Soc. 58 1195

Lankford, J. 1982, J. Mater. Sci. Letters 1 493-495

Laugier, M. 1981, Thin Solid Films 76 289-294

Lawn, B. R. and Fuller, E. R. 1975, J. Mater. Sci. 10 2016-2024

Mai, Y. M. and Atkins, A. G. 1980, J. Strain Analysis 15 63-74

Lueth, R. C. and Hale, T. E. 1970, Mater. Res. Standards 10 23-29

Munz, D. Fracture Mechanics of Ceramics 1983, Fracture Mechanics of Ceramics, Vol. 6, Plenum Press, New York, pp 1-26

Niihara, K., Morena, R. and Hasselman, D. P. H. 1982, J. Mater. Sci. Letters 1 13-16

Okada, T. and Sines, G. 1983 J. Amer. Ceram. Soc. 66 719-725

Pletka, B. J., Fuller, E. R. Jr. and Koepke, B. G. 1979, Fracture Mechanics Applied to Brittle Materials, ASTM STP 678; Ed. S. W. Freiman. ASTM Philadelpia USA, pp 19-37

Roebuck, B. 1979, J. Mater. Sci. 14 2837-2844

Roebuck, B. and Almond, E. A. 1982, J. Mater. Sci. Letters 1 519-522

Rooke, D. P. and Cartwright, D. J. 1976, Compendium of Stress Intensity Factors HMSO London

Santhanam, A. T., Mizgalski, K. P. and McCoy,W. C. 1985. This Conference

Shaw, M. C., Braiden, P. M. and De Salvo, 1975 J. Eng. Ind. 77-87 (Trans. Amer. Soc. Mech. Eng.)

Shih, T. T. 1981 J. Testing Eval. 9 50-55

Steinbrech, R., Khehans, R. and Schaarwachter, W. 1983, J. Mater. Sci. 18 265-270
Stout, M. G. and Petrovic, J. J. 1984, J. Amer.Ceram. Soc. 67 14-18
Upit, G. S. and Varchenya, S. A. 1973, Science of Hardness Testing. ASM, Metals Park, Ohio, 135-146
Wachtman, J. B. Jr., Capps, W. and Mandel, J. 1972, J. Mater. JMSLA 7 188-194
Warren, R. and Johannesson, B. 1984, Powder Metall. 27 25-29
Warren, R. 1978, Acta Metall. 26 1759-1769
Weibull, W. 1939, Ing. Vetenskaps, Akademiens Handligar, Stockholm 151 1-45

Discussion on Fundamental properties, mechanical testing and characteristics

Rapporteur: G GRIMVALL (1)
Session Chairman: C A BROOKES (2)

(1) Department of Theoretical Physics, The Royal Institute of
Technology, Stockholm, Sweden
(2) Department of Engineering Science, University of Exeter, UK

Predictably, the time available for discussion was far too short to encompass the points raised in this stimulating session. In the event, most of the discussion centred around the two papers which reported the effects of 'dopants' - i.e. on the velocity of dislocations (Professor Hirsch) and the strength of interfaces (Dr Hondros).

The paper by Professor Hirsch gave rise not only to very general questions and comments but also some very specific debate. Professor Minkoff, and others, commented on the possible relation between the strain energy considerations of Professor Cohen and the dynamical studies of Professor Hirsch. There remains a gap between these two approaches and it is clearly desirable that first principles calculations of lattice energies are extended to include crystals with defects such as dislocation kinks. Since atomic bonds are broken and reformed during plastic deformation, the band structure considerations should be such as to shed light on this dynamical process - including the effect of an applied stress. Professor Minkoff also asked for general information on the dislocation structure in hard materials which would contribute to an understanding of the hardness of these materials. Professor Hirsch replied that he had only considered diamond cubic and zinc blende structures so far, but the mechanism for the doping effect should apply to other covalently bonded materials with a band gap.

Professor Williams followed up with three specific questions on how the model may be used to explain the observations from internal friction studies that (a) the dislocation velocity at high temperatures increases with n or p doping; (b) there is a lack of internal friction at low temperatures and (c) there is a linear dependence of dislocation velocity on stress. Professor Hirsch replied that (a) had not previously been considered but, if internal friction is due to dissipation of energy by the movement of kinks over energy barriers, then dislocation concentration and mobility increases for Si with n or p type doping; (b) the model is based on the idea that dislocation cores are reconstructed and that the secondary Peierls potential is high.   (Activation energies for kink migration have been measured recently by in situ experiments in the TEM and by internal friction (for Si). The resultant activation energies are found to be ≈ 1 eV);   (c) the model is based on the Hirth-Lothe model for dislocation velocity which gives linear dependence with stress. However, deviations from linearity are observed and there are various mechanisms to account for this. It should be noted that the dislocations are dissociated into partials and, for example at low stresses, kink nucleation on the two partials can be coupled.

Furthermore, the resolved shear stress driving the two partial disloca- tions, and their mobilities, may be different. Under these conditions, the simplicity of the Hirth-Lothe model is lost. The doping effect may also be different for different partials - which also complicates the detailed interpretation. The theory we have formulated ignores these com- plications to concentrate on the essential physical mechanism as proposed.

A further specific point, from Dr Page, concerned the possibility of dislo- cations (revealed by etch pitting) having become rearranged as the result of annealing indentations formed at somewhat lower temperatures. Professor Hirsch explained that the Ge specimens were given a short (30 minutes) post indentation anneal at the nominal temperature of indentation, defined as the temperature of the bulk of the specimen. Thus the rosettes reflect the response to the stress fields acting during and after the indentation at the same nominal temperature. The hardness response is controlled by the total plastic flow under and around the indenter, i.e. by slip under and close to the indenter by dislocations with Burgers vectors inclined to the surface as well as by slip in the etch rosettes further away from the indentation by dislocations with Burgers vectors parallel to the surface. It is likely that the latter slip occurs to relieve the stresses in the compressed plastic zone under the indenter formed by slip with inclined Burgers vectors, and may occur in part after the load is removed. The slip in the etch rosettes away from the indentation centre can at most make a contribution to the hardness response; in the case of Si this contribu- tion appears to be negligible in the temperature range investigated, since it is found that doping affects the rosette diameters but not the hardness values.

The paper by Professor Hondros gave rise to a lively discussion on the concept of the practical work of adhesion.   In reply to questions by Drs Exner and Buckley, Professor Hondros stated that the practical work of adhesion was only a rough estimate, based on the disbonding of par- ticles in metal matrices. By Auger spectroscopy, it was established that rupture always occurred at the metal-nonmetal interface. Professor Fisch- meister made the comment that a 'technical' work of adhesion could be the work done in tearing apart, or in some other way separating, two pieces of a couple. The separation requires a crack to propagate, usually in the metal phase close to the interface. Therefore, at first sight, one would not expect the thermodynamic work of adhesion to have any relation to the technical work of adhesion. In fact, they often differ by a factor of ten, but within a given family of materials pairs, they still co-vary surprisingly well. The reason is that in the technical work of adhesion, the work of plastic deformation dominates.   The extent of deformation, however, is limited by the decohesion event. Decohesion in microregions is determined by the thermodynamic work of adhesion. Thus the thermo- dynamic quantity assumes a trigger role in controlling a process which involves a much larger energy. The controversial nature of this approach to the strength of interfaces remained a talking point throughout the duration of the Conference.

Inst. Phys. Conf. Ser. No. 75: Chapter 3
Paper presented at 2nd Int. Conf. Science Hard Mater., Rhodes

Mechanical and physical properties of diamond

J E FIELD

PCS, Cavendish Laboratory, Cambridge CB3 0HE, U.K.

ABSTRACT

The extreme rigidity of the diamond lattice is the key to many of its exceptional properties. The paper first discusses the main physical properties of diamond including its structure, stability, the impurities diamonds contain and their effect on its thermal, optical and electrical behaviour. With this as background, recent results and data are then presented on the elastic, cleavage, strength, fracture, hardness, frictional and wear properties of diamond.

INTRODUCTION

Diamond is an exciting material with many outstanding properties. It is pre-eminent as a gemstone, an industrial tool and as a material for solid state research. Since natural diamonds grow deep below the Earth's surface before their ejection to mineable levels, they also contain valuable information for geologists. The key to many of diamond's properties is the rigidity of its structure which explains, for example, its exceptional hardness and its high thermal conductivity. Since 1953 it has been possible to grow synthetic diamond. Techniques are now available to grow gem quality synthetic diamonds greater than 1 carat (0.2 gm). However, the costs are high, and the largest commercially available industrial diamonds are about 0.01 carat in weight or about 1 mm in linear dimension. The bulk of synthetic diamonds used industrially are 600 µm or less. Over 75% of diamond used for industrial purposes today is synthetic material. A recent development has been the production of composites based on diamond; these materials have a significantly greater toughness than diamond while still maintaining very high hardness and reasonable thermal conductivity. The facility to grow synthetic diamond of chosen size, morphology and impurity content has further added to the versatility of the material. For these and a variety of other reasons, it is an opportune time to review the mechanical properties of diamond.

It is certainly not possible to understand the mechanical behaviour of diamond without some knowledge of its growth, the impurities and defects which it contains and its other physical and chemical properties. The remainder of the introduction, therefore, deals with such topics and thus gives the background to the key role diamond plays in physics, geology and technology. Further information on diamond in general can be found in the books edited by Berman (1965a) and Field (1979a) and the monograph by Davies (1984). Field (1979a) has an Appendix which summarises data on the physical properties of diamond. A detailed review of the mechanical properties of diamond will appear in Field (1986).

Diamond Stability

Diamond and graphite are both forms of carbon. In 1955, Berman and Simon considered the thermodynamics of the phase change and obtained the diamond/ graphite equilibrium diagram. Subsequent recalculation, using improved data, for specific heats, molar volumes and compressibilities has modified the line slightly above pressures of 40 kbar (0.4 GPa) (Berman 1965b, 1979). It is important to appreciate that although the line marks the boundary between regions of thermodynamic stability it says nothing quantitatively about the rate of change from one phase to another. Diamond it will be noted is <u>not</u> the stable form under environmental conditions; experience suggests that the transformation to graphite must be a slow process. Similarly when synthesising diamond it is not sufficient to simply take graphite to a P/T point above the Berman/Simon line; if this were so many of the early attempts at diamond synthesis would have been successful.

Oxidation and Graphitization

Since oxygen assists graphitization and allows diamond to be burnt at ca. 970 K such experiments have to be performed under chemically inert conditions (i.e. in a good vacuum or in the flow of an inert gas). Since there is a 43% volume increase involved in changing from diamond to graphite, graphitization is usually confined to the surface. Internal graphitization only takes place if a diamond contains microcracks or other defects which allow space for the expansion to be accommodated. At very high temperatures, the stresses caused by internal graphitization at defects can propagate cracks and cause dislocation movement (Evans and James 1964). These processes can buckle thin flakes or disrupt nearby surfaces (Evans and James 1964, Seal 1958a). In the presence of oxygen, Evans and Phaal (1962a) found that between 970 and 1270 K the oxidation rate varies linearly with oxygen pressure in the range 0.08 and 0.65 mbar and that $\{111\}$ surfaces have the highest oxidation rate, followed by $\{110\}$ and with $\{100\}$ surfaces having the lowest. Above 1300 K experiments at 0.52 mbar showed that the oxidation rates for these low index faces became constant. In the absence of oxygen, graphitization starts at ca. 1900 K (Evans and James 1964). The octahedral and dodecahedral surfaces of diamond graphitize with activation energies of 1060 ± 80 and 730 ± 50 kJmol^{-1} respectively. The activation volumes suggest that single atom processes are involved.

The practical consequences of such activation energies are that at atmospheric (i.e. effectively zero) pressure and at temperatures of 298, 1273 and 1473 K graphitization could just be detected after times of 10^{100}, 7500 and 1 year respectively (Evans 1976). However, at 2400 K a 0.1 carat (0.02 gm) diamond would be totally converted to graphite in about 3 minutes.

Crystallographic structure

Diamond has tetrahedral, covalent bonds between each atom and its four nearest neighbours. In diamond, the commonest way of linking the tetrahedral units into a three dimensional structure is the cubic form. An alternative structure is the hexagonal wurtzite lattice which was predicted as a possible polymorph of diamond by Ergun and Alexander (1962). Bundy (1962) has synthesised this form in the laboratory and it has now also been produced in various shock loading experiments. The cubic form of diamond occurs most commonly in nature as octahedra with {111} planes or in dodecahedra with {110} planes as a dissolution form. The simple cube with {100} planes is relatively rare with natural diamonds, but is common synthetically if particular conditions of T and P have been used in the growth capsule. Diamonds invariably contain an individual and complex internal structure which reflects the growth conditions, impurity content and any deformation while at high P and T. A range of techniques can image this structure. A polished surface shows small differences in topographic relief since, for example, type II regions (see below for a discussion of types) are more abrasion resistant than type I (Lang, 1979). Etched sections give information on growth since the impurities incorporated during growth affect the etch rates (Seal, 1962, 1965; Harrison and Tolanksy, 1964). Particularly sensitive techniques are those of U-V, cathodoluminesence and X-ray topography which taken together give a wealth of detail. An instructive reference is that by Lang (1979); his figure 14.1, for example, shows the internal structures of a diamond as revealed by four of the techniques mentioned above.

Impurities; inclusions

The observed properties of diamonds show unambiguously that elements other than carbon must be present in the diamond lattice. Much of recent diamond research has had the objectives of identifying these elements, their chemical relationship, their effect on optical, electrical, thermal and mechanical properties and their geochemical significance. The major impurity in diamond is nitrogen (see below) but recent research has shown that there are significant amounts of oxygen and hydrogen both in the bulk and at the surface of diamond (Madiba et al, 1984). As Table 1 in Sellschop (1979) details, 58 elements have now been identified as occurring in diamond. Some of these have been found using mass spectrometry or optical spectroscopy, but most ($\sim$50) have been detected by nuclear probe techniques.

Some diamonds contain visibly quite large inclusions of up to millimeter dimensions. The inclusions may be gaseous, liquid or solid. Solid inclusions occur either as totally enclosed single crystals or as crystalline material lying within fracture systems in the diamond. Two broad mineral groupings are recognised by the geologist. The first is the syngenetic minerals, which crystallize under the same physical conditions as diamond and are then enclosed and protected by the surrounding diamond, and the epigenetic minerals which formed subsequent to diamond growth, either while the diamond was in the kimberlite or later during transport to an alluvial deposit. It is the first category which has had the more detailed study (for a review, see Harris and Gurney, 1979). The considerable interest in the minerology and chemistry of the syngenetic inclusions is due to their value in evaluating ideas on the origin of diamonds, xenoliths and kimberlites and their inter-relation.

Diamond Types; The Role of Nitrogen

Measurements of the absorption of diamond in the infrared, visible and
ultra-violet has lead to a classification into types I and II (Robertson et
al, 1934; Sutherland et al, 1954; Clark et al, 1956). Most diamonds are
type I and have their absorption edge at ca. 330 nm, while a much smaller
group, type II, have this edge at ca. 220 nm. The absorption edge at
220 nm (5.48 eV) corresponds to the fundamental edge from the valence to
the conduction band (Clark et al 1956), and hence the edge at 330 nm has
been attributed to defects. The classification is further strengthened by
the infrared absorption; both types have an absorption band between 2 and
6 μm but only type I between 6 and 10 μm. Of particular significance are
peaks at 7.3 and 7.8 μm (see below). Sutherland et al (1954) further noted
that various lines in the spectra of a large number of type I diamonds
appeared to be related. For example, lines at 7.8, 8.3, 9.1, 20.8 μm
(group A features) varied by the same amount from diamond to diamond, and
similarly with a second set which they labelled group B at 7.0, 7.3, 7.5,
8.5, 10.0, 12.9, 30.5 μm. Subsequent work (see, for example, Clark, 1965)
has shown that the line at 7.3 is not a B feature but the general breakdown
still holds and the current view is that the A and B features are caused by
particular segregations of nitrogen impurity.

As would be expected from the size of the band gap, most diamonds are good
insulators. However, Custers (1952, 1954) found that a small proportion of
type II diamonds are semi-conductors, and proposed classifying the non-
conducting type II as IIa and the semi-conductors as IIb.

The major factor in explaining the absorption spectra of diamond is the
presence of nitrogen in amounts up to ca. 0.3% (Kaiser and Bond, 1959). In
their experiments, Kaiser and Bond graphitized diamonds and measured the
gases evolved. The amount of nitrogen was several orders of magnitude
higher than that of any other known impurity of diamond. Another valuable
observation was that there is a direct correlation between the 7.8 μm peak
and the nitrogen content, thus providing a non-destructive technique for
measuring the amount of nitrogen in a particular diamond. This is forbid-
den in a perfect diamond lattice because of its symmetry (Lax and Burstein,
1955) but the presence of the nitrogen defect lowers the local symmetry and
provides coupling between the electromagnetic radiation and lattice
vibration.

It would be wrong to give the impression that there is a rigid distinction
between the various types of diamond. Some diamonds are 'mixed' in the
sense that they contain regions of both type I and II. This is almost
certainly associated with the growth history. Tables 1 and 2 summarise
some of the main points of classification. As shown later, it is also
possible to make correlations with various mechanical properties and the
diamond types.

TABLE 1   Diamond Types and Nitrogen Content

Type	Abundance	Nitrogen
Ia	ca. 98% of natural	N up to 0.3% (i.e. 10^{19} to 5×10^{20} atoms cm^{-3}) A and B aggregates; low levels of single N atoms ($<10^{16}$ atoms cm^{-3}); platelets.
Ib	ca. 0.1% of natural; main synthetic type	N in single substitutional sites; paramagnetic (10^{17} to 10^{20} atoms cm^{-3}).
IIa	ca. 2%	Very low N level.
IIb		Extremely low N level; semi-conducting due to B acceptors.

TABLE 2   Diamond Types and Absorption

Type	U-V Absorption edge	Infra-red	Comments
Ia	ca. 330 nm	Absorption 2.5 µm to 10 µm	A and B spectral features related to A and B N aggregates. Intensity of 7.8 µm peak relates to total N content. Intensity of 7.3 µm peak relates to X-ray spike intensity and to platelet size and concentration.
II	220 nm; at fundamental (5.48 eV)	Absorption 2.5 µm to 6 µm; transparent > 6 µm	---

Platelets

The nature of the platelike features found in type Ia diamonds is one of the most intriguing mysteries of diamond science. The main contenders are nitrogen impurity or carbon itself. In the last few years the concensus of opinion about the amount of nitrogen in the platelets has fluctuated wildly. There is only space here to comment in the briefest terms. Direct observation of platelets by electron microscopy was made in 1962 by Evans and Phaal. They showed that they formed on {100} planes in type I diamonds but were absent in type II. They found that the platelet size and density varied markedly from specimen to specimen and also in different regions of each specimen. Platelet densities were in the range (1 to 3.5) x 10^{15} cm^{-3} and diameters ranged up to ∿ 0.1 μm. More recent observations have found "giant" platelets of up to 5 μm diameter (Woods, 1976). James and Evans (1965) have shown that the platelet contains extra material which forces apart the diamond matrix (i.e. an extrinsic defect). Platelets have an effect on strength. For example, they pin dislocations and so affect the resistance of type I diamond to plastic deformation.

Thermal Properties

The very high thermal conductivity of diamond is a major factor in explaining its performance in many technological applications. As is well known, diamond is frequently used in processes such as machining, grinding, cutting and polishing which generate high temperatures at the work interface. It is paradoxical that a material which graphitizes in air at ca. 900 K should survive such conditions. The answer is the thermal conductivity which has a value at room temperature about 5 times greater than that of copper. The heat is conducted by phonons so that the rigidity of the lattice which gives diamond its great hardness also favours a high conductivity.

The nearest approach to ideal behaviour is with type IIa diamonds. As would be expected, Umklapp and boundary processes are factors and in addition there is an isotope effect (natural diamond has ca. 1% of C^{13}) but there are only small effects from impurities and dislocations. Good type IIs have a peak in their conductivity at about 80 K and a value of ca. 1.5 x 10^4 W m^{-1} K^{-1}. Other types of diamond, see Table 3, have lower conductivities because of dislocations, impurities, nitrogen aggregates and platelets.

TABLE 3   Thermal Conductivity of Diamond in W m^{-1} K^{-1}

Type	At 293 K	At 80 K
Ia	600 - 1000	2000 - 4000
IIa	2000 - 2200	up to 15,000

Electrical Properties

Pure diamond, with an energy gap of 5.48 eV is a good insulator. However, a very small number of diamonds (type IIb) are p-type semi-conductors, and recent work has shown that the acceptor centre is substitutional boron

(Chrenko, 1971, 1973; Lightowlers and Collins, 1976). Nitrogen is a common impurity in diamond and it has been suggested that substitutional nitrogen is the donor centre in semi-conducting diamond (Dean, 1965). Because n-type conductivity is not observed with natural diamonds it is thought that the donor ionization energy is much greater than the acceptor ionization energy. Messmer and Watkins (1970) have proposed that the unusual depth of the nitrogen donor level is a manifestation of the Jahn-Teller effect. For a detailed account of the electrical properties of diamond the reader is referred to Collins and Lightowlers (1979).

MECHANICAL PROPERTIES

Elastic Constants

The extreme rigidity of the diamond lattice is the key to many of its mechanical properties. Tables 4 and 5 summarize recent determinations of its elastic constants and values for the modulii and Poisson ratio. Poisson's ratio varies markedly with orientation, with values ranging from 0.01 to greater than 0.2. The υ_{21} value equals 0.104. The isotropic aggregate value is given in Table 5.

TABLE 4
Elastic constants, moduli and pressure and temperature coefficients

Source	C_{11}	C_{12}	C_{44}	$E_{11}/10^{11} Nm^{-2}$	$K/10^{11} Nm^{-2}$
(a)	10.76	1.25	5.76	10.50	4.42
(b)	10.79	1.24	5.78	10.53	4.42
(c)	10.764	1.252	5.774	10.50	4.42
(d)	-	-	-	-	5.6
dC/dP	5.98±0.7	3.06±0.7	2.98±0.3	-	-
dC/CdT	-1.37±0.2	-5.70±1.5	-1.25±0.1		

(a) McSkimin and Bond (1957) (c) Grimsditch and Ramdas (1975)
(b) McSkimin and Andreatch (1972) (d) Drickamer et al (1966)

TABLE 5
Isotropic aggregate values computed from single crystal data (Ruoff, 1979)

Material	$E/10^{11} Nm^{-2}$	$G/10^{11} Nm^{-2}$	$K/10^{11} Nm^{-2}$	υ
Ge	1.316	0.545	0.76	0.207
Si	1.629	0.666	0.979	0.223
Diamond	11.41	5.53	4.42	0.07

Cleavage Plane

The dominant cleavage plane is the (111) but others have been observed.
Sutton (1928) noted that (110) cleavage and conchoidal fractures are
possible. Ramachandran (1946) found (211) cleavage, while Ramaseshan
(1946) in a study of 15 crystal fragments from Raman's collection observed
(111), (211), (110), (322), (332), (221) and (322). The (111) was most
common while the (221) and (110) were not uncommon. These results are
consistent with the order given in Table 6 based on Ramaseshan's
calculations.

However, it is not clear from these figures why the (111) should dominate,
particularly if the largest tensile stresses are normal to other planes.
The answer is most likely to be due to defects on the (111) plane giving
preferential weakening. Since the (111) plane is the growth plane it seems
possible that it will contain layers of faulted growth. Evidence for this
has come from the studies of Pandya and Tolansky (1954), Ramage (1956) and
Wilks (1958). Tolansky (1955, 1965) has emphasized the irregular nature
of most diamond cleavage surfaces. The traditionally perfect cleavage is
usually only on a micro-scale. Type II diamonds in general produce
smoother cleavage surfaces than type I.

TABLE 6   Theoretical cleavage energy for diamond

Plane	Angle between plane and (111) plane	Cleavage energy/ $J\ m^{-2}$
111	0° and 70° 32'	10.6
332	10° 0'	11.7
221	15° 48'	12.2
331	22° 0'	12.6
110	35° 16' and 90'	13.0
322	11° 24'	13.4
321	22° 12'	14.3
211	19° 28'	15.0
320	36° 48'	15.3
210	39° 14'	16.4
311	29° 30'	16.6
100	54° 44'	18.4

Note: to obtain a fracture surface energy, γ, divide by 2.

Velocity of Cleavage

Not surprisingly there are few experimental observations on the velocity of
cleavage. Ramage (1956), while studying cleavage surfaces, noticed that
many of them contained Wallner lines. According to Wallner (1939) the
lines are caused when a fracture front passes through a defect and releases
a burst of stress waves. Since a distortional wave has tension associated
with it, it disturbs the crack front along the locus of intersection,
causing a line marking. From the geometry of the situation it is possible
to obtain the fracture velocity in terms of the distortional wave velocity.

From her study of ten cleavage surfaces, Ramage estimated that typical cleavage velocities could reach about one fifth the distortional wave speed, i.e. about 2000 m s^{-1}.

Field (1962, 1979b) has used high-speed photography at sub-microsecond framing intervals to observe crack growth in diamonds. Velocities up to 7200 m s^{-1} were recorded. Since relatively few specimens were fractured, this does not rule out the possibility of even higher crack velocities. The value of 7200 m s^{-1} is about 0.7 of the Rayleigh wave velocity (the theoretical limit for brittle fracture). Since diamond transmits stress waves faster than any other material it is no surprise that it also has the highest measured fracture velocity. However, under conditions similar to that used in diamond cleaving, velocities of 1000 to 3000 m s^{-1} are more typical. A fuller discussion of fast crack growth in brittle solids and data comparing diamond with other brittle materials can be found in Field (1971, 1979b).

Cleavage Energy

Several authors (Ewald,1914; Kraus and Slawson,1939; Ramaseshan,1946) have approached the problem of obtaining a theoretical cleavage energy value by calculating the number of bonds which cross unit area of a chosen plane and multiplying this by the C-C bond strength. Table 6 is similar to one drawn up by Ramaseshan (1946), but with a value of 83 kcal mol^{-1} (5.8 x 10^{-19} J) for the C-C bond (Pauling, 1960). Since a cleavage crack in advancing unit area produces two unit areas of fracture surface, the corresponding fracture surface energies are half the cleavage energies. The theoretical value for γ is therefore 5.3 J m^{-2}.

By monitoring the extent of crack growth beneath a loaded indenter Field and Freeman (1981) have been able to measure the fracture energy, γ, of diamond as 5.50 ± 0.15 J m^{-2} for (111) cleavage. Fig. 1 gives an example of a 'cone' crack propagating from an indented (100) surface. Loading a (100) surface gives the simplest situation. The 'ring' crack on the indented surface is usually four-sided (Seal, 1958b; Howes, 1956) as would be expected from the way the 111 planes intersect the (100) plane. If viewed from beneath, the 'cone' maintains a four-fold symmetry but with conchoidal fracture joining the sides.

Experimental values of γ for most materials are usually greater than, often very much greater than, the theoretical value. This is because of plastic processes at the crack tip. The fact that the measured value is only slightly greater than the theoretical value for diamond suggests that little, if any, dislocation movement occurs in the stress field at the crack tip. The factor most likely to cause an increase in γ for diamond is secondary cleavage or conchoidal fracture. The limited amount of data obtained by Field and Freeman for 'cones' produced from (110) surfaces supports this view since γ equals 7.1 ± 1.0 J m^{-2}. No significantly different values were obtained between the different diamond types.

The critical stress intensity factor K_{IC} is related to γ and G by

$$K_{IC}^2 = 2E\gamma = EG$$

For diamond, K_{IC} equals 3.4 MNm$^{-3/2}$ and this compares with 0.75 MNm$^{-3/2}$ for soda-lime glasses and sapphire, silicon carbides and silicon nitrides which all fall in the range 3 to 5 MNm$^{-3/2}$. Tungsten carbide can have a K_{IC} value three to five times the diamond figure. A value for diamond composites measured at the Diamond Research Laboratory, Johannesburg, was 7.3 and 8.6 MNm$^{-3/2}$ for specimens made with particles of average diameter of 10 μm and 25 μm respectively.

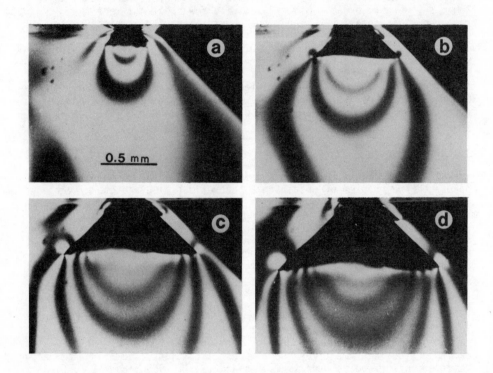

Fig. 1   A flat-ended diamond cone of tip diameter 0.27 mm is loaded against the (100) face of a diamond block and the cone crack development viewed from the side in polarised light. The dark region in the top right of each picture is caused by a facet on the diamond which impedes the view.

Fracture Strength

Rather different methods have been used for measuring the strength of macro-diamonds (those of millimeter dimensions) and micro-diamonds (sub-millimeter particles, grits, powders, etc...).

(a) Macro-diamonds

Obtaining fracture data by indentation is frequently the only approach available when specimen size is limited as, for example, with diamond. Fracture data has invariably been obtained using blunt indenters (Tolansky and Howes, 1956; Seal, 1958b; Howes, 1959, 1962; Bowden and Tabor, 1965; Lawn and Kamatsu, 1966; Lawn, 1968; Bell et al, 1975; Chrenko and Strong, 1975; Field, 1979b).

One of the most comprehensive studies was by Howes and Tolansky. Their study began in 1955 with the application of sensitive interferometric techniques to the observation of ring cracks in octahedral, cube and dodecahedral diamond surfaces. A later paper (Tolansky and Howes, 1956) showed that ring cracks could be induced in diamond with indenters softer than diamond, for example tungsten carbide and sapphire. Howes (1959) reported theoretically predicted and experimentally measured values of strengths for a variety of surfaces. He found about 25% <u>lower</u> values when tungsten carbide indenters were used. He attributed this to a size effect, but interfacial frictional stresses (see Field, 1979b) are also likely to be important. A summary of strength data is given in Table 7. p_c is the average stress obtained by dividing the load by the crack area, and p'_c is the average stress obtained from the load and the observed, or calculated, area. Most workers agree that the observed area, and that calculated from Hertz theory, are in close agreement. It should be emphasized that the data of Table 7 (taken from Field, 1979) are not necessarily values given in the original papers. To obtain consistency, the data was all recalculated using $E = 10.5 \times 10^{11}$ Nm^{-2} and $v = 0.2$ (Poisson's ratio depends markedly on orientation. It now appears that a mean value of $v = 0.1$ would be better. This would increase all the strengths by 33%).

TABLE 7   <u>Strength data for diamond. Data recalculated with $E = 10.5 \times 10^{11}$ Nm^{-2} and $v = 0.2$.</u>

Author	Face	Indenter radius/mm	p_c/Nmm^{-2}	p'_c/Nmm^{-2}
(a)	Octahedral		10,300	
	Cube	0.5, tungsten carbide	13,250	
	Dodecahedral		11,770	
(a)	Octahedral		10,790	13,730
	Cube	0.35, diamond	13,730	32,370
	Dodecahedral		17,660	
(b)	Cube	0.28, diamond	13,730	32,370
(c)	Cube	0.245, diamond		29,920
(d)	Cube	0.25, diamond		16,380-32,370
(d)	Synthetic octa-hedral and cube	0.25, diamond		21,580-32,370

(a) Howes (1965);   (b) Bowden and Tabor (1965);   (c) Bell et al (1977); (d) Chrenko and Strong (1975).

Bell et al (1975) used diamond indenters to compare different diamonds. They have found that diamonds with strong birefringence are usually weaker. Their results are given in terms of the critical loads applied, but if we take the mean load value for their best diamond (SC2B) and calculate an area of contact using Hertz theory the value of p'_c quoted in Table 7 is found. The work of Chrenko and Strong (1975) is of interest since they investigated both synthetic and natural diamonds. Highest strengths were obtained with high quality, low nitrogen content, colourless, synthetic gem stones. The presence of atomically dispersed substitutional nitrogen in synthetic diamond reduced the strength. The main conclusions from Table 7 are that the strength of diamond depends on the crystallographic face indented, and that for good quality diamonds the surface strength values are encouragingly consistent.

From the above data, it is possible to calculate the tensile strength of diamond (Field,1979). The value is 290 kg mm^{-2} (2.8×10^{9} Nm^{-2}) using $v = 0.2$ or 380 kg mm^{-2} (3.7×10^{9} Nm^{-2}) using $v = 0.1$. These compare with a theoretical strength (see Lawn and Wilshaw 1975) of 1.21×10^{11} Nm^{-2} which suggests that good quality diamonds have sharp ended defects of about 0.5 µm. Other results, to be discussed below, show that there is a strength-size effect with diamond so higher strength values would be expected if smaller areas are stressed.

(b)   Micro-diamonds

Various methods have been devised for assessing the strength of diamond grits and references are given in Field (1979b, 1985) and Field and Freeman (1981). An example of an approach used in industry is the Friatest machine developed by De Beers (Belling and Dyer, 1964; Belling and Baily, 1974) which involves putting a sample of two carats in a cylindrical steel capsule with a steel ball which is vibrated to and fro along the axis for a set number of 2400 cycles. The amount of particle fracture which occurs is measured by grading (screening) the grit. In the latest version of the test, the impact strength is expressed in terms of the number of cycles to produce the 'half-life' strength value.

Single grit compression is more laborious but allows a systematic study of particle strength and factors affecting it. It involves placing grits between anvils and loading in compression until failure occurs. Particles which fracture at low stress usually break into two fragments, but those which withstand a high load shatter into a fine micron-sized powder. In this latter situation there are often several small 'clicks' during loading as small high regions fracture off (Tsypin and Gargin, 1973; Field and Freeman, 1981). Fig. 2 gives results for a natural diamond grit product. It is clear that there is a strength/size relation since the vertical axis gives a measure of strength. If all sizes of particles had failed at the same stress the data points would have fallen on a horizontal line. The curve shows that smaller particles, though they withstand lower loads, have the largest strengths. Other workers have found a similar trend (see for example Bakul et al, 1969; Stupkina, 1971; Butuzov and Bezrukov, 1971; Shulshenko, 1969, 1973). The greater curvature with steel is due to the 'bedding in' of particles (Field et al, 1974, 1979), which relaxes the high local stresses at surface roughness. For a given particle size, the failure load is less the harder, material. It is not due to interfacial friction, which would have the opposite effect to that observed. If the same data is plotted in terms of load versus particle radius then an approximately linear relation holds (Field and Freeman, 1981). This suggests that Auerbach's law (1891) applies for this range of particle size of diamond.

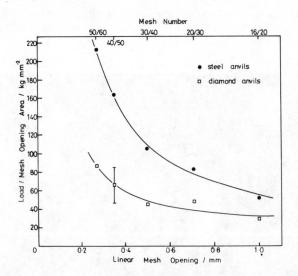

Fig. 2   Strength data for a natural diamond grit. The smallest particles (high mesh number) have the highest strengths (load/area). Steel anvils of hardness 9 GPa; diamond anvils of hardness about 90 GPa.

Synthetic material frequently has a pair of opposing flat crystal faces. This allows the contact area to be measured and the compressive strength calculated, from the recorded load in a crushing experiment. In experiments on synthetic material, Field et al (1974), Field (1979) showed that particles with a high magnetic impurity had only about 35% of the strength of similar material with a low magnetic impurity. Fig. 3 shows two examples of synthetic particles taken from a batch of Saw Diamond Abrasive (SDA) material. The individual particles are cubo-octahedral. As viewed here a (100) cube face is uppermost. In recent experiments, Field and Freeman (1981) measured the contact areas, and then loaded individual particles between hardened steel anvils. In the experimental arrangement, it was possible to photograph and identify the various stages of the crushing process. It is industrially useful to have SDA grits of different strengths and various grades are marketed. The Friatest, mentioned earlier, can give relative strength values, the single grit method gives absolute failure stresses.

A given batch of particles (i.e. same grit type, certain mesh size) can still contain a wide variety of particles. For example, surface condition, shape, defect concentration, birefringence, transparency, luminescence, diamond type (I or II), etc. All of these factors could have an effect on strength. In fact the very large variation in fracture loads in one series of tests is a strong indication that several factors are important. In the single particle crushing experiments by Battat (reported in Field, 1979) some of these variables were studied. With 20/30 mesh, De Beers Improved Dust (a product from good quality natural diamond) smooth particles were found to be ca. 43% on average stronger than rough and blocky shaped material was ca. 40% stronger than needle-shaped. In similar tests on 20/30 natural diamond grit (a boart, which has been surface and shape-treated), material which had few visible (to eye) defects was 38%

stronger than the 'opaque' particles with many visible inclusions. Field
and Freeman (1981) have shown that etching SDA material can increase its
strength by about 35%.

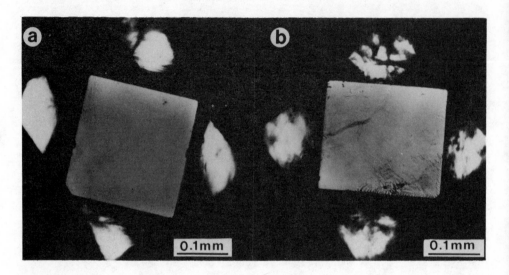

Fig. 3   Examples of SDA synthetic grit particles with cubic faces. It is
possible to measure the contact area of the faces accurately.

In his work Zsolnay (1971) considered crushed natural diamonds and divided
them into the following categories: pebbles, blocks, oblongs, flats and
needles. 'Pebble'-shaped material was not suitable for abrasive work since
the smooth shape meant that high-pressures were needed before cutting
started: these high pressures quickly loosened the diamond from the
matrix. Flats and needles were unsuitable for abrasion since they crushed
too easily. Zsolnay, therefore, eliminated these grits by carefully
sorting and obtained a better product for abrasive applications.
Bogatyreva (1972) successfully sorted grits of different roughnesses by a
floatation method in which 'smooth' grits move towards the surface and the
'rough' grits downwards. It was found that the 'bath' product (rough)
material was 25-35% weaker than the 'surface' (smooth) material. This is
of the order of magnitude mentioned above. The conclusion is that although
some sorting and selection is already performed, further sorting of
particles would be beneficial. Sizeable effects have already been
established based on the selection of some properties. However, several
other important parameters need to be investigated systematically.

Plastic Flow

Diamond has the greatest resistance to plastic flow of any material. Its
exceptional yield strength arises essentially for two reasons; the first
is that the theoretical strength is high because of the high density of
strong atomic bonds, and secondly because the strongly directional nature
of the bonding only allows the presence of narrow dislocations which are
difficult to move. There has been considerable interest in determining the
conditions under which flow is possible.

Natural diamonds give clear evidence from their internal structure of
dislocation motion and shear deformation while the material was at high T
and P in the Earth. In their high temperature experiments on bulk plastic
deformation, Evans and Wild (1965, 1966, 1967), Evans (1967), Wild et al
(1967) diamond plates were stressed in 3-point bending. The plates were of
approximate dimensions 5 x 3 x 0.5 mm. Fig. 4 is an example of a type I
specimen which has bent through 9°. The results showed that bulk deformation, in this loading situation, occurred at temperatures above 1900 K.
Further, type II diamonds deformed at significantly lower stresses than
type I. This was explained (Wild et al, 1967) by the different initial
defect distributions. Type I diamonds have a low dislocation density and a
high density of platelets. In contrast, type II diamonds have a high
dislocation density but with no platelets present (Fig. 5). Electron
micrographs of deformed type I diamonds showed dislocation pinning at the
platelets (Fig. 6). The platelets clearly play an important role in
hindering dislocation movement in type I diamonds and this explains the
higher stresses needed to obtain permanent deformation. The question of
plastic deformation at room temperature has been more controversial. It
now appears that it can take place, if there is sufficient constraint to
eliminate tensile fracture. Research to generate high pressures by
pressing a sample between diamond anvils has led to a deeper understanding
of the ultimate compressive and shear strength of diamond.

Ruoff (1979) has assessed the yield strengths of Ge, Si and diamond by
considering (i) dynamic yield strengths obtained from Hugoniot elastic
limits, (ii) yield strengths obtained from hardness data, and (iii) yield
strengths obtained by measuring the first sign of permanent deformation to
an indenter. Table 8 summarises the results obtained from an analysis of
(i). Note that the G and E used for diamond are the isotropic aggregate
values given in Table 5. For a ductile solid, the yield strength σ_0
is $\sim 1/3$ the hardness value, H (Tabor, 1951). According to Marsh (1964)
this fraction increases slowly as the ratio σ_0/E increases. The
hardness of diamond depends on the crystal orientation but if we take
H = 90 GPa then σ_0 equals ~ 30 GPa if $H = 3\sigma_0$ and ~ 36 GPa if
$H = 2.5 \sigma_0$.

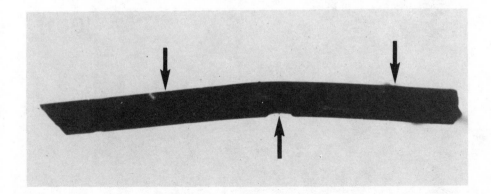

Fig. 4   Optical micrograph of the side view of a diamond plate which has
been bent through about 9°. The thickness of the plates is about 0.5 mm
(Evans, 1976).

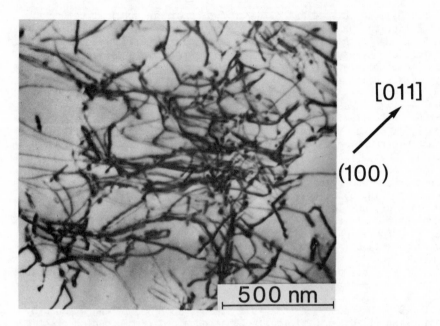

[011]

(100)

500 nm

Fig. 5 Electron micrograph of a diamond flake from the plastically deformed region of a bent type II diamond. The dislocations are shown to be tangled as a result of their movement during bend of the diamond (Evans, 1976).

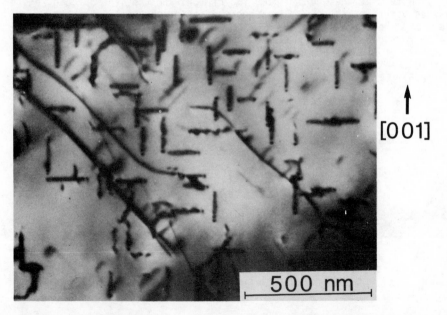

↑
[001]

500 nm

Fig. 6 Electron micrograph of a diamond flake from the plastically deformed region of a bent type I diamond. The dislocations whose creation and movement were responsible for the plastic deformation can be seen as well as the platelets which are present in this type of diamond (Evans, 1976).

TABLE 8   Dynamic yield stress (After Ruoff, 1979)

Material	σ_0/GPa	σ_0/G	σ_0/E
Ge	3.61	0.0662	0.0274
Si	4.33	0.0650	0.0266
Diamond	35*		0.0308+

* Computed from average value of σ_0/G for Ge and Si and E for diamond.
+ Computed from average value of σ_0/E for Ge and Si and E for diamond.

The maximum load diamond anvils can withstand is clearly of practical interest, since diamond cells are used in a variety of high pressure optical and X-ray work. The first objective of any design is to keep the tensile stresses below the tensile failure strength of diamond. Assuming this can be done the question then arises as to the ultimate pressures which can be achieved without extensive plastic yielding of the anvils. The relation between the maximum pressure p_0 in the applied pressure distribution and σ_0 the yield strength depends on the details of the pressure distribution. For uniform pressure $p_0 = 1.28 \sigma_0$ and for a spherical indenter (Hertz situation) $p_0 = 1.36 \sigma_0$ (Ruoff and Chan, 1976; Ruoff, 1979). For these pressure distributions plastic deformation would be expected to begin at $\sim$ 45 and $\sim$ 48 GPa respectively. Ruoff (1979) estimates that plastic dents would become visible at $\sim$ 70 GPa and that limiting pressures of up to $\sim$ 85 GPa could be achieved. Mao et al (1978a and b) have indeed observed plastic dents in diamond anvils and claim to have achieved pressures of 172 GPa. In view of Ruoff's work this pressure is about a factor of 2 too high, and Ruoff questions the calibration technique used. Further research will be needed to settle this issue. This question does not detract from the achievements of Mao and co-workers in designing a cell which reaches such extreme pressures and proving conclusively that diamond will plastically deform at room temperature.

Hardness of Diamond

Early measurements on the indentation hardness of diamond have been reviewed by Brookes (1971, 1979). Particularly noteworthy is the apparent lack of reproducibility in the quoted values ranging from H_K = 45 GPa, Knoop et al (1939), to H_B = 525 GPa, Kruschov and Berkovich (1951). It is now known that the hardness of diamond depends on a great many factors such as orientation, diamond type, impurity content, load and shape of indenter. Added to the fact that the indents themselves are small, thus giving relatively large experimental errors, the early scatter is understandable. Recent measurements, with greater control of the variables listed above, show good agreement. Table 9 from Brookes (1979) gives data on diamonds, indented at room temperature with a 1 kg load, on mechanical polished surfaces.

TABLE 9
Knoop hardness (H_K) of diamond - GPa (after Brookes, 1979)

Plane	Direction	Type I	Type II
(001)	[110]	81.0	89.0
(001)	[100]	96.0	101.0
(110)	[110]	86.0	92.0
(110)	[001]	106.0	113.0
(111)	[110]	55.0	75.0
(111)	[112]	62.0	108.0

Note that the type II values are consistently higher than those for type I and that there is a marked anisotropy by a factor of about 1.5. Few crystals in fact show such a large variation with orientation as diamond. Brookes et al (1971) and Pospeich and Gryziechi (1970) have developed models, based on the resolved shear stresses developed on the primary slip systems in the bulk of the crystal beneath the indenter which successfully explain and predict the anisotropy of hardness in most crystals. The fact that diamond follows the predictions of these modes, assuming a {111} <1$\bar{1}$0> slip system has been argued by Brookes (1970) that the indents are evidence of plastic flow at room temperature. Chrenko and Strong (1975) have published important data on General Electric laboratory grade diamonds, and a selection of results are given in Table 10.

Note that as the nitrogen content decreases the tensile strength (cleavage pressure), abrasion resistance and Vickers hardness all increase. Interestingly G.E. results on natural diamonds show high hardness but rather modest abrasion resistances. The different types of indenter and lower load explain the higher hardness found for natural diamond in the G.E. work compared with that of Brookes.

Data taken from Brookes (1970) Bakul et al (1973) show that there is a clear evidence of higher hardness values with decreasing load. At first sight this is surprising since geometric similarity implies that, for pyramidal indenters, the hardness of a crystal should be independent of normal load. However, this trend is not confined to diamond but has been found for a whole range of crystalline solids (Moxley, 1974; Brookes, 1979). An important additional observation is that the effect becomes more pronounced as the crystal's hardness increases. The true explanation of the hardness increase at low loads is most likely associated with the early stages of plastic deformation when small volumes are stressed. However, whether it is due to the small number of dislocation sources involved, or due to the initial stages of slip being on a fine scale with increased dislocation interaction, and thus greater hardening (a mechanism favoured by Brookes and his co-workers) needs further study.

TABLE 10
Strength, hardness, abrasion resistance of laboratory grade
diamond (after Chrenko and Strong, 1975)

Diamond	Colour	Nitrogen/ ppm atomic	Cleavage pressure[a]/kb	Hardness[b]/ GPa	Abrasion resistance[c]
70	yellow	70	203	95	0.84
12	yellow	20	---	106	1.31
82	pale green	∿2	245	---	1.72
15	colourless	∿0.3	---	110	3.05
87	pale green	∿0.1	∿300	131	3.11

(a) Indentation on a (001) face with a spherical indenter.
(b) Vickers indenter on a (001) face, with the diagonals of the indenter parallel to <100> direction.
(c) Grinding wheel test with 70 Al_2O_3 grit and with grinding ratio expressed as volume of wheel wear in m^3 divided by weight of diamond lost in gm.

There is no space here to review all the high temperature hardness work on diamond. However, mutual indentation studies with octahedra by Evans and Sykes (1974) showed that type I hardnesses above 1650°C were higher than type II. This result is consistent with Evans' other work on bulk plastic deformation (discussed above). It conflicts, however, with the room temperature indentation data given in Table 8. The Evans and co-workers' measurements on yield stresses and creep behaviour are consistent with the fact that type II has little nitrogen, but many dislocations, while type I has a relatively high level of nitrogen including platelets which hinder dislocation movement. The surprise is really the reversal of behaviour when diamonds are indented at room temperature, which suggests that when small volumes are stressed, and localised flow commences, that type I diamonds are softer than type II. This result taken with the dramatic increase in hardness at low load suggests that further research on indentation plasticity in hard crystals is needed.

The usual behaviour when a pure crystalline solid is irradiated by fast neutrons is to introduce point defects and dislocation loops which increase the density. With diamond Brookes and Vance (1972) have shown that both the crystal density and the Knoop indentation hardness decrease. These results are consistent with the view that any change from the perfect diamond lattice at room temperature, be it by the addition of nitrogen or point defects, decreases its hardness.

Wear

There is a considerable practical interest in the abrasion and wear properties of diamond (for reviews, see Wilks and Wilks, 1965, 1972, 1979). Firstly, there are the situations where diamond is itself ground and polished to make jewelry or industrial tools. Secondly, there are the many industrial applications where diamond is used for polishing, grinding, cutting and machining.

The work of Wilks and Wilks (1965, 1972) has demonstrated three main results about the abrasion resistance of diamond: (i) its great dependence on the orientation of the facet and the direction of abrasion; (ii) the very sharp changes in magnitude which occur near certain planes and directions. A change of a few ° can more than double the material removal rate; and (iii) the impossibility of representing the variation of abrasion resistance by a simple vector quantity. In other words, if the direction of traversal is reversed a different wear rate can result.

The technique developed by Wilks and Wilks involves rotating a cast iron or bonded wheel of 1 cm radius and allowing the U-shaped edge to be loaded against the diamond under test. The cuts produced are then measured using multiple-beam interferometry at spacings of $\lambda/2$, where λ is the wavelength of a monochromatic light source (546.1 nm). An important result is that each revolution of the wheel removes approximately the same amount of material over a wide speed range. This suggests that thermal effects at 'hot spots' are not important. An example from their extensive data is given in Table 11. Note that there is an order of 10 difference in the depth of cuts which implies a difference of over 100 in the rate of removal of material!

The first attempt at explaining the mechanism of material removal was that of Tolkowsky (1920). His model, subsequently refined and developed by Wentorf (1959) and Wilks and Wilks (1965, 1972), is based on micro-cleavage with cleavage on the {111} planes dominating. The model has proved amazingly successful in explaining the main features of the results. It is unfortunate that some authors have introduced the phrase 'abrasion hardness'. The 'hard' and 'soft' planes and directions for abrasion are not the same as with indentation hardness, which is not surprising since abrasion of diamond is primarily a brittle process. For this reason the phrase 'abrasion resistance' is to be preferred.

TABLE 11

Depths of cuts made by bonded wheels, normalized to a
depth of 10 ($\lambda/2$) in the softest direction (Wilks and Wilks, 1972)

Plane and direction						
Plane	Dodecahedron (011)		Cube (001)		Octahedron (111)	
Direction	[100]	[001]	[100]	[110]	[11$\bar{2}$]	[$\bar{1}\bar{1}$2]
Depth of cut ($\lambda/2$)	10.0	1.8	8.5	1.1	3.9	2.1

Friction

Tabor (1979) has reviewed research on the friction of diamond. The main results, in air, can be summarized as follows:

(a) The friction depends on crystal face and orientation. On the octahedral face the friction is low and there is practically no anisotropy. On the cube face, polished in the <100> direction there is four-fold symmetry in the friction. The lowest friction is along the <011> or cube diagonal direction: the highest along the <100> or cube edge direction.

(b) If the cube face is polished in the <110> direction the four-fold
symmetry is replaced by two-fold symmetry.
(c) In some situations repeated traversal over the same track on the cube
face of diamond reverses the order of frictional anisotropy. After a few
hundred traversals the friction is higher in the <011> direction and low in
the <100> direction.
(d) The frictional anisotropy on the cube face becomes less marked if a
conical stylus of large apical angle is used.
(e) The coefficient of friction may or may not depend on load depending on
the geometry of the sliding system.
(f) The friction is scarcely changed by the addition of a lubricant.
(g) Over a wide range of experimental conditions the coefficient of
friction lies between $\mu = 0.05$ and 0.15: it rarely exceeds $\mu = 0.2$.
(h) Finally we note that although the wear rate on the (100) and (110)
faces of diamond may depend on the sense of sliding the friction in such
situations does not depend on the sense of sliding.

For diamond sliding on diamond under vacuum conditions, the friction can
reach a high value ($\mu \sim 1$). For example, if the surfaces are cleaned by
heating or bombardment or alternatively if there are repeated traversals
over the same track. The high friction is accompanied by a very high wear
rate (Bowden and Hanwell, 1966). If a mono-molecular oxygen film is
allowed to reform the friction decreases dramatically. Buckley and co-
workers at NASA have recently made a detailed study of the adhesion and
friction of diamond in contact with various metals. The results showed
that μ was related to the relative chemical activity of the metals in high
vacuum. All the metals studied transferred to the surface of diamond
during sliding under a vacuum of 10^{-8} N m^{-2} (Miyoshi and Buckley,
1980; Pepper, 1981). Several authors have considered the dissipative
processes when diamond slides on diamond. There is still no entirely
convincing answer to explain the coefficients of friction which are
observed. The loads at the asperities are high but will they cause
fracture (which does not expend much energy) or plastic deformation
(which might)? The question is made even more intriguing, since estimates
of the stresses to cause tensile failure (if the area of stressing is
small), plastic yielding and the transition of diamond to metallic
behaviour are all close. More discriminating experiments and analysis
are clearly required.

High Modulus Coatings

There is a growing interest in using high modulus coatings to protect
brittle materials from contact and impact damage. Recently we have
completed a finite-element analysis on the effect of a thin hard coating on
the Hertzian stress field (van der Zwaag and Field, 1982). Of particular
interest is the reduction of the maximum (radial) tensile stresses. The
amount of reduction in the substrate solid increases with the Young's
modulus and thickness of the coating. The reduction is very large with
diamond-like coatings since the coating modulus, E_c, can be typically
ten times the substrate modulus, E_s. Further, films of a few micrometers thickness can have a beneficial effect (van der Zwaag and Field,
1983). Fig. 7 summarises some of the results. It is a non-dimensional
plot where P* is the maximum axial stress in the coating, d is the film
thickness, a is the contact radius and υ the Poisson ratio. The dashed
curve is the Hertz stress at the depth of the interface for different d
values. Note that the ordinate is a log scale.

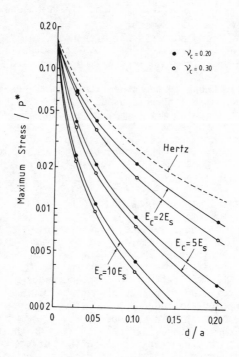

Fig. 7   Variation of the maximum tensile stress at the interface with the coating parameters (van der Zwaag and Field, 1982).

Acknowledgements

It is a pleasure to thank Professor T. Evans of Reading University for providing figures, and Dr F.A. Raal and Dr R.J. Caveney of D.R.L. Johannesburg for supplying specimens for the Cambridge experiments.

REFERENCES

Auerbach, F. 1891, Ann. Phys. Chem. 43 61-75
Bakul, V.N., Prinkhna, A.I. and Gobdanovich, M.G. 1969, Inst. Superhard
    Mat. Kiev Report 1-12
Bakul, V.N., Loshak, M.G. and Mal'nev, V.I. 1973, Sint. Almazy 6 16-19
Bell, J.G., Wilks, E.M. and Wilks, J. 1975, Ind. Diam. Rev. 135-138
Belling, N.G. and Bialy, L. 1974, The Friatest; Ten Years on, Ind. Diam.
    Info. Bureau, Ascot
Belling, N.G. and Dyer, H.B. 1964, Ind. Dia. Info. Bureau, Ascot
Berman, R. ed. 1965a, Physical Properties of Diamond. Clarendon Press,
    Oxford
Berman, R. 1965b, ibid. Ch 14
Berman, R. 1979, The Properties of Diamond. Ed. J.E. Field. Academic Press,
    London, pp 3-22
Berman, R. and Simon, F. 1955, Z Electrochem. 59 333-338
Bogatyreva, G.P. 1972, Sint. Almazy 3 35-38
Bowden, F.P. and Hanwell, A.E. 1966, Proc. Roy. Soc. Lond. A295 233-243
Bowden, F.P. and Tabor, D. 1965, Physical Properties of Diamond. Ed. R.
    Berman. Clarendon Press, Oxford, pp 184-220
Brookes, C.A. 1970, Nature Lond. 228 660-661
Brookes, C.A. 1971, Indus. Diam. Rev. 21-24
Brookes, C.A. 1979, The Properties of Diamond. Ed. J.E. Field. Academic
    Press, London, pp 383-402
Brookes, C.A. and Vance, E.R. 1972, Diamond Conference, Bristol
    (unpublished) reported in Brookes, 1979
Brookes, C.A., O'Neil, J.B. and Redfern, B.A.W. 1971, Proc. Roy. Soc.
    Lond. A322, pp 73-88
Bundy, F.P. 1962, J. Chem. Phys. 38 631-643
Butuzov, V.P. and Bezrukov, G.N. 1971, Sint. Almazy 3 11-14
Chrenko, R.M. 1971, Nature, Phys. Sci. 229 165-167
Chrenko, R.M. 1973, Phys. Rev. B7 4560-4567
Chrenko, R.M. and Strong, H.M. 1975, General Electric Report 75CRD089 1-46
Clark, C.D. 1965, The Physical Properties of Diamond. Ed. R. Berman.
    Clarendon Press, Oxford
Clark, C.D., Ditchburn, R.W. and Dyer, H.B. 1956, Proc. Roy. Soc. Lond.
    A234 363-381
Collins, A.T. and Lightowlers, E.C. 1979, The Properties of Diamond.
    Ed. J.E. Field. Academic Press, London, pp 79-106
Custers, J.F.H. 1952, Physica 18 489-496
Custers, J.F.H. 1954, Physica 20 183-184
Davies, G. 1984, Diamond. Adam Hilger, Bristol
Dean, P.J. 1965, Phys. Rev. 139 A588-602
Drickamer, H.G., Lynch, R.W., Clenenden, R.L. and Perez-Albuerne, E.A.
    1966, Solid State Physics. Eds. F. Seitz and D. Turnbull. Vol 19
    pp 135-228
Ergun, S. and Alexander, L. 1962, Nature Lond. 195 765-767
Evans, T. 1965, Physical Properties of Diamond. Ed. R. Berman, Clarendon
    Press, Oxford, pp 116-134
Evans, T. 1967, Science and Technology of Diamonds. Ed. J. Burls. Ind. Dia.
    Info. Bureau, London, pp 105-110
Evans, T. 1976, Contemp. Phys. 17 45-70
Evans, T. 1979, The Properties of Diamond. Ed. J.E. Field. Academic Press,
    London, pp 403-424
Evans, T. and James, P.F. 1964, Proc. Roy. Soc. Lond. A277 260-269
Evans, T. and Phaal, C. 1962a, Proc. Fifth Biennial Conf. on Carbon, Penn.
    State Univ. 147-153

Evans, T. and Phaal, C. 1962b, Proc. Roy. Soc. Lond. A270 538-552
Evans, T. and Rainey, P. 1975, Proc. Roy. Soc. Lond. A344 111-130
Evans, T. and Sykes, J. 1974, Phil. Mag. 29 135-147
Evans, T. and Wild, R.K. 1965, Phil. Mag. 12 479-489
Evans, T. and Wild, R.K. 1966, Phil. Mag. 13 209-210
Evans, T. and Wild, R.K. 1967, Phil. Mag. 15 447-451
Ewald, P.P. 1914, Ann. Phys. 44 257-282
Field, J.E. 1962, Proc. 6th Int. Conf. on High-Speed Photography.
    Willink and Zoon, The Hague, pp 514-521
Field, J.E. 1971, Contemp. Phys. 12 1-31
Field, J.E. ed. 1979a, The Properties of Diamond. Academic Press, London
Field, J.E. ed. 1979b, ibid, 281-324
Field, J.E. 1986, Rep. Prog. Phys. In press
Field, J.E., Hauser, H.M., Hutchings, I.M. and Woodward, A.C. 1974a,
    Ind. Diam. Rev. 255-259
Field, J.E. and Freeman, C.J. 1981, Phil. Mag. 43 595-618
Grimsditch, M.H. and Ramdas, A.K. 1975, Phys. Rev. B11 3139-3148
Harris, J.W. and Gurney, J.J. 1979, The Properties of Diamond.
    Ed. J.E. Field. Academic Press, London, pp 555-594
Harrison, E.R. and Tolansky, S. 1964, Proc. Roy. Soc. Lond. A279 490-496
Howes, V.R. 1959, Proc. Phys. Soc. Lond. 74 48-52
Howes, V.R. 1962, Proc. Phys. Soc. Lond. 80 78-80
Howes, V.R. and Tolansky, S. 1955, Proc. Roy. Soc. Lond. A230 287-293
Howes, V.R. and Tolansky, S. 1955, Proc. Roy. Soc. Lond. A230 294-301
James, P.F. and Evans, T. 1965, Phil. Mag. 11 113-129
Kaiser, W. and Bond. W. 1959, Phys. Rev. 115, L857-863
Knoop. F., Peters, C.G. and Emerson, W.B. 1939, J. Res. Nat. Bur. Stand.
    23 39-61
Kraus, E.H. and Slawson, C.B. 1939, Am. Min. 24 661-672
Kruschov, M.M. and Berkovitch, E.S. 1951, Indust. Diam. Rev. 11 42-44
Lang, A.R. 1979, The Properties of Diamond. Ed. J.E. Field. Academic Press,
    London, pp 425-472
Lawn, B.R. 1968, J. Appl. Phys. 39 4828-4836
Lawn, B.R. and Komatsu, H. 1966, Phil. Mag. 14 689-699
Lawn, B.R. and Wilshaw, T.R. 1975, Fracture of Brittle Solids, C.U.P.
Lax, M. and Burstein, E. 1955, Phys. Rev. 97 39-52
Lightowlers, E.C. and Collins, A.T. 1976, J. Phys. D: Appl. Phys. 9
    951-963
McSkimin, H.J. and Andretch, P. 1972, J. Appl. Phys. 43 985, 2944-2948
McSkimin, H.J. and Bond, W.L. 1957, Phys. Rev. 105 116-987
Mao, H.K., Bell, P.M., Shaner, J. and Steinberg, D.J. 1978a, J. Appl.
    Phys. 49 3276-3283
Mao, H.K. and Bell, P.M. 1978a, Science 2000 1145-1147
Marsh, D.H. 1964, Proc. Roy. Soc. Lond. A279 420-435
Madiba, C.C.P., Sellschop, J.P.F., Derry, T.E. and Fearick, R.W. 1984,
    Diamond Conference, Bristol (unpublished)
Messmer, R.P. and Watkins, G.D. 1970, Phys. Rev. Lett. 25 656-659
Miyoshi, K. and Buckley, D.H. 1980, Appl. Surf. Sci. 6 161-172
Moxley, B. 1974, PhD dissertation, Univ. of Exeter
Pandya, N.S. and Tolansky, S. 1954, Proc. Roy. Soc. Lond. A225 40-48
Pauling, L. 1960, The Nature of the Chemical Bond. Cornell University
    Press, p 85
Pepper, S.V. 1981, NASA Tech. Memo. 82741, 1-13
Pospeich, J. and Cryziecki, J. 1970, Arch. wun. Hutnictlia 15 267-277
Ramage, J.B. 1956, PhD thesis, University of London
Ramachandran, G.N. 1946, Proc. Ind. Acad. Sci. A221 163-198
Ramaseshan, S. 1946, Proc. Ind. Acad. Sci. A24 114-121

Ruoff, A.L. 1979, High Pressure Science and Technology.
    Eds. K.D. Timmerhaus and M.S. Barber. Vol. 2, Plenum, pp 525-548
Ruoff, A.L. and Chan, K.S. 1976, J. Appl. Phys. 47 5077-5080
Robertson, R., Fox, J.J. and Martin, A.E. 1934, Phil. Trans. R. Soc.
    A232 463-535
Seal, M. 1958a, Nature 182 1264-1267
Seal, M. 1958b, Proc. Roy. Soc. Lond. A248 379-393
Seal, M. 1962, Proc. First. Int. Cong. on Diamonds in Industry.
    Ed. P. Greene. Ind. Diam. Infor. Burea, London, pp 361-375
Seal, M. 1965, Am. Min. 50 105-123
Sellschop, J.P.F. 1979, Properties of Diamond. Ed. J.E. Field.
    Academic Press, London, pp 107-163
Shul'zhenko, A.A., 1969, Sint. Almazy 6 7-15
Shul'zhenko, A.A. and Kruk, V.B. 1973, Sint. Almazy 3 4-11
Sutherland, G.B.B.M., Blackwell, D.E. and Simeral, W.G. 1954,
    Nature, Lond. 174 901-904
Sutton, J.R. 1928, Diamond. Murby and Co. London
Strupkina, L.M. 1971, Soviet Phys: Crystal 15 728-730
Tabor, D. 1979, The Properties of Diamond. Ed. J.E. Field.
    Academic Press, London, pp 325-350
Tolansky, S. 1955, The Microstructure of Diamond Surfaces. NAG Press,
    London
Tolansky, S. 1965, Physical Properties of Diamond. Ed. R. Berman.
    Clarendon Press, Oxford, pp 135-173
Tolanksy, S. and Howes, V.R. 1956, Proc. Phys. Soc. B70 521-526
Tolkowsky, M. 1920, PhD thesis, University of London
Tsypin, N.V. and Gargin, V.G. 1973, Sint. Almazy 3 32-35
Wallner, H. 1939, Zeit. Phys. 114 368-378
Wentorf, R.H. 1959, J. Appl. Phys. 30 1765-1768
Wild, R.K., Evans, T. and Lang, A.R. 1967, Phil. Mag. 15 267-279
Wilks, E.M. 1958, Phil. Mag. 3 1074-1080
Wilks, E.M. and Wilks, J. 1965, Physical Properties of Diamond.
    Ed. R. Berman, Clarendon Press, Oxford, pp 221-256
Wilks, E.M. and Wilks, J. 1972, J. Phys. D: Appl. Phys. 6 1772-1781
Wilks, J. and Wilks, E.M. 1979, The Properties of Diamond. Academic
    Press, London, pp 351-382
Woods, G.S. 1976, Phil. Mag. 34 993-1012
Zsolnay, L.M. 1971, DWMI Symp. on Grinding and Cutting with Diamond Wheels.
    Paper 8, pp 1-7
Zwaag, van der S. and Field, J.E. 1982, Phil. Mag. 46 133-150

Inst. Phys. Conf. Ser. No. 75: Chapter 3
Paper presented at 2nd Int. Conf. Science Hard Mater., Rhodes

207

The mechanical properties of cubic boron nitride — a perspective view

C A BROOKES

Dept. of Engineering Science, Univ. of Exeter, Exeter EX4 4QF, U.K.

ABSTRACT

The ratio of the resolved shear modulus (G) to hardness (H) is low for cubic BN (i.e. 8 : 1) - in common with most covalent solids. A comparatively small dislocated zone is formed which then expands on annealing at temperatures above 1173 K. {111} <1$\bar{1}$0> slip is established from Knoop hardness anisotropy and slip traces. The effect of temperature on hardness is comparable with that of diamond and indentation creep is observed at temperatures as low as 673 K. Preliminary measurements of K_{IC}, from cracked indents, yield an average value of 2.8 MN m$^{-3/2}$.

INTRODUCTION

Whilst there is insufficient information on the mechanical properties of cubic BN to warrant a review at this time, it would appear appropriate to view the data which are available against the properties of other crystalline solids. This material has to be synthesized and single crystals are now available with well formed and reasonably flat 'as-grown' {001} and {111} surfaces having edge lengths of only 200 to 500 μm. The crystal structure of cubic BN is of the sphalerite (or zinc-blende) type. This might alternatively be considered as two interpenetrating face-centred cubic sub-lattices; one consisting entirely of boron atoms and the other entirely nitrogen atoms. Consequently, {111} planes consisting entirely of boron atoms alternate with {111} planes entirely of nitrogen throughout the crystal. Although the bonding is predominantly covalent, some degree of ionic bonding is known to be present. Synthesized crystals usually contain many metallic inclusions and growth twins (Pipkin, 1980) in many ways similar to diamond, but the effect of such defects on the mechanical properties has not yet been ascertained. Particles for aggregation, which generally tend to be approximately 10 μm in diameter and consist of more than one crystal, are subsequently subjected to direct contact sintering. The conditions are such as to promote secondary growth of cubic boron nitride which forms bridges between the primary particles (Hooper and Brookes, 1986). Thus, the aggregates contain a continuous skeleton of cubic boron nitride.

As a consequence of their small dimensions and tendency to brittle behaviour, indentation hardness measurements have formed the basis of studies on the mechanical properties of cubic boron nitride crystals. The most important experimental variables influencing hardness measurements were outlined at the first International Conference on the Science of Hard Materials (Brookes, 1983) and it is now fully appreciated that the effects of these variables are more pronounced in ceramics than in other crystalline materials. Proper attention to such considerations enable us to derive much more than a consistent hardness number where, otherwise, the data might serve more to confuse than to enlighten.

ANISOTROPY AND SLIP SYSTEMS

As with all ceramic crystalline materials, the indentation hardness of cubic boron nitride crystals is most accurately measured having due regard to crystallographic effects and using the Knoop indenter (Brookes, 1983). Room temperature values for cubic BN may be compared in Table 1 with diamond, the only harder solid, with the measurements having been based on a normal load of 4.9 N.

TABLE 1 A Comparison of the Knoop Indentation Hardness of Cubic Boron Nitride with Diamonds at Room Temperature

Specimen	Knoop hardness - HK0.5 [GN/m^2]			
	(001)		(111)	
	[100]	[110]	[11$\bar{2}$]	[1$\bar{1}$0]
Type I diamond	96.04	81.34	61.74	54.88
Type II diamond	100.9	89.18	102.9	71.54
Cubic boron nitride	43.12	29.89	41.50	39.45

It is worth noting that a typical Type I diamond could contain up to 0.3% of nitrogen impurities, usually considered to be in the form of platelets, whilst the Type II diamond would be less impure (<500 ppm) but more heavily dislocated (Field, 1986). Nakamura and Yazu (1983) have shown that the hardness of diamond decreases somewhat with increasing nitrogen content. It is therefore tempting to speculate that the presence of impurities in these materials tends to reduce the number of strong interatomic bonds in the parent lattice thereby decreasing its strength. Further support for this contention, in diamond, may be derived from the observed decrease in hardness with increasing damage due to fast neutrons which, ultimately, convert the crystalline structure to an amorphous solid of lower density (Brookes, 1979). Whilst we might anticipate similar effects to be observed in cubic BN, at sufficiently high defect concentrations, a significant variation in the hardness of crystals from different manufacturers has not been apparent from our work.

An extensive review of the ratio of the shear modulus (G), as determined for the relevant slip direction and slip plane (Kelly, 1973), to the Vickers or Knoop indentation hardness (H) has recently been carried out (Ross, 1985). The results demonstrate that a G/H of 100 is not as widespread as previous work (Tabor, 1984) has implied - other than for face-centred cubic metals. Nevertheless, a general pattern does emerge such that a similar value obtains for crystals having a common type of bonding and crystal structure. The G/H values for a number of covalent materials are enumerated in Table 2 and the average value, of about 10, may be contrasted with the much higher values for the predominantly ionic or metallic solids.

TABLE 2   A Comparison of the Ratio of the Shear Modulus (G)
to Room Temperature Indentation Hardness (H) for
Various Crystalline Solids

SPECIMEN	SLIP SYSTEMS	G/H
Covalent:		
Diamond	$\{111\}$ $\langle1\bar{1}0\rangle$	5
Cubic boron nitride	$\{111\}$ $\langle1\bar{1}0\rangle$	8
Silicon	$\{111\}$ $\langle1\bar{1}0\rangle$	6
Germanium	$\{111\}$ $\langle1\bar{1}0\rangle$	6
Aluminium oxide	$\{0001\}$ $\langle11\bar{2}0\rangle$	14
Silicon carbide	$\{0001\}$ $\langle11\bar{2}0\rangle$	7
Transition metal carbides (average)	$\{110\}$ $\langle1\bar{1}0\rangle$	9
Ionic:		
Various halides and oxides	$\{110\}$ $\langle1\bar{1}0\rangle$	14
(range)	and	to
	$\{100\}$ $\langle011\rangle$	200
Metallic:		
Face-centred cubic (average)	$\{111\}$ $\langle1\bar{1}0\rangle$	100
Body-centred cubic	$\{1\bar{1}0\}$ $11\bar{2}$	
(average)	$\{12\bar{3}\}$ $\langle111\rangle$	60
Close-packed hexagonal (average)	$\{0001\}$ $\langle11\bar{2}0\rangle$	75

The process of indentation in crystalline materials might best be considered in two parts. Initially, indentation will be resisted by the resolved shear stress necessary to activate or initiate dislocation sources. Here, the rate of strain; the shape of the indenter; and the orientation of the primary slip systems can be assumed to be of major importance. The continued penetration of the indenter in the second part reflects the ease of subsequent dislocation movement. Then, dislocation interactions - giving rise either to crack formation or work-hardening - availability of secondary/alternative slip systems; and creep processes are of greater significance. Hence, the actual behaviour for a given crystal will lie within a spectrum of possibilities governed by the varying contributions due to the above effects.   At one end of the spectrum will be the metals where dislocation movement is initiated easily, and at comparatively low stresses (say 10^{-5}G), but in which work-hardening of the highly dislocated zone makes the most significant

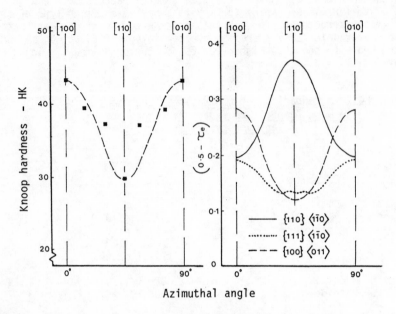

Fig. 1 The measured anisotropy in Knoop hardness of cubic BN
compared with predictions of the resolved shear stress model.

contribution to the hardness value. At the other end, and whilst work-
hardening may still occur, the principal form of resistance to indenter
penetration is the necessity for stresses close to the theoretical shear
strength (e.g. $10^{-1}G$) to initiate dislocation movement. On this basis,
it is clear that the covalent solids fall into the latter category but
that, equally, there can be little doubt that dislocations (or similar
defects) are produced during the indentation process - albeit in a zone
very close to the indenter (Humble and Hannink, 1978; Hill and Rowcliffe,
1979). Consequently it seems reasonable to assume that anisotropy in the
Knoop hardness of cubic BN can be used to identify the active slip systems
in the same way as demonstrated in earlier work - e.g. aluminium oxide
(Brookes, O'Neill and Redfern, 1971), transition metal carbides (Hannink,
Kohlstedt and Murray, 1972; Hollox and Rowcliffe, 1972). A comparison of
these results with the predictions of the resolved shear stress model
(Brookes, O'Neill and Redfern, 1971), as shown alongside the hardnesses in
Fig. 1, indicates that the slip systems controlling indentation are either
{100}<011> or {111}<1$\bar{1}$0>. The geometry of slip lines produced on the
surface adjacent to the indentations should enable a distinction to be made
between these two alternative systems but such features are not normally
observable in hard crystals deformed at room temperature. However, slip
lines may be developed by a subsequent anneal, as in silicon when heated
to approximately 673 K, or by dislocation etching. When indented crystals
of cubic BN were heated above a certain threshold temperature, in air,
etch features were developed on {100} surfaces which were thought to be
slip traces (Brookes, Hooper and Lambert, 1983). For example, the indentation shown in Fig. 2 was made, at room temperature, with a load of 4.9 N
and with the long diagonal of the Knoop indenter parallel to a [100] on

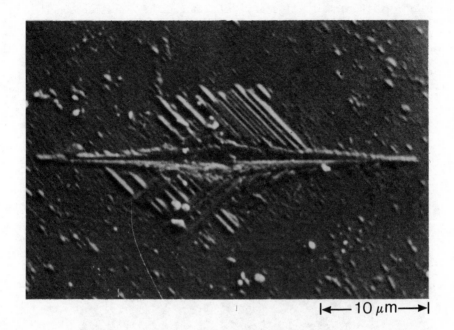

|←——10 μm——→|

Fig. 2 Etch features, thought to be slip traces, produced by annealing
a Knoop indentation in a cubic BN crystal at 1173 K in air. Note
that the traces are parallel to <110> directions on the (001) plane.

the (001) surface. Subsequently, the specimen was heated at 1173 K for
thirty minutes in air. This treatment served both to allow the relaxation
of residual stress, resulting in the expansion of the dislocated zone
beneath and around the indentation, and etching of the resultant slip
features. Slip traces due to {100}<011> systems on a (001) plane would
all lie in <100> whilst {111}<110> slip would give traces in <110>
directions. Since these particular features are all parallel to <110>,
then it may be concluded that the {111}<110> systems control the indentation process. There are two potentially important aspects of slip on
{111} planes in cubic BN. First, it is generally observed in other group
III-V compounds that dislocation loops lying in the {111} planes of the
group III element have different dislocation velocities to those lying in
the {111} planes of the group V element (Hirsch and co-workers, 1986).
Secondly, the surface of slip steps due to the {111}<1$\bar{1}$0> slip systems
would consist entirely of either boron or nitrogen ions/atoms. Consequently, significant environmental effects of a chemical:mechanical nature
can be predicted.

The expansion of a dislocated zone, as a result of annealing a crystal
which has been previously indented at room temperature, is not peculiar to
cubic BN. Etching techniques used in conjunction with annealing have been
used to demonstrate that the geometry and density of dislocation etch pits
about the expanded zone are very similar to those produced by actually
indenting at the annealing temperature. Therefore, it has been assumed
that, although constrained to the immediate vicinity of the indentation at
room temperature, the nature of defects produced and their habits are

compatible with normal dislocation behaviour. This treatment applied to
surfaces produced by fracture or polished by wear (see companion paper in
these proceedings by Hooper and Brookes) can yield further information on
the nature of deformation in the aggregate material. For example, Fig. 3
shows etch features, developed by annealing, in an individual particle
which has (presumably) been deformed during the initial direct sintering
process. The threshold temperatures that have been measured, e.g. diamond
>1423 K; cubic BN, 1173 K; Si, 673 K; Ge, 573 K; appear to be directly
related to the covalent bond energy (Brookes and co-workers, 1983; Ross,
1985).

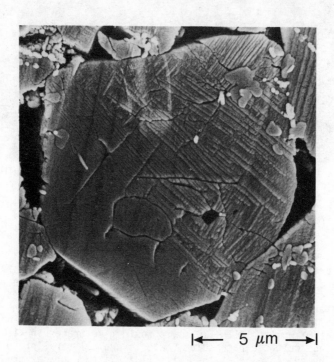

|◄— 5 μm —►|

Fig. 3 Etch features developed in an individual
particle of a cubic BN aggregate specimen.

THE LOAD INDEX 'n' AND THE DISLOCATED VOLUME

The use of the load index 'n' in the equation $P = Ad^n$ - where 'P' = the
normal load, 'A' = a material constant, and 'd' = the length of the
indentation diagonal - is now established as a means of comparing the
increase in measured hardness as a function of decreasing normal load.
(Note that 'n' = 2 for conditions where the measured hardness is truly
independent of the load.) Again, this effect is particularly pronounced in
the hard ceramic materials and 'n' is almost invariably significantly less
than 2 (Brookes, 1983). One explanation offered for this effect in
metallic materials, where the experiments were carried out in an electron
microscope with extremely low loads (10^{-6} N), was that the indenter was
sampling such small regions of the crystal as to need to produce its own

dislocation sources and therefore requiring stresses close to the theoretical level (Gane and Bowden, 1968). But for the covalent crystals
covered in this work it has already been demonstrated that the stresses
beneath the indenter are always, whatever the load, of the order of the
theoretical shear stress and yet the effect is most pronounced in these
solids. The possibility of an explanation based on surface characteristics and such environmental conditions as those associated with the
Rebinder type effects can be ruled out by the detailed work of Burnand
(1974) and Ross (1985). These researchers established that consistent
values of 'n'<2 were obtained irrespective of specimen surface preparation
(i.e. cleaved or chemically polished) and whether or not the crystals (LiF
and MgO) were immersed in a selection of electrolytes (Burnand, 1974).
Also, values of 'n'<2 remained reasonably constant for Si and Ge when
indented at temperatures up to 573 K - i.e. a temperature well above that
at which the Rebinder type effects can be maintained (Ross, 1985). An
alternative explanation has been investigated, generally using dislocation
etching techniques, by considering the dislocated zone beneath the indentation. It is now known that the geometry of the dislocation pattern
developed in this zone is similar whilst its overall volume increases with
increasing load (Brookes, 1983). A typical geometry consists of densely
packed etch pits in the immediate vicinity of the indentation, due to dislocation activity on a number of intersecting slip planes, with rows of
dislocations, generally on a few parallel slip planes, radiating away from
the indentation. The resultant etch pit arrays are commonly called dislocation rosettes and the extent of the rosette, i.e. the distance of the
furthermost etch pit from the centre of the indentation, has been shown to
be related to the critical resolved shear stress (τ_c) for a given crystal
(Hopkins, Miller and Martin, 1973; Hirsch and co-workers, 1986). However,
it should be noted that, at this stage, there is no direct and universal
relationship between τ_c and H. The classical expanding cavity model for
indentation (see review by Tabor, 1970) identified an outer hemispherical
boundary between the plastically deformed material due to the indentation
and an inner concentric heavily deformed 'hydrostatic' core immediately
beneath the contact region. This particular model was originally developed
for isotropic materials and there are obvious constraints on its application to the behaviour of single crystals with their discrete slip systems
enforcing anisotropic considerations and a certain preferred flow of
plastically deformed material. Nevertheless, in single crystal measurements, it seems reasonable to assume that the radius 'r' of a circle
centred on the indentation point and enclosing the whole of the dislocation rosette generated by the indenter will be related to the size of the
plastic:elastic boundary. Similarly, that the half of the diagonal length
of a square based Vickers indentation will be related to the radius of the
hydrostatic core ('a'). (Where a Knoop indenter is used then the
'equivalent' diagonal length for the same indentation pressure can be
used.) One obvious advantage of this particular exercise is that 'r' can
be measured with much greater precision for crystals where a dislocation
etchant can be applied, to identify the dislocations generated by the
indentation, than in polycrystalline materials. Even in the absence of a
suitable dislocation etchant it is easier to distinguish the extent of
plastic deformation from slip line observations on single crystals (such
as those shown in Fig. 2) than on the polycrystalline counterpart. The
results of carrying out such measurements for a fairly wide range of
materials (Ross, 1985) are summarised in Fig. 4 for single crystals of
those materials of most relevance to these proceedings, and it should be
stressed that r/a is independent of load. Here it can be seen that the
dislocated zone is very close to the indentation for diamond, cubic BN,

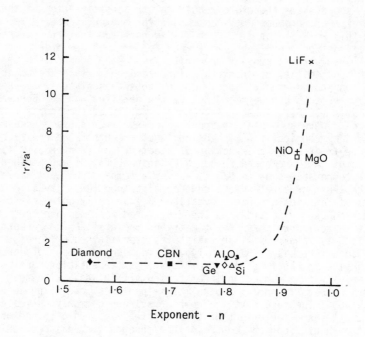

Fig. 4 The relationship between the geometry of the
dislocated zone (r/a) and the load index 'n' for a
selection of ceramic single crystals at room temperature.

Ge, Si and Al_2O_3 - i.e. those materials where the load effect is particularly pronounced. As the size of an indentation decreases, and whilst r/a
remains constant, then the ratio of the surface of the actual indentation
(proportional to d squared) to the dislocated volume (proportional to d
cubed) increases. This observation would appear to be consistent with, in
general, the surface of a given solid being harder than its bulk.

TEMPERATURE EFFECTS AND INDENTATION CREEP

The small dimensions of available single crystals of cubic BN ensure that
the effect of temperature on the indentation hardness of these specimens
is limited to low loads and temperatures that can be obtained by the use
of heated stages on hardness machines designed for microindentations.
Such measurements have been made, using a load of 4.9 N with a Berkovich
(triangular based pyramid) indenter aligned with its facets parallel to
<110> on an as grown (111) surface, and it has been shown that the hardness
falls immediately and continuously from 5400 HB 0.5 at room temperature to
2350 HB1 at 873 K (Brookes and Lambert, 1982). (It should be pointed out
here that Berkovich values are of the order of 10% to 20% greater than the
corresponding Knoop values (Brookes and Moxley, 1975). This continuous
and apparently high rate of softening is nevertheless comparable with that
of some covalent solids such as diamond and Al_2O_3 (Brookes, 1983) but
contrasts with others such as Si and Ge where, initially, the decrease in
hardness with temperature is relatively small.

Normally, the hardness of a given metal will be lower for single crystals than for its polycrystalline form. This behaviour is consistent with the predominant part of the indentation process being work-hardening due to disloaction interactions - in this case with grain boundaries. Equally, if the predominant component is that stress required to initiate dislocation movement, we would expect ceramic single crystals to be harder than their polycrystalline specimens. This is often found to be the case and represents another factor that should be borne in mind when comparisons are drawn. In the absence of sufficiently large single crystals for measurements above 873 K, we have measured the Knoop indentation hardness of a cubic BN aggregate tool material as a function of temperature up to 1700 K. The measurements up to 900 K were made on the hot stage of a microindentation hardness machine but using a load of 20 N whilst a load of 100 N, in a vacuum of about 10^{-5} mbar, was used for the measurements in a specially designed apparatus at higher temperatures (Harrison, 1974). It would appear, on the basis of the continuity of the values shown from one apparatus to the other, that the load effect was not significant under these particular experimental conditions (Fig. 5).

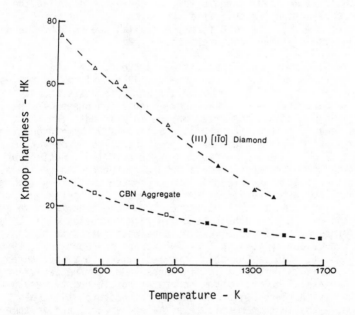

Fig. 5 The effect of temperature on the hardness of a cubic boron nitride aggregate cutting tool material, as a function of temperature, compared with that for a diamond under similar experimental conditions. Note that open and full symbols correspond to HK2 and HK10 values respectively.

It is clear, by comparing these results with those given earlier for single crystal cubic BN, that the single crystal form is harder than the aggregate - at least between room temperature and 1173 K. The results may also be compared with those for the (111) surface of a Type I diamond indented in a [110] direction and under, otherwise, precisely the same experimental conditions (see Fig. 5). Whilst there is inevitably some degree of convergence, the hardness of diamond at 1473 K, i.e. the temperature to be expected at the cutting tool:workpiece interface, is seen to be approximately twice that of the cubic BN aggregate. This observation appears to be in contradiction to the data presented in an earlier paper (Okada, 1976) where it was implied that a polycrystalline form of cubic BN was harder than diamond at temperatures above 1073 K. However, it is not clear whether both sets of results in that paper were from experiments carried out under the same conditions and therefore there may be some doubt about the validity of the comparison. Nevertheless, it is clear that those results for diamond at 1073 K were lower than ours at that temperature - i.e. approximately 3000 HV and 3400 HK respectively - whilst hardness values for the polycrystalline cubic BN aggregate were consistently some 60% higher over the temperature range from room temperature to 1273 K. A reason for the latter apparent inconsistency, may lie in the load effect previously mentioned in that the results of this work were based on loads of 20 N and above whilst those of Okada were with loads of 1 N. Furthermore, the diagonal of a Vickers indentation caused by 1 N in a material with a hardness of 3000 HV - i.e. the value at 1073 K - would be approximately 8 μm. Since the particle size of polycrystalline cubic BN is usually about 10 μm then there is further cause for doubting the validity of the earlier conclusion. It is hoped that current work, at temperatures up to 2500 K, will more firmly establish the relative hardness of these and other ultra-hard materials.

It may well prove that short term indentation hardness, i.e. where the dwell time for normal loading is of the order of 30 s maximum, is not as important in the evaluation of these materials as their resistance to time dependent plastic deformation as revealed by indentation creep measurements. Three types of indentation creep may be identified. The first is observed in the behaviour of all crystalline materials at homologous temperatures at and above 0.5 Tm and appears to be controlled by conventional creep mechanisms based on vacancy diffusion assisted climb. Activation energies determined from indentation data obtained at these temperatures, after Atkins, Silverio and Tabor (1966), are consistent with this mechanism and are generally in the region of 20 RTm - where 'R' is the gas constant and 'Tm' is the melting temperature. A second type is known as anomalous indentation creep, because the creep rate decreases with increasing temperature, and is observed in those materials with predominant bonding of covalent or ionic nature (Westbrook and Jorgensen, 1965; Westwood and Goldheim, 1965). But this form is limited to temperatures less than 473 K and is therefore of little consequence to the behaviour of cutting tool materials. The third type occurs at homologous temperatures less than 0.3 Tm, it increases with increasing temperature and it has been observed (Morgan, 1976) in some metals (e.g. Mo and V etc.) and in some ceramic solids (e.g. MgO and Al_2O_3).

During the metal cutting process there will be a temperature gradient, from ambient to a maximum of about 1500 K, measured at points remote from - and actually in - the tool:workpiece interface. Consequently, time dependent deformation under point loading could be an important aspect of cutting tool behaviour. We have previously reported that indentation creep was measured in a diamond at 1373 K (Brookes, 1979). Thus the hardness for a standard 12 s indentation was 2030 HK1 whilst a value of 1390 HK1 was obtained when the dwell time was increased to 1000 seconds. Similar experiments have now been carried out on specimens of cubic BN and some of the results are shown in Fig. 6. Again, it should be observed here that the results for temperatures below 673 K were obtained with a

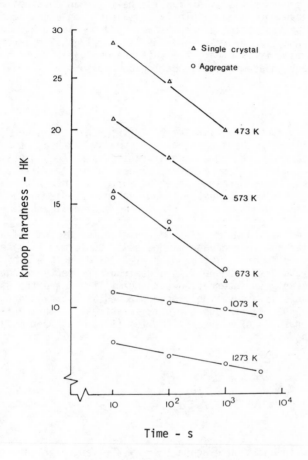

Fig. 6 Showing the effect of increasing the time of
loading ('dwell' time) on the measured hardness
of cubic BN crystals and aggregate material.

heated specimen stage mounted on a commercial microindentation hardness
machine and using a load of approximately 5 N or 10 N on single crystals;
whilst the results at higher temperatures were obtained with specially
designed apparatus and loads of about 100 N on aggregate material. Hence,
we should not place too much significance on the absolute hardness values
but, even so, it is quite clear that significant indentation creep is
taking place at very modest temperatures. Furthermore, the comparability
of results at 673 K for the aggregate and on the cubic BN crystal, with
the indenter aligned in [1$\bar{1}$0] on the (111) surface, verify that this time
dependent deformation is a fundamental property rather than a vagary of
sintered material - where it might have been suggested that localised
densification and reduction in internal porosity might be responsible for
the phenomenon. When these results were analysed in terms of the Atkins,
Silverio and Tabor (1966) model, activation energies in the region of 3
RTm (where R = the gas constant and Tm = the melting temperature) were

obtained for these crystals - similar results having been obtained for SiC and Al_2O_3. This represents a particularly low activation energy and does not appear to be indicative of any established mechanism but may reflect merely the need for another model at low homologous temperatures. Whatever the mechanism, it is clear that this phenomenon is worth further investigation for both academic and technological reasons.

ASPECTS OF FRACTURE

Brittle fracture in diamond, by cleavage on {111} planes, is a well known feature and is utilised in the shaping of gem stones. Although the aniso-tropy in certain wear rates has been explained in terms of the removal of blocks of crystal bounded by such cleavage planes (Wilks and Wilks, 1979), the experimental proof - in terms of facetted surfaces - has not been entirely convincing. In most industrial applications of diamond as a tool material the conditions are such as to develop more significant wear mechanisms - such as diffusion controlled solution and cratering. Further-more, cracks produced by indenting diamond (as also for Si and Ge) are not so clearly restricted to the cleavage planes as those formed in rather less brittle materials such as TiC (Brookes, 1966) and MgO (Key, Li and Chou, 1959). Nevertheless, toughness is a desirable property in shaping processes - particularly in interrupted cutting/milling - and the use of indentation techniques is increasingly acknowledged as a convenient method of evaluating this property through the critical stress intensity factor (K_{IC}). Thus, the analysis of Evans and Charles (1976) has been used for Vickers indentation data:

$$K_{IC} = 0.129 \left(\frac{c_v}{a_v}\right)^{-3/2} \frac{H_v\sqrt{a_v}}{\phi} \left\{\frac{E\phi}{H_v}\right\}^{2/5} \qquad 1$$

where a_v = half the indentation diagonal; c_v = the length of the cracks measured from the centre of the indentation; H_v = Vickers hardness; E = Young's modulus and ϕ is a constraint factor.

Typical K_{IC} values measured in this way include 2.2 MN m$^{-3/2}$ for Al_2O_3 (Chiang, Marshall and Evans, 1982); and 3.2 MN m$^{-3/2}$ for SiC (Henshall and Brookes, 1985). It would seem to be an obvious precaution to carry out such experiments using an indenter whose symmetry reflects that of the indented plane (Brookes, 1983). Therefore, Henshall and Brookes (1985) have applied a similar approach to the use of the Berkovich indenter on a (0001) plane of SiC. The modified equation proposed was:

$$K_{IC} = 0.0392 \left(\frac{c_B}{a_B}\right)^{-3/2} \frac{H_B\sqrt{a_B}}{\phi} \left\{\frac{E\phi}{H_B}\right\}^{2/5} \qquad 2$$

where a_B = height of the triangular indentation; c_B = the length of the cracks radiating from the centre of the indentation and H_B = Berkovich hard-ness. Their results showed that, whilst there was some degree of anisotropy in hardness, the values of K_{IC} were essentially independent of orientation with an average of 3.3 MN m$^{-3/2}$. Similar preliminary experiments on {111} planes of cubic BN have given an average of 2.8 MN m$^{-3/2}$, and, again, these measurements were not significantly anisotropic. However, the radial cracks produced by indentation had a pronounced preference for propagation on {110} planes. This is perhaps not surprising since, on the assumption of some ionic bonding, {110} surfaces would be electrically neutral whilst both {100} and {111} would be charged. Additionally, radial cracks tend to form on planes normal to the indented surface - as would be the case here for {110} cracking on a (111) plane of indentation.

ACKNOWLEDGEMENTS

The author would like to thank his colleagues J.D.J. Ross, W.A. Lambert and A.R. Parry for permission to use previously unpublished results, and De Beers Industrial Diamond Division Ltd for a grant to the laboratory.

REFERENCES

Atkins, A.G., Silverio, A. and Tabor, D. 1966, J. Inst. Metals, 94, 369
Brookes, C.A. 1966, Wear, 9, 103-115
Brookes, C.A. 1979, "Properties of Diamond", Ed. J.E. Field, Academic
    Press, London, 383-402
Brookes, C.A. 1983, "Science of Hard Materials", Ed. R.K. Viswanadham,
    D.J. Rowcliffe and J. Gurland, Plenum Publishing Co, New York, 181-199
Brookes, C.A., Hooper, R.M. and Lambert, W.A. 1983, Phil. Mag. A,
    Vol. 47, No. 5, L9-L12
Brookes, C.A., Hooper, R.M., Lambert, W.A. and Ross, J.D.J. 1983,
    "Ultrahard Materials Application Technology" Vol. 2, Ed. P. Daniel,
    De Beers Industrial Diamond Div., Ascot, 72-80
Brookes, C.A. and Lambert, W.A. 1982, "Ultrahard Materials Application
    Technology", Vol. 1, Ed. P. Daniel, De Beers Industrial Diamond Div.,
    Ascot, 128-136
Brookes, C.A. and Moxley, B. 1975, J. Phys. E., 8, 456-460
Brookes, C.A., O'Neill, J.B. and Redfern, B.A.W. 1971, Proc. R. Soc.
    A332, 73
Burnand, R.P. 1974, PhD dissertation, Univ. of Exeter
Chiang, S.S., Marshall, D.B. and Evans, A.G. 1982, J.Appl.Phys. Vol. 53,
    No. 1, 298-311
Evans, A.G. and Charles, E.A. 1976, J. Amer. Ceram. Soc., 60, 373-385
Field, J.E. 1986, These Proceedings
Gane, J. and Bowden, F.P. 1968, J. Appl. Phys. Vol. 39, No. 3, 1432-1435
Hannink, R.J., Kohlstedt, E. and Murray, M.J. 1972, Proc. R. Soc. A326,
    409-420
Harrison, P. 1974, PhD dissertation, Univ. of Exeter
Henshall, J.L. and Brookes, C.A. 1985, J. Mater. Sci. L4, 783-786
Hill, M.J. and Rowcliffe, D.J. 1979, J. Mater. Sci. 9, 1569-1576
Hirsch, P.B., Pirouz, P., Roberts, S.G. and Warren, P.D. 1986, These
    Proceedings
Hollox, G.E. and Rowcliffe, D.J. 1971, J. Mater. Sci. 6, 1261
Hooper, R.M. and Brookes, C.A. 1986, These Proceedings
Humble, P. and Hannink, R.J. 1978, Nature, Lond. 234, 112
Keh, A.S., Li, C.H. and Chou, Y.T. 1959, Acta. Metall. 7, 694
Kelly, A. 1966, "Strong Solids", Clarendon Press, Oxford, 12-28

Morgan, J.E. 1976, PhD dissertation, Univ. of Exeter
Nakaruma, T. and Yazu, S. 1983, Proc. High Pressure Conf. of Japan, Fukuoka, 136-137
Okada, S. 1976, Annals of the CIRP, Vol. 25, 1, 219-224
Pipkin, N. 1980, J. Mater. Sci. 15, Letters, 2651-2653
Ross, J.D.J. 1984, PhD dissertation, Univ. of Exeter
Tabor, D. 1970, Rev. Phys. Techol., 1, 145
Tabor, D. 1984, Special Technical Publication "Microindentation Hardness Testing", A.S.T.M.
Westbrook, J.H. and Jorgensen, P.J. 1965, Trans. AIME, 230, 613-625
Westwood, A.R.C., Goldheim, D.L. and Lye, R.C., 1967 Phil Mag. 16, 505-513
Wilks, J.E. and Wilks, E.M. 1979, "Properties of Diamond", Ed. J.E. Field, Academic Press, London, 351-380

Inst. Phys. Conf. Ser. No. 75: Chapter 3
Paper presented at 2nd Int. Conf. Science Hard Mater., Rhodes

221

Deformation substructure of diamonds subjected to high pressure and temperature

M LEE, R C DEVRIES AND E F KOCH

General Electric Company Corporate Research and Development
Schenectady, New York 12345

ABSTRACT

Twins are the prevailing microstructure of deformed diamonds and formation
of one set of twin bands create conditions favouring the formation of
secondary cross twins in some of the primary twin bands. Slip lines
representing $\{111\}$ $\langle110\rangle$ slip system are also observed along with
occasional observations of very dense dislocation forests.

INTRODUCTION

Because of the extremely rigid lattice structure of diamond, generating new
dislocations or moving existing dislocations in diamond by applying
mechanical stress at ambient temperature is very difficult. Analysis of
portions of diamonds subjected to indentation tests (Humble and Hannink,
1978) or of a diamond plate deformed under bending stress at elevated
temperature (Evans and Wild, 1965) has shown that diamond deforms plastically under suitable conditions and that its primary slip systems are
$\{111\}$ $\langle110\rangle$.

Until recently the plastic properties of diamond have not been of
significant engineering interest for either direct applications or for
processing. Recent improvements in diamond anvil technology, however,
have made it possible to produce sufficiently high static pressure to
cause significant plastic deformation in the anvils at ambient temperature
(Mao and co-workers, 1979). With this new improved design of the anvil
type presses, the upper bound of capable pressure for a system is limited
by the yield strength of the diamonds used as anvils.

Plastic deformation in diamond is more commonly observed during the high
temperature-high pressure sintering process of polycrystalline diamond
bodies (DeVries, 1975). The pressure and temperature conditions in the
sintering presses are sufficiently high (Wentorf, DeVries and Bundy, 1980)
that many diamond grains in the sintered bodies show deformed microstructures (Yazu and co-workers, 1983; Walmsley and Lang, 1983).

In this paper several commercially available polycrystalline diamond discs recommended for rock cutting applications were analysed for the deformation substructures in the diamond grains using transmission electron microscopy.

EXPERIMENTS

As noted above, individual diamond particles can be deformed plastically in a high pressure apparatus if the pressure and the temperature in the cell are sufficiently high (DeVries, 1975). However, preparing an appropriate sample for transmission electron microscopic analysis from deformed diamond particles is extremely difficult. Several different methods, such as direct examination of fine debris from a crushed sample or brazing deformed particles to a metal plate in order to secure them during a subsequent polishing and thinning process, were tried unsuccessfully.

A sintered diamond product commercially available for rock cutting applications was selected for this investigation. It has been reported (Wentorf, DeVries and Bundy, 1980) that the ranges of temperature and pressure used for processing this type of material are well within the pressure and the temperature range required for plastic deformation of diamond and that some of the diamond grains in this material are severely deformed (Hibbs and Lee, 1979). After optical examination of polished surfaces of the selected samples, the materials were mechanically thinned by rough grinding and polishing to less than 100 micrometer thick plates. These thin plates were thinned further by ion beam sputtering until a portion of the sample became transparent to the electron beam. Diffraction patterns from the selected areas were analysed to determine the crystal orientation. The materials used have some metallic inclusions such as the FCC phase of Co, which shows diffraction patterns virtually indistinguishable from the diffraction patterns of diamond. Chemical compositions of the uncertain areas were determined using the energy dispersive X-ray spectrum analysis method.

A metallurgical microscope equipped with a differential interference contrast stage was used for optical examinations. The differential interference contrast (Nomarski) technique improves visibility of a small difference in height on the object surface. A transmission electron microscope, equipped with a scanning system, was used for selective area energy dispersive X-ray spectrum analysis.

RESULTS

The polished surface of a large heavily deformed grain is shown in Fig. 1. Minute difference in hardness of the deformed region and the surrounding matrix cause formation of steps on the surface during polishing (DeVries, 1975). Other inhomogeneities in the diamond causing local difference in hardness also show ridges on polished surfaces. It is possible that the observed lines on the surface could be the scratch marks left by diamond particles in the polishing wheel. However, the photograph shows three distinctive sets of intersecting ridges (indicated by the letter A) corresponding to the active slip systems in the crystal, and at least two sets of those ridges are in significantly different orientations from the direction of polishing. It is also important to note that all marks on the surface are protruding out of the surface.

Low magnification optical micrographs of larger areas of two surfaces are shown in Fig. 2. Both photographs show a number of grains with many fine ridges. Some of the marks are from internal defects that existed prior to pressing such as the growth twins in the grain (A). But most of the ridges on the surface of grains are indicators of underlying deformed regions. As shown in the photographs, most of the diamond grains are not deformed. It should also be noted that the distribution of the deformed grains is entirely random. Deformed grains are located close to the surface (B) of the sample as well as the interior of the sample (C). Size of the grains also does not seem to have any relationship with the susceptibility of individual grains to deformation. One very interesting feature in photograph 2(b) is the partially deformed large grain located at the centre of the photograph (D). This grain shows a growth twin boundary across the grain (arrow). Only one half of the twinned crystal is severely deformed, while on the other side of the twin boundary there are no visible deformation marks. It seems that either the deformation characteristics of diamond are very sensitive to the orientation of the lattice, or the properties of deformed zones in diamond causing resistance to wear are very much orientation dependent.

The most prominent feature of the substructure of deformed diamonds is the abundance of twins and their structural complexity. A transmission electron microscope (TEM) photograph from a heavily deformed zone of diamond is shown in Fig. 3. In order to confirm that the area examined is indeed the section of a diamond, a slightly larger area including the area shown in Fig. 3 is examined using a scanning transmission electron microscope (STEM). The STEM micrograph of the area is shown in Fig. 4(a) along with a Co distribution map of the same area, Fig. 4(b). The X-ray energy dispersion spectrum from the central area of the photograph is shown in Fig. 4(c), and the spectrum from the dark triangular spot is shown in Fig. 4(d). The total counts (intensity) of the spectrum shown in Fig. 4(c) took six times longer to accumulate than the one shown in Fig. 4(d). A low level of Co was detected in the spectra from all over the area under examination, but the intensity of Co signal from the dark spot of the area is much stronger than the surrounding area, as indicated by the spectra and the Co map of the area. When a sample is thinned by an impingent ion beam, material removed from one area can be deposited on the surface of surrounding area; it is believed that elements from the metallic inclusions in the sample containing Co are thus evaporated and redeposited on the surface of diamond during the sputter thinning process. The low level of the Co spectrum detected from the central area of Fig. 4(a) is an indicator of this surface deposit (contaminant). The absence of any other strong signal supports the observation that the bulk of material in the region is diamond.

An electron diffraction pattern representing approximately a [110] orientation of diamond, as shown in Fig. 5, was obtained from the sample. A dark field micrograph obtained by placing the aperture of the microscope around the twin diffraction spot (Eddington, 1975) indicated by the arrow in Fig. 5 is shown in Fig. 6. A closely spaced set of twin lamellae inside the primary twin bands is clearly visible in this micrograph. Another bright field picture and corresponding diffraction pattern from the same area after rotating the sample slightly is shown in Fig. 7(a). This small change of the relative orientation of the crystal to the electron beam increased the intensity of streaks (arrow) around the (111) spot (Fig. 7b). A dark field micrograph associated with this streak is shown in Fig. 8. This time a different set of twins, having much finer scale than the primary or the secondary twin bands shown in Fig. 7, is seen.

A similar set of pictures from the deformed zone of another diamond particle is shown in Fig. 9. Again, a set of closely-spaced twins are shown in the picture (Fig. 9c). The severely twinned substructure observed from the two examples discussed was common in most of heavily deformed grains of all of the samples examined. It is quite possible that all of the deformed grains would have shown the twinned substructure if the appropriate orientation of all of the crystals had been examined.

Another very common internal structure of deformed diamonds frequently observed along the contact points of diamond grains is a zone showing a very high concentration of dislocations (Fig. 10). This high dislocation density zone can be seen more frequently when a (111) axis of the crystal is oriented parallel to the electron beam. However, as the two photographs of the same general area observed from two slightly different orientations (Fig. 11 a and b) show, zones located inside diamond grains and having minor dependence on the direction of observation were also present.

Fig. 11(a) also shows intersecting sets of slip lines. These slip bands are very similar in appearance to the bands of very fine scale twins. But the diffraction pattern (Fig. 12) obtained from the area shown in Fig. 11(a) shows no spots or streaks indicating twins in the structure. Some heavily deformed slip lines are shown also in Fig. 9(a), which did not match with the images on any of the dark field pictures corresponding to all of the twin spots in Fig. 9(b).

Another common feature observed in many grains examined is the wavy pattern shown in Fig. 13. The edges of twin plates shown in Fig. 6 also show this wavy pattern clearly. Regularity of the spacings between fringes and their distribution over a broad area may indicate that these are not the thickness contours generated during the milling process. No attempt was made to identify the origin of this pattern.

DISCUSSION

Since one of the authors of this paper reported (DeVries, 1975) that the deformation band in diamonds causes relief on the polished surfaces which is distinctively visible under an optical microscope, the underlying microstructure providing the higher wear resistance for these bands compared with the surrounding matrix has been the subject of several interesting speculations. Hibbs and Lee (1979), in their explanation of the wear characteristics of sintered diamonds during cutting of sandstones, proposed that the bands are multiples of twins and that interference of propagating crack tips by these twin bands strongly influences the degradation processes of diamond edges. They cited unpublished results of a transmission electron microscopic analysis of plastically deformed diamonds showing the presence of multiple twins in the deformation bands, but the paper did not disclose any further details of the microstructure of deformed bands.

Yazu and co-workers (1983) reported observations of three characteristic internal microstructures of diamonds, i.e. slip bands, dislocation forests, and strain fringes, from their TEM analysis of four types of sintered diamonds. These authors did not investigate the microstructure of the "slip bands", and also did not report observation of any twins. More recently, Walmsley and Lang (1983) also reported a TEM investigation of a "synthetic diamond compact", concluding that the microtwin is one of the prevailing features of the deformed microstrucrure of diamonds. Walmsley and Lang also discussed the presence of some intersecting twins, but their report did not reveal the nature of large twins containing a very large number of secondary cross twins as observed in this investigation.

Formation of very fine scale intersecting twins in the primary twin bands also changes the orientation of the exposed crystal faces of the deformed band from the surrounding matrix, when the particles are polished. This orientation difference between two adjacent areas can cause different degrees of polishing resistance for the two regions.

According to the results of this and previous investigations of the microstructure of deformed diamonds, the mechanism of diamond deformation can be described with three basic modes: slip in $\{111\}$ $\langle1\bar{1}0\rangle$ primary slip system; twinning, and localized, very severe deformation under extremely high stress. Steps describing the formation of the severely deformed zones, such as the ones frequently observed at the contact points of diamond grains, are difficult to define. It is certain that this type of zone appears only at locations where the applied stress is very highly localized and that the zones are composed of very small crystallites. The orientations of these crystallites seem to be quite random.

The distribution of applied stresses in a single particle in a mass of diamond particles is complex due to the multi-faceted shape and broad range of size distribution of the particles. The magnitude and the orientation of the applied stress on a particle will depend mainly on the location of contact points with the surrounding particles and the stress on the neighbouring particles. As deformation proceeds, the configurations of stresses in the grains will also change and the favoured mode of deformation can also change. At locations such as the contact points, where a localized stress of high magnitude prevails, a great deal of deformation can proceed without influencing the stresses in surrounding regions.

Inside the press, where the applied pressure is approximatley hydrostatic, distribution of stresses in the diamond grains will gradually approach an isotropic state. Several actions such as rearrangement, fracture, and plastic deformation of the particles contribute to this stabilization process. The plastic deformation seems to occur towards the end of the process after the mass of powder is well packed. At this latter stage of packing, only slight distortion in the lattice of diamond crystals seems to be sufficient to complete the final densification process and to achieve sufficiently uniform stress distribution in the pressed mass. The formation of a very large number of twin bands of various orientations and different width provides sufficient degree of distortion to accomplish the final stage of densification.

REFERENCES

DeVries, R.C. 1975, Mat. Research Bull. 10, pp 1193-1200
Eddington, J.W. 1975, Electron Diffraction in the Electron Microscope, Number 2 of Monographs in Practical Electron Microscopy in Materials Science, Macmillian-Phillips Technical Library
Evans, T. and Wild, R.K. 1965, Phil. Mag. 12 pp 479-489
Hibbs Jr., L.E. and Lee, M. 1979, Wear of Materials-1979, Ed. K.C. Ludema, W.A. Glaser and S.K. Rhee, ASME, pp 485-491
Humble, P. and Hannink, R.H.J. 1978, Nature 273 pp 37-39
Mao, H.K., Bell, P.M., Dunn, K.J., Chrenko, R.M. and DeVries, R.C. 1979, Rev. Sci. Instr. 50 (8) pp 1002-1009
Walmsley, J.C. and Lang, A.R. 1983, J. Mat. Science Letters, 2, pp 785-788
Wentorf, R.H., DeVries, R.C. and Bundy, F.P. 1980, Science, 208 pp 873-880
Yazu, S., Nishikawa, T., Nakai, T. and Doi, Y. 1983, Speciality Steels and Hard Materials, Proc. of Int. Conf. on Recent Development in Speciality Steels and Hard Materials, Ed. N.R. Comins and J.B. Clark, Pergamon Press, pp 449-456

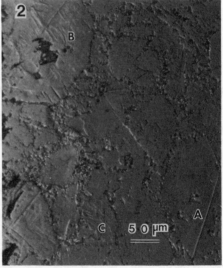

(1) Fig. 1 Polished surface of a heavily deformed large diamond particle. Three intersecting sets (A1, A2, A3) of deformation bands are seen.

(2) Fig. 2 Examples of polished diamond surface showing various internal defects, a large grain with growth twins (A in 2a) and a small deformed grain inside the body (C in 2a).

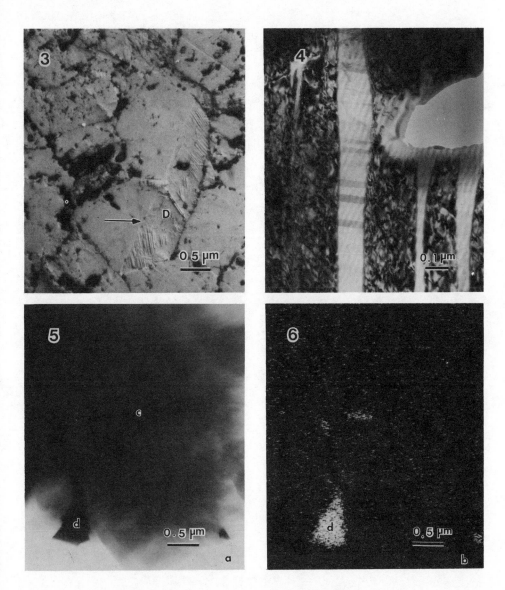

(3)   Fig. 2   A deformed grain close to the surface (B in 2a), and a
      partially deformed twin crystal (D in 2b).

(4)   Fig. 3   A TEM photograph of the deformed zone of a diamond showing
      several large twin bands.

(5,6) Fig. 4   A TEM photograph (a) of area including the section shown in
      Fig. 3, and an x-ray map of the area for cobalt (b).

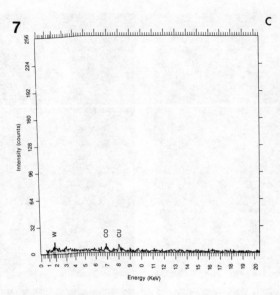

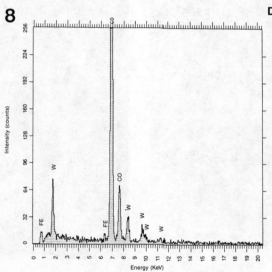

(7,8) Fig. 4   An x-ray emission spectrum from the central region (c) and
the dark inclusion (d) of the sample shown in Fig. 4a and b.
The weak cobalt peak of the spectrum (c) was accumulated
after six times longer exposure time than the spectrum (d).

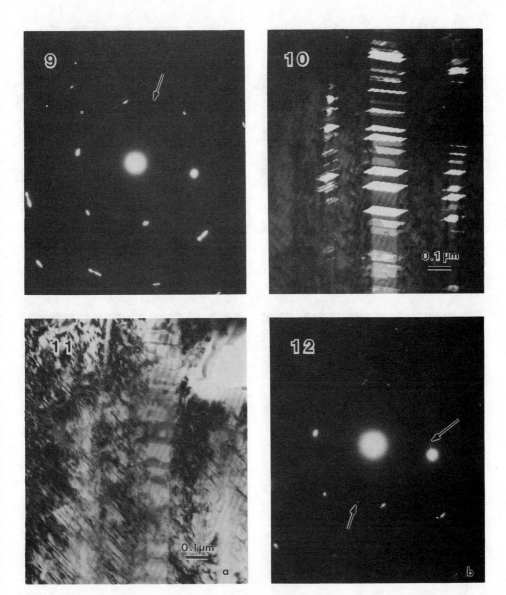

(9) Fig. 5 An electron diffraction pattern obtained from sample shown
            in Fig. 3. Weak dots and streaks (arrows) indicating pre-
            sence of twins in the crystal.

(10) Fig. 6 A dark field photograph related to the twin spot in Fig. 5.
            A large number of twins are seen inside the primary twin
            bands.

(11,12) Fig. 7 A bright field picture of the area shown in Fig. 3 observed
            from slightly different orientation (a) and the electron
            diffraction pattern from the area (b). The intensity of
            twin spots and streaks are increased by tilting the sample.

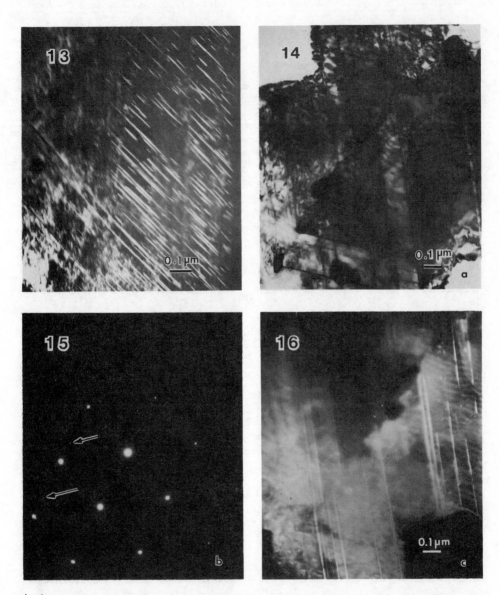

(13) Fig. 8 A dark field micrograph associated with the twin streak on
            Fig. 7 (b). Different set of secondary twins from the set
            seen in Fig. 6 is lighted in this picture.

(14,15,16) Fig. 9 A bright field TEM photograph of another deformed dia-
            mond (a), a diffraction pattern showing twin spots (b)
            obtained from the sample, and a dark field micrograph
            of the same area showing twins (c). The bright field
            picture (a) also shows set of lines not related to any
            of the twin spots.

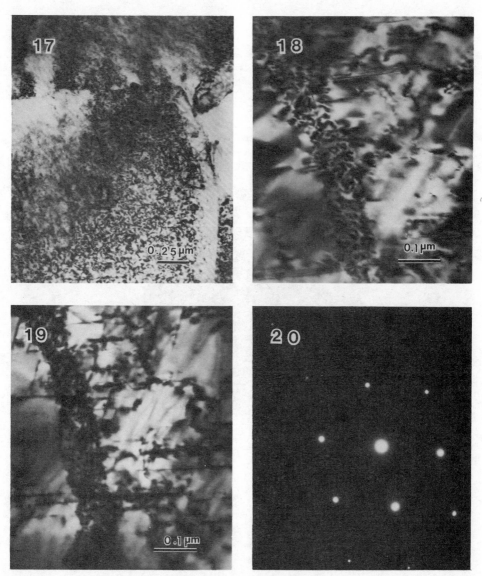

(17)   Fig. 10   A bright field picture of deformed diamond showing a
                    dense dislocation forest.

(18,19) Fig. 11   Two bright field TEM pictures of a generally same area of
                    a sample taken from slightly different orientations.
                    Both pictures show a high dislocation density band
                    stretching across the field. The pictures also show
                    multiple slip lines.

(20)   Fig. 12   The diffraction pattern from the area shown in Fig. 11
                    (a). The pattern indicates that the orientation of elec-
                    tron beam is [011] and that twin spots are not observed
                    from this orientation.

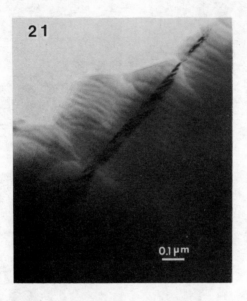

(21) Fig. 13 A thin section of diamond showing many wavy lines on the
TEM picture. It also shows a large deformation band containing many finer bands. The wavy patterns are not seen
inside the band.

Inst. Phys. Conf. Ser. No. 75: Chapter 3
Paper presented at 2nd Int. Conf. Science Hard Mater., Rhodes

233

Hardness of synthesized diamond crystals

Y DOI, S SATO, H SUMIYA AND S YAZU

Itami R & D Division, Sumitomo Electric Industries Ltd,
Itami City, Hyogo, Japan

ABSTRACT

This work establishes that the Knoop hardness on the (100) and the (110) planes of synthesized diamond crystals is isotropic for all orientations of the indenter on these planes. This behaviour is different from that reported for natural diamonds which exhibit marked anisotropy in Knoop hardness.

In this paper we have calculated the effective resolved shear stress for the {111} <110> slip systems and discussed the difference between the predicted and the measured hardness values.

INTRODUCTION

There are several published results on the hardness of natural diamonds using various kinds of indenters, such as the square-based Vickers pyramidal indenter, the triangular-based Berkovich indenter and the rhombohedral-based Knoop indenter. In these results, the Knoop indentation measurements are shown to be repeatable and consistent.

Brookes, O'Neill and Redfern (1971) established that the hardness of natural diamond crystals are of an anisotropic nature, and they explained that this behaviour was due to plastic deformation which occurs on {111} slip planes in <1$\bar{1}$0> directions (Brookes, 1970, 1971, 1979). They calculated the effective resolved shear stress developed during the indentation of a (001) surface, assuming these were the active slip systems, and their calculations agreed with the measurements.

Whilst there have been several reports on the hardness of natural diamond crystals, there have been few on synthesized diamonds. This work establishes the nature of the Knoop hardness of such crystals.

EXPERIMENTAL RESULTS

The nitrogen content of the specimens used is given in Table 1. This was determined for natural diamond crystals by utilizing infra-red absorption bands at 7.3, 7.8, and 8.5 micrometers (Davies, 1980). In the case of synthesized crystals, nitrogen corresponds to the 8.85 micrometer absorption (Chrenko, Strong and Wentorf, 1971). The specimens were between 0.1 to 0.2 carat.

TABLE 1

Specimen	Nitrogen Content (ppm)
SD-01	30
SD-02	51
SD-03	35
SD-04	57
ND-01	>1000
ND-02	410

Note. SD: Synthesized Diamond, Type Ib
       ND: Natural Diamond, Type Ia

The Knoop indenter was used on finely polished (001), (110) and (111) surfaces using a normal load of 4.9N applied for 15 seconds. The resultant indentation was measured at a magnification of x1000 using a TV monitor. Note that the included angles between the facets of the Knoop indenter are 172°30' and 130° degrees and that this shape, under conditions of ideal plasticity with negligible elastic recovery, would produce an indentation whose long diagonal was approximately seven times the length of the short diagonal.

Our measurements are shown in Figs. 1 and 2. Fig. 1 also includes the data for natural diamond crystals (Brookes, 1979) and illustrates the marked anisotropy in their Knoop hardness. The measurements of the ND-01 specimen on the (110) plane differ from the same plane of other specimens in that the [110] direction is harder than the [001]. This is the opposite of the nature of anisotropy reported previously for natural diamonds and we are currently investigating the reason for this apparent anomaly. On the other hand, and except for the (111) plane, the Knoop hardness is virtually isotropic for the synthesized crystals (Fig. 2).

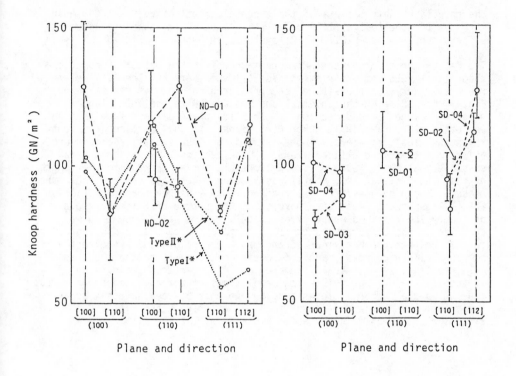

Fig. 1   Fig. 2

Anisotropy in the Knoop hardness of natural diamonds (Fig. 1)
and synthesized diamonds (Fig. 2). (* Brookes, 1979)

DISCUSSION

The theoretical effective resolved shear stress was calculated using
equation 1. This equation was derived by Brookes, O'Neill and Redfern
(1971) to compare Knoop hardness measurements with the resolved shear
stresses beneath the indenter. When the effective resolved shear stress
is at a maximum, according to this model, the Knoop hardness is a minimum.

$$\tau e = \frac{F}{A} \cos\phi\cos\lambda \frac{1}{2}\{\cos\psi + \sin\gamma\} \qquad 1$$

where:

τe = effective resolved shear stress.
F = applied force.
A = area supporting F.
λ = angle between axis of F and the slip direction.
ϕ = angle between axis of F and the normal to the slip plane.
ψ = angle between each facet on the indenter and the axis of
rotation for a given slip system.
γ = angle between each facet on the indenter and the slip direction.

The results of calculations of the theoretical effective resolved shear stresses for the (001), (110) and (111) planes of indentation, assuming slip is confined to {111} <110> systems, are shown in Fig. 3.

The measurements for natural diamonds agree with the calculations of the theoretical resolved shear stress for all specimens but the (111) plane where they were the opposite of the predicted values. As can be seen in Fig. 4, the shape of the indentations produced with the long diagonal parallel to the [112̄] direction on the (111) plane differs from that on the other planes and/or in other directions. In this particular orientation, new features are developed, around the four facets of the indentation, and these are aligned along <110> directions. It is considered possible that these features are responsible for a decrease in the length of the long diagonal and, therefore, a reduced hardness value. Thus, the measured Knoop hardness for <112̄> directions is higher than for <11̄0> on the (111) surface - i.e. the converse of the nature of anisotropy predicted by the resolved shear stress model. The same phenomenon was observed in the measurements on (111) surfaces of synthesized diamonds and in this respect the behaviour of the two kinds of diamond are directly comparable. In contrast, Knoop hardness measurements on the (001) and (110) surfaces were virtually independent of orientation, i.e. were isotropic, and did not follow the predictions of the resolved shear stress model. We are currently investigating the possibility that this apparently anomalous behaviour is due to slip not only on the {111} <110> systems but also on an alternative system - e.g. {110} <11̄0>.

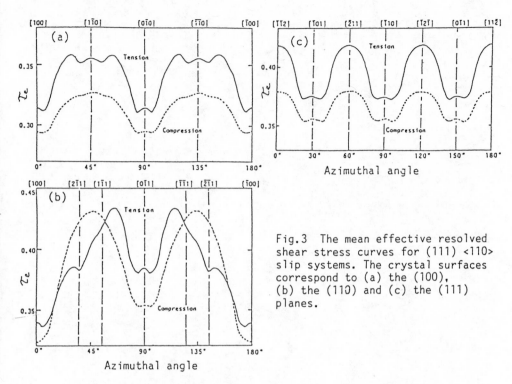

Fig.3 The mean effective resolved shear stress curves for (111) <110> slip systems. The crystal surfaces correspond to (a) the (100), (b) the (110) and (c) the (111) planes.

Fig. 3 The mean effective resolved shear stress curves for (111) <110> slip systems. The crystal surfaces correspond to (A) the (100), (b) the (110), and (c) the (111) planes.

Fig. 4 Knoop indentation in the <112> direction on the (111) plane of natural diamond.

REFERENCES

Brookes, C.A. 1970, Nature <u>228</u> 660
Brookes, C.A. 1971, Diamod Research 12
Brookes, C.A., O'Neill, J.B. and Redfern, B.A.W. 1971, Proc. Roy.
    Soc. Lond. A. <u>322</u> 73
Brookes, C.A. 1979, The Properties of Diamond, edited by J.E. Field
Chrenko, R.H., Strong, H.M. and Wentorf, R.E. 1971, Phil Mag. <u>23</u> 313
Davies, G. 1980, Industrial Diamond Review. 446

Inst. Phys. Conf. Ser. No. 75: Chapter 3
Paper presented at 2nd Int. Conf. Science Hard Mater., Rhodes

239

The problem of electronic transition in the conversion of graphite into diamond

HU EN-LIANG (1) AND CHEN ZHAO-WEI (2)

(1) General Machinery Research Institute, Hefei, China
(2) Tianjin University, Tianjin, China

ABSTRACT

The purpose of this paper is mainly to discuss the electronic transition from catalyst metal to the carbon atom of graphite, i.e. M---e--►C, thus causing the loosening and disintegration of π bonds. We also show that carbon ions are advantageous to the formation of the diamond structure. Calculating results obtained by ΔG - T curve under a pressure of 5.5 GPa shows that the oxidization conversion temperature of Fe and C is 2028°C. But that of Ni and C is 1732°C.

INTRODUCTION

The problem concerning the mechanism of formation of synthetic diamond has attracted the attention of many researchers throughout the world. Much of the current literature contains postulates that the conversion of graphite into diamond is due to the electronic transition C (graphite)---e--►M (catalyst). An alternative mechanism, concerning the role of carbon and carbonium ions, has been proposed by Hu En-liang (1982). Here we shall discuss the electronic transition from catalyst metal to carbon atom of graphite (M---e--►C) which causes the loosening and disintegration of the π bond thus enabling the transformation from SP^2 to SP^3 hybridization during the conversion of graphite to diamond.

ELECTRONIC TRANSITION OF CATALYST METAL TO CARBON ATOM OF GRAPHITE

The relation between $\Delta G°$: Temperature (T) of metal oxides at high temperature and pressure can be obtained by the Gibbs-Helmholtz equation as follows:

$$\Delta G° = RT \ln P_{O_2} \qquad\qquad 1$$

where $\Delta G°$ is the standard free energy and P_{O_2} is the partial pressure of the oxides.

Eqn. 1 is important in the study of electron gain and loss. It is clear that the lower the energy level the more stable the oxide, i.e. it is easiest for the metal to lose electrons. Conversely, the higher the energy level the easier the electrons will be gained.

We can now compare the relative ease of electron gain or loss, between metal and carbon, under high temperature and pressure. There is a certain amount of O_2 gas in the reaction cell during the synthesis of diamond and some O_2 will have been absorbed on the surface of the catalyst metal. Experimentally, it is shown that O_2 can be quickly, irreversibly and chemically absorbed on the surface if the clean metal (except Au) is exposed to the air - then it is difficult to remove. According to Eggleton and Tompkins (1957) absorbed O_2 on iron wire is difficult to re-move completely even if it is heated to 1473K in a vacuum or reduced by H_2. Additionally, there is a certain amount of O_2 and N_2 in the pores of graphite. However, some porosity is desirable and experiments have shown that graphite with a porosity of 25%-30%, in which there is a certain amount of O_2 and N_2, is entirely suitable.

Such chemical changes as $2M + O_2 \rightleftharpoons 2MO$ and $2C + O_2 \rightleftharpoons 2CO$ will usually occur at high temperatures and pressures in the reaction cell. Through calculating ΔG we can analyse the electronic transition in the synthesis of diamond. Thus, we use the following chemical reaction equation at constant temperature:

$$\Delta G = G^O + RT \ln Q_p \qquad\qquad 2$$

The argument of the logarithm (Q_p) is called the proper quotient of pressure. Ordinarily, the quotient shows proportionality in the product of the partial pressure of the resultants to that of the reactants. In a gas reaction:

$$aA + bB \rightleftharpoons cC + dD$$

we obtain:

$$Q_p = \frac{p_C^{'c} \cdot p_D^{'d}}{p_A^{'a} \cdot p_B^{'b}}$$

For a chemical reaction which contains the condensed state substance we may not calculate Q_p for the condensed phase. For example, in the follow-ing reaction:

$$2\,Fe + O_2 \rightleftharpoons 2FeO$$

$$Q_p = 1/P'_{O_2}$$

where P'_{O_2} is the partial pressure of O_2:

$$P'_{O_2} = P$$

Therefore we obtain:

$$\Delta G = RT \ln P_{O_2} - RT \ln P \qquad\qquad 3$$

It has been shown that Eqn. 3 is only suitable for ideal gases. If the corresponding algebraic equation were derived for real gases at higher pressure, using Eqn. 3, the fugacity (f) should replace the pressure and K replaces the pressure quotient:

$$K_f = Q_p \cdot K_r$$

$$f = r \cdot p$$

where r is the fugacity coefficient and K_r is the quotient of fugacity coefficient.

Hence we can obtain:

$$\Delta G = RT \ln P_{O_2} + RT \ln K_f$$

$$= RT \ln P_{O_2} + RT \ln Q_p + RT \ln K_r$$

$$\Delta G = RT \ln P_{O_2} - RT \ln P + RT \ln K_r \qquad 4$$

There are no values of the fugacity coefficient K_r for real gases, therefore we can only calculate the approximate value of ΔG at each temperature by Eqn. 3. Some calculated results are given in Table 1.

TABLE 1 $\underline{\Delta G \text{ Values of Some Oxides at a Pressure of 5.5 GPa}}$

Oxidation Reaction	T(K)	P_{O_2} atm	P atm	G (kJ)
$2Fe + O_2 \rightleftarrows 2FeO$	1000	2.0×10^{-22}	55000	-506.561
	1600	2.8×10^{-11}		-468.796
$2Ni + O_2 \rightleftarrows 2NiO$	1000	8.4×10^{-20}	55000	-456.277
	1600	1.2×10^{-9}		-418.764
$2Mn + O_2 \rightleftarrows 2MnO$	1000	3.0×10^{-33}	55000	-713.933
	1600	2.46×10^{-18}		-685.128
$2C + O_2 \rightleftarrows 2CO$	1000	1.2×10^{-21}	55000	-309.991
	1600	2.2×10^{-17}		-365.298
$2Cu + O_2 \rightleftarrows 2Cu_2O$	1000	1.5×10^{-11}	55000	-298.184
	1600	1.8×10^{-4}		-260.126

From such data as that shown in Fig. 1, and at a pressure of 5.5 GPa, the oxidation conversion temperature of Fe and C is 2301K but that for Ni and C is 2005K. Above these temperatures it will be easy for the relevant metal to lose electrons and for the carbon to gain them, i.e. M---e--▸C. Conversely, below these temperatures C---e--▸M is favoured. If the pressure is increased then electrons are lost more readily from the catalyst metals (e.g. Fe, Co, Ni and Mn) with a corresponding gain in electrons by the carbon.

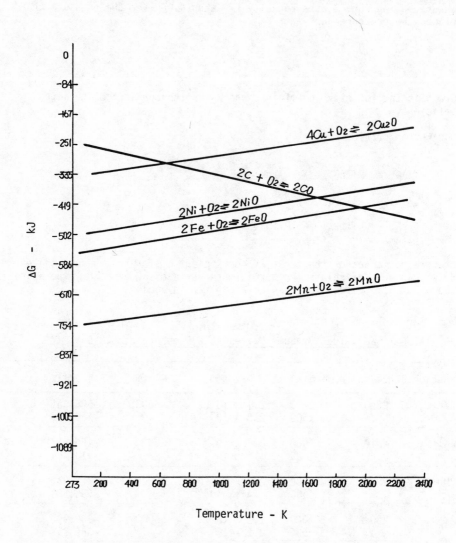

Fig. 1 ΔG: Temperature curves of some oxides at a pressure of 5.5 GPa.

A suitable temperature range or synthesizing diamond in the Ni:C system, at a pressure of 5.4 GPa, is from 1667K (1394°C) to 1728K (1455°C). But for the Fe:C system, at a pressure of 5.25 GPa, it is from 1830K (1557°C) to 1688K (1415°C). Here, the growth of diamond is restricted. All of the above temperature ranges for synthesizing diamond are below the oxidation conversion temperatures given above and, for this reason, we may conclude that the conversion of graphite to diamond is due to M---e--►C.

Microscopic observation of metal films by Feng Chu-de (1983) has shown
that some of the thicker films covering the (111) crystal surface consist
of three sub-layers. The carbon in the first and second layers is of
graphite-like structure, while that in the third layer is of diamond-like
structure. Ni^{2+} exists in the same layer. This indicates that the Ni^{2+}
has been formed by electronic transition of catalyst metal Ni to C in the
metal thin film while the graphite-like structure of carbon has been
immediately and quickly transformed to a diamond-like structure.

It is known that copper readily dissolves carbon, but it is impossible to
synthesize diamond using copper as a catalyst. Generally, this is thought
to be because the d shell of copper is filled with electrons and cannot
receive π-electrons from graphite. However, we believe that the main
reason is that it is impossible to obtain electronic transfer from the
catalyst metal to the carbon atom of graphite under the synthesis
conditions of high temperature and pressure.

From Table 1 we can see that C---e-->M when the temperature exceeds 923K,
under 5.5 GPa pressure, and it is difficult to synthesize diamond under
these conditions. If a small amount of either Ni or Si is added to the Cu
the electronic transition to the carbon atom of graphite is affected and
then it may be used as a solvent for diamond synthesis. Alternatively, an
electric field for electro-donating to the carbon atom of graphite can be
used. This has been verified, by Wakatsuki's (1975) experiment using
direct current heating under high temperatures and pressures. Thus pure
copper has been used as a solvent-catalyst in the synthesis of a small
quantity of diamond.

It is believed that the parent rock of natural diamonds in the Kimberlite
pipe contains 8%-13% of Fe_2O_3 and FeO, with a small quantity of NiO,
CoO_2 etc., and that the growth temperature is in the range 1470-1570K.
Let us suppose that, just as C---e-->M where the pressure does not affect
ΔG values, growth conditions are the same for high temperature - high
pressure as for high temperature - atmospheric pressure. The data shown
in Fig. 1 indicates that the reaction C---e-->M should occur for Cu but
not for Fe, Co, or Ni. The presence of elemental Cu but not Fe, Co, or Ni
in the Kimberlite pipe is consistent with these observations.

TRANSFORMATION FROM SP^2 TO SP^3 HYBRIDIZATION

We conclude that the process of diamond formation depends on M---e-->C.
Then the breaking of the π bond and twisting the planar hexagonal ring
structute into the chain regular tetrahedral configuration will accompany
the transformation from SP^2 into SP^3 hybridization.

In organic chemistry reactions, when the group connected with benzene is
electron-donating, the electron cloud density may be transferred to the
benzene ring making it more reactive. In conjugation, there is an alternating maximum and minimum in electron cloud density in the chain of the
benzene ring and this is advantageous to the tetrahedral configuration of
the SP^3 hybridization state. Conversely, when the substitutive group is
electron-accepting, the activity of the benzene ring is reduced and
reaction is made more difficult. In other words, for the transformation
of the SP^2 to the SP^3 hybrids, it is advantageous for it to lose an electron (carbanion) and disadvantageous for it to lose an electron (carbonium
ion).

We consider that the conversion of graphite into diamond is accomplished by the following sequence of events. A group of carbon atoms of graphite is dissolved by the catalyst metal, under conditions of high temperature and pressure, because M---e-->C and causes loosening of the π bond. This develops alternating minima and maxima in the electron cloud density leading to the complete rupture of the π bond. Then, coordinate covalent bonding is made between P and d orbitals. The disintegration of the π bond is accompanied by an increase in the distance between adjacent atoms, from 1.42A to 1.54A, and changes in the angles between them from 90° to 109° 28' (for the P - d bond and the graphite ring) and from 120° to 109° 28' (for the atoms within the original graphite ring). The distance between adjacent layers of graphite decreases with increasing pressure and this may be considered as assisting the disintegration of the π bond leading to the layer by layer transformation of SP^2 to SP^3 hybridization. A schematic representation of this transformation is shown in Fig. 2.

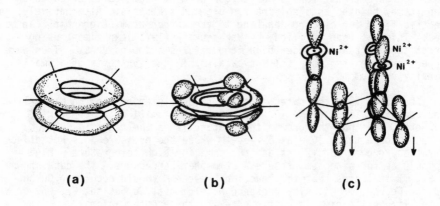

(a)   (b)   (c)

Fig. 2   Schematics of the transformation of SP^2 hybrid
         into SP^3 hybrid in the graphite ring.

(a) = π bond of graphite ring.
(b) = alternating phenomena of maximum and minimum
        electron cloud density.
(c) = coordinating covalent bonding (P - d bonding).

THE INFLUENCE OF AN ELECTRIC FIELD ON DIAMOND SYNTHESIS

Carbanions are negatively charged carbon atoms and, when derived from saturated carbon atoms, are believed to have a tetrahedral configuration. The electron pair is assumed to occupy one lobe of an SP^3 hybrid orbital.

The electron pair is assumed to occupy one lobe of an SP^3 hybrid orbital. Carbanions move to the positive, electrode under the influence of an external electric field. They quickly collect and supersaturate positive electrodes, when heated by a direct current, and diamond is readily synthesized. A carbonium ion is a positively charged carbon atom. Most carbonium ions have a three-fold coordination in a planar geometry with SP^2 hybrid orbitals in the plane and, normal to this, an empty (Pz) orbital. Carbonium ions move towards the negative electrode, under the influence of an external field, and form graphite.

In the electric field:

$$\vec{E} = -(du/dn)\ \vec{n} \qquad\qquad 5$$

where $\vec{E}$ = the strength of the electric field at any given point,

   $\vec{n}$ = unit vector,

   $\frac{du}{dn}$ = (negative) value of the potential gradient vector at this point·

When the resistance (R) is 1/1000 ohms and the direct current for heating is 1000 amps, the difference in diamond formation at both electrodes is not obvious. For a resistance of 1/100 ohms and a current of 500 amps, the rate of diamond transformation at the positive electrode is increased and that at the negative electrode is decreased.

Many experiments have shown that the presence of an external electric field has a significant effect on the nucleation process. One element, common to all these experiments, is that the essence of diamond formation is nucleation on ions under these conditions of thermodynamic instability. This process makes it possible to form small but extremely stable crystals. The critical value of P for an electrically charged nucleus is lower than that for the neutral case. There is not only a practical significance to the electronic transition in the conversion of graphite to diamond, but also it is important to diamond crystallisation under conditions of reduced pressure.

REFERENCES

Eggleton, H. and Tompkins, P. 1957, Contact Catalysis, Ed. R.H. Griffith, Oxford University Press, p 63
Feng Chu-de, 1983, Journal of Silicate, Vol. II, No. 1, pp 105-110
Hu En-liang, 1982, Latent Science, No. 4, pp 29-30 (in Chinese)
Kaschiev, 1972, Journal of Crystal Growth, 13/14, pp 129-130
Orlov, Yu. L. 1977, Mineralogy of Diamond, p 146 (in Chinese, translated from Russian by Huan Chao-en)
Sokolowski M. 1982, Journal of Crystal Growth, 57, pp 185-188
Tomashov, N.D. 1964, Theory on Corrosion and Protection of Metal, p 28
Wakatsuki, M. 1975, Proc. 4th High Pressure Conf, Japan, p 413
Wentorf, R.H. Jr. and Ber Ban Senges, 1966, Physical Chemistry 70, p 975
A Writing Collective of Jilin Poly-technical Institute and Jilin Engineering Institute 1979, Physical Chemistry, pp 166-169 (in Chinese)
A Writing Collective of Dept. of Organic Chemistry 1978, Tianjin University, Organic Chemistry, pp 106-108

Discussion on Superhard materials

Rapporteur: I MINKOFF (1)
Editor: C A BROOKES (2)

(1) Department of Materials Engineering, Israel Institute of
Technology, Haifa, Israel
(2) Department of Engineering Science, University of Exeter, UK

Discussion following the first session dedicated to the superhard
materials, in this series of Conferences, was principally of a general and
diverse nature (mostly arising from Dr Field's comprehensive review) but
the nature of slip and behaviour of dislocations elicited some rather more
detailed comments and discussion.

Professor Hirsch asked Dr Brookes if anisotropy in the hardness of cubic
BN and diamond could be used to differentiate between {111} <1$\bar{1}$0> or {111}
<11$\bar{2}$> slip because this might then allow a definition of whether nucle-
ation of dislocations is a rate controlling step for deformation at low
temperature since the energy of the critical nucleus is controlled by the
dislocation energy - i.e. Gb^2. In addition, he commented on the possibility that, since cubic BN is of the zinc blende structure type, then (111)
should have a different hardness to ($\bar{1}\bar{1}\bar{1}$) planes. Professor Heuer also
asked if the stacking fault energy for cubic BN had been determined.
Dr Brookes explained that the resolved shear stress model he used could
not differentiate between <1$\bar{1}$0> and <11$\bar{2}$> slip and that he did not know of
any data on the stacking fault energy for cubic BN. However, as grown
{111} planes were indented at random and the hardness values were consistent to ±5%, he therefore believed that this 'polarity' effect was not
significant in hardness measurements - but could well be important in wear
processes. Dr Santhanam expressed surprise at the amount of indentation
creep observed in cubic BN at temperatures as low as 200°C since creep in
commercial WC-Co cemented carbides was negligible below 800°C. He asked
if the stress exponents had been measured. Dr Brookes replied that creep
under indentation conditions had now been observed at temperatures <0.3 T_m
in a number of materials which did not creep under uniaxial uniform stress
conditions. Consequently, he believed that this particular form of indentation creep was not amenable to analysis on the basis of existing creep
models and, therefore, had not attempted to determine the stress exponents.

The spectacular twinning micrographs shown by Dr Lee excited some discussion on the relative ease of twinning in diamond. Professor Heuer
contrasted copper and diamond in this respect and pointed out that the
ratio of the shear modulus to the stacking fault energy was important (the

Burgers vector is similar for the two materials). Professor Hirsch agreed
that twinning is governed by the movement of partial dislocations and
thought that the critical parameter (Gb/ɤ) of diamond was about half that
of copper. Dr Lee was in full accord with this point but pointed out that
this work was more concerned with the cause of differences in the wear
characteristics of the pre-existing deformed zones, and their neighbour-
hood, rather than a model for their original formation.

Dr Y Doi reported that the nature of hardness anisotropy for synthetic
diamonds was contrary to that predicted by the resolved shear stress model
- if slip only on {111} <1$\bar{1}$0> was assumed. He therefore suggested that
other slip systems could be involved.   Dr Brookes observed that this
apparent anomaly was often reported on {111} surfaces for a number of
cubic crystals and maintained that it was due to the assymetric 'pile-up'
of material against the Knoop long diagonal. However, he did agree that
slip might be possible on other systems - e.g. {001} <110> - but felt that
{111} would predominate. Professor Hirsch asked if nitrogen platelets were
observed in synthetic diamonds. In natural diamonds, dislocation loops on
{100} planes with Burgers vectors of [100] or 1/2 [110] had been observed.
They appeared to be the result of transformation of N_2 platelets into
perfect dislocations. Dr Doi replied that no experimental work had as yet
been carried out to determine whether, or not, nitrogen platelets were
present in these particular diamonds.

Inst. Phys. Conf. Ser. No. 75: Chapter 4
Paper presented at 2nd Int. Conf. Science Hard Mater., Rhodes

249

Phase equilibria and structural chemistry in ternary systems: transition metal–boron–nitrogen

I SMID AND P ROGL

Institut für Physikalische Chemie der
Universität Wien
A-1090 Wien, Währingerstraße 42, Austria

ABSTRACT

The phase equilibria in the ternary systems Fe-B-N, Co-B-N and Ni-B-N
have been established from X-ray powder diffraction data and isothermal
sections at 900°C and 1 atm Ar have been constructed. In all three
systems the binary transition metal borides form two-phase equilibria
with hexagonal boronnitride.

The iron-boron binary system has been reinvestigated by means of X-ray
powder, metallographic and melting point analysis. "FeB_2" and "Fe_3B"
are not stable in thermodynamic equilibrium and "FeB_{19}" rather represents
a solid solution phase of iron in β-rh boron than a binary compound.
A complete constitutional T-X diagram is presented.

INTRODUCTION

The so-called superhard materials (diamond, cubic boron nitride and boron-
carbide) with their superior properties in hardness and wear resistance
have attracted considerable attention; this is particularly true for cubic-
boronnitride due to its rather attractive fracture-toughness values. For
effective production of c-BN, a knowledge of the exact phase equilibria
and mutual solid solubilities in ternary systems, transition metal-boron-
nitrogen (T-B-N) is of importance.

In a systematic study of the phase equilibria and structural chemistry of
such ternary systems we have investigated in particular combinations with
the iron metals, T= Fe, Co and Ni.

EXPERIMENTAL

The samples were prepared by powder-metallurgical reaction of compacted
powder-mixtures of the elements and boronnitride in a high frequency
furnace using a tungsten susceptor and Mo(W) or boronnitride substrate
materials. Heat treatments were usually 3 days at 900(+5)°C under either
1 bar of argon or high purity nitrogen atmosphere. The thermodynamic
phase equilibria and solid solubilities were derived by means of X-ray
powder diffraction analysis.

* Dedicated to Prof.Dr.K.Schlögl on occasion of his 60th birthday.

RESULTS

Isothermal sections for the three systems Fe-B-N, Co-B-N and Ni-B-N were established at 900°C revealing the formation of two-phase equilibria between the transition metal borides and boronnitride (Fig.1). The mutual solid solubilities at 900°C were found to be rather small and no ternary compounds formed under high purity conditions. As far as the Fe-B-N ternary is concerned, the phase equilibria derived confirm the earlier information obtained from partial phase equilibria studies by Kiessling and Liu (1951), and by Fountain and Chipman (1962). Technical conditions however - using graphite crucible material and technical argon - prompted the stabilization of a compound $Fe_3(B,N,C)$ with Fe_3C-type structure. A comparison of the lattice parameters for $Fe_3(B,C,N)$ (a= 4.459(1), b= 5.347(3), c= 6.664(3), V= 158.9(1)) with the variation of the unit cell dimensions along the homogeneity range $Fe_3B_xC_{1-x}$ (see Borlera and Pradelli, 1967) revealed a rather small concentration of C and N.

The iron-boron binary system has been reinvestigated with respect to the confusing literature reports about the existence of a congruently melting "FeB_2" which furthermore was claimed to be isostructural with the $A\ell B_2$-type (Voroshnin and coworkers, 1970). X-ray powder and metallographic analysis as well as the determination of the melting points (microoptical pyrometer) were not compatible with the existence of an "FeB_2" compound. Similarly, we did not observe a compound "Fe_3B" which was recently proposed by Khan, Kneller and Sostarich (1982) to be thermodynamically stable within a small temperature range 1150<T<1250°C. All our samples with a boron concentration of 20 to 30 at% B prepared by arc melting, were found to be melted after a heat treatment was applied at 1200 (±15)°C followed by quenching.

Concerning the binary phase "FeB_{19}", earlier proposed by Portnoi, Levinskaya and Romashov (1969) with a peritectic formation at 1980°C, $\ell + B \rightleftharpoons$ "$FeB_{\sim 19}$", we observed that their X-ray powder intensities and d-values are practically identical with the corresponding data calculated for a solid solution of ca. 3-5 at% Fe in β-rhombohedral boron.

In FeB-samples containing excess boron and at temperatures below T $\sim$ 1000°C a transposition type FeB $\longrightarrow$ CrB was observed in agreement with earlier findings (for a recent review see Kanaizuka, 1982). X-ray powder data of these alloys reveal an interesting pattern consisting of sharp and diffuse reflections. Despite the close resemblance of the sharp subset of reflections with the CrB-type, indexing is possible on the basis of an FeB-type subcell. From a microdensitometric study and applying Scherrers formula the undisturbed regions were estimated to be $\sim$ 80 Å. The relationship between the FeB and CrB-type structures in terms of a simple shift operation has been pointed out by Hohnke and Parthé (1966). Based on this transposition we have derived a simple model to calculate the X-ray powder pattern in fine agreement with the observed data (Fig.2). A complete constitutional diagram: temperature versus concentration (T>800°C) is presented for the Fe-B binary, Fig.3, which essentially confirms the diagram selected and calculated by Chart (1981). The binary systems cobalt-boron and nickel-boron were also reinvestigated at the temperature of interest (900°C, Figs.4,5) in respect to the corresponding ternary boronnitride-systems. In both cases the phase equilibria derived are in excellent agreement with the phase diagram data as earlier proposed by Rundqvist (1958a,b, 1959) and by Schöbel and Stadelmaier (1965 and 1966). This is particularly true for the retarded formation of the

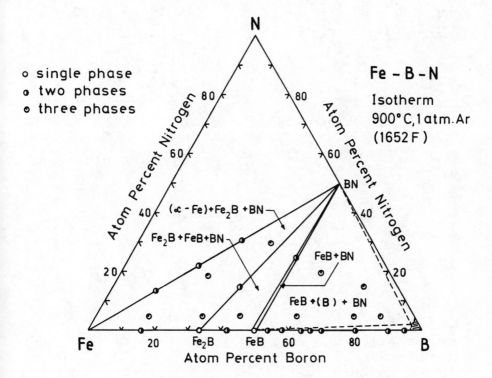

Fig. 1a)   Phase equilibria in the ternary system Fe-B-N,
           isothermal section at 900°C and 1 atm argon.

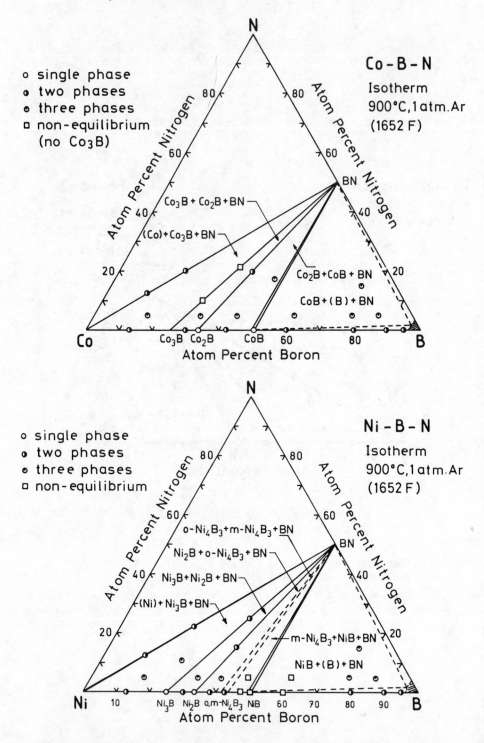

Fig. 1b) and c)   Phase equilibria in the ternary systems Co-B-N and Ni-B-N, isothermal sections at 900°C and 1 atm argon.

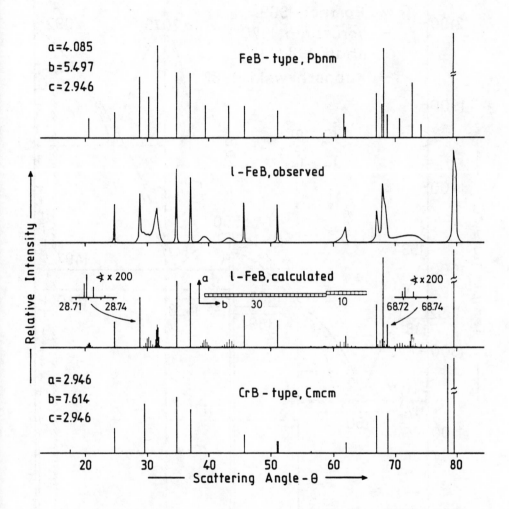

Fig. 2 X-ray powder diagram of iron monoboride; observed and calculated intensities and comparison with the calculated patterns for the FeB-type and CrB-type structure (for details see text).

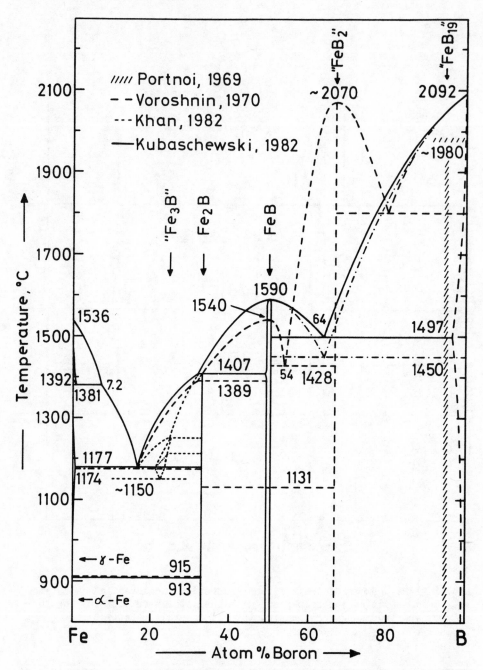

Fig. 3 Constitutional diagram of the binary system iron-boron (T > 800°C),
and comparison with earlier results obtained by Portnoi et al.
(1969) (dashed area, "FeB$_{19}$"), Voroshnin and coworkers (1970)
(dashed lines), and by Khan, Kneller and Sostarich (1982) (dotted
lines). The solid line refers to a critical assessment work by
O.Kubaschewski-von Goldbeck (1982). This work is shown by the
dashed-dotted curve.

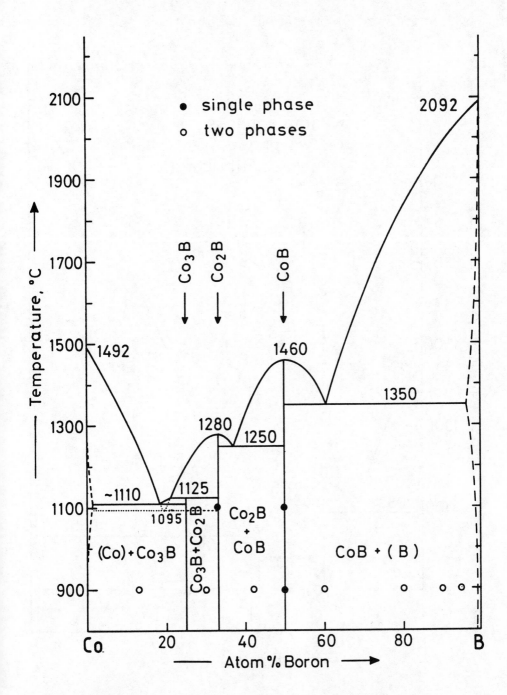

Fig. 4   Constitutional diagram of the binary system cobalt-boron;
phase equilibria based on Schöbel and Stadelmaier (1966).

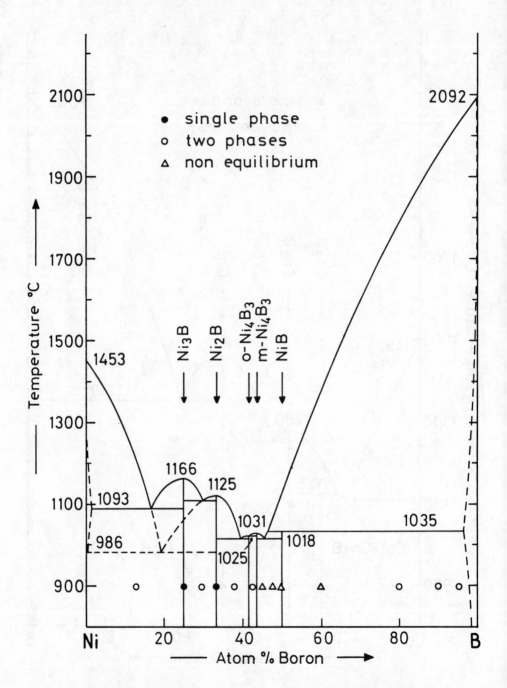

Fig. 5 Constitutional diagram of the binary system nickel-boron; phase equilibria based on Schöbel and Stadelmaier (1965).

cementite-type phases Co_3B and Ni_3B, which were only obtained after prolonged heat treatments. There was no indication in our samples for the formation of diborides "CoB_2" or "NiB_2" as earlier described by Kolomytsev (1959). "Ni_3B_2" (Kolomytsev, 1959) very likely corresponds to orthorhombic Ni_4B_3 (Rundqvist, 1959). There are furthermore considerable doubts about the existence of a compound "CoB_{12}" (Avlokhashvili et al., 1979). Even from their X-ray powder data it is evident, that "CoB_{12}" more likely corresponds to an extended solid solubility of cobalt in β-rhombohedral boron as described by Andersson and Callmer (1974). The so-called τ-borides "$Fe_{23}B_6$", "$Co_{23}B_6$" and "$Ni_{23}B_6$" with the $Cr_{23}C_6$-type of structure were not observed in thermodynamic equilibrium. These phases are metastable and were reported to appear as recrystallization products in rapidly quenched melts. Complete constitutional diagrams Co-B and Ni-B are presented in Figs.4,5.

ACKNOWLEDGEMENTS

The research reported herein has been sponsored by the U.S.A. through its European Research Office.

REFERENCES

Andersson, S. and Callmer, B. 1974, J. Solid State Chem. 10 219-231
Avlokhashvili, F. A., Tavadze, F. N., Tavadze, G. F., Tsikaridze, D. N., Gabunia, D. L., and Tsomaya, K. P., 1979, J. Less Common Met. 67 367-373
Borlera, M. L., and Pradelli, G. 1967, La Metallurgia Italiana 11 907
Chart, G. T. 1981, Comm. Cummunautes Europ. CECA No 7210-CA/3/303
Fountain, R. W., and Chipman, J. 1962, Trans. Met. Soc. AIME 224 599
Hohnke, D., and Parthé, E. 1966, Acta Crystallogr. 20 572-582
Kanaizuka, T. 1982, J. Solid State Chem. 41 195
Khan, Y., Kneller, E. and Sostarich, M. 1982, Z. Metallkde. 73 624-626
Kiessling, R. and Liu, Y. H. 1951, J. Metals N. Y. 3 639-643
Kolomytsev, P. T. 1959, Dokl. Akad. Nauk. SSSR 124 1247-1250
Kubaschewski-von Goldbeck, O. Iron-Binary Phase Diagrams, Springer-Verlag 1982
Portnoi, K. I., Levinskaya, M. and Romashov, V. M. 1969, Sov. Powder Metall, Met. Ceram. 8 657
Rundqvist, S. 1958a, Nature 181 259
Rundqvist, S. 1958b, Acta Chem. Scand. 12 658
Rundqvist, S. 1959, Acta Chem. Scand. 13 1193
Schöbel, J. D. and Stadelmaier, H. H. 1965, Z. Metallkde. 56 856
Voroshnin, L. G., Lyakhovich, L. S., Panich, G. G. and Protasevich, G. F. 1970, Metallov. i. Term. Obrab. Met. 9 14

Inst. Phys. Conf. Ser. No. 75: Chapter 4
Paper presented at 2nd Int. Conf. Science Hard Mater., Rhodes

259

High-temperature x-ray investigations of Nb$_2$C and some transition metal disilicides

B LÖNNBERG, I ENGSTRÖM AND T LUNDSTROM

Institute of Chemistry, Box 531, S-751 21 Uppsala
Sweden

ABSTRACT

The low temperature, orthorhombic modification of Nb$_2$C transforms at 1468$\pm$ 20 K into a hexagonal modification, crystallizing in the ε-Fe$_2$N type structure. The thermal expansion of the orthorhombic modification is strongly anisotropic while that of the hexagonal modification is nearly isotropic.

The thermal expansions of the group IV-VII disilicides have been studied. No phase transformations were observed in the investigated temperature interval, 300 K to 1500 K. The thermal anisotropy varies considerably among the disilicides. The volume expansion of the disilicides broadly reflects the order of their melting points.

INTRODUCTION

Composites and hard materials are composed of phases with large differences in chemical and physical properties. Knowledge of these properties is essential when designing new materials. For materials to be used at elevated temperatures it is important to use constituents of similar thermal properties. In particular, differences in thermal expansion can create large stresses in the materials. The same problem is encountered when the materials are used as intermediate or protective layers in electronic devices. As a part of more detailed high-temperature property studies of the group V-VI carbides and the group IV-VII disilicides, the thermal expansion of two Nb$_2$C modifications (Lönnberg and Lundström, 1984) and eleven disilicides (Engström and Lönnberg, 1984) have been investigated. Some of the results are presented in the present paper.

EXPERIMENTAL

The samples were prepared by arc-melting mixtures of metal and carbon or silicon under an argon atmosphere. The niobium carbide samples were subsequently heat-treated in evacuated and sealed silica capsules at 1073 K for 14 days. The powder diffraction patterns of the arc-melted silicides displayed sharp lines, indicating homogeneous samples. No further heat treatments of the silicides were therefore performed. The linear thermal expansion coefficients were obtained by measuring the lattice parameters versus temperature, using a focusing high-temperature X-ray powder diffraction camera (Hägg, Ersson, Rudenholm and Sellberg, 1979) with CuKα radiation.

Al_2O_3 was used as internal calibration standard. The specimens were heated in vacuum (10^{-4}Pa) and the temperature measured with a Pt/Pt-10 % Rh thermocouple. The lattice parameters at ambient temperature were determined using a Philips XDC 1000 Guinier-Hägg focusing camera with CuKα radiation. The samples were examined from room temperature to 1573 K. The room temperature cell dimensions were redetermined after each run. No significant changes were observed, which indicates that no changes as regards composition and impurities had occurred in the specimens.

RESULTS AND DISCUSSION

Thermal Expansion of Two Nb_2 Modifications

Three samples with different nominal compositions were prepared in the Nb-C system. The results of the chemical analysis and phase analysis are given in Table 1.

TABLE 1   Chemical Analysis of the Samples after Heat Treatment at 1073 K for 14 Days (Samples Quenched in Water) and Phase Compositions after Arc-melting

	$NbC_{0.30}$	$NbC_{0.54}$	$NbC_{0.68}$
Nb (at. %)	76.6	64.9	62.8
C (at. %)	22.7	35.0	37.1
O (at. %)	0.7	0.1	0.1
Phase composition after arc-melting	Nb + Nb_2C (ortho)	Nb_2C (hex)	Nb_2C + NbC (hex)

Nb_2C occurs in three modifications: an orthorhombic low temperature modification, a hexagonal modification with the ε-Fe_2N type structure, stable at intermediate temperatures, and a hexagonal modification with the L'3 type structure stable above 2770 K. The main difference between the three modifications is the different ordering of the carbon atoms. The hexagonal → orthorhombic transformation is more sluggish in C-rich samples than in niobium-rich samples. As can be seen from Table 1, only the sample containing excess metal transformed into the orthorhombic modification on arc-melting. Even a heat treatment at 1073 K for 14 days was not enough to obtain a transformation of Nb_2C in the C-rich sample. A similar composition dependence of the transformation rate was observed by Yvon, Nowotny and Kieffer (1966) for the transformation of the ε-Fe_2N type structure to the ζ-Fe_2N type structure in V_2C. It was suggested that epitaxial growth of the ε-Fe_2N type structure on close-packed (111) planes of the cubic monocarbide occurs and that epitaxial growth of the orthorhombic structure on the close-packed (110) planes of the metal occurs for the metal-rich sample. By analogy a similar explanation seems reasonable for the Nb_2C transformation. The difficulty in obtaining the orthorhomic modification on cooling carbon-rich specimens could thus be ascribed to the absence of a substrate displaying a low interfacial energy towards the orthorhombic phase, leading to nucleation problems. In metal-rich specimens the (110) plane of niobium might be favourable for coherent nucleation. The lattice mismatch towards orthorhombic Nb_2C is, however, considerable.

The transformation temperature was found to be 1468 ± 20 K. The transformation of Nb_2C in niobium-rich specimens was highly reversible and only a small hysteresis was observed. A non-linear variation of the lattice parameter with temperature was obtained. Second order polynomials were fitted to the experimental data using a least squares procedure. The polynomials for hexagonal Nb_2C is given in eqn. (1) and for orthorhombic Nb_2C in eqn. (2) (T in K):

$$a = 5.4089 + 2.377 \times 10^{-5} \times T + 1.005 \times 10^{-8} \times T^2 \text{ Å} \qquad \text{1a}$$
$$c = 4.9662 + 1.465 \times 10^{-5} \times T + 1.498 \times 10^{-8} \times T^2 \text{ Å} \qquad \text{1b}$$

$$a = 10.9001 + 3.057 \times 10^{-5} \times T + 1.065 \times 10^{-8} \times T^2 \text{ Å} \qquad \text{2a}$$
$$b = 3.0867 + 2.774 \times 10^{-5} \times T + 4.801 \times 10^{-9} \times T^2 \text{ Å} \qquad \text{2b}$$
$$c = 4.9667 + 2.445 \times 10^{-5} \times T + 4.868 \times 10^{-9} \times T^2 \text{ Å} \qquad \text{2c}$$

The average linear thermal expansion coefficients were determined by inserting eqn. (1) and (2) into eqn. (3):

$$\bar{\alpha} = 1/a_0 \times (a-a_0)/(T-T_0) \qquad 3$$

The lattice parameters refer to T K and T_0 K, respectively. In the present study T_0 is 298 K. The variation of $\bar{\alpha}$ with temperature is given in Figs. 1a and 1b.

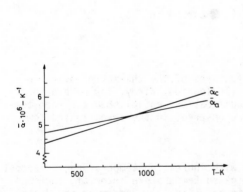

Fig. 1a Variation of the average linear thermal expansion coefficients with temperature for hexagonal Nb_2C.

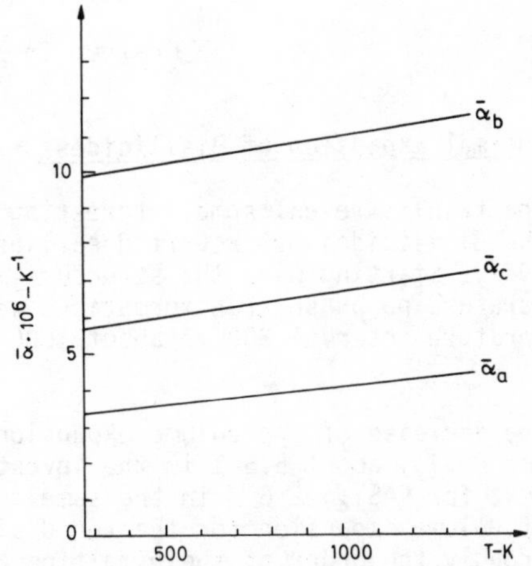

Fig. 1b Variation of the average linear thermal expansion coefficients with temperature for orthorhombic Nb_2C.

It is seen that hexagonal Nb_2C is almost isotropic while the orthorhombic modification is very anisotropic. The change in the expansion behaviour takes place in the basal plane, while the expansion along the c-axis remains the same during the transformation. The orthorhombic distortion of the hexagonal structure is very small, i.e. the Nb-Nb and the Nb-C bond strengths do not change very much. The main change lies in the basal plane while the c-axis remains unchanged. The main difference between the two Nb_2C structures lies in the ordering of the carbon atoms as demonstrated in Fig. 2. Although no C-C contacts occur, the carbon atoms in the ξ-Nb_2C type structure form double rows parallel to the b-axis, while in the ε-Fe_2N type structure they are arranged in alternating more densely and less densely packed carbon "layers". It is probable that the change in carbon ordering is responsible for the large thermal expansion anisotropy of orthorhombic Nb_2C.

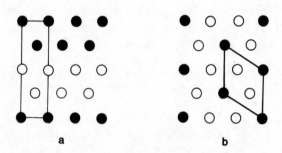

a   b

Fig. 2 The carbon arrangement in Nb_2C crys-
       tallizing in the ξ-Nb_2C type struc-
       ture (a) and ε-Fe_2N type structure
       (b).
       ○ carbon in Z=0, ● carbon in Z=½.

Thermal Expansion of Disilicides

The results reveal some interesting features of the thermal behaviour of the disilicides not reported earlier (Touloukian, Kirby, Taylor and Lee, 1977). Starting with the structure modifications established for room temperature no phase transformations were observed in the investigated temperature interval 300 to about 1500 K.

The increase of the volume expansion with increasing temperature is largest for $CrSi_2$, about 5.5 % in the investigated temperature interval, and smallest for $ReSi_2$, 2.6 % in the same interval (Fig. 3). Comparing the degree of volume expansion for the ten disilicides investigated the order reflects broadly the order of their melting points.

Structurally, most of these disilicides belong to the $TiSi_2$ structure family, exceptions being the isomorphous $ZrSi_2$ and $HfSi_2$. A member of the $TiSi_2$ family is characterized by a specific stacking sequence of a close-
-packed $MeSi_2$ layer of atoms. Adjacent layers are displaced in a specific way for each member. The structures of $TiSi_2$ (orthor), $CrSi_2$ (hex) and $MoSi_2$ (tetr) are characterized by the stacking sequences ABCD, ABC and AB, respectively.

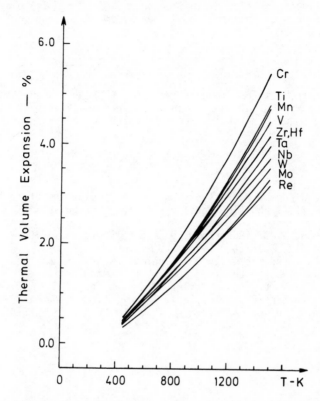

Fig. 3 Disilicide volume expansion versus temperature.

Comparing thermal expansion data for the disilicides belonging to the TiSi$_2$ family shows that the thermal anisotropy varies considerably. While VSi$_2$ and MoSi$_2$ representing different stacking sequences exhibit only very small differences in thermal expansion for different crystallographic directions, CrSi$_2$ shows a very anisotropic behaviour. In this compound the expansion in the close packed layer is, at 1500 K, almost twice the expansion between the layers.

Comparing disilicides with the same stacking sequence as CrSi$_2$ shows that the other representatives (VSi$_2$, NbSi$_2$, TaSi$_2$) have a very isotropic thermal behaviour. Thus the same stacking sequence does not necessarily imply the same thermal properties. CrSi$_2$ is the only disilicide of the TiSi$_2$ family which is a semiconductor.

The thermal expansion behaviour of the two silicides representing the ZrSi$_2$ structure-type, ZrSi$_2$ and HfSi$_2$, is very similar. Comparison with the other disilicides shows that the expansion in one of the crystallographic directions (b-axis) is remarkably small. The main part of the thermal expansion thus occurs in the a- and c-directions of this structure.

The thermal expansion coefficients calculated from the consecutive observations of the expansion are generally 8-9 ($\times 10^{-6} \cdot K^{-1}$) at room temperature, increasing to around 11-15($\times 10^{-6}K^{-1}$) at 1300 K. Exceptions being here on one hand the almost constant expansion coefficient 8 for the b-axes of $ZrSi_2$ and $HfSi_2$ and on the other hand the extreme expansion coefficient of 22 for $CrSi_2$ at 1500 K.

ACKNOWLEDGEMENT

We are most grateful to Dr. Bengt Allan Roos, AB SANDVIK, Hard Materials, Stockholm for carrying out the carbon and oxygen analyses. The financial support of the Swedish Natural Science Research Council is gratefully acknowledged.

REFERENCES

Engström, I. and Lönnberg, B. 1984. Thermal Expansion Measurements on Group IV to VII Transition Metal Disilicides. To be submitted to J. Appl. Phys.
Hägg, G., Ersson, N.-O., Rudenholm, G. and Sellberg, B. 1979. J. Appl. Crystallogr. 12, 221-224.
Lönnberg, B. and Lundström, T. 1984. Thermal Expansion and Phase Analytical Studies of Nb_2C. Submitted to J. Less-Common Met.
Touloukian, Y.S., Kirby, R.K., Taylor, R.E. and Lee, T.Y.R. 1977. Thermophysical Properties of Matter, Vol. 13, IFI/Plenum, New York-Washington.
Yvon, K., Nowotny, H. and Kieffer, R. 1966. Mh. Chem. 97, 689-694.

Inst. Phys. Conf. Ser. No. 75: Chapter 4
Paper presented at 2nd Int. Conf. Science Hard Mater., Rhodes

265

Deformation mechanisms of polycrystalline tungsten hemicarbide W$_2$C

J DUBOIS, G FANTOZZI, T EPICIER, C ESNOUF

Groupes d'Etudes de Métallurgie Physique et de Physique des Matériaux
LA 341 - INSA DE LYON - Bât. 502 - 69621 VILLEURBANNE CEDEX FRANCE

ABSTRACT

A precise thermodynamic analysis of the mechanical test results carried out between 800 and 2200°C on polycrystalline W$_2$C allows us to determine the terms of the constitutive law. Particularly, the stress dependence of the preexponential factor is determined following the analysis of Surek and co-workers. The preceding results are compared with those obtained by measurement of carbon diffusion coefficient, electron microscope observations and crystallographic determinations by neutron diffraction. All the results are consistent with the proposed deformation mechanisms.

INTRODUCTION

Hitherto, the mechanical behaviour of transition metal hexagonal carbides is poorly known; only a few studies have been made on the plasticity of these carbides. We shall pay attention to studies that have been performed on pure metallic carbides, works on cemented carbides being mentioned in a recent review (Almond (1983)).

Most publications concern the tungsten monocarbide WC, and particularly the study of plasticity using hardness test (Corteville, Monier and Pons, 1963-1965; French and Thomas, 1965; Takahashi and Freise, 1964). These works reveal extreme hardness anisotropy of WC crystals that persists over the entire range of temperatures (up to 900°C (Lee, 1982)). The hardness test reveals slip lines surrounding the microhardness indentations and reveals the slip systems at room temperature :
$$\{1\bar{1}00\} \ \langle0001\rangle \ \text{and} \ \{1\bar{1}00\} \ \langle11\bar{2}0\rangle$$
This last system would be the most probable because it is activated more easily under an applied stress. These observations have mostly been confirmed by electron microscopy analysis (see review in Almond (1983)). Kelly and Rowcliffe (1967) have studied the plastic deformation of hot-pressed polycrystalline samples of WC in four-point bending and under compression up to 2100°C. The stress vs displacement curves for WC were different from those of cubic carbides : they exhibit work-hardening, rates much higher than for cubic carbides. Kelly and Rowcliffe did not observe a recovery effect and the transition from brittle to ductile behaviour occured at approximately half the absolute melting point. The apparent activation energy for plastic deformation is approximately 0.5 10^{-19} J/atom from 1220°C to 2070°C. On the other hand Lewis and Porter (1969), studying wear pieces (tool tips and single crystals)

observed by X-ray techniques, deformation stacking faults involving slip on the basal plane (0001). These samples having been deformed up to 1300°C, the authors suggest that slip an the basal plane may occur at high temperature. Hibbs, Sinclair and Rowcliffe (1983) have shown that plastic deformation induced by indentation at room temperature and 1000°C occurs by motion of 1/3 $\langle 1123 \rangle$ dislocations split into 1/6 $\langle 1123 \rangle$ on $\{1\bar{1}00\}$ planes; they also observe interactions between the leading partials of such extended defects, which create a new partial dislocation type (1/6 $\langle 1\bar{2}10 \rangle$) which causes the configuration to become sessile. The view of the authors is that this mechanism could hinder plastic flow, and also promote nucleation of cracks.

A few studies have been made on hemicarbides. Vahldiek, Mersol and Lynch (1966) have studied Mo_2C single crystals by microhardness (Knoop and Vickers indentation) to verify dependence of hardness as a function of crystal axes and produce plastic deformation. Optical and electron microscopy revealed slip and twin traces on all planes studied. Basal (0001) $\langle 2\bar{1}10 \rangle$ slip was determined to be the primary slip system and a secondary slip was also observed : $\{10\bar{1}0\}$ $\langle 2\bar{1}10 \rangle$. The latter also produced twinning of the $\{10\bar{1}2\}$ $[000\bar{1}]$ type. The tungsten hemicarbide has been studied by Koester and Moak (1966) pointing out a "knee" in the hardness curve at approximately 1000°C. This behaviour is entirely different from the cubic carbides that show a regular decreasing as a function of temperature. Orange and co-workers have worked with polycrystalline samples (large grains) and have observed slip lines surrounding Vickers microhardness indentations. The slip system has been identified by the focus method : (0001) $\langle 1\bar{2}10 \rangle$ at room temperature. This result has been confirmed by Epicier and co-workers (1981).

A preliminary investigation of the microstructure and plasticity of W_2C (by TEM observations and bending tests) has produced interesting results concerning deformation mechanisms. As in metallic monocarbides, a process involving the glide of narrowly dissociated dislocations is thought to control the plastic deformation of the material in a middle temperature range (1100° to about 2000°C); at higher temperatures, evidence for dislocation climb has been obtained .

The aim of this paper is to show that a precise thermodynamic analysis of the mechanical test results (between 800 up to 2200°C) allows the determination of the terms of the constitutive law, more particularly the pre-exponential factor that is generally considered as constant. In order to obtain a consistent analysis of the plastic deformation of W_2C, these preceding results are compared with those obtained by measurement of self diffusion parameters, TEM observations and crystallographic determinations.

RESULTS

Material

To study the intrinsic plastic deformation behaviour of tungsten hemicarbide it is necessary to prepare a material with high purity and a negligible amount of porosity, since the presence of impurities (such as oxygen and nitrogen which occupy similar atomic positions to the carbon atoms) and pores considerably modify the mechanical behaviour of carbides. The specimens

are polycrystalline with a large grain size (about 300 μm) and a carbon
content near stoichiometry (Tréheux, Dubois and Fantozzi, 1981).
The study of the plastic deformation of tungsten hemicarbide requires a
good knowledge of this material, particularly from a physico-chemical
point of view. The chemical analysis has been carried out (O$\langle$30 ppm,
N$\langle$10 ppm) and published elsewhere (Tréheux, Dubois and Fantozzi, 1981).

Carbon Diffusion

There is a lack of data on the diffusion coefficient of carbon in tungsten
hemicarbide. Therefore, we have carried out the measurements of the bulk
and grain boundary diffusion of ^{14}C in W_2C by the radioactive tracers
technique (Tréheux, Dubois and Fantozzi, 1981).

The apparent grain boundary diffusion coefficient is represented by the
expression:

$$P_{GB} = 1.8 \ 10^{-10} \ \exp - (287,918/RT) \ m^3 \ s^{-1} \qquad 1$$

whereas the bulk diffusion coefficient is given by the relation :

$$D_V = 18.3 \ 10^{-4} \ \exp - (382,470/RT) \ m^2 \ s^{-1} \qquad 2$$

Neutron Diffraction

The crystal structure of most of the metal carbides may be described on
the basis of a close-packed metal sublattice, the carbon atoms occupying
all or part of the octahedral interstitial sites. Concerning the structure
of hemicarbides, several crystallographic studies have been performed
(Butorina and Pinsker, 1960 ; Morton and co-workers, 1972 ; Telegus,
Gladyshevskii and Kripyakevich, 1968 ; Rudy and Windish, 1967 ; Yvon,
Nowotny and Benesovsky, 1968 ; Bowman and Arnold, 1971 ; Epicier and
co-workers, 1983a). Different structures have been observed (Fig. 1) : an
ordered hexagonal phase α - W_2C (C6 type) at low temperatures (only a
carbon plane is occupied). At middle temperatures two types of ordered
phases have been reported called β - W_2C (orthorombic ζ - Fe_2N and
hexagonal ϵ - Fe_2N). At high temperatures (above about 2400°C), all the
authors observe only a disordered hexagonal phase (L'$_3$ type) called γ -
W_2C. Their results are not consistent and so it is interesting to perform
crystallographic determinations of structure by neutron diffraction.
Indeed, neutron diffraction is better than X-ray diffraction because it is
sensitive to heavy metals and light atoms. Furthermore the scattering
amplitudes for neutrons are approximately equal and neutron
diffraction is carried out by transmission through the sample (Bowman and
Arnold, 1971).

Hitherto, only two studies have been made of W_2C using neutron diffraction
: Yvon, Nowotny and Benesowsky (1968) observed at room temperature, after
quenching from 2100°C up to 2400°C, an ordered structure of ϵ-Fe_2N type.
Bowman and Arnold (1971) noted that the high temperature phases of W_2C
cannot be quenched and therefore they studied the neutron diffraction as
function of temperature. They observed a transformation phase between the
random L'$_3$ structure (γ - W_2C) and on ordered trigonal form at about
1750°C in the presence of WC. It was very difficult to study this phase
change because of severe interference from the graphite background. These
non-conclusive results have led us to perform neutron diffraction as a
function of temperature in order to specify the carbon atom distributions

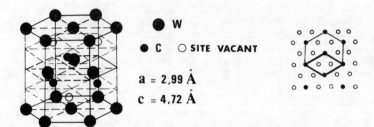

Fig. 1 W$_2$C structure, hexagonal cell ε Fe$_2$N-type. The parameters for the metallic sub-lattice are a = 2.99 Å, c = 4.72 Å (unit cell shown). The β -W$_2$C cell is the triple cell, with a = 5.18 Å, c = 4,72 Å. It is worth noting that no interstitial (0001) c type plane is empty of carbon atoms.

(ordered, partially ordered or random) in the structure of W$_2$C between 800 up to 2200°C. Neutron diffraction was performed with the High Flux Reactor of the Institut Laue-Langevin (Grenoble, France), on the D1B multidetector ; the details of these experiments as well as the complete results, will soon be published.

The study has shown that only two structures are observed between room temperature and 2200°C : at low temperatures, we observe the ε - Fe$_2$N type structure, probably mixed with an L'$_3$ structure (ε - Fe$_2$N is a

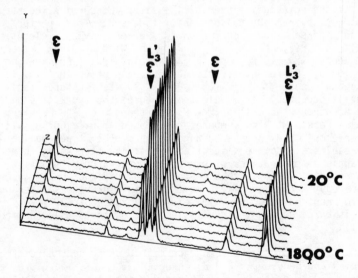

Fig. 2 Neutron powder diffraction patterns as a function of temperature

hexagonal ordered structure whereas L'$_3$ is a hexagonal disordered structure) as illustrated in Fig. 2. Above 1800°C we observe only the L'$_3$ structure. This experiment shows equally that the ordered phase α - W$_2$C (C$_6$ type) is never observed.

The study of the evolution of the intensity of the spectrum lines :

$$(01\bar{1}1) \text{ and } (02\bar{2}1) \text{ for}$$
$$(11\bar{2}1) \text{ and } (11\bar{2}2) \text{ for L'}_3 \text{ and } \mathcal{E}$$

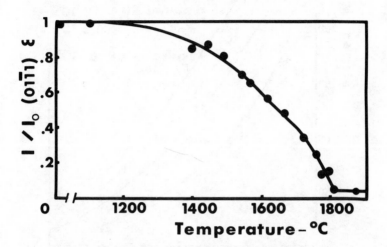

Fig. 3 Order-disorder transition of W_2C shown in terms of the relatively intensity of the (0111) neutron diffraction peak of $\mathcal{E}$

as a function of temperature has allowed us to show that an order-disorder transition takes place between 1500 and 1800°C (Fig. 3) and that this transition is perfectly reversible. Above 1800°C and up to 2200°C we observe only the disordered structure. These first results are very important because they confirm on the one hand the results obtained with many difficulties by Bowman and Arnold (1971) and the other hand that all carbon planes are at least partly occupied at low temperatures.

Mechanical Behaviour

The mechanical tests (Dubois, Orange and Fantozzi, 1980 ; Dubois and co-workers, 1983) were performed with carbide samples (25 x 4 x 0.3 mm^3), previously predeformed ($\mathcal{E}_p = 10^{-3}$ at 2100°C) for relieving the great brittleness. They were plastically deformed up to $\mathcal{E} = 10^{-2}$ at different strain rates (from 10^{-7} up to 10^{-5} S^{-1}) on a four point bending micromachine described previously (Orange and co-workers 1980). Fig. 4 shows the variation of the yield stress as a function of temperature and strain rate at a plastic strain $\mathcal{E} = 5 \cdot 10^{-4}$. It is worth noting that the form of the curves is typical of thermally activated deformation. Activation parameters (volumes V_a and enthalpies ΔH_a) were determined using the data of the stress-strain curves, relaxation tests or strain rate changes (Dubois and co-workers 1983). We also measured by relaxation tests or dip-tests the evolution of interval stress, σ_i, as a function of temperature (Fig. 5). The modulus-reduced internal stress varies strongly as a function of temperature (E is the modulus corrected for temperature).

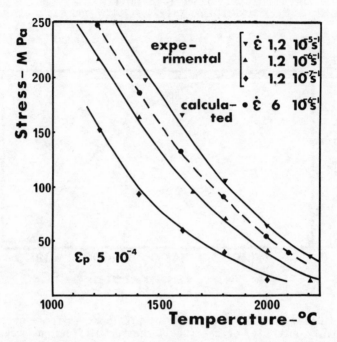

Fig. 4 Variation of the yield stress at 0.05 % plastic strain
as a function of temperature and strain rate.

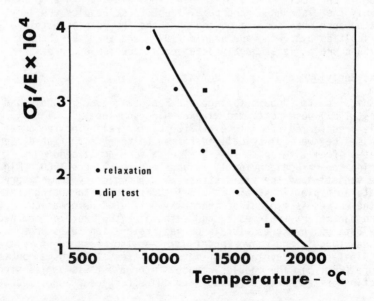

Fig. 5 Modulus-reduced internal stress as a function of temperature.

Electron microscope observations

The TEM analysis required the preparation of thin plates of W_2C bent up to 1% at different temperatures. After deformation tests, the samples were cooled down under stress in order to freeze ($1°C\ s^{-1}$) the dislocation arrays obtained at high temperature. The method of thinning (Epicier and co-workers, 1981) uses classical electropolishing and ion beam thinning in order to obtain a thin foil very near the surface, where the strain is the highest. TEM observations were made at 100 and 200 kV (Jeol 100 C and CX respectively) and 1 MV (E.M. Cameca -CEN Grenoble France). It should be noted that the use of a 200 kV microscope seems to be particularly suitable for observations of materials such as metal carbides especially for weak beam images necessary to reveal narrow dissociations. Only the basal slip system $\langle 11\bar{2}0 \rangle$ (0001), is activated in the temperature range up to 2200°C (Orange and co-workers, 1978; Epicier and co-workers, 1981). Many dislocations have been studied by weak beam dark field (WBDF). Up to 1750°C the microstructure shows mainly mixed dislocations gliding on the basal plane sometimes with interactions in the same plane (Dubois and co-workers, 1983). The accurate analysis of these dislocations shows that they are nearly all dissociated (Fig. 6).

Fig. 6 Weak beam Dark Field image showing the dissociation of a near edge dislocation in W_2C deformed at 1500°C.

The dislocations of type 1/3 1120 are dissociated into two Shockley partials following the reaction :

$$1/3\ [\bar{2}110] \rightarrow 1/3\ [\bar{1}010] + 1/3\ [\bar{1}100] \qquad\qquad 3$$

The stacking fault energy γ , as well as the elastic parameter ν were deduced from measurements of the partial separation of many dislocations as a function of their character (3 to 5 nm from the screw to the edge orientation). These yielded : γ = 92 mJ/m^2, and ν = 0.23 (Epicier and Esnouf 1984).

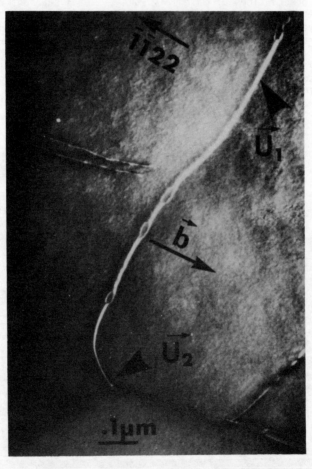

Fig. 7 Weak Beam Dark Field image showing a climbing near-edge dislocation (the line directions U$_1$, U$_2$ are out of the basal glide plane) in W$_2$C deformed at 2000°C (sample cooled under stresses). Note that the defect is mostly undissociated with some small portions, probably lying in different (0001) planes, which are split (the image was taken after a short irradiation in the microscope, which has been shown to promote spreading of the dissociated dislocations; the undissociated ones remaining unaffected (Epicier and co-workers, 1984a)).

Above 1750°C dislocation networks and rows of loops (Dubois and co-workers, 1983) are frequently observed in the basal planes. These manifestations of dislocation climb are confirmed by the analysis of non-basal dislocations (mixed in character) as shown by Fig. 7 where the segments of dislocation are dissociated into the basal plane and recombined out of the basal plane or are dissociated with jogs indicating "jumps" from one (0001) plane to another (Epicier and co-workers, 1984).

DISCUSSION

The method of deformation, by flexion of a thin plate is not convenient to reach the steady state, as the compression test, because it's necessary to verify the Nadaï's approximation (Martin, 1972). Nevertheless, the thermodynamic analysis of the mechanical test results can be applied.

For thermally activated deformation processes the stress and temperature dependence of the strain rate is usually given by an Arrhenius equation as in :

$$\dot{\mathcal{E}}(T,\sigma^*) = \dot{\mathcal{E}}_0 (T,\sigma^*) \cdot \exp\left[-\Delta G (T,\sigma^*/ kT)\right] \qquad 4$$

where $\dot{\mathcal{E}}_0$ is the pre-exponential factor, ΔG the activation free energy of the process and σ^* the effective stress : $\sigma^x = \sigma_a - \sigma_i$ with σ_a : applied stress and σ_i : long range internal stress.

Surek and co-workers (1975) have shown that it is normally impossible to separate the stress and temperature dependence of the strain rate into the two components $\dot{\mathcal{E}}_0$ and ΔG without an additional supposition. These authors propose an analysis of experimental results which allows the separation of the stress and temperature dependence of $\dot{\mathcal{E}}$ into those of the pre-exponential factor and the activation free energy. This method gives the true values of activation volume, V, and energy, ΔH, of the plastic deformation process.

We have shown previously that σ_i/ E strongly depends on the temperature. This variation means that there is a significant evolution of the microstructure and therefore of the pre-exponential factor $\dot{\mathcal{E}}_0$ which depends strongly on the microstructure. Consequently, we have applied the method of Surek and co-workers (1975) to determine the temperature and stress dependance of $\dot{\mathcal{E}}_0$ and ΔG.

Thermodynamic analysis for linear elastic obstacles shows that the true values of the activation parameters are given by the following equations (Surek and co-workers 1975) :

$$\Delta G = \left[\Delta H + V \sigma^* (T/E \cdot dE/dT)\right] / \left[1 - T/E \cdot dE/dT\right] \qquad 5$$

$$\Delta H = \Delta H_a - k T^2 \cdot (\partial \ln \dot{\mathcal{E}}_0/ \partial T)_{\sigma^*} \qquad 6$$

$$V = V_a - k T \cdot (\partial \ln \dot{\mathcal{E}}_0/ \partial \sigma^*)_T \qquad 7$$

ΔH_a and V_a represent the experimental values of an activation enthalpy and volume calculated from the relation :

$$\Delta H_a = k T^2 \cdot (\partial \ln \dot{\mathcal{E}} / \partial T)_{\sigma^*} = - TV \cdot (\partial \sigma^*/ \partial T)_{\dot{\mathcal{E}}} \qquad 8$$

$$V_a = k\,T\,.\,(\partial \ln \dot{\mathcal{E}}\,/\,\partial \sigma^*)_T \hspace{5cm} 9$$

The values of these apparent activation parameters are given in other papers (Dubois and co-workers 1983). When the pre-exponential factor is a function of σ^*/E only, Surek and co-workers have shown from eqs. 6 and 7 that the true values of the activation parameters are given by :

$$\Delta H = \Delta H_a + k\,\sigma^* \,.\, T^2/E^2 \,.\, dE/dT \,.\, F' \hspace{3cm} 10$$

$$V = V_a - k\,.\,T/E\,.\,F' \hspace{4.5cm} 11$$

$$\Delta G = [\Delta H_a + V_a\,.\,\sigma^*\,.\,T/E\,.\,dE/dT]\,/\,[1 - T/E\,.\,dE/dT] \hspace{1cm} 12$$

where $\qquad F' = dF/d\,(\sigma^*/E)$ and $\ln \dot{\mathcal{E}}_0 = F\,(\sigma^*/E)$.

Very often, it is observed that the dependence of $\dot{\mathcal{E}}_0$ is of the form : $\dot{\mathcal{E}}_0$ = $(\sigma^*/E)^n$, thus eqs 10 and 11 became respectively :

$$\Delta H = \Delta Ha + n\,k\,T\,.\,T/E\,.\,dE/dT \hspace{4cm} 13$$

$$V = V_a - n\,k\,T/\sigma^* \hspace{5cm} 14$$

We have verified that the stress and temperature dependence of the activation volume is consistent with the relation 14 as shown in Fig. 8. In this way, we deduce : stress exponent n = 2.90 and constant true volume $V = 0.6\ 10^{-27}\ m^3$.

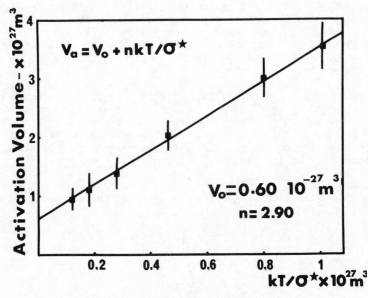

Fig. 8 Stress and temperature dependence of the
activation volume.

The knowledge of n and and dE/dT allows us to correct the experimental values and to determine the true activation energy by applying eq. 13. The corrected values of energy ΔH give, at $\sigma^* = 0$, an activation energy of about 3.4 10^{-19} J instead of 6.4 10^{-19} J for experimental values as shown in Fig. 9. If we compare this value of ΔH ($\sigma^* = 0$) to these obtain previously by measurement of carbon grain boundary diffusion coefficients we can say that they are no very different.

For cubic carbides (Toth 1970), authors have generally found an apparent activation energy for plastic deformation similar to that measured for bulk diffusion of carbon atoms. Rigorously, only the trace activation energy must be considered when making comparison with diffusion energy. The experimental results do not allow calculation of the true activation energy from experimental values. The preceding comparison is therefore not really accurate and the analysis must be re-examined.

Our analysis of the experimental results of mechanical tests seems to show that the process controlling the plastic deformation of W$_2$C is linked to the carbon diffusion along the dislocations. Rowcliffe (1965) has proposed a synchro-shear process to explain the deformation behaviour of TiC.

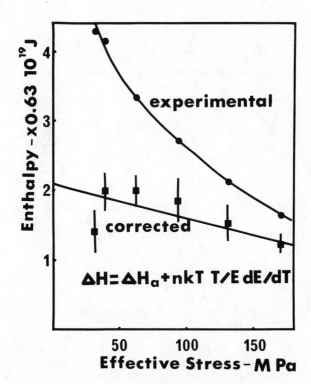

Fig. 9 Apparent and true activation enthalpies as a
function of effective stress.

This synchronized shear model, proposed initially by Kronberg (1957) for sapphire, requires the presence of zonal dislocations and the carbon diffusion is not necessary. Hollox (1968) has proposed that the synchronized shear of the carbon plane takes place by simultaneous diffusion of carbon atoms. In the case of W$_2$C, this "shear-diffusion" mechanism supposes that dislocations are dissociated into Shockley partials in the basal plane (Epicier and co-workers, 1981 ; Dubois and co-workers, 1983). This dissociation has been effectively observed by TEM analysis at least below 1750°C so the shear diffusion mechanism involving glide of dissociated dislocations and simultaneous diffusion of carbon atoms into the core of dislocations (Fig. 10) is thought to be the predominant process controlling the plastic deformation. This model which requires the presence of carbon atoms in all the carbon planes is consistent with the results obtained by neutron diffraction which shows that all carbon planes are at least partly occupied throughout the temperature range (α - W_2C structure is never observed).

Above 1800°C evidence of dislocation climb has been obtained through observations of hexagonal dislocations, dipole trays, rows of loops resulting from dipole annealing by diffusion (Dubois and co-workers, 1983 ; Epicier and co-workers, 1984a). Furthermore, this phenomenon is confirmed by the appearence of non-basal dislocations which are not dissociated (Epicier and co-workers, 1984). The recombination process is not well defined.

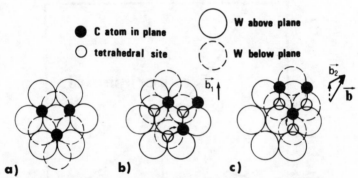

Fig. 10 Mechanism of shear-diffusion in the basal plane of W_2C (the carbon layer is schematically represented filled with intersticial atoms), Epicier and co-workers (1981).
a) stacking of tungsten and carbon planes perpendicular to the C - Axis.
b) Shear b_1 = 1/3 $[\bar{1}010]$ of the low tungsten layer and simultaneous diffusion of C-atoms from previous sites (become tetrahedral) to new octahedral sites.
c) Shear b_2 = 1/3 $[\bar{1}100]$ and repeated diffusion of carbon atoms; passing of the dislocation b = 1/3 $[\bar{2}110]$.

The occurence of climb at high temperature may explain the deviation of the true activation enthalpy curve from the straight line but this point must be confirmed by experiments in compression in order to obtain larger plastic deformation.

In conclusion, the mechanical behaviour of W_2C can be described by the constitutive law :

$$\dot{\mathcal{E}} = K \left(\sigma^*/E \right)^n \exp - \left(\Delta H_0 - \sigma^* V_0 \right)/kT \qquad 15$$

with $K = 1.5 \ 10^9 \ s^{-1}$ $n = 2.9$
          $\Delta H_0 = 3.4 \ 10^{-19} \ J$ $V_0 = 0.6 \ 10^{-27} \ m^3$

This law is well verified over the whole temperature range studied as seen in Fig. 4 where the calculated $\sigma^* $– T curve coincides with the experimental one.

However, at high temperatures the proposed model is not rigorously verified, climb processes taking place in this temperature range.

REFERENCES

Adda, Y. and Philibert, J. 1966, "La diffusion dans les solides". P.U.F., Paris pp. 717-729.
Almond E.A., 1983, Science of Hard Materials, Ed. R.K. Viswanadham, D.J. Rowcliffe and J. Gurland, Plenum Press, New-york, pp. 517-561
Bowman, A.L. and Arnold, G.P. 1971, Advan. High Temp. Chem. 4 243-264.
Butorina, L.N. and Pinsker, Z.G. 1960, Soviet. Phys. Cryst. 5 560-563.
Corteville, J. and Pons, L. 1963, C.R. Acad Sc. Paris, 257 1915-1918.
Corteville, J., Monier, J.L. and Pons, L. 1965, C.R. Acad. Sci. Paris 260 2773-2776.
Dubois, J., Orange, G. and Fantozzi, G. 1980, Scripta Met. 14 107-111.
Dubois, J., Epicier, T., Esnouf, C. and Fantozzi, G. 1983, Science of Hard Materials Ed. R. K. Viswanadham, D.J. Rowcliffe and J. Gurland, Plenum Press, New-York, pp. 201-218.
Epicier, T., Esnouf, C., Dubois J. and Fantozzi, G. 1981, Scripta Met. 15 1279-1283.
Epicier, T., Dubois, J., Esnouf, C. and Fantozzi, G. 1983a, C.R. Acad. Sc. Paris 297 215-218.
Epicier, T., Esnouf, C., Dubois, J. and Fantozzi, G. 1984a, "Deformation of Ceramics II", Plenum Press, pp. 73-86.
Epicier, T. and Esnouf, C. 1984, J. Microsc. Spectrosc. Electron. 9 17-33.
Epicier, T., Dubois, J., Esnouf, C. and Fantozzi, G. 1984b, Yamada Conference IX "Dislocations in Solids", to be published.
French, D.N. and Thomas, D.A. 1965, Trans. Met. Soc. AIME 233 950-952.
Hibbs, M.K., Sinclair, R. and Rowcliffe, D.J. 1983, Science of Hard Materials Ed. R.K. Viswanadham, D.J. Rowcliffe and J. Gurland, Plenum Press, New-York, pp. 121-135.
Hollox, G.E. 1968 Mater. Sci. Eng. 3 121-137.
Kelly, A. and Rowcliffe, D.J. 1967, J. Amer. Ceram. Soc., 50 251-256.
Koester, R.D. and Moak, D.P. 1967, J. Amer. Ceram. Soc. 50 290-296.
Kronberg, M.L. 1957, Acta Met. 5 507-524.
Lee M. 1983, Met. Trans. 14A 1625-1629
Lewis, D. and Porter, L.J. 1969, Appl. Crys. 2 249-252.
Martin, J.L. 1972, Thèse Univ. Paris-Nord.
Morton, N., James, B.W., Wostenholm, G.H. and Hepburn, D.C.B. 1972, J. Less-Common Metals 29 423-426.
Orange, G., Dubois, J., Fantozzi, G. and Gobin, P.F. 1978, Mat. Sc. Eng. 34 291-294.
Orange. G., Dubois, J., Fantozzi, G. and Gobin, P.F. 1980, Mem. Sci. Rev. Met. 2 131-143.
Rowcliffe, D.J. 1965, Ph. D. dissertation "The Mechanical Properties of Transition Metal Carbides", University of Cambridge.
Rudy, E. and Windisch, S.J. 1967, J. Amer. Ceram. Soc. 50 272-273.
Surek, T., Kuon, L.G., Luton, M.J. and Jonas, J.J. 1975, Proceed. John E. Dorn Symposium, pp. 629-655.
Takahashi, T. and Freise, E.J. 1965, Phil. Mag., 12 1-8.

Telegus, V.S., Gladyshevskii, E.I. and Kripyakevich, P.I. 1968, Soviet. Phys. Cryst. 12 813-815.

Toth, L.E. 1971, Transition Metal Carbides and Nitrides, Academic Press New-York, pp 163-166.

Tréheux, D., Dubois, J. and Fantozzi, G. 1981, Ceram. Inter. 7 142-148.

Vahldiek, F.W., Mersol, S.A. and Lynch, C.T. 1966, Trans. Met. Soc. AIME 236 1490-1496.

Yvon, K., Nowotny, H. and Benesovsky, F. 1968, Mh. Chem. 99 726-729.

Inst. Phys. Conf. Ser. No. 75: Chapter 4
Paper presented at 2nd Int. Conf. Science Hard Mater., Rhodes

Microchemistry and high temperature properties of sintered SiC

R HAMMINGER, G GRATHWOHL, F THÜMMLER

University and Nuclear Research Center Karlsruhe, Post Box 3640,
D-7500 Karlsruhe 1, Federal Republic of Germany

ABSTRACT

The present paper relates on the microchemistry of differently doped sintered SiC materials, analysed by classical and very modern microanalytical methods. Elucidating statements were achieved by Auger electron spectroscopy using a high resolution equipment in combination with an ultra high vacuum fracture device. Results on long term high temperature properties are summarized and correlations between properties and microstructure as well as recommendations for future-developments are given.

INTRODUCTION

Ceramic components for Diesel engines and automotive gas turbines are supposed to become industrial realities in the near future. Today, sintered SiC (SSiC) can be considered as a leading candidate in the group of non-oxide ceramics for such high performance applications at low as well as high temperatures.

Sintering of pure non-oxide ceramics with covalent bonding, such as BN, SiC and Si_3N_4, without external pressure is not possible to high densities, for several reasons. In order to achieve products of high density and reasonable homogeneity by normal sintering, the addition of small amounts of sintering aids becomes necessary. The additives enhance densification by forming multi-phased rather than single-phased ceramic compounds, most obvious in the case of Si_3N_4.

Since the pioneer work of Prochazka (1974) numerous investigations have been published concerning the manufacturing process of SSiC. For sintering, the customary utilized additives are B and C or more recently Al and C. Although different explanations have been offered these developments have surpassed complete understanding of the operative sintering mechanisms so far. The densification behaviour may also be affected by certain impurities of the starting powders, even at very low concentrations.

The physical and mechanical properties of SSiC and of other covalently-bonded ceramics, on the other hand, are influenced by such microstructural features as presence and nature of second phases and by the state of the grain boundaries. The distribution of the sintering additives and impurities thus plays an important role in influencing the mechanical behaviour of the materials, especially at elevated temperatures. Though many deve-

lopment activities were initiated in order to improve the homogeneity of
SSiC, the microstructures of available materials are still heterogeneous.
This is true with respect to the necessary sintering additives, i.e. B
and C or Al and C, resp., and different impurities, incorporated during
the various steps of processing and by the primary SiC powders. Detailed
microstructural and microchemical analysis is therefore required.

EXPERIMENTAL METHODS

Characteristic features of the microstructure of a high-quality SSiC material are the homogeneity, the very small grain size and minimized values
of porosity and second-phase inclusions. Impurities and additives can be
present in solid solution with SiC, but also in the form of second phases.

In order to achieve knowledge about the interesting microstructural and
microchemical effects, the combined use of complementary investigation methods has been successfully performed. In the case of SSiC containing
(Al,C) or (B,C), analytical problems arise from the detectability of extremely small concentrations, thus requiring high sensitivities and both
high lateral and spatial resolutions of the used analytical methods. Another problem exists in the nature of the sintering aids themselves: they
are all elements with very low atomic numbers, which generally cannot be
analysed in a simple manner. The analytical work of this study was performed by the following methods:

> - Wavelength dispersive analysis of X-rays (WDX)
> - Auger electron spectroscopy (AES) and
>   high resolution AES (HRAES)
> - α-microautoradiography

WDX, AES and α-microautoradiography were used to identify the inhomogeneities in the form of inclusions and precipitations, whereas high resolution
Auger electron spectroscopy (HRAES) enabled the study of UHV-exposed grain
boundaries. The degree of uniformity of the B-distribution of (B,C)-doped
materials was determined by α-microautoradiography.

MATERIALS

Different high-quality α-SSiC materials as delivered from international
suppliers were examined in the present work, as listed in **Table 1**. These
materials have been characterized by chemical analysis and ceramography.
The concentration values of the intentional sintering additives Al and B
were determined to be 0.40 - 0.58 wt.% and 0.22 - 0.34 wt.%, resp. While
SSiC I and SSiC V are (Al,C)-doped, SSiC II, III and IV are (B,C)-doped
materials. However, as shown in **Table 1** the (Al,C)-doped qualities are
not really free of B. The content of impurities is low and at nearly the
same level for all materials. Differences can be seen in the lower content
of O in materials II and IV, but with a higher content of N in material V,
originating in AlN which was added as a sintering aid.

All materials have high densities, as shown in **Table 1**. The grain growth
during sintering was controlled; exaggerated grain growth was not detected.
Grain sizes not larger than 12 μm were found. The average grain size differed by a factor of less than 2 for the materials investigated, the values are given in **Table 1**. Both transgranular and intergranular fracture

modes were detected with a higher degree of the latter in the case of (Al,C)-doping.

TABLE 1 <u>Characteristics of the Investigated α-SSiC Materials</u>

Material	O	N	C(free)	Si(free)	Fe	Ca	Mg	Al	B	C(tot)
SSiC I	0.10	0.07	0.51	<0.20	<0.1	<0.05	<0.01	0.40	<0.1	29.5
SSiC II	<0.01	0.055	0.50	<0.50	<0.05	<0.05	<0.1	0.05	0.34	30.4
SSiC III	0.08	0.03	0.54	<0.2	<0.1	<0.05	<0.01	<0.1	0.25	30.9
SSiC IV	0.024	0.01	0.5	n.d.*	0.065	<0.02	<0.01	0.015	0.225	29.95
SSiC V	0.07	0.29	1.02	<0.2	<0.1	<0.05	<0.01	0.58	<0.1	31.1

*not determined

Material	Density g/cm³	T.D. %	Grain size, av. μm
SSiC I	3.11	97.1	3.4
SSiC II	3.15	98.3	5.6
SSiC III	3.13	97.7	3.7
SSiC IV	3.10	96.4	4.0
SSiC V	3.09	96.5	3.5

MICROANALYTICAL RESULTS

Inhomogeneities

The results of our measurements give clear evidence of heterogeneous distribution of the sintering additives Al resp. B and C in both types of materials. Considerable amounts of the additives and the impurity elements are concentrated within differently sized inclusions which were found in all materials. Many foreign phases and inclusions with very different chemical compositions have been analysed by WDX and AES as well, both methods verifying each other. It was impossible to detect Si and/or C in some inclusions, where extremely high concentrations of O have been measured.

A general characteristic of SSiC seems to be the existence of particles of free C. The dark inclusions found in all materials were investigated by HRAES-point analyses to consist of excess or residual pure graphite, as shown for (B,C)-doped SSiC in Fig. 1.

Excess C is considered to be a remainder of the production process, which involves the usage of C-additions. Very characteristic differences were revealed with respect to accompanying elements of these particles. A demonstration with (B,C)-doped SSiC is given in Fig. 2, showing also a secondary electron image and an appropriate distribution map of N. The coincidence of the dark inclusions with the map is conclusive. The only detectable elements in the spectrum are Si (from the bulk material) C, B and N beside a slight contamination of O, originating from residual molecules of the UHV-system. Quantitative measurements gave evidence for stoichiometric BN being present. The atomic concentrations were calculated to be 4.9 at-% B and 5.3 at-% N. Thus, the material reveals BN inside the gra-

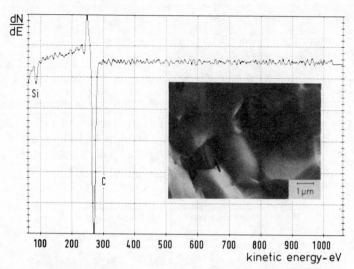

Fig. 1   HRAES-point-analysis of (B,C)-doped SSiC, indicating pure graphite

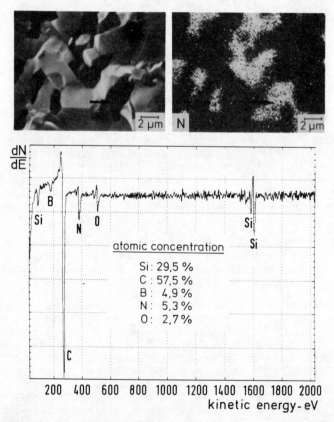

atomic concentration

Si: 29,5 %
C : 57,5 %
B : 4,9 %
N : 5,3 %
O : 2,7 %

Fig. 2   HRAES-point-analysis of (B,C)-doped SSiC, indicating the existence
of stoichiometric BN-compounds beside graphite

phitic inclusions. Generally, B-rich inclusions were also observed by WDX in the (B,C)-doped qualities.

Small particles of SiO_2 were found in the (B,C)-doped materials only by the application of both WDX and AES. This is demonstrated in Fig. 3 by absorbed electron imaging and the line profiles of Si and O.

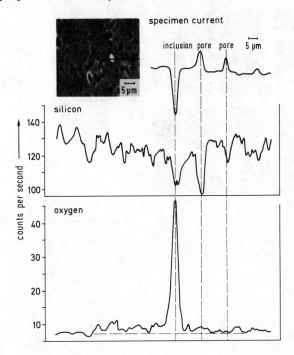

Fig. 3 WDX-line profiles of Si and O in (B,C)-doped SSiC

Other O-containing inclusions were found to exist as silicates in the systems Si-O-Al, Si-O-K, Si-O-Ca and Si-O-Na. An example is presented in Fig. 4 by Auger distribution maps and the spectra from two point analyses of a fracture surface of a (B,C)-doped sample. Beside the presumed existence of silicates the heterogeneous distribution of B is clearly revealed in this figure.

The (Al,C)-doped materials exhibit Al-rich areas of different sizes with a heterogeneous distribution. These areas are generally connected with enrichments of C and O. One very large area (2.0 x 0.5 mm^2) of this type was investigated in depth by bombarding with Ar-ions. During sputtering, (12 nm/h), the high O- and C-contents are very rapidly reduced, the Si-concentration increases, while the Al-content remains nearly unchanged up to a depth of 1.4 μm where the investigation was stopped. These effects are demonstrated in Fig. 5, also showing the contamination-induced O-profile, because this specimen was fractured in air a short time before the analysis.

The B-distribution of (B,C)-doped SSiC, evaluated by α-microautoradiography, generally differs strongly in the investigated materials, as shown in Fig. 6.

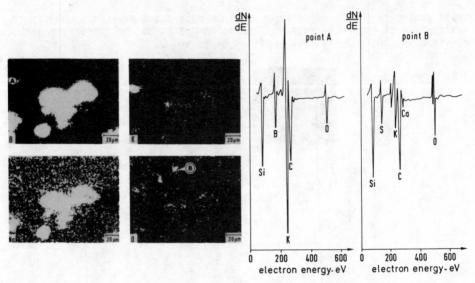

Fig. 4   Fracture surface of (B,C)-doped SSiC analysed by AES; distribution
maps and spectra at the point A and B

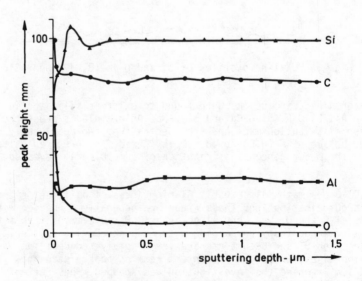

Fig. 5   AES-depth analysis of a selected Al-rich area in a
fracture surface of (Al,C)-doped SSiC

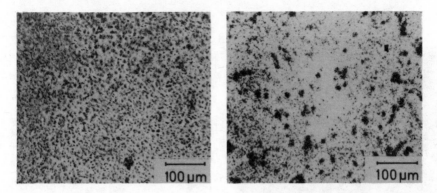

Fig. 6 Microautoradiograms of polished specimen surfaces

Grain Boundaries

The analysis of the grain boundaries was performed by the application of HRAES in combination with an UHV specimen fracture attachment. The lateral resolution offered here was as high as 50 nm in contrast to the value of the conventional AES being lower by a factor of ca. 100. Thus, the main advantage of using HRAES can be seen in its excellent resolution limit which enables illustration and analysis of single grains and grain boundaries, as demonstrated by Hamminger, Grathwohl and Thümmler (1983a, 1983b, 1983c).

Neither B nor impurity elements, including O, could be detected along the grain boundaries of different (B,C)-doped materials. These grain boundaries are absolutely free of any intergranular layer within the monolayer-sensitive depth resolution of the method. B could only be measured in the matrices of (B,C)-doped specimens and in enriched spots as mentioned above. Fig. 7 shows the typical secondary electron image of an intergranular fractured region of (B,C)-doped SSiC. The Auger spectrum of this figure was taken by a point analysis from the marked intergranular area. Besides the Si- and C-peaks there is no further element visible.

In the case of (Al,C)-doping, the grain boundaries exhibit extremely thin segregation films of Al and B. The chemical analyses gave evidence that these materials contain some 100 ppm B. Typical HRAES-results for two (Al,C)-doped materials are presented in Fig. 8, showing the secondary electron image with marked transgranular and intergranular fractured regions and the appropriate representative Auger spectra of point analyses in these regions. Sputtering by Ar-ions revealed the thickness of the Al- and B-enriched grain boundary layers to be less than 1 nm. Besides the spectra in Fig. 8 the grain boundary film is also represented by taking up the appropriate low energy Al-distribution map to the given secondary electron image. The transgranular and intergranular fractured regions can be clearly distinguished and the correspondence between the Al-map and the intergranular areas is possible in spite of the high level of noise. Traces of N were also detected at the grain boundaries of this material as AlN was used as intentional sintering additive.

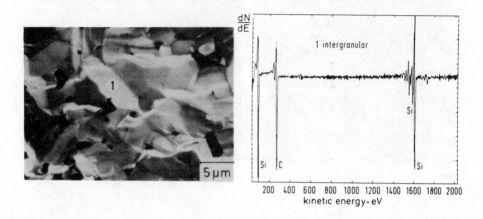

Fig. 7   Secondary electron image and HRAES-spectrum of an intergranular
fractured region of (B,C)-doped SSiC

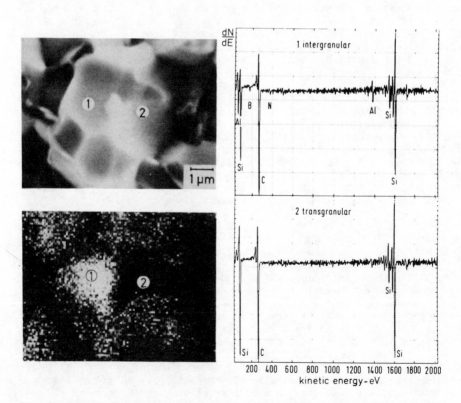

Fig. 8   Secondary electron image, appropriate Al-distribution map and
representative HRAES-spectra of (Al,C)-doped SSiC

RESULTS ON HIGH TEMPERATURE PROPERTIES

Oxidation

SSiC normally has a very good resistance against oxidation in ambient air. This is based on the formation of protective SiO_2-layers, being glassy or partly crystalline. The protection is limited at very high temperatures ($\sim 1700^{\circ}C$) by the low viscosity of SiO_2. Schlichting (1980) found the oxidation kinetics to be controlled by the penetration of O_2-molecules through the SiO_2-layer. The oxidation may be influenced by the partial pressure of O and the sintering additives used, which may be enriched in the protective layer. Fig. 9 summarises a few oxidation experiments, demonstrating principally very low values and a somewhat better behaviour of (Al,C)-doped SSiC than (B,C)-doped materials, measured at 1300ºC. After an oxidation treatment of 500h at 1300ºC and 500h at 1500ºC, these differences disappear. It was found, that (Al,C)-doped materials form somewhat thicker SiO_2-layers than (B,C)-doped SiC, the values being about 4-8 μm and 3-5 μm, resp. The concentration of Al in the SiO_2-layer is about 10 times higher than in the bulk material, as measured by microanalytical methods. Enrichments of B in the layer have not been observed presumably due to the high volatility of B-oxide.

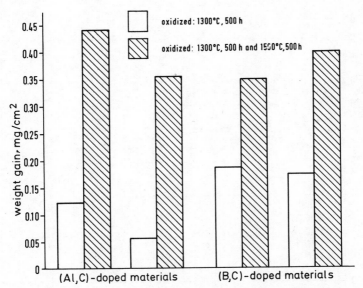

Fig. 9 Weight gain of differently doped SSiC materials after oxidation

Time to Failure

The strength degradation with time of SSiC at high temperatures has been investigated by several authors. A review of important experimental work has been given by Grathwohl and co-workers (1983) considering static as well as dynamic fatigue tests. There is experimental evidence, that SSiC exhibits a relatively high sensitivity to slow crack growth. Especially the influence of Al as sintering additive has been mentioned in this respect.

by Grellner, Schwetz and Lipp (1981). The state of the grain boundaries (e.g. their strength) should exert a decisive influence since slow crack growth normally occurs intergranularly.

The materials investigated in this work were ground and preflawed by Knoop indentation (50 N), heated 2h in air and the specimens of 3.5 x 4.5 x 45 mm were loaded in 4-point-bending with 20 mm inner span at 1.25 N/s. Flexural stress rupture time was measured at $1300^{\circ}C$ in air and vacuum (1×10^{-3} Pa); some specimens were also preoxidized at $1300^{\circ}C$, 500 h. Additionally, an injection molded material, (B,C)-doped SSiC VI was tested.

The measured stress rupture data are presented in Fig. 10, the stress given as stress ratio between the applied stress and the measured short time strength at $1300^{\circ}C$. All data points relate to specimens which failed from a growing crack introduced by the Knoop-indentation. The calculated range of the n-values for the five materials is from n = 64 to n = 23 (n = slow crack growth parameter, $v \sim K_I^n$, v = crack growth rate, K_I = stress intensity factor). In vacuum, SSiC showed a high resistance against slow crack growth. The room temperature strength of "vacuum-survivors" was considerably higher than the strength of the preflawed specimens and increased with increasing load sustained in the rupture test.

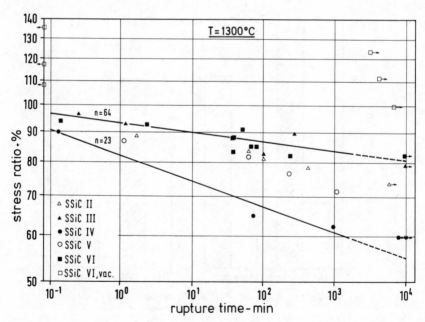

Fig. 10   Results of SSiC-rupture tests in air and in vacuum. Fitting lines
through extreme slope values of SSiC III (n = 64)
and SSiC IV (n = 23)

The SEM-micrographs in Fig. 11 illustrate the fracture surfaces of (Al,C)-doped SSiC V and (B,C)-doped SSiC VI. The semicircular Knoop-flaw in Fig. 11a is surrounded by a ring-like region of slow crack growth. A clear distinction between this region and the fracture surface of the catastrophic failure can be made for the (B,C)-doped material. As can be seen from Fig. 11b, the fracture mode in the region of slow crack growth is intergranular, while the fast fracture occurs mainly transgranularly (Fig. 11c). In contrast, the (Al,C)-doped SSiC does not allow identification of the transition between the slow and the fast crack growth regions, because the fracture mode is intergranular in both regions. This is illustrated with the fracture surface of SSiC V in Fig. 11d.

Fig. 11   SEM-micrographs of (B,C)-doped SSiC VI (a-c) and
(Al,C)-doped SSiC V

Creep

The creep behaviour was investigated using specimens of 45 x 4.5 x 3.5 mm in a four-point bending mode with 20 mm inner span. Isothermal tests as well as load changing and temperature changing experiments have been performed up to 1600°C.

SSiC is generally a very creep resistant material. After a primary range the creep curves in vacuum, e.g. at 1500-1600°C, exhibit a steady state (Grathwohl, Reetz and Thümmler, 1981) with a stress exponent n*∿ 1. However, there are considerable differences in the creep rate of SSiC with different dopings, as shown in Fig. 12. (B,C)-doped materials generally tend to show higher creep strength than (Al,C)-doped qualities. The creep strength increases with increasing density, possibly as a result of porosity effects as well as from the larger grain size of the high-density materials. The creep rate in vacuum seems to be higher than in air, but only a few experiments have been performed in air for comparison. This is consistent with the observation, that the creep rate at 1400°C in air can be reduced after long term preoxidation of the samples (1300°C, 500 h + 1500°C, 500h), which is especially true for (Al,C)-doped materials.

Beside the higher creep resistance a considerable reduction in the achiev-
able creep rupture strain to values below 10^{-3} was found for the preoxidized samples at a stress level of 190 MN/m².

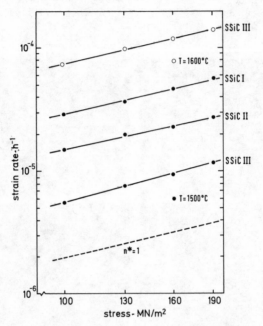

Fig. 12   Log strain rate - log stress curves
of SSiC in vacuum

The fact that the creep rate sensitivity in vacuum was described under all
used conditions by a linear stress function (Grathwohl, Reetz and Thümmler,
1981) leads to the assessment of a stress-induced diffusion process as the
deformation controlling mechanism. As a result of the temperature-change-
experiments it was found that the dependence of the creep rate on temperature is different for the individual materials. Under simplifying assumptions the conclusion can be drawn from the available diffusion data (Hong,
Hon and Davis, 1979; Hon, Davis and Newbury, 1980) and from the apparent
creep activation energies that in (Al,C)-doped SSiC a grain boundary diffusion process is rate determining, while in (B,C)-doped SSiC lattice diffusion becomes more important.

DISCUSSION

A detailed and fairly comprehensive knowledge about the microstructure of
sintered silicon carbide materials has been achieved by the combined use
of different microanalytical methods. The microstructural features reveal
a large number of inclusions and precipitations with high concentrations
of the additives and impurities in all investigated materials. Although
the exact chemical compositions of the very differently sized inclusions
have as yet not been determined in detail, the enrichment of Al ($\sim$ 30 %)
and B in some of them is obvious. These inhomogeneities lead to the ques-

tion whether the amounts of the sintering additives are chosen and admixed properly with respect to optimal effectivity. Although the microstructural inhomogeneity detected in the sintered samples may not be identical with the state in the green compact before sintering, the presence of inhomogeneities in the green compacts is obvious and often underestimated.

The analysis of the grain boundaries by HRAES offering a lateral resolution of 50 nm in combination with a UHV fracture device provided an excellent tool for investigating uncontaminated fracture surfaces including grain boundaries and single grains. The difference in the grain boundary state of the (B,C)- and (Al,C)-doped qualities was evident: The grain boundaries of the former are free of B or any other element in the sintered state. Clean grain boundaries in a covalently bonded ceramic may be very favourable with respect to high temperature long term properties, as shown in this investigation. The extremely thin intergranular films found in (Al,C)-doped materials can affect the measured properties. Further investigations have to be done in order to understand these phenomena including formation and stability of the grain boundary films.

Considering the (passive) oxidation behaviour in ambient air, no strong influence of the different dopings can be stated. The oxidation rates measured at 1300 and 1500°C are extremely small, so that the influences of chemical composition are irrelevant, at least in practice. On the other hand, the influence of long term oxidation, leading to reduced creep rates but at the same time also to reduced creep fracture strain is more evident. These effects could be a consequence either of the diffusion of the dopants or of the penetration of O along grain boundaries. The latter was indicated by HRAES in (AlN,C)-doped SSiC after heating in vacuum 2000°C, 30 min.

A similar statement is possible with respect to the strength degradation effects. An important fact in this respect is the different fracture mode of the two types of material: The (B,C)-doped SSiC shows intergranular fracture in the region of slow crack growth and mainly transgranular fracture during catastrophic failure, while (Al,C)-doped SSiC fractures intergranularly in both regions. This again is a consequence of the different grain boundary states in the two materials. Another point of interest is the very small susceptibility of (B,C)-doped SSiC to slow crack growth in vacuum, which is due to the high intrinsic grain boundary strength of this materials, i.e. without environmental influences during the experiment. It seems, that this property of "strong" grain boundaries will decrease somewhat in ambient air. These experiments, on the other hand, cannot give an unambiguous correlation between the stress-rupture behaviour and the kind of doping used, as long as the rupture events are not known in more detail. Thus, the question of weakening of the grain boundaries and, consequently, enhancement of slow crack growth by the Al-films cannot be definitively decided. It can be mentioned, however, that Al-segregation at grain boundaries increased considerably after long term high temperature treatment, as revealed by HRAES-investigations.

With respect to creep, it seems that the highest creep strength can be reached in the material with extremely clean grain boundaries, i.e. (B,C)-doped SSiC, in which a classic creep mechanism is active, namely a lattice diffusion process. Materials, whose grain boundaries are impure, i.e. (Al, C)-doped SSiC, behave in a more complex manner and show a somewhat lower creep strength. Although small differences in grain size, porosity and content of free C are neglected in this interpretation, the conclusion seems to be obvious.

The inhomogeneities and inclusions found in these materials can be expected to be detrimental also for properties other than those measured in this work, e.g. for strength. It is not surprising, that the level of strength of SSiC has reached only 400-450 MN/m^2 so far, although the potential of this material seems to be much higher. The difference in strength between SSiC and HIPSiC (600-650 MN/m^2) is not only a question of residual porosity but very probably also of chemical inhomogeneity.

Finally, these investigations demonstrate that the manufacturing process of SSiC is far from being fully optimized, hitherto. The processing should be improved, mainly in different steps of powder technology. These are e.g. purer starting powders, lower content of O or improved deoxidation, improved homogeneity, mainly with respect to the sintering additives, adjusting state and amount of additives, etc. The results of this work suggest that improvements in the properties of SSiC can be expected in the future as a result of improved homogeneity.

REFERENCES

Grathwohl, G., Reetz, T. and Thümmler, F. 1981, Sci. of Ceramics 11 425-431

Grathwohl, G., Hamminger, R., Iwanek, H. and Thümmler, F. 1983, Sci. of Ceramics 12 583-590

Grellner, W., Schwetz, K.A. and Lipp, A. 1981, Proc. Brit. Ceram. Soc. 31 27-36

Hamminger, R., Grathwohl, G. and Thümmler, F. 1983a, J. Mat. Sci. 18 353-364

Hamminger, R., Grathwohl, G. and Thümmler, F. 1983b, J. Mat. Sci. 18 3154-3160

Hamminger, R., Grathwohl, G. and Thümmler, F. 1983c, Sci. of Ceramics 12 299-305

Hong, J.D., Hon, M.H. and Davis, R.F. 1979, Ceramurgia Intern. 5 155-160

Hon, M.H., Davis, R.F. and Newbury, D.E. 1980, J. Mat. Sci. 15 2073-2080

Prochazka, S. 1974, Ceramics for High Performance Applications. Ed. J.J. Burke, Brook Hill, Chestnut, pp. 239-252

Schlichting, J. 1980, Energy and Ceramics. Ed. P. Vincenzini, Elsevier, Amsterdam, Oxford, New York, pp. 390-398

Inst. Phys. Conf. Ser. No. 75: Chapter 4
Paper presented at 2nd Int. Conf. Science Hard Mater., Rhodes

293

Quantitative electron energy loss spectroscopy of non-stoichiometric titanium carbide

C ALLISON(1) AND W S WILLIAMS(2)

Department of Physics and Materials Research Laboratory
University of Illinois at Urbana-Champaign, Urbana, Illinois 61801

(1) Present address: Department of Physics, Oklahoma State University
Stillwater, OK 74074
(2) Also with the Department of Ceramic Engineering

ABSTRACT

We report an analysis of carbon/metal ratio, x, at three points within an individual 1 μm grain of TiC_x in a Ni-Mo bonded composite. The value of x in the center of the grain was 0.81, considerably below the practical limit of 0.97. Values of x in the TiC_x grain nearer the grain boundary indicated the presence of Mo (but no Ni). The micro-analytical technique employed was electron energy loss spectroscopy (EELS) with a dedicated scanning transmission microscope (STEM).   Experimental cross sections for inner shell ionization were obtained earlier from evaporated elemental films and tested successfully on foils of ordered vanadium carbide for which the carbon/metal ratio was known.

INTRODUCTION

The mechanical properties of the refractory compound titanium carbide, TiC_x, which is non-stoichiometric, depend strongly on the value of the carbon/metal ratio, x. For example, Williams showed that the micro-hardness at room temperature and the critical resolved shear stress at 1200K both increase approximately linearly for TiC_x with increasing value of x (Figs. 1-3) (Williams, 1967; Williams, 1964). The parameter x has an astonishing range of values, from 0.5 to 0.97, and the corresponding variations in microhardness and yield stress are substantial. The micro-hardness increases 40% as x ranges from 0.8 to 0.97, and the critical resolved shear stress at high temperatures for the same single crystal specimens increases by approximately the same amount over the same range of x.

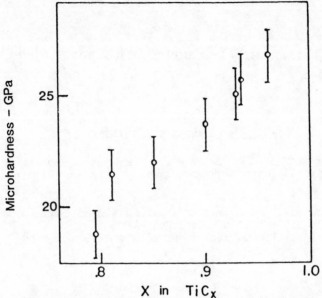

Fig. 1 Dependence of room-temperature microhardness of TiC$_x$ single
crystals on carbon/metal ratio, X. (From Williams, 1967).

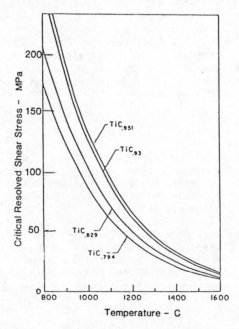

Fig. 2 Temperature dependence of the critical resolved shear stress
for TiC$_x$ single crystals at high temperatures. Curves for various
values of carbon/metal ratio, x, are similar but displaced,
indicating dependence of yield stress on x.   (From Williams, 1964).

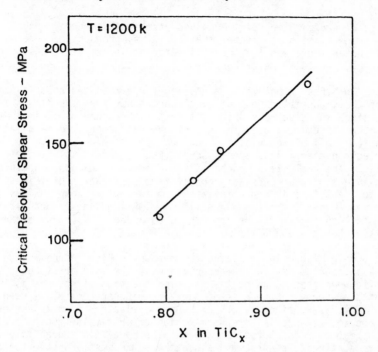

Fig. 3 Dependence of critical resolved shear stress of TiC_x single
crystals on carbon/metal ratio, x, presented explicitly from data
in Fig. 2. Yield stress increases with increasing value of x.
(From Williams, 1964).

Thus both for fundamental studies of the relation between the chemical
composition and mechanical properties of titanium carbide, which is noted
for its hardness at room temperature, and for developing practical
strategies to improve its resistance to plastic deformation failure at
high temperatures, where it finds practical application in cutting tools,
methods for measuring its carbon/metal ratio are essential. If sample
volume is not limited, chemical analysis is satisfactory, or existing
curves relating lattice constant to x can be used (Storms, 1967; Toth,
1971). However, if the microstructure has a submicron scale, and if
variations are expected in the value of x within the microstructure, a
more localized method of analysis (e.g., an electron beam technique) is
required.

The problem of electron beam microanalysis in this system is complicated,
however, by the presence of a light element, carbon, as a major constituent. Carbon exists as a potential contaminant in experimental systems,
and large spurious signals for carbon can be obtained. Even if the
specimen can be protected from contamination, carbon cannot be analyzed by
its characteristic X-radiation in most of those scanning-transmission
electron microscopes that have sufficiently small electron beam focal
spots, for three reasons: (1) the typical energy-dispersive X-ray
detector has a window which absorbs soft X-radiation from carbon; (2) the

probability of deexcitation of the carbon atom in this energy range by
Auger electron emission is greater than for X-ray production, leading to
low X-ray photon flux; and (3) the SiLi detector is not very sensitive in
the low energy region where carbon X-radiation occurs.

However, all of these problems in analyzing for carbon can be by-passed by
the use of the transmitted beam of electrons rather than the X-rays it
produces, since the electrons which interact with carbon atoms in the foil
lose an amount of energy equal to that of the characteristic X-radiation,
and essentially all of the electrons can be collected and counted. This
method, electron energy loss spectroscopy (EELS), should in principle be
applicable to the analysis of carbon in thin foils of carbon-containing
compounds such as TiC_x. Indeed, a qualitative application of this
approach to identify TiC in precipitate form in another compound has
been reported (Mochel, Allison and Williams, 1981) and a semiquantitative
study of stacking faults in the isomorphic compound, TaC_x, was carried
out (Allison, Hoffman and Williams, 1982).

The other problem faced in analyzing for carbon in TiC_x in sub-micron
structures is spatial resolution. Fortunately, the most outstanding
feature of a dedicated scanning transmission electron microscope is the
small spot diameter. In the case of the Vacuum Generators HB-5, used
here, the spot size claimed is 0.5 nm. As the electron beam traverses
the foil specimen, there is some degradation of this value due to scattering of the electrons, but since the foil is made thin enough to ensure
only single scattering events, and since these are mostly in the forward
direction, the effective sampling diameter is still only a few nanometers.
By contrast, the effective sampling diameter for X-ray generation in
transmission through a thin foil can be many times larger because of
fluorescence in the foil, and the extraordinary spatial resolution
of the probe is not realized.

In the present work, EELS has been used to study the microchemistry of
TiC_x in TiC/Ni-Mo cemented carbides--systems in which variations of x over
submicron dimensions might be expected from diffusion. This project
utilized the quantification of energy loss spectroscopy for Ti and C
developed earlier from experimental inner-shell ionization cross sections
obtained from evaporated elemental films and tested in an EELS study of
ordered vanadium carbide (Allison, Hoffman and Williams, 1984).

EXPERIMENTAL PROCEDURE AND RESULTS

Specimens of TiC/Ni-Mo, that is, titanium carbide grains liquid-phase sintered with a nickel-molybdenum binder, were obtained from D. Moskowitz of
the Ford Motor Company. The carbide grains were 1-2 μm in diameter.
Disks of these cemented carbides 3mm in diameter were made with a diamond
coring tool, polished mechanically, and thinned in a Gatan ion thinner for
analysis in the HB-5 scanning transmission electron microscope. Microscope operating conditions were: an accelerating voltage of 100 kV; a
medium objective aperture and large collector aperture, giving a collection angle β of 2.49 m rad; and a slit width, which defined the energy
resolution, of 16 eV. The slit width was chosen to correspond with the

conditions used earlier to obtain experimental inner shell ionization cross sections for Ti and C. Energy resolution was sacrificed for signal intensity, as in this exploratory study, maximum spatial resolution was desired. Data were taken in the spot mode, which focuses the electron beam to 0.5 nm.

The carbon contamination problem was controlled by lightly baking the specimen and halting the data collection when the carbon edge showed a change in size from one sweep of the energy range to the next. The stability of the carbon edge over the sweeps that were stored gives us confidence that contamination had little effect on the data taken.

Electron energy loss spectra were taken from several TiC grains at various distances from the grain boundary. Three representative spectra are shown in Figs. 4, 5, and 6. The spectrum in Fig. 4 was taken 0.5 μm from the edge of a grain; i.e., in the center; that in Fig. 5 was taken 0.1 μm from the edge; and that in Fig. 6 was taken about 0.2 μm from the edge.

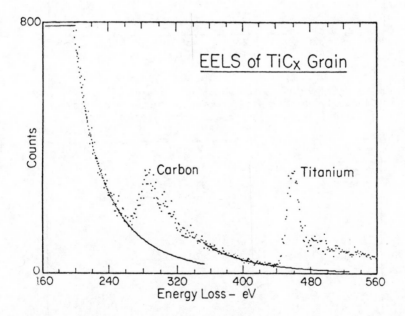

Fig. 4   Electron energy loss spectrum of TiC_x grain in TiC/Ni-Mo cemented carbide 0.5 μm from grain edge, i.e., in center of grain. Solid lines are background extrapolations.

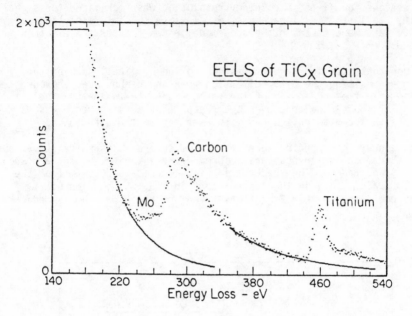

Fig. 5 Electron energy loss spectrum of TiC_x grain 0.1 μm from grain
edge showing molybdenum M edge as well as carbon and titanium.
Solid lines are background extrapolations.

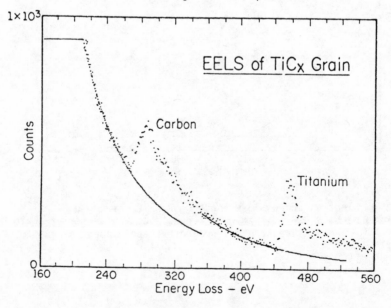

Fig. 6 Electron energy loss spectrum of TiC_x grain 0.2 μm from grain
edge. Solid lines are background extrapolations.

The carbon-to-titanium ratio for these spectra was calculated using the areas under the Ti and C peaks and energy "windows," which limit these areas, of 38 eV, 58 eV, and 78 eV. Experimental cross-sections for inner shell ionization, needed to reduce the raw data, were taken from the companion study of elemental films by EELS reported by Allison, Hoffman and Williams. Background subtraction using a function of the form E^{-r} was employed. The results for the spectra nearest the center of the grain (Fig. 4) are shown in Table 1.

TABLE 1   <u>EELS of TiC_x Grain</u>

ΔE	σ_C(barns)	σ_{Ti}(barns)	$A_C(\Delta E)$	$A_{Ti}(\Delta E)$	C/Ti
38 eV	384	292	7490	7300	0.78
58 eV	481	363	9410	8790	0.81
78 eV	572	434	10980	10980	0.82

DISCUSSION

Taking the average of the C/Ti ratio for energy windows 58 eV and 78 eV gives $x = 0.81$, a reasonable result for the carbon-to-metal ratio in TiC_x, where x can vary from 0.50 to 01.97. The uncertainty is estimated to be ±0.02.

The spectrum in Fig. 5 includes a feature on the low energy loss side of the carbon edge. The shape and location of this feature are consistent with the known presence of molybdenum in this specimen. Molybdenum has an M_{45} edge at 230 eV, which is at the observed position, and calculations indicate that the intensity should start very low and increase the energy loss (Leapman, Rez and Mayers, 1978). This shape can appear as a change in background slope, as observed here. The presence of Mo was confirmed by energy-dispersive X-ray analysis in the HB-5, but the location of the Mo could not be determined with with this method because of fluorescence in the foil which degrades the spatial resolution.

If we analyze with EELS a grain of titanium carbide at a point midway between those of the two previous examples (0.5 μm and 0.1 μm from the edge of the grain) we would expect to find an intermediate amount of molybdenum and titanium. Fig. 6 shows a spectrum taken 0.2 μm from a TiC_x grain edge. There is no obvious sign of molybdenum M edge in this spectrum and there is relatively less carbon than in Fig. 5. The carbon-to-metal ratio has been found using the experimental cross sections and gives the (non-physical) value of 1.11. However, if we assume that 19% of the carbon lattice sites remain vacant, as was found near the center of a grain, it can be calculated that an <u>apparent</u> C/Ti ratio of 1.11 implies that 10% of the titanium atoms have been replaced by molybdenum. Although a molybdenum M edge was not visible in this spectrum, we found the background fit to be noticeably worse than the fit obtained for Fig. 5. An amount of molybdenum below the limit of detectability may be responsible for this bad fit.

In previous work we saw that thickness had a large effect on the carbon-to-vanadium ratio. Because of the risk of accumulating carbon contamination, we were not able to take spectra of both the low and higher loss regions to determine whether the samples exceeded the thickness requirements imposed by multiple scattering. However, we later took zero-loss spectra of nearby locations in order to estimate the thickness of the sample where quantitative analysis was done. We found no case in which first plasmon to zero-loss ratio exceeded the limit of 0.2 established by Zaluzec (Zaluzec, 1980); in fact, this ratio was always less than 0.1. Also, we may note that the spectrum in Fig. 4 was from the thinnest region (as determined by the number of counts per second collected at the carbon edge) as well as the lowest in carbon content. Samples that are too thick should underestimate the carbon-to-titanium ratios, yet our thicker samples show an increase in carbon content. Thus it seems clear that we are seeing a decrease in titanium near the grain boundaries and not a spurious fluctuation due to changing thickness.

This decrease in Ti is not accompanied by an increase in Ni, however--only Mo. The absence of nickel in the TiC_x grain despite its presence in the binder phase is explained by its inability to form a solid solution with titanium carbide.

The absence of molybdenum in the center of the grain shows that the grains are not completely dissolved in the sintering process, and that diffusion of molybdenum in solid TiC is too slow at the sintering temperature to allow penetration beyond the liquid/solid interface.

The relatively low value of x, 0.81, in the center of a TiC_x grain, has implications for mechanical behavior: adding carbon to the system during processing could raise x to the limit observed for this material, 0.97, and thereby increase its resistance to plastic deformation. The benefit to a high temperature structural material or a cutting tool containing TiC_x grains and operating at high temperatures might be significant, as the yield stress of TiC/Ni-Mo falls to low values at the range of temperatures produced by frictional heating during use in metal machining (Wolfe, 1982).

SUMMARY

We have shown that the critical parameter, x, the carbon/metal ratio in TiC_x, which controls its mechanical properties, can be measured at different points in individual micrometer-size grains of titanium carbide using EELS. We have applied the technique to a sample of unknown stoichiometry -- titanium carbide cemented with a molybdenum-nickel binder -- and we have found a C/Ti ratio of 0.81 in the center of a grain. The proportion of titanium decreases as one approaches the grain boundary, and we attribute this change to the replacement of titanium by molybdenum during the sintering process.

ACKNOWLEDGEMENTS

This work was supported by the Department of Energy, Division of Materials Sciences, under Contract DE-ACO2-76ERO1198 through the University of Illinois Materials Research Laboratory. The STEM/EELS work was performed in the Center for Microanalysis of that laboratory. The assistance and cooperation of P. Mochel and M. Hoffman are gratefully acknowledged.

REFERENCES

Allison, C., Hoffman, M., and Williams, W.S. 1982, J. Appl. Phys. <u>53</u>, 6757-6761.
Allison, C., Hoffman, M. and Williams, W.S. 1984, Ultramicroscopy, <u>13</u>, 253-264.
Hren, J.J., Goldstein, J.I., and Joy, D.C. 1979, Introduction to Analytical Electron Microscopy, Plenum, New York,
Leapman, R., Rez, P. and Mayers, D.F. 1978, Proc. of Int. Cong. on Electron Microscopy, Vol. I, 526-527.
Mochel, P., Allison, C., and Williams, W.S. 1981, J. Am. Ceram. Soc. <u>64</u>, 185-187.
Storms, E.K. 1967. The Refractory Carbides. Academic Press, New York.
Toth, L. 1971. Transition Metal Carbides. Academic Press, New York.
Williams, W.S. 1964. J. Appl. Phys. <u>35</u>, 1329-1338.
Williams, W.S. 1967. Proprietes Thermodynamiques Physiques et Structurales des Derives Semi-metalliques (Editions du Centre National de la Recherche Scientifique, Paris) 181-189.
Wolfe, C. 1982. Plastic Deformation of TiC/Ni-Mo Cemented Carbide and Model Binder Phase. M.S. Thesis, University of Illinois, Urbana, IL.
Zaluzec, N. 1980. Proc. Electron Microsc. Soc. Am., San Francisco, (38th meeting), p. 112.

Inst. Phys. Conf. Ser. No. 75: Chapter 4
Paper presented at 2nd Int. Conf. Science Hard Mater., Rhodes

303

The presence of cobalt at WC/WC interfaces

A HENJERED(1), M HELLSING(2), H-O ANDRÉN(2) AND H NORDÉN(2)

(1) Surahammars Bruks AB, Box 201, S-735 00 Surahammar, Sweden
(2) Department of Physics, Chalmers University of Technology,
S-412 96 Göteborg, Sweden

ABSTRACT

WC-WC boundaries in WC-Co type cemented carbides with 6-20 wt% Co were
analysed with STEM/EDS and atom-probe. All boundaries were found to
contain cobalt, localized to the boundary plane. The amount of cobalt
corresponds to about half a monolayer only, so the boundaries can be
described as grain boundaries with a cobalt segregation. In a material
containing 0.5% of chromium carbide, some of the cobalt in the boundaries
was replaced by chromium. The results support the skeleton model of WC-Co.

INTRODUCTION

WC-Co type cemented carbides generally exhibit a high hardness, a high
rupture strength, and a low ductility. In order to explain these
properties, it was suggested long ago that a continuous carbide "skeleton"
forms during liquid phase sintering of WC-Co with 10% or less cobalt
(Dawihl, 1940).

Chemical dissolution experiments performed by Dawihl and Hinnüber (1943)
and Sandford and Trent (1947) supported the skeleton model. However, the
skeleton idea was opposed by Gurland and Norton (1952), who from
electrolytic extraction, wettability, dilatometry, and infiltration
experiments concluded that <u>all</u> WC grains were surrounded by binder phase
after sintering.

A compromise between these two models was developed by Gurland and Norton
(1956) and Gurland (1963): At low cobalt contents, WC-WC junctions exist,
whereas at higher cobalt contents, thin cobalt layers separate the WC
grains. These layers were predicted to be very thin, 2-40 nm (Dawihl and
Frisch, 1962, 1964; Dawihl and Altmeyer, 1963). The supposedly very high
yield strength of these thin cobalt layers would then explain the high
rupture strength of WC-Co.

During the last three decades a large number of mechanical and
microstructural investigations of WC-Co have been performed, and the
results have often been interpreted as supporting either the skeleton or
the cobalt layer theory.

As modern methods of microanalysis gradually became available, they were used in the search for thin cobalt layers between WC grains, but it was not possible, either with transmission electron microscopy (Bernhard, 1956; Persson, 1968; Hara, Nishikawa and Nishimoto, 1970) or with electron microprobe analysis (Fischmeister 1963), to detect any cobalt between what appeared as WC grain boundaries.

Recently, however, the presence of cobalt between WC grains has been observed by several workers. Lea and Roebuck (1980), using Auger spectroscopy, found cobalt "of monolayer proportions" in WC-WC boundaries, whereas Sharma and co-workers (1980), who studied thin foils of WC-Co with STEM/EDS (scanning transmission electron microscopy/ energy dispersive X-ray spectrometry), found a 2 nm thick layer between the WC grains.

Friedrich (1983) also detected cobalt at WC grain boundaries using STEM/EDS and interpreted the observed concentration profiles in terms of a cobalt diffusion profile, some 50 nm wide. Using lattice imaging in the transmission electron microscope, Jayaram and Sinclair (1983) found several WC-WC boundaries which were contiguous at an atomic level but also some which contained a 1 nm layer of cobalt.

We have used STEM/EDS and also the microanalytical instrument that has the highest spatial resolution available today, the atom-probe, to study WC-WC boundaries in WC-Co type materials. Our results show that cobalt is present in WC-WC boundaries in the form of a submonolayer grain boundary segregation.

EXPERIMENTAL

Material

Four WC-Co type cemented carbides with cobalt contents between 6 and 20% were studied (Table 1). Material B also contained 2% of cubic carbides, and Material C contained a small amount of chromium carbide, Cr_3C_2. All four materials were produced by the normal liquid sintering process used commercially.

Needle shaped specimens were used both for atom-probe microanalysis and

TABLE 1 Cemented Carbide Materials Studied

Material	Composition wt%			
	Co	WC	Cr_3C_2	(Ta,Nb,Ti)C
A	6	94	-	-
B	6	92	-	2
C	10	89.5	0.5	-
D	20	80	-	-

STEM/EDS. Starting from thin rods, cut out from the cemented carbide material with a low speed diamond wafer saw, sharp needles were prepared by electropolishing in sulphuric acid (Henjered and Nordén, 1983). After inspection by transmission electron microscopy, the specimens were further electropolished in small steps until a WC-WC boundary appeared at the specimen tip (Fig. 1).

Atom-Probe Analysis

The atom-probe instrument used in this investigation has been described before (Andrén and Nordén, 1979). A specimen temperature of 90 K was used, and field-ion imaging was made with 30 mPa of neon. Analyses were performed in ultra high vacuum (a pressure less than 100 nPa) using an analysing aperture size (projected diameter) of between 1.5 and 4.4 nm and an electrical evaporation pulse amplitude of 15-25% of the standing voltage. Three specimens with a WC-WC boundary at the tip were analysed with the atom-probe.

STEM/EDS

For the STEM/EDS analysis a JEOL 200CX instrument equipped with a Link model 860 X-ray spectrometer was used, operated at 200 kV and using an incident electron probe diameter of 6 nm. The recorded spectra were treated with the RTS 2/FLS software system, utilizing standard spectra from pure elements.

As will be described below, cobalt was recorded in the STEM/EDS analyses of WC-WC boundaries. In order to quantify the amount of cobalt present, the incident beam was assumed to have a gaussian lateral intensity distribution, and all cobalt was assumed to be localized to the boundary plane. (The justification for this assumption is given below.) Further it was assumed that the intensity distribution of the beam, as it traverses the specimen, retains its gaussian nature but broadens according to the single scattering model (Goldstein and co-workers, 1977; Reed, 1982). Integrating the beam intensity distribution over the analysed volume then gave an estimate of what composition ought to be measured for a given specimen thickness and cobalt content in the boundary (Hellsing, 1985). Using this method, measured cobalt compositions could be converted to actual WC-WC boundary coverages.

RESULTS

Localization of Cobalt

The exact localization of cobalt in the WC-WC boundary region was determined with atom-probe microanalysis. A specimen of Material C, which contained an axial WC-WC boundary, is shown in Fig. 1. Atom-probe analysis of the boundary, using a projected aperture diameter of 1.5+0.5 nm (Fig. 2), gave 5.8 at% cobalt and 2.6 at% chromium, in addition to tungsten and carbon. (Material C contained chromium carbide.) After moving the aperture a distance of 2 nm in either of the two directions perpendicular to the boundary, only tungsten and carbon was recorded. Thus, cobalt (and chromium) is present in the boundary, and must be confined to a zone not wider than 2 nm.

An even higher spatial resolution was obtained when analysing a specimen of

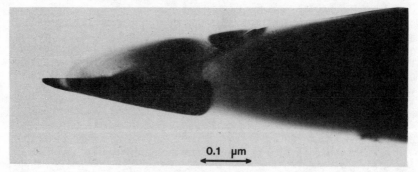

Fig. 1   Needle shaped specimen containing an axial WC-WC boundary.
Transmission electron micrograph.

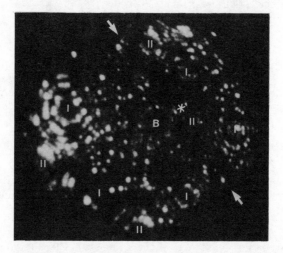

Fig. 2   Neon field-ion micrograph of the specimen in Fig. 1.
The 2 nm aperture (asterisk) has been moved just outside the boundary
(arrows). In the two WC grains, one basal plane (B) and several
pyramidal planes of type $(10\bar{1}1)$ (I) and $(11\bar{2}1)$ (II) are seen.

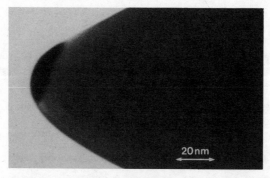

Fig. 3   A specimen containing a transverse WC-WC boundary close to
the tip apex. Transmission electron micrograph.

Material C which contained a transverse boundary (Fig. 3). Atom-probe analysis was performed through the boundary, at an angle of 65⁰ to the boundary plane, using an aperture of 2.6+0.3 nm. Cobalt (and chromium) were recorded over a distance of 1.0+0.3 nm. This distance should be compared to the distance over which the boundary plane was actually being analysed, namely 2.6 cot 65⁰ or 1.2+0.1 nm. Obviously, the cobalt and chromium must have been localized to a layer at the boundary approximately one atomic layer thick.

Amount of Cobalt

From the atom-probe data it is straightforward to determine the amount of cobalt (and chromium) present in the WC-WC boundaries, since the geometry of the analysis is accurately known. The results from the four analyses that were made with the atom-probe are given in Table 2. In all four cases, the cobalt (or the cobalt plus chromium in Material C) was present to an amount that corresponds to about half a monolayer coverage. (We define one monolayer as a 0.2 nm thick layer having the same density of atoms as h.c.p. cobalt.) In Material B (containing cubic carbides) no

TABLE 2 Atom-Probe Analysis of WC-WC Boundaries

Material	Recorded composition Co at%	Cr at%	Aperture size nm	Corrected composition Co (monolayers)	Cr (monolayers)
B	3.1+1.0	-	4.4+0.4	0.56+0.23	-
C	5.8+1.2	2.6+1.0	1.5+0.5	0.36+0.19	0.16+0.11
C	4.2+1.3	1.7+1.0	2.2+0.5	0.38+0.20	0.18+0.15
C	2.9+0.8	0.7+0.5	2.6+0.3	0.32+0.13	0.08+0.07

TABLE 3 STEM/EDS Analysis of WC-WC Boundaries

Material	Corrected composition Co (monolayers)	Cr (monolayers)
A	0.38+0.24	-
C	0.42+0.14	0.26+0.14
D	0.49+0.15	-

titanium or niobium was recorded in the WC-WC boundary, and due to the limited mass resolution it was not possible to decide whether any tantalum was present in the boundary.

Three of the cemented carbide materials were also studied with STEM/EDS. In each material two to four WC-WC boundaries were analysed. About ten analyses were made along each WC-WC boundary at positions of different specimen thickness. (It should be noted that the use of needle shaped specimens makes the determination of specimen thickness simple and accurate - it can be measured directly after having tilted the specimen 90° around its axis.) When the results of EDS analysis were corrected for the beam broadening, using the method described above, it turned out that all boundaries contained approximately half a monolayer of cobalt (cobalt plus chromium in Material C), see Table 3.

DISCUSSION

Cobalt

The presence of cobalt in the WC-WC boundaries is discussed first, and the presence of chromium afterwards. The small amount of cobalt found in the WC-WC boundaries, less than one atomic layer, and the fact that all cobalt was localized to a zone at the boundary, the thickness of which is of atomic dimensions, rule out the possibility that there could exist a separate layer of binder phase forming two phase boundaries with the adjoining WC grains. Instead, the WC-WC boundaries must be considered as grain boundaries, to which cobalt has segregated. Materials with different cobalt contents were studied, and grain boundaries with different relative orientations were analysed, but the result - using two independent methods - was the same: cobalt to an amount that corresponds to half a monolayer was present in all boundaries. Our results thus support the old skeleton model of WC-Co.

Chromium

The presence of so much chromium in the WC grain boundaries of Material C is surprising - about 1/3 of the cobalt segregation was replaced by chromium in these boundaries. It is known that chromium carbide has about the same high solubility in cobalt as tungsten carbide at the sintering temperature (Hollek and Kleykamp, 1982), and a STEM/EDS analysis of the binder phase of Material C gave a chromium content of 2.5+0.7 wt%. Thus there is ample supply of chromium in the binder. The fact that chromium replaces some of the cobalt in the grain boundaries suggests that chromium and cobalt are bound in a similar way in the boundary and compete for the same positions.

CONCLUSIONS

1. WC-WC boundaries in four WC-Co type cemented carbides containing between 6 and 20% cobalt were analysed with STEM/EDS and atom-probe.

2. Cobalt was detected in all WC-WC boundaries that were analysed.

3. In one of the materials, which contained 0.5 wt% Cr_3C_2, chromium was also detected in the boundaries.

4. All cobalt (and all chromium) was localized to a zone in the boundary, approximately one atomic layer thick.

5. The amount of cobalt (or cobalt plus chromium) found in the boundaries corresponds to approximately half a monolayer coverage.

6. Thus grain boundaries form between the WC grains in WC-Co cemented carbides, and cobalt (and chromium) is present in these grain boundaries in the form of a submonolayer segregation.

7. These results support the skeleton model of WC-Co.

ACKNOWLEDGEMENTS

This work was supported by AB Sandvik Hard Materials, Stockholm, and by the National Swedish Board for Technical Development (STU).

REFERENCES

Andrén, H-O. and Nordén, H. 1979, Scand. J. Metall. 8 147-152
Bernard, R. 1956, Proc. 2nd Plansee Seminar, ed. Benesovsky, F., Pergamon, London, pp. 41-9
Dawihl, W. 1940, Z. techn. Phys. 21 336-345
Dawihl, W. and Altmeyer, G. 1963, Z. Metallkunde 54 645-650
Dawihl, W. and Frisch, B. 1962, Arch. Eisenhüttenw. 33 61-66
Dawihl, W. and Frisch, B. 1964, Cobalt 22 22-30
Dawihl, W. and Hinnüber, J. 1943, Kolloid Z. 104 233-236
Exner, H.E. and Fischmeister, H. 1966, Arch. Eisenhüttenw. 37 417-426
Fischmeister, H. 1963, Jernkont. Ann. 147 200-217
Friedrich, K.M. 1983, Metal Sci. 17 456
Goldstein, J.I., Costley, J.L., Lorimer, G.W., and Reed, S.J.B. 1977, Scanning Electron Microscopy, ed. Johari, O., IIT Research Institute, Chicago, Ill., 1 pp. 315-324
Gurland, J. 1963, Jernkont. Ann. 147 4-21
Gurland, J. and Norton, J.T. 1952, J. Metals 4 1051-1056
Gurland, J. and Norton, J.T. 1956, Proc. 2nd Plansee Seminar, ed. Benesovsky, F., Pergamon, London, pp. 99-110
Hara, A., Nishikawa, T., and Nishimoto, T. 1970, Sumitomo Electric Techn. Rev. 14 106-109
Hellsing, M. 1985, Metall. Trans. 16A 686-689
Henjered, A. and Nordén, H. 1983, J. Phys. E 16 617-619
Hollek, H. and Kleykamp, H. 1982, Refractory and Hard Materials, 1 112-116
Jayaram, V. and Sinclair, R. 1983, J. Am. Ceram. Soc. 66 C137-C139
Lea, C. and Roebuck, B. 1980, Metal Sci. 15 456-458
Persson, G. 1968, Nature 218 159-160
Reed, S.J.B. 1982, Ultramicroscopy 7 405-410
Sandford, E.J. and Trent, E.M. 1947, Spec. Rep. No. 38, Iron Steel Inst., London, pp. 84-91
Sharma, N.K., Ward, I.D., Fraser, H.L. and Williams, W.S. 1980, J. Am. Ceram. Soc. 63 194-196

Inst. Phys. Conf. Ser. No. 75: Chapter 4
Paper presented at 2nd Int. Conf. Science Hard Mater., Rhodes

311

Determination of the binder phase composition in WC-Co cemented carbides

M HELLSING AND H O ANDRÉN

Department of Physics, Chalmers University of Technology,
S-412 96 Göteborg, Sweden

ABSTRACT

The composition of the binder phase in a 75W -5C -20Co (wt %) cemented carbide was determined with two microanalytical methods of high spatial resolution, STEM/EDS and Atom Probe. The material was studied both as-sintered and after equlibrium treatment at 1250°C at fixed carbon activities, a_c =0.4 and a_c =1.0. The results indicate that a macroscopic decarburized zone exists near the surface, and that tungsten depleted zones approximately 50 nm wide are present around the WC grains in the as-sintered material. In the interior of the material, the binder phase composition corresponded to beta-WC equilibrium at about 1250°C.*

INTRODUCTION

The importance of the binder phase composition in determining the performance of cemented carbides is well documented. Cobalt binders, which dominate the market, principally dissolve tungsten and carbon even when mixed carbides such as (W,Ti)C or (W,Ti,Ta,Nb)C are present in the material. In the present study we have focussed on the Co-W-C system, and in particular, on the composition of the binder phase (beta phase) in WC-Co cemented carbides.

Information on the Co-W-C system is scarce compared to the wealth of literature avaliable on corresponding Fe based systems. Isothermal sections and basal projections of liquidus surfaces based on experimental determinations have been presented by several authors (Takeda, 1936; Rautala and Norton, 1952; Grüter, 1959; Pollock and Stadelmaier, 1970; Hoffmann and Mohs, 1974). However, there is substantial disagreement between the various results especially in the areas concerning the existence and/or composition of the carbides and the solubility of C + W in the beta phase.

A new approach to the problem was adopted by Uhrenius, Carlsson and Franzén (1976), who applied a thermodynamic model developed by Hillert and Staffansson (1970) to an experimental study of the Co-W-C system at liquidus temperatures. With this model, thermodynamic parameters, such as partial free energies and interaction energies are evaluated from both binary system data and experimental data on the ternary system. Consequently, fewer experimental determinations need be obtained. The parameters thus calculated for the three component system can be used to compute its equilibrium data e.g. phase eqilibria, tie lines and iso-activity lines. Modelling more complicated systems i.e. with four or more components, is substantially facilitated when the parameters for the ternary system are known.

Several thermodynamic studies of the Co-W-C system based on the above model have been published in recent years. Isothermal sections at the following temperatures have been presented:

    1260 $^{\circ}$C (Uhrenius, Carlsson and Franzén, 1976)
    1440 $^{\circ}$C (Johansson, 1977)
    1150 $^{\circ}$C (Johansson and Uhrenius, 1978)
    1000,1100 and 1200 $^{\circ}$C (Tüma and Ciznerova, 1981)
    1200,1275,1350 and 1425 $^{\circ}$C (Åkesson, 1982)

The reliability of such calculated equilibrium diagrams depends on the accuracy of the experimental determinations on which the thermodynamic parameter evaluations are based. Here, the problem associated with conventional experimental techniques is to determine the composition, including the carbon content, of one phase, e.g. the beta phase, when other phases are present, e.g. WC, graphite or eta phase. The equilibrium compositions are usually therefore estimated either by extrapolation from single phase regions or by interpolation between the composition in a single phase region close to a two or three phase equilibrium and the average composition of the material when precipitates of other phases can be detected.

On the other hand, even if an accurate Co-W-C equilibrium diagram is constructed it cannot be used directly to predict the final composition of sintered and slowly cooled WC-Co materials.

What is needed therefore, is a means of measuring the composition of the beta phase, including the carbon content, in the presence of other phases. Direct measurments of the binder phase composition and possible segregation in the as-sintered state would serve as a link in understanding what actually happens during the cooling process. This information would be of potential value in the further development of cemented carbide materials.

In our investigation we have used two high resolution techniques to approach the problems discussed above: Firstly, Atom Probe time-of-flight spectrometry which has a spatial resolution of less than 1 nm and equal detection efficiency for all elements, and secondly, Scanning Transmission Electron Microscopy / Energy Dispersive X-ray Spectrometry (STEM/EDS) which is a convenient and less time consuming technique. However, STEM/EDS could not be used to determine the carbon content.

The purpose of the present investigation is threefold:

    a) To show that Atom Probe can be used to make accurate analyses of the binder phase in WC-Co cemented carbides.

    b) To establish the positions of some phase equilibria in the cobalt rich corner of the Co-W-C system in order to test the validity of and refine existent equilibrium diagrams.

    c) To determine the composition of the binder phase in the as-sintered cemented carbide material.

EXPERIMENTAL

Two materials were investigated:

1. A homogenous, single phase alloy (5.2W-0.6C-94.2Co (wt %)).

2. A commercially manufactured WC-Co cemented carbide grade (75W-5C-20Co (wt %)).
Material 1 was produced from a pure homogenous Co-W alloy which was carburized at a fixed carbon activity at 1250°C. The composition of this material was established using X-ray fluorescence for the tungsten content and with a combustion technique to determine the carbon content.

Material 2 was vacuum sintered in bar form (0.5 X 0.4 X 2.0 cm) at 1360°C. Some of the bars were then studied in the as-sintered condition and some after equilibration. The bars intended for equilibrium studies were cut longitudinally in 0.5 mm thick plates, before heat treatment, to ensure instant quenching throughout the material. Heat treatments were made at 1250°C in equilibrium with graphite, i.e. at a_c =1.0, and at a_c =0.4, where a_c is the carbon activity. Further details on the heat treatment procedures are given by Åkesson (1982).

Electropolishing in 8% sulphuric acid in methanol was used to prepare needle shaped specimens with a tip radius in the order of 50 nm, which is required to undertake Atom Probe analysis. Only the outermost tip of the needle can be analyzed in the Atom Probe, a fact which presents a major difficulty for specimen preparation. In order to get a binder phase domain at the tip we used stepwise electropolishing together with Transmission Electron Microscopy (TEM), see Fig. 1. Another method which has been successfully applied is ion beam etching from behind the tip along the shank of the needle, similar to the way a pencil is sharpened with a knife.

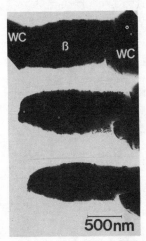

Fig. 1 The effect of stepwise electropolishing of a cemented carbide specimen (Material 2). The WC grain to the left in the top micrograph (TEM) was removed by applying a 5 ms/15 V pulse. A second pulse sharpens the tip further (bottom).

The specimens were examined and analysed in a STEM (JEOL 200CX) equipped with an EDS system (Link Systems 860 Analyser) and in a combined Field Ion Microscope (FIM) - Atom Probe. The latter instrument is described in a paper by Andrén and Nordén (1979). The recorded EDS spectra were treated with the RTS 2/FLS software system, utilizing standard spectra from pure elements.

RESULTS

a. Homogenous Test Alloy (Material 1)

A homogenous beta phase alloy (Material 1) with a composition well within the single phase region at 1250 °C was analysed with both STEM/EDS and Atom Probe in order to demonstrate the reliability of the experimental techniques in determining the composition of the binder in WC-Co cemented carbide alloys. The results of these measurements are shown in Table 1 together with the results obtained using chemical analytical techniques for bulk material analysis. The Atom Probe results show very good agreement with the results of the bulk analysis in both the tungsten and carbon content. A typical Atom Probe mass spectrum from a binder phase analysis is shown in Fig. 2. A minor disagreement is however noted for the tungsten content as determined by EDS analysis of the test alloy. Wirmark and Nordén (1984) found that this deviation can be explained in terms of a somewhat too high value of the sensitivity factor for the W L-line with respect to the Co K-line.

b. Equilibrium Composition (Material 2)

The composition of the binder phase in the sintered WC-Co grade (Material 2) at 1250 °C was determined at the three phase equilibrium beta-WC-graphite (a_c =1) and at one point along the beta-WC equilibrium line (at a_c =0.4). The results are presented in Table 2 and plotted in the calculated isothermal section shown in Fig. 3.

c. The As Sintered State. (Material 2)

Three needle shaped specimens were prepared from a sintered bar (Material 2). The needles were cut out from two depths below the surface: one at 0.25 mm and two at 2 mm. The Atom Probe results are shown in Table 3. While the tungsten content did not vary significantly with depth, a notable

TABLE 1   Analysis of a Homogenous Beta Phase Test Alloy
(Material 1)

	W wt.%	C wt.%
Bulk analysis	5.20	0.60
Atom Probe	5.2 $\pm$ .3	0.57 $\pm$ .05
EDS	5.6 $\pm$ 0.2	----

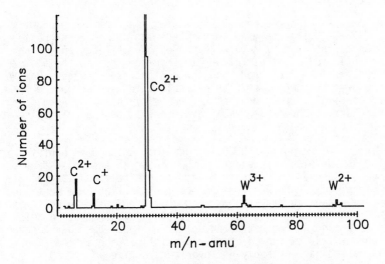

Fig. 2 A typical Atom Probe mass spectrum
obtained in a binder phase analysis.

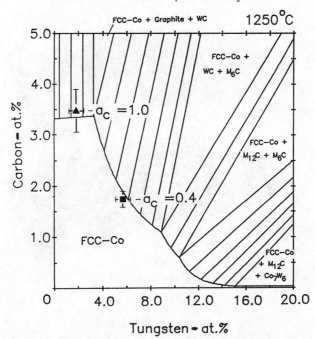

Fig. 3 The cobalt rich corner of a calculated isothermal section
of the Co-W-C system at 1250 °C (Uhrenius, 1984). The beta phase
composition at the three phase equilibrium beta-WC-graphite (triangle)
and at the two phase equilibrium at a_C =0.4 (square), as determined
in the present study, are also shown.

difference between these two depths lies in their carbon content - the
binder phase in the specimen taken at 0.25 mm from the surface (Fig. 4) was
depleted of carbon.

Within the specimen at a depth of 0.25 mm (Fig. 4) from the surface a
series of STEM/EDS point analyses were made in the same binder phase domain
at increasing distances from a beta-WC phase boundary (Fig. 5). The
boundary was oriented parallel to the electron beam during the analyses.
The contribution to the measured tungsten content from the adjacent WC
grain due to beam broadening effects was calculated and accounted for. A
model based on single Rutherford scattering as discussed by Reed (1982) was
applied to integrate the estimated X-ray intensity generated by electrons
scattered beyond the phase boundary.

These STEM/EDS measurements at 0.25 mm show that a tungsten depleted zone
approximately 50 nm wide is present in the binder at the phase boundary.
The presence of such a zone also at greater depth below the surface is
plausible since Atom Probe analysis of one specimen (at 2 mm depth and 20 -
30 nm from the phase boundary) shows a tungsten depletion in agreement with
the STEM/EDS measurements at 0.25 mm (Fig. 5). Furthermore, in this Atom
Probe measurement no carbon was detected. This indicates that a carbon
depleted zone similar to that of tungsten exists in the interior parts of
the material which have probably not been decarburized.

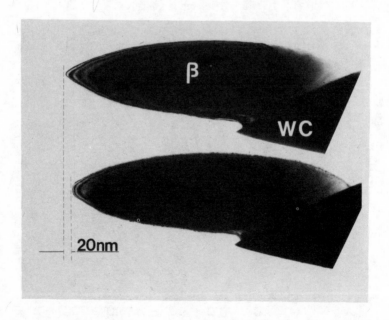

Fig. 4   Two TEM micrographs showing the tip of a specimen before (top)
and after (bottom) Atom Probe analysis. This specimen was prepared
from Material 2 in the as-sintered condition 0.25 mm from the surface.
The results of the Atom Probe analysis is shown in Table 3 and
together with STEM/EDS measurements in the same binder phase domain
in Fig. 5.

TABLE 2   The Equilibrium Composition of the Binder Phase in a 75W-5C-20Co Cemented Carbide Grade at 1250°C. (Material 2)

	W wt.%	C wt.%
i. In equilibrium with WC and graphite ($a_c = 1$)		
Atom probe	5.7 ± .7	0.70 ± .12
EDS	5.9 ± .4	----
(Hoffmann and Mohs (1974)	6.0	0.95 experimental)
(Uhrenius (1984)	9.9	0.66 calculated,
		see Fig 3.)
(Åkesson (1982)	7.3	0.70 calculated)
ii. in equilibrium with WC at $a_c = 0.4$.		
Atom Probe	16.3 ± .8	0.33 ± .03
EDS	16.4 ± 1.2	----

TABLE 3   The As-Sintered Composition of the Binder Phase in a 75W-5C-20Co Cemented Carbide Grade. (Material 2)

	W wt.%	C wt.%
i. 0.25 mm from the surface of the material.		
> 100 nm from phase boundary:		
Atom Probe	8.8 ± 1.5	0.03 ± .02
ii. 2 mm from the surface of the material.		
> 100 nm from phase boundary:		
Atom Probe	8.0 ± 1.3	0.46 ± .07 [*]
20 - 30 nm from phase boundary:		
Atom Probe	3.1 ± 0.8	<0.03

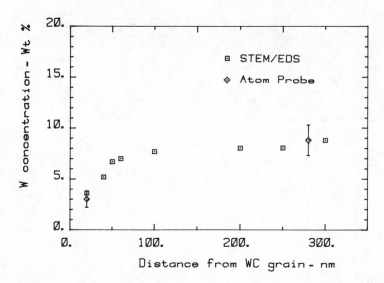

Fig. 5   The tungsten content in the binder phase domain shown in Fig. 4
(Material 2, as-sintered, 0.25 mm from surface) as a function of
distance from the WC grain. The Atom Probe analysis close to the
WC/binder phase interface was however made in another specimen
(2 mm from the surface), see Table 3.

DISCUSSION

It has been shown above that the Atom Probe can be used to make correct
determinations of both the tungsten and the carbon content of the binder in
fine grained cemented carbides. One could therefore predict that Atom
Probe microanalysis will become a standard tool in constructing phase
diagrams in the future. It is important however, to realize that the
technique is relatively complicated and time-consuming and should therefore
be used only to solve problems which cannot be solved satisfactorily with
conventional techniques.

Equilibrium Composition

i. Beta - WC - graphite - equilibrium (a_C =1)

At this three phase equilibrium point our measured tungsten content is in
agreement with the corresponding point in the isothermal section presented
by Hoffmann and Mohs (1974).These authors performed an extensive study of
the cobalt rich corner of the Co-W-C system at 1250 $^{\circ}$C using optical
microscopy and X-ray diffraction. However, Atom Probe analysis at this
point indicates that the carbon content in their work has been
overestimated. The equilibrium composition calculated by Uhrenius (1984)
(Fig. 3), agrees with the present study in carbon content, but shows higher
tungsten levels.

We believe that the present Atom Probe determination of the composition of
the beta phase in equilibrium with WC and graphite at 1250 $^{\circ}$C is more
accurate than those previously published as none of these result from
direct measurements at the three phase equilibrium.

ii. Beta - WC equilibrium at a_C =0.4.

The measured composition at this lower carbon activity is in good agreement
with both the calculated (Fig. 3) and the experimentally determined
(Hoffmann and Mohs, 1974) solubility line.

As-Sintered Composition

The results concerning the composition of the binder in the as-sintered
material is perhaps the most interesting part of the present study as, to
our knowledge, no direct measurements of the tungsten and carbon contents
of the binder in WC-Co cemented carbides have previously been published.

The tungsten content of the binder in the as-sintered condition seems to be
fairly constant througout the material, except for a thin depleted zone
near the WC/Co phase boundary. On a more macroscopic scale, a decarburized
zone exists near the surface of the material. The extent of this zone has
not yet been quantified, further investigation is needed.*

A possible explanation of the observed composition of the binder in the
as-sintered condition will now be discussed. During sintering it is
preferable to reach WC-liquid equilibrium without preciptitation of lower
carbides or graphite. In order to avoid precipitation of graphite or eta
phase the material has to contain approximately equimolar quantities of
tungsten and carbon. The measured composition of the binder at 2 mm from
the surface expressed in molar fractions is 0.024 C and 0.026 W (compare
Table 3). These contents are equal to within the error limits, and they
correspond fairly well to the beta-WC equilibrium composition at 1250°C,
which is about 30°C below the temperature at which the material is
completely solidified. Thus, equilibrium conditions seem to be retained in
the material down to this temperature. The decarburization of the surface
probably occurs below this temperature as a result of a decreasing carbon
activity of the furnace atmosphere. Since the mobility of tungsten in the
binder is much lower than that of carbon, the tungsten content does not
change correspondingly, except in a narrow zone around the WC grains.

Exner (1979) suggested the existence of an equilibrium constant [W]·[C] of
about 8 ·10^{-4}, based on the data of Hoffmann and Mohs (1974) at 1250 °C.
W and C are the atomic fractions of tungsten and carbon in the beta
phase. The present data gives the following values of the solubility
product:

$$[W] \cdot [C]$$

Equilibrium at 1250 °C

upper carbon limit (a_C =1)   6.4· 10^{-4}
a_C =0.4   10.3· 10^{-4}

As-sintered   6.2· 10^{-4}

Thus, the equilibrium constant concept is also approximately valid in the
as-sintered condition.

CONCLUSIONS

1. Atom Probe analysis can be used to make accurate determinations of the binder phase composition in fine grained cemented carbides.

2. The composition of the binder phase at the beta-WC-graphite equilibrium (a_c =1.0) and at the beta-WC equlibrium at a lower carbon activity (a_c =0.4) have been determined at 1250 C.

3. In the sintered and slowly cooled cemented carbide material a tungsten depleted zone approximately 50 nm wide is present in the binder at the beta-WC phase boundary.

4. The Atom Probe measurements indicate that a macroscopic decarburized zone exists near the surface of the as-sintered material.

5. The composition of the binder 2 mm from the surface corresponds approximately to a beta-WC equilibrium composition at 1250 $^{\circ}$C.*

ACKNOWLEDGEMENTS

Dr. L. Åkesson is greatfully acknowledged for performing the heat treatments and for numerous helpful discussions. We also wish to express our gratitude to Dr. B. Uhrenius for providing the calculated isothermal section and for sharing his profound knowledge with us. This work was supported by AB Sandvik Hard Materials, Stockholm, and by the National Swedish Board for Technical Development (STU).

REFERENCES

Andrén, H-O. and Nordén, H. 1979, Scand. J. Metall. 8 147-152
Exner, H.E. 1979, Int. Met. Rev. 4 149-173
Grüter, H. 1959, Thesis, Universität Münster.
Hillert, M. and Staffansson, L.I. 1970, Acta Chem. Scand. 24 3618-3626
Hoffmann, A. and Mohs, R. 1974, Metall. 28 661-666
Johansson, T. and Uhrenius, B. 1978, Met. Sci. 12 83-94
Johansson, T. 1977, Thesis, University of Uppsala
Pollock, C.B. and Stadelmaier, H.H. 1970, Met. Trans. 1 767-770
Rautala, P. and Norton, J.T. 1952, Trans. AIME 194 1045-1050
Reed, S.J.B. 1982, Ultramicroscopy 7 405-410
Takeda, S. 1936, Sci. Rep. Research Inst. Tohoku Univ. (1st ser.) 864-881
Tüma, H. and Ciznerová, M. 1981, Kov. Mater. 19 389-401
Uhrenius, B., Carlsson, B. and Franzén, T. 1976, Scand. J. Metall. 5 49-56
Uhrenius, B. 1984 Calphad XIII (Conference), Villard de Lans, May 14-18
Wirmark, G. and Nordén, H. 1984, to appear in proceedings of Analytical Electron Microscopy Workshop, Lehigh Univ.
Åkesson, L. 1982, Thesis, Kungl. Tekn. Högsk., Stockholm.

*
Note added in proof:

A more thorough investigation of the composition of the as-sintered material showed that the carbon content of the binder phase is lower than 0.02 wt % throughout the material.

Inst. Phys. Conf. Ser. No. 75: Chapter 4
Paper presented at 2nd Int. Conf. Science Hard Mater., Rhodes

Analytical electron microscopy of phase separated Ti/Mo cemented carbides and carbonitrides

A H HEUER(1), J S SEARS(1) and N J ZALUZEC(2)

(1)Case Western Reserve University, Department of Metallurgy and
Materials Science
Cleveland, Ohio, USA
(2)Argonne National Laboratory, Materials Science Division
Argonne, Illinois, USA

ABSTRACT

Analytical electron microscopy has been used to characterize the
microchemistry and microstructures of two commercial Ti/Mo cemented
carbonitrides, a commercial Ti/Mo cemented carbide, and several
experimental carbonitrides. All samples contained a core/rim structure,
the Mo being very much enriched in the rim relative to the core, and a
continuous Ni or Ni/Co binder phase. Substantial N was found in the rims
of the carbonitrides, although not as much as in the cores. The
commercial carbonitride alloys were not in chemical equilibrium, and our
results suggest that revision in the currently accepted Ti-Mo-C-N and
Ti-Mo-C phase diagrams may be necessary.

INTRODUCTION

Alloys within the Ti-Mo-C-N system represent an important class of Ni- (or
Ni/Co) bonded cemented carbonitride cutting tools (Rudy et al., 1974).
The carbonitride grains within such materials are two-phase, and
posess a characteristic core-rim structure (Rudy et al., 1974). The phase
separation within this quaternary system is attributed to the tendancy for
reciprocal salt systems to phase separate whenever the end members (in
this case TiC, "MoC", TiN, and MoN) have different free energies of
formation (Rudy, 1973). Mo-rich and Mo-poor phases, both with the NaCl
structure, form in the present system, and the preferential wetting of the
Mo-rich phase by the Ni (or Ni/Co) binder during liquid phase sintering
(Parikh and Humenik, 1957) causes the characteristic core-rim
microstructure. (Following earlier useage, the Mo-poor and Mo-rich phases
are referred to here as α' and α''.) We have characterized the
microstructure and microchemistry of two such commercial cutting tools
using analytical electron microscopy (Hren, Goldstein and Joy, 1979), and
compared them with a series of experimental alloys sintered for various

times, and also with a commercial N-free Ti/Mo Ni-bonded cemented carbide. The microstructures of all materials are very similar, although the microchemistry of the carbide is of course different from that of the carbonitrides; furthermore, the microchemistries of the experimental alloys vary with sintering time. In this paper, we describe the general features of the microstructures of the carbide and carbonitrides; the specific differences in the microchemistries, and their variation with sintering time, are deferred to a separate publication.

EXPERIMENTAL

A Phillips EM400T electron microscope equipped with an EDAX 9100-60 X-ray energy dispersive detector was used for the XEDS analysis (for Ti, Mo (and W and V when present), Ni and Co), and a Gatan 607 electron energy loss spectrometer was used for the EELS analysis (for C and N); conventionally prepared ion-thinned foils were employed for all analyses. The EELS analyses were conducted at 77K to minimize buildup of contamination layers, which otherwise would obscure the C analysis. The software provided with the EDAX 9100-60 system was used to reduce the intensity of characteristic X-ray peaks and the intensities of energy loss absorption edges to atom percent (metal analyses) or atom ratios (C and N analyses), as appropriate. In all cases, chemical analyses were preformed at the thinnest regions (< 20nm) of the electron microscope foils; this was necessary to prevent multiple scattering in the case of the EELS analyses and to prevent preferential absorption of the softer Mo L (and W M) X-rays relative to the Ti K X-rays. As the EELS data analysis routines are difficult and fairly involved, we discuss them at some length in the text.

RESULTS

Microstructural Analysis

Figs. 1 and 2 show general microstructures of one of the commercial cemented carbonitrides and the commercial carbide, respectively. The grain size for both materials is approximately 1 micrometer, and the core-rim structure (α' and α'', respectively) is unmistakable. In conformity with previous work, and as will be discussed below, the α' core is almost devoid of Mo relative to the α'' rim. The lighter appearance of the α' cores arises from absorption contrast, as they have a lower mean atomic weight than the α'' Mo-rich rims. The binder phase is visible as discrete particles in Figs. 1 and 2, but the specific morphology of the binder phase is illustrated below.

Individual grains of the carbonitride and carbide phases are shown in Figs. 3 and 4, respectively. The dislocations and inclusions (which subsequent analysis showed to be Ni binder) in Fig. 3 are common in many grains, usually at the core-rim interface. We believe that binder inclusions were trapped during the early stages of sintering, when the core-rim structure was being established, and do not represent an equilibrium aspect of the microstructure. The difference between the

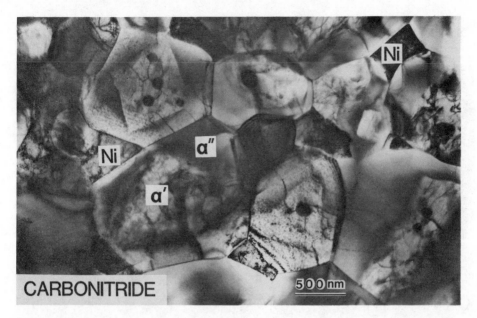

Fig.1 Microstructure of commercial cemented Ti/Mo carbonitride tool.
Bright field electron micrograph. α' and α" refer to core and rim,
respectively, and Ni to the binder.

Fig.2 Microstructure of cemmercial cemented Ti/Mo carbide tool. α', α"
and Ni have the same meaning as in Fig.1. Bright field electron
micrograph.

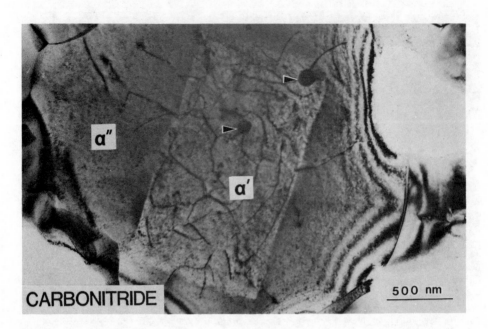

Fig.3   Core/rim microstructure of a single grain in a commercial cemented Ti/Mo carbonitride. A substantial dislocation content, in addition to entrapped binder particles (arrowed), are common in this material. Bright field electron micrograph.

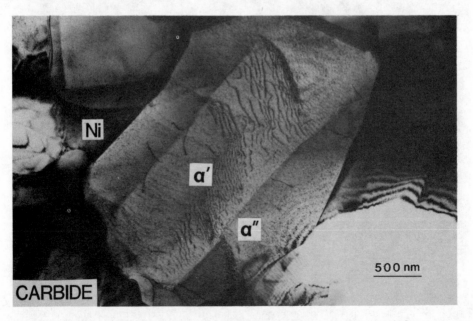

Fig.4   Core/rim microstructure of a single grain in a commercial cemented Ti/Mo carbide. Bright field electron micrograph.

commercial carbide and commercial carbonitride in this regard are not understood at this time.

The dislocations present in both materials may arise from the thermal expansion mismatch between the core and rim of the carbonitride (or carbide) phases, or they may be misfit dislocations present to accommodate lattice parameter differences between core and rim. However, any such lattice parameter difference must be very small, as no spot splitting or other diffraction evidence pointing to two discrete phases was found on selected area diffraction patterns taken from combined core/rim areas.

The binder phase is most conveniently imaged by making a dark field image with a unique Ni reflection, as shown in Figure 5. As seen in this figure, the binder appears to be single crystal over many carbonitride grain diameters, a result of the difficulty of nucleation during the solidification of Ni that follows liquid phase sintering. This is somewhat surprising, as it might be expected that carbonitride-liquid metal (or carbide-liquid metal) interfaces should provide good nucleation sites for crystallization. This presence of mm size binder grains in WC-Co alloys has been reported previously (Willbrand and Wieland, 1972), and Fig. 5 shows the same is probably true in alloys based on (Ti/Mo)(C/N) - Ni.

Dark field images of this sort are also useful for determining whether or not the binder phase is continuous and therefore completely wetting, i.e. if there are any $\alpha''-\alpha''$ (rim-rim) contacts; actually, a continuous binder phase might be expected because of the good wetting of Ni alloys in contact with the Mo-containing Ti carbonitride grains. The continuity question is most effectively demonstated by tilting a "grain boundary" so that it is accurately parallel to the electron beam (Fig. 6). When this is done, the boundaries are seen to have thicknesses of approximately 5 nm. (Our results do not preclude the existence of "special" $\alpha''-\alpha''$ boundaries of particularly low energy (less that between α'' and binder) but none were found.)

Microchemistry

Fig. 7 shows the XEDS spectra of a core and rim of a commercial carbonitride grain. There is almost complete segregation of the Mo to the rim; the Ti-content of this phase is ~85 at % Ti, while that of the core is ~98 at % Ti. Comparable data for the carbide and the second carbonitride are shown in Fig. 8. (These latter spectra have been scaled so that all curves have the same Ti intensity.) The carbide contains W in addition to Mo; the carbonitride also contains V.

Real variability in the Ti content of core and rim was found. For example, the cores contained between ~99 and ~95 at % Ti. Similar variability was found in the experimental alloys. No correlation between core and rim microstructure and the variable microchemistry was apparent. When W was present in the alloy, the Mo/W ratio was similar in the core and rim, but V segregated almost entirely to the rim (see Fig. 8). This chemical variability is under further investigation and will be reported on in a separate publication.

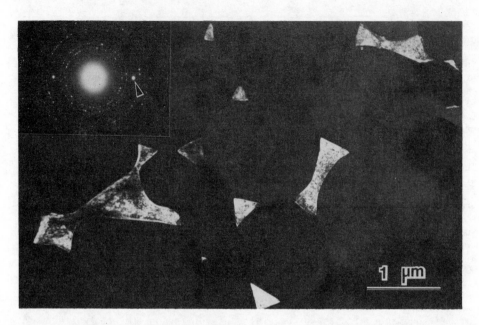

Fig. 5 Dark field electron micrograph showig Ni binder phase (bright regions) in Ni-bonded cemented carbonitride. The inset is a selected area diffraction pattern from a large area; a discrete Ni reflection is arrowed.

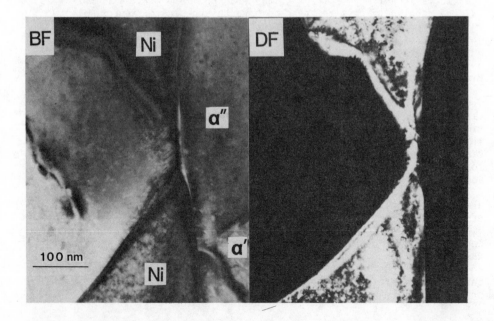

Fig. 6 Bright field (BF)/dark field (DF) pair of micrographs showing that the Ni binder phase is continuous, and some tens of nanometers wide, in the carbonitride of Fig. 5.

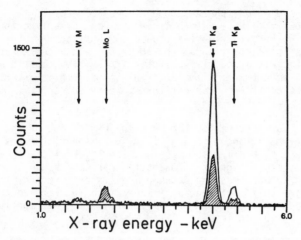

Fig. 7 EDS spectra of core (hatched) and rim (open) of a single grain in a commercial carbonitride. The energies of the W M, Mo L, and Ti Kα and Ti Kβ x-rays are indicated.

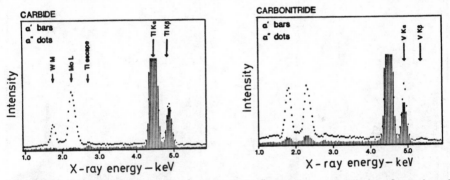

Fig. 8 EDS spectra of α' core and α" rim in an individual grain of a carbide and carbonitride. The carbide contains W as well as Mo; the carbonitride contains W and V as well as Mo.

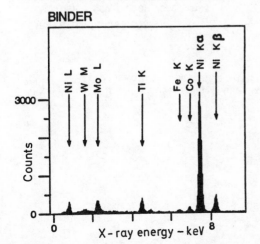

Fig. 9 EDS spectrum of binder phase in a commercial Ni-bonded carbonitride.

An XEDS spectrum of the binder phase in one of the commercial carbonitrides is shown in Fig. 9. Both Ti and Mo have dissolved to an appreciable extent in the binder during liquid phase sintering; the intensity of the Ni signal corresponds to ~60 at %. The binder phase in the other carbonitride was a roughly equimolar Ni/Co alloy, but it also contained comparable amounts of Ti and Mo. This finding of substantial solubility of Ti in the binder phase confirms an earlier report (Moskowitz and Humenik, 1966).

The electron energy loss spectra for the core and rim of one of the commercial carbonitrides is shown in Fig. 10. The absorption edges corresponding to the C K, N K and Ti L_{32} edges are labeled; the α' and α'' spectra are very similar. In order to quantify this data, the rapidly decaying pre-edge background must be extrapolated into the energy loss range of interest, and then stripped, as is described in standard texts on electron microscopy (Hren, Goldstein and Joy, 1979). The extrapolated background is shown by the solid black lines in Fig. 10.

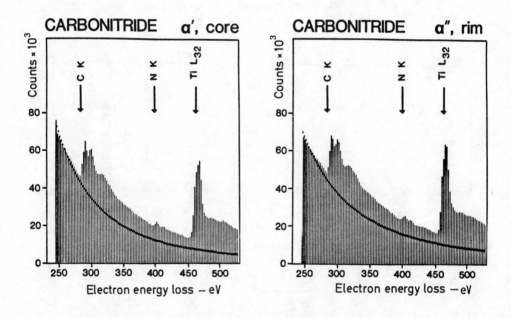

Fig. 10 EELS spectra of core and rim in a grain of a commercial carbonitride. The extrapolated backgrounds under the C K, N K and Ti L_{32} edges are indicated.

An α' spectrum after background subtraction is shown in Fig. 11a.
(Figs. 10 and 11 are actually from different specimens.) At this stage,
the integrated intensity under the lowest energy absorption edge, in this
case the C K edge, can be summed; the total number of counts is
proportional to the C concentration in the area illuminated by the
electron beam. One would normally use an integration window of 100 volts
in such analyses but the proximity of the N K edge at 401 volts to the Ti
L_{32} edge at 453 volts made this impossible; a 50 volt integration window
was used for all quantification. Following the integration of the C
signal, the background just below the onset of the N edge must also be
stripped; the counts above the dashed line in Fig. 11a correspond to the N
signal. Integration, again using a 50 volt window, of the intensity in
the N edge was then conducted; a C to N ratio of ~4 was determined from
this spectrum using Egerton's SIGMAK program (Egerton, 1978). Replicate
core analyses were very variable, but the C/N ratio was always greater
than one. In principle it would be possible to continue this analysis to
determine the Ti concentration, and thus the Ti/C+N ratio. However, the
cross section of the Ti L edge is not known with the same certainty as the
K edges, and we contented ourselves with the C/N analyses.

The spectrum of Fig. 11b is from the rim of the same grain and looks very
similar to that of Fig. 11a except for the intensity just below the
C edge. This intensity is due to the Mo content of the rim, the Mo M
absorption edge being very broad and starting at about 240 volts, as shown
in the pure Mo spectrum of Fig. 11c. Unfortunately, the carbonitride rim
spectrum includes the Mo M_{32} edge at 390 and 410 volts, which overlap the
N edge at 401 volts and is a serious interference for determining the N
content of the rim. To cope with this problem, we assumed that the Mo M
edge in the Ti/Mo carbonitride had the same shape as the Mo edge in pure
Mo, and subtracted counts from the rim spectrum equivalent to pure Mo and
scaled at 283 volts, the onset of the C K edge. The resulting stripped
spectrum is shown in Fig. 11d and indicates that most of the signal at
400V in this spectrum is due to N and not to Mo. Proceeding with the
analysis as just described, we found a greater C/N in the rim than the
core, so that the N core /N rim ratio is >1. Great variability was found
in the ratio, due in some measure to the difficulty of the C/N analysis,
but the basic conclusion that N is present in the rim but at a lesser
concentration than in the core seems unassailable.

Comparable core/rim spectra for an experimental alloy sintered for an
extended period are shown in Fig. 12 (we show the raw spectra with the
fitted background for convenience). Careful inspection of the data
suggests that the N core/N rim ratio is greater than for the commercial
alloys. The lower N rim concentration appears to be the major
microchemical change induced by the longer sintering time. (The
microstructure did not vary appreciably with sintering time.)

Our finding of N in the rims of this material is contrary to the phase
diagram published by Rudy for this system (Rudy, 1973); his diagram
suggests very small solubility of N in the rim. We thus were concerned
that the Mo content of the rim might have given rise to a spurious N K
edge due to the Mo M_{32} edge at 390 and 410 volts. To check this point, we
acquired EELS spectra of the commercial carbide (Fig. 13). As seen in the

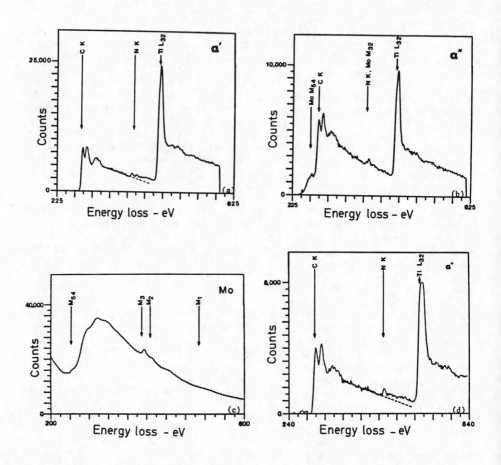

Fig. 11   EELS spectra of α' core (a) and α" rim (b) after background
subtraction. (This is not the same grain as shown in Fig. 10.) The
spectrum in (c) is a stripped spectrum of pure Mo, whose shape was used to
strip the Mo contribution from the rim spectrum, as described in the text.
The spectrum in (d) is the final spectrum used to determine the C/N ratio
of the rim.

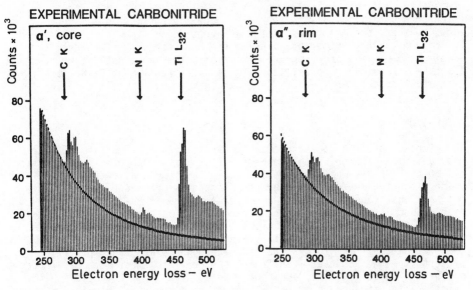

Fig. 12   EELS spectra of core and rim in an experimental carbonitride alloy sintered for long times.

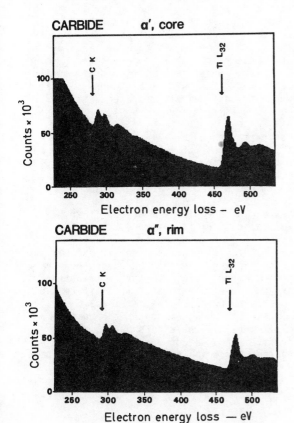

Fig. 13   EELS spectra of core and rim in a commercial carbide.

rim spectrum of this material, there is a small edge visible at 400 volts, which goes hand in hand with the Mo M_5 edge present just below the C K edge; however, the magnitude of the 400 volt edge is lower than in any of the carbonitrides. Given that the Mo rim concentration is roughly the same in the carbide and the various carbonitrides under study, we conclude that the intensity in the rim spectra at 400 volts in the carbonitrides is real and indicates an appreciable N concentration.

DISCUSSION

The microstructure of these commercial Ti/Mo carbides and carbonitrides provide a vehicle for which the benefits of analytical electron microscopy can be exploited very effectively. The microstructures obtained by liquid phase sintering are dominated by two factors -- phase separation, which has previously been reported (Rudy, 1973) in the Ti-Mo-C-N quaternary system, but also appears to occur in the Ti-Mo-C ternary system, at least at high TiC compositions (see below), and the fact that the liquid Ni binder phase preferentially wets the high-Mo carbide or carbonitride phase. The capillary forces are strong enough to almost completely prevent any α''-α'' contacts, while all the core/rim interfaces are coherent and have a low energy. In addition, these forces lead to rapid densification and the rapid establishment of the continuous Ni binder/core-rim microstructure during sintering. In fact, densification and the establishment of the wetting-dominated microstructures must occur so rapidly that chemical equilibrium is not assured during the sintering, in spite of the ready transport path through the liquid binder. This is manifested by the variable Ti content in both core and rim in the various materials studied, and the changing C/N ratio in the carbonitrides during extended liquid phase sintering. These latter changes can occur by solid state diffusion within an individual two-phase carbonitride grain, while chemical equilibration on the metal sublattice within a two-phase grain is impeded because of the much lower metal diffusivity compared with non-metal diffusivity in such carbides (Sarian, 1968, 1969).

It might seem that the rim concentration shoud be able to equilibrate by transport within the binder. That this does not occur must be due to the difficulty of changing the core concentration readily -- deviating from core/rim equilibria with regard to Ti/Mo concentration will not occur to eliminate chemical inhomogeneity in the rims of various grains.

The implications of our work for the phase equilibria of the Ti-Mo-C-N system are profound and their full discussion is deferred to a later paper. Two points are worth emphasizing here, however. The TiC-"MoC" binary has long been thought (Rudy, 1973) to exhibit extensive solubility at the TiC end. Core/rim microstructures of the type discussed here have been previously observed (Rudige and Exner, 1976) but were attributed to a lack of equilibration rather than to the existence of a miscibility gap. However, the similarity of the microstructures of the carbide and carbonitrides, accepting that a miscibility gap exists in the quaternary system, casts doubt on this explanation. To check this point, we determined a Mo concentration profile across a cored grain in the commercial carbide, reasoning than an appreciable gradient would indicate

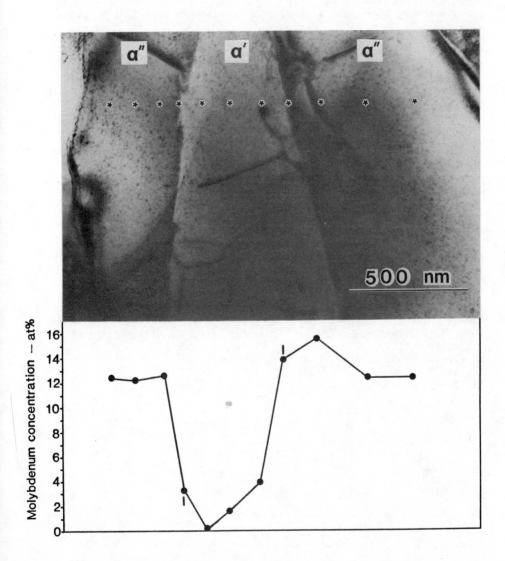

Fig. 14   EDS Mo concentration profile across a two-phase grain
in the commercial carbide.

an attempt to equilibrate a non-equilibrium "case", (Rüdige and Exner, 1976), whereas a sharp gradient would confirm a miscibility gap. The result is shown in Fig. 14, where the symbol 1 in the profile shows the data when the probe, ~20 nm in diameter and shown by the points in the upper micrograph, was placed directly on the core/rim interface. The data point strongly to a sharp interface rather than to a diffuse diffusion zone.

The second point concerns our finding N in the rim of the carbonitride, even after long sintering times. While the C/N ratio in the rim and the N core/N rim ratio both increase with sintering times, they never reach the values predicted by considering this system an ideal higher order solid solution (Rudy, 1973). CALPHAD calculations (Kaufman, 1983) using reasonable interaction parameters yields a computed phase diagram in reasonable agreement with our best estimates of the core/rim microchemistry, as will be discussed elsewhere.

REFERENCES

Egerton, R. F., 1978, Ultramicroscopy 3, 243.
Hren, J. J., Goldstein, J. I. and Joy, D. C., 1979, Introduction to
    Analytical Electron Microscopy.
Kaufman, L. 1983, Private Communication.
Moskowitz, D. and Humenik, M. Jr. 1966, Modern Developments in Powder
    Metallurgy, Vol 3, Plenum Press, N.Y.
Parikh, N. M. and Humenik, M. Jr. 1957, J. Amer. Ceram. Soc. 40, 315-320
Rüdige, O. and Exner, H. E. 1976, Powder Metall. Int. 8, 7-13
Rudy, E. 1973, J. of Less-Common Metals, 33, 43.
Rudy, E., Worcester, S. and Elkington, W. 1974, 8th Plansee Seminar
    [30], 18.
Sarian, S. 1968, J. Appl. Phys. 39, 3305-3310
Sarian, S. 1969, J. Appl. Phys. 40, 3515-3520
Willbrand, J. and Wieland U. 1972, Inter. J. Powder. Met. 8, 89-93

Inst. Phys. Conf. Ser. No. 75: Chapter 4
Paper presented at 2nd Int. Conf. Science Hard Mater., Rhodes

335

Study of recarburised eta-phase by transmission electron microscopy

H LE ROUX

University of the Witwatersrand, Johannesburg 2001, R.S.A.

ABSTRACT

On resintering, eta-phase changes chemically to Co and WC sometimes retaining a rosette-like appearance. Large Co contents remaining localised, consist of an allotropic mixture of Co and, at times of Co_7W_6, indicating a residual C-deficiency. Neighbouring small lakes of f.c.c. Co are C-sufficient. It is postulated that C-deficient regions attract Co which flows until eta-phase has formed and remains trapped until a C balance is restored. Carbide grain growth occurring in enlarged Co lakes is ascribed to the recrystallisation of W from solution during recarburisation of eta-phase.

INTRODUCTION

Undesirable eta-phase forming in cemented carbides can be converted into WC-Co by careful resintering under carburising conditions. As the size of components as well as the degree of recarburisation have a large influence

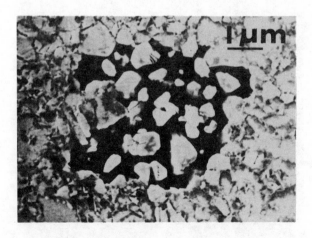

Fig. 1 is a scanning electron micrograph of an
eta-phase rosette (dark region) in coarse-
grained WC-5wt% Co.

on the final product a range of undercarburised to carburised states can result. It has been observed that recarburised material often shows features with the rosette-like morphology of eta-phase although the etching reaction is that of normal WC and Co in the standard etchant (Murakami's reagent). This ghost-like phase is called 'phantom eta-phase' herein.

Investigation by optical microscopy reveals that the phantom phase is dominated by a large localised concentration of Co. In conventional hardmetals, enlarged WC grains are packed within these large 'lakes' of Co. As the morphology of eta-phase is retained in the phantom phase a study by optical, scanning (SEM) and conventional transmission electron

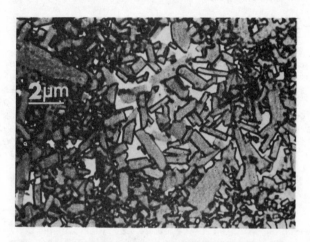

Fig. 2 is an optical micrograph of 'phantom eta-phase' (white region) in recarburised fine-grained WC-5wt% Co.

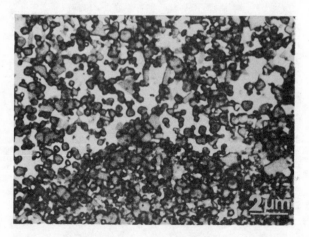

Fig. 3 is an optical micrograph of a portion of 'phantom eta-phase' in recarburised WC-(15wt% TiC-TaC)-8wt% Co.

microscopy (TEM) has been combined with electron diffraction (TED) to determine the component phases of these phantoms. The cause of the pronounced localisation of the Co as well as the carbide grain growth require elucidation.

A typical eta-phase rosette is shown in Fig. 1 which is a scanning electron micrograph of coarse-grained WC-5wt% Co. Tungsten carbide grains which are distributed in the eta-phase have the same grain size as the neighbouring, normal carbide regions.

In Fig. 2 an optical micrograph of a part of the phantom eta-phase in a recarburised, fine-grained, conventional WC-5wt% Co specimen shows the large localised concentrations of Co (white areas) in which enlarged WC grains occur. The boundary of the phantom with that of apparently normal WC-Co is quite sharp as it is between eta-phase and its boundary in Fig. 1.

In Fig. 3 an optical micrograph shows part of a phantom rosette in a tool grade of WC-(15wt% TiC-TaC)-8wt% Co with still larger Co lakes but exhibiting little or no grain growth. The edges of these lakes can be as large as 4 μm. The boundary between this phantom and the normal carbide region is very sharp.

Differences in etch contrast between Fig. 1 and Figs. 2 and 3 should be noted.

EXPERIMENTAL

Specimen surfaces illustrated in Figs. 2 and 3 were subjected to a detailed examination in a JEM 100C electron microscope. Thin films (~50 μm thick) were obtained by slicing these surfaces off the specimens with a diamond saw and then etching to transparency by ion milling. About 30 selected area diffraction patterns (SADPs) of different areas were obtained for each specimen. Dark field micrographs were also obtained and these helped to identify superimposed spot electron diffraction patterns of the different phases by locating the various phases separately in the micrographs. The diffraction aperture used was 1 μm in diameter. All studies were made at room temperature.

The mechanical properties of the phantom eta-phase could not be determined because test specimens were too small to retain the phantom on resintering the undercarburised material.

RESULTS

In Fig. 4 a set of TEM micrographs of an area in recarburised WC-5wt% Co is shown. In Fig. 4a and 4a' an SADP of this area and the analysis of the SADP, respectively, are shown. The SADP has been correctly oriented with respect to the images. The analysis of the SADP shows a superposition of the (110) f.c.c. Co (a = 0.354 nm), the (8 22.1) Co_7W_6 (hexagonal: a = 0.473 nm; c = 2.55 nm) and (12.0) h.c.p. Co (a = 0.251 nm; c = 0.408 nm) reciprocal lattice planes in an orientation relationship with each other. The respective zone axes lie along [110] f.c.c. [8 22.1] Co_7W_6 and [12.0]

h.c.p. directions. The dark field micrograph of Co in Fig. 4b was

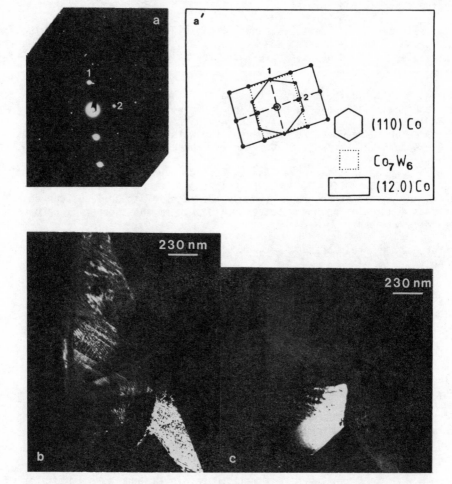

Fig. 4 is a set of TEM micrographs of recarburised, conventional
WC-5wt% Co. a) is an SADP of the area shown in b and c and has
been corrected for rotation with respect to the images. a') is
a diagram of the SADP shown in a. The (110) f.c.c. Co (8 22.1)
Co_7W_6 and (12.0) h.c.p. Co reciprocal lattice planes are shown.
The zone axes are, respectively, [110] f.c.c. Co, [8 22.1] Co_7W_6
and [12.0] h.c.p. Co. At point "1" there are three coincident
reflections: $\bar{1}11$ f.c.c. Co, $2\bar{1}.6$ Co_7W_6 and 00.2 h.c.p.Co
b) is a dark field micrograph of these three phases imaged by
spot "1". The triangular region is ~1.3 μm^2. c) is a dark
field micrograph of Co_7W_6 imaged by the $\bar{1}0.8$ reflection marked
"2" lying on the small rectangle.

imaged by nearly coincident $\bar{1}11$ f.c.c. Co, $2\bar{1}.6$ Co_7W_6 and 00.2 h.c.p. Co reflections marked "1" in Fig. 4a and 4a'. In Fig. 4c, Co_7W_6 is shown in dark field as imaged by the $\bar{1}0.8$ reflection marked "2" in Fig. 4a and 4a'. This phase has a rhombohedral structure but the reflections are given in hexagonal indices obtained from a transformed unit cell.

The orientation relationships are as follows:

[12.0]	h.c.p. Co	//	[110]	f.c.c. Co
(00.1)	h.c.p. Co	//	$(\bar{1}11)$	f.c.c. Co
[12.0]	h.c.p. Co	//	[8 22.1]	Co_7W_6
$(2\bar{1}.0)$	h.c.p. Co	//	$(\bar{1}0.8)$	Co_7W_6

The size of the triangular Co lake in Fig. 4b is approximately 2.8 μm long by 0.9 μm wide and has an area of ~1.3 μm^2.

In Fig. 5a an SADP of a different lake in the fine-grained WC-5wt% Co specimen is shown. This is the (111) reciprocal lattice plane of f.c.c. Co where the foil normal is the [$1\bar{1}1$] zone axis. In Fig. 5b a dark field micrograph of this lake imaged by a 220 type reflection is shown. Although a (111) reciprocal lattice plane is illustrated, the most usual f.c.c. plane observed in this specimen was the (110) reciprocal lattice plane in lakes which are ~0.3 μm^2 in area. The surface area of the lake in Fig. 5b is also ~0.3 μm^2.

The size of these small lakes corresponds to those Co lakes lying outside the phantom while the larger lake has the larger dimensions of the Co inside the phantom.

In this specimen only 10% of the observations were of the allotropic mixture and these all had surface areas larger than 1 μm^2. Forty per cent of the observations were of carbide grains and of other f.c.c. orientations. All f.c.c. lakes had surface areas much smaller than 1 μm^2. These results imply that lakes containing a mixture of the allotropes are larger than lakes containing only the f.c.c. allotrope.

In Fig. 6a the SADP of a similar area in recarburised WC-(15% TiC-TaC) -8% Co (tool grade) is shown. This is also the (111) f.c.c. Co reciprocal lattice plane where the foil normal is the [111] zone axis. In Fig. 6b a dark field micrograph imaged by a 220 type reflection is shown. The Co lake has a surface area of ~0.4 μm^2. This area is comparable with that seen in the conventional hardmetal specimen. Both figures show evidence of stacking faults.

In Fig. 7 a set of micrographs of recarburised WC-(15% TiC-TaC)-8% Co is shown. An SADP of the Co lake in Fig. 7b is shown in Fig. 7a. The (00.1) reciprocal lattice plane of h.c.p. Co is in an orientation relationship with the (123) reciprocal lattice plane of f.c.c. Co as analysed in Fig. 7a'. The zone axes are, respectively, [00.1] and [123]. Spot "1" is a composite not only of nearly coincident f.c.c. and h.c.p. Co reflections but also of stacking fault streaks which are due to thin platelets of either f.c.c. or h.c.p. Co or of both. The diffraction pattern in Fig.7a occurs in 60% of the observations made on this specimen. In Fig. 7b a dark field micrograph imaged by the nearly coincident $\bar{1}\bar{1}1$ f.c.c. and

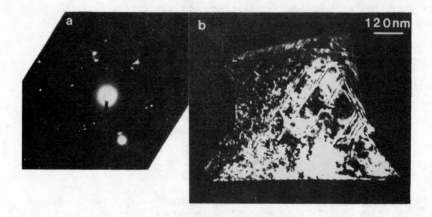

Fig. 5   a) is an SADP of the (111) reciprocal lattice plane of
f.c.c. Co in recarburised WC-5wt% Co. The foil normal is the
[111] zone axis.   b) is a dark field micrograph of f.c.c. Co
imaged by a 220 type reflection. The surface area of this lake
is ~0.3 μm².

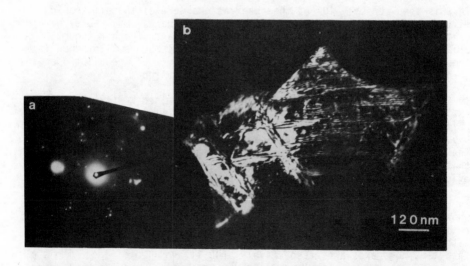

Fig. 6   a) is an SADP of the [111] zone axis of f.c.c. Co in
recarburised WC-(15wt% TiC-TaC)-8wt% Co shown in b. b) is a
dark field micrograph of f.c.c. Co imaged by a 220 type
reflection. This lake has a surface area of ~0.4 μm².

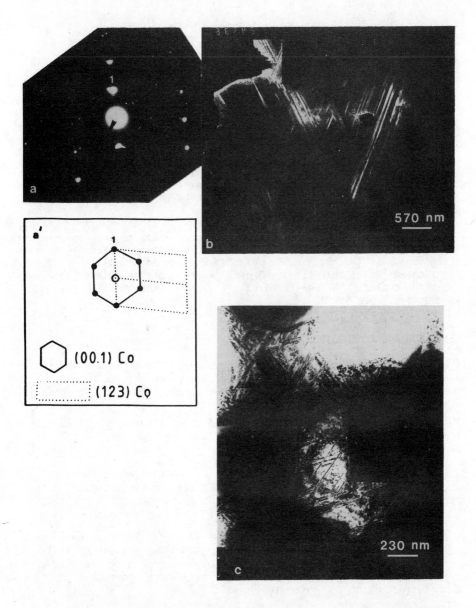

Fig. 7 shows an SADP and TEM micrographs of recarburised WC-(15wt% TiC-TaC)-8wt% Co. a) is an SADP of the area in b). a') is the analysis of a) and shows the (00.1) and (123) reciprocal lattice planes ([00.1] and [123] zone axes respectively) of h.c.p. and f.c.c. Co. b) is a dark field micrograph of h.c.p. and f.c.c. Co imaged by spot marked "1" in a and a' which is a combined $\bar{1}0.0$ h.c.p. and $\bar{1}\bar{1}1$ type f.c.c. Co reflection. This lake has a surface area of ~9 μm². c) is a bright field micrograph of h.c.p. and f.c.c. Co having the same SADP as a and shows the wide separation of carbide grains in the 'phantom eta-phase'.

$\bar{1}0.0$ h.c.p. Co reflections, with streaks, (marked "1" in Fig. 7a and 7a') is shown.

The orientation relationship is as follows:

$$[00.1] \quad \text{h.c.p. Co} \quad // \quad [123] \quad \text{f.c.c. Co}$$
$$(\bar{1}0.0) \quad \text{h.c.p. Co} \quad // \quad (\bar{1}\bar{1}1) \quad \text{f.c.c. Co}$$

This micrograph is typical of the lakes in this specimen. The lake size is ~4 μm by ~2.3 μm with a surface area of ~9 μm^2.

In Fig. 7c a bright field micrograph of a different area in the same specimen which has the same SADP as illustrated in Fig. 7a is shown. The large separation of the carbide grains by thick Co lakes is evident in this micrograph.

The average surface area of all the measured Co lakes in the tool grade is ~2.3 μm^2. This average was obtained from large lakes of mixed allotropes making up 60% of the total number of observations and small lakes of f.c.c. Co making up 10% of this total. Thirty per cent of the observations were of carbide grains. The mean area of the larger lakes containing the allotropic mixture is ~3 μm^2.

DISCUSSION AND CONCLUSIONS

In both specimens large and small Co lakes were investigated. The large lakes consisted of allotropic mixtures of h.c.p. and f.c.c. Co which sometimes also contained the C-deficient mu-phase, Co_7W_6. This phase is concentrated in one corner of the lake suggesting that the W atoms have clustered in the Co. The surface areas of these lakes were all larger than 1 μm^2. In contrast the small Co lakes of average area = ~0.3 μm^2 consisted of f.c.c. Co only.

It became necessary to correlate the spatial relationship between the lakes seen in the TEM micrographs and those seen in the optical micrographs and to show that the TEM micrographs represent views either of the phantoms or of the neighbouring areas.

In recarburised, conventional WC-5wt% Co foils, 50% of the observed areas were small lakes of f.c.c. Co mainly in the [110] orientation. It is assumed, therefore, that the areas examined formed part of a single crystal of f.c.c. Co. The mean surface area of the larger lakes constituting 10% of the observations was ~1.5 μm^2.

In the tool grade mainly large lakes were observed. The smallest lakes consisting of f.c.c. Co in the [111] orientation compare well with the small lakes of the same orientation in recarburised WC-5% Co. The largest lake observed was ~9 μm^2 in area. In 60% of the observations the lakes consisted of the [00.1] h.c.p. and [123] f.c.c. Co zone axes in an orientation relationship.

There is a correlation between the size of the small Co lakes observed in the TEM micrographs and the small Co lakes in the regions exterior to the phantoms as seen in the optical micrographs of both specimens. It is therefore inferred that the small lakes represent this exterior region. There is a similar correlation between the sizes of the larger lakes in the phantom seen in TEM and optical microscopy. A similar inference is made that the large lakes lie within the phantoms.

The spot diffraction patterns of both h.c.p. and f.c.c. Co appearing together in the SADPs is evidence that single crystals of both allotropes exist in the large Co lakes.

Single crystals of h.c.p. Co can be present in hardmetals after certain heat treatments e.g., at ~700°C (Giamei co-workers, 1968; le Roux, 1981) because dissolved W increases the h.c.p. to f.c.c. transition temperature. Hexagonal Co can also be present when the carbon content is low. This is certainly true of the metal and it has been inferred to be true for WC-Co from the mutually exclusive solubility of C and W in Co (Rüdiger co-workers, 1971) and the stabilisation of f.c.c. Co by C. The solubility of W was shown by these authors to vary from 1 to 20% for a variation in the dissolved C content from 1.3% to 0.12%, respectively.

The presence of the mu-phase, Co_7W_6, which forms in the binary Co-W system in the absence of C, confirms not only that the enlarged lake in which it was observed is C-deficient but that it is rich in W. A composition of approximately 50 at% W is necessary for the mu-phase to form (Hansen and Anderko, 1958).

It is unlikely that deformation of specimens during preparation was responsible for the presence of h.c.p. Co crystals as it should not then occur only in localised regions but be distributed randomly throughout the materials.

The results therefore show that there is a correlation between Co lake size and their allotropic content.

The relative quantities of Co in enlarged lakes in the two grades examined appear to be different. The mean surface area calculations support this conclusion as the areas in the two specimens differed by a factor of two whereas the Co content varies in the ratio of 2 to 3. The two grades will be compared and contrasted.

Different lake sizes can be expected as the nominal Co contents are not the same. The more usual distribution of Co in WC-Co is such that large grained material induces the formation of thick Co layers and fine grained material much thinner layers. This is not true of the phantoms where the opposite appears to be the case. Both materials are essentially fine grained but, in the phantoms, the conventional material has become coarse grained whereas the tool grade has remained fine grained. The Co lakes on the other hand are larger for the smaller than for the larger grains.

It has been shown that large quantities of W are dissolved in the phantoms. Some of this dissolved W has crystallised, while some W remains bound to Co as Co_7W_6, causing enlargement of the carbide grains in the conventional grade. On the other hand no grain growth has occurred in the tool grade. It is assumed that TiC-TaC, which usually has an inhibiting effect on grain growth is responsible for this lack of growth. However, noticeable clustering of carbide grains in the tool grade's phantom has occurred thereby simulating large grains. Consequently larger lakes may have formed. Considering the enlarged carbide grains in the conventional material and the simulated enlargement of carbide grains in the tool grade there is some justification for making a direct comparison between lake sizes in the two specimens.

For equally large carbide grains in the two specimens (one real and one simulated) equally large Co lakes should form. Since the lake areas differ by a factor of two and the Co contents only in the ratio of 2 to 3 it has been inferred that there is proportionately more Co in the tool grade than in the conventional grade.

Clustering of WC grains is very pronounced in hardmetals with Co-rich binder phases, particularly in WC-30wt% Co (le Roux, 1984) and the distribution of Co in Fig. 3 is similar to that occurring in Co-rich materials. However, a distinction must be made between large Co lakes in a Co-rich grade and localised concentrations of Co in a Co-lean grade.

It is also assumed that the phantom was preferentially thinned in the tool grade because the Co lakes are larger than in the conventional material where more of the neighbouring areas were thinned. Larger lakes make a softer material which can be more readily thinned.

Several conclusions may be drawn from the foregoing.

There is a compositional disturbance within the hardmetal such that, in the phantom regions, the quantity of Co and W is representative of equiatomic compositions i.e., Co = W = 50at% which contrasts with the composition in the material external to the phantoms where the normal W-rich/Co-lean composition holds. Resintering of the material has increased the C content of the phantoms so that a chemical change from eta-phase to WC-Co has been able to take place. However, a residual C-deficiency still exists in the phantoms where the Co remains trapped within the original boundaries of the eta-phase.

A C imbalance evidently induces the migration of Co away from the C-sufficient to C-deficient regions, the quantity migrating depending on the degree of C-deficiency and this Co is held in these regions until the proper C balance is achieved.

It is therefore postulated that both the excess Co required for eta-phase formation and the non-return flow of the Co from the phantoms, on resintering, are caused by localised C-deficiencies.

The temperatures of formation of Co_3W_3C (M_6C) and Co_6W_6C ($M_{12}C$) are different (Sarin, 1976). The former is found to occur at 1425°C while the latter occurs at 1000°C. In one case the Co is liquid and in the other solid. Since $M_{12}C$ is the most C-deficient eta-phase known and the amount of Co (and W) necessary correspondingly greater, larger quantities of Co must be able to migrate in the solid (for $M_{12}C$) than in the liquid state (for M_6C) to supply the needs of eta-phase. It is known that solid state reactions do occur at high temperatures e.g., between 700°C and 1040°C, involving extensive rearrangements and chemical changes of component phases, albeit different phases from those discussed herein (le Roux 1976; le Roux 1985).

Because of the apparent difference in Co content of the phantoms in the two grades, differences in the degree of C-deficiency may exist in the two materials and therefore two different eta-phases may have been present. It is not possible to confirm this as no observations on the original material were made.

SUMMARY

The correlation observed between the Co lake size and the presence of h.c.p. Co and Co_7W_6 in the phantom eta-phase was a necessary step in the process of interpreting the observations.

The result obtained from this investigation that Co does not flow from C-deficient regions raises the additional question of how Co comes to be so concentrated in these localised regions. A logical consequence of the finding that the Co is trapped in the phantom by a residual C-deficiency is that a C-deficient state, existing regionally, attracts Co towards it preparatory to the formation of eta-phase. Thus the driving force for the reaction is a C-deficiency. The quantity of Co attracted must therefore be dependent on the degree of C-deficiency such that the greatest quantity of Co would be attracted by the least quantity of C to form, e.g. $M_{12}C$.

The concentration of such large quantities of Co has a secondary effect. Tungsten, which has a high solubility in Co at or near sintering temperatures (1000° C to 1450°C) becomes regionally concentrated at the same time. It participates in the formation of eta-phase but on recarburisation is able to crystallise as WC. The existing carbide grains act as nuclei so that enlargement of these grains occurs. When grain growth inhibitors are present (e.g., TiC-TaC) the excess W evidently remains in solution.

In the course of the formation of eta-phase the normal distribution of Co and W is disturbed so that distinct compositional ranges exist in different parts of the material. In the normal region the WC-rich/Co-lean composition occurs whereas in the phantom regions a W-rich/Co-rich composition develops under the influence of C-deficiency.

The inability to obtain small specimens which retain the phantom indicates not only that the effect of the phantom on the mechanical properties of the material cannot easily be measured, because test specimens are usually small, but confirms that it is possible to recarburise the phantom until the reverse Co flow has occurred.

ACKNOWLEDGEMENTS

The author gratefully acknowledges discussions with Mr. E. Lardner and Mrs. A. Paxton, very pertinent comments by Mr. R. Cooper, Dr. C.T.Peters and Mr. N. Dent, technical assistance by Mrs. S.S.Taylor and sponsorship by Boart Research Centre, Krugersdorp.

REFERENCES

Giamei, H.F., Burma, J., Rabin, S., Cheng, M. and Freise, E.L. 1968, Cobalt (Eng. Edn) (40) 140

Hansen, M. and Anderko, K. 1958, Constitution of Binary Alloys. McGraw-Hill, New York

le Roux, H. 1976, Acta metall. 24 299-305

le Roux, H. 1981, High Temp. High Press. 13 503-506

le Roux, H. 1984, Int. J. Refractory and Hard Metals, 3 99-100

le Roux, H. 1985, Acta metall. 33 309-315

Rüdiger, O., Hirschfeld, D., Hoffmann, A., Kolaska, J., Ostermann, G. and Willbrand, J. 1971, Tech. Mitt. Krupp Forsch. Ber. 29 1-14

Sarin, V.K. 1976, Modern Developments in Powder Metallurgy. Eds. H.H. Hausner, P.W. Taubenblatt, Princeton, NJ, American Powder Metallurgy Institute, 1-13

Inst. Phys. Conf. Ser. No. 75: Chapter 4
Paper presented at 2nd Int. Conf. Science Hard Mater., Rhodes

347

A study of microstructural development during the liquid-phase sintering of WC-Co alloys

R DESHMUKH (1) AND J GURLAND (2)

(1) ATT Technology, Engineering Research Center, Princeton, NJ
(2) Division of Engineering, Brown University, Providence, RI

ABSTRACT

The interrelationship of the contiguity of tungsten carbide (WC) with other microstructural parameters is defined and the effects of contiguity on properties are briefly discussed. Some experimental results related to the development of contiguity are described, namely, the change of contact parameters with sintering time, and the crystal orientation relationships across grain boundaries. The results suggest that special contact boundaries in the tungsten carbide may persist during the liquid phase sintering of WC-Co.

INTRODUCTION

The microstructure of sintered WC-Co alloys consists of the carbide phase WC and the cobalt-rich binder phase. The state of aggregation of the carbide phase varies as a function of cobalt content and processing history, ranging from more or less isolated WC particles dispersed in the binder phase of alloys of high cobalt content to a highly connected skeleton of contiguous WC grains in low cobalt alloys. The state of aggregation of the carbide phase is expressed by the degree of contiguity which is a measure of the extent of the carbide grain boundary area relative to the total surface area of the carbide grains.

The study of contiguity in multi-phase structures such as cemented carbides is of interest for two reasons: (1) contiguity may affect the mechanical properties which relate to tool performance, and (2) the change of contiguity during sintering may give insight into the sintering mechanism. In this paper, the interrelationship of contiguity with other microstructural parameters is defined, the largely undocumented effects of contiguity on properties are briefly discussed, and some experimental results on the development of contiguity during sintering are described.

Contiguity as a Microstructural Parameter

The carbide phase WC and the cobalt-rich binder phase are labeled α and β, respectively, in accordance with ASTM Method B657 (1983). The binder phase will be referred to as "cobalt" for simplicity.

Contiguity is defined by the ratio of grain boundary area shared by WC grains, $S_{\alpha\alpha}$, to the total surface area of the carbide grains, the latter consisting of WC/WC grain boundary area, $S_{\alpha\alpha}$, and WC/cobalt interfacial area, $S_{\alpha\beta}$ (per unit volume of alloy).

$$C_\alpha = \frac{2\,S_{\alpha\alpha}}{2\,S_{\alpha\alpha} + S_{\alpha\beta}} = \frac{2\,(N_L)_{\alpha\alpha}}{2\,(N_L)_{\alpha\alpha} + (N_L)_{\alpha\beta}} \qquad 1$$

where the average number of intercepts per unit length of test lines with traces of WC/WC grain boundaries is $(N_L)_{\alpha\alpha}$ and with traces of WC/cobalt interfaces, $(N_L)_{\alpha\beta}$. As shown by the last term of the above equality, contiguity is obtained operationally from boundary intercepts with test lines on planar sections. The contiguity ranges from 0 to 1 as the dispersion of the α phase changes from a completely dispersed to a completely agglomerated structure.

The interrelationship between contiguity and other primary structural parameters is obtained by combining the following two stereological relations:

$$\frac{\bar{L}_\beta}{\bar{L}_\alpha} = \frac{(V_V)_\beta}{[1-(V_V)_\beta]} \qquad \text{Underwood (1970)}$$

and

$$\bar{L}_\alpha = (\bar{L}_3)_\alpha/(1-C_\alpha) \ , \qquad \left|\begin{array}{l}\text{Exner and Fischmeister (1966)}\\ \text{Gurland (1966)}\end{array}\right.$$

giving

$$\bar{L}_\beta = \frac{(\bar{L}_3)_\alpha}{(1-C_\alpha)}\ \frac{(V_V)_\beta}{[1-(V_V)_\beta]} \qquad 2$$

where $\bar{L}_\alpha$ and $\bar{L}_\beta$ are the mean free paths (between α/β interphase boundaries) in the carbide phase and binder phase, respectively; $(\bar{L}_3)_\alpha$ is the mean-intercept grain size of the carbide (i.e. particle size); and $(V_V)_\beta$ is the volume fraction of binder.

When $C_\alpha = 0$, Eq. 2 reduces to the expression given by Underwood (1970) for the matrix mean free path between dispersed non-contiguous particles. As Eq. 2 shows, the primary structural parameters are not independent of each other. Any three of the four parameters are sufficient to characterize fully the degree of dispersion of the carbide phase in WC-Co alloys.

Effect of Contiguity on Deformation and Fracture

Little is known about the effects of contiguity on the strength of cemented carbides. In a 1970 review, Exner and Gurland cite limited results which indicate that hardness increases and transverse rupture strength decreases, with increasing contiguity. A practical difficulty is that of separating the effects of contiguity from those of the other variables in Eq. 2. In general, cobalt content and WC grain size are the independent variables controlled or selected by the manufacturer, which makes $\bar{L}_\alpha(1-C_\alpha)$ the dependent term. In practice the mechanical properties are often interpreted as a function of L_β alone, with C_α neglected as a contributing variable.

Nevertheless, a recently proposed theory of hardness of WC-Co alloys features the contiguity explicitly, together with the mean free path of cobalt, and the grain size of WC. In the model of Lee and Gurland (1978), the hardness load (i.e. the indentation pressure) is carried in parallel by the binder phase, with embedded discontinuous portions of WC, and the continuous portion of WC, defined by $[(V_V)_\alpha . C_\alpha]$. The alloy hardness (DPH) is given by a modified law of mixtures,

$$H_C = H_\alpha (V_V)_\alpha C_\alpha + H_\beta [1 - V_V)_\alpha C_\alpha] \qquad 3$$

The "in-situ" hardnesses of the carbide and matrix phases, H_α and H_β, respectively, are represented by empirical equations of the Hall-Petch type, namely:

$$H_\alpha = 1382 + 23.1 \ [\bar{L}_3)_\alpha^{-1/2} \ kg/mm^2$$

$$H_\beta = 304 + 12.7 \ \bar{L}_\beta^{-1/2} \ kg/mm^2$$

The agreement between theory and experiment is very good for WC-Co alloys. Also, when applied to several other carbide-binder systems, Warren (1983) found that Eq. 3 correctly estimates the actual hardness in most cases.

Contiguity enters the hardness model through a consideration of the role of grain contact boundaries in transmitting plastic deformation through the connected portion of the carbide constitutent. Indeed, in a few cases, well developed slip lines were observed in WC particles, originating or ending at WC-WC grain boundaries (Lee and Gurland, 1978), presumably associated with high local stress concentrations. Also, there is strong evidence that the WC particle contacts have an important role in fracture (Luyckx, 1981). Hong and Gurland (1983) showed that WC/WC boundary debonding and WC intragranular cleavage are the precursors of cobalt rupture in the fracture of WC-Co, with WC/WC intergranular fracture predominating at low cobalt contents and small WC particle sizes.

The Development of Contiguity During Liquid Phase Sintering

It has been repeatedly shown that the contiguity of WC decreases with sintering time, although the path by which the microstructure evolves is not yet clear. Exner and Fischmeister (1966) attribute the decrease in contiguity mainly to grain agglomeration and grain boundary migration in the agglomerates, whereas Lee, Jaffrey and Browne (1980) interpret their results to show that the decrease of contiguity is due mainly to continuous cobalt penetration of WC-WC grain boundaries.

In this study, stereological measurements of contiguity, grain size, contact number and contact length are combined with crystal orientation determinations in an attempt to elucidate the processes responsible for the observed changes. As will be seen, the presence of various competing processes makes it difficult to arrive at a simple interpretation.

EXPERIMENTAL STUDIES OF MICROSTRUCTURAL EVOLUTION

Liquid Phase Sintering

The composition considered here is 75 wt % WC - 25 wt % Co. The powders
were milled in acetone, pressed at room temperature and sintered in
hydrogen at various temperatures. The microstructures were characterized
by scanning electron microscopy (SEM) and optical microscopy.

The state of agglomeration at the WC constitutent is shown in Figs. 1 - 5
at various early stages of the sintering process. Figs. 1 and 2 compare,
respectively, the appearances of the WC powder in the as-received condition
and after milling with 25 w/o cobalt. It is seen that the particle agglomerates initially present have not been completely broken up during milling.
These WC agglomerates are also still present during the next stage (Fig.
3) namely, after sintering at 1250°, i.e. below the solidus temperature
which is approximately 1320°C. At this stage there is evidence of solid-
state sintering of cobalt. However, after sintering at 1450°C, appreciably
above the solidus temperature, the particle agglomerates have been replaced by large WC grains of characteristic shapes and limited contact
with each other (Fig. 4). What occurs between the last two stages is
shown in Fig. 5, which refers to a specimen sintered at 1300°C, i.e. very
near the quasi-eutectic temperature, and which exhibits partial melting
within the channel indicated on the figure. Visibly, the liquid binder
phase has disintegrated some of the WC agglomerates. Interpretation of
this sequence of photomicrographs leads to the conclusion that penetration
of contact boundaries of the solid constituent by the liquid phase is an
essential step in the densification process since it facilitates the
particle rearrangement typical of liquid-phase sintered systems.

Quantitative Characterization of Sintering Effects on the Microstructure

The major effects of liquid phase sintering on WC were determined by means
of the evolution of the following parameters: (a) average grain size
(Fig. 6), (b) contiguity (Fig. 7), (c) average number of contacts per
grain (Fig. 7), and (d) average contact length per contact (Fig. 8). As
shown by our results, there is considerable grain growth, but decrease of
contiguity and grain contacts during isothermal sintering. These findings
are in agreement with those previously reported in the literature (Exner
and Fischmeister, 1966, and Lee and co-workers, 1980). For related
studies of the effect of cobalt content on contact parameters, see Lee and
co-workers (1980) and Luyckx (1981).

Orientation Relationship Between Contiguous WC Crystals

The contact between carbide grains occurs at grain boundaries of complex
fine structure. Detailed studies by means of scanning transmission
electron microscopy (Sharma and co-workers, 1980) and field-ion atom probe
(Hellsing and co-workers, 1983) show elevated cobalt concentrations at
the boundaries, which could indicate the presence there of cobalt-rich
layers with thickness of the order of 20Å (Sharma and co-workers, 1980).
However, there is also evidence, obtained by Hagège and co-workers (1980)
by transmission electron microscopy that at least some of the grain
boundaries provide a direct transition between contiguous WC crystals
which, pairwise, exhibit relatively good lattice coincidence and are able
to accommodate a small mismatch of low-index atomic planes by the defect
structure of the grain boundary.

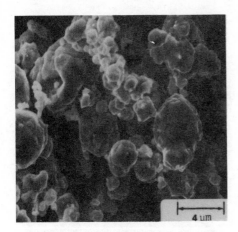

Fig. 1 SEM micrograph of as-received WC powder, 4 μm particle size.

Fig. 2 SEM micrograph of WC-25 **wt%** Co powder milled 24 hours, washed with alcohol.

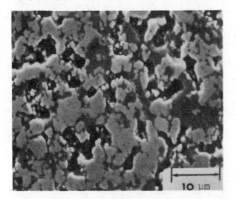

Fig. 3 Microstructure of WC-25 **wt%** Co sintered at 1250°C for 1 hour.

Fig. 4 Microstructure of WC-25 w/o sintered at 1450°C for 1/2 hour.

Fig. 5 Microstructure of WC-25 **wt%** Co sintered at 1300°C for 15 minutes. Area indicated by the channel A_1A_2 shows local melting of the cobalt-rich binder.

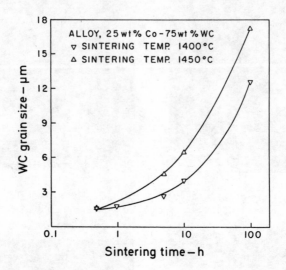

Fig. 6 The change in grain size of WC (mean linear intercept) with time in WC-25 **wt%** Co sintered at 1400°C and 1450°C.

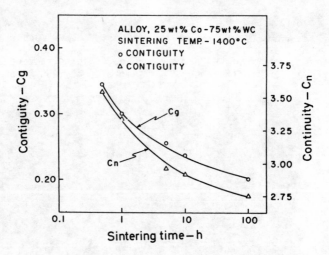

Fig. 7 The change of contiguity and continuity (i.e. mean number of contacts per WC grain) with sintering time at 1400°C in WC-25 **wt%** Co.

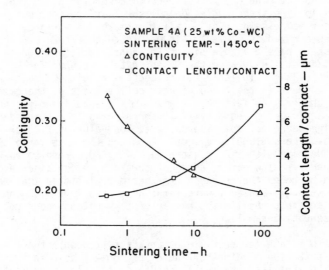

Fig. 8 The change of contiguity and length of contact bewteen WC grains with sintering time at 1450°C in WC-25 wt% Co.

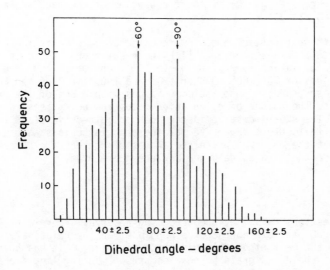

Fig. 9 Measured dihedral angles of contact between WC grains in WC-25 wt% Co sintered 100 hours at 1450°C.

In the present study, the orientation relationship between adjacent WC grains was investigated on ten contact boundaries, in a specimen sintered 100 hours at 1450°C. The crystal orientations were obtained from the analysis of slip line trace angles on polished surface sections. The method is described elsewhere in detail (Deshmukh and Gurland, 1982). The results are summarized in Table 1. It was found that seven of the ten pairs of crystals have possible high coincidence relations, on the basis of the coincidence site lattice theory and data developed for WC by Hagège and co-workers (1980). In Table 1, the degree of coincidence is represented by Σ, where Σ is the ratio of unit cell volumes of the coincidence site lattice and crystal lattice. A given misorientation may represent a special high-angle grain boundary if it corresponds to a coincidence site lattice of high degree of coincidence, i.e. low value of Σ. For instance, Hagège and co-workers (1980) previously reported the observance of a high coincidence boundary with Σ=2 in sintered WC-Co. Our results as well, with Σ<14, appear to indicate a tendency toward high coincidence contacts, but it should be noted that because of experimental limitations the sampling was strongly biased toward large WC grains in a high-cobalt composition, and that the limited number of observations does not constitute a representative sampling.

In an effort to relate the particle orientations to contact angles, the frequency distribution of dihedral angles of contact between faces of contiguous WC grains was obtained. The dihedral angles were measured on SEM photomicrographs of sections of a sintered WC-Co alloy with 25 w/o cobalt, after short and long time sintering. The resulting frequency distributions (for example, Fig. 9) show pronounced peaks at 60° and 90° and, perhaps, less easily distinguishable peaks at other values. Assuming that the observed frequency distribution results from the superimposed distribution expected from true dihedral angles of 30°, 60°, 90° and 120° one finds fair agreement between the observed and calculated cumulative distribution curves (Fig. 10) if the frequencies of the four angles are in relative ratios 2:4:3:1.

Theoretical dihedral contact angles were calculated for coincidence boundaries with Σ values from 2 to 14. The results are listed in Table 2. While the angular values range from 8° to 157°, one or more of the particular values of 30°, 60°, 90° or 120° are found in 7 of the 13 orientation cases, and, if one includes values within ±5° of the precise angular values, the number of orientation cases including these angles rises to 10. The experimental evidence does not directly relate the observed dihedral contact angles to particular orientations across grain boundaries, but the observed preferred angles of contact are compatible with several of the high coincidence contact geometries.

DISCUSSION

One of the questions raised by the preceding observations relates to the origin of the WC particle contacts responsible for the relatively high contiguity at the beginning of liquid phase sintering. It has been suggested that the WC/WC grain boundaries present during liquid phase sintering are survivors of boundaries already present in the poly-crystalline powder (Lardner, 1973) or formed during cold pressing (Lee and co-workers, 1980) or during the solid-state sintering stage (Nelson and

TABLE 1 Comparison of Experiment and Theoretical* Axis-Angle
Values for Pairs of Contiguous WC Crystals

Pair No.	Experimental		Theoretical	
	Rotation Axis	Rotation Angle, degrees	Possible Coincidence, Σ	Rotation Axis at Angle
1	[10 2 9]	163	14b	[10 2 9] 164.641°
2	[102]	105	8	[102] 104.478°
3	[001]	38	7a	[001] 38.213°
4	[423]	85	4	[423] 82.819°
5	[625]	152	14b	[625] 153.234°
6	[403]	139	7b	[403] 141.787°
7	[001]	27	13a	[001] 27.796°
8	[627]	160	-	
9	[203]	78	-	
10	[10 8 3]	122	-	

* Reference: Hagège and co-workers (1980)

TABLE 2 Calculated Dihedral Angles for Pairs of
Contiguous WC Crystals of High Coincidence

Coincidence Σ	Possible Values of Dihedral Angles Degrees						Within ±5° of Angles			
							30°	60°	90°	120°
2	30,	60,	90,	120			x	x	x	x
4	30,	60,	120				x	x		x
5	30,	60,	90,	120			x	x	x	x
7a	22,	38,	82,	90,	98				x	
7b	8,	82,	98							
8	40,	53,	75,	134,	145					
10a	30,	60,	90,	120			x	x	x	x
10b	16,	75,	106,	130						
11	32,	62,	70,	74,	110,	140	x	x		
13a	28,	32,	88,	90,	92		x		x	
13b	32,	58					x	x		
13c	30,	60,	90,	120			x	x	x	x
14a	26,	84,	95,	136,	197		x		x	

Milner, 1972). Considering the thermodynamic condition for particle adherence in a liquid, namely, that

$$\gamma_{S_1S_2} < 2\,\gamma_{SL} \qquad\qquad 4$$

where $\gamma_{S_1S_2}$ is the grain boundary energy and γ_{SL} is the solid-liquid interfacial energy, it is seen that only boundaries that satisfy the above condition could be present in the liquid in a stable or quasi-stable manner. Generally, high coincidence boundaries are associatd with low boundary energies, and it is therefore possible that only the boundaries of special orientations (as schematically shown in Fig. 11) would satisfy the condition of Eq. 4, especially if it is assumed that for random orientations across grain boundaries

$$\gamma_{S_1S_2} > 2\,\gamma_{SL} \qquad\qquad 5$$

According to a relationship developed by Warren (1976), the contiguity of spherical particles relates to the number of interparticle contacts and the ratio of interphase and interface boundary energies, to a close approximation, by the equation

$$\frac{\gamma_{SS}}{\gamma_{SL}} \quad \frac{2\,[1-3(C/\bar{n})]}{[1-(C/\bar{n})]} \qquad\qquad 6$$

where C is the contiguity and $\bar{n}$ is the average number of contacts per particle. Substitution of experimental values of C and $\bar{n}$ from Fig. 7 into Equation 6 gives values of γ_{SS}/γ_{SL} which vary from 1.6 to 1.7 as sintering time increases from 1 to 100 hours. These are in good agreement with the value of $\gamma_{SS}/\gamma_{SL} = 1.72$ reported by Warren (1976) for WC-Co. Although the conditions of Equation 4 (i.e. $\gamma_{SS}/\gamma_{SL} < 2$) is obeyed, in this case it does not strictly apply because of the shape anisotropy of WC grains. Nevertheless, the relatively small change in the values of the energy ratio with sintering time indicates that perhaps a degree of equilibrium is attained at the contacting interfaces.

These considerations suggest that special WC contact boundaries may be thermodynamically stable in the microstructure during the dynamic evolution of the structure towards equilibrium. While many of the WC agglomerates are disintegrated by the liquid binder phase when it first appears, it is not clear whether the special boundaries found after long-time sintering are survivors from the solid state stage, or were formed during liquid phase sintering. More work is required not only on the topology after long-time sintering, but also on the contact development in the very earliest stage of liquid-phase sintering.

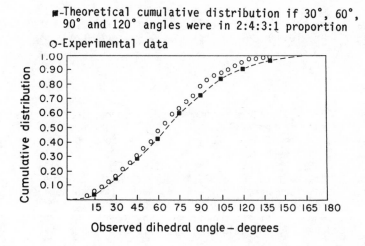

Fig. 10 Experimental and theoretical cumulative distribution of dihedral angles of contact between WC grains in WC-25 **wt%** sintered 100 hours at 1450°C.

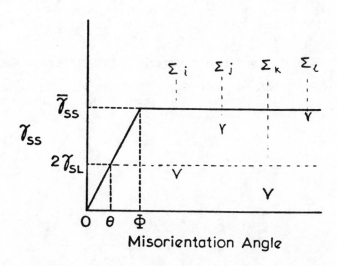

Fig. 11 Schematic variation of grain boundary energy with misorientation angle and for special high angle boundaries (indicated by Σ_i, Σ_j, Σ_k, Σ_l). The two contacting particles will adhere only if $\gamma_{ss} < 2\gamma_{SL}$, i.e. if the angle of misorientation is less than θ, or corresponds to the special angles for Σ_i and Σ_k.

ACKNOWLEDGEMENT

Financial support by Brown University Materials Research Laboratory is
gratefully acknowledged, as is the invaluable technical assistance of
Mr Herbert J. Stanton.

REFERENCES

ASTM 1983, B657-79, Standard Method for Metallographic Determination of
    Microstructure in Cemented Carbides, 1983 Annual Book of ASTM
    Standards, Volume 02.05 (1983), p. 537, Am. Soc. for Testing and
    Materials, Philadephia, PA
Deshmukh, R. 1982, PhD. Thesis, Brown University
Deshmukh, R. and Gurland, J. 1982, Metallography 15 383-390
Exner, H.E. and Fischmeister, H.F. 1966, Arch. Eisenhuttenwesen 37 417
Exner, H.E. and Gurland, J. 1970, Powder Metallurgy 13 13 -31
Gurland, J. 1966, Trans. Met. Soc. AIME 236 642-646
Hagège, S., Nonet, G. and Delanquette, P. 1980, physica status solidi
    61 97-107
Hellsing, M., Henjered, A., Nordén, H. and Andrén, H.O. 1983, in Science
    of Hard Materials, R.K. Viswanadham et al, eds. Plenum Press, New York
    931 (discussion)
Hong, J. and Gurland, J. 1983, in Science of Hard Materials,
    R.K. Viswanadham et al, eds. Plenum Press, New York, 649-666
Lardner, E. 1973, Powder Metallurgy 16 107 (discussion)
Lee, H.C. and Gurland, J. 1978, Materials Science and Engineering 33 125-133
Lee, J.W., Jaffrey, D. and Browne, J.D. 1980, Powder Metallurgy, 23 57
Luyckx, S.B. 1981, in Proceedings, 5th Int. Conf. on Fracture, Cannes,
    France, 1981, D. Francois, ed. Pergamon Press, New York, Vol. 2, 1075
Nelson, R.J. and Milner, P.R. 1972, Powder Metallurgy 15 346-363
Sharma, N.K., Ward, I.D., Fraser, H.L. and Williams, W.S. 1980,
    J. Am. Ceramic Soc. 63 194-196
Underwood, E.E. 1970, Quantitative Stereology, Addison Wesley, Reading,
    MA, Ch. 4
Warren, R. 1983, in Deformation of Multi-Phase and Particle Containing
    Materials, J.B. Bilde-Sorensen et al, eds. RISO, Roskilde, Denmark, 575
Warren, R. 1976, Metallography 9 183-191

Discussion on Fundamentals of microstructure and deformation

Rapporteurs: M G GEE (1), T LUNDSTROM (2) AND S B LUYCKX (3)
Session Chairman: H FISCHMEISTER (4)

(1)   National Physical Laboratory, Teddington, Middlesex, UK
(2)   University of Uppsala, Sweden
(3)   University of the Witwatersrand, Johannesburg, RSA
(4)   Max Plank Institute, Stuttgart, FRG

In the discussion on the paper by Smid and Rogl, Dr Telle asked about the
purity of the initial boron powder used in the syntheses and for the
reason for selecting such a low temperature as 900°C for the ternary
sections. Mr Smid answered that the use of beta-boron or amorphous boron
did not result in any differences and that an increase of the c axis of BN
was observed at temperatures above 900°C, indicating a pick-up of impurities. Dr Lundström commented that a few years ago, attempts were made at
Uppsala to prepare an iron diboride using arc-melting as well as solid
state reaction techniques. No diboride was obtained however, and this
supports the present results that an iron diboride does not exist. He also
said that it is also worthwhile noting that the diffraction intensities of
the solid solution $FeB_{\sim 29}$ deviate strongly from those of pure beta-boron,
some of these lines might therefore erroneously have been ascribed to the
FeB_2 phase.

In the discussion of the contribution by Lönnberg, Engström and Lundström
Prof. Williams commented on the similarity of the basal plane (Nb) in Nb_2C
to the cube plane in NbC. The latter material also utilizes the similarity
to accommodate stacking faults between dissociated dislocations in NbC.
The faults must be hexagonal in the cubic structure of NbC but NbC has no
hexagonal modification. Nb_2C has, however, hexagonal symmetry, and so it
can satisfy this need. The fault is then of lower carbon/metal ratio than
the unfaulted material. We have confirmed this said Prof Williams, for
the isomorphous system TaC/Ta_2C using electron energy loss spectroscopy
(EELS). The implication for deformation is that dislocation mobility will
be limited by the ability of this low-carbon region to migrate, which, in
turn, will be limited by the mobility of carbon atoms (diffusion). The
migration energy of carbon in carbides is very large ($\sim 5eV$) so one
expects a high yield stress at high temperature, as observed. Prof Fischmeister asked for the degree of correlation between the thermal expansion
coefficient and melting point. Dr Lönnberg said that within a group of
materials with similar chemical bonding a relatively strong correlation
occurs but if materials of a deviating bond type are included the correlation decreases (see Professor Krikorian's contribution to the present
conference).

The paper presented by Dr Fantozzi on tungsten-hemicarbide attracted the
comment from Dr Lundström that neutron diffraction and chemical analysis

by Horsta and co-workers, Acta Chem Scand, 1977, had shown that the
hemicarbide was invariably substoichiometric when prepared at 1600-1700°C.
In his reply, Dr Fantozzi explained that his materials had been prepared
by coating plates of tungsten with carbon and then heating to 2600°C to
promote the formation of a hemicarbide which was found to be fully stoichiometric by neutron diffraction. Professor Williams asked whether the
speaker had tried to model the order-disorder reaction in these compounds,
and whether the carbon content of the stacking faults had been measured
since these would limit carbon diffusion. However, neither had been
attempted.

The preparation of sinterable SiC powder requires intensive milling, and
Dr Komac asked whether this had introduced metallic impurities into the
powder used by Dr Hamminger in the study of microchemistry and high temperature properties of SiC. The author confirmed that this was true. To
a further question from Professor Williams he said that there was no proof
that the Auger-analysed carbon was graphite, and consequently its role as
a sintering aid was not certain. However, Dr Schwetz explained that carbon
was introduced in an amorphous form as a phenolic resin, which is converted to glassy carbon in the heating cycle and transformed to graphite
during the high-temperature sintering of SiC-C doped compacts. The graphite contents of 0.5 to 3.0 wt% in sintered SiC can be easily detected by
X-ray diffraction. Dr Schwetz was also able to answer a question about the
role of boron and carbon, and aluminium and carbon in the sintering of
SiC. It appears that carbon acts similarly in SiC and BC (Schwetz and
co-workers, this conference). Boron and aluminium enhance grain boundary
and volume diffusion by entering into solid solution to as much as 0.2 wt%
and 0.6 wt% respectively in SiC during sintering.

Dr Hellsing and his co-workers used field-ion microscopy, atom probe
analysis and electron diffraction analysis in their investigation of the
composition and disposition of the binder phase in WC-Co cemented carbides. Dr Hellsing responded to a question from Dr Exner, that the application of these techniques had established that there was no evidence of
phases other than WC and β-Co at WC/Co interfaces in the 150 specimens
they had examined. To a second question, from Dr Exner on the W-C-Co phase
diagram, Dr Hellsing said that they had examined the equilibrium composition at 1250°C (25°C below the pseudo-eutectic according to Åkesson´s
thesis, 1982) and found that the solubility of tungsten near the Co-WC-C
boundary was lower than previously reported. Materials containing equilibrium contents of W and C could be cooled from the liquidus without
passing through a graphite or eta-phase region, according to the isothermal sections at 1200, 1275, 1350 and 1425°C obtained by Åkesson.

Dr Viswanadham remarked that the detection of Co at all WC/WC boundaries
was at variance with the high resolution election microscopy results of
Jayaram and Sinclair (J Amer Ceram Soc 1983, C-137). Dr Norden pointed out
that the quoted authors did not directly detect Co: in fact, they assumed
that Co was present whenever there was a lattice mismatch between grains.
In the present work, it was not possible to determine whether the boundaries were coincident or their relative orientation because the atom
probe tips were exposed for the minimum time needed for imaging in the
transmission electron microscope without risking contamination. In reply
to a question from Dr le Roux about the results of Loretto and Greenwood
who found relatively large amounts of Co at the WC/WC boundaries, Dr
Norden thought their preparation technique had been at fault. Forward and

back-sputtering occurs during ion milling and the formation of surface
layers of non-typical compositions can lead to ambiguous results.

The questions to Professor Williams on the paper on quantitative electron
energy loss spectroscopy of non-stoichiometric carbides, mainly concerned
the experimental technique. To Dr A Doi's question on the spatial resolution relative to the material's grain size, he replied it was 10-15 nm
which enabled analyses to be performed at many points within the 3 μm
grains. There was no evidence that specimen preparation affected the
results. Dr Sarin asked whether lattice parameters had been measured to
establish the degree of non-stoichiometry but this had not been attempted
yet.

Dr Chatfield opened the discussion on Dr le Roux' paper on recarburization
of eta-phase by asking about the recarburization temperature, and the
relevance of the allotropy of the cobalt to changes in the phantom eta
phase. The resintering temperature was $1450^{\circ}C$; the process was not understood but a correlation existed between the size of the lakes and the
allotropic composition. Further questions by Dr Exner about the process,
brought forth the comment from Professor Fischmeister that the relative
slow diffusion of W during sintering may have been responsible for the
phantom eta-phase. In reply to a question from Dr Sarin on the structure,
Dr le Roux said that the eta-phase was equally likely to be M_6C, $M_{12}C$ or
$M_{12}C_4$. Dr Sarin further suggested that the disappearance of the eta-phase
did not occur during sintering since there was no driving force for
diffusion, but Dr Chatfield thought that the volume change should serve
this function. Professor Williams wondered whether sub-stoichiometric
carbides could act as sources of carbon but Dr le Roux did not think this
was likely.

In his presentation on microstructural developments during sintering of
WC-Co hardmetals, Professor Gurland included a micrograph showing WC
crystals on the surface of a pore. When questioned about the growth
mechanism by Dr Kny, he had no explanation but suggestions had beed made
that it was a result of the non-uniform distribution of Co. There was some
discussion with Dr Exner and Professor Fischmeister about the strength of
coincidence boundaries. Professor Fischmeister commented that although
such boundaries should be strong because of their low energy, this may not
be true if interface segregation of Co occurred.

Dr Chatfield was critical of the used of the Lee and Gurland expression
for the hardness of cemented carbides. He thought that it overemphasised
the binder phase contribution and the good agreement was mainly due to an
overestimate of the binder phase hardness as a result of using a Hall-
Petch type equation. Thus for an equation:

$$1/H_c = f_\alpha/H_\alpha + f_\beta/H_\beta$$

where
$$H_\alpha = 1382 + 23.1 \, (d_\alpha)^{1/2}$$

then using Lee and Gurland's results, values for H_β can be obtained. The
spread was large but a mean Hardness of 440 was obtained in a Hall-Petch
type relationship of

$$H_\beta = 377 + 1.45 \, (\ell_\beta)^{-1/2}$$

He thought the effective Hall-Petch parameter was probably in better
agreement with that expected for Co-binder materials than that used by Lee
and Gurland. Using this expression, equally good agreement was obtained
between predicted and measured hardness as with Lee and Gurland's model.
Good fit was also obtained with Warren's results for other carbides, even
for VC-Co materials, where the agreement with the Lee and Gurland model
was poor. Dr Chatfield stated that he was not suggesting that his explanation was correct, he simply wished to show that more work was needed to
find a better model for the deformation of hardmetals.

Professor Gurland replied that he was surprised that the equivalent stress
model worked very well, as Chermant and his co-workers had already
noted. He thought that a combined criteria might be needed:

$$\sigma = \sigma_\alpha f_\alpha + \sigma_\beta f_\beta \tag{1}$$

$$\varepsilon = \varepsilon_\alpha f_\alpha + \varepsilon_\beta f_\beta \tag{2}$$

where equation 2 leads to Chermant's model. He agreed that more research
was needed. In conclusion, Dr Warren commented that whilst the series
model equation may give a better fit to the results, the parallel model of
Gurland was more physically plausible.

Inst. Phys. Conf. Ser. No. 75: Chapter 5
Paper presented at 2nd Int. Conf. Science Hard Mater., Rhodes

363

SiAlON hardmetal materials

K H JACK

Wolfson Research Group for High-Strength Materials
University of Newcastle upon Tyne, U.K.

ABSTRACT

Sialons are phases in the Si-Al-O-N and M-Si-Al-O-N systems where M
includes Li, Mg, Ca, Sc, Y and the rare earth elements.   They are
comparable in variety with the mineral silicates and are built up of one-,
two-, and three-dimensional arrangements of $(Si,Al)(O,N)_4$ tetrahedra in the
same way that the structural units of the silicates are SiO_4 tetrahedra.
These new materials include N-containing ceramics, glasses and glass-
ceramics that are being explored for their thermal, chemical and physical
properties.

β'-sialon, isostructural with β-Si_3N_4, is a successful engineering ceramic
and cutting tool for machining metals.   The equally promising α' and $0'$
sialons are based respectively on α-Si_3N_4 and the oxynitride Si_2N_2O.
Their dual-phase composites with β'-sialon have advantages over single-
phase materials.

INTRODUCTION

Silicon Nitride

Si_3N_4 has a unique combination of properties: it is strong, hard, wear-
resistant, stable to higher than 1800^0C, oxidation-resistant and, because
of its low coefficient of thermal expansion, has excellent resistance to
thermal shock.   It is also less than half as dense as steel.   These
features were pointed out by Parr, Martin and May (1959) at the Admiralty
Materials Laboratory twenty-five years ago and several organisations and
research groups in Britain became involved in attempts to produce silicon
nitride engine components.   The hotter an engine runs the more efficient
it becomes.   Ni-based superalloy gas turbines run at not much more than
1050^0C which is about the limit for a metallic alloy.   An engine that
operates at 1350^0-1400^0C would not only be more efficient but could use
poorer fuels and cause less pollution of the environment.   Thus, in 1971
the U.S. Department of Defense selected silicon nitride as the material for
the ceramic gas turbine.

The Densification of Silicon Nitride

After thirteen years and an expenditure of more than \$400 million in the
U.S.A., Germany and Japan on development programmes, there is still no

Fig. 1   Silicon nitride bar hot-pressed with
15wt%Y_2O_3 and oxidized at 1000ºC for 120h

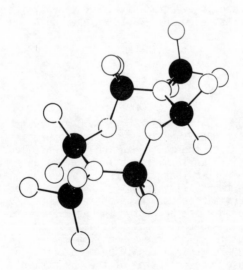

Fig. 2   The crystal structure of β-Si_3N_4
and β'-$(Si,Al)_3(O,N)_4$
● metal   ○ non-metal

commercial ceramic engine. One reason is that the desirable properties of Si_3N_4 are achieved only in fully dense material and the technological difficulty is in fabricating dense, precisely shaped components. The strong interatomic covalent bonding means that self diffusivity is small and the temperature at which the atoms begin to move is so high that the Si_3N_4 starts to decompose by volatilisation of nitrogen. It does densify at 2100°C under a high nitrogen pressure, but then excessive grain growth weakens the product. Thus, Si_3N_4 cannot be densified like an ordinary ceramic merely by firing; it must be hot-pressed in graphite dies at about 30MPa and 1800°C. The process is limited to simple shapes and the hot-pressed product is so hard that the final shape must be obtained by expensive diamond grinding. Moreover, so-called "fluxing agents" that are usually metal oxides, e.g. MgO or Y_2O_3, must be added to the Si_3N_4 powder to achieve densification. Their function is to react with a little Si_3N_4 and a surface layer of SiO_2 that is always present on each particle of nitride to give, at the high hot-pressing temperature, a metal oxynitride liquid. The latter allows mass transport and densification by "liquid-phase sintering".

The liquid that is necessary for densification cools and, in the case of MgO additive, forms a Mg-Si-O-N glass in the grain boundaries. Although the properties of the product are good at room temperature, the inter-granular glass softens above 1000°C and so the high-temperature creep resistance is poor.

With Y_2O_3 additive, the Y-Si-O-N liquid cools to give one or more crystalline oxynitrides. These oxidise at about 1000°C with an increase of up to 30% in specific volume. Oxidation therefore opens up grain boundaries, exposes fresh surfaces for further attack, and so the oxidation becomes catastrophic; see Fig. 1. Almost all additives necessary for densifying Si_3N_4 degrade its properties in some way.

THE Si-Al-O-N SYSTEM

β'-Sialon

Si_3N_4 is built up of tetrahedral SiN_4 units (see Fig. 2) just as the fundamental structural unit in all mineral silicates is the SiO_4 tetrahedron. Indeed, the atomic arrangement in β-Si_3N_4 is the same as that in Be_2SiO_4 and Zn_2SiO_4. Since Al can always partially replace Si in the wide variety of silicates that make up the lithosphere, it is not too surprising that up to two-thirds of the Si of β-Si_3N_4 may be replaced by Al without change of structure provided that an equivalent concentration of nitrogen is replaced by oxygen (Jack and Wilson, 1972) as in the equation:

$$Si^{4+} N^{3-} \longrightarrow Al^{3+} O^{2-} \qquad\qquad 1$$

In the Si-Al-O-N phase diagram at 1800°C shown in Fig. 3, Al content is plotted on the x-axis (the balance being Si) and O content on the y-axis (balance, N). The β'-sialon phase, in which the β-Si_3N_4 crystal structure is retained, extends over a range of composition $Si_{6-z}Al_zO_zN_{8-z}$ with z between 0 and 4. Because the sialon has the same atomic arrangement, its mechanical and physical properties are similar to those of Si_3N_4. Chemically, however, it has some of the features of Al_2O_3. It is a solid solution and, like all solutions, its vapour pressure is lower than that of the pure solvent, i.e. it has a lower melting temperature and a

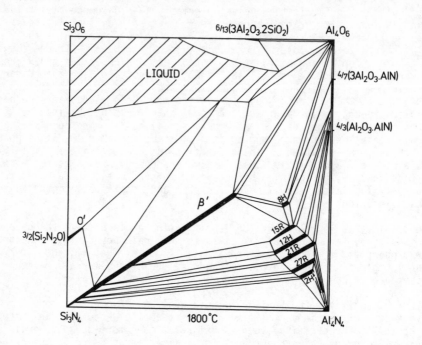

Fig. 3   The Si-Al-O-N behaviour diagram at 1800°C

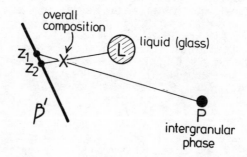

Fig. 4   Post-preparative heat-treatment
of β' + glass to give β' + YAG

higher boiling point. Thus, compared with Si_3N_4, β'-sialon will readily form more liquid at lower temperatures with an oxide additive such as Y_2O_3. This, in turn, means that the ceramic can be fully densified by pressure-less sintering without the need for hot-pressing. The lower temperature of densification avoids excessive grain growth and so the fine-grain strength of the material is retained. Finally, the lower vapour pressure reduces any volatilisation and decomposition at high temperatures.

Post-preparative Heat-treatment

The Y-sialon liquid necessary for densification at high temperature cools to give an intergranular glass but by post-preparative heat-treatment at about 1400°C, or by controlled cooling, this can be reacted with the matrix to give Y-Al-garnet (YAG, $Y_3Al_5O_{12}$) in the grain boundaries and a slightly changed β'-sialon composition:

$$Si_5AlON_7 + Y\text{-}Si\text{-}Al\text{-}O\text{-}N \longrightarrow Si_{5+x}Al_{1-x}O_{1-x}N_{7+x} + Y_3Al_5O_{12}$$

$\quad\beta'$-sialon $\qquad$ glass $\qquad\qquad\qquad\beta'$-sialon $\qquad\qquad$ "YAG" $\qquad\qquad$ 2

These relationships are shown schematically in Fig. 4. Since the grain-boundary phase is an oxide, the product has excellent oxidation resistance (Fig. 5), and because there is no inter-granular glass it has good creep properties (Fig. 6).

COMMERCIAL SIALON CERAMICS

Process Development

Based upon the above principles, several years of process development (see Lewis and co-workers, 1980) has resulted in the present commercial production of β'-sialons. This is a family of materials consisting of two groups with different main microstructural constituents: (i) β'-sialon plus glass; and (ii) β'-sialon plus YAG. The strength of the glass-containing materials is high at room temperature but decreases above 1000°C. The YAG materials have a somewhat lower room-temperature strength but retain strength and creep resistance up to 1400°C. Properties of a type (i) β'-sialon are listed in Table 1; it is used for extrusion tooling, drawing dies and plugs, welding components and industrial wear parts.

Forming, Sintering and Machining

The appropriate mixes of nitride, oxide and oxynitride powders can be cold or warm formed prior to sintering by using all the shaping methods normally employed for oxide ceramics. These include: (a) isostatic pressing and, for large numbers of components, uniaxial pressing with and without binders and lubricants; (b) warm or cold extrusion with addition of plasticisers to produce continuous sections of pre-formed material; (c) injection moulding for intricate shapes; and (d) slip casting in aqueous media.

Pre-forms may be machined in the green state and, after debonding where necessary in air at up to 500°C, are then sprayed with a protective coating of refractory oxides before sintering in a nitrogen atmosphere at 1750°-1850°C. Reproducible linear shrinkage occurs and, because this is allowed for, the final diamond machining of the hard, fully dense component to precise dimensional tolerances is reduced to a minimum.

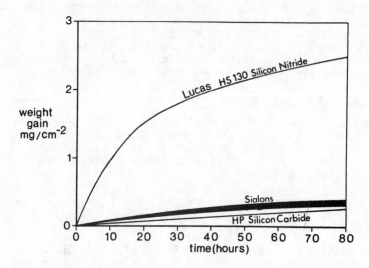

Fig. 5   Oxidation of β'-sialon at 1400°C in flowing dry air
(after Arrol, 1974)

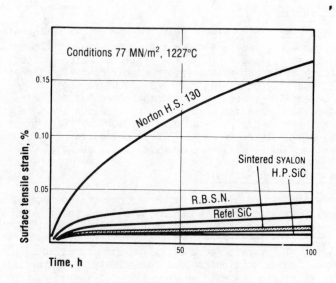

Fig. 6   Creep of β'-sialon at 1127°C and 77MPa
(after Lumby, North and Taylor, 1978)

TABLE 1   Typical Physical Properties of Type (i) β'-Sialon

Property	Value	Units
3-point Modulus of Rupture, $20^{\circ}C$	945	MPa
Weibull Modulus	15	-
Tensile Strength, $20^{\circ}C$	450	MPa
Compressive Strength, $20^{\circ}C$	3,500	MPa
Young's Modulus of Elasticity, $20^{\circ}C$	300	GPa
Hardness, $20^{\circ}C$	1,800	HV0.5
Fracture Toughness (K_{1C})	7.7	MPa m$^{\frac{1}{2}}$
Poisson's Ratio	0.23	-
Density	3,250	kg m^{-3}
Thermal Expansion Coefficient (0°-$1200^{\circ}C$)	3.04×10^{-6}	K^{-1}
Thermal Conductivity, $20^{\circ}C$	22	Wm^{-1}K^{-1}
Electrical Resistivity, $20^{\circ}C$	10^{10}	ohm m
Permittivity at 10GHz, $20^{\circ}C$	8.2	-
Loss Tangent at 10GHz, $20^{\circ}C$	0.002	-
Thermal Shock Resistance (quenched in cold water)	900	$T^{\circ}C$
Coefficient of Friction (Sialon on Sialon in 10W40 oil at $80^{\circ}C$)	0.04	-

TABLE 2   Cutting Performances of Co-bonded WC, Al_2O_3, and a β'-Sialon

		Cast Iron	Hardened Steel EN31	Incoloy 901
WC	Cutting speed, m/min	250	5	20
	Depth of cut, mm	6.5		
	Feed rate, mm/rev	0.50		
Al_2O_3	Cutting speed, m/min	600	Impossible	300
	Depth of cut, mm	6.5	to cut	No second
	Feed rate, mm/rev	0.25		entry
β'-SIALON	Cutting speed, m/min	1100	120	300
	Depth of cut, mm	10.0	0.5	2.0
	Feed rate, mm/rev	0.50	0.25	0.25

Engineering Applications

Although the main motivation for the development of engineering ceramics was, and still is, the prospect of the ceramic gas turbine, its commercial realisation might well take another decade and meanwhile the unique properties of β'-sialons and their ease of fabrication are being applied in other directions.

So-far, the most successful application has been as a cutting tool for machining metals. Fig. 7 shows a selection of tool tips, and their performances in cutting cast iron, hardened steel and a nickel-based alloy are compared with those of Co-bonded WC and of Al_2O_3 in Table 2. The "lead time" for machining some of the RB211 aeroengine turbine discs is said to have been reduced to less than one-quarter by using sialon inserts. They are being manufactured under licence by two of the world's major tool companies, one in Sweden and one in the U.S.A.

Excellent thermal shock resistance, high-temperature mechanical strength and electrical insulation combine to make sialon unexcelled in welding operations. Fig. 8 shows location pins for the resistance welding of captive nuts on vehicle chassis. The usual hardened steel pins in Al_2O_3 insulating sleeves last for 7000 operations i.e. a shift; sialon pins have completed 5 million operations - i.e. a year - without signs of wear.

Ten years ago, hot-pressed Si_3N_4 was predicted to be an ideal material for ball and roller bearings but the cost of machining from the simple hot-pressed shapes was prohibitive. With a β'-sialon sintered to almost the final required dimensions, and with even better wear resistance, hardness and tribological properties, these applications are again possible; see Fig. 9.

Sialon die inserts used in the extrusion of brass, copper, bronze, aluminium, titanium and steel have given remarkable improvements in surface finish, dimensional accuracy and higher extrusion speeds. The material also copes with a wide range of wear environments in contact with metals with or without lubrication and the tube-drawing dies and mandrel plugs shown in Fig. 10 are examples of where there are marked increases in productivity compared with the conventional use of tungsten carbide.

Other applications depend on the resistance of sialons to most molten metals, including steel, although they are attacked by slag. It is being used in casting and metal spraying. The centre-piece of Fig. 10 is a coracle used in pulling single crystals from a GaP melt that attacks most other materials; it is a slip-cast β'-sialon.

PROCESSING ROUTES AND RAW MATERIALS

Two of the Si-Al-O-N elements are the most abundant in the earth's crust and the other two make up its atmosphere so that there is no possibility of a raw materials shortage in sialon production. In what is now a large field of sialons, the β'-phase is the only one that has been explored in detail for its technological potential; others with at least equal promise are derived from α-silicon nitride and silicon oxynitride. Moreover, the production of β'-sialon by different methods over its full range of composition has not yet been fully investigated.

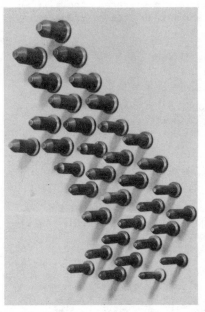

Fig. 8 Captive weld nut location pins

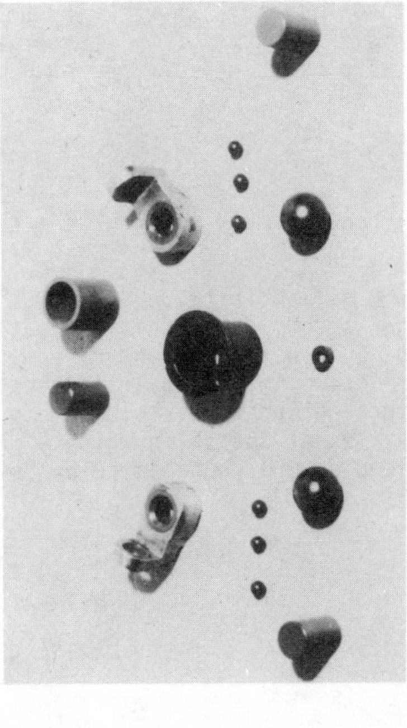

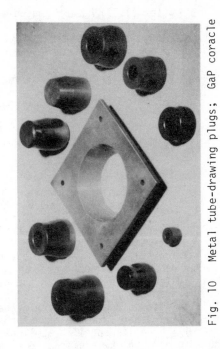

Fig. 10 Metal tube-drawing plugs; GaP coracle

Fig. 7 Selection of sialon cutting tool tips

Fig. 9 Ball, roller and trunnion bearings

Lee and Cutler (1979) have produced sialon powder according to equation 3 by heating pelletised mixtures of clay and coal in nitrogen, and similar production of nitrogen ceramics by carbothermal reduction of oxides in nitrogen is now being extensively explored.

$$3Al_2 \left[Si_2O_5 \right] (OH)_4 \ + \ 15C \ + \ 5N_2 \longrightarrow 2Si_3Al_3O_3N_5 \ + \ 6H_2O \ + \ 15CO$$

$$\text{clay} \qquad\qquad \text{coal} \quad \text{nitrogen} \qquad \beta'\text{-sialon, z=3}$$

3

Then, Umebayashi (1977) in Japan has produced β' by reaction in nitrogen of volcanic ash (impure silica) and aluminium powder; see equation 4.

$$2SiO_2 \ + \ 4Al \ + \ 2N_2 \ \longrightarrow \ Si_2Al_4O_4N_4$$

$$\text{volcanic} \quad \text{aluminium} \qquad \beta'\text{-sialon, z=4}$$
$$\text{ash} \qquad \text{powder}$$

4

Rice husks are mainly cellulose and silica. Their pyrolysis gives "black ash", an intimate mixture of carbon and silica, which is a useful starting material for silicon nitride and sialon production. Clay and coal, volcanic ash and rice husks are not going to produce pure β'-sialons for sophisticated applications like the ceramic turbine but they will provide useful refractory bricks, furnace linings and materials resistant to molten metals. Just as there are many grades of alumina ceramics and refractories, there will be many forms and grades of β'-sialon each of which will have its specific applications and its appropriate methods of manufacture. Unlike alumina, β'-sialon has excellent thermal shock properties and will replace alumina, with advantage, in many of its uses.

α'-SIALONS

α' Structures based on the $Si_{12}N_{16}$ unit cell of α-silicon nitride occur in M-Si-Al-O-N systems where M includes Li, Ca, Y and all the rare earth elements from Nd to Lu. Appropriate mixtures of the nitrides, or of nitrides plus oxide, are heated without pressure at 1750°C in nitrogen or argon:

$$0.5Ca_3N_2 \ + \ 3Si_3N_4 \ + \ 3AlN \ \longrightarrow \ Ca_{1.5} \left[Si_9Al_3N_{16} \right]$$

5

$$\text{or} \quad CaO \ + \ 3Si_3N_4 \ + \ 3AlN \ \longrightarrow \ Ca \left[Si_9Al_3ON_{15} \right]$$

6

Unlike β-silicon nitride, the α unit cell has two large interstitial sites that can accommodate additional atoms; α' is derived by partial replacement of Si^{4+} with Al^{3+} and valency compensation is effected by cations such as Ca^{2+} occupying these interstices in the Si-Al-N network. The structural principle is similar to that in the formation of "stuffed" derivatives of quartz in which Al^{3+} replaces Si^{4+} and positive valency deficiencies are compensated by "stuffing" Li^{1+} or Mg^{2+} into interstitial sites.

$$Si_2O_4 \ \longrightarrow \ Li \left[SiAlO_4 \right]$$

$$\text{quartz} \qquad \beta\text{-eucryptite}$$

7

$$Si^{4+} \ \longrightarrow \ Li^{1+} \ Al^{3+}$$

8

When α' is synthesised entirely from nitrides the valency compensation is due solely to the additional modifier cations and the limiting compositions contain not more than two of these per unit cell, e.g. $Ca_2 \left[Si_8 Al_4 N_{16} \right]$ or $Y_2 \left[Si_6 Al_6 N_{16} \right]$, although these limits have not been achieved experimentally. Where a modifier oxide is used, oxygen replaces nitrogen in the network but its limit is probably about one oxygen atom per unit cell. The α'-phases are stable in inert, nitriding and carburizing atmospheres up to 1750°C. Like β'-sialons, they can be prepared by carbothermal reduction of mixed oxides in nitrogen. They have good oxidation resistance up to 1350°C and have coefficients of thermal expansion similar to that of silicon nitride ($\sim 3 \times 10^{-6}$/°C). Prospects for their technological application are as good as those for β'-sialon (Hampshire and co-workers, 1978) and mixed phase $\alpha' : \beta'$ composites with controlled microstructures are already being used in preference to β' in some cutting tool applications.

O'-SIALONS

Fig. 3 shows that there is a limited solubility, about 10 mole% Al_2O_3 at 1800°C, of alumina in silicon oxynitride, Si_2N_2O, to give an O'-sialon solid solution without change of structure. A suitable additive such as Y_2O_3 lowers the melting temperature and, by increasing the volume of liquid, allows densification by pressureless sintering in the same way as in the processing of β'-sialons. Thus, reaction and densification of mixed powders of Si_3N_4, SiO_2, Al_2O_3 and Y_2O_3 produce O'-sialon ceramics with an inter-granular glass phase that can be devitrified by post-preparative heat-treatment to give $Y_2Si_2O_7$. Once again the products have low coefficients of thermal expansion and hence good thermal shock properties. They are oxidation resistant to 1350°C and should be as useful as β' and α' in engineering applications (Trigg and Jack, 1984).

Because its structure consists of parallel planes of covalently-bonded silicon-nitrogen atoms linked by Si-O-Si bonds, O'-sialons can accommodate additional interstitial cations between the planes when some of the linking oxygens are replaced by nitrogen:

$$O^{2-} \longrightarrow N^{3-} \quad Li^{1+} \qquad\qquad 9$$

It is suggested that Li-O'-sialon materials might provide ceramics with fast ion-transport of lithium and hence the possibility of solid electrolytes for battery applications.

NITROGEN GLASSES

Glasses occur in all the sialon systems so far studied and are important because the mechanical properties of the nitrogen ceramics, particularly their high-temperature strength and creep resistance, depend on the amount and characteristics of the grain-boundary glass. The glasses are also of interest in their own right, and a systematic study by Drew, Hampshire and Jack (1981) of Mg, Ca, Y and Nd sialon systems shows that by fusing powder mixtures of SiO_2, Al_2O_3, Si_3N_4 and AlN with the appropriate metal oxide at 1600°-1700°C in a nitrogen atmosphere and then furnace cooling, glasses containing up to 15 at%N are obtained.

The Y-Si-Al-O-N glasses are typical and, as shown by Figs. 11 and 12,

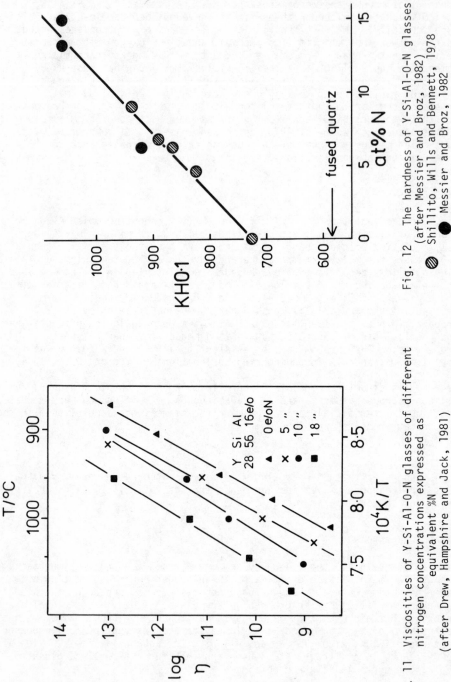

Fig. 11 Viscosities of Y-Si-Al-0-N glasses of different
nitrogen concentrations expressed as
equivalent %N
(after Drew, Hampshire and Jack, 1981)

Fig. 12 The hardness of Y-Si-Al-0-N glasses
(after Messier and Broz, 1982)
Shillito, Wills and Bennett, 1978
Messier and Broz, 1982

their viscosity and hardness both increase with increasing nitrogen content. Where more refractory and erosion-resistant glasses are required, e.g. in radome applications, these nitrogen glasses are possible candidates.

CONCLUSIONS

Simple principles of "ceramic alloying" have produced an expanding field of new crystalline and vitreous metal-Si-Al-O-N materials. The first to be commercially developed, β'-sialon, has wide applications in engines, cutting tools, bearings and welding. It promises to be a leading contender for the ceramic gas-turbine.

Other sialons, notably α' and O' based upon the structures of α-silicon nitride and silicon oxynitride respectively, show equal promise. They have the same low coefficients of thermal expansion and, by suitable processing, materials containing mixtures of two or more of the compatible phases β', α' and O' may be obtained. By varying the proportions of the constituent phases and their rates of formation, microstructures can be tailored to give desirable mechanical and physical properties more easily than with a single-phase sialon.

ACKNOWLEDGMENT

Publication of this paper has been possible through the award of an Emeritus Fellowship to the author by the Leverhulme Trust.

REFERENCES

Arrol, W.J. 1974, "Ceramics for High Performance Applications", Proc. 2nd Army Mat. Tech. Conf., Hyannis. Eds. J.J. Burke, A.E. Gorum and R.N. Katz. Brook Hill Publishing Company, Chestnut Hill U.S.A. pp 729-738

Drew, R.A.L., Hampshire, S. and Jack, K.H. 1981, "Special Ceramics 7", Proc. Brit. Ceram. Soc. 31 119-132

Hampshire, S., Park, H.K., Thompson, D.P. and Jack, K.H. 1978, Nature 274 880-882

Jack, K.H. and Wilson, W.I. 1972, Nature Phys. Sci. (London) 238 28-29

Lee, J.G. and Cutler, I.B. 1979, Ceramic Bull. 58 869-871

Lewis, M.H., Bhatti, A.R., Lumby, R.J. and North, B. 1980, J. Mat. Sci. 15 103-113

Lumby, R.J., North, B. and Taylor, A.J. 1978 "Ceramics for High Performance Applications - II", Proc. 5th Army Mat. Tech. Conf., Newport R.I. Eds. J.J. Burke, E.N. Lenoe and R.N. Katz. Brook Hill Publishing Company, Chestnut Hill U.S.A. pp 893-906

Messier, D.R. and Broz, A. 1982, J. Am. Ceram. Soc. 65 C-123

Parr, N.L., Martin, G.F. and May, E.R.W. 1959, "Study and Application of Silicon Nitride as a High Temperature Material", Admiralty Materials Laboratory Report No. A/75(s)

Shillito, K.R., Wills, R.R. and Bennett, R.B. 1978, J. Am. Ceram. Soc. <u>61</u> 537

Trigg, M.B. and Jack, K.H. 1984, Proc. 1st Int. Symposium on Ceramic Components for Engine, Hakone, Japan.   Eds. S. Somiya, E. Kanai and K. Ando.   KTK Scientific Publishers, Tokyo pp 199-207

Umebayashi, S. 1977, "Nitrogen Ceramics", Proc. NATO Advanced Study Inst. Ed. F.L. Riley.   Noordhoff, Leyden pp 323-328

Inst. Phys. Conf. Ser. No. 75: Chapter 5
Paper presented at 2nd Int. Conf. Science Hard Mater., Rhodes

377

High temperature strengthening of zirconia toughened ceramics

N CLAUSSEN

Technische Universität Hamburg-Harburg, D-2100 Hamburg 90, FRG

ABSTRACT

Transformation-toughened (i.e. ZrO_2-toughened) ceramics represent a new class of high performance ceramics with spectacular strength properties at low and intermediate temperatures. However, at temperatures above about 700°C, most of these tough oxide-base ceramics can no longer be used as load-bearing engineering parts because of characteristic deficiencies. The aim of the present paper is to provide and discuss microstructural design strategies which may enable ZrO_2-toughened ceramics to be applied at higher temperatures. From the various strategies suggested, three appear to show good prospects, namely (a) the prevention of glassy inter-granular films, (b) the addition of hard high modulus particles and (c) whisker or fibre reinforcement. Experimental approaches are presented from some ZrO_2-toughened ceramics, e.g. tetragonal ZrO_2 polycrystals and ZrO_2-toughened cordierite, spinel and mullite.

INTRODUCTION

The transformation of ZrO_2-toughened ceramics has recently received increasing interest in its application to high performance structural parts, mainly because of some rather spectacular strength data (Heuer and Hobbs, 1981; Claussen, Rühle and Heuer, 1984). For instance, values of over 2500 MPa for tetragonal ZrO_2 polycrystals (TZP) have been reported (Tsukuma and co-workers, 1985), values which 10 years ago would have been unthinkable for a bulk polycrystalline ceramic material. Table 1 shows some of the strength and toughness improvements to various engineering ceramics obtained by using ZrO_2 in the form of tetragonal or monoclinic particles dispersed in other ceramic matrices. The results for partially stabilized ZrO_2 (PSZ) (Table 1, second row) and for TZP (Table 1, third row) are compared with those for fully stabilized cubic ZrO_2 (c-ZrO_2) (Table 1, first row). Different mechanisms, e.g. stress-induced trans-formation, microcracking, crack deflection, inhibition of grain growth, contribute to the increase in toughness and strength to various extents (Claussen, 1984). Compared with the two currently available main competi-tors for high performance engineering ceramics (SiC and Si_3N_4), ZrO_2--toughened ceramics have a clear advantage in the high applied stress regime (Fig. 1), while the two covalent representatives exhibit a much

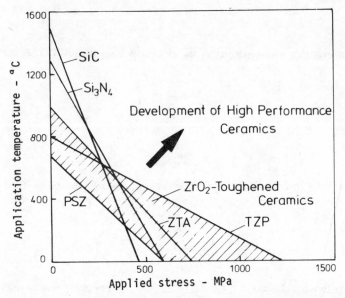

Fig. 1 Schematic diagram indicating the potential application range of some engineering ceramics (ZTA is zirconia toughened alumina)

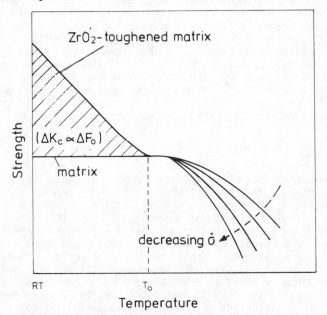

Fig. 2 Characteristic dependence of strength on temperature for a zirconia toughened ceramic. At $T > T_o = 1000\,°C$ (equilibrium temperature $T_o = 0.5(M_s + A_s)$ where $M_s = 850\,°C$, $A_s = 1150\,°C$, and M_s and A_s are defined in the text), stress-induced transformation no longer occurs.

TABLE 1   <u>Room Temperature Toughness K_{Ic} and Strength σ_f
of ZrO_2-toughened Ceramics</u>

Ceramic material	Matrix alone		Matrix + ZrO_2	
	K_{Ic} MPa m$^{1/2}$	σ_f MPa	K_{Ic} MPa m$^{1/2}$	σ_f Mpa
c-ZrO_2	2.4	180	2-3	200- 300
PSZ			6-8	600- 800
TZP			7-12	1000-2500
Al_2O_3	4	500	5-8	500-1300
Mullite	1.8	150	4-5	400- 500
Spinel	2	180	4-5	350- 500
Al_2TiO_5	0.8	40	2.5	120
Cordierite	1.4	120	3	300
Sintered Si_3N_4	5	600	6-7	700- 900

greater high temperature potential. There are two principal reasons why
the present ZrO_2-toughened ceramic types must be considered to be low
temperature ceramics; firstly, the creep rates in the usually oxide-based
ZrO_2-toughened ceramics (especially PSZ and TZP) are high relative to those
of SiC and Si_3N_4 and, secondly, the most important toughening mechanism,
which results from the stress-induced phase transformation, decreases
drastically towards the equilibrium transformation temperature T_o (which
is equal to about 1000°C for bulk ZrO_2; $T_o = 0.5 (M_s + A_s)$ where M_s
($\approx$ 850°C) is the temperature at which the transformation of tetragonal
ZrO_2 (t-ZrO_2) to monoclinic ZrO_2 (m-ZrO_2) starts during cooling and A_s
($\approx$ 1150°C) is the reverse transformation temperature) as schematically
depicted in Fig. 2. This is due to the decreasing driving force for the
transformation ΔF_o with increasing temperature. A typical example for the
strength decrease at $T < T_o$ is given in Fig. 3 for a ZrO_2-toughened Al_2O_3
(Lange, 1982). For $T > T_o$ the mechanical properties are further degraded
depending on the type of matrix and on the stress situation, e.g. with
decreasing stress rates $\overset{\bullet}{\sigma}$ the critical failure stress is reduced.

This disappointing characteristic feature limits the load-bearing application of ZrO_2-toughened ceramics to low or, at the most, medium
temperatures. Therefore it represents a materials science challenge to
search for ways of improving the high temperature mechanical properties of
these materials. The typical low temperature toughness of ZrO_2-toughened
ceramics remains an important property even when they are used at higher
temperatures, e.g. under thermal shock conditions the maximum stresses
usually occur at rather low temperatures. However, a better strength performance up to 1000-1200°C would considerably widen their technical use.

The purpose of the present paper is to discuss some microstructural design strategies which will, on the one hand, retain at least some of the
strength and toughness at low temperatures imparted by the stress-induced
transformation and, on the other hand, yield improved mechanical behaviour
at elevated temperatures. In order to understand these strategies, the
high temperature deficiencies associated with the microstructures of ZrO_2-
toughened ceramics are outlined first.

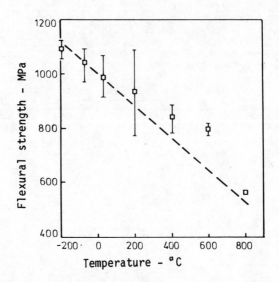

Fig. 3 Temperature dependence of strength of an alumina- 29.5 vol%
zirconia with 2 mol% yttria composite, hot pressed at 1600°C for 1h,
annealed at 1500°C. (Compare with Fig. 2, for T < T$_0$): ------,
predicted from K$_c$ data. (After Lange, 1982)

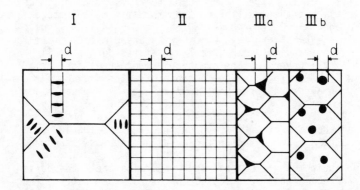

Fig. 4 Schematic diagram of the three basic zirconia toughened ceramic
microstructures: I, PSZ with cubic grains of 50-100μm containing
tetragonal precipitates with d = 0.2μm; II, TZP with predominantly
tetragonal grains with d = 0.2-1μm; IIIa, dispersion-type
zirconia-toughened ceramics with intergranular t-zirconia (0.5-1μm) or
m-zirconia particles; IIIb, dispersion-type zirconia-toughened
ceramics with intragranular zirconia dispersion. A great variety of
combinations of these microstructures is possible (Claussen, 1984)

HIGH TEMPERATURE DEFICIENCIES OF PRESENT ZrO_2-TOUGHENED CERAMICS

The general criteria of high temperature failure in ceramics naturally
also apply to ZrO_2-toughened ceramics, i.e. crack nucleation and propagation, both phenomena being associated with either diffusive cavitation or
viscous hole crack growth (Evans and Rana, 1980). Additionally, ZrO_2-
toughened ceramics are prone to destabilization of c- or t-ZrO_2.

Some characteristic microstructural features of ZrO_2-toughened ceramics
closely related to the high temperature failure can be summarized using
the schematic diagram in Fig. 4. It shows the three basic microstructures
of these materials, and these microstructures can be combined with each
other and modified in various ways (Claussen, 1984).

Type I microstructure. Conventional PSZ alloyed with 8-10 mol. % MgO
(Mg-PSZ) or CaO (Ca-PSZ) is usually sintered in the cubic solid solution
range, i.e. at relatively high temperatures ranging between 1650 and
1850°C. With appropriate cooling, the microstructure can be made to
consist of large (50-100 μm) cubic grains within which are dispersed
coherent precipitates of tetragonal symmetry. The precipitates are then
coarsened or modified by aging at temperatures between 1100 and 1450°C in
order to optimize their transformability under stress. The shape and the
size d of these precipitates depend on the species of the stabilizer, i.e.
lens shaped in Mg-PSZ, cube shaped in Ca-PSZ and plate shaped in Y-PSZ
(Swain and Hannink, 1981).

Type II microstructure. TZP is a fine-grained (d = 0.2-1 μm) and
predominantly tetragonal material doped with rare earth oxides, e.g.
2-4 mol. % Y_2O_3 (Y-TZP) or 9-14 mol. % CeO_2. The critical grain size d_c*
depends on alloying, e.g. Y_2O_3 content, Y_2O_3 distribution, amount and
distribution of glassy phase.[3] In order to enhance densification at relatively low temperatures (1350-1500°C) as required by the tetragonal phase
field boundary (Ruh and co-workers, 1984; Rühle, Claussen and Heuer,
1984), impurities such as silicates and Al_2O_3 have been tolerated and
result in a continuous amorphous film between the grains (Rühle, Claussen
and Heuer, 1984).
[* It is convenient to define a critical size d_c above which, at room
temperature, spontaneous transformation will occur and below which it will
not.]

Type III microstructure. Dispersion-type ZrO_2-toughened ceramics encompass non-ZrO_2-based matrices in which ZrO_2 particles are incorporated
either intergranularly (type IIIa microstructure) or intragranularly (type
IIIb microstructure) by a variety of processing techniques (Claussen,
1984). Type IIIa composites are the most common and most advanced ZrO_2-
toughened ceramics because (a) they usually result from conventional
processing and (b) the toughening effect, i.e. the transformability of
irregularly shaped t-ZrO_2 particles, is enhanced relative to that of
intragranular normally rounded particles (Heuer and co-workers, 1982).
The critical ZrO_2 particle size is similar to those of the other
microstructures (Fig. 4, types I and II) i.e. it is between 0.2 and 1 μm
depending on matrix properties, impurity content etc. For instance,
Claussen and Rühle (1981a) have shown that in ZrO_2-toughened Al_2O_3, $d_c \approx$
0.6 μm. For geometrical reasons, the grain size of type IIIa ZrO_2-
toughened ceramics is of the same order of magnitude (0.5 - 5 μm).

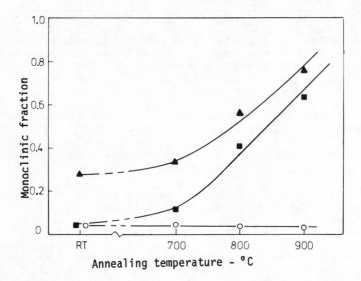

Fig. 5 Change in monoclinic fraction at the surface of Mg-PSZ (▲), plasma-sprayed Mg-ZrO$_2$ (■) and Y-TZP (o) with annealing temperature (annealed for 100h in a slightly reducing atmosphere). (Anderson and Hermansson, 1983)

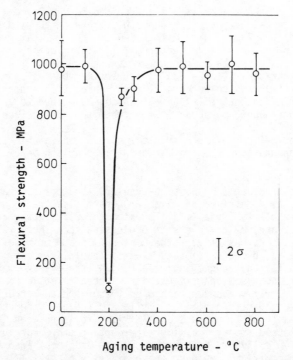

Fig. 6 Strength vs aging temperature of 3 mol% yttria-TZP (sintered for 1h at 1400°C; 2000h anneal in air)(Matsui, Soma and Oda, 1984)

PSZ ceramics seem to be rather suitable for high temperature applications because the grain sizes are large compared with those in TZP. However, they usually contain a substantial amount of silicate impurities, leading to an amorphous film at the grain boundary. Grain boundary sliding and viscous crack nucleation and growth will become a problem at T $\gtrsim$ 700°C depending on the glass composition. Another severe deficiency of Mg-PSZ (this type is already used at high temperatures) is the diffusive decomposition to m-ZrO$_2$ plus MgO at T $\gtrsim$ 700°C (Hannink, 1983). Fig. 5, (Anderson and Hermansson, 1983), shows the increase in monoclinic fraction at the surface of a conventional Mg-PSZ with annealing temperature in a slightly reducing atmosphere (the decomposition in an oxidizing atmosphere was actually the same). There is probably no means of overcoming this handicap for materials with MgO doping alone, but the use of CaO- or Y$_2$O$_3$-doped PSZ materials at T > 700°C may represent an alternative.

Y-TZP is thermodynamically stable under the conditions shown in Fig. 5 and is the reason why this material is at present preferred for engine applications as liners, hot-plates etc. (Larsen and Adams, 1984). However, a deficiency of TZP arises from the small grain size and silicate-wetted "grain boundaries", both of which will produce severe creep problems at T $\gtrsim$ 800-900°C.

A further drawback of Y-TZP is its mechanical property degradation in the temperature range 150-300°C. This is due to a transformation to monoclinic symmetry starting from the surface (Claussen, 1984; Schubert, Claussen and Rühle, 1984a), and is particularly enhanced in humid atmospheres (Kobayashi and Masaki, 1982; Tsukuma, Kubota and Tsukidate, 1984) and hot aqueous solutions (Nakajima, Kobayashi and Murata, 1984). The destabilizing mechanisms are not yet fully understood but may be associated with hydrothermally induced surface nucleation of m-ZrO$_2$, e.g. hydroxide formation or Y^{3+} depletion (Schubert, Claussen and Rühle, 1984a). The extent of the surface transformation depends on grain size and Y$_2$O$_3$ solute content, i.e. with increasing grain size at a given Y$_2$O$_3$ level or decreasing Y$_2$O$_3$ content at a given grain size both the conversion rate and the toughness increase (Claussen, 1984). For instance, 3 mol. % Y$_2$O$_3$-TZP with a grain size of 0.4 µm shows no surface transformation after a 50 h anneal at 200°C, while the same material with a 0.67 µm grain size has transformed to about 70 vol. % m-ZrO$_2$ (Sato and Shimada, 1984). For extended air annealing (2000 h) at 200°C, 3 mol. % Y$_2$O$_3$-TZP with a grain size of about 0.5 µm exhibits a drastic decrease in strength as depicted in Fig. 6 (Matsui, Soma and Oda, 1984). The results of bend creep tests at a constant outer-fibre stress of 400 MPa at temperatures between 150 and 300°C are shown in Fig. 7, which demonstrate that creep is most rapid at 250°C, resulting in failure after about 2 h, while for temperatures less than 150°C and greater than 300°C, up to about 1000°C, no measurable creep was observed within 5 h test time. X-ray examination of the failed samples revealed m-ZrO$_2$ on the tensile but not on the compressive surface, indicating that tensile stresses drastically enhance the transformation from tetragonal to monoclinic phase. This short-term stability test of TZP ceramics is more realistic than autoclave tests in water vapour (Rühle, Claussen and Heuer, 1984; Schubert, Claussen and Rühle, 1984a) for simulating behaviour under service conditions.

Most other non-ZrO$_2$ matrix ZrO$_2$-toughened ceramics (Fig. 4, type IIIa) will exhibit high temperature failures that are characteristic for the respective fine-grain-size matrix materials. Depending on processing technique and impurity contents, intergranular phases are expected to degrade the mechanical properties at elevated temperatures. However, the chances of producing more refractory dispersion-type ZrO_2-toughened ceramics are in principle higher than for PSZ or TZP if matrices that are more creep resistant are developed, e.g. Al_2O_3, spinel, mullite, with controlled processing to try to eliminate wetting intergranular films.

HIGH TEMPERATURE MICROSTRUCTURAL DESIGN STRATEGIES

General microstructural design criteria also apply to ZrO_2-toughened ceramics, i.e. on the assumptions that the creep rate for viscous flow is proportional to $\sigma/\eta d$, the creep rate for plastic flow is proportional to σ^n (n = 2-20), and the creep rate for diffusional creep is proportional to $\sigma D/d^m$ (m = 2 or 3) then the viscosity η should ideally be high, the applied stress σ should be low, the grain size d should be large, and the diffusivity D should be low. The dihedral angle formed in the dispersed phase should be large but may be uncontrollable (Evans and Rana, 1980). However, if the stress-induced transformation toughening mechanism is to be partly active, compromises have to be made in microstructure design for high temperature applications. This is because some microstructural requirements for transformation toughening are opposed to those for high temperature creep resistance. In TZP for instance, a small grain size is required to retain the tetragonal phase and a large grain size to reduce the creep rate. In dispersion types of ZrO_2-toughened ceramics a narrow size distribution of small $t-ZrO_2$ particles at the grain boundaries (triple points) is impossible in a matrix of large grain size with 10-20 vol. % ZrO_2. Hence the strategy for microstructural developments will be rather specific for the different ZrO_2-toughened ceramics.

Table 2 summarizes a number of possible measures to improve the high temperature mechanical properties of ZrO_2-toughened ceramics; some are being tested, and other tests are planned. Some of the strategies have already been demonstrated with the materials given in the third column. Other suggestions will also be discussed further below.

Prevention of Glassy Intergranular Films

Thick continuous films wetting the grain boundaries (Fig. 8(a)) of low viscosity will promote rapid failure (at high temperatures). Thus, refined and clean processing is essential, especially for PSZ and TZP. For TZP, however, the extent to which the amorphous grain boundary phase contributes to the high toughness and strength is as yet unclear, i.e. sintering strong TZP may even be a requirement. Annealing procedures, similar to those applied to Si_3N_4 alloys to crystallize the intergranular phase (Fig. 8(c)), will be advantageous. For instance, the improved high temperature toughness and thermal shock behaviour of "subeutectoid" heat-treated Mg-PSZ (Swain and co-workers, 1982) may be due to crystallization (possible formation of forsterite). Much more exploratory effort must be devoted to this aspect of improving the mechanical behaviour of ZrO_2-toughened ceramics at high temperatures.

Another possibility for retarding creep is to reduce the wetting of the grain boundaries by composition changes (Fig. 8(b)), e.g. by using controlled dopants. In attrition-milled fused mullite, ZrO_2 has been found to affect the wetting, producing small isolated islets of the amorphous phase which is located at ZrO_2-mullite interfaces (but not at mullite-mullite grain boundaries) as characterized in Fig. 9(a) (Prochazka,

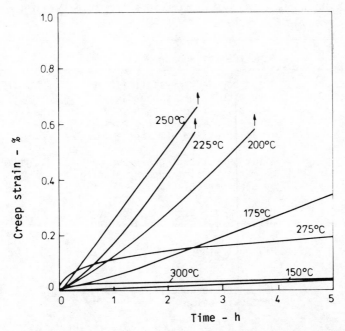

Fig. 7 Creep behaviour for a constant stress of 400MPa of the material shown in Fig. 6. At 250°C, failure has already occurred after about 2h because of an enhanced t→m zirconia transformation at the tensile surface of the sample: ↑, failed (Matsui, Soma and Oda, 1984)

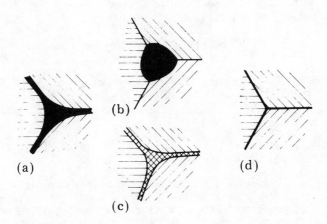

Fig. 8 Grain boundary configurations in ceramics: (a) amorphous phase wetting the grain boundaries; (b) non-wetting grain boundary; (c) crystallized grain boundary phase; (d) clean grain boundary

TABLE 2 Underline: High Temperature Microstructural Design Strategies
of ZrO$_2$-toughened Ceramics

Strategy	Experimental approaches	Examples
(1) Prevention of glassy inter-granular phases	Grain boundary crystallization; reduction of grain boundary wettability; hot forging of single crystals; hot isostatic pressing in an oxidizing atmosphere	PSZ, TZP, spinel, Al$_2$O$_3$, mullite
(2) Special grain boundary design	Sliding-resistant grain boundary geometry; grain boundary pinning by hard particles	PSZ, TZP
(3) Directional grain growth	Controlled crystallization of plastically deformed material	PSZ
(4) Increase in T_0 (A_s)	HfO$_2$ alloying	Al$_2$O$_3$
(5) Retention of compressive surface stresses	HfO$_2$ surface anneal; surface destabilization	Al$_2$O$_3$, mullite
(6) Precipitation hardening	Solution sintering followed by aging	Spinel, PSZ, Al$_2$O$_3$
(7) Dispersion strengthening	Addition of hard and high Young's modulus phases	TZP, mullite, cordierite
(8) Fibre reinforcement	Introduction of fibres or whiskers of SiC, Al$_2$O$_3$ etc.	Mullite, TZP cordierite
	Directional solidification	Al$_2$O$_3$-ZrO$_2$-MgO
(9) Enhancement of thermal stability	Surface alloying with CeO$_2$, Y$_2$O$_3$, MgO etc.	PSZ, TZP

Wallace and Claussen, 1983). In "pure" mullite (Fig. 9(b)), thin films continuously wetted the grain boundaries.

Elimination of a siliceous intergranular phase (Fig. 8(d)), e.g. in TZP and PSZ, may be difficult and expensive. It is claimed that this goal can be achieved by hot pressing or hot isostatically pressing ultrapure ZrO$_2$ powders, but air annealing at T $\geq$ 600°C has led to severe strength degradation (Schubert, Claussen and Rühle, 1984b) due to reoxidation, causing for example Co-pressure-induced grain boundary porosity (Bennison and Harmer, 1983). A possible way out of this dilemma would be hot pressing with oxide dies or hot isostatically pressing in an oxidizing atmosphere. A reliable but expensive way of preventing grain boundary phases would be to hot forge PSZ single crystals followed by a recrystallization anneal.

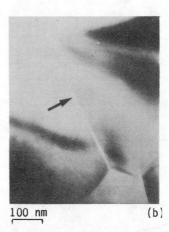

100 nm   (a)   100 nm   (b)

Fig. 9 Dark field TEM photograph of (a) sintered mullite-zirconia and (b) zirconia-free mullite (Prochazka, Wallace and Claussen, 1983). The glassy phase in (a) is mainly associated with the zirconia particles while the mullite grain boundaries are essentially glass free (Fig. 8(b)). In (b) all grain boundaries are wetted with an amorphous film (Fig. 8(a))

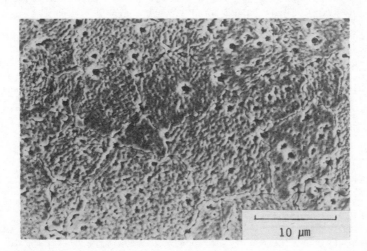

10 μm

Fig. 10 Meandering grain boundaries in the near-surface region of 2 vol% alumina-TZP with 2.2 mol% yttria sintered in a magnesia powder bed at 1500°C. The grain boundaries were probably pinned by a reaction products such as spinel and forsterite (micrograph courtesy of H Schubert, Max Planck Institute, Stuttgart)

Special Grain Boundary Design

In superalloy technology the dispersion of hard phases (e.g. TiC, TaC) is applied to retard high temperature creep (grain boundary sliding). Similar steps should be explored with PSZ and TZP ceramics. Additions of Al_2O_3 or spinel particles, but also of non-oxides (e.g. ZrC, SiC, Si_3N_4) where reactions during processing and use are limited, may be active in grain boundary pinning. An example of grain boundary pinning is demonstrated by the scanning electron microscopy (SEM) photograph in Fig. 10, showing the meandering grain boundaries in the surface region of a 2 vol. % Al_2O_3-TZP with 2.2 mol. % Y_2O_3. This material was sintered in an MgO powder bed at 1500°C; the MgO reacts with Al_2O_3 to form spinel, probably reacts with the intergranular phase to form forsterite for example and also dissolves in the ZrO_2, hence resulting in a PSZ-type microstructure.

Directional Grain Growth

The controlled grain growth of ZrO_2-toughened ceramics represents an interesting opportunity for developing more creep-resistant micro-structures. This should especially apply to PSZ types in which large grains are typical. Again, techniques used in the area of high temperature metal alloys may be applicable, e.g. plastic deformation by hot forging followed by controlled recrystallization in a temperature gradient as at present performed with turbine blades (Sahm and Speidel, 1974). Instead of utilizing a high dislocation density (created by hot forging) a liquid phase or a solute gradient could be used with a temperature gradient. Utilization of a liquid phase may achieve oriented crystal-lization and, at the same time, remove that phase from the remaining grain boundaries. Solute gradients, e.g. caused by the diffusion of Y_2O_3 or MgO into a ZrO_2 body, may also result in oriented grain growth. The gradient may be produced by annealing or sintering in an atmosphere (powder bed) containing the diffusing species (Schubert, Claussen and Rühle, 1984a) or by locally doping the ZrO_2 sample. From sintering ZrO_2-Y_2O_3 in the tetragonal-cubic two-phase field it is known that the Y_2O_3-rich cubic grains grow at a much higher rate than the tetragonal grains (Rühle, Claussen and Heuer, 1984), i.e. high temperature gradient annealing in the cubic single-phase field may result in directional growth of large grains. Epitaxial growth on seeds could be examined as a pre-liminary study to test for growth anisotropy.

Increase in Transformation Temperature

An enhancement of stress-induced transformation toughening at higher tem-peratures can be achieved by dispersing HfO_2-alloyed ZrO_2 particles which would shift T_o to higher temperatures. For instance, in an Al_2O_3-15 vol. % $(Zr_{0.5}, Hf_{0.5})O_2$ composite, A_s is shifted to 1460°C (Claussen, Sigulinski and Rühle, 1981). The technological problem associated with HfO_2 alloying is that the tetragonal phase becomes more difficult to retain to room temperature, i.e. d_c is reduced (e.g. about 0.3 μm for the above-mentioned Al_2O_3-$(Zr, Hf)O_2$ composite). For HfO_2 contents of less than 50 mol. %, normal sintering at T greater than the austenite finish temperature A_f ($\approx 1550°C$) would probably result in increased grain sizes and dispersed particle sizes, thereby opposing the retention of tetragonal symmetry. Furthermore amorphous intergranular phases should be avoided to make full use of the transformation. Special processing techniques, e.g. low temperature sintering of very fine powders or direct hot isostatic pressing of encapsulated compacts, may represent a feasible way of producing HfO_2-alloyed dispersion ZrO_2-toughened ceramics.

Retention of Residual Compressive Surface Stresses

Residual compressive surface stresses originating for instance from grinding-induced transformation of the near-surface region will usually anneal out when the material is heated to $T > A_s$ ($\approx 1100°C$) and the strength is consequently reduced (Claussen, 1982). In order to retain the stresses, the surface region either should be HfO_2 alloyed (see Section 3.4), e.g. by annealing in an HfO_2 powder bed, or should be destabilized locally by coarsening the ZrO_2 particle size (e.g. by laser heating) or annealing under conditions that deplete the surface of stabilizer solute (Claussen and Rühle, 1981). In the latter case the compressive stresses would be relieved at $T > A_f$; however, on cooling they will reappear because of spontaneous transformation.

This strategy is, of course, only useful for short-term application at high temperatures, i.e. the creep rates must be about zero.

Precipitation Hardening

Precipitation hardening is a standard method applied to many metal alloys, the strengthening resulting from blockage of dislocation movement which leads to an increase in the yield strength. At high temperatures, and also in ceramic materials (especially with cubic symmetry), interactions of precipitates with dislocations and growing cracks may be effective. The influence of precipitates in a number of ceramic systems has been studied mainly with respect to room temperature hardness and strength (Fine, 1972). In Y-PSZ single crystals with tetragonal precipitates, the strength remained constant (600 MPa) at $T > 800-1600°C$; since transformation is excluded, direct crack-precipitate interaction is claimed to be the cause of the strengthening (Ingel and co-workers, 1984).

A potential system for combining ZrO_2 toughening at low and intermediate temperatures with precipitation hardening at elevated temperatures is Al_2O_3-rich spinel ($Al_2O_3 \cdot MgO$) with dispersed t-ZrO_2 particles. Strengthening at room temperature by precipitates in 3.5 $Al_2O_3 \cdot MgO$ single crystals (Bansal and Heuer, 1974) and in hot-pressed polycrystalline 1.5 $Al_2O_3 \cdot MgO$ (Kauzaki and co-workers, 1980) has been demonstrated. Also ZrO_2 toughening in stoichiometric and Al_2O_3-rich spinel is highly effective (Claussen, 1982), i.e. no MgO stabilization (to the cubic phase) of the ZrO_2 particles takes place for Al_2O_3-to-MgO ratios greater than unity. Hence, some experimental prerequisites are given. However, the practical combination of the two strengthening mechanisms seems to be feasible only if specific microstructural conditions are met: (a) the ZrO_2 particles should be located at the grain boundaries and retained in the tetragonal phase; (b) the spinel grains should be unwetted; (c) the spinel grains should be large enough that mixed intergranular and transgranular fracture would occur at the application temperature; (d) the size and shape of the ZrO_2 particles should be stable during precipitation aging; (e) α-Al_2O_3 precipitates which have been aged at a temperature above the application temperature should nucleate and grow within the spinel grains.

To study this ZrO_2-toughened ceramic system, several compositions marked in the Al_2O_3-MgO phase diagram in Fig. 11 (Claussen, 1982) are currently being investigated. The preliminary results show a number of features.

(a) The strength at 1000°C of the 1.5 Al_2O_3·MgO composite (Fig. 11, composition 2) is considerably higher than that of the 1.0 Al_2O_3·MgO composite (Fig. 11, composition 1), for samples both with and without ZrO_2. The system is not yet understood because, even after a 70 h anneal at 1300°C, no α-Al_2O_3 precipitates could be observed, indicating that the the Al_2O_3 solute content was too low. However, the ZrO_2 particle size and morphology were unchanged.

(b) In 3.5 Al_2O_3·MgO composites (Fig. 11, composition 4) sintered in nitrogen at 1800°C, ZrO_2 particles dispersed at the grain boundaries grew too large to be retained in tetragonal symmetry. Aging for short periods of time at 1300°C caused thin plate-like precipitates, sometimes fully crossing the spinel grains and penetrating each other. Because of the high sintering temperature as required by the single-phase field boundary, short-time hot isostatic pressing may result in a dense material with tetragonal particles.

(c) 2.0 Al_2O_3·MgO composites (Fig. 11, composition 3) appear to be most promising. Samples prepared by attrition milling of reactive stoichiometric spinel with the appropriate amount of γ-Al_2O_3 (with 15 vol. % ZrO_2) and by sintering at 1650°C for 1 h resulted in dense composites with predominantly t-ZrO_2 particles. An aging treatment at 1300°C produced α-Al_2O_3 platelets similar to those in the 3.5 Al_2O_3·MgO material (Fig. 12).

Dispersion Strengthening

Toughening of ceramics by dispersing second-phase particles has been widely examined because of technological simplicity. However, most experiments have been directed towards improving the low temperature toughness or thermal shock resistance, and very little is known about the high temperature behaviour of these particulate ceramic composites. Furthermore, second phases will facilitate crack nucleation and growth (Evans and Rana, 1980). This is especially likely for high strength, high modulus materials, e.g. SiC and Si_3N_4. However, for some "softer" ceramics, the incorporation of hard high modulus second-phase particles may become effective at elevated temperatures. Al_2O_3-containing t-ZrO_2 seems to be an encouraging example (Tsukuma and co-workers, 1985).   As shown in Fig. 13, the strength of 2 mol. % Y_2O_3-TZP with 40 vol.% Al_2O_3 at 1000°C is nearly four times higher than that of the Al_2O_3-free 2 mol. % Y_2O_3-TZP. A partial contribution to the strength of enhancement possibly arises because Young's modulus for Al_2O_3 is higher and decreases less rapidly with temperature than that of TZP. A further contribution may result from the hindering of grain boundary sliding.   A comparable improvement has been observed with ZrO_2-toughened cordierite containing 10-40 vol. % Si_3N_4 and Mg-PSZ with dispersed spinel particles (Rice, 1981).

Fibre Reinforcement

Fibre reinforcement of brittle ceramic matrices is now widely assumed to be the best option for effectively enhancing the high temperature properties, especially creep resistance. In a recent review article (Rice, 1981)

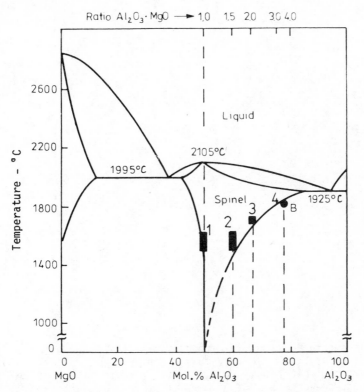

Fig. 11 Alumina-magnesia phase diagram with spinel compositions used for precipitation-hardening experiments. On aging at 1300°C in air, alpha-alumina precipitation was possible only with compositions 3 and 4

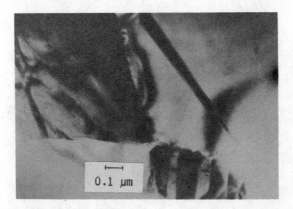

Fig. 12 Alumina-magnesia spinel with 15 vol% zirconia sintered at 1650°C in air and aged for 30 min at 1300°C in air. Alpha-alumina platelets appear in the grains. The zirconia particles have been transformed by the sample preparation for the TEM analysis

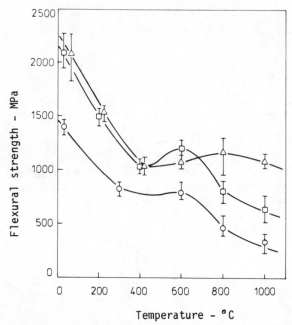

Fig. 13 Temperature dependence of bend strength of hot isostatically pressed (1500°C; 100MPa; 0.5h) 2 mol% yttria-TZP(⊖) 2 mol% yttria-TZP with 20 wt% alumina(⊟) and 2 mol% yttria-TZP with 40 wt% alumina(⊿) (Tsukuma and co-workers, 1984)

Fig. 14 Silicon carbide whiskers in zirconia

the general toughening mechanisms in ceramic fibre-ceramic matrix composites has been discussed. For relatively soft low modulus ceramics, most
of the strengthening mechanisms found in metal-matrix fibre composites
become valid at high temperatures. The preferred fibre strengthening
mechanism for ZrO_2-toughened ceramics at high temperatures is modulus
load transfer, to reduce the matrix stress and hence to inhibit creep and/
or premature crack nucleation. The essential design criteria for fibre-
reinforced ceramics are (Rice, 1981): (a) a high fibre modulus and
tensile strength (preferably more than twice that of the matrix), (b) a
small fibre diameter, i.e. of the order of the matrix grain size, (c) a
homogeneous distribution of fibres, i.e. no fibre-free areas or fibre
clustering (ideally aligned and continuous), (d) no, or little, chemical
bonding of fibres to the matrix (to prevent singularities within the fibre
in the presence of a matrix crack), (e) no reactions that cause the fibre
properties to deteriorate, (f) large fibre volume fractions and (g)
comparable thermal expansion coefficients α (but rather with the thermal
expansion coefficient α_f of the fibres greater than the thermal expansion
α_m of the matrix so that compressive pre-stressing of the matrix is
developed).

These requirements limit the number of useful fibre-ZrO_2-toughened ceramic
combinations. For PSZ and TZP, Al_2O_3 fibres or whiskers are the best
choice. The thermal mismatch with SiC and Si_3N_4 fibres will have a negative effect. Furthermore, reactions would take place during sintering,
especially with Si_3N_4. Further restrictions in using SiC or Si_3N_4 fibres
may result from the oxygen permeation through ZrO_2 at high temperatures.
Hence, ZrO_2-toughened ceramics with ceramic matrices compatible with the
fibre or whisker materials at present available seem to offer a better
opportunity. Mullite, cordierite or aluminium titanate (Al_2TiO_5) which
have low expansion coefficients make good choices for ZrO_2 toughening.
Since these ZrO_2-toughened ceramics (except for Al_2TiO_5) usually have small
grain sizes, whiskers are to be preferred to fibres because of the smaller
diameters available (Table 3). Furthermore, the strength of the available
whiskers is almost an order of magnitude higher than that of the respective fibres (but they are not continuous).

To explore the potential of fibre reinforcement of ZrO_2-toughened ceramics,
the combinations given in Table 4 have been chosen. The properties of the
single components are given in Table 3. The SiC whiskers used in the
experiments are shown in Fig. 14.

(a) ZrO_2-toughened cordierite has been composed such that the matrix
coefficient of expansion matches that of the α-Si_3N_4 whiskers. Since
reactions between Si_3N_4 and ZrO_2 during sintering become considerable
at T > 1350°C, the composites were hot pressed (at 1200°C for 20 min).
However, even under these processing conditions the reaction between ZrO_2
and Si_3N_4 seems to have caused the Si_3N_4 whiskers to degrade. Hence a
preliminary conclusion would be to use Si_3N_4 whiskers only for a cordierite matrix without ZrO_2 addition.

(b) The combination of ZrO_2-toughened cordierite (as in (a)) with SiC
whiskers would result in matrix compressive stresses, a desirable stress
state especially when modulus load transfer is to act as the strengthening
mechanism. Samples with 20 vol. % of randomly dispersed SiC whiskers were
hot pressed at 1300°C for 20 min. In this case, no detectable reactions
took place. The four-point bending strength of the ZrO_2-toughened cordierite matrix was increased from 180 MPa at room temperature (and 170 MPa

TABLE 3 <u>Properties of ZrO_2-toughened Ceramic Matrices and Whiskers (and Fibres) used for Exploratory Studies</u>

Material	σ_f MPa	K_{IC} MPa	α $\times 10^{-6} K^{-1}$	E Gpa	Grain or fibre diameter μm
ZrO_2-toughened ceramic matrix					
80 vol.% cordierite- 20 vol.%(3 mol.%Y_2O_3-TZP)	180	5[a]	2.5	135	<0.5
3 mol.%Y_2O_3-TZP	705	10[a]	10.5	200	<0.4
ZrO_2-toughened mullite	400	4	5	220	0.6
Whiskers					
β-SiC (SCW 1)[b]	21000	(5)	4.7	490(581)[c]	0.05-0.2[d]
α-Si_3N_4 (SNW 1)[b]	14000	(4)	2.5	385	0.2-0.5[e]
Fibres					
Al_2O_3 (FP Al_2O_3)[f]	1380	(4)	8.5	380	20

[a] Value obtained when the grain size is optimized.
[b] Supplier's property data.
[c] Petrovic and Roof, 1984.
[d] Length, about 30 μm.
[e] Length, about 100 μm.
[f] Supplier's property data.

at 1000°C) to 390 MPa at room temperature (and 306 MPa at 1000°C) by the whisker reinforcement. At 1000°C, some debonding and pull-out of the whiskers was obvious from SEM analysis, indicating "favourable" weakening of the whisker-matrix interface at elevated temperatures. Preliminary tests with centrifugally slip-cast mixtures of cordierite powder with SiC whiskers and Si_3N_4 powder with SiC whiskers have shown that an almost perfect two-dimensional alignment of the whiskers can be achieved. A further improvement of the room temperature strength of the ZrO_2-toughened cordierite-SiC whisker composite can be obtained by reducing the Y_2O_3 content in the ZrO_2. In the present samples the sizes of the t-ZrO_2 particles were too small (less than 0.2 μm) to be transformed by stresses.

(c) The conditions for the system consisting of ZrO_2-toughened mullite with β-SiC whiskers are similar, i.e. the property ratios given in Table 4 are such that they would favour the desired strengthening mechanism of modulus load transfer.

(d) The system 3 mol. % Y_2O_3-TZP with SiC whiskers was chosen to study the effect of a thermal mismatch that would produce tensile stresses in the matrix. It is expected that the high shear stresses at the whisker-matrix interface may be relaxed by local stress-induced transformation to m-ZrO_2. This microstructural configuration would then favour debonding and pull-out as the strengthening mechanism because of microcracking

TABLE 4 <u>ZrO$_2$-toughened Ceramic Matrix-Whisker Systems Chosen for Exploratory Studies</u>

Matrix-whisker (fibre system)	Volume fraction of fibres	Ratios of a specific fibre property to the corresponding matrix property				Microstructure and stress state at room temperature	Potential strengthening mechanisms at high temperatures
		E_f/E_m	α_f/α_m	σ_f/σ_m	d_f/d_m		
ZrO$_2$-toughened cordierite + Si$_3$N$_4$ whiskers	0.2	2.85	1	76	1	Stress-free interface; two matrices (cordierite + TZP)	Modulus load transfer, pull-out, debonding
ZrO$_2$-toughened cordierite + SiC whiskers	0.2	3.63	1.4	114	1	Compressive matrix stresses; two matrices (cordierite + TZP)	Modulus load transfer, debonding, pull-out
ZrO$_2$-toughened mullite + SiC whiskers	0.1-0.2	2.22	1	52	1	Stress-free interface; t- and m-ZrO$_2$ particles	Modulus load transfer, debonding, pull-out
3 mol.%Y$_2$O$_3$-TZP + SiC whiskers	0.2	2.45	0.5	30	1	Tensile matrix stresses	Pull-out, debonding, microcracking, load transfer
3 mol.%Y$_2$O$_3$-TZP + Al$_2$O$_3$ fibres	0.15	1.9	0.8	2.0	50	Low tensile matrix stresses; fibre diameter similar to critical flaw size	Debonding, pull-out, crack deflection

adjacent to the SiC whiskers at the interface during cooling. The strength of composites containing 20 vol. % SiC whiskers, hot pressed at 1400°C for 20 min, was 645 MPa at room temperature (and 300 MPa at 1000°C) compared with 705 MPa at room temperature (and only 165 MPa at 1000°C) without SiC whiskers. As in example (b) the grain size of the TZP has not been optimized, i.e. it was too small to yield any stress-induced transformation. However, it is interesting to note that the whiskers changed the fracture mode at 1000°C from a nearly 100% intergranular type to a predominantly transgranular type as demonstrated in Fig. 15.

(e) TZP with chopped Al_2O_3 fibres was chosen to investigate the TZP-fibre interface and the fibre microstructure and morphology after hot isostatic pressing. This specific combination is certainly not ideal, i.e. a more realistic system should contain submicron Al_2O_3 whiskers. Even though a strength increase at room temperature cannot be expected, improvements should be gained at T > 900°C, especially as a result of the increasing ratio of E_f/E_m with temperature.

The preliminary results with whisker-reinforced ZrO_2-toughened ceramic show that this strengthening strategy will probably be the most effective one.

Enhancement of Thermal Stability

There is probably no way of preventing the thermal decomposition of Mg-PSZ at T > 700°C. The thermal stability of dispersion ZrO_2-toughened ceramics, e.g. with Al_2O_3, mullite or spinel matrices, for temperatures up to about 1200°C should be a lesser problem, especially when unstabilized ZrO_2 particles are used.

The low temperature degradation specific to Y-TZP, as discussed above, merits intensive research. Parts used at high temperatures are cycled through the critical temperature range (150-300°C), and it is desirable that the lifetime of the part is unaffected by temperature cycling. Increasing the Y_2O_3 content (to more than about 4 mol. %) and reducing the grain size will prevent the surface transformation but with some loss of toughness. Further protection can result from using a fine-grained PSZ type of surface layer that can be produced by sintering or annealing in a Y_2O_3-bearing atmosphere (Claussen, 1984; Schubert, Claussen and Rühle, 1984). However, care has to be taken to prevent excessive strength reduction because of this layer. A highly inhomogeneous Y_2O_3 distribution among the different grains of ≳ 3 mol. % Y_2O_3-TZP has been found to be surprisingly stable (Rühle, Claussen and Heuer, 1984). The stability is attributed to the fact that the overstabilized grains at the surface remain tetragonal while the low Y_2O_3 grains transform to monoclinic symmetry, and compressive surface stresses develop that inhibit surface cracking. The addition of Al_2O_3 particles to Y-TZP has also been found to retard the surface transformation as shown in Fig. 16 (Sata and Shimada, 1984). The exact cause, however, is not known; the average tetragonal grain size and also the transformation zone may have been reduced by the Al_2O_3 additions. Furthermore, Al_2O_3 solute may be effective in stabilizing the tetragonal phase.

Fig. 15 TEM of hot-pressed 3 mol% yttria-TZP with 20 vol% silicon
carbide whiskers

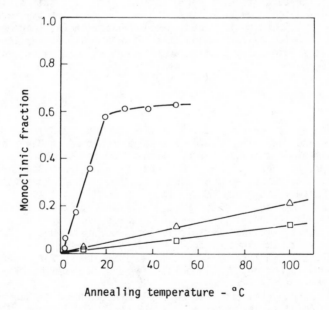

Fig. 16 Change with annealing time of monoclinic fraction of 3 mol%
yttria-TZP (o), 3 mol% yttria-TZP with 5 wt% alumina (△) and 3 mol%
yttria-TZP with 10 wt% alumina (□) sintered at 1500°C; annealed at
200°C (Sato and Shimada, 1984)

CONCLUDING REMARKS

A number of promising strategies exist to develop microstructures for ZrO_2-toughened ceramics in order to use this class of high performance ceramics at high temperatures. However, for further clarification of micro-structural design criteria, it is important to investigate the high temperature thermomechanical behaviour of the present ZrO_2-toughened ceramic types to a greater extent, i.e. too little is known about the failure mechanisms, especially under static and dynamic stress conditions at elevated temperatures.

From the various strategies given in Table 2, the following measures seem to offer the best prospects for strength improvements at temperatures in the tetragonal stability range: (a) prevention or control of amorphous intergranular phases, i.e. by using high purity processing, by crystalliz-ing the glass or by preventing the wetting of the grain boundaries; (b) addition of hard and high modulus second phases; (c) whisker (fibre) reinforcement, especially for TZP, ZrO_2-toughened mullite and ZrO_2-toughened cordierite.

ACKNOWLEDGEMENTS

The author is grateful to R.L. Coble and G. Petzow for stimulating dis-cussions and critical review of the manuscript. Part of the work was supported by the Deutsche Forschungsgemeinschaft. Acknowledgement is made to Elsevier Sequoia for permission to publish a modified version of this paper which was originally published in Materials Science and Engineering, 1985, 71, 23-28.

REFERENCES

Anderson, P. and Hermannson, L. 1983, Silikat-Rapp, 83-3
Bansal, G.K. and Heuer, A.H. 1974, Fracture Mechanics of Ceramics. Eds.
    D.P.H. Hasselman and F.F. Lange, Plenum, New York, p 667
Bennison, S.J. and Harmer, M.P. 1983, Character of Grain Boundaries,
    Advances in Ceramics, Vol. 6, Eds. M.F. Yan and A.H. Heuer, American
    Ceramic Society, Columbus, OH, p 171
Claussen, N. and Ruhle, M. 1981, Proc. Int. Conf. on the Science and
    Technology of Zirconia, Advances in Ceramics, Vol. 3, Eds. A.H. Heuer
    and L.W. Hobbs, American Ceramic Society, Columbus, OH, p 137
Claussen, N. 1982, Z. Werkstofftech. 13 138 and 185
Claussen, N., Sigulinski, F. and Rühle, M. 1981, Proc. Int. Conf. on the
    Science and Technology of Zirconia, Advances in Ceramics, Vol. 3,
    American Ceramic Society, Columbus, OH, p 164
Claussen, N. 1984, Proc. 2nd Int. Conf. on the Science and Technology of
    Zirconia, Advances in Ceramics, Vol. 12, Eds. N. Claussen, M. Rühle
    and A.H. Heuer, American Ceramic Society, Columbus, OH, p 325
Claussen, N., Rühle, M. and Heuer, A.H. 1984, Eds. Proc. 2nd Int. Conf. on
    the Science and Technology of Zirconia, Advances in Ceramics, Vol. 12,
    American Ceramic Society, Columbus, OH
Evans, A.G. and Rana, A.S. 1980, Acta Metall. 28 129
Fine, M.E. 1972, Amer. Ceram. Soc. Bull. 51 510

Hannink, R.H.J. 1983, J. Mater. Sci. 18 457
Heuer, A.H. and Hobbs, L.W. 1981, Eds. Proc. Int. Conf. on the Science and
    Technology of Zirconia, Advances in Ceramics, Vol. 3, American Ceramic
    Society, Columbus, OH
Heuer, A.H., Claussen, N., Kriven, W.M. and Rühle, M. 1982, J. Amer.
    Ceram. Soc. 65 642
Ingel, R.P., Lewis, D., Bender, B.A. and Rice, R.W. 1984, Proc. 2nd Int.
    Conf. on the Science and Technology of Zirconia, Advances in Ceramics,
    Vol. 12. Eds. N. Claussen, M. Rühle and A.H. Heuer, American Ceramic
    Society, Columbus, OH, p 408
Kanzaki, S., Hamano, K., Nakagawa, Z. and Saito, K. 1980, Yogyo Kyokai Shi
    88 59
Kobayashi, K. and Masaki, T. 1982, J. Jpn. Ceram. Soc. 17 427
Lange, F.F. 1982, J. Mater. Sci. 17 225
Larsen, D.C. and Adams, J.W. 1984, Final Tech. Rep. AFWAL-TR-83-4141, Air
    Force Wright Aeronautical Laboratories
Matsui, M., Soma, T. and Oda, I. 1984, Proc. 2nd Int. Conf. on the Science
    and Technology of Zirconia, Advances in Ceramics, Vol. 12, Eds.
    N. Claussen, M. Rühle and A.H. Heuer, American Ceramic Society,
    Columbus, OH, p 371
Nakajima, K., Kobayashi, K. and Murata, M. 1984, Proc. 2nd Int. Conf. on
    the Science and Technology of Zirconia, Advances in Ceramics, Vol. 12,
    Eds. N. Claussen, M. Rühle and A.H. Heuer, American Ceramic Society,
    Columbus, OH, p 382
Petrovic, J.J. and Roof, R.B. 1984, J. Amer. Ceram. Soc. 67 C219
Prochazka, S., Wallace, J.S. and Claussen, N. 1983, J. Amer. Ceram. Soc.
    66 C125
Rice, R.W. 1981, Ceram. Eng. Sci. Proc. 2 661
Ruh, R., Mazdiyaasni, K.S., Valentine, P.G. and Bielstein, H.O. 1984,
    J. Amer. Ceram. Soc. 67 C190
Rühle, M., Claussen, N. and Heuer, A.H. 1984, Proc. 2nd Int. Conf. on the
    Science and Technology of Zirconia, Advances in Ceramics, Vol. 12,
    Eds. N. Claussen, M. Rühle, and A.H. Heuer, American Ceramic Society,
    Columbus, OH, p 352
Sahm, P.R. and Speidel, M.O. 1974, Eds. High Temperature Materials in Gas
    Turbines, Elsevier, Amsterdam
Sato, T. and Shimada, M. 1984, J. Amer. Ceram. Soc. 67 C212
Schubert, H., Claussen, N. and Rühle, M. 1984a, Proc. Brit. Ceram. Soc.
    34 157
Schubert, H., Claussen, N. and Rühle, M. 1984b, Proc. 2nd Int. Conf. on
    the Science and Technology of Zirconia, Advances in Ceramics, Vol. 12,
    Eds. N. Claussen, M. Rühle and A.H. Heuer, American Ceramic Society,
    Columbus, OH, p 766
Swain, M.V. and Hannink, R.H.J. 1981, Proc. 5th Int. Conf. on Fracture,
    Advances in Fracture Research, Vol. 4, Ed. D. François, Pergamon Press,
    Oxford, p 1559
Swain, M.V., Garvie, R.C., Hannink, R.H.J., Hughan, R. and Marmach, M.
    1982, Proc. Brit. Ceram. Soc. 32 343
Tsukuma, K., Kubota, Y. and Tsukidate, T. 1984, Proc. 2nd Int. Conf. on the
    Science and Technology of Zirconia, Advances in Ceramics, Vol. 12, Eds.
    N. Claussen, M. Rühle and A.H. Heuer, American Ceramic Society,
    Columbus, OH, p 382
Tsukuma, K., Ueda, K., Matsushita, K. and Shimada, M. 1985, J. Amer.
    Ceram. Soc. 68 C56

Inst. Phys. Conf. Ser. No. 75: Chapter 5
Paper presented at 2nd Int. Conf. Science Hard Mater., Rhodes

401

High temperature properties of ceramics in the SiAlON system

D A BONNELL(1), T Y TIEN(1), M LEE(2) AND M K BRUN(2)

(1) University of Michigan, Ann Arbor, MI   41109
(2) General Electric Corporate Research and Development, PO Box 8,
Schenectady, NY 12301

ABSTRACT

A series of SiAlON materials with a cordierite-based matrix were annealed for different lengths of time to cause crystallization of the glass
phase. Their fracture toughness, hardness, and elastic modulus were
measured from room temperature up to 1100°C. The fracture toughness
generally decreased with temperature. Short time annealing raised
toughness at lower temperatures, while further annealing lowered it
back to the value for as-hot pressed materials. At higher test temperatures annealing had no effect on toughness. This annealing behavior
is significantly different from that previously reported in the system
SiAlON-YAG. Hardness decreased monotonically with temperature for
all samples. Both it and the elastic modulus were not affected by
the annealing treatment. At elevated temperatures appreciable scatter
of modulus results allowed only a rough trend of decrease with temperature to be observed.

INTRODUCTION

Strength, toughness and hardness are the most important mechanical
properties of the metalcutting tools. Hardness and toughness are
necessary to prevent rapid abrasion of the tool. The tool must also
have sufficient strength and toughness to sustain cutting action.
When compared to traditional carbide and high speed steel tools, ceramic tools have generally better hardness and high temperature deformation resistance. However, their strength and toughness are only mar-
ginal.

Recently stronger and tougher ceramic materials have been introduced
in cutting tool markets, and some, most notably from the silicon nitride
family, have shown significant success in machining high temperature
superalloys. Superalloy cutting involves very high temperatures,
on the order of 900°C, at the cutting edge. As cutting forces in
machining superalloys are relatively high, the properties of particular importance are the strength, and particularly the toughness of
the material, in the temperature range up to 900°C. The toughness
of SiAlON materials has been shown in previous work (Lee, Brun and
Tien, 1983) to remain relatively high up to 900°C, where it is roughly
twice that of alumina-based tools. It is believed that this higher

toughness is largely responsible for their success, as their bend strength is not substantially higher than that of alumina-TiC tools (900MPa vs. 950MPa for typical SiAlON and typical alumina-TiC, respectively).

Silicon nitride-based tools, including SiAlONs, are generally two-phase materials, consisting of elongated grains of beta-silicon nitride (or beta' SiAlON) with silicate glass matrix. The most popular additives for sintering of silicon nitride to date have been oxides of yttrium, magnesium, aluminum, etc. Silicon oxide is always present as an impurity in silicon nitride. If samples are cooled relatively rapidly from the sintering temperature, all of the oxide phase will be glassy. It is possible, however, to crystallize some or all of the glass by post-sintering annealing treatment. In the case of SiAlONs with yttrium aluminum garnet (YAG) type matrix material, it was shown (Lee, Brun and Tien, 1983) that the toughness of the material is appreciably decreased by crystallization of the YAG phase. There are several possible explanations for the observed decrease of toughness after annealing, such as the stresses resulting from the differences in the thermal expansion of YAG and silicon nitride (Bonnel and Tien, 1983), or the decrease in the specific volume upon crystallization and resultant formation of voids. It was believed that changing to a different matrix which would have better thermal expansion match and would not exhibit a substantial volume change on crystallization might alter that behavior.

The aim of this work was to measure toughness and elastic modulus of SiAlON materials with a cordierite-based matrix at temperatures up to 900°C, for a series of different compositions and different heat treatments. The results would be compared to results obtained on a YAG-based system.

EXPERIMENTAL PROCEDURE

Sources of the starting materials and purities are summarized in Table 1. Starting powders were ballmilled with alumina balls using alcohol as the fluid medium. The compositions were compensated for the pickup of alumina during milling. Consolidation was accomplished by hot pressing in graphite dies in flowing nitrogen atmosphere. Compositions and hot pressing conditions are shown in Table 2. It can be seen that hot pressed materials consisted of beta' SiAlON and glass, with the exception of composition 30-10, which exhibited some x-ray diffraction lines which could not be interpreted. Samples of these four compositions annealed for different lengths of time were chosen for this study. Annealing was done in nitrogen at 1000°C.

Elastic modulus was measured by an ultrasonic technique. At room temperature, the time-of-flight and the resonance method gave similar values. At elevated temperatures only the time-of-flight technique was used.

The fracture toughness was measured by the indentation technique of Evans and Charles (1976). It is an empirical technique, based on correlation of the fracture toughness with several other properties for a variety of different materials. Since its derivation takes

TABLE 1 Sources and Purities of Starting Materials

	Sources	Purities	Particle Size, μm
Si_3N_4	H. Starck LC 12	90%-α Si_3N_4 1.59% O_2 0.01% Fe 0.01% Ca	0.5
AlN	H. Starck	2.87% O_2	3
Al_2O_3	Alcoa A16	99.8%	1
Y_3O_3	Molycorporation 5600	99.99%	3

into account limited plasticity under the indenter, it was assumed that the technique will apply at temperatures up to the point where massive plastic deformation becomes possible.

A high temperature Vickers hardness tester, with 5Kg weight, was used for all tests. Samples were first heated to the highest temperature used, indented, and then cooled to next lower test temperature and the procedure repeated. In this manner, the amount of time the indented sample spent at elevated temperature was minimized, and the indented sample was never exposed to temperatures higher than the one at which the indentation was made. Four indentations were made at each temperature. Indentation and crack lengths were measured at room temperature with an optical microscope. There is a possibility that crack length may change during cooling to room temperature. It was shown previously (Lee, Brun & Tien, 1983), however, that performing optical measurements after cooling, as opposed to doing them at the test temperature, did not change the results.

RESULTS AND DISCUSSION

The values of the elastic modulus at room temperature were 279.9 and 284.1 GPa for the hot pressed and for the annealed sample 10-5, respectively. Although an increase of about 2% on annealing is indicated, the difference is within the possible variation in the measurement. Thus, the effect, if any, of annealing on the modulus is minor.

The effect of temperature on the elastic modulus of the hot pressed sample 10-5 is shown in Fig. 1. It is immediately apparent that there is a significant scatter of data, particularly at intermediate temperatures. At temperatures above about 700°C the data are reasonably well behaved and a rough trend of decrease with temperature could be seen.

It was initially assumed that the scatter is due to the measurement

TABLE 2 Composition, Hot Pressing Conditions and Crystalline Phases Present In Hot Pressed Samples

| Composition Identification | Composition wt.% | | | | Hot Pressing | | Phases* |
	$Beta_5$	$Beta_{10}$	$Beta_{30}$	Corderite ($Mg_2Al_4Si_5O_8$)	h	Temp.	
5-10	90			10	2	1680	B
10-10		90		10	2.5	1690	B
30-10			90	10	3.5	1690	B, U
5-20	80			20	2	1680	B

*$Beta_5$, $Beta_{10}$ and $Beta_{30}$ are Beta'-SiAlONs with general formula

$Si_{6-x}Al_xO_xN_{8-x}$ where $x = 0.39$, 0.77 and 2.18, respectively

**B = beta' SiAlON
U = unidentified

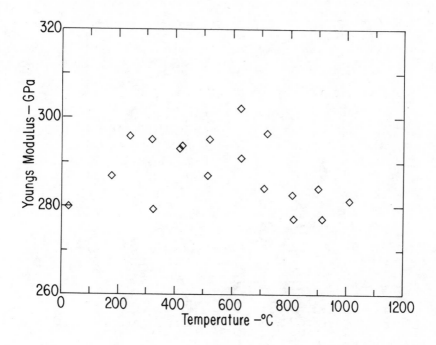

Fig. 1 Elastic modulus of composition 10-5 vs temperature.

problems. Similar scatter, however, has been encountered with every
type of silicon nitride-based material tested to date. To illustrate
the point, data obtained on a number of different materials on the
same experimental set-up are shown in Fig. 2. The materials shown
are two hot pressed SiAlON materials with YAG and cordierite matrices,
respectively, a sintered SiAlON material, a hot pressed silicon nitride with yttria and a zirconia-toughened alumina. Showing all
of the experimental points would be confusing, due to the overlap
of different materials. Therefore, only the range over which the
experimental data were scattered for each material are shown.

Several observations can be made about the data. The data for the
alumina-based material showed very little scatter throughout the temperature range investigated. Somewhat higher scatter was obtained for
the hot pressed silicon nitride material, followed by the two hot
pressed SiAlONs and the sintered SiAlON material. For all materials
there was less scatter at higher temperatures. With the exception
of sintered SiAlON, the scatter also decreased as temperatures approached room temperature. With the exception of alumina-based material,
the room temperature values of modulus were lower than would be obtained by extrapolation of the rest of the data.

The lack of scatter of the alumina-zirconia data clearly demonstrates
that the scatter is not caused by the equipment. (Good data have
also been obtained on a number of metallurgical samples.)

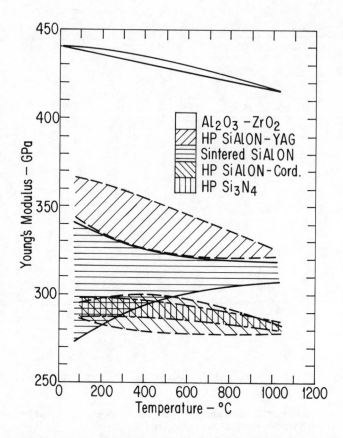

Fig. 2 Elastic moduli of hot poressed SiAlON-YAG, SiAlON-
Cordierite, and $Si_3N_4-Y_2O_3$, sintered SiAlON and HIPped
alumina-zirconia, as a function of temperature.

A completely satisfactory explanation for the scatter of data cannot
now be offered. It has been suggested by Dole and co-workers (1978),
that scatter of data at intermediate temperatures can be caused by
microcracking. Their model predicts less scatter at high temperature,
and increase in the scatter as the microcracks open on cooling due
to differential thermal expansion. The room temperature value of
modulus is expected to be low. The model describes the behavior of
silicon nitride ceramics reasonably well. However, microcracking
should be confirmed by an independent test to confirm such an explanation.

The result of fracture toughness and hardness measurements are shown
in Table 3. For discussion purposes, the data for the composition
10-10 have also been shown in Fig. 3. The values for the hot pressed
sample and samples annealed for 6 and 68h are shown as diamonds, squares
and triangles, respectively. While there were some minor differences
in behavior of other compositions, the composition 10-10 is representative of typical behavior. It can be seen that the fracture toughness

TABLE 3 Fracture Toughness and Hardness as a Function of Temperature, Composition and Anealing Treatment

Composition	Annealing Time[1]	Room Temp. ΔK_{IC}[2]	Room Temp. HV5[3]	300°C K_{IC}	300°C HV5	600°C K_{IC}	600°C HV5	900°C K_{IC}	900°C HV5	1100°C K_{IC}	1100°C HV5
β5-10C[5]	0	4.84 0.15	16.31 0.12	4.87 0.06	14.81 0.19	4.69 0.20	13.67 0.19	4.40 0.15	12.66 0.24		
	95	5.51 0.14	16.80 0.08	5.14 0.11	14.97 0.13	4.68 0.16	14.01 0.17	4.26 0.07	12.77 6.35		
β10-10C	0	4.55 0.21	15.86 0.19	4.46 0.21	14.28 0.40	4.17 0.12	13.31 0.29	4.07 0.27	12.65 0.21	3.85 0.11	11.30 0.17
	68	4.77 0.19	16.40 0.34	4.50 0.21	14.70 0.17	4.19 0.13	13.57 0.52	3.91 0.16	12.38 0.17	3.45 0.20	10.42 0.05
	6	5.50 0.20	16.19 0.23	4.87 0.23	14.54 0.28	4.41 0.12	13.37 0.17	3.78 0.07	12.38 0.14	4.24 0.04	11.09 0.21
β30-10C	0	3.25 0.18	15.23 0.28	2.84 0.30	13.86 0.89	3.03 0.05	12.11 0.95	2.79 0.11	11.72 0.11		
	1000	3.15 0.23	14.93 0.19	3.09 0.42	13.87 0.18	2.95 0.13	12.75 0.18	2.91 0.17	12.31 0.03		
	54	3.18 0.45	15.54 0.39	3.01 0.20	14.01 0.17	2.98 0.20	13.17 0.32	2.69 0.34	12.17 0.25	2.78 0.21	11.43 0.29
β5-20C	0	5.36 0.13	15.69 0.36	5.14 0.11	14.14 0.22	4.95 0.09	12.80 0.64	4.58 0.09	11.25 0.14		
	46	4.71 0.27	15.18 0.34	4.60 0.33	13.87 0.29	4.94 0.17	12.64 0.19	4.51 0.18	11.94 0.11		
	6	5.47 0.12	15.11 0.26	5.37 0.09	13.93 0.22	4.85 0.21	12.61 0.29	4.38 0.07	11.92 0.17		

Table 3 continued

Material		Room Temp.			300°C		600°C		900°C		1100°C	
Composition	Annealing Time[1]	K_{IC}[2]	ΔK_{IC}[4]	HV5[3]	K_{IC}	HV5	K_{IC}	HV5	K_{IC}	HV5	K_{IC}	HV5
β5-5Y[6]	0	5.19		17.80	4.79	16.17	4.30	15.31	3.92	14.01		
	24	4.57		17.31	4.35	15.85	3.99	14.66	2.98	14.04		
β5-10Y	0	5.26		17.36	5.08	16.25	4.64	14.77	4.53	13.99		
β5-20Y	0	5.48		17.95	5.31	16.25	4.84	15.35	4.21	14.39		
β10-5Y	0	4.95		17.28	4.58	15.88	4.19	14.99	3.72	14.06		

1 In hours
2 MN/m$^{(3/2)}$
3 GPa
4 ΔK_{IC} indicates the standard deviation in measured data
5 Cordierite
6 YAG

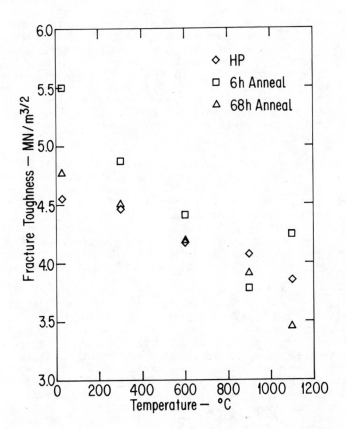

Fig. 3 Fracture toughness vs. temperature for samples of
       composition 10-10 as hot pressed, and after 6- and 68-h anneal.

decreases monotonically with temperature, with the exception of the
value at 1100°C for the sample annealed 6h. At room temperature,
the fracture toughness for sample annealed for short period of time
was appreciably higher than the value for the hot pressed sample.
The increase of toughness thus correlates with the initial crystallization of cordierite. At longer annealing times the fracture toughness
decreased to approximately the same value as the hot pressed sample.
This type of behavior was not as pronounced in compositions with higher
glass content or higher level of substitution. At temperatures of
600°C and higher, there was no appreciable difference in toughness
of hot pressed and annealed samples. One likely explanation for uniformity of high temperature toughness values may be the possibility
of annealing out some of the stresses introduced by previous annealing
treatments. Also, if there were residual stresses present in the
material due to the thermal expansion mismatch of different phases,
they would decrease as temperature is raised.

Comparing compositions 5-10 and 5-20, it can be seen that an increase in glass content caused an increase in the toughness which remained even at elevated temperatures. When comparing samples with 10 vol.% glass and different level of aluminum substitution in SiAlON phase, it can be seen that toughness of hot pressed samples at room temperature decreased slightly by increasing substitution level from 5 to 10 and dramatically by increasing it to 30 eq% Al. Toughness values for all the samples of composition 30-10 were consistently much lower than for the other compositions. As has been shown in Table 2 the composition 30-10 contained an unidentified crystalline phase in addition to beta' SiAlON, and toughness has been apparently adversely affected by its presence.

An interesting point is the behavior of fracture toughness at 1100°C. It was reported previously (Lee, Brun and Tien, 1983) that fracture toughness of materials in the SiAlON-YAG system decreased up to 900°C. The glass transition temperature (Tg) for glasses of composition similar to the grain boundary glass-phase are in the range $900-1000^\circ$C (Mosher, Raj and Kossowski, 1976). It can be expected that the fracture toughness may increase above Tg due to the possible plastic deformation of the glass phase. In the present experiments, with the one exception already mentioned, toughness continued to decrease at 1100°C.

Evans and Wiederhorn (1976) measured fracture toughness of silicon nitride between 1000 and 1400°C. They found that in the regime at low crack velocity, the toughness decreased with temperature. In the regime at higher crack velocities, which should be comparable with present work, the toughness was found to increase with temperature. One possible reason for the apparent discrepancy of present results and their work may be that the increase in K_{IC} that they measured was small between 1000 and 1200°C and more pronounced at higher temperatures, so that there was not sufficient overlap of the temperature ranges to make a valid comparison. The differences may also be due to the difference in the techniques used. Although the indentation technique assumes limited plasticity of the material around the indenter, it may yield misleading results when used at temperatures where plastic or viscous deformation become extensive.

When comparing the results reported here to results obtained on SiAlON-YAG systems, it can be seen that the toughness of hot pressed samples of comparable compositions was not appreciably different at any of the test temperatures. The conclusion that could be drawn is that the level of substitution in SiAlON phase and the amount of glass phase are more important in determining the toughness than the composition of the glass phase. The annealing behavior in the two systems, however, is very different. While the low temperature toughness could be increased by proper annealing treatment in the SiAlON-cordierite system, in SiAlON-YAG system crystalization of glass phase invariably led to a toughness decrease. As already discussed in the introduction, the differences in density between the parent glass (~ 3000 Kg/m^3) and YAG (4500 Kg/m^3) are relatively large, possibly leading to microcracking or void formation while the density of cordierite (2500 Kg/m^3) more closely approximates that of the parent glass (~ 2400 Kg/m^3). Also, the thermal expansion coefficient of YAG is appreciably higher than that of silicon nitride, while cordierite has a low ther-

mal expansion coefficient. The residual stresses could therefore
lead to toughening in the case of cordierite, and have the opposite
effect in the case of YAG.

Hardness results are shown in Table 3. Results for composition 10-
10 are also shown in Fig. 4. In all cases hardness decreased monotonically with temperature. Annealing caused essentially no changes
in hardness, as can be seen from Fig. 4, where all the samples fall
in one band. Such behavior was exhibited by all of the compositions.
The hardness of different compositions, on the other hand, was different, and the differences were maintained at all temperatures. As
expected, the hardness was highest for composition 5-10, and the hardness decreased with increase of the amount of glass phase and the
level of substitution. These results are consistent with trends observed previously in SiAlON-YAG system.

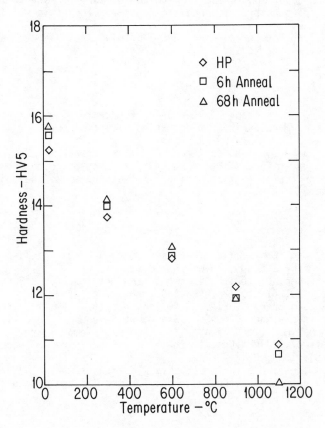

Fig. 4 Hardness vs. temperature for samples of composition 10-
       10 as hot pressed, and after 6- and 68-h anneal.

CONCLUSIONS

Fracture toughness and hardness in the temperature range from room
temperature up to 1100°C were measured for a series of SiAlON-cordierite compositions, annealed for different lengths of time to crystallize the cordierite phase. It was shown that toughness generally
decreased monotonically with temperature for all samples. Short time
annealing raised toughness at lower temperatures, while further annealing lowered it again to the value for the hot pressed samples. At
higher temperatures previous annealing had essentially no effect on
toughness. This behavior is significantly different from that reported for the system SiAlON-YAG, where annealing always resulted in
lowering of the toughness.

Hardness decreased monotonically with temperature for all samples.
It was not affected by the annealing treatment. Elastic modulus was
not affected by annealing. At elevated temperatures significant scatter of modulus results was observed, and only a roughly decreasing
trend could be observed.

ACKNOWLEDGEMENT

The authors wish to thank P. L. Dupree for performing the indentation
test and C. A. Markowski for preparation of the manuscript.

Two of the authors (D. Bonnell and T. Y. Tien) thank the U.S. Department of Energy, Office of Basic Energy Sciences, Division of Materials
Sciences, for providing the funding for the program at The University
of Michigan.

REFERENCES

Bonnell, D. and Tien, T.Y., "The System SiAlON-Cordierite", Progress
    Report DOE/ER/10619-3, November 1983.
Dole, S., Hunter, O., Jr., Calderwood, and Bray, D.J., 1978 J. Am.
    Ceramic Soc., 61 (11-12), 486-490.
Evans, A.G. and Charles, E.A., J.Am.Cer.Soc. 1976, 59,(7-8), 371-372.
Evans, A.G. and Wiederhorn, S., 1976 J.Mat.Science, 9, 270.
Lee, M., Brun, M.K., and Tien, T.Y.: "The High Temperature Fracture
    Toughness of SiAlON",Ceramic Engineering and Science Proceedings,
    4, (9-10), 1983, 864.
Mosher, D.R., Raj, R., and Kossowsky, R., "Measurement of Viscosity
    of the Grain Boundary Phase in Hot Pressed Silicon Nitride", J.
    of Mater. Sci., 1976, 11, 49-53.

Inst. Phys. Conf. Ser. No. 75: Chapter 5
Paper presented at 2nd Int. Conf. Science Hard Mater., Rhodes

413

Mechanical properties of HIP treated sintered boron carbide

K A SCHWETZ (1), W GRELLNER (1) AND A LIPP (2)

(1) ESK-GmbH, D-8960 Kempten, F.R.G.
(2) ESK-GmbH, D-8000 Munich, F.R.G.

ABSTRACT

Pressureless sintered boron carbide materials, doped with 1 and 3 wt% carbon, were subjected to HIP without encapsulating at 2000 °C under a gas pressure of 200 MPa argon. Materials, with sintered densities, D_S, in the range of 2.35 to 2.45 g/cm^3 (93 - 97 % of theoretical density, TD) allowed a post densification to a final density, D_{HIPS}, of $\geqslant 2.50$ g/cm^3 ($\geqslant 99$ % TD). In addition to elimination of residual porosity significant improvements were observed in flexural strength, elastic constants and wear resistance of sintered and posthipped boron carbide. The properties data, together with corresponding microstructures, are also compared with those of axially hot-pressed B_4C.

INTRODUCTION

According to the classification of hard materials by Kieffer and Bene-sovsky in "Hartstoffe" (1963), boron carbide belongs to the group of non-metallic hard materials, as e.g. Si_3N_4, SiC, cubic BN and diamond. Though it was prepared the first time a century ago (Joly, 1883), the formula B_4C was given only in 1934 (Ridgway, 1934). Today a homogeneity range of compositions from $B_{4.0}C$ to $B_{10.4}C$ is established (Thevenot and Bouchacourt, 1979; Bouchacourt and Thevenot, 1981; Schwetz and Lipp, 1985). The composition of the commercially available boron carbide is usually close to a boron : carbon atomic ratio of 4 : 1 (Adlassnig, 1958; Lipp, 1965 a, b, c; Lipp, 1966 a), which represents the stoichio-metric limit on the high carbon side.
Structurally boron carbide can be considered as the prototype of so-cal-led "interstitial compounds of alpha-rhombohedral boron" (Economy, Matko-vich and Giese, 1965), which includes $B_{12}C_3$ (= B_4C, formula referred to the unit cell, Lipp, 1966 b), $B_{12}C_2Al$ (Lipp and Roeder, 1966), $B_{12}O_2$ (Petrak, Ruh and Atkins, 1974), $B_{3-4}Si$ (Ettmayer, Horn and Schwetz, 1970) etc. The crystal structure for these materials consists of boron icosahedra which occupy the points of a rhombohedron, plus a chain of atoms (i.e. C-C-C, Duncan, 1984; or C-B-C, Werheit, De Groot and Malkemper, 1981), which runs along the c-axis of the rhombohedron. There is no chain at all in alpha-rhombohedral boron. The rigid framework of polyatomic units of closely bonded atoms is responsible for the refractory nature, high melting point, great hardness etc.

In contrast to silicon carbide crystals which in their purest form are transparent and colourless, the pure B_4C crystals which are obtained by the electrothermic production process

$$2 B_2O_3 + 7 C = B_4C + 6 CO \qquad (\Delta H = 1812 \text{ kJ/mole, 9.1 kWh/kg } B_4C)$$

show a deep black, shiny appearance (Lipp, 1965 a).

Congruent melting of B_4C occurs at 2450 °C (Elliot, 1961; Kieffer and co-workers, 1971), the specific weight is 2.52 g/cm³, the Knoop Microhardness Number HK 0.1 is about 3000 (Lipp and Schwetz, 1975), as compared with 2000 for alpha-Al_2O_3, 2600 for silicon carbide and elemental boron, 4700 for cubic boron nitride and 7000 - 8000 for diamond. Thus, except for diamonds, the hardest materials known all contain boron (Schwetz, Reinmuth and Lipp, 1981).

However, B_4C is vulnerable to oxidation at about 600 °C and reacts with metals that form borides or carbides at 1000 °C (Thompson, 1971). The latter property, in particular, explains why so many attempts to bind it with a metal to produce a dense cermet similar to the cemented carbides have failed. Therefore, dense B_4C articles, which are used in the abrasive wear industry e.g. as sand blasting nozzles (Lukschandel, 1972), dressing tools, and light-weight armour plates hitherto have been fabricated exclusively in self-bonded form via hot-pressing (Lipp, 1981).

As shown by Fig. 1, sand blasting nozzles made of dense boron carbide show a "non plus ultra" wear behaviour as compared with that of alumina, Si-infiltrated SiC, WC-Co hardmetal and sintered silicon carbide, respectively.

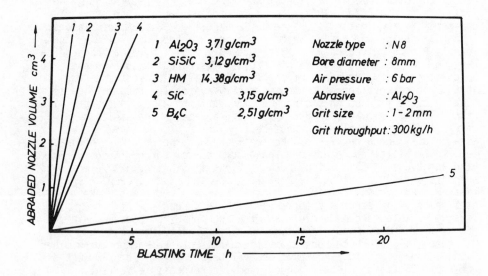

Fig. 1   Wear behaviour of blasting nozzles made of different hard materials

In the late seventies a process was developed (Schwetz and Vogt, 1977) for pressureless sintering of boron carbide, which uses submicrometre particle sized B_4C powder together with small additions of amorphous carbon as a sintering aid. Since sintering shrinkage is uniform and predictable, close tolerances can be met, i.e., near net shape B_4C parts are produced without the need for expensive diamond machining. However, the residual porosity of $\leqslant 2$ % affects the mechanical and wear properties of sintered boron carbide (SB_4C), e.g. the flexural strength is only 75 % of that of hot-pressed B_4C (Kislyi and Grabchuk, 1975; Champagne and Angers, 1979; Schwetz and Grellner, 1981; Osipov and co-workers, 1982; Beauvy, 1982; De With, 1984; Reinmuth and co-workers, 1984).

It was, therefore, the object of the present work to investigate the effect of post-hipping, i.e. post-densifying by hot isostatic pressing without cladding, on the mechanical properties of sintered boron carbide and to compare properties data of HIPS-B_4C with those of the axially hot pressed boron carbide (HP-B_4C).

RESULTS OF SINTERING AND POST-HIP-TREATMENT

The sintered B_4C materials, doped with 1 and 3 wt% C, were prepared by a standard isostatic cold-pressing and sintering process described elsewhere (Schwetz and Vogt, 1977; Schwetz and Grellner, 1981). The boron carbide powder used has a purity of B + C = 97.2 % (total carbon, 20.5 wt%), a specific surface area of 22 m^2/g and a maximum particle size of 1 μm. The impurities contained are as follows (ppm): oxygen (23000), nitrogen (2900), Fe (200), Si (1300), Al (500). The carbon containing additive was used in form of a phenolformaldehyde resin and introduced in methanolic solution. During drying of the boron carbide - methanol dispersion a thin resin film is deposited on the carbide particles. This film firstly serves as a pressing aid and later pyrolyses to amorphous vitreous carbon coatings during the heating up cycle of pressed powder compacts. Sintering of compacts (plates of dimensions 53 x 23 x 7 mm) was carried out under reduced pressure (0.1 mbar), at a temperature ranging from 1900 to 2150 °C (hold time, 30 min) in graphite crucibles using a graphite-tube resistor furnace.

The role of carbon as a sintering aid is attributed to two major effects (Schwetz, Reinmuth and Lipp, 1983):

(1)   Chemical effect (temporary); the removal of surface B_2O_3-contamination layers from the carbide particles by the chemical reaction 2 B_2O_3 + 7 C = B_4C + 6 CO

(2)   Sintering activation effect (permanent); a certain amount of carbon is necessary to enhance grain boundary and/or lattice diffusion by inhibiting the surface-to-surface matter transport processes. The carbon migrates with the boundary and finally is agglomerated at grain corners ("Solid-State-Diffusion"-sintering mechanism). This implies that the sintered bodies exhibit a significant concentration of residual carbon.

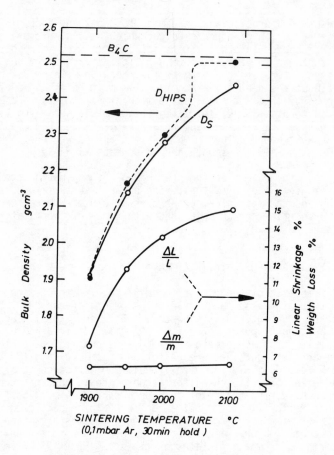

Fig. 2   Weight loss $\Delta m/m$, linear shrinkage $\Delta L/L$,
sintered (D_S) and posthipped (D_{HIPS}) density
of B_4C (22 m²/g) + 1 wt% C compacts
(green density, 59 % of theoretical)

Subsequent to sintering the sintered plates were subjected to hot
isostatic pressing (Hunold, 1983) without encapsulating, at 2000 °C in a
high-pressure autoclave under a gas pressure of 200 MPa argon (120 min
hold).
Fig. 2 illustrates - for the case of B_4C doped with 1 wt% carbon - the
linear shrinkage, sintered density and weight loss increase with
increasing firing temperature from 1900 °C up to 2100 °C. At temperatures
of 2050 to 2100 °C (30 min hold) sintered densities, D_S, of 2.35 to 2.45
g/cm³ were achieved, which allow a post-HIP-densification to a final
density, D_{HIPS}, of $\geqslant 2.50$ g/cm³: the photomicrographs Figs. 3 and 4
clearly show elimination of porosity by posthipping of sintered boron
carbide doped with 3 wt% C. Note that black points in Fig. 4 (HIPS-B_4C)
are not pores but residual carbon.

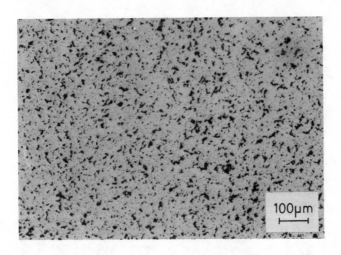

Fig. 3 Polished and unetched section of S-B$_4$C (3 % C doped),
C_{total} = 24.8 %, D_S = 2.46 g/cm³; $\sigma_{B,4pt}$ = 353 MPa;
sintering conditions: 2100 °C, 0.1 mbar, 30 min

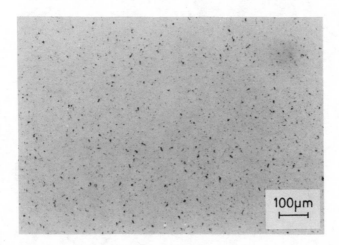

Fig. 4 Polished and unetched section of HIPS-B4C (3 % C doped)
C_{total} = 24.8 %, D_{HIPS} = 2.50 g/cm³; $\sigma_{B,4pt}$ = 429 MPa
same material as shown in Fig. 3 after HIP treatment
(2000 °C, 200 MPa, 2 h)

TABLE 1   Sintering Response and Mechanical Properties of HP-, S- and HIPS-B$_4$C Materials

B$_4$C-C addition wt% C	Firing Temp. a) °C	Sintering condit.	Final b) Density g/cm³	Relative Density c) % TD	Mean Grain Size d) μm	Flexural Strength 4pt, MPa	Young's Modulus GPa	Shear Modulus GPa	Poisson's Ratio
B$_4$C (21.7) e)	-	HP	2.51	99.6	5.4	480 ± 40	441	188	0.17
B$_4$C-1 C (22.5) e)	1900	S	1.91	75.8		224 ± 13			
		HIPS	1.90	75.4		241 ± 16			
	1950	S	2.14	84.9		340 ± 20			
		HIPS	2.17	86.1		293 ± 17			
	2000	S	2.28	90.5		350 ± 33			
		HIPS	2.30	91.3		323 ± 19			
	2100	S	2.44	96.8	7.7	351 ± 40	390	166	0.17
		HIPS	2.51	99.6	8.1	401 ± 19	433	183	0.18
B$_4$C-3 C (24.8) e)	2000	S	2.32	92.1		320 ± 37			
		HIPS	2.33	92.5		359 ± 39			
	2050	S	2.35	93.3		349 ± 46			
		HIPS	2.50	99.2		428 ± 58			
	2100	S	2.46	97.6	7.1	353 ± 30	372	158	0.17
		HIPS	2.50	99.2	7.8	429 ± 63	405	171	0.16
	2150	S	2.46	97.6	10.0	340 ± 38			
		HIPS	2.50	99.2		416 ± 50			

a) hold time, 30 min (0.1 mbar)

b) by sample immersion in H$_2$O at 20 °C

c) TD ... theoretical density, based on 2.52 g/cm³ for pure B$_4$C (C$_{total}$: 21.7 wt%)

d) by linear intercept method

e) content of C$_{total}$ (wt%) after sintering

PROPERTIES OF HIP-TREATED BORON CARBIDE (HIPS-B₄C), COMPARISON WITH
AXIALLY HOT PRESSED BORON CARBIDE (HP-B₄C)

In Table 1 the final density, mean grain size, flexural strength and
elastic constants (Young's modulus, shear modulus and Poisson's ratio) of
sintered (S) and posthipped (HIPS) boron carbide, doped with 1 and 3 wt%
C, are summarized and compared with corresponding data of dense single
phase hot-pressed boron carbide (HP).

For property measurements small prismatic bars (2 x 4 x 34 mm) were
prepared from the sintered and post-hipped plates by cutting and wet
grinding with a plastic bonded 90 μm diamond disc (maximum surface
roughness, 1.3 μm). Flexural strength values ($\sigma_{B,4pt}$) are mean values of
10 measurements, the relative standard deviation, $(s/\sigma) \times 100$, was in
the range of 8 to 12 %. Details of strength measurements have already
been described (Schwetz and Grellner, 1981). Young's and shear modulus,
as well as Poisson's ratio (E, G and ν, respectively), were deducted
from the given resonance frequencies. The experimental set-up induces
mechanical vibration into the prismatic bars. From resonance frequencies
of transversal and torsional harmonics the values of E and G can be
calculated. Poisson's ratio   is then, according to following equation:

$$\nu = (E / 2G) - 1 \qquad\qquad 1$$

Using appropriate specimen preparation (satisfactory homogeneity of the
materials used, rectangular bars, good surface quality) the resolution is
better than 0.5 %. Comparison examinations with Young's moduli of static
loaded samples showed excellent agreement.

Figs. 6 and 8 are photomicrographs with 500 fold magnification and show
the fine-grained and uniform microstructures of pure HP-B₄C and HIPS-B₄C
(1 % C doped) materials. The microstructures shown were revealed via
electrolytic etching of polished specimen-sections with dilute sulphuric
acid.

The comparison of Figs. 6 and 8 clearly shows the small difference in
average grain size and the more frequent presence of twins in HIPS-B₄C.
Residual carbon can just be detected as a second phase in form of small
graphite particles, ≤ 2 μm in diameter, homogeneously dispersed (inter-
and transgranular).

The SEM-micrographs, Fig. 5 and 7, show almost 100 % transgranular
fracture and substantial freedom from porosity for both boron carbide
materials. As can be seen from Figs. 9 to 12, in addition to elimination
of residual porosity significant improvements were observed in flexural
strength, elastic constants and wear resistance of sintered and
posthipped B₄C.

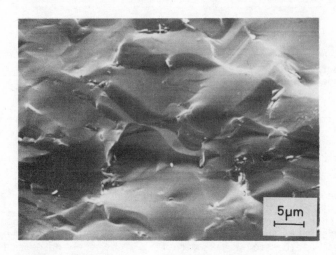

Fig. 5 Fracture surface (SEM, 2000 fold) of HP-B$_4$C
D_{HP} = 2.51 g/cm^3; $\sigma_{B,4pt}$ = 480 MPa
C_{total} = 21.7 wt%

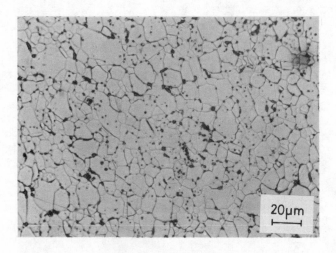

Fig. 6 Microstructure of HP-B$_4$C, C_{total} = 21.7 wt%
D_{HP} = 2.51 g/cm^3, $\sigma_{B,4pt}$ = 480 MPa
Mean grain size = 5,4 μm

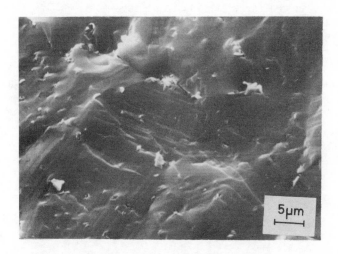

Fig. 7   Fracture surface (SEM, 2000 fold) of HIPS-B$_4$C
(1 % C doped), D$_{HIPS}$ = 2.51 g/cm^3;
$\sigma_{B,4pt}$ = 401 MPa, C$_{total}$ = 22.5 wt% C

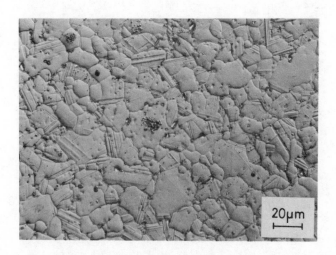

Fig. 8   Microstructure of HIPS-B$_4$C (1 % C doped),
C$_{total}$ = 22.5 wt% C, D$_{HIPS}$ = 2.51 g/cm^3;
$\sigma_{B,4pt}$ = 401 MPa, Mean Grain Size: 8.1 μm

Since high gas pressure of approximately 200 MPa was used, the post densification could be carried out at a 50 to 100 °C lower temperature than used for pressureless sintering to closed density (93 - 95 % TD[x]). Therefore the porefree post-hipped bodies underwent no grain growth and had substantially the same average size as the predensified bodies (~ 8 μm). Containerless hot isostatic pressing did not result in elimination of pores in the case of sintered densities less than 2.35 g/cm³; the HIP-treatment of the porous bodies showing inter-communicating porosity produced strength degradation (see Figs. 9 and 10).

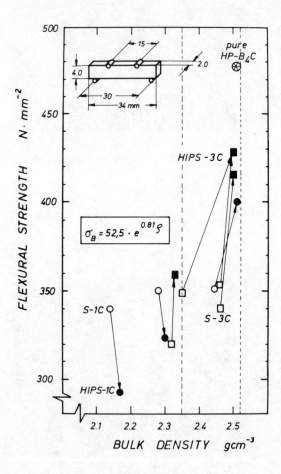

Fig. 9   Flexural strength as a function of sintered density for sintered (S), hot-pressed (HP) and HIP-treated sintered (HIPS) boron carbide materials.

---

[x]TD ... of theoretical density, based on 2.52 g/cm³ for pure B₄C
    (C_{total} = 21.7 wt%)

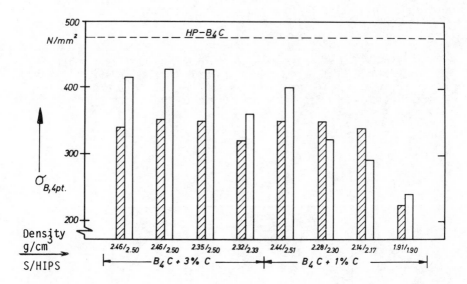

Fig. 10   Flexural strength of carbon doped sintered (hatched)
           and post-HIPed (unhatched) boron carbide materials

Within the grain size range of 5 to 10 μm the strength properties may be
described by the exponential expression

$$\sigma_B = 52.5 \times \exp (0.81 \times \rho) \qquad\qquad 2$$

where σ_B is the 4-point flexural strength in MPa and ρ is the final
density in $10^3 \times kg/m^3$.

The mean flexural strength of the samples that have been post-densified
to densities of ≥ 2.50 g/cm³ is increased from 345 to 415 MPa, i.e. by
more than 20 % (see Fig. 10) with respect to samples which were only
pressureless sintered.It is rather surprising that flexural strength as
well as grain size of S- and HIPS-boron carbide are almost unaffected by
the amount of free carbon in the sintered materials (up to total carbon
contents of ~ 25 wt%).

As shown by Fig. 11, Young's (E) and shear (G) moduli are increased with
respect to pressureless sintered samples towards the maximum values of
pure, hot-pressed B4C. The obtained E/G dependence is strictly linear and
from the slope 2.3529, a Poisson's ratio of 0.176 is computed.

Wear testing of HIPS-B4C components clearly demonstrates the potential
for future utilization of this new material in abrasive wear industry
(see Fig. 12).

Wear behaviour of porefree HIPS-B4C sand blasting nozzles actually
matches convential hot-pressed boron carbide nozzles (HP-B4C). However,
this is not the case for pressureless sintered boron carbide (i.e.
without HIP-post-densification) mainly due to the residual porosity of
≤ 2 %.

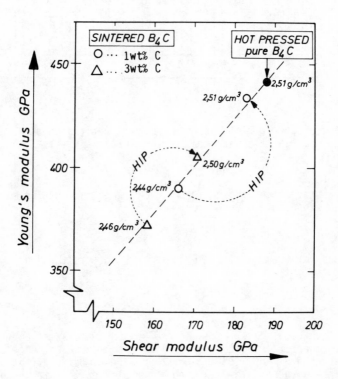

Fig. 11   Influence of post-HIP-treatment on elastic constants
          of pressureless sintered B$_4$C

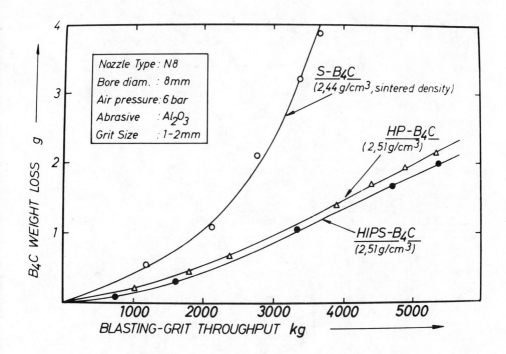

Fig. 12   Wear characteristics of boron carbide sand blasting
nozzles made by different fabrication methods

ACKNOWLEDGEMENTS

The authors thank H. Knoch and J. Kracker for conducting the wear
measurements of various sand blasting nozzles, Mrs. Petra Sacher for
typing and J. Lukschandel for critical review of the manuscript.

REFERENCES

Adlassnig,K. 1958, Planseeber. Pulvermetall. 6, 92-103
Bouchacourt,M. and Thevenot,F. 1981, J. Less-Common Met. 82, 227-235
Beauvy,M. 1982, Rev. Int. Hautes Temper. Refract. 19, 301-310
Champagne,B. and Angers,R. 1979, J. Am. Ceram. Soc. 62 (3-4), 149-153
De With,G. 1984, J. Mat. Sci. 2 (19), 457-466
Duncan,T.M. 1984, J. Am. Chem. Soc. 106 (8), 2270-5
Elliot,R.P. 1961, IIT Research Inst., ARF-2200-12, Final Report,
    US At. En. Comm., Contract At. (11-1), 578, Project. Aggr. No. 4, 445
Economy,J., Matkovich,V.I. and Giese,R.F. 1965, Z. Kristallogr. 122, 248
Ettmayer,P., Horn,H. Ch. and Schwetz,K.A. 1970, Mikrochimica Acta,
    Suppl. IV.,87-95

Hunold, K. 1983, cfi/Ber. d. DKG, 60, 182-189

Joly, A. 1883, C. R. Acad. Sci. 97, 456-458

Kieffer, R. and Benesovsky, F. 1963, "Hartstoffe", Springer Verlag, Wien/New York

Kieffer, R., Gugel, E., Leimer, G. and Ettmayer, P. 1971, Ber. Dt. Keram. Ges. 48, 385

Kislyi, P. S. and Grabchuk, B.L. 1975, Proc. 4th Eur. Symp. on Powder Metallurgy, Vol 3, Paper 10-2, INIS-mf-2082

Lipp, A. 1965a, Techn. Rundschau 57 (14), 5

Lipp, A. 1965b, Techn. Rundschau 57 (28), 19

Lipp, A. 1965c, Techn. Rundschau 57 (33), 5

Lipp, A. 1966a, Techn. Rundschau 58 (7), 3

Lipp, A. 1966b, Ber. Dt. Keram. Ges. 43, 60-62

Lipp, A. and Roeder M. 1966, Z. Anorg. Allgem. Chem. 343, 1

Lipp, A. and Schwetz K.A. 1975, Ber. Dt. Keram. Ges. 52, 335-338

Lipp, A. 1981, Silikattechnik 32, 373-375

Lukschandel, J. 1973, Fachberichte f. Oberflächentechnik OT 11, 273-275

Osipov, A.D., Ostapenko, I.T., Slezov, V.V., Tarasov, R.V., Podtykan, V.P. and Kartsev, N.F. 1982, Porosh. Met. 229, 63-67

Petrak, D.R., Ruh, R. and Atkins, G.R. 1974, Am. Ceram. Soc. Bull. 53, 569

Reinmuth, K., Lipp, A., Knoch, H. and Schwetz, K.A. 1984, J. Nuclear Mat. 124, 175-184

Ridgway, R.R. 1934, Trans. Electrochem. Soc. 66, 117-133

Schwetz, K.A. and Vogt, G. 1977, DE-2751998 (US-4195066, "Process for the Production of dense sintered shaped Articles of polycrystalline Boron Carbide by Pressureless Sintering")

Schwetz, K.A. and Grellner, W. 1981, J. Less Common Met. 82, 37-47

Schwetz, K.A., Reinmuth, K. and Lipp, A. 1981, Radex Rdsch. Nr. 3, 568-85

Schwetz, K.A., Reinmuth, K and Lipp, A. 1983, Sprechsaal 116 (No. 12) 1063-1070

Schwetz, K.A. and Lipp, A. 1985, Chapter "Boron Carbide, Boron Nitride and Metal Borides" in ULLMANN's Encyclopedia of Industrial Chemistry, Verlag Chemie GmbH, D-6940 Weinheim

Thevenot, F. and Bouchacourt, M. 1979, Ind. Ceram. 732, 655-61

Thompson, R. 1971, in Progress in Boron Chemistry" (Brotherton J., Steinberg H., Eds.) Vol 2, Pergamon Press, 173-230

Werheit, H., De Groot, K., and Malkemper, W., 1981, J. Less-Common Met. 82, 153-62.

Inst. Phys. Conf. Ser. No. 75: Chapter 5
Paper presented at 2nd Int. Conf. Science Hard Mater., Rhodes

427

Particulate titanium carbide–ceramic matrix composites

J G BALDONI, S T BULJAN, AND V K SARIN
GTE Laboratories, Inc., Waltham, MA 02254

ABSTRACT

Mechanical properties of Si_3N_4 and Al_2O_3 matrix composites containing 30 vol % TiC were evaluated. The room and elevated temperature K_{IC} of the Si_3N_4 composite were found to be essentially unaffected by TiC additions, while the strength exhibited a maximum at 1000°C. The addition of TiC to Al_2O_3 was found to increase K_{IC} at room temperature and, as opposed to monolithic Al_2O_3, K_{IC} was found to increase above 900°C. The strength was observed to decrease with increasing temperature. The results are discussed and compared to crack-dispersoid interaction toughening mechanisms.

INTRODUCTION

The development of composite materials is one of the focal points of ceramic research in recent years. Composites allow more radical changes and tailoring of materials properties and new property tradeoff directions which are often fixed or limited in typical "single phase" ceramic materials. Silicon nitride ceramics are materials with high strength, thermal shock resistance, and relatively high fracture toughness and as such represent an appealing base for the development of either fiber or particulate modified microstructures.

One of the more recently developed materials, silicon nitride-titanium carbide dispersed particulate phase-composites, is successfully being utilized in metal cutting-tool and wear applications. In a number of specific applications, this silicon nitride-based composite performs substantially better than aluminum oxide-particulate titanium carbide ceramics (Sarin and Buljan, 1983).

This study examines and compares the room and elevated temperature mechanical properties of silicon nitride and alumina-based composites containing dispersed, particulate titanium carbide.

EXPERIMENTAL PROCEDURE

Hot-pressed billets of Al_2O_3 + TiC and Si_3N_4 + TiC, both nominally 30 vol % TiC and fully dense, were used in this study. The Al_2O_3-based material was obtained from a commercial source, while the Si_3N_4-based material was fabricated using Si_3N_4 and TiC powders with 6 wt % Y_2O_3 as a densification aid. The Al_2O_3 content from milling pickup was determined by emission spectroscopy to be 2.2 wt %. MOR samples, which were also used for controlled surface flaw samples, were rectangular bars, 0.127x0.259x2.5 cm, machined from the billets with the tensile face perpendicular to the hot-pressing direction. The machine lay was parallel to the length of the bars, and the edges were chamfered to prevent notch effects. The tensile face of each sample was wet-polished to a 1 μm diamond paste finish.

A randomly selected group of test bars was precracked with a single Knoop indentation. A 2.5-kg mass was used, with the long axis of the indentation aligned perpendicular to the tensile stress direction of the bars. Precracked samples were annealed in vacuum at 1200°C and slowly cooled (60°C/h) to room temperature in order to relieve residual stresses.

Test bars of each material, both precracked and non-precracked, were broken in a four point loading fixture placed in a tungsten mesh furnace. Inner and outer loading points were 1.016 and 2.286 cm, respectively. Room temperature testing was performed in air and high temperature testing in flowing argon. A crosshead speed of 0.05 cm/min was used for all tests. The fracture surfaces of each broken precracked test bar were examined by optical microscopy. Each individual precrack size was measured from photomicrographs. Fracture toughness, K_{IC}, was calculated using the relationship (Petrovic and co-workers, 1975):

$$K_I = \sigma M(\pi a/Q)^{1/2} \qquad 1$$

Where σ is the maximum outer fiber tensile stress, M is a numerical factor related to flaw and specimen geometry, a is the flaw depth, and

$$Q = \Phi^2 - 0.212(\sigma/\sigma_y)^2 \qquad 2$$

where σ_y is the tensile yield stress and Φ is the elliptic integral, which is tabulated in standard mathematical tables.

The value of M for a semicircular flaw is 1.03, and this value was used for all calculations. The plastic zone correction factor 0.212 $(\sigma/\sigma_y)^2$ is considered negligible for all calculations except for those temperatures at which nonlinear force-time response in MOR testing is observed.

Samples for microstructural characterization and fracture toughness determination by indentation (indentation fracture toughness, IFT) were polished sections cut from the same hot pressed billets used for test bars. The polished surface was perpendicular to the hot-pressing direction.

IFT was determined by indenting the sample with a Vickers indenter using a 10 kg mass and measuring the total length of cracks emanating from the corners of the Vickers indentation. Based on the work of Evans and Charles (1976), fracture toughness values were calculated using the following relationship:

$$IFT = 0.113HD^{0.5}/(1+C_L/2D)^{1.5} \qquad 3$$

IFT=Indentation Fracture Toughness, MPa•m$^{1/2}$
H=Microhardness, GPa
D=Diagonal of Vickers Indentation, μm

$$C_L = \text{Total Crack Length} = \sum_{n=1}^{4} c_n \equiv \text{sum of crack lengths emanating from the four corners of the Vickers indentation}$$

Microhardness values were obtained using both Vickers and Knoop indenters. The microstructure of each material was analyzed quantitatively from optical micrographs.

RESULTS

Microstructure

The mean size of the dispersed phase in both composites was found to be approximately 1.5 μm. The calculated average nominal mean-free path in the matrix of both composite materials was approximately 3.0 μm A higher concentration of TiC grains larger than 4 μm was observed in the Si_3N_4-based material (Fig. 1).

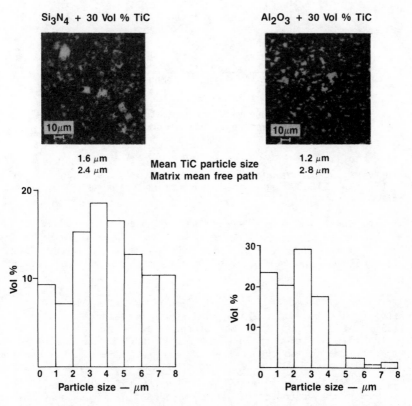

Fig. 1: Quantitative microstructural analysis of the Al_2O_3-TiC and Si_3N_4-TiC composites.

Modulus of Rupture

The values of modulus of rupture of materials measured up to 1200°C are shown in Fig. 2. The observed dependence of modulus of rupture on temperature for the Al_2O_3-TiC composite was similar to that which has been reported in the literature for Al_2O_3 materials (Lynch and co-workers, 1966).

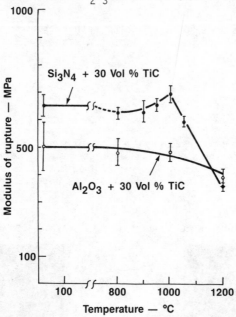

Fig. 2: Modulus of rupture versus temperature for two 30 vol % TiC dispersed phase-ceramic matrix composites.

The strength of the Al_2O_3 + 30 vol % TiC material remained constant up to approximately 1000°C, experiencing a gradual loss of strength in the range 1000-1200°C. At 1200°C, non-linear force-time response prior to fracture was observed.

Relative to room temperature, the strength of the Si_3N_4 + 30 vol % TiC material appeared to decrease slightly at 800°C and then increase to maximum at 1000°C. Above 1000°C, a rapid loss in strength was observed. Nonlinear force-time response was not observed in this material during testing up to 1200°C.

The strength increase near 1000°C is not typical for simple binary Si_3N_4 materials where the modulus of rupture is constant to approximately 1000-1100°C, after which point the strength decreases at higher temperatures. The rate of strength loss at high (>1000°C) temperature is dependent upon the amount and composition of the glass phase. The presence of Al_2O_3 was found to affect the elevated temperature strength and relates to the intergranular SiO_2 glass viscosity reduction through Al_2O_3 additions as well as through the volume of glass increase (Smith and Quackenbush, 1980). The observed strength behavior in the temperature range 800-1200°C for the Si_3N_4 + TiC composite is more typical of Al_2O_3-based ceramics containing appreciable amounts of SiO_2 (10 wt %). The increase in strength at elevated temperature has been associated with plasticity of the glass phase, where it is postulated that at the maximum in strength the viscosity of the glass is such that localized crack blunting occurs by viscous flow, relieving stresses at the crack tip. Above this temperature the viscosity of the glass phase is reduced and the strength of the material decreases rapidly (Davidge and Evans, 1970).

The lower room temperature strength of Si_3N_4 containing TiC compared to Si_3N_4 is attributed to the agglomeration of TiC grains during processing, creating larger size critical flaws in the composite material.

Indentation Fracture Toughness

Literature for cemented materials has shown that fracture toughness values determined by indentation, Palmquist fracture toughness, can be affected by the residual strain state of the polished sample surface. Residual compressive strain produced by grinding and polishing can inhibit crack propagation during indentation, which reduces the total crack length produced and results in a higher apparent fracture toughness (Exner, 1969). Fig. 3 shows the measured crack length as a function thickness of surface removed for a Si_3N_4 + TiC sample. From these data it is apparent that the initial

grinding and polishing produced a damaged layer approximately 10 μm deep, which remained after the initial final polishing step using 1-μm diamond paste. Based upon results such as this, it was determined that automatic polishing with frequent diamond paste replenishment (every 5-10 min) yielded reproducible minimum damage surfaces and consistent crack length measurements.

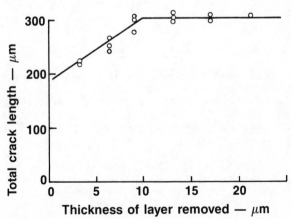

Fig. 3: Measured total crack length produced by a 10 kg Vickers indenter as a function of thickness of damage layer (produced by rough grinding & polishing) removed.

The measured crack lengths and average diagonals produced by the 10 kg Vickers indentation are shown in Table 1 for Si_3N_4 and Al_2O_3-based materials. Also included are the average Vickers and Knoop microhardness values for each material. For a given load, the diagonal of a Vickers indenter is about one third of the length of the Knoop indenter due to the differences in indenter geometry. This affects the calculated values of microhardness with the Knoop values being approximately 15% less than the Vickers values for the same material. Consequently, the value of fracture toughness calculated using the expression previously given, which contains a hardness term, is dependent upon microhardness technique; i.e., the IFT values calculated using Knoop microhardness are approximately 15% lower than those calculated using Vickers microhardness.

TABLE 1:   Indentation Data and Calculated Indentation Fracture Toughness (IFT)

	Si_3N_4*	Si_3N_4 + 30 Vol % TiC	Al_2O_3**	Al_2O_3 + 30 Vol % TiC
Vickers Diagonal, µm	109.5 ± 0.6	104.4 ± 0.6	102.1 ± 0.5	95.4 ± 0.7
Total Crack Length, µm	298.1 ± 9.0	341.4 ± 18.2	528.6 ± 37.4	475.9 ± 19.6
Microhardness, GPa				
Vickers, HV10	15.2 ± 0.3	16.7 ± 0.3	17.4 ± 0.2	20.0 ± 0.4
Knoop, HK1	13.6 ± 0.8	14.6 ± 0.5	15.2 ± 0.8	16.8 ± 0.7
IFT, MPa • $m^{1/2}$				
using HV10	5.0 ± 0.1	4.5 ± 0.1	2.9 ± 0.1	3.5 ± 0.1
using HK1	4.4 ± 0.1	4.0 ± 0.2	2.6 ± 0.2	3.0 ± 0.2

*Composition; 6 w/o Y_2O_3, 2 w/o Al_2O_3, Balance Si_3N_4, Sintered
**Fine grained Al_2O_3 cutting tool

Fracture Toughness by Controlled Surface Flaw Technique

Fracture toughness measurements using controlled surface flaws showed that, for both composite materials, the flaws introduced by the 2.5 kg Knoop indentations were semielliptical in shape, the ratio of flaw depth to half crack length was 0.60-0.85. The induced flaw size was observed to vary appreciably for each composite, the flaw depth was 95-120 µm for the Al_2O_3 + 30 vol % TiC material and 60-100 µm for the Si_3N_4 + 30 vol % TiC material.

The samples which were annealed at 1200°C and slowly cooled to relieve residual tensile strain produced by the indentation process had the same range of induced flaw sizes as the unannealed, as indented samples indicating that flaw healing during annealing did not occur.

Values of induced flaw depth using 2.6-kg Knoop indentations reported in the literature for hot-pressed Si_3N_4 not containing a dispersed hard second phase are approximately 70 µm with a standard deviation of 3% (± 2.0 µm) (Petrovic and co-workers, 1975). The observed larger range in induced flaw sizes in Al_2O_3 + TiC and Si_3N_4 + TiC materials is attributed to the nature of the composite. Crack growth during indentation may be influenced by the TiC distribution and degree of agglomeration. The measured strength of the precracked samples, as illustrated in Table 2, was found to be inversely proportional to the induced flaw size.

TABLE 2: <u>Measured</u> <u>Data</u> <u>and</u> <u>Calculated</u> <u>Fracture</u> <u>Toughness</u> <u>of</u>
<u>Si$_3$N$_4$+TiC</u> <u>and</u> <u>Al$_2$O$_3$+TiC</u> <u>Composites</u>
<u>At</u> <u>Room</u> <u>Temperature</u>

Sample	Measured Crack Depth μm	Modulus of Rupture GPa	Calculated K_{IC} MPa·m$^{1/2}$
Si$_3$N$_4$ + TiC			
#1	77	409.6	4.7
#2	54	473.8	4.8
#3	80	408.6	4.5
#4	99	351.7	4.6
Al$_2$O$_3$			
#1	95	285.5	3.3
#2	120	280.1	3.6
#3	95	295.6	3.4
#4	95	303.9	3.4

The fracture toughness values at room and elevated temperatures of the two composite materials containing 30 vol % TiC dispersed phase obtained using precracked and annealed specimens are shown in Fig. 4. The fracture toughness of the Al$_2$O$_3$ + TiC composite decreased between 900 and 1000°C in a

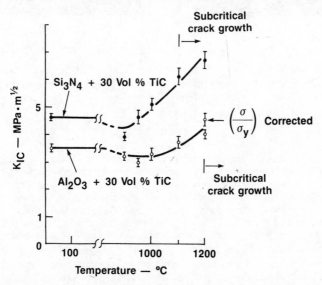

Fig. 4: Fracture toughness versus temperature for two 30 vol % TiC dispersed phase-ceramic matrix composites.

manner similar to that reported in literature for dense Al_2O_3 (Claussen, 1975). However, unlike Al_2O_3, above 1000°C the measured fracture toughness of the Al_2O_3 + 30 vol % TiC increases, and at 1200°C limited subcritical crack growth prior to the fast fracture is observed (Fig. 5b). Nonlinear force-time behavior was not observed for this composite as was observed during modulus of rupture measurements.

The fracture toughness of the Si_3N_4 + TiC composite was observed to decrease slightly at 900°C and then increase with increasing temperature in an analogous manner to data reported for hot pressed Si_3N_4 (Petrovic and co-workers, 1975). Subcritical crack growth was observed in the Si_3N_4 composite at 1100°C (Fig. 5a), and became extensive at 1200°C. Fracture toughness was calculated using the increased flaw size prior to fast fracture at the higher temperatures. Subcritical crack growth prior to fast fracture has not been previously reported to occur in hot pressed Si_3N_4 containing Y_2O_3 sintering aid below 1400°C (Kris, 1983) but has been observed in Si_3N_4 containing MgO at 1200-1250°C (Petrovic and co-workers, 1975; Govila, 1979; Kris, 1983).

(a)   (b)

Fig. 5: Subcritical crack growth observed for a) Si_3N_4 + 30 vol % TiC at 1100°C and b) Al_2O_3 + 30 vol % TiC at 1200°C. (Black arrows indicate original crack produced by Knoop indenter and white arrows ind᷃ e crack growth prior to fast fracture.)

Crack Trajectory

Fully propagated cracks, induced at room temperature using a Vickers indenter, (Fig. 6) were observed to pass both around and through TiC grains in both composites. Composite theory predicts that crack trajectory is controlled by residual stresses produced by dispersoid-matrix property mismatches, i.e., thermal expansion coefficient (α)(Binds, 1962) and elastic modulus mismatch (E) (Erogen, 1974). For a composite in which the dispersed phase has a higher thermal expansion coefficient and elastic modulus than the matrix, i.e., Si_3N_4 matrix-TiC dispersoid, the resultant thermal expansion mismatch strain and modulus mismatch under load tend to divert an advancing crack around the particle (Binds, 1962, Faber, 1982). For the case in which the thermal expansion coefficient of the dispersoid is less than that of the matrix but the elastic modulus is higher, i.e., Al_2O_3 ma-

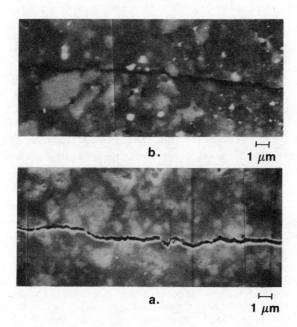

Fig. 6: Crack trajectory in a) Si_3N_4 + 30 vol % TiC and b) Al_2O_3 + 30 vol % TiC composites.

trix-TiC dispersoid, the thermal expansion mismatch stress tends to divert the advancing crack toward the dispersed second phase, while the modulus mismatch stress tends to divert the crack around the particle and the crack trajectory is difficult to predict. No dominance of either crack-dispersoid interaction, i.e., deflection around or through the TiC grains, was observed in either composite.

DISCUSSION

The Si_3N_4 composite containing 30 vol % TiC had a value of fracture toughness at room temperature which was lower, approximately 10%, than that measured for the matrix material alone. Published data pertaining to this composite system are conflicting: the room temperature fracture toughness of Si_3N_4 containing TiC (8.0 μm average particle size) and a CeO_2 sintering aid exhibits a maximum at 20 vol % dispersoid and then decreases with increasing TiC content (Mah, 1981), while the room temperature K_{IC} decreases with increasing dispersoid content for Si_3N_4 composites containing fine TiC (2.0 μm average particle size) and a Y_2O_3 sintering aid (Buljan and Sarin, 1983). Neither of these behaviors is explained by crack interaction mechanisms, crack deflection, line tension, or surface roughening, which predict increased fracture toughness with increased dispersoid content (Faber and co-workers, 1983).

Recent literature (Faber, 1982; Faber and co-workers, 1983) predicts toughening by crack deflection, irrespective of the size of the dispersoid or the sign of the thermal expansion/elastic modulus mismatch stress between dispersoid and matrix. However, this has been confirmed only in composite systems which do not contain matrix phase grain boundaries, i.e., glass matrix dispersed particulate phase composites (Faber, 1982; Swearanger, 1978). The TiC dispersoid was observed to have little influence on the trajectory of a propagating crack, which further indicates that crack interactive toughening mechanisms are not operative in this composite system at room temperature.

The room temperature fracture toughness of TiC ($\approx$3.9 MPa$\bullet$m$^{1/2}$) (Chermant and co-workers, 1978; Wunschmann, 1978) is lower than that for hot pressed Si_3N_4 containing Y_2O_3 ($\approx$4.5 MPa$\bullet$m$^{1/2}$), and it appears that the addition of this more penetrable dispersoid lowers the fracture toughness of the Si_3N_4-TiC composite, since propagating cracks do penetrate the dispersed particulate phase.

At elevated temperature, the fracture toughness of the Si_2N_4 + 30 vol % Tic composite was observed to increase with increasing temperature in a manner similar to hot pressed Si_3N_4, a material whose high temperature mechanical properties are controlled by the amount and viscosity of the intergranular glass phase (Smith and Quackenbush, 1980). The observed maximum in modulus of rupture at 1000°C and the detection of substantial subcritical crack growth at 1100°C for this composite indicate that the intergranular glass phase is substantially altered, compared to hot-pressed Si_3N_4, by the introduction of the TiC dispersoid. ESCA (Electron Spectroscopy for Chemical Analysis) analysis of as-received powders has shown the presence of TiO_2 on the surface of the TiC particles. STEM (Scanning Transmission Electron Microscopy) coupled with energy dispersive x-ray analysis of the glass phase at triple points in Si_3N_4-TiC composite microstructures has shown the presence of titanium in the glass (Zilberstein and Buljan, 1983). Apparently the glass softening point is appreciably lowered by the addition of TiO_2 or TiC, which allows slow crack growth to occur at a temperature 200-400°C lower than previously reported for Si_3N_4 with MgO or Y_2O_3 sintering aids, respectively.

In the Al_2O_3-TiC composite system in which the dispersoid has a higher room temperature fracture toughness than the matrix, the addition of TiC increases the fracture toughness of the composite, and the fracture toughness has been observed to increase with increasing dispersoid content (Buljan and Sarin, 1983; Wahi and Ilseher, 1980). Crack penetration of the higher K_{IC} phase appears to increase the fracture toughness of the composite. Although crack deflection toughening cannot be ruled out, the thermal expansion coefficient mismatch and elastic moduli mismatch are predicted to divert the crack in opposite ways (Faber, 1982) and the effects may cancel. Very little crack deflection was observed in room temperature IFT testing (Fig. 6).

As opposed to alumina, at elevated temperature the fracture toughness of the Al_2O_3-TiC composite increases with increasing temperature. Since TiC undergoes a brittle-ductile transition (Williams and Schaal, 1962) at 800-900°C, crack interaction with the "ductile" dispersoid, crack blunting or line tension (Lange, 1970; Evans, 1974) may account for this increase in

K_{IC} and simultaneous drop in modulus of rupture. Weakening of grain boundaries due to the presence of TiO_2 could also contribute to the increase in K_{IC} and decrease in strength.

SUMMARY AND CONCLUSION

The addition of TiC to alumina increases the room temperature fracture toughness of the composite relative to the matrix, and the high temperture K_{IC} increases with increasing temperature. At room temperature, TiC additions to Si_3N_4 decrease K_{IC}, and at elevated temperature, slow crack growth is observed at temperatures much lower than has been reported for monolithic Si_3N_4. A maximum in modulus of rupture is observed at 1000°C for the Si_3N_4-TiC composite.

Observation of the trajectories of cracks generated at room temperature has indicated that the cracks propagated both through and around TiC grains with little apparent deflection. While toughening by crack deflection has been predicted and confirmed in glass matrix composites, this mechanism does not appear to be operative in either Si_3N_4-TiC or Al_2O_3-TiC composites. Weak grain boundaries in the matrix may control crack propagation and reduce the probability of deflection by thermal expansion coefficient/ moduli mismatch stresses.

REFERENCES

Binds, D.B., 1962, in Science of Ceramics, Vol. 1 (ed. G.H. Stewart, Academic Press, London, 315-334.

Buljan, S.T. and Sarin, V.K., 1984, in Proc. Third International School on Sintered Materials (ed. G.S. Upadhyaya), Elsevier Science Publishers, Amsterdam, 455-468.

Chermant, J.L., Deschanvres, A. and Osterstock, F., 1978, in Fracture Mechanics of Ceramics, Vol. 4 (ed. R.C. Bradt, D.P.H. Hasselman and F.F. Lange), Plenum Press, New York, 891-901.

Calussen, N., Pabst, R. and Lahman, C.P., 1975, Br. Ceram. Soc. Proc., No. 25, 139-149.

Davidge, R.W. and Evans, A.G., 1979, **Mater.** Sci. Eng., 6, 5, 281-298.

Erogen, F., 1974, in Fracture Mechanics of Ceramics, Vol. 1 (ed. R.C. Bradt, D.P.H. Hasselman and F.F. Lange), Plenum Press, New York, 245-267.

Evans, A.G., 1974, J. Mater. Sci., 9, 1145-1152.

Evans, A.G. and Charles, E.A., 1976, J. Amer. Ceram. Soc., 59, 7-8, 371-372.

Exner, H.E., 1969, Trans. Metall. Soc. AIME, 245, 677-683.

Faber, K.T., 1982, Ph.D. Thesis, Univ. of Calif., Berkeley.

Faber, K.T., Evans, A.G. and Drory, M.D., 1983 in Fracture Mechanics of Ceramics, Vol. 6 (ed. R.C. Bradt, A.G. Evans, D.P.H. Hasselman and F.F. Lange), Plenum Press, New York, 72-99.

Govila, R.K., Kinsman, K.R., and Beardmore, P., 1979, J. Mat. Sci., 14, 1095-1102.

Kriz, K., 1983 in Progress in Nitrogen Ceramics (ed. F.L. Riley) Matinus Nijhoff Publishers, Boston, 523-528.

Lange, FF., 1930, Phil. May,1 22, 179, 983-992.

Lynch, J.F., Ruderer, C.G. and Duckworth, W.H., 1966, Engineering
    Properties of Selected Ceramic Materials, Amer. Ceram. Soc., Inc.,
    Publisher, Columbus, Ohio, 5.4.0-1--5.4.1-28.
Mah, T., Mendiratta, M.G. and Lipsitt, H.A., 1981, Bull. Amer. Ceram. Soc.,
    60, 11, 1299-1231.

Petrovic, J.J., Jacobson, L.A., Talty, P.K. and Vasudevan, A.K., 1975,
    J. Amer. Ceram. Soc., 58, 3-4, 113-116.
Sarin, V.K. and Buljan, S.T., 1983, SME Technical Paper MR83-189.
Smith, J.T. and Quackenbush, C.L., 1980, Bull. Amer. Ceram. Soc.,
    59, 5, 529-533.
Swearengen, J.C., Beauchamp, E.K., Eagen, R.J., 1978, in Fracture
    Mechanics of Ceramics, Vol. 4 (ed. R.C. Bradt, D.P.H. Hasselman
    F.F. Lange), Plenum Press, New York, 973-987.
Wahi, R.P. and Ilschner, B., 1980, J. Mat. Sci., 15, 875-885.
Williams, W.S. and Schaal, R.D., 1962, J. Appl. Phys., 33, 3,
    955-962.
Wunschmann, M., 1978, Diplom. Thesis Univ. of Erlange-Nurnberg,
    Erlaangen, West Germany.
Zilberstein, G. and Buljan, S.T., 1983, presented at Second Conference
    on Characterization, Alfred Univ., Alfred, NY (in print).

Inst. Phys. Conf. Ser. No. 75: Chapter 5
Paper presented at 2nd Int. Conf. Science Hard Mater., Rhodes

439

Physico-chemical model of destruction of ceramics under high speed deformation

V B LAZAREV, V Ya SHEVCHENKO, N M ZHAVORONKOV, A D IZOTOV

N.S. Kurnakov Institute of General and Inorganic Chemistry,
USSR Academy of Sciences, Moscow 117071

ABSTRACT

The problem of failure of ceramics under a rapidly changing dynamic load has been studied from the point of view of physical chemistry. It is shown that, when the velocity of deformation is greater than the velocity of the crack propagation, the energy of dissociation of matter and the velocities of elastic waves are the main factors in the dissipation of the energy of loading.

INTRODUCTION

The calculation of strength for solids of different physico-chemical nature (metals, plastics, ceramics etc.) is an important stage in the study of their fundamental properties and in estimating their suitability to applied problems (Physical Metallurgy, 1965; Esposito and co-workers, 1980; Borin and co-workers, 1982; Kozachuk and co-workers, 1983; Osipov, 1983). Modern quantum theory of solid state enables one, when an appropriate interatomic interaction potential is chosen, to make reliable calculations whose accuracy is conditioned by the adequacy of the potential model chosen. However, when applying the estimates to a practical end, one often meets considerable (up to several orders of magnitude) divergences between theoretical strength values and experimentally observed data.

The reason for this is, in our opinion, mostly the complexity of obtaining the experimental data, and the difficulty of accounting for all the manifold details accompanying loading and deformation of solids. Until recently, describing the process of material failure, researchers regarded the speed of deformation as the most essential of the process factors and the corresponding behaviour of the material was regarded as elastic or as plastic viscous (or as a combination of both). The physico-chemical characteristics of the materials sustaining deformations were therefore neglected. However, studying the processes of polymer material failure, researchers came to the conclusion that new concepts on the breakdown of these materials were needed. The conclusion resulted in creation of the fluctuation theory of polymer failure under static as well as dynamic loading. The process of failure of such materials as ceramics has also some characteristic features which become most apparent at certain loading conditions and, in particular, in high-speed deformations.

THEORETICAL CONSIDERATION

The theory of failure under a static load is well developed. It states that the failure occurs due to the formation and development of Griffiths cracks due mostly to nucleation and movements of dislocations (Finkel, 1970). The study of materials failure at deformation velocities, V_d, which are greater, than the velocity of cracks spreading, V_c, and the velocity of sound in a solid, V_s, is far less advanced. For this case, some notions based on a "hydrodynamic" model have been developed (Birkhoff and co-workers, 1948; Lavrentyev, 1957). These notions, however, are not applicable to the so called brittle bodies, such as many of the ceramics. We call brittle, bodies whose deformations under load are elastic up to the breaking-point (Physical Metallurgy, 1965). These notions do not take into account some mechanical and physical chemistry notions of the failure processes and do not give a basis for quantitative studies of ceramic and of other solids (Wilkins, 1978).

It is advisable then to consider some other models of failure of solid materials at high deformation velocities ($V_d > V_c$). It is to be stressed that when $V_d < V_c$ then failure is due to both crack formation and to non-elastic distortions; elastic distortions also occur. The whole bulk of material is then involved in the deformation process. This is due to the propagation of longitudinal waves of elasticity. Thus, at $V_d < V_c$, the failure is a macroscopic process. At $V_d > V_c$, however, it is necessary to use microscopic notions, such as interatomic interactions. To describe the failure process the notions of physical chemistry are necessary. Indeed, the upper limit possible of the deformation velocity in a solid is reached when the atomic bonds break during the period of one vibration of an atom. It yields an estimate of the order of magnitude of the velocity waves of the transfer of energy between the neighbouring atoms (Physical Metallurgy, 1965). The order of magnitude of the velocity of waves of elasticity is approximately 10^5 cm/s, and the estimates of different authors of the velocity of crack propagation vary from 0.5 - 0.6 V_t (Bateson, 1960) to 0.5 - 0.7 V_l (Kozachuk and co-workers, 1983), where V_t and V_l are, respectively, the transverse and longitudinal velocities of sound in a solid. The velocities of dislocation movements are also not greater than V_t (Dynamics of dislocations, 1975), the theory permitting, however, the dislocation velocities which are equal or near to the velocity of sound. This occurs when the glide plane is able to transmit the energy to a moving dislocation. It depends on the magnitude of the friction stress in a given case (Hirth and Lothe, 1970). Thus, at $V_d > V_s$ (which often occurs in reality (Kurzzeitphysik, 1967)) interatomic bonds break during time which is less than the period of one atomic vibration, i.e. the process of failure being non-cooperative event (wave process behind the shock-wave front (Zeldowich and Raizer, 1966)). It is a chain of successive breaks of atomic bonds. It seems reasonable to describe the process as a dissociation of a "totally frozen" solid using bipartial interatomic potentials.

The stress, σ_p, at the breaking-down point of a solid is one of the most widespread characteristics of strength; σ_p depends on properties of the material itself, on the temperature and on speed of deformation. The behaviour of a solid at the moment of failure is described by the magnitude of the maximum relative deformation, $\mathcal{E}_p$, which is determined by the crystalline atomic structure and by physico-chemical properties of the material as well as by the state of the surface of a given sample, by the temperature, by the kind of deformation and its speed.

To calculate the ultimate strength of material when $V_d > V_c$ we make use of two-particle potentials of interatomic interactions. Two-particle interatomic potential $U(r)$ can be rewritten in the general form (Fig. 1 (a)), where r is the distance between the atoms, r_m is the equilibrium distance between the atoms (in a solid, the smallest distance between the atoms), r_0 is the distance between the atoms at $U(r) = 0$, r_1 is the interatomic distance at $d^2U(r)/dr^2 = 0$, i.e. with the potential curve passing through the bend point, D_b is the lattice energy of the material per bond.

Fig. 1(b) shows the general form of the dependence of the force of two-atom interaction on r: $F(r) = +dU(r)/dr$. At $r = r_1$ the force of inter-action reaches a maximum, F_{max}, application to an atom of the force F_{max} is sufficient to bring the atoms far apart, i.e. to break the molecule. The intersection of the $F(r)$ axis by the tangent to the curve $F(r)$ at the point $F(r) = 0$ is denoted F_F, its magnitude being proportional to the elastic modulus, E, of the material

$$E = (dF/dr)_{r=r_m} r_m \qquad 1$$

We consider now some of the simpler forms of the two-particle potential of interatomic interaction which can be applied to the solution of the problem (Fig. 2).

As a first approximation in many theoretical models the solid sphere potential is used (Fig. 2(a)). It is, however, unsuitable for the solution of our problem, since an infinitely small force is sufficient to bring about material failure for such an interaction.

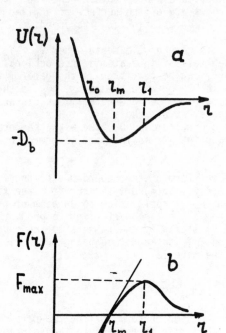

Fig. 1   The generalised form of the two-particles potential of
interatomic interaction (a) and the dependence of interaction
forces between two atoms on the distance between them (b).

" δ " - potential, i.e. a narrow and infinitely deep potential well
(Fig. 2(b)) is better suited to its role as a first approximation; in a
slightly changed form it corresponds to the case of the absolutely rigid
link of atoms with the energy D_b (Fig. 2(c)), the failure being due in
this case to dissociation only.   The rectangular well potential
(Fig. 2(d)) is similar to the latter. The Sutherland potential (Fig. 2(e))
is nearer to reality, its application being, however, difficult because
it cannot be differentiated at the point τ_o.

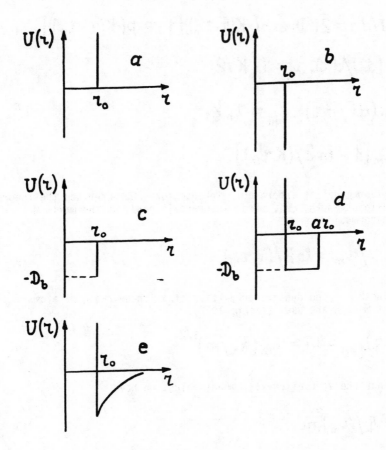

Fig. 2   The forms of some two-particle potentials of interatomic
interaction: solid sphere potential (a); "δ" - potential (b);
"δ_D" - potential (c); rectangular well potential (d);
Sutherland potential (e)

We make use of the Morse potential

$$U(r) = D_b \exp(-K(r-r_m))[\exp(-K(r-r_m)) - 2] \qquad 2$$

$$K = (k_r/2D_b)^{1/2} \qquad 3$$

where D_b - dissociation (atomization) energy, r_m - equilibrium
bond length, k_r - force constant characterising bond lengthening
(Bartenev, 1974).

We have:

$$F = dU/d\tau = 2KD_b \exp(-K(\tau - \tau_m))[1 - \exp(-K(\tau - \tau_m))] \quad 4$$

$$F_{max} = (dU/d\tau)_{\tau = \tau_1} = D_b K/2 \quad 5$$

$$E = \tau_m (dF/d\tau)_{\tau = \tau_m} = \tau_m k_\tau \quad 6$$

$$\tau_1 = \tau_m [1 + \ln 2/(K\tau_m)] \quad 7$$

F max in given model of interatomic interactions characterises the theoretical tensile strength, σ_p, of an interatomic bond, at the maximum relative deformation:

$$\varepsilon_p = \Delta\tau/\tau_m = \ln 2/(K\tau_m) \quad 8$$

To determine k_τ , the dynamic properties of a linear chain of atoms having mass m each are used (Ziman, 1972):

$$(d\nu/dq)_{q=0} = \upsilon_l = \tau_m (k_\tau/m)^{1/2} \quad 9$$

Assuming that the virtual crystal model holds, we have

$$k_\tau = (\upsilon_l/\tau_m)^2 \bar{m} \quad 10$$

where $\bar{m}$ is the average mass of the atoms. For one bond we obtain:

$$F_{max} = 0.25 \frac{\upsilon_l}{\tau_m} (2\bar{m} D_b)^{1/2} \quad 11$$

$$E = \upsilon_l^2 \frac{\bar{m}}{\tau_m} \quad 12$$

$$\varepsilon_p = \frac{\ln 2}{\upsilon_l} (2D_b/\bar{m})^{1/2} \quad 13$$

or per unit square:

$$\sigma_P = 0.25 \, v_l \, \rho \, (2D/M)^{1/2} \tag{14}$$

$$E = \rho v_l^2 \tag{15}$$

$$\varepsilon_P = \frac{\ln 2}{v_l} \, (2D/M)^{1/2} \tag{16}$$

where ρ is the density of the solid (kg/m^3), M - the mass of one g-mol of matter (kg), D - the dissociation (atomization) energy (J/mol).

For a given case, when calculating σ_p and ε_p for a monocrystal subjected to a rapidly changing load, it is necessary to allow for crystallographic direction as well as for the situations and character of chemical bonds along which loading occurs. Table 1 gives some data for certain ceramics. In order to calculate σ_p and ε_p for the amorphous, polycrystalline and fine-dispersed ceramics it is necessary to estimate the average number of interatomic bonds per unit area which is perpendicular to the loading direction.

TABLE 1   Characteristics of Theoretical Strength of Interatomic Bonding under High-Speed Loading along the Bonds

Compound	z_m^*, nm	D^*, kJ/mol	V^{**}, km/s	$F_{max} \cdot 10^9$ N/bond	$E \cdot 10^8$, N/bond	ε_p, arb. units
diamond	0.15445	707.73	18.6	4.61	4.46	0.286
SiC	0.1887	1227.70	11.2	2.73	2.21	0.342
SiO_2 (cristobalite)	0.162	1844.95	6.75	1.92	0.93	0.569
Al_2O_3 (corundum)	0.192	3058.604	10.84	2.62	2.07	0.350
α-Si_3N_4	0.1738	3993.197	7.85	2.00	1.18	0.471

* Experimental (Wyckoff, 1963; Pearson, 1958, 1967; Thermodynamic properties of original substances, 1978-1982);
** sound velocity.

The dynamic modulus of Young can be calculated from the known formula

$$E_d = 3 v_l^2 \rho - v_t^4 \rho / (v_l^2 - v_t^2) \tag{17}$$

when V_1 and V_t are known. In our model, $E = \rho V_1^2$ corresponding to the maximum of E_d determined by 17.

If, σ_p is calculated using Hooke's law:

$$\sigma_p^{el} = E \varepsilon_p \tag{18}$$

the σ_p value will be greater than the value given by 14 by the quantity

$$\sigma_p^{el}/\sigma_p = 4\ln 2 \tag{19}$$

being in good agreement with the estimates of Esposito and co-workers (1980).

CONCLUSION

Using the above approach, the well-known problem of high-speed collisions of solids can be solved (Birkhoff and co-workers, 1948; Lavrentyev, 1957).

Let us consider a long thin undeformable cylindrical rod (with density $\rho_\imath$, its length and area, $1_\imath$ and S) which moves along its axis with speed $V_\imath$ and meets a half-infinite solid body A (with the density ρ), their interaction being directed along the normal to the surface of the body, and all mechanical energy W of the rod being spent on dissociation (atomization) of the matter of A:

$$W = \frac{1}{2} \rho_\imath l_\imath S v_\imath^2 = \mathfrak{D} S l_c / V \tag{20}$$

where 1_c is the depth of the cavity formed in A, V is the volume of 1 g-mole of the matter of A. Expressing D from 16 and assuming

$$\varepsilon_p = l_c / l_\imath \tag{21}$$

we obtain

$$l_c = a l_\imath \left(\rho_\imath v_\imath^2 / \rho v_1^2 \right)^{1/3} \tag{22}$$

where $a = (\ln 2)^{2/3}$. Eqn. 22 differs from the following formula based on hydrodynamic considerations (Birkhoff and co-workers, 1948; Lavrentyev, 1957; Sagomonyan, 1974; Zlatin and Kozhushko, 1982):

$$l_c = l_\imath \left(\rho_\imath / \rho \right)^{1/2} \tag{23}$$

This is because different deformation mechanisms are considered. Unlike the hydrodynamic model, the notions developed in this paper imply that at high-speed loading both the dissociation energy of the matter and the velocity of sound in the matter are the factors which determine the dissipation of the energy of loading (Shevchenko and co-workers, 1984).

According to the present model, the dissociation mechanism is responsible for the failure of ceramic materials at speeds $V_d > V_c$ and, also under certain loads, at other speeds. The conclusions obtained also are valid for other classes of materials under the loading and deformation conditions considered above.

REFERENCES

Bartenev, G.M. 1974, Superstrong and highly strong unorganic glass. Stroyizdat, Moscow (Russian)

Bateson, S. 1960, Phys. a. Chem. Glassen. 1 139-141

Birkhoff, G., McDougall, D.P., Pugh, E.M., Taylor, G.I. 1948, J. Appl. Phys. 19 563-582

Borin, I.P., Novikov, S.A., Pogorelov, A.P., Sinicyn, V.A. 1982, DAN SSSR (Soviet Academy Doklady) 266 1377-1381 (Russian)

Dynamics of disclocations. 1975, Kiev, Naukova Dumka (Russian)

Esposito, E., Carllson, A.E., Ling, D.D., Ehrenreich, H., Gellat, C.D. 1980, Phil. Mag. A 41 251-259

Finkel, V.M. 1970, The Physics of Failure. Metallurgiya, Moscow (Russian)

Hirth, J.P., Lothe, J. 1970, Theory of Dislocations. McGraw - Hill Book Company, N.Y. - St.Louis - San Fransisco - Toronto - London - Sydney

Kozachuk, A.I., Solnceva, I.Yu., Stepanov, V.A., Shneizman, V.V. 1983, Fiz. tverd. tela. 25 1945 (Russian)

Kurzzeitphysik. 1967, Springer-Verlag, Wien, New York.

Lavrentyev, M.A. 1957, Uspekhi Mat. Nauk. 12 41-56 (Russian)

Osipov, K.A. 1983, DAN SSSR (Soviet Academy Doklady) 271 657-661 (Russian)

Pearson, W.P. 1958, 1967, Handbook of lattice spacing and structures of metals and alloys, 1 - 2 Pergamon Press, London

Physical Metallurgy, 1965, Ed. R.W. Cahn. Northon-Holland Publishing Company, Amsterdam

Sagomonyan, A.Ya. 1974, Penetration. Izdat. Moscow University, Moscow (Russian)

Shevchenko, V.Ya., Izotov, A.D., Lazarev, V.B., Zhavoronkov, N.M. 1984, Izv. AN SSSR ser. Neorg. Materialy. 20 1047-1052 (Russian)

Thermodynamical properties of original substances. 1978-1982. Ed. Gloushko V.P. Nauka, Moscow (Russian)

Wilkins, M.L. 1978, Int. J. Eng. Sci. 16 793-796

Wyckoff, W.G. 1963, Crystal structures. Intersci. Publ. New York

Zeldovich, Ya.B., Raizer, Yu.P. 1966, Physics of shock-waves and high-temperature hydrodynamical phenomena. Nauka, Moscow (Russian)

Ziman, J.M. 1972, Principles of the theory of solids. University Press, Cambridge

Zlatin, N.A., Kozhushko, A.A. 1982, Zh. Tekhn, Phys. 52 330-334 (Russian)

Inst. Phys. Conf. Ser. No. 75: Chapter 5
Paper presented at 2nd Int. Conf. Science Hard Mater., Rhodes

449

Indentation cracking in ceramics and cermets

M T LAUGIER
MTL Materials Testing Laboratories
9, Nova Croft, Coventry, CV5 7FJ, U.K.

ABSTRACT

Indentation crack geometry is determined by serial sectioning in a range of ceramics and WC-Co cermets. Cracks are generally of approximately rectangular profile; half-penny cracks are not observed. However, indentation fracture in ceramics is successfully described by current analyses based on half-penny crack geometry. An explanation for this surprising behaviour is suggested from the results of simple stress intensity calculations which demonstrate the insensitivity to quite marked changes in crack geometry of the surface stress intensity factor which controls (surface) crack extension.

INTRODUCTION

Indentation methods provide a low cost, rapid alternative to the standard methods of toughness determination in hard materials. These methods currently find application in materials development and will increasingly be employed in quality control, particularly in the developing structural ceramics industry. Indentation techniques are highly surface sensitive and special care is needed in materials preparation if reproducible results representative of the bulk are sought. However, these methods provide a direct measure of surface toughness (this may differ appreciably from the bulk value) which in some applications may control performance.

This work looks in detail at the form of the cracking induced by a Vickers diamond indenter in a range of WC-Co cermets and in a selection of ceramics (sialon, 'mixed' Al_2O_3 - TiN - TiC ceramic, 'pure' Al_2O_3). Palmqvist cracking only is found in all cases. It is shown how current indentation fracture analyses (based on half-penny crack geometry) may be modified to incorporate these findings.

INDENTATION CRACKING IN GLASSES, CERAMICS AND CERMETS

Cracks induced by a loaded sharp indenter in glasses have been observed to nucleate beneath the indenter near the plastic/elastic boundary (Lawn and Swain, 1975); above a threshold load, well formed half-penny cracks are produced which extend without change in geometry (Lawn and Fuller, 1975). It is well known that the radial surface traces of these cracks obey $P/c^{3/2} = k = $ const. for a given material (P is the applied load, c is the crack length: see Fig. 1). It has been widely accepted that the radial surface traces associated with indentation cracking in ceramics (which are opaque) also correspond to the intersection with the surface of cracks of half-penny geometry. Although no direct evidence that cracks are of half-penny geometry appears to be available, bend specimens that fail at indentation sites show evidence of approximately semi-circular fracture origins (Petrovic and Mendiratta, 1979) and additionally the quantity $P/c^{3/2} = k$ which has the dimensions of toughness is also found to be constant (for a given material).

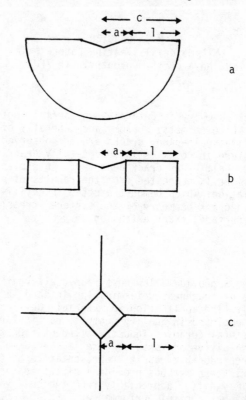

Fig. 1   Illustration of half-penny and Palmqvist crack geometries
a) Half penny crack profile b) Palmqvist crack profile
c) Surface traces corresponding to a and b

Indeed the stress intensity factor for a centre loaded penny crack is of the above form and working by analogy, Evans and Charles (1976) succeeded in developing a 'universal' relationship linking toughness K_c and k although it was necessary to introduce a normalising factor in $(E/H)^n$, with $n \sim 2/5$ (E is Young's modulus, H is hardness). Later work by Lawn, Evans and Marshall (1980) based on a physical model, linked the factor in E/H to the residual plastic driving force for crack extension which was found to be of the form $P_r \sim (E/H)^{n_p}$. They showed semi-empirically that $n = 1/2$ provides a suitable exponent for (E/H), leading to an expression for toughness of the form

$$K_c \sim \frac{P_r}{c^{3/2}} = (E/H)^{1/2} P / c^{3/2} \qquad\qquad 1$$

This form has been validated by Anstis, Chantikul, Lawn and Marshall, ACLM (1981) for a wide range of brittle materials. The ACLM formula is

$$K_c = 0.016 (E/H)^{1/2} P / c^{3/2} \qquad\qquad 2$$

Cermets (composites containing ceramic and ductile phases e.g. WC-Co) on the other hand are known to exhibit a different cracking behaviour. In these cases, the surface radial traces have been found by serial sectioning to correspond to relatively shallow cracks of the Palmqvist type (see Fig. 1). Here too bend specimens that fail at indentation sites show evidence of approximately semi-circular fracture origins, (Ingelstrom and Nordberg, 1974) although the original indentation cracks are of the Palmqvist type. This suggests that the semi-circular cracks observed at indentation sites after failure are a result of the bending process.

CRACK PROFILES IN WC-Co CERMETS AND CERAMICS

The sequence of Palmqvist crack formation obtained by serial sectioning in some WC-Co materials under increasing load is illustrated in Fig. 2 (surfaces were carefully lapped and polished using successively finer diamond paste prior to indentation, in order to remove grinding damage). Material properties are listed in Table 1. The serial sectioning procedure was straightforward. Material was removed from the surface by careful diamond lapping and polishing and crack traces were measured optically. A reliable depth scale was provided by the geometry of the Vickers indentation. Thickness of material removed is

$$d = (a_1 - a_2) (\sqrt{2}/2) \cot\alpha = 0.286 (a_1 - a_2) \qquad\qquad 3$$

where $\alpha = 68^{o}$ is the semi-angle between pyramid faces and a_1, a_2 are impression half diagonals before and after material removal. Cracks are observed to nucleate near the surface and not beneath the indenter as has previously been thought (Ogilvy, Perrott and Suiter, 1977). This suggests that even a relatively thin toughened surface layer may be effective in controlling crack nucleation in WC-Co materials, leading to improved wear behaviour. Little change is found in the separation of the inner portions of the cracks with depth, although slight penetration into the indentation zone is sometimes observed initially. Crack length l decreases mainly from the outer portions with depth, so that in some cases the final observed crack length l is only about 50% of its original value. No analytic relation was apparent between the length of surface traces of cracks and their depth. Cracks become relatively deeper with decreasing Co content but no tendency towards half-penny crack formation has been observed.

TABLE 1   WC-Co and Ceramic Material Properties

Material	Heyn's grain size μm	Co vol fraction	Hardness HV30
WC-Co (10M)	0.9	0.16	1350
WC-Co (6M)	1.0	0.10	1600
WC-Co (3F)	0.6	0.05	1990
sialon			1330
'mixed' ceramic	2		1960
Al$_2$O$_3$ + TiN + TiC			
'pure' Al$_2$O$_3$	2		1620

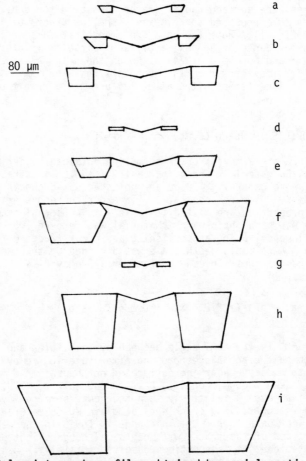

80 μm

a
b
c
d
e
f
g
h
i

Fig. 2   Palmqvist crack profiles obtained by serial sectioning in WC-Co cermets at several indentation loads, P.
For 10M: a) 196N; b) 294N; c) 490N
For 6M: d) 98N; e) 294N; f) 490N
For 3F: g) 34N; h) 294N; i) 490N

Crack profiles were also determined by the same serial sectioning technique for several ceramic tool materials. Surface preparation was the same as for WC-Co. The materials investigated were a sialon, a 'pure' Al_2O_3, and a 'mixed' Al_2O_3 - TiN - TiC ceramic (see Table 1). Profiles of indentation cracks produced under relatively high load conditions (294N) are shown in Fig. 3. Indentations were also performed at a number of

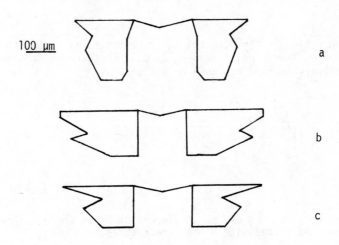

Fig. 3   Crack profiles obtained by serial sectioning in ceramics. Applied load 294N.   a. sialon,   b. 'mixed' ceramic Al_2O_3 - TiN - TiC,   c. 'pure' Al_2O_3

different loads; the general form of the cracks remained unchanged. Cracks are generally quite similar to the Palmqvist cracks found in WC-Co materials; again no tendency to half-penny crack formation was observed. Interestingly and in contrast to findings for WC-Co materials, it is for the toughest material - the sialon - that the relative crack depth was greatest. If confirmed, this result could be significant in the context of wear behaviour.

STRESS INTENSITY FACTORS FOR PALMQVIST CRACKS

Although indentation induced cracking in ceramics has not been found to be of the half-penny type assumed in previous analyses, the predictive capabilities of the ACLM formula will remain unaffected. It is the aim of this section to provide an explanation for the somewhat surprising success of the ACLM formalism.

Recently, Oore and Burns (1980) proposed an integral approach enabling the determination of the stress intensity factor K for an irregular planar crack under arbitrary normal loading. Oore and Burns give a general expression for the weight function

$$W_{QQ'} = \frac{K_{PQQ'}}{P_Q} = \frac{\sqrt{2}}{\pi} \frac{1}{l^2_{QQ'} \left(\int_S \frac{ds}{\rho_Q^2} \right)^{1/2}} \qquad 4$$

in the notation if Fig.4, where $K_{PQQ'}$ is the stress intensity factor at Q' resulting from a point load P_Q at Q. The corresponding expression for arbitrary loading may then be written

$$K_{Q'} = \int_A \frac{\sqrt{2}\, \sigma_Q\, d A_Q}{\pi\, l^2_{QQ'} \left(\int_S \frac{ds}{\rho_Q^2} \right)^{1/2}} \qquad 5$$

where $P_Q = \sigma_Q\, dA_Q$ and dA_Q is an element of crack surface area at Q, and A is the area of one side of the crack surface.

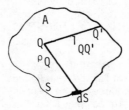

Fig.4   Irregular planar crack loaded by normal point forces at Q.

The method of Oore and Burns may be used to estimate the stress intensity factors associated with the Palmqvist cracks and these estimates may then be compared with the stress intensity factor for a centre loaded penny crack, the form of which appears to satisfactorily describe toughness, even when cracking is of the Palmqvist type. The driving force for crack extension is generated in the plastic region surrounding the indentation where the material displaced by the indenter is accommodated, and for a point loading approximation to be reasonable, the relative crack size c/a should considerably exceed b/a where b is the plastic zone size. This requirement is fairly well satisfied for ceramics where the ratio is about 3:1, but this is not so for WC-Co materials where the ratio only ranges from about 1:1 to 2:1.

A first estimate for the stress intensity factor associated with Palmqvist cracks may be obtained when cracks are represented by semi-circles of diameter l loaded by point forces as shown in Fig.5. In the notation of

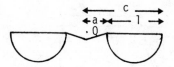

Fig. 5   Semi-circular Palmqvist crack representation.
Crack opening force P_r acts as Q.

the diagram, the expression for the stress intensity factor for this
semi-circular representation is (Laugier, 1985)

$$K^{semi-circular} = 2 \left(\frac{\pi}{2 + \pi}\right)^{1/2} \left(\frac{1}{a}\right)^{-1/2} K^{CLP} = 1.6 \left(\frac{1}{a}\right)^{-1/2} K^{CLP} \qquad 6$$

where $K^{CLP} = 1/_\pi 3/2 . \ P_r/_c 3/2$ 7

is the stress intensity factor for a centre loaded penny crack and P_r
is the residual plastic driving force.

A second and possibly better estimate may be obtained by treating the
crack profiles as rectangles of length l and depth d, to which they more
closely approximate (Fig. 6.).   An expression for the stress intensity
factor for a rectangular crack may be obtained by direct application of
equation 4, noting the integral $\int \frac{ds}{\rho_Q^2}$ along a line AB is the product

of the angle subtended at Q by AB, with the perpendicular distance from
Q to AB (See Fig. 6). The expression for the stress intensity factor is

$$K^{rectangular} = \left(\frac{A}{2 \pi}\right)^{-1/2} \left(\frac{1}{a}\right)^{-1/2} K^{CLP} \qquad 8$$

where A is a geometrical term which cannot be evaluated analytically;

$$A = \left(\frac{c}{a}\right)\left(\frac{a}{1}\right)\theta + \phi\left(\frac{a}{1}\right) + \frac{1}{m}\left(\frac{c}{1}\right)\left(\frac{a}{1}\right)(\theta - \phi) + 1 \qquad 9$$

with m = d/1.

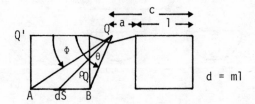

Fig. 6   Rectangular Palmqvist crack representation.
Crack opening force P_r acts at Q. $\rho_{QQ'} = c$

Values of A have been determined using the Palmqvist crack profiles
obtained in both WC-Co and ceramic materials. Values of the geometrical
term A together with necessary crack profile parameters are listed in
Table 2. The geometrical term $A^{-1/2}$ is found to remain almost constant

TABLE 2   Values of Geometrical Term A, together with Crack Profile
Parameters

Material	load P N	l/a	c/l	θ deg	φ deg	m	A
10M	196	0.50	3.0	7	4	0.4	2.3
	294	0.50	3.0	15	10	0.6	3.0
	490	0.54	2.9	20	13	0.7	3.3
6M	98	0.80	2.2	8	5	0.15	2.4
	294	1.22	1.8	26	13	0.4	2.8
	490	1.44	1.5	40	19	0.6	3.3
3F	34	1.00	2.0	18	8	0.33	2.8
	294	2.00	1.5	60	31	0.9	3.7
	490	2.18	1.4	59	28	0.8	3.7
Sialon	294	1.9	1.5	63	35	1.05	3.4
'mixed' ceramic Al_2O_3-TiN-TiC	294	3.2	1.3	60	24	0.6	2.8
'pure' Al_2O_3	294	2.6	1.4	53	22	0.6	2.9

for these crack profiles with a mean value $\bar{A}^{-1/2} = 0.58$ and a coefficient
of variation v of only 6%. The numerical term $(A/2\pi)^{-1/2} = 1.5$ and this
is very nearly equal to the corresponding value 1.6 for the half-penny
case. These stress intensity factors in turn differ in form from the
stress intensity factor for the centre loaded penny only by the factor
$(1/a)^{-1/2}$ which has a mean value $(1/a)^{-1/2} = 0.92$ and a coefficient of

variation v = 34%. This variation is rather more than might be expected experimentally and originates with the WC-Co data. However, it may be remarked that the coefficient of variation c of the calibration coefficient 0.016 in the ACLM approach is 25%. When the ceramics are considered alone $(\overline{1/a})^{-1/2}$ = 0.63 and the coefficient of variation v is only 11%.

These simple calculations indicate the relative insensitivity of the surface stress intensity factor which controls observed crack extension to quite marked changes in crack geometry, and show quite clearly why the earlier half-penny crack analyses led to valid toughness predictions in ceramics.

CONCLUSION

Indentation crack profiles have been determined by serial sectioning for a range of WC-Co cermets and ceramics, and in all cases cracks have been found to be of the relatively shallow Palmqvist variety; half-penny cracks were not observed. Fracture mechanics calculations based on the observed crack profiles have shown that the stress intensity factor which controls surface crack extension only differs from that of a centre loaded penny crack by a factor $(1/a)^{-1/2}$. This factor is almost constant for ceramics (coefficient of variation $\sim$ 11%), showing clearly why previous analyses based on half-penny crack geometry could provide a reliable description of indentation crack extension.

ACKNOWLEDGEMENTS

Much of the background to this work was performed whilst the author was at Sandvik UK Hard Materials Research Centre.

REFERENCES

Anstis, G.R., Chantikul, P., Lawn, B.R. and Marshall D.B., 1981   J. Am. Ceram. Soc. 64   533-538.
Evans, A.G. and Charles E.A., 1976   J. Am. Ceram. Soc. 59   371-372.
Ingelstrom, N., Nordberg, H., 1974   Eng. Fract. Mech. 6   597-607.
Laugier, M.T., 1985   J. Am. Ceram. Soc. 68   C51-C52.
Lawn, B.R., Evans, A.G. and Marshall, D.B., 1980   J. Am. Ceram. Soc. 63   574-581.
Lawn B.R. and Fuller E.R., 1975   J. Mater. Sci. 10   2016-2024.
Lawn, B.R. and Swain, M.V., 1975   J. Mater. Sci. 10   113-1122.
Ogilvy, M., Perrott, C.M. and Suiter, J.W., 1977   Wear 43   239-252.
Oore, M. and Burns D.J., J. Press. Vessel Tech. Trans. 1980 ASME 102   202-211.
Petrovic, J.J. and Mendiratta, M.G., 1979   Fracture Mechanics Applied to Brittle Materials, edited by S.W. Freiman, ASTM STP 678, 1979, 83-102.

Inst. Phys. Conf. Ser. No. 75: Chapter 5
Paper presented at 2nd Int. Conf. Science Hard Mater., Rhodes

459

Mode III fracture toughness of glassy and crystalline silicates

D A TOSSELL, J P SARGENT AND K H G ASHBEE

H H Wills Physics Laboratory, Tyndall Avenue, Bristol BS8 1TL, England

ABSTRACT

An optical interferometry technique has been developed to study the tearing mode fracture (mode III) of slabs of brittle material rigidly supported by epoxy resin through which the fracture stress is applied. Discontinuity of fringes facilitates measurement of the displacement field around the crack tip. Linear elastic fracture mechanics formulae have been used to convert these measurements into values for fracture toughness. K_{III} data over the velocity range 10^{-8} to 10^{-2} ms^{-1} is presented for three materials, namely soda-lime glass, borosilicate glass and natural low quartz.

INTRODUCTION

In his first paper A A Griffiths (1920) assumes an energy criterion for the fracture of glass. In his second paper (1924) he assumes a stress criterion. By using the idea that strain energy released during fracture can be calculated from the work which imaginary tractions applied to a crack's surfaces would have to do in order to close a crack, Frank and Lawn (1967) have shown that the two Griffith criteria are differentially related. Griffith's second paper concerns fracture in a plate subjected to two finite principle stresses, and he finds that the largest tangential stress acting in the surface of an elliptic crack is not quite at the crack tip. Physically, this means that, if it has any shear acting on it, an existing crack will change its direction. This is illustrated schematically in Fig. 1. Although most cracks in nature are predominantly mode I, deviations of crack orientation are sufficiently commonplace to be worth exploring.

The linear fracture mechanics approach for evaluating stresses and displacements associated with fracture follows the Griffith-Irwin theory based on an elastic solution derived by Westergaard (1939). In this approach the stress field near a crack tip can be divided into three basic types, each associated with a kinetic movement of two crack surfaces relative to each other. These three types are illustrated in Fig. 2, and are denoted as the opening mode(I), the edge sliding mode(II) and the tearing mode(III). Modes I and II can be regarded as plane extensional problems of the theory of elasticity and mode III the pure shear problem. Using this approach Eshelby (1971) has shown that for mode III deformation, characterized by the displacement (0,0,w) in cartesian coordinates, the elastic equations reduce to one, namely

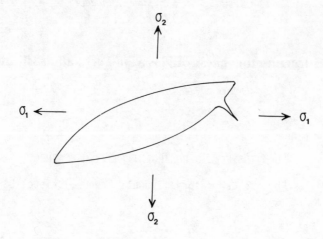

Fig. 1 Crack propagation from a location of maximum tensile stress on an elliptic cavity in a plate subjected to biaxial tension.

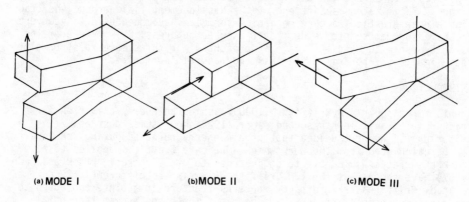

Fig. 2 Basic modes of crack displacements.

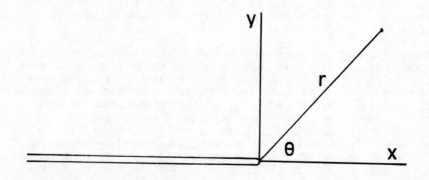

Fig. 3 Crack tip coordinates.

$$\nabla^2 w = 0 \qquad\qquad 1$$

which shows that the surviving displacement component is a harmonic function of x and y. The only non-zero stress components are

$$\sigma_{zx} = \mu \frac{\partial w}{\partial x} \quad , \qquad \sigma_{zy} = \mu \frac{\partial w}{\partial y} \qquad\qquad 2$$

where μ is the shear modulus. With reference to the cartesian and polar coordinates at the crack tip shown in Fig. 3. The displacements can be expanded in terms of the basic sets of harmonic functions,

$$r^n \sin n\theta \quad , \qquad r^n \cos n\theta \qquad\qquad 3$$

and since w is odd in θ,

$$w = \sum a_n r^n \sin n\theta \qquad\qquad 4$$

n cannot be integral, since then w would be continuous across the crack and the crack faces would not suffer a relative displacement. Again Eshelby (1971) has shown that subject to the condition that the crack faces must be free of stress, i.e. $\sigma_{zy} = 0$ for $\theta = \pm \pi$, then n must be plus or minus half an odd integer. Due to energy considerations the negative values of n must be excluded. Close enough to the crack tip the term n = 1/2 will be dominant and using standard notation the displacement w in the z direction is

$$w = \frac{2 K_{III}}{\mu (2\pi)^{1/2}} r^{1/2} \sin 1/2 \, \theta \qquad\qquad 5$$

and the stresses

$$\sigma_{zy} = \frac{K_{III}}{(2\pi)^{1/2}} r^{1/2} \cos 1/2 \, \theta \quad , \quad \sigma_{zx} = \frac{- K_{III}}{(2\pi)^{1/2}} r^{1/2} \sin 1/2 \, \theta \qquad\qquad 6$$

In particular, ahead of the crack tip ($\theta = 0$)

$$\sigma_{zy} = \frac{K_{III}}{(2\pi)^{1/2}} r^{-1/2} \qquad\qquad 7$$

and the relative displacement of the faces just behind the tip ($\theta = \pm \pi$)

$$\Delta w = \frac{4 K_{III}}{\mu (2\pi)^{1/2}} r^{1/2} \qquad\qquad 8$$

where K_{III} is the mode III stress intensity factor. The latter equation predicts a parabolic relationship between the relative displacement, Δw, and the distance, r, directly behind the crack tip. So if the displacement field around a mode III crack can be measured with sufficient accuracy it is possible to deduce a value for the fracture toughness, K_{III}, of the material which contains the crack. The experiment described here enables Δw to be measured for propagating mode III cracks.

EXPERIMENTAL

The technique for measuring the displacement field of tearing mode cracks originated in a study of water induced swelling of epoxy resin adherends by Sargent and Ashbee (1981). With a multi-ply laminate as one adherend and a microscope cover slip as the other adherend, it was found that the swelling stress was of a magnitude sufficient to fracture the cover slip. Moreover, these cracks were found to be pure mode III shear and, by using an optical flat in order to create a cavity within which optical interference can take place, some preliminary values were obtained for the fracture toughness of the cover slip glass.

Fig. 4 is a schematic diagram of the specimen block devised to produce controlled mode III fracture and so exploit the Sargent and Ashbee experiment. A thin slab of brittle material is bonded to two metal plates using a partially cured epoxy resin and these in turn are bonded to a metal block using the same resin. The specimen slab is pre-cracked and oriented so that misorientation of the planes of the metal plates produces differential uplift either side of the crack. This misorientation (anti-plane shear) is achieved using the apparatus shown in Fig. 5. The block is securely clamped to an optical bench by a vice and a hardened steel wedge is advanced into the resin using a micrometer screw driven by a constant torque motor through a gear box. Different gear ratios give wedge velocities in the range 10^{-5} ms^{-1} to 10^{-5} ms^{-1} which in turn realised crack velocities in the range 10^{-7} ms^{-1} to 10^{-2} ms^{-1}. Curiously it was sometimes observed that slow wedge velocities produce fast cracks and vice versa, although, in general, the procedure outlined above gives sufficient control to generate cracks with the velocity in a desired range.

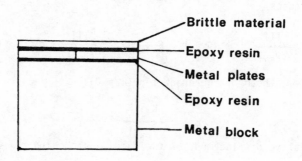

Fig. 4 Specimen mounted on metal block.

The displacement field around the crack tip is determined from optical
interference fringes created between the free surface of the specimen slab
and a $\lambda/10$ optical flat kinematically supported on centre points above the
specimen. The optical arrangement is shown schematically in Fig. 6.
Illumination is provided by a helium-neon laser emitting light of
wavelength (λ) 632.8 nm. Due to the coherent nature of laser light it is
possible to have sufficient distance between the specimen surface and the
optical flat so that deformation of the test piece does not bring the two
surfaces into contact. An example of a typical fringe pattern is shown in
Fig. 7, the specimen being so oriented that the crack marks the boundary of
the underlying metal plates, Fig. 4. The region of most interest is that
around the crack tip and Fig. 8 is an enlarged image of the fringe pattern
from that vicinity. Using standard numerical techniques the position of the
fringes in relation to the crack tip permits measurement of the
displacement to an accuracy of $\sim\lambda/20$.

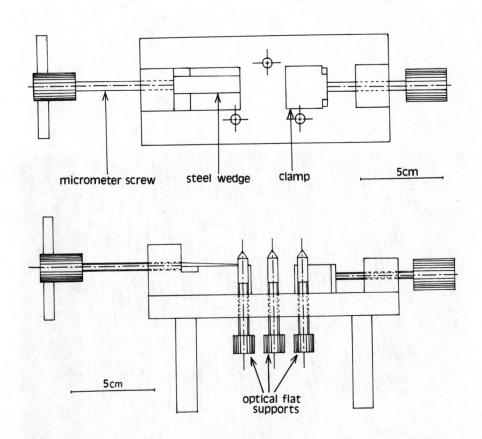

micrometer screw   steel wedge   clamp   5cm

optical flat
supports

Fig. 5 Apparatus used to advance metal wedge and propagate a mode III crack.

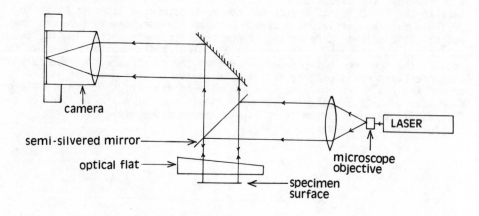

Fig. 6 Optical system for generating and recording interference fringes.

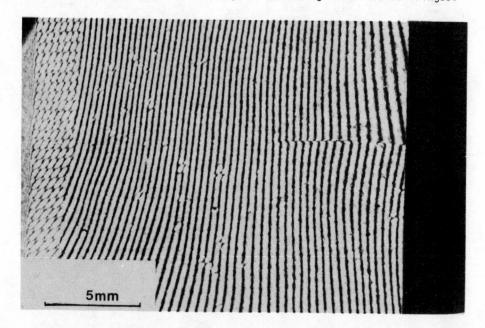

Fig. 7 Typical fringe pattern, the crack marks the boundary of the
    underlying metal plates.

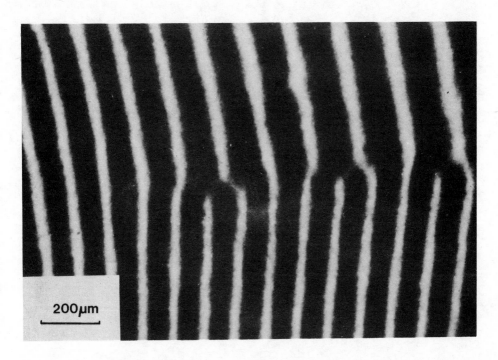

Fig. 8 Fringe pattern in the vicinity of the crack tip.

In some experiments aluminization of the surfaces of both the specimen and the flat has been used in order to obtain sharp multiple beam interference fringes. Tolansky (1948) gives a full treatment of the advantages and limitations of multiple beam interferometry for the determination of surface topography of solids. One of the conditions stipulated is that the interfering surfaces be separated by no more than a few wavelengths of light. This conflicts directly with the need to keep the surfaces apart in order to accommodate specimen deformation. This difficulty was overcome to some extent firstly by limiting the reflectivity of the coating, and secondly by minimising the inclination of the optical flat to the specimen surface. Fig. 9 shows the effect of the angle, θ, upon multiple reflections between the two interfering surfaces, and how the image at the point of interference can contain information from the specimen surface some distance away. If however the reflectivity of the coating is limited to 75% then by the twelth reflection the intensity of that light beam has fallen to 3% of the incident beam intensity. Using the geometry of Fig. 9 it has been calculated that for an initial separation, t_o, of 0.5mm (the smallest practically possible) and an angle of 0.002 radians between the surfaces, the distance X_n for twelve reflections is of the order of 100 microns. It is evident that the relative magnitudes of the contributions need to be carefully considered since the intensities fall off quickly after the first few reflections. A number of experimental tests were carried out to determine the effect of using the multiple beam technique. The first of these is based on the moiré effect.

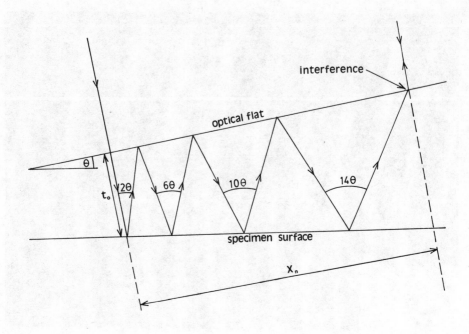

Fig. 9 Multiple reflections between highly reflective surfaces.

Moiré fringes are generated when patterns with nearly identical periodic
structures are made to overlap. These fringes might be expected to
faithfully follow any changes in the shape of the specimen surface as it is
deformed. However, if the specimen is left undeformed and the optical flat
is raised or lowered with a small tilt the moiré fringes formed between the
interference patterns recorded consist of evenly spaced straight lines.
Fig. 10 shows the moiré pattern generated from two interference images of
the same surface, one photographed when the optical flat was nearly
touching the specimen, the other when the flat had been raised, with
unavoidable tilt, some 4mm above the specimen. The moiré fringes can be
seen as grey bands across the image running from the top left hand corner
to the bottom right, and are found to be straight and evenly spaced. This
test which was repeated for a cracked specimen, gave an identical result.
That is, the moiré fringes traversed the cracked area with no detectable
deviation. A second experimental test consisted of deriving values for K_{III}
from different images of the same crack, where the height of the optical
flat was varied. The values obtained were in good agreement within the
experimental and numerical errors.

The explanation for the above dilemma lies in the nature of the surface
being examined, in this case smooth and continuous, the only feature being
the crack itself. Tolansky warns of a loss of definition of minute detail
if the surfaces are too far apart. It is evident that any deviations in the
immediate vicinity of the crack are swamped by the continuity of the
surrounding material.

Fig. 10 Moiré pattern generated between two interference patterns.

A typical set of multiple beam fringes in the region of the crack tip is shown in Fig. 11, and are to be compared with those of Fig. 7, the non-aluminized specimen. In general, the use of multiple beam interference is limited due to the need to constantly adjust the orientation of the optical flat so that it is parallel or very nearly parallel to the region of the crack tip thereby complying with the need to keep θ (Fig. 9) small. This meant there were few fringes generated by the area from where they were most needed, and also made constant re-alignment of the rest of the optical system necessary. Of the results obtained to date for the two glass compositions the aluminizing technique was used in 80% of the experiments with no systematic difference evident when compared with results for the non-aluminized specimens. None of the quartz crystal specimens were aluminized due to the fact that it has so far proved impossible to polish the surface to the necessary degree of optical flatness.

RESULTS

Equation 8 is a truncation of a power series solution and is therefore an approximation valid only for small r. For the materials studied here, the parabolic relationship was found to hold for distances of between 0.5mm to 1.0mm behind the crack tip. The area immediately around the tip can be further investigated using the moiré technique. By superimposing a photograph of an image from an uncracked specimen on a subsequent photographic image during crack growth, and comparing the moiré fringes in equivalent positions either side of the crack, it is possible to derive the

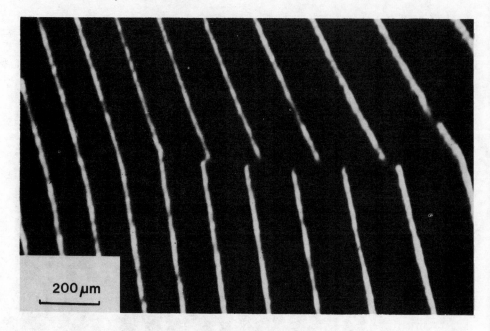

Fig. 11 Multiple beam fringe pattern in the vicinity of the crack tip.

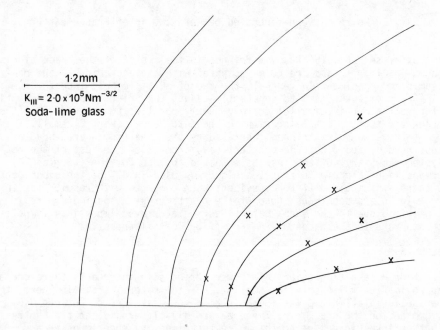

Fig. 12 Theoretical and measured displacements in the crack tip region.

displacement field due solely to the mode III stress. In Fig. 12 the solid lines are derived from equation 5 and represent lines of constant displacement in the z direction. The experimentally derived displacements are plotted for comparison and good agreement is found from a distance of 0.75mm behind the crack to a distance several millimetres in front of it.

The results obtained for soda-lime glass, borosilicate glass and {10Ī0} cleavage in α quartz are presented graphically in Fig. 13. Following Wiederhorn (1974) the data were fitted to a power function,

$$V = A \, K_{III}{}^{n} \qquad\qquad 9$$

where v is the crack velocity. The values obtained for the constants A and n are summarised in Table. 1. Often a crack propagates without any resolvable discontinuity in the interference fringes, i.e. apparently with zero mode III component. Also, predominantly mode III cracks would sometimes rotate into mode I orientation. Neither of these deviants produced useful results.

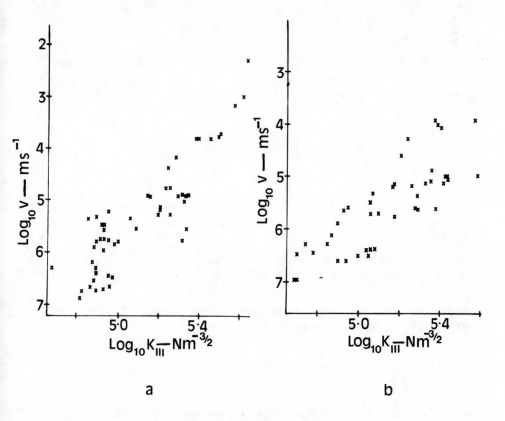

Fig. 13 v-K_{III} relationships for
(a) Borosilicate glass, thickness 0.5mm
(b) Soda-lime glass, thickness 1.1mm

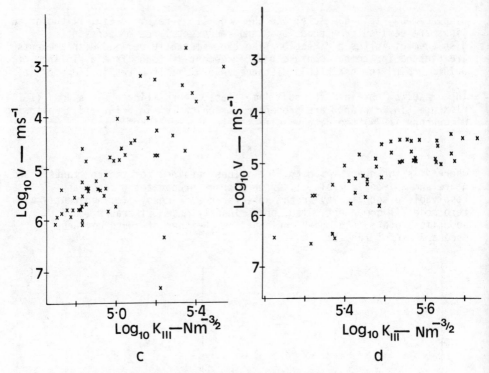

Fig. 13 (c) Soda-lime glass, thickness 0.3mm
        (d) α-quartz {10I0} cleavage
            observed for 50% relative humidity at 20°C

In nature, crystalline quartz is characterised by its concoidal fracture. However it has been found that mode III tests on thin (300µm) basal (0001) slabs produces cleavage on the prismatic planes {10I0}, Fig. 14. This observation of cleavage on {10I0} planes is independent of the orientation of the slab. Even with the most unfavourable orientations, cleavage on {1010} is maintained by switching from, say, {10I0} to {01I0}, see Fig. 14. The total absence of cleavage on {1120} planes is noteworthy since the bond density count is indistinguishable from that for {10I0}.

A physically more meaningful parameter than the fracture toughness is the mode III strain energy release rate,

$$G_{III} = \frac{K_{III}^2}{2\mu} = 2\gamma \qquad\qquad 10$$

where γ is the specific fracture surface energy. Lawn and Wilshaw (1975) have shown that 2γ can be computed from a knowledge of molecular dimensions and for soda-lime glass is found to be 2.4Jm^{-2}. Table 1 summarises the known G_I and G_{III} data for soda-lime and borosilicate glasses.

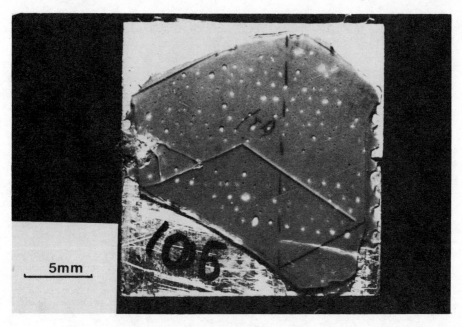

Fig. 14 {10Ī0} changing to {01Ī0} cleavage in α-quartz.

TABLE 1 <u>Summary of Crack Propagation Data</u>

Material	n	ln A	G(Jm^{-2})	Velocity range (ms^{-1})
Soda lime glass 0.3mm	3.0+0.3	-43.8	0.045 - 2.25	10^{-6} - 10^{-3}
Soda lime glass 1.1mm	2.3+0.3	-40.2	0.045 - 1.79	10^{-7} - 10^{-4}
Soda lime Mode I	Ī5	-30.0	1.75 - 4.48	10^{-7} - 10^{-4}
Borosilicate	5.0+0.3	-54.0	0.077 - 4.83	10^{-7} - 10^{-2}

REFERENCES

Eshelby, J.D. 1971, Sci. Prog. , Oxf. <u>59</u> 161-179
Frank, F.C. and Lawn, B.R. 1967, Proc. Roy. Soc. Lond. <u>A299</u> 291
Freiman, S.W. 1974, J. Amer. Ceram. Soc. <u>57</u> 350-353
Griffith, A.A. 1920, Phil. Trans. Roy. Soc. Lond. <u>A221</u> 163-198
Griffith, A.A. 1924, Proc. First. Int. Congr. Appl. Mech. (ed. C.B. Bienzo
    and J. M. Burgers) J Waltman Jr. Delft, p.55-80
Lawn, B.R., Wilshaw, T.R. 1975, Fracture of Brittle Solids. Cambridge
    University Press, Cambridge.
Sargent, J.P. and Ashbee, K.H.G. 1981, J. Comp. Mater. <u>15</u> 492-501
Tolansky, S. 1948, Multiple-Beam Interferometry. Claredon Press, London.
Westergaard, H.M. 1939, J.Appl. Mech. Trans. Am. Soc. Mech. Engrs. <u>6</u> A49-53
Wiederhorn, S.M., Evans, A.G., Fuller, E.R., and Johnson, H. 1974, J. Amer.
    Ceram. Soc. <u>57</u> 319

Inst. Phys. Conf. Ser. No. 75: Chapter 5
Paper presented at 2nd Int. Conf. Science Hard Mater., Rhodes

473

A model for the influence of a small amount of ductile second phase on the crack resistance shown for Al_2O_3 ceramics

H G SCHMID

MPI für Metallforschung, Institut für Werkstoffwissenschaften,
Seestr. 92, D-7000 Stuttgart 1

Present address: Institut de génie atomique, EPF-Ecublens
PHB-Ecublens, 1015 Lausanne

ABSTRACT

When testing ceramic materials containing a glassy phase, plastic effects
occur at high temperatures and influence the stress-strain relationship.
The mechanical behaviour does not only depend on temperature but also significantly on the loading rate. A general model is developed which explains
the non-linear elastic behaviour in terms of adhesive effects produced by
unbroken ductile islands behind the crack tip. Using typical values for the
glassy phase, the model agrees qualitatively well with the experiment.

INTRODUCTION

Ceramic materials are typically used in high temperature applications. One
of the biggest problems with regard to their technical application is their
low crack resistance. The most important reason for this is that most ceramic materials do not have the **capacity for plastic deformation** , and their
behaviour is brittle. When ceramic materials containing a glassy phase are
tested at high temperatures plastic effects influence the crack resistance
behaviour (Kromp and Pabst, 1980; Bornhauser and co-workers, 1983; Schmid and
co-workers, 1984; Haug and co-workers, 1984). With decreasing displacement
rate $\dot{\delta}$ the crack resistance increases to a higher value and the behaviour
of the ceramic becomes more and more plastic. The motivation for this work
was to explain these higher crack resistance values by means of a two dimensional model.

EXISTING EXPERIMENTS

Testing procedure

Bornhauser (1983) and Haug (1985) carried out bend tests of three point
bend specimens having an artificial notch of width 70 μm. These specimens
were tested in a displacement controlled manner at temperatures **up to**

1100°C in air. The displacement rate was varied between 0.3 and 30 μm/min. The specimen dimensions were 35 x 7 x 2.3 mm and the span width was 30 mm. The displacement of the specimens was directly measured with an inductive displacement transducer which controls the testing machine. Some material parameters are listed in the following table:

TABLE 1 **Material Parameters**

Material	grain size	K_{IC} (20°C)	Young's Modulus (20°C)	D (1100°C)	n
	μm	$MPam^{1/2}$	GPa	$MPa^{-1}h^{-1}$	
Al_2O_3 + 3 wt% glassy phase	10-15	4.2 +/- 0.4	365	1×10^{-6}	1

The testing machine used, permits direct observation of the crack tip by a video system and continuous in-situ measurements of crack propagation. The crack length was measured with an inductive displacement transducer which was fixed at the cross-table of the observation microscope. Investigations with this material have shown that it is only necessary to measure the crack tip at the surface because crack front curvatures were negligible (Wieninger, 1984). The optical tracking of the crack tip was possible up to crack velocities of 10 mm/min. By using this method the actual crack length a and crack velocity å is known at each point of the load-displacement curve. The crack velocity å was nearly constant over a range of about 1.5 mm within the specimen thickness of 7 mm and proportional to the displacement rate δ̇.

Results

The crack resistance R was derived from the applied stress σ_a supposing linear elastic behaviour, according to the well known relation (Haug and co-workers 1984):

$$R = \frac{K_I^2}{E} = \frac{\sigma_a^2 a \pi}{E} \qquad\qquad 1$$

where K_I is the stress intensity factor of the fracture mode I and E is the Young's modulus.
A general way for calculation of the crack resistance is the measure of the energy ΔU (Bornhauser, 1983) necessary to extend a crack with a straight crack front over the distance Δa. This leads to the definition of the crack resistance (Schwalbe, 1980):

$$R = \frac{\Delta U}{\Delta a} \frac{1}{B} \qquad\qquad 2$$

where B is the specimen thickness.

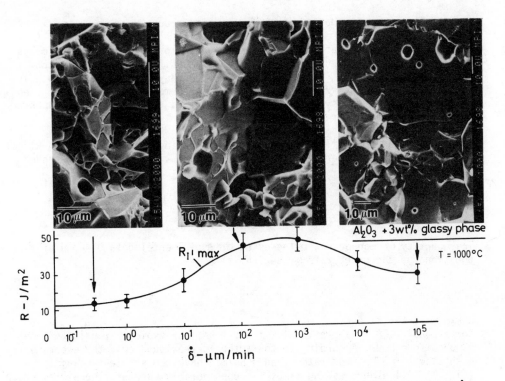

Fig. 1 The crack resistance as a function of the displacement rate $\dot{\delta}$
    (Haug and co-workers 1984) and crack surfaces observed by SEM

The crack resistance R of the tested alumina (Tab. 1) at low temperature is
about 30 N/m according to both methods (see eq. 1 and 2) and hardly depends
on the displacement rate $\dot{\delta}$ (Bornhauser, 1983). The fracture mode is trans-
crystalline at all displacement rates and is similar to the behaviour of
pure alumina (Kromp and Pabst, 1980).

The behaviour of alumina at high temperature and high displacement rates is
similar (Fig. 1). But with decreasing displacement rate the evaluated crack
resistance shows a maximum (Haug and co-workers, 1984, see **Figs. 1 and 9**).
The tendency of both curves **(Figs. 1 and 9)** is qualitatively the same. But
the maximum value according to both the K-concept and the crack resistance
concept (eq. 1 and 2) are 50 N/m and 150 N/m, respectively. The reason for
this discrepancy is that the application of equation 1 is limited by a linear
elastic behaviour of the material. In the range where the discrepancy of R
occurs the shape of the load displacement curves becomes typical for plastic
deformation (Bornhauser, 1983). Therefore, only the values of R according to
eq. 2 are valid. With decreasing displacement rate the fracture mode becomes
increasingly intercrystalline and the area fraction of the plastically deformed glassy phase on the crack surface f_p increases up to a maximum in
the range of 30% (Fig. 1). The reason for the large area fraction of the
glassy phase in the case of intercrystalline fracture is its distribution
in the material. The glassy phase is distributed at triple points or holes

between the grain-boundaries in different size groups.

Microscopical investigations of the environment of the crack tip by means of the in situ method, discussed above, and scanning electron microscopy have been done by Haug (1985). He has not found any plastic deformation in front of the crack tip regardless of the displacement rate and testing temperature. But the glassy phase increasingly links the crack surfaces behind the crack tip with decreasing displacement rate.

To estimate the possibility of creep effects during crack propagation Riedel (1981) gives a characteristic transition time t_R which has to be exceeded in the test:

$$t_R = \frac{K_I^2}{(n+1)\ C^*\ E} \qquad\qquad 3$$

The quantity C^* can be evaluated using the experimental data D and n of the equation for the creep rate $\dot{\epsilon}$ (see Tab. 1):

$$\dot{\epsilon} = D\sigma^n \qquad\qquad 4$$

where σ is the stress.
For n equal 1, C^* is approximatly K_I^2 D. Thus, the typical transition time is about one hour. According to the author's knowledge only the experiments with slower displacement rates than 1 µm/min have comparable testing times (Fig. 1). All other testing times are considerably shorter. Therefore, creep mechanisms as suggested by the work of Marion and co-workers (1983) and Raj and Ashby (1975) seem to be negligible in the present experiments and the mechanical behaviour of the continuum in front of the crack tip must be a linear elastic one. A non-linear elastic mechanism can be located only between the crack surfaces behind the crack tip.

It seems therefore that adhesive effects behind the crack tip are responsible for the elastic-plastic behaviour of the tested alumina for displacement rates greater than 1 µm/min and temperatures below 1100°C. For a better understanding of these effects we will develop in the next section a general model for the description of an adhesive zone in a brittle matrix with ductile inclusions in terms of fracture mechanics.

MODEL

General model

Kristic (1985) explained that the crack resistance of WC-Co hard metals was due to adhesive forces behind the crack tip (ligament rupture mechanism) using the linear elastic continuum mechanics approach of Dugdale based on the K-concept.
Such a model works under the condition that the matrix is brittle and continuous (a skeleton) and that the adhesion between the ductile Co-phase and the brittle WC-matrix has to be high enough (Kristic,1983). But alumina with glassy phase is, of course, not a continuum. The behaviour of the material is plastic at high temperatures (Bornhauser,1983). It is therefore

not clear whether a continuum mechanical approach is useful or not.
Schmid (1984) proposed a model for description of the crack resistance of
a brittle matrix with ductile inclusions which is based on the calculation
of the energy ΔU_p for plastic deformation of the adhesive zone during crack
propagation.
By means of this model we will show in the following that one can define an
average crack resistance of multiphase materials and we demonstrate that
the models of the continuum mechanics for description of adhesive effects
are special cases of the proposed theory.

Let us assume a two dimensional model (Fig. 2), where the ductile phases
are distributed in layers of different size groups with an average distance
from each other of Δa embedded in a brittle matrix. The geometry of one
class of layers including any ductile phases with a plastic behaviour $\sigma_i(h)$
is a rectangular one with l_{ai} layer width and h_{ai} layer thickness, where
σ_i is the stress working on the layer surface perpendicular to the x-axis.
If the crack is growing in a brittle manner through the matrix and the
crack tip touches the ductile layers, then the local stress relaxes by plastic deformation of the ductile phase, such that the fracture stress of the
ductile phase is not reached. This is because the fracture deformation of
the ductile phase is much higher than the fracture deformation of the brittle material. The ductile phase remains as an unbroken island behind the
crack tip. Due to crack propagation the distance between the crack surfaces
h increases. Therefore, the unbroken island has to be deformed until a maximum separation distance h_{bri} where the island breaks. By this, a so called
adhesive zone is created.

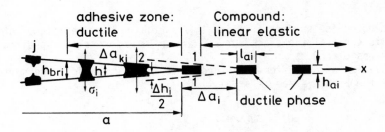

Fig. 2 Model for crack propagation in a brittle matrix with n
ductile inclusions (thickness of the body, B)

Let us now calculate the energy ΔU_{pi} which is necessary for deformation
of a class of islands when the crack propagates over the distance Δa_i.
Before the crack starts, the adhesive zone does not yet exist. This zone
is generating while the crack propagates and produces new islands. Therefore the length of the adhesive zone increases at first until a maximum
Δa_{ki}, where the separation distance h has reached an elongation h_{bri} and
the first generated island breaks. Therefore the adhesive zone cannot increase further and a stable zone length Δa_{ki} is established. For a stable
zone length we can calculate the energy ΔU_{pi}^{ki} to deform j ductile islands
of a single size group over the separation displacement Δh_i if the crack

propagates over distance Δa_i:

$$\Delta U_{pi} = B\, l_{ai} \left(\int_{h_{ai}}^{h_{ai}+\Delta h_i} \sigma_i(h)\,dh + \int_{h_{ai}+\Delta h_i}^{h_{ai}+2\Delta h_i} \sigma_i(h)\,dh + .. + \int_{h_{ai}+(j-1)\Delta h_i}^{h_{bri}+h_{ai}} \sigma_i(h)\,dh \right) \qquad 5$$

where $l_{ai} B \sigma_i(h)$ is the force working on the islands perpendicular to x. Obviously, ΔU_{pi} is equal to the energy necessary to deform an undeformed island until fracture.

$$\Delta U_{pi} = B\, l_{ai} \int_{h_{ai}}^{h_{bri}+h_{ai}} \sigma_i(h)\,dh \qquad 6$$

The plastic crack resistance of a single size group of layers can be evaluted (see eq. 2) to:

$$R_{pi} = \frac{l_{ai}}{\Delta a_i} \int_{h_{ai}}^{h_{bri}+h_{ai}} \sigma_i(h)\,dh = f_i R_{mi} \qquad 7$$

where f_i denotes the surface fraction $l_{ai}/\Delta a_i$ of the ductile islands of a single size group at the crack surface and R_{mi} is the so called specific plastic crack resistance.
For calculation of ΔU_i of one size group, we have to consider the surface energy of the phase γ_i too (Friedel, 1964):

$$\Delta U_i = \Delta U_{pi} + 2\,\gamma_i\, l_{ai}\, B \qquad 8$$

Thus, the crack resistance of a single size group can be solved by:

$$R_i = f_i\,(R_{mi} + 2\,\gamma_i) = f_i R_{si} \qquad 9$$

where R_{si} is the so-called specific crack resistance.
The crack resistance R_i defines only an average value within an interval of crack propagation Δa_i.

The only possibility to find an expression for the crack resistance R of the whole material including all size groups is to define an average value R that does not scatter very much. Therefore, the chosen interval Δa, where we will define an average crack resistance has to be large enough (see eq.2):

$$\Delta a = k\Delta a_i + arest_i \qquad 10$$

where k is integer and $arest_i$ is less than Δa_i.
Under the condition that Δa is much greater than $arest_i$, we may calculate:

$$\Delta U = k\Delta U_1 + 1\Delta U_2 + \ldots + m\Delta U_n + 2\gamma_{n+1} B\Delta a \ f_{n+1} \qquad 11$$

where k, 1, m are integers.
n+1 denotes the contribution of the matrix to the crack resistance,
$B \ \Delta a \ f_{n+1}$ is the surface of the brittle matrix within the interval of the
surface Δa B, γ_{n+1} is the surface energy of the matrix and f_{n+1} is the surface fraction of the brittle matrix at the crack surface.
We obtain for the whole crack resistance according to eq. 2:

$$R = \frac{\Delta U}{\Delta a B} = \sum_{i=1}^{n+1} f_i \ R_{si} \qquad 12$$

We would like to point out that the scatter of the measured crack resistance
$\Delta R(\Delta a)$ should be a characteristic value of the composite material too. The
crack resistance defined for a continuum (Rice, 1968):

$$R(\Delta a \rightarrow 0) = J = \frac{\partial U}{\partial a} \frac{1}{B} \qquad 13$$

(see eq. 2) describes only the local behaviour of one phase and should be
equal to the specific crack resistance R_{si} within this phase.
It seems therefore not meaningful to describe the average crack resistance
of multiphase materials by means of one J-integral. But the J-integral could
be a special case of the proposed theory.

Let us assume an elastic continuum (Fig. 3). Within the zone length Δa_k behind the crack tip exist adhesive effects. The restraining stress σ may be
a function of the separation distance h. The maximal separation is denoted
by h_{br}. Rice (1968) has shown that if one shrinks down the contour S to the
lower and upper surface of the adhesive zone then J is given by:

$$J = \int_0^{h_{br}} \sigma(h) dh \qquad 14$$

Rice (1968) has furthermore shown that the description of the adhesive zone
by means of the J-concept includes the models of Barenblatt and Dugdale,
which are based on the K-concept.

For comparing the proposed model with the results of continuum mechanics,
we have to define the following conditions (see eq. 7)
1. Inclusions in the continuum do not exist; therefore h_{ai} has to be zero.
2. There is only one continuum; therefore the index i is useless.
3. For the area fraction of the adhesive effects within the adhesive zone,
   f is equal to 1
These conditions for a continuum lead, according to eq. 7, to:

$$R = \int_{0}^{h_{br}} \sigma(h)\,dh \qquad\qquad 15$$

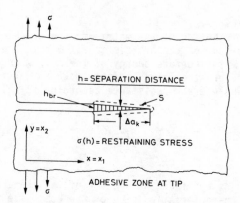

Fig. 3 Model of an adhesive zone behind the crack tip in the continuum (Rice 1968) with restraining stress dependent on separation distance h

We may conclude for description of adhesive effects that the J-integral is a special case of the proposed theory (see eq. 14, 15) and therefore not valid for two- or multiphase materials.

With respect to the above discussion, we would like to point out once more the validity and the limits of the models of fracture mechanics. The concept of the crack resistance (eq. 2) is applicable to all materials regardless of their composition and mechanical behaviour. This concept includes all con- **tinuum mechanics models.**
The concept of the J-integral is applicable to materials that one can describe with linear or nonlinear elastic continuum mechanics. This concept includes the K-concept. The K-concept (eq. 1) describes only a linear elastic continuum.

According to these explanations, it is clear that models based on the R concept are most suitable for **describing multiphase materials.**

Application of the Model to Glassy Phase Inclusions

For a calculation of the crack resistance R, it is necessary to find out the load displacement relation $\sigma_i(h)$ for the viscous glassy phase which is located in a layer. It is however very difficult to give a description for the viscous behaviour of a glass close to the transition to the linear elastic behaviour with respect to the deformation rate, especially in very thin layers. The main problems are the following:

i) Under high pressures the viscosity is a function of the flow rate (e.g. Non Newtonian viscous flow, Simmons and co-workers, 1982).
ii) The impurity content is not well known. Its influence on the mechanical

behaviour can be very strong (Greil,1982).
iii) The concentration of the impurities can depend on the local geometry.

Therefore it seems to be only possible to give a qualitative description of
the stress-strain behaviour of a glassy phase.

A model to understand the plastic behaviour of the glassy phase is shown in
Fig. 4 which describes the situation for a viscous liquid between two flat

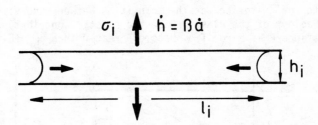

Fig. 4 Model of viscous liquid flow between two moving plates

plates. If these plates are displaced with the opening rate $\dot{h}$, viscous
flow in the centre must take place. The plates correspond to the crack surfaces, $\dot{h}$ to the crack opening rate, h_i to the layer thickness and the viscous liquid to a glassy island. The stress σ_i to induce the viscous flow is
usually described by the relation (Drucker,1960; Raj and Dang,1975):

$$\sigma_i \sim \frac{\eta \dot{h}}{h_i^3} \qquad\qquad 16$$

where η is the viscosity. l_i is the width of the glassy phase.
In our model we assume the following stress displacement relation which has
been given by Drucker (1960) with the condition that $\dot{h}$ is proportional to $\dot{a}$:

$$\sigma_i = \frac{A \eta \beta \dot{a} \, l_i^2}{h_i^3} \qquad\qquad 17$$

where A is a shape factor and h_i is given by:

$$h_i = h + h_{ai} \qquad\qquad 18$$

This description establishes a criterion to understand the transition between
the brittle and the viscous behaviour of a glassy phase as a function of the
crack propagation rate $\dot{a}$. If at a given $\dot{a}$ the resulting σ_i is greater than
the stress σ_{br} for brittle fracture of the glassy island then the behaviour
is brittle. If σ_i is smaller than σ_{br} it is possible to induce viscous flow
and the behaviour of the glassy island is viscous.

For an estimate of the qualitative behaviour of the crack resistance R_i it seems to be sufficient to consider the geometry variable with the strongest influence h_i. For the following calculation l_i is replaced by the value for the average layer width $\bar{l}$. Under the condition that h_{ai} is much greater than h_{bri}, R_i can be calculated using eq. 6 and 17:

$$R_i = f_i \frac{A \eta \beta \dot{a} \bar{l}^2}{2h_{ai}} \qquad\qquad 19$$

According to the criterion for the transition between the viscous and the brittle behaviour of the glassy phase the plastic contribution R_i to the crack resistance R is zero if σ_i becomes greater than σ_{br} (Fig. 5).

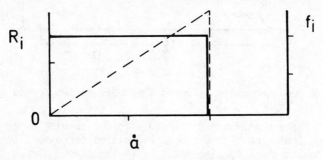

Fig. 5 The crack resistance R_i --- and the area fraction f_i —— as a function of the crack propagation rate $\dot{a}$.

If we assume, for example, three different size groups with a different layer thickness and a different area fraction (Fig. 6) then the crack resistance behaviour as a function of the crack propagation rate becomes more realistic (Fig. 7). If $\dot{a}$ is very fast then σ_i for all glassy islands is greater than σ_{br}. If $\dot{a}$ decreases the area fraction of the deformed glassy phase f_p increases due to the fact that more size groups of glassy islands become viscous and R_p increases, where f_p is given by:

$$f_p = \sum_{i=1}^{m} f_i \qquad\qquad 20$$

where m denotes the number of size groups of viscous islands.
Due to the proportionality of R to $\dot{a}$, R_p must decrease if f_p approaches its limit f_{max}, where f_{max} is the maximum possible area fraction of glassy islands on the crack surface.

COMPARISON OF THE MODEL WITH EXPERIMENT

Test of the Length of Adhesive Zone

The experiments of Bornhauser (1983) have shown that at a given displacement rate the crack resistance increases as a function of the crack length (Fig. 8). At the crack initiation (a=4.2mm) glassy islands do not exist and the crack resistance value should only be determined by the linear elastic behaviour of

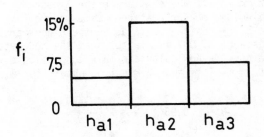

Fig. 6 Histogram of three size groups of glassy islands with different layer thickness h_{ai} and different area fraction f_i.

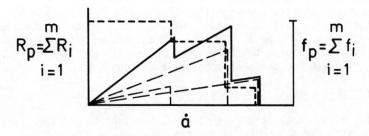

Fig. 7 The plastic contribution of three size groups to the crack resistance R_p ——, the crack resistance of each size group R_i —— and the area fraction f_p --- as a function of the crack propagation rate $\dot{a}$.

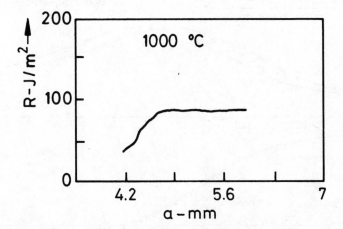

Fig. 8 The crack resistance of alumina with 3wt% glassy phase as a function of the crack length a. (Bornhauser 1983). The notch depth was 4.2 mm, $\dot{\delta}$ = 3 µm/min and the maximum crack length 7 mm.

the whole material. The measured value of the crack resistance at the starting crack length of 4.2 mm is already similar to the linear elastic crack resistance of the same alumina measured at room temperature. The increase from this linear elastic crack resistance value to the higher crack resistance level occurs over a distance of 0.7 mm. In terms of the model the adhesive zone should have this length. We now calculate this zone using eq. 7 and eq. 17 and assuming that h_{bri} is less than 0.2 h_{ai} we obtain for R_i:

$$R_i \geqslant 0.96 \, f_i \, \frac{A \, \eta \, \beta \, \dot{a} \, \bar{1}^2}{2h_{ai}^2}$$
21

This calculated value of R_i lies in the scatter of the experiments. Therefore a crack opening of a factor five corresponds to a full developed zone behind the crack tip.

On this basis the order of magnitude of the adhesive zone can be calculated, where the average layer thickness of the glassy phase in the alumina $\bar{h}$ is 0.3 μm and β is 10^{-3} see eq. 17. The average adhesive zone length Δa_k is given by 1.5 mm. With respect to the rough simplification of the model this value agrees well with the measured adhesive zone length.

Test of the Crack Resistance Behaviour

The calculation of the crack resistance as a function of the crack velocity was done with the following parameters (Fig. 9):

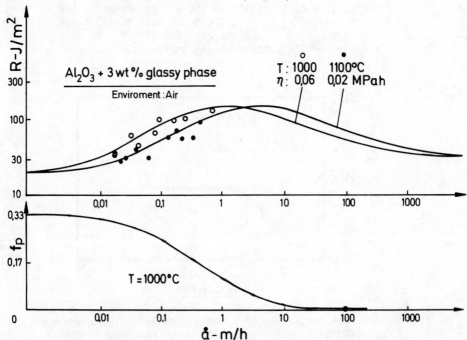

Fig. 9 The crack resistance and the average fraction of plastically deformed glassy phase f_p as a function of crack velocity calculated using the model and experimental values using Born-hauser's method (Haug and co-workers, 1984)

f_{max} = 30%, $\bar{l}$ = 2μm, $\bar{h}$ = 0.3μm

$\eta(1000^{o}C)$ = 0.06 MPa, σ_{br} = 3000 MPa

f_p and $\bar{l}$ were measured with a scanning electron microscope. The histogram of the size groups of glassy islands (Fig. 10) and $\bar{h}$ were derived from an analysis of the microstructure (Schmid 1984).

The viscosity η was estimated from the measured glassy point and melting point and σ_{br} for a silicon glass according to Scholze (1977). A is about ten.

Using all these measured, estimated, and fitted parameters the model agrees well with the observed crack resistance as a function of the crack propagation rate $\dot{a}$ (Haug and co-workers, 1984) (Fig. 9). $\dot{a}$ was measured directly using a travelling microscope. This method is very complicated and is only applicable in a small range of crack velocity (Bornhauser, 1983; Haug and co-workers, 1984).

The calculated dependence of the area fraction of the plastically deformed glassy phase on the crack surface f_p is in good agreement with the observations.

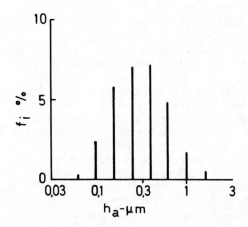

Fig. 10 Histogram of the distribution of the derived size groups
of glassy islands (h_a is the layer thickness).

The relation between the displacement rate $\dot{\delta}$ and the crack velocity $\dot{a}$ is at present not known. Here we assume that $\dot{\delta}$ is proportional to $\dot{a}$. This allows a comparison of the theory with experiment over a wide range of the displacement rate (Fig. 1) and leads to a fair agreement between experiment and model.

CONCLUSIONS

The model shows that the crack resistance of alumina with a glassy phase at high temperatures could be controlled by adhesive effects of the unbroken glassy phase behind the crack tip. The model agrees qualitatively well with the crack resistance behaviour reported in the literature (Kromp and co-workers,1980; Bornhauser and co-workers 1983; Schmid and co-workers,1984; Haug and co-workers,1984). However a reliable test of the model would require more experimental data.

Two practical implications of the observed crack resistance behaviour at elevated temperatures are:

i) At high displacement rates the behaviour of the material is linear elastic. In the range of increasing crack resistance the adhesive effect may be helpful if a crack is produced. Therefore the mechanical strength of the material may be the same as at room temperature.

ii) At slower displacement rates the crack resistance decreases since the deformation of the glassy phase becomes unimportant with respect to the crack resistance. Furthermore macroscopic creep has to be taken into account. Accordingly the long term mechanical stability of alumina with glassy phase at high temperatures is poor.

REFERENCES

Bornhauser, A.C. 1983, Phil. thesis, Universität Stuttgart

Bornhauser, A.C., Kromp, K., Schmid, H.G., Pabst, R.F. 1983, Proc. 4th Europ. Conf. on Fracture, Leoben/Austria, Vol. II, 655-664

Drucker, D.C. 1960, in "High Strength Materials", Ed. F. Zackay, J. Wiley & Sons, New York, pp. 795-833

Friedel, J. 1964, "Dislocations", Pergamon Press, New York, pp. 320-347

Greil, P. 1982, Phil. thesis, Universität Stuttgart

Haug, T., Bornhauser, A.C., Schmid, H.G., Gerold, V. and Pabst, R.F. 1984, 2nd Int. Conf. on Creep and Fracture of Engineering Materials and Structures, Swansea, England, pp. 473-484

Haug, T. 1985, Phil. thesis, Universität Stuttgart

Kristic, V.D. 1983, Phil. Mag. A., 48, No.5, 695-708

Kristic, V.D. and Komac, M. 1985, Phil. Mag. A., 51, No.2, 191-203

Kromp, K. and Pabst, R.F. 1980, Materialprüf., 22, 241-245

Marion, E.J., Evans, A.G., Drory, M.D. and Clarke, D.R. 1983, Acta Metall., 31, 1445-1457

Raj, R. and Ashby, M.F. 1975, Acta Metall., 23, 653-666

Raj, R. and Dang, C.H. 1975, Phil. Mag., 32, No.5, 909-922

Rice, J.R. 1968, J. Appl. Mech., 35, 379-386

Riedel, H. 1981, J. Mech. Phys. Solids, 29, 35-49

Schmid, H.G. 1984, in "Beiträge zum Festigkeitsseminar November 1983", Ed. R.F. Pabst, MPI Stuttgart, pp. 70-107

Schmid, H.G., Haug, T., Bornhauser, A., Gerold, V. and Pabst, R.F. 1984, in "Deformation of Ceramics II", Ed. R.E. Tessler, R.C. Bradt, Plenum Pub. Cop. New York, pp. 631-641

Scholze, H. 1977, "Glas", Springer-Verlag, New York, pp. 123-141

Schwalbe, K.H. 1980, "Bruchmechanik metallischer Werkstoffe", C. Hanser Verlag, München, pp. 17-93

Simmons, J.H. Mohr, R.K., Montrose C.J. 1982, J. Appl. Phys., 53, 4075-4080

Wieninger, H. 1984, Phil. thesis, Universität Wien

Discussion on Ceramics

Rapporteur: A H HEUER (1)
Session Chairman: R WARREN (2)

(1) Case Western Reserve University, Cleveland, Ohio, USA
(2) Volvo Flygmotor AB, Trollhättan, Sweden

Referring to Professor Jack's presentation on SiALON materials, Professor Minkoff asked about the nature of precipitation in the glass systems described and whether, for example, spinodal decomposition occurs. Professor Jack said that precipitation on crystallization had been studied by X-ray diffraction, electron microscopy, optical microscopy and DTA. In the β'-SiAlONs containing intergranular glass, crystallization of the glass does not involve spinodal decomposition. Nevertheless this process is likely in some other systems. As an example, Professor Jack mentioned Mg-Si-Al-O-N glasses, with compositions rich in Mg, Si and O, from which a phase near Mg_2SiO_4 crystallizes out below 1000°C.

Dr Page asked Professor Jack whether there was any evidence as to the stress state of the glassy material in SiAlONs and whether this stress state could affect behaviour. Professor Jack said that all the grain boundary phases in the Y-Si-Al-O-N system had lower thermal expansions than β'-SiAlON and therefore would be expected be in tension after cooling. There is however no information on the detailed stress state in these materials although this is undoubtedly necessary in understanding their properties.

Professor Popper questioned the use of the term "hardmetal" to describe SiAlONs since they are presumably non-metallic in nature. Professor Jack said that it has become usual to use the term to describe cutting tools and wear parts generally.

On the subject of TZP (tetragonal zirconia polycrystal) Dr Page asked Dr Claussen whether he had meant by the term "clean" grain boundaries, boundaries free of particles and films, free of segregants or free of amorphous layers. Dr Claussen confirmed that he had referred to amorphous films. While such films will worsen creep properties their effect on room-temperature toughness was still unclear; they could be an essential feature of tough TZP.

Dr Heuer mentioned recent Japanese work showing that additions of up to 20% Al_2O_3 to Y-TZP can lead to strengths of around 2500 MPa, he asked what strengthening mechanisms could be operating. Dr Claussen said that a small contribution would arise from the increased overall Young's modulus but that dissipation mechanisms such as microcracking and crack deflection would also be likely with a fine Al_2O_3 dispersion. He also assumed that the grain size in the ceramic is extremely small (< 0.3 μm).

Dr Warren asked what the main hurdles were that had to be overcome before fibre and whisker reinforced ceramics could be introduced as viable materials. Dr Claussen said that he believed that obtaining a homogeneous dispersion of the phases was the main problem. Precisely controlled slip-casting followed by hot pressing or HIP was a suggested route. Dr Ahmad considered that HIP and high temperature sintering would be expected to damage whiskers and in particular their surface perfection, the source of their strength. He suggested that in oxide ceramics it might be advantagous to use α-Al_2O_3 or other oxide whiskers rather than SiC whiskers which might react chemically with the material during HIP. Dr Claussen felt that some whisker damage was unavoidable but that sufficient strength could remain to reinforce most oxide-based, zirconia-toughened materials at high temperatures. At low temperatures the whiskers could lead to weakening by creation of increased flaw size. SiC whiskers are probably well suited to reinforce ZrO_2-toughened mullite because a silicate interface phase would reduce chemical reactions. For PSZ or TZP, an aluminia whisker might be preferable.

Dr Heuer asked Dr Bonnel if, in her work on SiALON materials, any volume change had been observed upon crystallization of the glassy phase. She replied that, though some change might be expected, none was observed in the density measurements.

In response to a question by Dr Sheinberg, Dr Schwetz stated that a value of 2.25 g/cm^3 had been used for the density of carbon in their calculations of the theoretical densities of the composites.

Responding to the paper by Baldoni et al. on mixed TiC-ceramic particulate composites, Dr Page asked whether plastic zone residual stress effects had been corrected for when obtaining K_c by the Evans and Charles method. Dr Buljan replied that this method had only been used for the room-temperature measurements and for that no correction had been applied. To an enquiry by Professor Shveikin he replied that he had no results relating to Al_2O_3-Ti(CN) composites.

Commenting on mixed ceramics of this type, Dr Claussen said that the notion that, to be effective, the second phase particles should be tougher than the matrix should not be applied without reservation. There are many examples in which less-tough particles have been found to improve toughness of the matrix. In these cases, the thermo-elastic properties of the dispersed phase are more significant through their influence on the microstress state of the composite.

Inst. Phys. Conf. Ser. No. 75: Chapter 6
Paper presented at 2nd Int. Conf. Science Hard Mater., Rhodes

489

Advanced TiC and TiC-TiN base cermets

H DOI

Central Research Institute, Mitsubishi Metal Corporation
Omiya, Saitama, Japan

ABSTRACT

The present paper reviews the progress and the present state of researches
carried out in relation to the microstructure and properties of TiC and
TiC-TiN base cermets to be used for high performance cutting. Before
discussing specific materials, reference is made to the important properties
required for a cutting tool material. Then, general features of the
microstructure of TiC and TiC-TiN base cermets and the relation between the
microstructure and the mechanical properties are outlined. Some mention is
also made on studies relating to the improvement of the cutting performance
of these cermets through the control of the various properties. Finally,
a prospect of more refractory Al_2O_3 base ceramics and composites as cutting
tool materials is briefly discussed.

INTRODUCTION

Increasing industrial needs for the more efficient cutting of modern materials,
which are sometimes difficult to cut, and higher reliability of tool
performance give strong incentives for developing, besides conventional WC
base hardmetals, new tool materials which possess improved properties such
as high wear resistance and toughness, and greater heat and oxidation
resistance, etc.

In reviewing, the developments of tool materials in the last decade are
characterised by the gradual encroachment of the so-called non-conventional
tool materials such as coated carbides, cermets and ceramics, which had been
developed for higher performance cutting, upon classic WC base hardmetals.
Nearly a half of the conventional WC base hardmetal inserts has been replaced
with the inserts made of such non-conventional hard materials of which coated
carbide inserts occupy a substantially large proportion. This means that
the last decade was not the age of innovation, but one of technical
improvements. A closer scrutiny reveals further that these technical
improvements have been made mostly on an empirical basis. However, research
and development based only on empiricism do not lead to a breakthrough.
New concepts having sound theoretical foundations are always highly in need,
and time is now ripe to expect such concepts, especially in the field of new
ceramics.

One important aspect in developing new tool materials or drastically improving
existing ones is the multiplicity of the properties required as a tool
material. An effective guideline for improving the tool materials is obtained

through the analysis of various types of tool damage.   The relative
importance of the strength and toughness of a tool material can be best
understood through the analysis of tool failure in interrupted cutting of
steels.   Therefore, before going into a general survey of the tool materials,
recent studies on the tool edge failure due to chipping or breakage of
hardmetal tools in interrupted cutting of steels will first be referred to
and discussed in some detail in relation to the tool material properties.
Then, the progress and the present state of advanced TiC and TiC-TiN base
cermets will be reviewed, and other promising, more heat resisting materials
will also be mentioned briefly in order to provide a sound basis for their
further advance.   All chemical compositions of the tool materials covered
in this review are expressed in wt% unless otherwise stated.

GENERAL PROPERTIES REQUIRED OF TOOL MATERIALS AND SIGNIFICANCE OF STRENGTH
AND TOUGHNESS

A brief mention will first be made of the types of tool damage which develop
in the course of machining. Fig.1(a) features the case of normal tool wear
of hardmetals. In Fig.1(b), plastic deformation of the tool nose due to
heat generation under large cutting forces is shown. Machining of steels
at high feed rates often causes such deformation and wear. Fig.1(c) shows

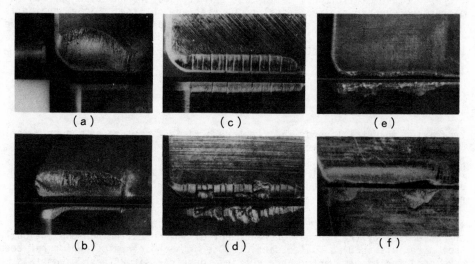

(a)   (c)   (e)

(b)   (d)   (f)

Fig.1 Typical examples of various types of tool damage.

another example of tool damage which is characterised by the generation of
parallel thermal cracks perpendicular to the tool edge. Severe thermal
stresses which develop in the course of milling of steels at high speeds
bring about this type of tool damage. In Fig.1(d), these thermal cracks
have already been generated extensively and developed side branches which
have interconnected and caused spalling of several small blocks from the
tool edge. Finally, Fig.1(e) and (f) show tool damage due to microchipping
and breakage of the tool edge. When the tool material is fairly hard and
brittle, the tool edge is liable to chip off upon repeated application of
excessive mechanical shocks on the tool edge due, for example, to severe
interrupted cutting or cutting of an unpremachined workpiece of non-circular
cross section. The analysis of the origin of the various tool damage shown

in Fig.1 is anticipated to give an effective guideline for developing new, improved tool materials. General requirements of the cutting tool for high efficiency machining are listed below.

Wear resistance
Resistance against plastic deformation of the tool nose
Less chipping and breakage of the tool edge under repeated impacts
Thermal fatigue resistance
Oxidation resistance
Less welding and adhesion

Now, strength and wear resistance of a material increase as the hardness increases. However, the toughness, which plays a vital role in preventing tool failure due to chipping of the tool edge, necessarily declines. Since relatively little is known in regard to the relative contributions of the strength and toughness of a tool material in preventing tool failure, it appears worthwhile here to examine in some detail essential features of the tool failures in interrupted cutting of steels.

According to the recent extensive studies on the tool failure due to microchipping or breakage of WC base hardmetal tools in interrupted cutting of steels (Negishi and Aoki,1976; Negishi, Aoki and Sata,1980a; Negishi, Aoki and Sata,1980b; Negishi, Aoki and Sata,1981), the probability distribution of the occurrence of chipping failure of a tool edge obeys a Weibull distribution, and a characteristic tool life curve shown in Fig.2 is obtained. Here, a tool life is defined by the total number of the cutting impacts which produce either flank wear of 0.3mm, crater wear of 0.1mm, or in the case of

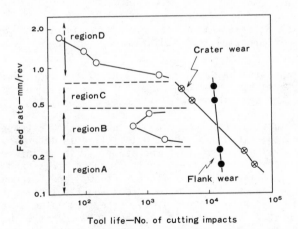

Fig.2 Tool life of WC base hardmetal (P10) in interrupted cutting
plotted against feed rate.

Cutting condition
workpiece: chromium molybdenum steel
(C:0.4%, HB 286)
tool geometry: -5; -5,5,15,15,0.8mm
cutting speed: 100m/min
depth of cut: 1.5mm

chipping failure, a mean value of the Weibull distribution of the tool edge chipping. It is interesting to note that in some medium feed rate region (referred to as region C), the tool edge becomes free from chipping, which means that high reliability of the tool performance can be expected in this region. Outside this region, i.e. in regions B and D, the tool life is limited by the occurrence of chipping. Fig.3 shows representative photographs of the three kinds of the tool failures ocurring in regions B, C, and D, respectively. The tool failure in region B (referred to as type B failure) is characterised by the breakage of the tool corner; moreover, the fractured surface is remarkably flat (see Fig.3(a)). In comparison, in the type D failure, the fractured surface is very irregular and several hair cracks are visible on the tool surface. Such hair cracks are rare in type B failure. Now, in the case of type C failure, chipping does not occur as already mentioned, even though cracks are seen on the tool face. Short time cutting tests carried out to elucidate the behaviour of hair cracks indicate that in region B, the tool edge chips off within a very short time once a hair crack has been formed.

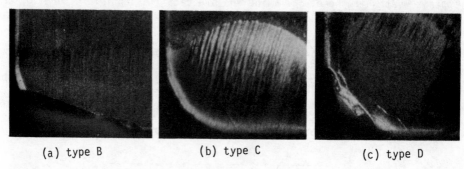

    (a) type B   (b) type C   (c) type D

    Fig.3 Representative photographs showing three kinds of
        hardmetal tool failures.

On the other hand, in the region D, chipping of the tool edge occurs after several hair cracks have been generated and the tool edge has survived for some time unbroken against repeated impacts.

In interrupted cutting, the tool edge is subjected to a temperature rise and high normal cutting stress. Since the tool life is determined by the combined effect of the cutting stress and temperature, it is of interest to investigate the effect of cutting conditions upon the cutting stress and tool surface temperature at the time of tool failure. Figs. 4 and 5 show the normal cutting stress acting on the   tool face and the tool surface temperature plotted against feed rate, respectively. The mean value of the normal cutting stress is calculated from the cutting force acting on the tool edge and chip contact length, while the tool temperature is evaluated at a specific point beneath the tool surface, as illustrated in Fig.5, by the use of FEM analysis. Both of these parameters increase with an increase of the feed rate. Moreover, an increase of the cutting speed or workpiece hardness causes an increase in the normal cutting stress and tool surface temperature. Now, the characteristic of the tool failures shown in Fig.3 and the relations between cutting speed or workpiece hardness and feed rate shown in Figs.4 and 5 can be combined to give a map of the tool failure in interrupted cutting, see Fig.6. The horizontal and vertical axes correspond to the tool surface temperature and normal cutting stress, respectively. When the tool surface temperature is below 660°C, the tool failure belongs to the chipping type (type B failure). Another kind of

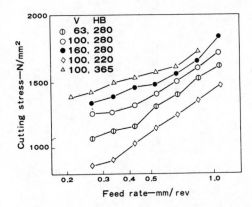

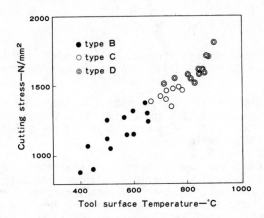

Fig.4 Normal cutting stress acting
on WC base hardmetal (P10)
tool face plotted against
feed rate。

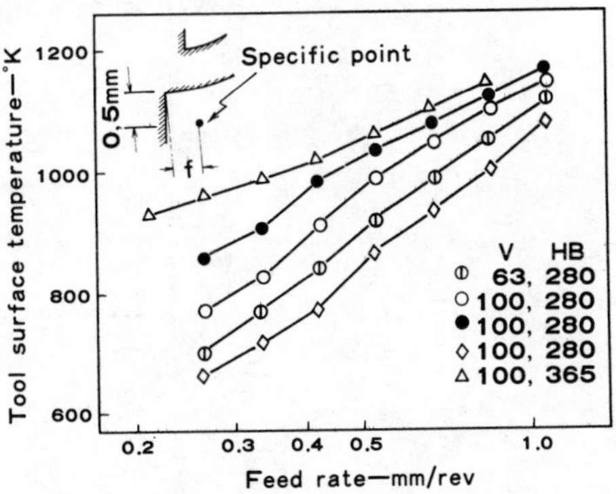

Fig.5 Tool surface temperature of WC
base hardmetal (P10) plotted
against feed rate.

Fig.6 A map of WC base hardmetal (P10) tool failure
in interrupted cutting.

chipping type failure occurs under normal cutting stresses $\geq$1.5kN/mm^2 (type
D failure). Only when the tool surface temperature is 660~800°C and the
normal cutting stress is$\leq$1.5kN/mm^2 does tool failure belong to the wear
type (type C failure).

To correlate the observed tool failures in interrupted cutting with the tool
material properties, measurements of the TRS (transverse rupture strength)
were carried out at ambient as well as elevated temperatures with unnotched
and notched WC base hardmetal specimens of the same grade (P10) used for the
interrupted cutting tests. From the strength of the notched specimen,
fracture toughness (K_{1c}) can be calculated.. As shown in Fig.7, above 500°C,
strength decreases, whereas fracture toughness increases. The observed
variations of TRS with increasing temperature correlate well with the features
of the tool failure map shown in Fig.6. The correlation can be summarised
in **Table** 1 in which cutting test parameters, tool material properties and

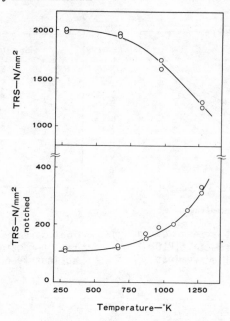

Fig.7 Transverse rupture strength of unnotched and notched
WC base hardmetal (P10) plotted against temperature.

crack growth velocity are listed with respect to the three representative
feed rates which characterise the three kinds of the tool failure.
Conclusions drawn from **Table** 1 with respect to the mechanisms of the tool
chipping failures are as follows.

(1) The conditions for type B failure are characterised by the largest
crack growth velocity. Therefore, although the bulk strength of the
tool edge is much larger than the normal cutting stress, once a hair-
like crack is generated at some notch-like irregularity on the tool
edge surface, it rapidly propagates, leading to a catastrophic
failure of the tool edge.

(2) Type C failure can be correlated with the fairly large bulk strength
(which is larger than the normal cutting stress) and fracture
toughness of the material at the tool failure temperature. Moreover,
microscopic observations revealed that growth of hair cracks once
generated is very soon arrested owing to the diffusional influx of
Co towards the crack front. This explains the high toughness of the
tool edge against chipping failure.

(3) In the case of type D failure, the temperature rise at the tool
surface is already considerably large and consequently, the normal
cutting stress which acts on the tool edge becomes comparable or
larger than the bulk strength of the tool material. Therefore,
although the fracture toughness has become conspicuously high and
the crack growth velocity has been much reduced as in the case of
type C failure, the tool edge can no longer bear the large cutting
stress acting on it and as a result, chipping of the tool edge is
initiated from several crack sources in such a way that the fractured
surface becomes very irregular.

TABLE 1 <u>Summary of Interrupted Cutting Test Data and the Relevant</u>
<u>Material Properties of WC Base Hardmetal Tool (P10)</u>

Type of tool failure	Feed rate region	Limiting factor of tool life	Morphology of fractured surface	Cutting test parameters at tool failure		Tool material properties at tool failure temperature		Crack growth velocity
	typical feed rate mm/rev			tool surface temperature °C	normal cutting stress N/mm²	TRS N/mm²	strength (notched) N/mm² ($\propto K_{IC}$)	
Type B	B (0.33)	Chipping	Flat	660	800-1400	1800-2000	130(170)	very large
Type C	C (0.53)	Wear	——	660-800	1300-1500	1600-1800	170-220	$\cong 0$ *
Type D	D (1.06)	Chipping	Irregular	>700	>1500	<1700	>220	>0

* Growth velocity is initially >0; crack growth is soon
arrested owing to the diffusional influx of Co.

The tool life curve characteristics for interrupted cutting are affected
by (a) cutting speed, (b) hardness of the workpiece, and (c) the grade of
the tool material. An increase of the cutting speed or workpiece hardness
causes the tool life curve (see Fig.2) to shift downwards along the vertical
axis. The effect of the tool material grade is shown in Fig.8.. General
features of the tool life curve of TiC base cermet are the same as those of
the WC base hardmetal (P10) although it has been shifted as a whole towards
a shorter tool life direction and the width of the region B becomes wider
and that of the region C narrower. The reason for these variations can be
attributed to the lower strength, fracture toughness, and thermal conductivity
of TiC base cermet. In the case of a WC base hardmetal of a lower grade (M20),
the tool life curve is shifted along the vertical axis downwards in comparison
with P10 hardmetal, and moreover, the region B is missing; chipping failure

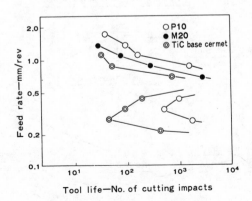

Fig.8 The effect of tool material grade on the relationship
between tool life due to chipping and feed rate.

does not occur in this region. The reason for this must be due to the larger thermal softening and larger fracture toughness of M20 hardmetal in comparison with those of P10 hardmetal.

The above discussion on the mechanisms of the tool failure in interrupted cutting has disclosed the following as highly desirable characteristics for a tool material for high performance interrupted cutting.

(a) High bulk strength at ambient temperature, and small thermal softening
(b) Large fracture toughness at ambient as well as elevated temperatures
(c) Operation of some built-in crack arresting mechanisms
(d) Large thermal conductivity to avoid accumulation of heat at tool surface and the resulting generation of large thermal stresses

It is to be noted that (b) and (c) are closely related each other; operation of a built-in crack arresting mechanism leads to an improvement in fracture toughness.

ATTAINMENT OF DESIRABLE PROPERTIES OF CERMETS AS HIGH PERFORMANCE TOOL MATERIALS THROUGH MICROSTRUCTURAL CONTROL

Evolution of TiC Base Cermets

The evolution of the TiC base cermet family is illustrated in Fig.9. The root of the family can be traced back to 1929 when cermet type hardmetals consisting of TiC base solid-solution (TiC-Mo$_2$C) as the hard phase and 10~15% Ni, Ni-Cr or Ni-Mo$_2$C as the binder phase were first commercially produced in Germany (Kieffer and Benesovsky,1965). However, this type of cermet was too brittle to be used even in a limited field of machining. The real emergence of TiC base cermets for machining started with the announcement of a TiC-Ni-Mo cermet in USA in 1959. Humenik and Parikh (1956) demonstrated how Mo additions to TiC-Ni cermet improve wetting of carbide towards binder metal by forming a mixed carbide shell (Ti,Mo)C round TiC grains thereby inhibiting carbide coalescence and grain growth; the resulting microstructure gives rise to improved hardness and impact resistance (toughness). As a

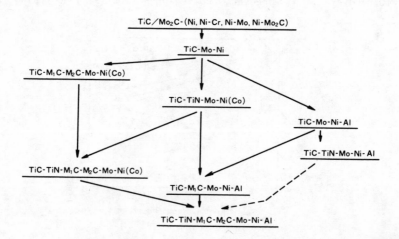

Fig.9 Evolution of TiC base cermet family.

result, TiC-Ni-Mo cermets found a certain field of use in finish machining of steels.

Modifications of TiC-Ni-Mo cermets have been driven on the basis of the following three concepts. First, improvement of TiC-Ni-Mo cermets through the additions of the second and the third carbides was pursued in 1960's and early 1970's in an attempt to strengthen and toughen the carbide shell round TiC core and the binder phase. The approach was mostly made in a trial and error manner. Notwithstanding, a better tool performance resulted from such continual efforts. The field of the application was accordingly widened to include semi-finishing and roughing of steels.

In the meantime, Kieffer and coworkers investigated for some years around 1970 whether nitrides and carbides of transition metals are suitable hard materials for cemented carbides (Kieffer and Fister,1970; Kieffer, Ettmayer and Freundhofmeier,1971a; Kieffer, Ettmayer and Freundhofmeier,1971b); they found that Ti carbonitrides cemented with a 70%Ni-30%Mo alloy binder exhibit optimum hardness and TRS, which are comparable to those of TiC-Ni-Mo cermets, and improved tool performance. On the other hand, Rudy demonstrated that by the use of the miscibility gap in the system Ti-Mo-C-N, a novel cermet which is highly tough and wear resistant displays much improved cutting performance (Rudy,1973; Rudy,Worcester and Elkington,1974). Rudy maintains that through an appropriate selection of the composition and solution temperature, a single solid-solution phase (Ti,Mo)(C,N) will, upon cooling, enter into the spinodal two-phase range, and Ti and N rich α'-phase surrounded by Mo and C rich α''-phase should result. This spinodal reaction, when utilized for making hard alloy phases of cemented hardmetals, brings about an extremely fine structure, excellent stability of the structure towards grain growth, and an oxygen scavenging effect. The above desirable characteristics are believed to originate in this. Although Rudy's concept of making spinodal cermets is theoretically self-consistent, questions may arise as to the significance of this concept as will be discussed later. However, Rudy's work aroused worldwide interests in TiC-TiN base cermets and has contributed greatly to intensify research activities, especially in Japan, on this type of cermet. Other important aspects of the incorporation of TiN into TiC base cermets are the improved thermal shock resistance and chemical stability which, together with the grain refinement effect, significantly suppresses cratering and grooving wear and increases resistance against mechanical and thermal shocks. As a result of these improvements in material properties the field of use of TiC-TiN base cermets for finishing and roughing of steels has been considerably widened.

The third concept of the modification of TiC-Ni-Mo cermets is to precipitate fine intermetallic compound Ni_3Al (γ') particles in the binder phase for improving the heat resistivity. This was proposed by Doi and Nishigaki in late 1970's (Doi and Nishigaki,1977; Nishigaki, Oosawa and Doi,1979) and shortly after by Moskowitz and Humenik (1980). The idea came from the strengthening concept in Ni base superalloys; the combined effect of solid-solution strengthening and precipitation hardening of the binder phase is anticipated to improve the creep resistance of TiC-Ni-Mo cermets. Originally, Doi and Nishigaki (1977) aimed to strengthen TiC base cermets with relatively high binder concentrations for use in machining operations hitherto restricted to HSS. Optimum precipitation of fine γ' in the binder phase considerably improved wear resistance in cutting while keeping toughness at a comparatively high level. However, the effect of γ' precipitation hardening of the binder phase on the toughness-hardness relationship of the cermet has not been well understood especially in regard to the high temperature properties; therefore this aspect should be further investigated for its use in cutting and wear resisting tools.

Microstructural Features of TiC and TiC-TiN Base Cermets

TiC-Ni-Mo(or Mo_2C) cermets evolved, as already mentioned, from the finding
that Mo added tó Ni enhances wetting of TiC to the binder metal and brings
about refinement and a homogeneous distribution of carbide grain-size
(Humenik and Parikh,1956; Parikh and Humenik,1957; and Parikh,1957). Mo which
has been added to the binder phase diffuses into the carbide phase from the
grain periphery to form a Mo rich mixed carbide shell. A considerable amount
of work has been undertaken on the formation of the mixed carbide shell
(Moskowitz and Humenik,1966; Yamaya and Sadahiro,1969; Suzuki, Hayashi and
Terada,1971a,b;Snell,1974; Lindau and Stjernberg,1976). This was motivated
through the observations that for use as a cutting tool material, too extensive
formation of the mixed carbide shell round TiC core during sintering is
undesirable since such a tool is susceptible to chipping and breakage (Fukatsu
and Yamaya,1963; Suzuki,1980). Growth of the thickness of the mixed carbide
shell is enhanced when the sintering temperature is relatively high ($\geq 1400°C$)
(Snell,1974). Another means of regulating the thickness of the mixed carbide
shell is the control of the carbon content; the thickness of the shell rapidly
grows and the shape of the carbide grains changes from spherical to angular
forms with decreasing carbon content (Yamaya and Sadahiro,1969; Snell,1974).

The growth of the mixed carbide shell is believed to occur through dissolution
and reprecipitation processes (Moskowitz and Humenik,1966; Suzuki, Hayashi
and Terada,1971a;Snell,1974; and Lindau and Stjernberg,1976). Suzuki and
coworkers attempted to elucidate, through the analysis of the phase relations
relevant to $TiC-Ni-Mo_2C$ system, the mechanism of the formation of the mixed
carbide shell. They éoncluded that the shell is formed in two or three steps,
the initial precipitation of Mo_2C-25%TiC on the surface of TiC grains, followed
by the dissolution and reprecipîtation, and the final precipitation during
cooling from the sintering temperature. They confirmed the existence of such
layered structure utilizing very coarse grained $TiC-N-Mo_2C$ cermets. Lindau
and Stjernberg (1976) investigated the kinetics of coarsening of a TiC-Ni-Mo
cermet by quantitative metallography and concluded that grain growth occurs
by an Ostwald ripening process, and the rate of coarsening is determined by
a second-order interfacial reaction.

Controlling the carbon content of TiC-Ni-Mo cermets is of utmost importance
in the production technology of the cermets. According to Stover and Wulff's
ternary Ti-Ni-C phase diagram (Stover and Wulff,1959), there is a two-phase
region (δ(TiC) + γ); small deviations from this region give rise to precipitatio
of either free graphite or γ phase (Ni_3Ti). In the case of WC-Co hardmetals,
the width of the corresponding two-phase is rather small, i.e. 0.1~0,5%C (Suzuki
1972). In comparison with this, the width of the two-phase region of TiC-Ni
cermets amount to as much as 1~2%C. This is related to the fact that TiC
has a wide range of non-stoichiometric composition (Storm,1967). The essential
features of the two-phase region for TiC-Ni cermets do not change in the case
of TiC-Ni-Mo cermets. The formation of the mixed carbide shell, however, affect
the width in carbon content of the two-phase region, which can be determined
by methods such as saturated magnetization and X-ray lattice parameter
measurements of γ and carbide phases (Snell,1974; Suzuki and coworkers,1983).
The width of the two-phase region increases with increasing Mo content (Suzuki,
Hayashi and Terada,1971b).

The importance of controlling the carbon content of TiC-Ni-Mo cermets is two-
fold. First, it determines the amounts of Ti and Mo in solution in the binder
phase. Secondly, as mentioned earlier, it has a considerable influence on the

growth and characteristics of (Ti,Mo)C shell around TiC grains. Increased solution of Ti and Mo in the binder phase is anticipated to contribute to solid-solution strengthening as in the case of dissolved W in the binder phase of WC-Co hardmetals (Suzuki and Kubota,1966; Rudiger and coworkers,1971; Ueda and coworkers,1977; Suzuki, Hayashi and Taniguchi,1978). Therefore, exact characterization of the binder phase composition is needed. A large amount of work on the determination of the binder phase composition of TiC-Ni-Mo cermets has been carried out (Moskowitz and Humenik,1966; Suzuki, Hayashi and Terada,1971b; Snell,1974; Nishigaki and Doi,1980; Suzuki and coworkers, 1983). Table 2 summarizes the results of measurements of Ti and Mo contents

TABLE 2 Analysis of the Binder Phase Composition of TiC-Ni-Mo Cermets

Composition of cermet	Composition of the binder phase						Method of analysis of the binder phase	Reference
	Ti content (%)			Mo content (%)				
	low C	medium C	high C	low C	medium C	high C		
TiC-0.8%Mo-12.3%Ni	—	2.4	—	—	0.4	—	wet chemical	Moskowitz and Humenik (1966)
TiC-4.3%Mo-12.3%Ni	—	6.0	—	—	0.5	—	"	
TiC-7.7%Mo-12.3%Ni	—	8.6	—	—	0.9	—	"	
TiC-11.2%Mo-12.3%Ni	—	11.7	—	—	2.1	—	"	
TiC-14.7%Mo-12.3%Ni	—	15.7	—	—	6.6	—	"	
TiC-24%Mo-15%Ni	10.5	—	3	6.5	—	2	EPMA	Snell (1974)
TiC-11%Mo₂C-24%Ni	—	12.7	—	—	0.2	—	wet chemical	Suzuki and coworkers (1983)
TiC-19%Mo₂C-24%Ni	16.3	12.0	4.9	1.8	1.1	0.3	"	
TiC-27%Mo₂C-24%Ni	—	11.9	—	—	2.1	—	"	
TiC-15%Mo-30%Ni	8.4	—	3.7	1.0	—	0.6	EPMA	Nishigaki and Doi (1980)
TiC-30%Ni	9	—	3	0	0	0	EPMA	Suzuki, Hayashi and Terada (1971)
TiC-10%Mo-30%Ni	11	—	5	2	—	0	"	
TiC-20%Mo-30%Ni	10	—	4	4	—	1	"	
TiC-30%Mo-30%Ni	9	—	4	6	—	3	"	

in the binder phase of TiC-Ni-Mo cermets of low, medium and high carbon contents. As shown in the Table, for a TiC-24%Ni-19%Mo cermet, the binder Ti content for low carbon contents amounts to as high as 16%; it fairly rapidly decreases to 5% with increasing the carbon content in the two-phase region. On the other hand, the binder Mo content is much less than the Ti content; a similar dependence on the carbon content is noticed. Moreover, the binder Mo content gradually increases with increasing the Mo/Ni ratio (Suzuki and coworkers,1983).

The volume fraction of the binder phase at constant Ni content in general increases with decreasing carbon content owing to the increased solubility of Ti and Mo in the binder phase (Komac and Lange,1982). Increasing Mo additions at constant Ni content also give a similar effect. This effect is worthy of notice when investigating the effect of carbon content of TiC and TiC-TiN base cermets on the mechanical properties.

The addition of various carbides to TiC-Ni-Mo cermets does not change essential features of the microstructure. Suzuki and coworkers investigated the effect of the additions of WC, TaC, NbC, HfC, Cr_3C_2 or V_4C_3 (Suzuki and Hayashi,1971; Suzuki, Hayashi and Terada,1978; Suzuki, Hayashi and Yamamoto,1979). The following is a summary of the observations.

(1) The lattice parameter of the carbide phase of $TiC-Ni-Mo_2C$ cermets increases with the addition of ZrC, HfC, NbC or TaC; the effect of ZrC or HfC addition is especially conspicuous. The addition of WC or Cr_3C_2, on the other hand, causes a reduction.

(2) The carbide addition, however, hardly affects the lattice parameter of the binder phase except the case of Cr_3C_2 addition; Cr dissolves fairly copiously in the binder phase and causes it to increase.

(3) The carbide (MeC) added to $TiC-Ni-Mo_2C$ cermets predominantly diffuses into the carbide shell to form a multi-component solid-solution consisting of TiC, Mo_2C and MeC.

(4) The effect of carbide additions upon the average grain-size of a cermet is in most cases relatively small (VC, TaC, WC, ZrC and NbC) except the case of Cr_3C_2 addition. The latter contributes considerably to coarsen the carbide particles.

The addition of TiN to TiC-Ni-Mo cermets drastically changes the microstructure and the composition of the binder phase (Suzuki and coworkers,1976; Suzuki, Hayashi and Yamamoto,1977; Nishigaki and coworkers,1980; Nishigaki and Doi, 1980; Fukuhara and Mitani,1982; Suzuki and coworkers,1983). As mentioned in the preceding section, the addition of TiN to TiC-Ni-Mo cermets brings about a fine and stable microstructure. The example is shown in Fig.10. According to Suzuki and coworkers (1976), the addition of TiN to $TiC-20\%Ni-20\%Mo_2C$ by as much as 15% leads to the finest microstructure. On the other hand, Nishigaki and Doi found that the grain size and thickness of the (Ti,Mo)(C,N) shell of

$TiC_{1-x}N_x-15\%Ni-30\%Mo_2C$ cermet in general decrease with increasing N content ($x \leqq 0.4$). Very recent data shown in Fig.10 ($TiC_{1-x}N_x-15\%Ni-15\%Mo$, $0\leqq x\leqq 0.5$) qualitatively support this view. Fukuhara and Mitani (1982) investigated the effect of N content upon coarsening rate of Ti carbonitride grains in $TiC_{1-x}N_x$

10%Ni-10%Mo cermet over a wide range ($0\leqq x\leqq 1$) and found that the cermet with composition of x=0.5 exhibits the minimum growth rate. They concluded that the thinning of (Ti,Mo)(C,N) shell with increasing N content reflects the effect of decreasing Ti content and increasing Mo content in the binder phase.

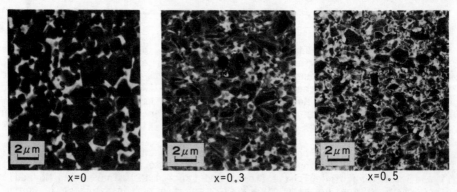

x=0   x=0.3   x=0.5

Fig.10 Effect of N content on the microstructure of
$TiC_{1-x}N_x-15\%Ni-15\%Mo_2C$ cermet.

With regard to the mechanism of grain refinement of TiC base cermet due to
TiN additions, there is a considerable disagreement in opinion among different
investigators. Rudy's concept of a spinodal decomposition reaction of
(Ti,Mo)(C,N) into Ti and N rich α'-phase and Mo and C rich α''-phase leading to
a grain refinement has already been mentioned. Several criticisms were raised
against this concept. Before discussing this matter in some detail, the
relevant experimental data should first be cited. Fig.11 and **Table** 3 give the
information on the processes of conventional sintering of TiC-TiN-Ni-Mo cermets
of various TiN and Mo additions as revealed through measurements of the lattice
parameters of the constituent phases (Nishigaki and Doi,1980). Important
points in connection with the sintering reaction are summarised below.

(a) Mo added to TiC-TiN-Ni cermets reacts with C to form Mo_2C at $600°C$;
    the amount of Mo_2C thus formed starts to decrease from the sintering
    temperature of $900{\sim}1000°C$, through a dissolution and reprecipitation
    mechanism, to move into the TiC lattice to form TiC-Mo_2C solid-solution.
    On the other hand, TiN remains unreacted until $1000°C$. Above $1000°C$,
    TiN is transferred into the (Ti,(Mo))C lattice, again through a
    dissolution and reprecipitation mechanism, to form a solid-solution
    of (Ti,(Mo))(C,N). The temperatures for the complete disappearance
    of Mo_2C and TiN are $1200{\sim}1300°C$ and $\geqq1400°C$, respectively.

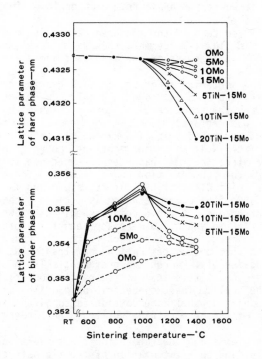

Fig.11 Lattice parameters of hard particle
       phase and binder phase for TiC-(0-20)
       %TiN-30%Ni-(0-15)%Mo cermets.

TABLE 3 X-ray Diffraction Analyses of the Constituent Phases of
TiC-30%Ni-15%Mo-2%C and TiC-20%TiN-30%Ni-15%Mo-2%C
Hardmetals Sintered at Various Temperatures for 1 h

Composition of cermet	Constituent phase	Intensities of X-ray diffraction peaks for each phases of cermet					
		600	800	1000	1200	1300	1400
TiC-30%Ni-15%Mo-2%C	TiC	S	S	S	S	S	S
	Ni	S	S	S	S	S	S
	Mo_2C	S	S	W	—	—	—
TiC-20%TiN-30%Ni-15%Mo-2%C	TiC	S	S	S	S	S	S
	Ni	S	S	S	S	S	S
	TiN	S	S	S	M	W	V W
	Mo_2C	S	S	M	W	—	—

(b) The amount of Mo in solution in the binder phase increases with
increasing the temperature up to 1000°C; however, above 1000°C, it
decreases with temperature. This corresponds to the transfer of Mo
into the Ti carbonitride phase from each grain periphery to form a
(Ti,Mo)(C,N) shell.
(c) The binder Mo content of the sintered cermet is in general much
larger with TiC-TiN-Ni-Mo cermets than with TiC-Ni-Mo cermets; it
increases with an increase of the amount of added TiN.

Another example of the sintering reaction of a TiN bearing cermet, TiC-20%TiN
-15%WC-10%TaC-5.5%Ni-11%Co-9%Mo is shown in Fig.12. Here, WC and TaC were
added as the second and the third carbides to a TiC-TiN-Ni-Mo cermet, and
2/3 of the Ni binder was replaced with Co. The general characteristics on
disappearance of various carbides and TiN during sintering are essentially
the same as that of TiC-TiN-Ni-Mo cermet mentioned above. In this case
also, a fine grained microstructure has been obtained due to the addition
of TiN. Fig.13 shows the concentration profiles within the binder phase and
the carbonitride shell (Ti,W,Ta,Mo)(C,N) of Ti, W, Mo and Ta in a TiC-12%TiN-
9%WC-6%TaC-16.5%Ni-33%Co-5.5%Mo cermet which is a higher, coarse-grained
version of the cermet used for the sintering experiment shown in Fig.12; the
ratios of TiC/TiN/WC/TaC/Mo and Ni/Co were kept equal in both hardmetals.
The important point to be noted with respect to the structure of the
carbonitride shell is that the concentration profiles for W, Mo, and Ta inside
the shell differ considerably; Mo resides preferentially in the periphery of
the shell, and W resides a little inside. Ta exists most richly in the
central region of the shell. These concentration profiles of the refractory
elements in the carbonitride shell are considered to give a favourable effect
in enhancing wetting of the carbonitride shell by the binder phase.

The sequence of the events outlined above for the conventional sintering of
TiN bearing cermet leading to the formation of the microstructure with a
carbonitride shell round each TiC(N) grain core appears to bear no direct
relation to the spinodal decomposition reaction which Rudy considers as an
essential step in the development of a fine microstructure. In short, an
essentially similar, fine microstructure easily develops in TiN added TiC
base cermets through the conventional sintering process, i.e. without
recourse to a preliminary heat treatment causing the spinodal reaction.
Moreover, the duplex $\alpha' + \alpha''$ structure which Rudy claims to have developed

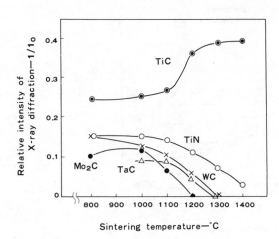

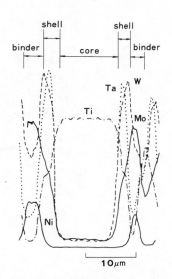

Fig.12 Relative intensities of X-ray
diffraction peaks for TiC, TiN,
WC, TaC and Mo₂C phases plotted
against sintering temperature.

Fig.13 Concentration profiles of Ti,
W, Mo, Ta and Ni in TiC-12%
TiN-9%WC-6%TaC-16.5%Ni-33%Co
-5.5%Mo cermet.

through the spinodal reaction is a kind of coaxial or shell structure which
is very commonly observed with the conventionally sintered TiN bearing cermet.
It has been confirmed in the author's laboratory, shortly after the
publication of Rudy's work, that the microstructure of a conventionally
sintered TiC-TiN base cermet is hardly distinguishable from that of the
so-called spinodal alloy of the same composition produced in exactly the
same way as Rudy has described.

As mentioned previously, TiN additions to TiC-Ni-Mo cermet have a significant
effect on the binder phase composition. Fig.14 shows the effect of TiN
additions on the binder phase composition of TiC-30%Ni-15%Mo cermet (Nishigaki
and Doi,1980). Binder Ti and Mo contents were determined by means of EPMA.
On examining the result, several features become apparent.

   (a) **Binder Mo content increases rapidly with TiN addition.** The addition
       of TiN by only 5% increases the binder Mo content by a factor of 4.
   (b) In contrast, binder Ti content markedly decreases with TiN additions.
       However, the reduction rate appears to diminish rapidly, and about
       a half of the Ti content originally in solution in the binder phase
       still remains after the addition of as much as 20%TiN.
   (c) The dependence of binder Mo content upon carbon content is small,
       whereas that of binder Ti content is much larger.

Table 4 tabulates the binder phase composition of various kinds of TiN bearing
cermet (Nishigaki and Doi,1980; Fukuhara and Mitani,1982; Suzuki and coworkers,
1983). Binder Ti and Mo contents are in reasonably good agreement among
different investigators. The effect of carbon content on binder Mo content
is rather conspicuous with a $TiC_{0.7}N_{0.3}$-24%Ni-19%Mo₂C cermet (Suzuki and
coworkers,1983). This is in contrast to the result obtained by Nishigaki
and Doi (1980) with EPMA. The difference may be attributed to the difference
in the method of the analysis.

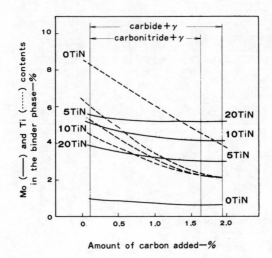

Fig.14 Binder Mo and Ti contents of TiC-(0 20)%TiN-30%Ni-15%Mo cermet plotted against the amount of carbon added.

TABLE 4 <u>Analysis of the Binder Phase Composition of TiC-TiN Base Cermets</u>

Composition of cermet	Composition of binder phase						Grain size μm	Method of analysis of binder phase	Reference
	Ti content %			Mo content %					
	low C	medium C	high C	low C	medium C	high C			
TiC-24%Ni-27%Mo$_2$C	—	11.9	—	—	2.1	—	1.2	Wet chemical method	Suzuki and coworkers (1983)
TiC$_{0.7}$N$_{0.3}$-24%Ni-11%Mo$_2$C	—	10.6	—	—	3.4	—	1.1		
TiC$_{0.7}$N$_{0.3}$-24%Ni-19%Mo$_2$C	9.3	8.4	2.7	10.0	3.4	—	LC 0.9 MC 1.0 HC 1.1		
TiC$_{0.7}$N$_{0.3}$-24%Ni-24%Mo$_2$C	—	5.9	—	—	8.2	—	0.9		
TiC-20%Ni-10%Mo		7.5			1.6		9.4	Wet chemical method	Fukuhara and Mitanl (1982)
TiC$_{0.8}$N$_{0.2}$-20%Ni-10%Mo		7.2			3.7		4.4		
TiC$_{0.8}$N$_{0.2}$-20%Ni-10%Mo		6.2			5.7		4.0		
TiC$_{0.8}$N$_{0.2}$-20%Ni-10%Mo		5.0			7.0		4.1		
TiC-30%Ni-15%Mo	8.6	6.2	4.7	0.9	0.8	0.6	Coarse grained	EPMA	Nishigaki and Doi (1980)
TiC-5%TiN-30%Ni-15%Mo	5.7	3.7	2.5	3.9	3.3	3.0			
TiC-10%TiN-30%Ni-15%Mo	5.1	3.3	2.4	5.1	4.4	3.0			
TiC-20%TiN-30%Ni-15%Mo	4.3	3.1	2.2	5.6	5.2	5.1			

Relation between the Microstructure and the Mechanical Characteristics of TiC and TiC-TiN Base Cermets

Parikh (1957) pointed out fairly early an analogy in the characteristics of fracture between WC base hardmetals and TiC-Ni-Mo cermets; fine grained TiC-Ni-Mo cermets fracture intergranularly, and with an increase in carbide grain-size ($\gtrsim 2\sim4$ μm), the mode of fracture changes to a transgranular type. Therefore, it was anticipated that the relation between various strengths and microstructural parameters derived with WC base hardmetals, such as TRS (or hardness) vs. binder volume fraction (or mean free path), would similarly hold with TiC-Ni-Mo cermets. This turned out to be true (Moskowitz and Humenik,1966; Suzuki, Hayashi and Terada,1972). On the other hand, in TiC-Ni-Mo cermets, the (Ti,Mo)C shell is much less hard than TiC core, but substantially harder than the binder phase, see **Table 5**. Moreover, the shell was found to be very brittle; Fig.15 shows that (Ti,Mo)C phase around the Vickers impression has swollen and the surface looks roughened and crumbling. This suggests that although a TiC-Ni-Mo cermet appears to be a simple two-phase composite with hard carbide grains embedded in a ductile binder phase like WC base hardmetals, a characteristic feature of the former which is in contrast with the latter is the existence of this considerably thick and relatively soft and brittle (Ti,Mo)C shell round TiC core; this significantly affects the mechanical properties of TiC-Ni-Mo cermets as will be discussed below.

Various factors such as starting powder characteristics, carbon content,etc. were found to affect the mechanical and cutting properties of TiC-Ni-Mo cermets; several important points will be briefly mentioned. In regard to the starting powder characteristics, use of TiC powder with a higher combined carbon content and a lower oxygen content is found to display better mechanical and cutting properties in the final product (Suzuki,1980). A small amount of impurity in the starting TiC powder leads to the formation of anomalous phases such as Ti_xS ($x\gtrsim1$) in the sintered state; this lowers significantly the room temperature TRS (Suzuki, Hayashi and Yamamoto,1978a).

The room temperature mechanical behaviour of TiC-Ni-Mo cermets in relation to carbon content was studied by several investigators (Nishigaki and coworkers,1974; Snell,1974; Suzuki, Hayashi and Yamamoto,1978; Moskowitz and Humenik,1978). Fig.16 shows representative data; carbide grain-size, hardness and compressive strength and fracture strain of TiC-15%Ni-8%Mo$_2$C are plotted against carbon content (Nishigaki and coworkers,1974). As can be seen, hardness and compressive strength monotonically increase with increasing the carbon content, while TRS and compressive strain have a maximum at relatively high carbon content in the two-phase region. The

TABLE 5 Hardness of the Constituent Phases of a
TiC-40%Ni-20%Mo Cermet

Composition of cermet	Hardness HV		
	TiC core	(Ti, Mo)C shell	binder phase
TiC-40%Ni-20%Mo	2900 ± 200	1500 ± 300	250 ± 50

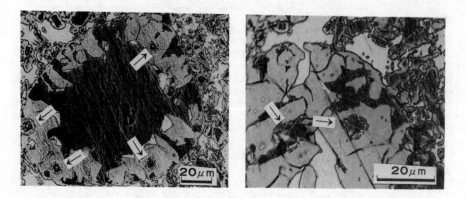

Fig.15 Effect of Vickers impression on (Ti,Mo)C shell
round TiC core of coarse grained TiC-40%Ni-20%
Mo cermet.

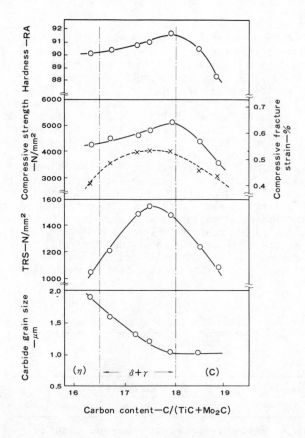

Fig.16 Effect of carbon content of TiC-15%Ni-8%Mo$_2$C cermet upon
hardness, compressive strength and strain, TRS and carbide
grain size.

observed effect of carbon content upon room temperature hardness and
compressive strength is the opposite of that for WC base hardmetals (Doi,
Fujiwara and Oosawa,1972), whereas the effect of carbon content on the TRS
is apparently similar (Suzuki and Kubota,1966; Nishigaki and coworkers,1974;
Ueda and coworkers,1977; Moskowitz and Humenik,1978; Suzuki, Hayashi and
Yamamoto,1978). It is well known that hot isostatic pressing (HIP) of WC-Co
hardmetals gives rise to a monotonically increasing TRS with decreasing the
C-content (Ueda and coworkers,1977). This suggests that WC-Co hardmetals of
lower carbon contents usually contain considerable amounts of residual
micropores which may act as internal crack sources (Suzuki and Hayashi,1975),
and their elimination brings about a pronounced increase of TRS (Doi, Fujiwara
and Oosawa,1972; Suzuki, Tanase and Hayashi,1975). The observed dependence
of TRS on carbon content for HIP'ed WC-Co hardmetals suggests that room
temperature strength depends predominantly on a solid-solution strengthening
effect of dissolved W in the binder phase whenever the microstructural
features are similar to each other. HIP of as-sintered TiC-Ni-Mo cermets
causes a rise in TRS similar to the case of WC-Co hardmetals. On the other
hand, as already mentioned, the structure of the hard particle phase of the
former is characterised by the extensive formation of (Ti,Mo)C shell round
TiC core, an aspect which is highly different from that of the latter. The
relation between TRS and carbon content with HIP'ed TiC-Ni-Mo cermets must
therefore be different from that with HIP'ed WC-Co hardmetals. Unfortunately,
the pertinent experimental data are lacking.

The effect of the replacement of the Ni in TiC-Ni-Mo cermets with Co upon
the TRS and hardness was investigated by Suzuki, Hayashi and Yamamoto (1977,
1978b). The room temperature TRS of $TiC-20\%Ni-20\%Mo_2C$ was found to be
slightly higher than in $TiC-20\%Co-20\%Mo_2C$ whereas at temperatures above
600°C, the reverse was observed. On the other hand, Co-bonded cermets
exhibit higher hardness at all temperatures (RT to 1000°C) in comparison
with Ni-bonded cermets. The observed higher strength of Co-bonded cermets
appear to be due to the larger high temperature strength of Co-binder(Suzuki,
Hayashi and Terada,1978; Roebuck and Almond,1981). Moreover, a comparison
of the strength properties between $TiC-20\%Ni(Co)-20\%Mo_2C$ and WC-8%Co hardmetals
both having the same binder volume fraction revealed that although at
temperatures below 900°C, WC-Co hardmetals exhibit a much higher TRS, above
900°C, $TiC-20\%Ni(Co)-20\%Mo_2C$ cermets become stronger than WC-8%Co hardmetals
(Suzuki, Hayashi and Yamamoto,1978 b). The authors ascribed the observed
reversal in strength to the larger skeleton strength of the cermets at
elevated temperatures.

A partial replacement of TiC in a $TiC-20\%Ni(Co)-20\%Mo_2C$ cermet with VC, Cr_3C_2,
NbC, TaC, or WC by 10% in volume of the cermet was carried out by Suzuki,
Hayashi and Terada (1978b) in order to investigate its effect upon the
mechanical properties. As mentioned previously, addition of VC, TaC, or WC
does not introduce an appreciable change in the microstructure, whereas
Cr_3C_2 or NbC additions lead to considerable coarsening of the carbide grains
(2.2-2.6~3.0μm). The effect of the second carbide addition upon microhardness
of the cermet and the binder phase was examined and the result is shown in
Table 6. It is interesting to note that the effect of the addition of each
carbide is different. WC or NbC additions to Co-bonded cermet contributes
to both cermet and binder phase hardening, whereas a TaC addition contributes
only to cermet hardening. This indicates that TaC has almost wholly
transferred into the carbide phase during sintering. The case of the Cr_3C_2
addition is just the reverse. Here, Cr_3C_2 has dissolved copiously into
the binder phase and hardens it conspicuously; whereas a combination of the
effect of coarsening due to Cr_3C_2 addition and that of the binder hardening

TABLE 6 Effect of Addition of the Other Kinds of Carbide by 10vol% upon Microhardness of TiC-20%Ni(Co)-20% Mo_2C Cermets and the Binder Phase

Target of observation	binder phase of cermet	Hardness —HV20					
		Standard cermet *	Carbide added				
			VC	Cr_3C_2	NbC	TaC	WC
Cermet	Ni	1340	1360	1360	1340	1380	1350
	Co	1440	1455	1460	1530	1520	1590
Binder phase	Ni	230	210	300	250	250	270
	Co	340	380	480	410	350	440

*   TiC-20%Ni(Co)-20%Mo_2C

leads to almost no hardening of the cermet as a whole. On the other hand, the Ni-bonded cermet has not shown such a pronounced change. Hardening of the cermet and the binder phase is rather slight and much less than that of Co-bonded cermet. This indicates that the solubility of refractory metals in Co is considerably larger than that in Ni. The effect of the addition of the second carbide upon TRS was found to be rather discouraging. In Co-bonded cermets, TRS generally decreases with such an addition except for WC additions for which TRS remains almost unchanged. On the other hand, the TRS slightly increases with a WC or TaC containing cermet which has been bonded with Ni; the addition of the second carbide considerably lowers the TRS. Post-sintering treatment by HIP does not essentially change the situation. Although HIP is effective in raising the TRS, its effect upon the cermet with the added second carbide is always somewhat lower than that on the standard cermet (TiC-Ni(Co)-Mo_2C). Suzuki, Hayashi and Terada (1978b) maintain that the addition of WC to TiC base cermet should be noted because the hardness of the cermet is raised by as much as 10% while the TRS remains unchanged with the addition.

The effect of the addition of a second carbide to TiC base cermet was further investigated at 900°C (Suzuki, Hayashi and Yamamoto,1979; Suzuki, Hayashi and Kubo,1980). Among various carbides added to a standard cermet (TiC-20%Ni-20%Mo_2C), i.e. V_4C_3, Cr_3C_2, NbC, TaC, WC, ZrC and HfC, only ZrC and HfC were found effective in improving high temperature strength. Addition of ZrC or HfC to replace 1 mole% TiC increases the microhardness of the cermet by 13%, moreover, as shown in Fig.17, the TRS is raised by 23% (ZrC addition) or 19% (HfC addition). Fig.17 also indicates that the addition of ZrC or HfC increases the TRS fairly rapidly with increasing the additions up to 1~2 mole%; therefore, the TRS more or less drops with further addition. The decrease in TRS was found to originate in micropores. Therefore, the strength drop due to excessive additions of ZrC or HfC can be largely recovered by HIP. The effect of carbon content upon the TRS of ZrC (or HfC) containing cermet at 900°C was found to be very small in contrast with the case of the standard cermet which shows almost linear increase of TRS with increasing the carbon content. The investigation on bend deformation and micro-cracking of a ZrC containing cermet at 1000°C reveals that the ZrC addition gives much improved

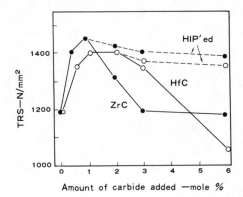

Fig.17 Effect of addition of ZrC or HfC to TiC-20%Ni
-20%Mo$_2$C upon TRS at 900 C.

deformation resistance, and the generation of surface micro-cracking is
considerably suppressed. The observed strengthening due to the addition of
ZrC (or HfC) is attributable to either the increase of the strength of the
carbide skeleton or the retardation of the dynamic recovery of dislocations
in the binder during high temperature deformation.

The effect of TiN addition upon the strength of TiC-Ni(Co)-Mo$_2$C cermets was
extensively investigated mainly by Suzuki and coworkers (Suzuki and coworkers,
1976; Suzuki, Hayashi and Yamamoto,1977; Suzuki, Hayashi and Yamamoto,1978**b**;
Nishigaki and Doi,1980; Nishigaki and coworkers,1980; Suzuki, Hayashi and Kubo,
1980; Suzuki and coworkers,1981). The grain refining effect of TiN additions
was found to persist with **increasing the TiN addition up to 15%; a further**
addition, however, leads to the appearance of TiN particles and this causes
an increase in the average grain-size (Suzuki and coworkers,1976; Suzuki,
Hayashi and Yamamoto,1977). The hardness of the TiC-TiN-Ni-Mo$_2$C cermet
considerably increases accordingly with increasing the TiN addition up to
30%. As shown in Fig.18, HIP brings about a pronounced increase in the TRS
of the TiN containing cermet; the rate of the increase becomes more conspicuous
with increasing TiN (Suzuki, Hayashi and Yamamoto,1977). Replacement of the
Ni binder with Co significantly raises hardness and lowers the TRS as in the
case of a TiC base cermet. Examination of the fracture-inducing origins
reveals that the average defect size of as-sintered TiC-(0~15)%TiN-20%Mo$_2$C
cermet rapidly increases with TiN addition. Since HIP of the cermet eliminates
residual micropores, the average defect size becomes even smaller than in a
TiN-free cermet.

The effect of variation of the binder volume fraction is shown in Fig.19.
The TRS of the Ni-bonded cermet reaches the maximum value of 2.7kN/mm^2 at
the binder content of 20%; further increases in volume fraction does not raise
the TRS, but slightly decreases it. Another point to be noted is that the
fracture-inducing defect size steadily decreases with increasing the binder
content of as-sintered cermet.

Fig.20 shows a comparison of the relation between hardness and TRS among
various hardmetals (Suzuki, Hayashi and Yamamoto,1977). It indicates the
superiority of a Co-bonded TiN bearing cermet to a Ni-bonded one and even

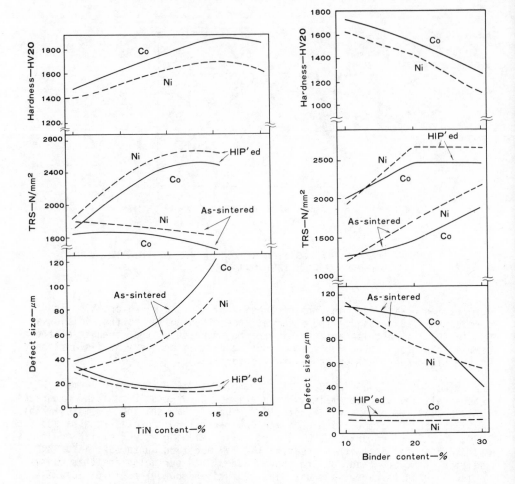

Fig.18 Effect of TiN content of TiC-
(0-20)%TiN-20%Ni(Co) -20%Mo₂C
cermets upon defect size, TRS
and hardness.

Fig.19 Effect of binder content of TiC-
15%TiN-(10-30)%Ni (Co)-20%Mo₂C
cermets upon defect size, TRS and
hardness.

to WC-Co hardmetals in realizing a combination of a high hardness and a large
TRS.

As mentioned before, binder phase strengthening through precipitation of γ'
is a possible way of improving heat resistance of TiC base and TiC-TiN base
cermets. The addition of Al to TiC or TiC-TiC-TiN base cermets increases
yield strength and hardness in the as-sintered state (Doi and Nishigaki,1976;
Nishigaki, Oosawa and Doi,1979; Nishigaki, Yoshimura and Doi,1980). The
improvement in hardness due to γ' precipitation persists up to ~1000°C. On
the other hand, in regard to the effect of γ' precipitation upon the TRS, there
are cases in which an optimum amount of γ' precipitation significantly raises
the TRS (Doi and Nishigaki,1976; Nishigaki, Oosawa and Doi,1979), whereas
in other cases γ' precipitation slightly lowers the TRS. When the binder phase
of the cermet is wholly converted into γ' mono-phase, the TRS in general

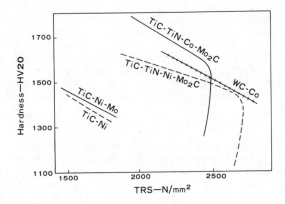

Fig.20 A comparison of hardness-TRS relation among
various kinds of hardmetal.

considerably decreases. This has partly been attributed to the effect of
residual micropores (Nishigaki, Oosawa and Doi,1979; Nishigaki, Yoshimura and
Doi,1980). Fig.21 shows an example of a transmission electron micrograph
showing precipitated γ' particles in the binder of TiC-10%WC-47.7%Ni-2.3%Al-
15%Mo-0.15%C cermet acting as barriers for dislocation movements. Although
an age-hardening treatment is effective for further strengthening of the
as-sintered cermet with high binder concentrations (Doi and Nishigaki,1976;
Doi and Nishigaki,1984), it is not practical to apply the treatment to a
cermet intended for use as a high grade cutting tool material owing to the
inevitable introduction of thermal cracks during rapid cooling from solution-
treatment temperature. The choice of the optimum amount of Al to be added
and the form of the Al addition is a technologically important problem.
AlN is regarded as a desirable form due to its excellent chemical stability
(Nishigaki, Yoshimura and Doi,1980).

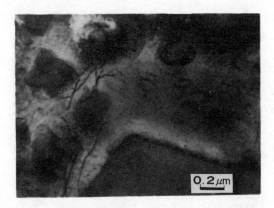

Fig.21 Dislocation-γ' particle interactions in TiC-
10%WC-47.7%Ni-2.3%Al-15%Mo-0.15%C cermet.

In spite of the technological importance, studies on high temperature deformation behaviour of TiC and TiC-TiN base cermets have hitherto been comparatively scarce. Recently, however, Suzuki and coworkers carried out a series of investigations on the bend deformation and bend creep deformation of cermets (Suzuki, Hayashi and Kubo,1980; Suzuki and coworkers,1981(a); Suzuki and coworkers,1983). **Fig.22 shows bend** creep deformation curves for TiC and $TiC_{0.7}N_{0.3}$ base cermets of various carbon contents and Mo_2C contents at 1000°C under the stress of 400N/mm^2 (Suzuki and coworkers,1983). It can be seen that the steady-state creep rate is much smaller for a $TiC_{0.7}N_{0.3}$

base cermet in comparison with a TiC base cermet. Moreover, it is shown that creep deformation is more suppressed with increasing Mo_2C content or decreasing carbon content. Suzuki and coworkers found that the effect of nitrogen or carbon content in the cermet upon the steady-state creep rate can consistently be expressed by the concentration of Mo in the binder phase; increasing nitrogen content and the amount of Mo_2C addition, and decreasing carbon content contribute to raising the binder Mo content, and **causes a decrease in creep rate.** It is interesting to note that the effect of dissolved Ti, which is also a potent solid-solution strengthener like Mo, appears to be comparatively slight. The effect of the grain size of Ti carbide or carbonitride particles was also investigated in the size range of 0.8-2.4μm at 1000°C and under the stresses of 300~400N/mm^2. It is indicated that a fine grained TiC base cermet exhibits a smaller creep rate than a coarse grained one, although the difference rapidly diminishes with decrease in stress and when the bend stress becomes as low as 300N/mm^2, the reverse tendency is sometimes noticed.

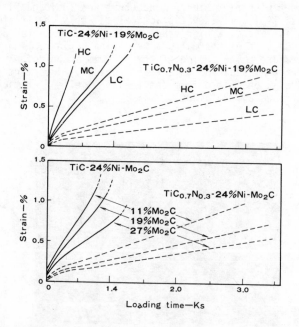

Fig.22 Creep deformation curves of TiC and $TiC_{0.7}N_{0.3}$ base cermets of various carbon and Mo_2C contents at 1000°C, 400 N/mm^2.

A comparison of the deformation behaviour of TiC-24%Ni-19%Mo$_2$C and TiC-14%TiN -24%Ni-19%Mo$_2$C cermets with WC base hardmetals (WC-10%Co, and WC-33%βt-12%Co) all with the same binder volume fraction indicates that at temperatures as high as 1000°C, TiC-TiN base cermets exhibit the largest flow stress, followed by TiC base cermets, WC-βt-Co and WC-Co hardmetals (Suzuki, Hayashi and Kubo, 1980). Here, βt signifies a solid-solution of WC, TaC and TiC. The activation energy of creep deformation was calculated from **the creep data taken at** 900-1000°C under the stress of 500N/mm^2. The result is as follows: TiC-TiN base cermet, 500kJ/mol; TiC base cermet, 460kJ/mol; WC-βt-Co hardmetal, 420kJ/mol; and WC-Co hardmetal, 340kJ/mol. Creep rupture times at 1000°C become about the same for the various hardmetals at stress levels as high as 600N/mm^2; the difference becomes increasingly large at lower stresses. In such a low stress region, the TiC-TiN base cermet exhibits the longest rupture time, followed by the TiC base cermet, WC-βt-Co and WC-Co hardmetals.

Suzuki and coworkers attribute the excellent creep deformation resistance of TiC-TiN or TiCN base cermets to the effect of the dissolved Mo in the binder in suppressing dynamic recovery of the binder phase (Suzuki and coworkers,1983).

As mentioned previously, improvement in fracture toughness of a tool material is an important property requirement for high performance cutting. Measurements of fracture toughness of TiC and TiC-TiN base cermets are comparatively few (August and Kalish,1981; Sadahiro,1982; Ohtsuki and coworkers,1983). K_{1C} and the limiting strength (strength of a hardmetal free from fracture-inducing defects) of TiC-Ni cermets are considerably smaller than those of WC-Co hardmetals (Ohtsuki, Kito and Masuda,1981; Ohtsuki and coworkers,1983); K_{1C} of a TiC-Ni cermet is approximately 50~60% of that for WC-Co hardmetals with the same amount of binder phase. The effect of the second carbide addition to TiC-Ni cermets was investigated by Ohtsuki and coworkers; the result is shown in Fig.23; here, K_{1C} for various hardmetals is replotted against hardness. It is apparent that a second carbide addition in general increases hardness and decreases K_{1C} except for WC additions. It is to be noted that although the TiC-Ni cermet exhibits considerably lower K_{1C} than that of WC-Co hardmetals as mentioned above, a comparison of G_{1C} (fracture energy release rate) indicates that G_{1C} for both kinds of hardmetal is nearly the same. Since G_{1C} is expressed by the equation:

$$G_{1C} = (1-\nu^2)K_{1C}^2/E$$

where ν and E are Poisson's ratio and Young's modulus, respectively, Ohtsuki and coworkers conclude that a comparatively low Young's modulus of the cermet is responsible for low K_{1C} value. The lowering of K_{1C} of TiC-Ni cermet with the addition of the second carbide must therefore be due to a decrease in G_{1C} caused by the embrittlement of Ni binder through the partial dissolution of the added carbide into the binder. The nature of the conspicuous improvement in mechanical properties due to the addition of Mo$_2$C (or Mo) discovered by Humenik and Parikh (1956) is not directly related to toughening, but to the reduction of fracture-inducing defects. Unfortunately, in the above discussion, the effect of the addition of the second carbide upon interfacial bonding energy between the carbide and the binder has not adequately taken into account. This appears to play an important role in affecting the fracture toughness of the cermet.

Sadahiro,1982, very recently studied the effect of TiN additions on K_{1C} of TiC-9%WC-9%TaC-16%Ni-16%Mo$_2$C cermet. Here, WC and TaC have been added to improve heat resistance and toughness of the cermet. Fig.24 shows an example

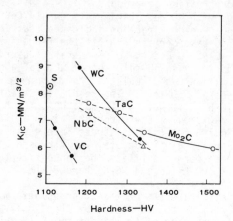

Fig.23 Effect of second carbide addition on K_{1C}-hardness relation with TiC-17vol%Ni cermet.

of the result. The relation between fracture toughness (K_{1C}) and hardness (HV) of the TiN containing cermet is shown. Although K_{1C} for TiC-Ni-Mo$_2$C cermet is in general much lower than that of WC-Co hardmetals when a comparison is made at some fixed hardness level, the addition of TiN is highly effective in improving K_{1C} of the cermet; the addition of TiN by as much as 7% appears to be sufficient to increase the K_{1C} to the level of WC-Co hardmetals. The effect of carbon content was also investigated; increasing the carbon content of the above cermet in the two-phase region slightly increases K_{1C}, although in the free graphite precipitation region, K_{1C} falls considerably.

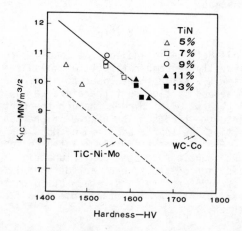

Fig.24 Relation between K_{1C} and hardness with TiC-(5∼13)%TiN -9%WC-9%TaC-16%Ni-16%Mo$_2$C cermets.

Improvement in Cermet Tool Performance through the Control of Mechanical, Thermal and Other Properties

On account of its high hardness, thermal stability, and other favourable properties, TiC was originally introduced into WC-Co hardmetals in order to suppress crater wear in machining of steels. The idea of utilizing TiC base cermets has resulted from the success of such an incorporation. However, the thermal conductivity of TiC base cermets is considerably inferior to WC base hardmetals, and thermal shock and fatigue resistance is also poorer. On account of these drawbacks, a TiC base cermet tool is susceptible to nose deformation in continuous turning of steels at high speeds or feed rates, and to thermal cracking in milling steels at high speeds. Moreover, insufficient toughness of TiC base cermets frequently leads to chipping and breakage in interrupted cutting of steels at high feeds. The desirable tool material properties for the prevention of these types of tool damage or failure were already mentioned. As outlined in the preceding section, the evolution of the TiC base cermet family reflects the history of improvement in cutting properties through the control of microstructure and properties of the cermet. In the following, attention will be focussed on the relation between cutting properties and tool material properties, and the various approaches to improve the cutting properties are surveyed.

Excellent cutting properties of the proto-type cermet TiC-Ni-Mo were first reported by Moskowitz and Humenik (1966). The TiC base cermet exhibits about a four fold increase of tool life in **semi-finish** and finish machining of steels; optimum tool life of TiC-12.3%Ni-Mo cermets is obtained for a composition containing 10~11%Mo.

An increase of carbon content of TiC-Ni-Mo cermets in the two-phase region in general leads to a considerable increase of wear resistance in cutting (Nishigaki and coworkers,1974; Snell,1974) in contrast with WC base hardmetals which show an increase of wear resistance with decreasing the carbon content due to the increased dissolution of W in the binder phase. In the case of TiC-Ni-Mo cermets, although solid-solution hardening due to increased dissolution of Ti and Mo in the binder phase increases with decreasing the carbon content (see TABLE 2), this does not lead to an increase in wear resistance since thickening of the relatively brittle, and less tenacious (Ti,Mo)C phase adversely affects the tool life.

A few years later, Moskowitz and Humenik (1978) attempted to explain tool life in terms of the effect of solid-solution strengthening by dissolved Ti in the binder phase of TiC-22.5%Ni-(1~16)%Mo cermets. They maintain that the tool life maximum occurs at a Mo addition of 10% and binder Ti content of 6~8% which corresponds to the highest Ti content in the two-phase region. They further pointed out that the resistance of plastic deformation at the nose of the cutting edge increases with decreasing carbon content, whereas abrasion resistance improves with increasing carbon content. As already mentioned above, an interpretation of the tool performance of TiC-Ni-Mo cermets in terms of only solid-solution strengthening of the dissolved Ti in the binder phase appears to be an over-simplified view since the effect of carbide coarsening and thickening of (Ti,Mo)C shell has not been taken into account. In the meanwhile, heat resistance of the cermet must depend on the amount of Ti and Mo dissolved in the binder phase; the excellent resistance of low carbon cermets against tool nose deformation is in accord with the characteristics of high temperature creep deformation of TiC-Ni-Mo cermets (see Fig.22).

γ' precipitation hardening of the Ni binder of TiC base cermets was shown to improve wear resistance in cutting (Doi and Nishigaki,1976; Nishigaki, Oosawa and Doi,1979). In the case of a TiC-15%Ni-8%Mo cermet,γ' precipitation by as much as ~50% in volume exhibits the highest wear resistance. A further increase in volume of precipitated γ' tends to decrease the wear resistance, the use of the binder phase consisting exclusively of γ' phase leads to an almost complete disappearance of the improvement in wear resistance due to Al addition. The possible interpretation for this is as follows. Since γ' binder is characterised by a positive temperature coefficient of the strength properties, the strength of this type of cermet becomes much lower at temperatures below 700°C than $\gamma+\gamma'$ bonded cermets and this causes the wear resistance to decrease considerably.

The effect of TiN additions upon microstructure and mechanical properties of TiC base cermets has already been discussed in considerable detail. On the other hand, improvements in thermal properties are anticipated to improve the cutting properties of TiC base cermets. Studies on this aspect are comparatively rare (Nishigaki and coworkers,1980). As shown in Fig.25, increasing nitrogen content in a TiCN base cermet significantly increase the thermal conductivity. Thermal conductivity is an important factor in affecting the thermal shock resistance of a tool material. However, in regard to the thermal shock resistance of the cermet at ambient as well as elevated temperatures, useful data are lacking to the author's knowledge. TABLE 7 shows the effect of nitrogen content of a TiCN base cermet upon thermal shock resistance; here, only newly measured data were used in the calculation of thermal shock (Nishigaki and coworkers,1980). It is apparent that an increase of nitrogen content considerably enhances thermal shock resistance.

Figs.26 and 27 show wear resistance in continuous turning and chipping resistance in interrupted cutting of steels, respectively (Nishigaki and coworkers,1980). An increase in chipping resistance with increasing nitrogen content of a TiCN base cermet is shown to be in excellent agreement with the trend of the thermal shock resistance (Table 7). Although crater wear resistance is improved with increasing nitrogen content, wear resistance in general decreases with increasing nitrogen content as shown in Fig.26. Therefore, there must be an optimum nitrogen content in a TiCN base cermet

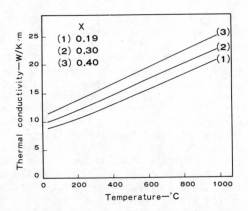

Fig.25 Temperature dependence of thermal conductivity of $TiC_{1-x}N_x$-13%Ni-30%Mo_2C cermet ($0.19 \leq x \leq 0.40$).

TABLE 7 Thermal **Shock** Resistance (R) of $TiC_{1-x}N_x$-13%Ni-30% -Mo_2C Cermet ($0.19 \leqq x \leqq 0.40$) Calculated from the Relation: $R = \sigma_B \theta / \alpha E$, where σ_B, θ, α and E are TRS, Thermal Conductivity, Thermal Expansion Coeff., Young's Modulus, respectively

Alloy	T(°C) N/(C+N) Ratio	R.T.	200	400
A	0.186	4.7	3.1	3.0
B	0.297	5.8	3.6	3.5
C	0.396	5.9	4.1	3.9

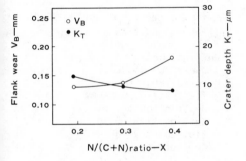

Fig.26 Effect of N/(C+N) ratio on wear resistance of a $TiC_{1-x}N_x$-13%Ni-30%Mo_2C cermet.

> Cutting condition
> work:SNCM-8(HB230)
> speed:200m/min
> feed rate:0.36mm/rev
> depth of cut:1.5mm
> cutting time:5min
> tip geometry:SNP432
> dry cutting

Feed rate N/(C+N) Time	0.256		0.310		0.330		0.360	
	1	2 min	1	2	1	2	1	2
0.19					X0.30			
				X0.50		X1.75		
0.30								X1.50
						X1.70		
								X.90
0.40								
								X1.50

Fig.27 Effect of N/(C+N) ratio on chipping resistance of a $TiC_{1-x}N_x$-13%Ni-30%-Mo_2C cermet in interrupted cutting of steel.

> Cutting condition
> work:SNCM-8(HB275)
> tool holder:NIIR44
> tip geometry:SNP432,-25 x0.08 honed
> speed:140m/min
> depth of cut:2mm
> No. of cutting impact:280times/min
> dry cutting

for exhibiting the best cutting performance. The process of γ' precipitation hardening of the binder phase of TiC base cermets due to Al additions in enhancing heat and wear resistance is anticipated to work also with TiCN or TiC-TiN base cermets. Fig.28 shows that this is the case (Nishigaki, Yoshimura and Doi,1980). Additions of 0.7% Al appear to give the highest wear resistance in cutting; this roughly corresponds to 40% in volume of precipitated γ' particles in the binder. Moreover, chipping resistance in interrupted cutting is also shown to be considerably improved, see Fig.29. It is to be noted that the addition of 0.7% Al significantly improves both the hardness and TRS. The observed improvement in both wear and chipping resistance is consistent with this.

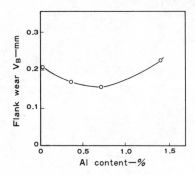

Cermet \ Feed rate	0.23	0.27	0.33	0.37
$TiC_{0.7}N_{0.3}$-15%Ni-8%Mo		-x		
			-x	
$TiC_{0.7}N_{0.3}$-15%Ni-8%Mo -0.7%Al				-x
				-x

Fig.28 Effect of Al addition to a $TiC_{0.7}N_{0.3}$-15%Ni -8%Mo$_2$C cermet on wear resistance.

    Cutting condition
      work:SNCM-8(HB220)
      speed:140m/min
      depth of cut:2mm
      cutting time:2min
      tip geometry:SNP432
      dry cutting

Fig.29 Effect of Al additions to $TiC_{0.7}N_{0.3}$-15%Ni-8%Mo cermet on toughness in interrupted cutting of steels.

    Cutting condition
      work:SNCM-8(HB220)
      speed:180m/min
      feed rate:0.36mm/rev
      depth of cut:1.5mm
      cutting time:15min
      tip geometry:SNP 432
      dry cutting

Simultaneous additions of a carbide, TiN and Al to a TiC base cermet are anticipated to improve mechanical, thermal and cutting properties of the cermet. According to Moskowitz and Humenik (1980), additions of 5%Al and 5%VC to TiC-22.5%Ni-10%Mo$_2$C cermets markedly improves nose deformation resistance; the effect is comparable to that of a TiN addition. Triple additions of 5%Al, 5%VC and 10%TiN further improve node deformation resistance (see Fig.30).

In regard to the effect of these additions upon thermal cracking, a 10%TiN addition was found to give considerably greater thermal crack resistance than additions of 5%Al and 5%VC. Moskowitz and Humenik are of the opinion that a selection of the compositions based on the TiC-VC-TiN-Ni-Mo system gives rise to a superior tool performance relative to WC base hardmetals in milling and rough machining of steels; sufficiently large deformation resistance can be obtained without the addition of Al.

Yoshimura and coworkers (1981) demonstrate that multiple additions of 20%TiN, 15%WC, and 10%TaC to a TiC-Ni-Mo cermet markedly improve the hardness, TRS, oxidation resistance and thermal conductivity at 900-1000 C as shown in Table 8. As a result of such a conspicuous improvement in the material properties, grooving wear resistance in continuous turning of alloy steels, toughness of the tool edge in intermittent turning of alloy steel and chipping wear resistance in milling carbon steel have greatly increased, and this kind of tailor-made cermet is believed to exhibit a much superior tool performance to TiC-Ni-Mo cermets and conventional WC base hardmetals for finish cutting.

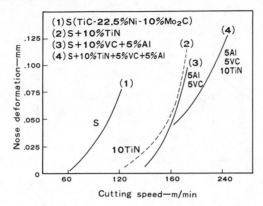

Fig.30 Nose deformation of TiC-22.5%Ni-10%Mo$_2$C containing no additive, 10%TiN, 5%Al+5%VC, and 5%Al+5%VC+10%TiN. 4340 steel (HB300) was used as the workpiece.

TABLE 8 A Comparison of High Temperature Hardness, TRS, Oxidation resistance and Thermal Conductivity of TiC-TiN-WC-TaC-Ni-Co-Mo Cermets

Composition of cermet	HV (1000°C)	TRS(900°C) N/mm²	Oxidation resistance wt gain for 1000°C/1h—mg/cm²	Thermal conductivity at 1000°C—W/k·m
TiC-16.5%Ni-9%Mo	500	1050	11.8	24.7
TiC-20%TiN-15%WC-10%TaC-5.5%Ni-11%Co-9%Mo	650	1360	1.66	42.3

Fig.31 shows another example of the tool performance of a tailor-made cermet plotted against fracture toughness (Sadahiro,1982); chipping resistance in milling of WC, TaC and TiN containing TiC-Ni-Mo cermet increases with increasing fracture toughness. As already mentioned fracture toughness of such a cermet doped with multiple additives can favourably compare with that of WC-Co hardmetals (see Fig.24).

The desirable mechanical, thermal and chemical properties of the newest cermets as high performance tool materials leading to improved thermal shock and thermal cracking resistance, grooving wear resistance, chipping resistance and nose deformation resistance originate in multiple additions of materials, such as TiN, WC, TaC, VC and Al, etc. As mentioned in the introduction, such development work has mostly been carried out on the basis of empiricism. That is to say, fundamental studies utilizing modern analytical tools of the

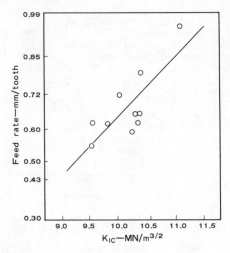

Fig.31 Relation between chipping resistance in milling and K_{1C} with TiC-9%TiN-9% WC-9%TaC-16%Ni-16%Mo$_2$C cermet.

Cutting condition
  work:S55C(HB250-300)
  speed:100m/min
  depth of cut:1.5mm
  1 pass:100x200mm
  tool geometry:150 mm face milling
                 cutter with single blade
                 (A.R.=5, R.R.=10, C.A.=30)

newest, complex cermet are highly necessary for elucidating problems relating to phase analysis, phase stability, chemical bonding, nature of fracture toughness, etc.

Improvement of wear resistance of cermets for high speed cutting may also be realized through an increase of the melting point of the binder phase. Precious metals appear to be interesting as a potential binder phase owing to the high melting point, low chemical reactivity and good thermal conductivity (Jackson, Warren and Waldron,1974). TiC-(Ru,Ni) system was investigated as the most promising tool material. **Small additions of Ni were beneficial for** the full densification during sintering. Although preliminary machining tests of TiC-12%Ru cermet were reported to be encouraging (Jackson, Warren and Waldron,1974), a question may arise as to whether the apparent disadvantage of high cost of Ru could be sufficiently offset by the anticipated improvement in tool performance. On the other hand, relatively small additions of Ru caused marked hardening of the Ni binder phase due presumably to a precipitation hardening process. It appears that investigations on the effect of small additions of Ru to advanced TiC-TiN base cermets on the mechanical and cutting properties are of considerable interest.

PROSPECT FOR MORE REFRACTORY HARD MATERIALS

Al$_2$O$_3$ ceramic is an outstanding tool material owing to the excellent heat resistance, chemical inertness and stability. As a ceramic tool material, it has a fairly long history. During the 1950's, lack of sufficient toughness and tool reliability motivated investigations for further strengthening of Al_2O_3; the grain refining effect of additives such as MgO was discovered and fairly high strength (TRS:500N/mm^2) was obtained ("white ceramic"). Towards the end of the 1960's, Al_2O_3 ceramic tool failure due to thermal cracking became a serious problem (Sugita and Yamada,1963a; Sugita and Yamada,1963b; Sugita and Yamada,1963c; Sugita and Yamada,1966) and consequently, intensive studies were undertaken to improve the **thermal shock resistance.** For this purpose, incorporation of TiC(TiN) was found effective and as a result, the Al_2O_3-30%TiC(TiN) composite ceramic ("black ceramic") was born. Application of^{2} ^{3}pressure sintering techniques (hot pressing, HIP) was found necessary to achieve sufficiently high densification (99%T.D.). Wahi and Illschner (1980) maintain that since fracture energy ($\propto K_{1C}^2$) of TiC is much larger than that of Al_2O_3, TiC dispersion in the matrix of Al_2O_3 significantly contributes to increase K_{1C} in addition to the improvement in thermal properties.

Al_2O_3 ceramic can also be toughened by a ZrO_2 dispersion. This concept was advocated **by Claussen;** a fine dispersion of unstabilized ZrO_2 grains in the in a matrix of Al_2O_3 gives rise to transformation-induced microcracks which absorb elastic energy in front of a running crack thereby giving an increase of K_{1C} (Claussen,1976; Claussen, Steeb and Pabst,1977). Balancing of high hardness and large toughness is a critical factor in utilizing this novel ceramic for cutting application (Doi,1983).

There is another possibility for enhancing the toughness of Al_2O_3 base ceramics. Unidirectional solidification of oxide-metal eutectic composites^{3}has been studied extensively (Briggs and Hart,1976). If in situ eutectic composite powder is available, the dense compact made from it can be said to have "built-in" crack arresting mechanism. Claussen and Petzow carried out the exploratory work (Claussen, Petzow and Yahn,1974; Petzow and Claussen,1977). They found that in situ eutectic structure is effective for increasing work of fracture. Lux and coworkers (Schnitt, Banik and Lux,1980; Banik, Schnitt Weiss and Lux,1980) investigated eutectic Al_2O_3-Cr_2O_3-Cr(Mo) and other systems; hot pressing of a composite powder made from eutectic melt by crushing and milling gave results which were not very satisfactory. Further work on processing, especially on refinement of eutectic composite powder seems to be necessary.

REFERENCES

Banik,G. Schmitt,T. Wruss,W. and Lux,B. 1980, Radex-Rundschau 4 337-343
Brigg,J. and Hart,P.E. 1976, J. Am. Ceram. Soc. 59 530-531
Claussen,N. Petzow,G. and Jahn,J 1974, 8th Plansee Seminar, Preprint II, No.45
Claussen,N. 1976, J. Am. Ceram. Soc. 59 49-51
Claussen,N. Steeb, J and Pabst,R.F. 1977, Am. Ceram. Soc. Boll. 55 559-562
Doi,H. Fujiwara,Y. and Oosawa,Y. 1972, Proc.Int. Conf.on Mechanical Behavior
     of Materials,1971, V 207-219
Doi,H. and Nishigaki,K. 1977, Modern Development in P/M 11 525-542
Doi,H. 1983, New Ceramic Powder Handbook, Science Forum, 536-551
Doi,H. and Nishigaki,K. 1984, This conference
Elliot,J.F. and Gleiser,M. 1960, Thermochemistry for Steelmaking, Addison-
     Wesley, London,1 160
Fukatsu,T. and Yamaya,S. 1963, Japan Soc.of Powder and Powd. Met.10 167-171
Fukuhara,M. and Mitani,H. 1982, Powder Met. Intern. 14 196-200
Humenik,M. and Parikh,N.M. 1956, J. Am. Ceram. Soc.39 60-63
Kieffer,R. and Benesovsky,F. 1965, Hartmetalle, Springer Verlag, Wien

Jackson,J.S. Warren,R. and Waldron,M.B. 1974, Powder Met. 17 255-270
Kieffer,R. and Fister,D. 1970, Planseeber. Pulvermet.18 246-253
Kieffer,R. Ettmayer,P. and Freundhofmeier,M. 1971a, Metall,25 1335-1342
Kieffer,R. Ettmayer,P. and Freundhofmeier,M. 1971b, Modern Development in
    P/M 5 201-214
Komac,M. and Lange,D. 1982, Int.J. Powder and Powd. Techn.18 313-321
Lindau,L. and Stjernberg,K.G. 1976, Powder Met. (4) 210-213
Moskowitz,D. and Humenik,M. 1966, Modern Development in.P/M 3 83-94
Moskowitz,D..and Humenik,M. Int.J. Powder and Powd. Techn. 14 39-45
Moskowitz,D. and Humenik,M. Modern Development in P/M 14 307-320
Negishi,N. and Aoki,K. 1976, Proc.4th Int.Conf.Prod. Eng. 263-269
Negishi,N. Aoki,K. and Sata,T. 1980a, Annals of CIRP 29 57-60
Negishi,N. Aoki,K. and Sata,T. 1980b, Prod. 4th.Int.Conf. Prod.Eng. 480-485
Negishi,N. Aoki,K. and Sata,T. 1981, Annals of CIRP 30 43-46
Nishigaki,K. Ohnishi,T. Shiokawa,T. and Doi,H. 1974, Modern Development in
    P/M 8 627-643
Nishigaki,K. Oosawa,Y. and Doi,H. 1979, Japan Soc. of Powder and Powd. Met.
    26 169-173
Nishigaki,K. and Doi,H. 1980, Japan Soc.of Powder and Powd. Met.27 130-136
Nishigaki,K. Doi,H. Shingyoji,T. and Oosawa,Y. 1980, Japan Soc.of Powder and
    Powd. Met. 27 160-165
Ohtsuki,E. Kito,T. and Masuda,Y. 1981, Trans. JIM 45 432-437
Ohtsuki,E. Fujita,H. Kito,T. and Masuda,Y. 1983, Trans. JIM 47 567-574
Parikh,N.M. and Humenik,M. 1957, J.Am.Ceram.Soc.40 315-320
Parikh,N.M. 1957, J.Am.Ceram.Soc.40 335-339
Petzow,G. and Claussen,N. 1977, Radex-Rundschau 2 110-116
Rudy,E. 1973, J.less-Common Metals,33 43-70
Rudy,E. Worcester,S. and Elkington,W. 1974, 8th Plansee Seminar, II, No.30
Rudiger,O. Hirschfeld,D. Hoffmann,A. Kolaska,J. Ostermann,G. Willbrand,J.
    1971, Mitt.Krupp.Forsch.Ber.29 1-14
Sadahiro,T. 1982, Japan Soc. of Powder and Powd. Met. 29 266-270
Schmitt,T. Banik,G. and Lux,B. 1980, Ber.Dt.Keram.Ges.57 80-83
Snell,P.O. 1974, Planseeber.Pulvermet.22 91-106
Storms,E.K. 1967, The Refractory Carbides, Academic Press, N.K. 44
Stover,J.F. and Gleiser,M. 1960, Thermochemistry for Steelmaking, Addison-
    Wesley, London,1,160
Sugita,T. and Yamada,I. 1964a, J. Japan Precision Eng. 30 662-670
Sugita,T. and Yamada,I. 1964b, J. Japan Precision Eng. 30 727-732
Sugita,T. and Yamada,I. 1964c, J. Japan Precision Eng. 30 779-784
Sugita,T. and Yamada,I. 1966, J. Japan Precision Eng. 32 357-363
Suzuki,H. and Kubota,H. 1966, Planseeber.Pulvermet.16 96-109
Suzuki,H. Hayashi,K. and Terada,O. 1971a, Trans. JIM 35 936-942
Suzuki,H. Hayashi,K. and Terada,O. 1971b, Trans. JIM, 35 146-150
Suzuki,H. and Hayashi,K. 1971, Japan Soc of Powder and Powd. Met. 17 262-266
Suzuki,H. 1972, Trans. JIM 11 125
Suzuki,H. Hayashi,K. and Terada,O. 1972, Trans. JIM, 36 514-518
Suzuki,H. and Hayashi,K. 1975, Trans. JIM 16 353-360
Suzuki,H. Tanase,T. and Hayashi,K. 1975, Planseeber. Pulvermet.23 121-130
Suzuki,H. Hayashi,K. Yamamoto,T. and Wan Jae Lee, 1976, Japan Soc. of Powder
    and Powd. Met. 23 224-229
Suzuki,H. Hayashi,K. and Yamamoto,T. 1977, Trans. JIM, 41 432-437
Suzuki,H. Hayashi,K. and Yamamoto,T.1978a,Planseeber.Pulvermet. 26 42-50
Suzuki,H. Hayashi,K. and Yamamoto,T.1978b, Japan Soc. of Powder and Powd.Met.
    25 136-140
Suzuki,H. Hayashi,K. and Terada,O. 1978, Japan Soc.of Powder and Powd.Met.
    25 132-135
Suzuki,H. Hayashi,K. and Taniguchi,Y. 1978, Japan Soc.of Powder and Powd.
    Met. 25 94-98

Suzuki,H. Hayashi,K. and Yamamoto,T. 1979, Japan Soc. of Powder and Powd.Met.
    26 22-26
Suzuki,H. Hayashi,K. Kubo,Y. and Jin Yukang 1981a, Japan Soc. of Powder and
    Powd. Met. 28 67-69
Suzuki,H. Hayashi,K. Kubo,Y. and Matsubara,H. 1981b, Japan Soc. of Powder and
    Powd. Met.28 147-151
Suzuki,H. Hayashi,K. Matsubara,H. and Tokumoto,K. 1983, Japan Soc. of Powder
    and Powd. Met. 30 106-111
Suzuki,H. Hayashi,K. and Kubo,Y. 1980, Japan Soc. of Powder and Powd.Met.
    27 266-270
Yamaya,S. and Sadahiro,S. 1969, Japan Soc. of Powder and Powd.Met. 16 190-196
Suzuki,S. 1980, Japan Soc. of Powder.and Powd.Met.27 39-44
Ueda,F. Doi,H. Fujiwara,Y. and Masatomi,H. 1977, Powder Met. Intern. 9 32-35
Yoshimura,H. Sugisawa,T. Nishigaki,K. and Doi,H. 1981, 10th Plansee Seminar
    2 HM16 727-741
Wahi,R.P. and Illschner,B. 1980, J. Mater. Sci. 15 875-885

Inst. Phys. Conf. Ser. No. 75: Chapter 6
Paper presented at 2nd Int. Conf. Science Hard Mater., Rhodes

525

New interstitial alloy hardmetals

A J CARR, P KORGUL AND K H JACK

Wolfson Research Group for High-Strength Materials
University of Newcastle upon Tyne, U.K.

ABSTRACT

β^m-Ni_2Mo_3N with the metal-atom arrangement of β-Mn shows remarkable chemical and thermal stabilities and promising hardmetal properties. It can be bonded with Ni by pressureless-sintering to give dense compacts with hardnesses up to 1300 HV30 and excellent oxidation resistance. The Ni can be partly or wholly replaced in the binder and in the β^m-hardmetal by Fe, Co and other metals. Two-thirds of the Mo is replaceable by chromium and less extensively by Nb and Ta, and one-half of the N is replaceable by C, all without change of structure. These extensive substitutions give prospects of controlling the chemical and mechanical properties.

INTRODUCTION

The aim is to develop new hardmetals in which the strategic materials WC and Co are not used. The properties of a hard metal require (i) a high interatomic bond strength and (ii) a crystal structure in which easy glide is impossible. Some hard materials, e.g. WC and Fe_3C, also have structures in which the environment of the interstitial atom is a trigonal prism of metal atoms. On the basis of these simple crystal chemistry considerations it is thought that several interstitial alloy phases should have suitable hardmetal properties. For example, δ-MoN and the corresponding carbonitride Mo(C,N) have the same metal-atom arrangements as WC but, despite this and their high hardnesses, they have not been fully explored as potential tool materials. Again, the Z-phase CrNbN might be expected to be hard, strong and chemically stable with limited solubility in Fe and Cr. Thus, the production of sintered Z-phase compacts with Fe or Cr as a binder seems worth exploring.

In the present work, it was thought that the most promising new hardmetals might be found among interstitial nitrides that are iso-structural with β-Mn metal. The structures are complex, making easy glide impossible, and each N is coordinated by a distorted trigonal prism of metal atoms. In addition, the phase β^m-Ni_2Mo_3N has a remarkable stability - much greater than that of Ni or Mo nitrides - and so it has been the first to be investigated.

PREPARATION AND CHARACTERIZATION OF β^m-Ni$_2$Mo$_3$N

Interstitial alloys with the metal-atom arrangement of β-Mn were first
reported by Kuo (1956) in the Fe-W-C system, by Goldschmidt (1957) in
Fe-Cr-W-C, by Evans and Jack (1957) in Fe-Mo-N and by Grieveson and Jack
(1960) in the systems Fe-Mo-N, Co-Mo-N and Ni-Mo-N.   Subsequent work
(Jeitschko, Nowotny and Benesovsky, 1964) showed β^m-phases in T$_3$M$_2$X systems
where T is a transition metal, M is a B-sub-group element, and X is interstitial C or N, but these are expected to be less hard and strong. The
cubic unit cell of the β^m-M$_2$Mo$_3$N phases (M = Fe, Co, Ni) has $a \sim 6.7$Å
(1Å = 10^{-1}nm = 10^{-10}m) and contains 8M and 12Mo atoms in two different
sites of space group P4$_1$32.   Each of the four N-atom sites is at the
centre of a twisted trigonal prism of Mo atoms that is mid-way between the
N environments observed respectively in δ-MoN and γ-Mo$_2$N.

$\underline{\beta^m\text{-Ni}_2\text{Mo}_3\text{N}}$

β^m nickel-molybdenum nitride was obtained by Nutter, Grieveson and Jack
(1969) by the reduction of nickel molybdate in hydrogen containing trace
impurities of nitrogen.   The mixed metals act as a "getter" for nitrogen,
and as little as 0.05 vol% in otherwise pure H$_2$ will give the β^m-nitride
at 800°C.   In the present work, differential thermal analysis (DTA) of the
phase shows that its decomposition temperature under one atmosphere of
nitrogen is 1360°C.   These two observations allow the calculation of the
free energy of formation which, as shown by Fig. 1, is much more negative
than that of either Ni$_3$N or Mo$_2$N.   The reason for this remarkable thermodynamic stability of β^m-Ni$_2$Mo$_3$N is unknown but it means that the interatomic bond strength is high.

A variety of preparative methods has been explored.   On a laboratory scale,
the most convenient is to mix Ni and Mo powders ($<$ 300 mesh; $<$ 53μm) by
dry-milling and then cold-press the mixture at $\sim$140MPa to give cylindrical
compacts 10mm diameter x 8mm high.   These pellets are nitrided in a
horizontal tube furnace at 1000°C in cracked ammonia (25N$_2$:75H$_2$) for 4h.
The reaction is followed by weight changes and X-ray diffraction analyses.
No changes in unit-cell dimensions ($\underline{a}$ = 6.6344 $\pm$ 0.0005Å) are observed for
reaction times varying from 1h to 16h, or varying reaction temperatures
900°-1200°C, or for different partial pressures of N$_2$ up to one atmosphere.
Unit-cell dimensions of β^m in nitrided mixes containing excess of either
Ni or Mo are also identical showing that its range of homogeneity is
negligible.

COMPACTION OF β^m-Ni$_2$Mo$_3$N WITH Ni BINDER

Processing

All β^m compacts densified with Ni show X-ray diffraction patterns in which
the Ni reflexions shift to lower angles due to the dissolution of Mo.
These shifts correspond with about 25 wt%Mo in the binder and there are no
reflexions due to free Mo.   Estimates of the binder volume by optical
microscopy show that it is approximately twice that expected from the
amount of added Ni powder, because of this solution of Mo, and so is in
reasonable agreement with the X-ray observations.

In initial experiments, β^m-Ni$_2$Mo$_3$N was crushed, milled with carbonyl Ni
powder and cold-pressed.   The green compacts were hot-pressed at 27MPa

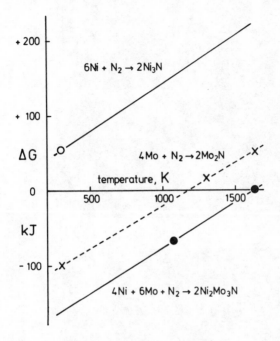

Fig. 1   Free energies of reactions to form
Ni_3N, Mo_2 and β^m-Ni_2Mo_3N

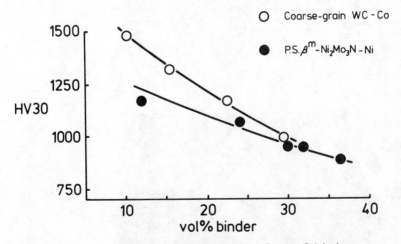

Fig. 2   Hardness variation with volume of binder
for pressureless sintered β^m:Ni and WC:Co

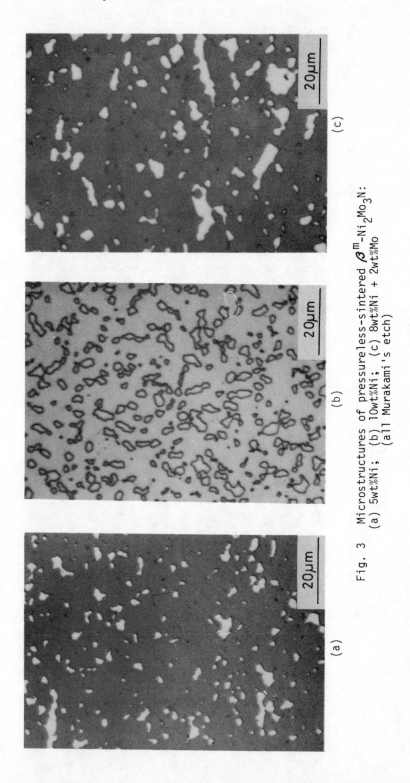

Fig. 3   Microstructures of pressureless-sintered β^m-Ni$_2$Mo$_3$N:
(a) 5wt%Ni;   (b) 10wt%Ni;   (c) 8wt%Ni + 2wt%Mo
(all Murakami's etch)

and $1200^{\circ}C$ for 20min but, due to NiO formation, the density of the compacts was not more than 98% theoretical. Improved processing, involving wet-milling the β^m and Ni powders in isopropyl alcohol and pressureless sintering in $50N_2:50H_2$ gave 99.8 ± 0.1 theoretical density with hardnesses approaching those of coarse-grain WC:Co. Hardnesses are shown in Fig.2 and typical microstructures by Figs.3(a) and 3(b). The microstructure consists of a skeleton of hardmetal grains, fused together, with pores of the skeleton filled with a Ni binder. Fig.4 shows a crack in β^m:15.5wt%Ni induced by a Vickers hardness indenter. The crack does not propagate through the binder which, instead, is plastically deformed. Moreover, the binder is not pulled away from the hardmetal interface, indicating good coherency.

"One-step" nitriding and sintering was explored by processing a mix of Ni and Mo containing 10wt% Ni in excess of that required for β^m-Ni_2Mo_3N formation. However, only 86% theoretical density was obtained and it seems that a rigid hardmetal skeleton is formed during nitriding that prevents subsequent sintering.

With the two-stage process, full densities are obtained by pressureless sintering, even with only 5wt% Ni-binder, and the composites have acceptably fine microstructures. Although hardnesses increase with decreasing binder contents it seems unlikely, even with improved processing leading to finer microstructures, that the hardness will exceed 1300 HV30.

Ni-Mo and Ni-Fe Binders

To reduce the solubility of the β^m hardmetal in the binder, a powder mix of 8wt%Ni, 2wt%Mo and 90wt% β^m-Ni_2Mo_3N was pressureless-sintered. The final binder volume was only about 1.5 times that of the initial binder volume showing, as expected, that increasing the molybdenum concentration in the nickel binder reduces the solubility of the β^m-phase. However, the final composite density is slightly reduced (see Table 1), the microstructure is a little coarser, and the hardness (1090 $\pm$ 10 HV30 for 16vol% binder; compare Fig. 2) is lower than expected. These differences, compared to composites processed with pure Ni binders, are small and require more detailed investigation. 8wt%Ni:2wt%Fe seemed as effective a binder as pure Ni and also reduced the solubility of the β^m phase.

TABLE 1   Composition, Density and Hardness of
Pressureless-Sintered β^m-Ni_2Mo_3N:Ni

Initial binder addition wt%	Final binder volume vol%	Theoretical density %	Hardness HV30
5Ni	12	99.7	1170 $\pm$ 5
10Ni	24	100.0	1070 $\pm$ 7
8Ni + 2Mo	16	98.0	1090 $\pm$ 10
10Ni excess in Ni_2Mo_3 mix	-	86.0	-

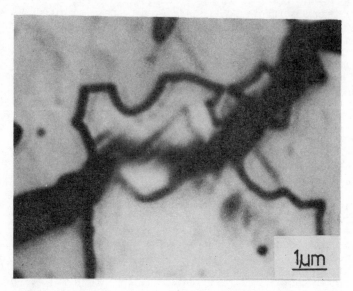

Fig. 4   Crack propagation in β^m-Ni$_2$Mo$_3$N:Ni

Fig. 5   Microstructure of β^m-Ni$_2$Mo$_3$N:Ni
oxidised in air at 1000°C for 24h

TABLE 2   Oxidation Penetration in mm after 24h for Pressureless-Sintered β^m-Ni$_2$Mo$_3$N:15.5wt%Ni-binder

800°C	900°C	1000°C
± 0.002	± 0.026	± 0.022

PROPERTIES OF β^m-Ni$_2$Mo$_3$N:15.5wt%Ni-BINDER COMPOSITES

Oxidation Resistance

The microstructure of β^m-Ni$_2$Mo$_3$N hot-pressed with 15.5wt%Ni and then oxidised in air for 24h at 1000°C is shown by Fig. 5.   Under these severe conditions the weight gain was less than 0.5% and, as shown, the oxidation occurs preferentially at the exposed Ni binder surfaces.   For similar pressureless-sintered composites, Table 2 shows that the oxidation penetration depths after 24h at 800°, 900° and 1000°C are of the same order as the associated errors of measurement and so no absolute values for oxidation resistance can be given.   The materials are, however, much more oxidation resistant than, for example, TiC bonded with Co, Mo or W.

Mechanical Properties

Preliminary measurements of fracture toughness on a sample with hardness 1000 HV30 gave 9.5MPa m$^{0.5}$ which is about one-half of that expected for WC:Co with the same binder content ($\sim$25vol%).

The wear number, defined in a standard rock test as the ratio of the weight of rock removed to the weight of tool lost, was 0.4x10^6.   For WC:Co and hardened steel the corresponding values are respectively 10x10^6 and 0.8x10^6.   Nevertheless, these results for the first samples of β^m:Ni prepared are promising in that modified processing is expected to improve the figures.

SUBSTITUTION IN β^m-Ni$_2$Mo$_3$N

Replacement of Ni by Fe and Co

Ni was replaced by increasing amounts of Fe and Co in the initial powders and the pelletised mixes were nitrided as previously at 1000°C in N$_2$:H$_2$. Under these conditions, 25at% of the Ni was replaced by Fe, and 75at% by Co in the β^m structure.   The results are in agreement with previous observations (Evans and Jack, 1957;   Grieveson and Jack 1960) that the pure β^m-phases Fe$_2$Mo$_3$N and Co$_2$Mo$_3$N are obtained by nitriding with NH$_3$:H$_2$ mixtures or with molecular nitrogen at high pressures.   Again the relative high stability of β^m-Ni$_2$Mo$_3$N is demonstrated.

Substitution of Mo by Cr

Since Cr is nearer in atomic radius to Ni than to Mo it seemed possible that Ni in β^m-Ni$_2$Mo$_3$N might be replaceable by Cr.   However, X-ray

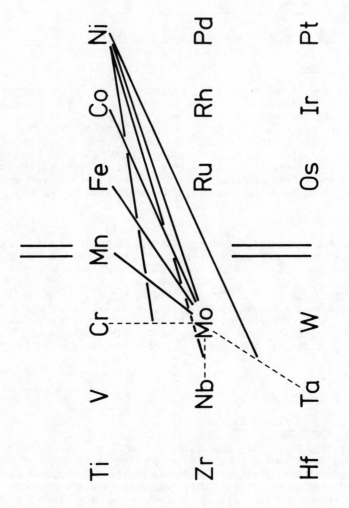

Fig. 6   The "Effective Periodic Group Number" of β^m structures

diffraction patterns of $Cr_2Ni_6Mo_{12}$ nitrided in $N_2:H_2$ at $1000^\circ C$ showed shifts of β^m reflexions to higher angles (i.e. a smaller unit cell) together with reflexions from b.c.c. Mo(Cr).   These observations suggest replacement of Mo by Cr and are in agreement with earlier speculations concerning the "Effective Group Number" and the valence electron concentration of β^m structures (Jack, 1967).   As shown by Fig. 6, the "EGN" is approximately 7, as for β-Mn itself.

Kikuchi, Wakita and Tanaka (1973) reported β-Mn-type precipitates in austenitic alloy steels containing nitrogen, and in one steel containing no Mo the precipitate was apparently an (Fe,Ni)-Cr nitride.   In the present work, a Mo-free β^m phase has not been found.   To determine the limit of Mo replacement, mixed Ni-Cr-Mo compositions were nitrided at $1200^\circ C$ in $67N_2:33H_2$.   The X-ray pattern from nitrided Ni_2Cr_2Mo gave, in addition to β^m, very weak additional reflexions from f.c.c. Ni(Cr.Mo) with $\underline{a} \sim 3.57\text{Å}$; see Fig. 7.   The expected change in the intensities of some β^m reflexions with increasing chromium replacement is accompanied by appreciable decrease in unit-cell dimensions (see Fig. 8), and on the basis of these observations a maximum of two-thirds of the Mo of β^m-Ni_2Mo_3N can be substituted by Cr.   Attempts to prepare a Mo-free β^m-nitride by using only Ni and Cr were unsuccessful.

Substitution of Mo by W

Although W is in the same Periodic Group VI as Cr and Mo, and has about the same atomic radius as Mo, no change in the unit-cell dimensions of the β^m-phase was observed when increasing amounts of Mo were replaced by W in the nitrided mixture and only additional reflexions of b.c.c. W and f.c.c. Ni(Mo,W) were obtained.   W has a lower affinity than Mo for N and so replacement of Mo by W might occur only at temperatures lower than $1200^\circ C$ or at N potentials higher than one atmosphere.   Even so, the thermal stability of the product would be expected to be too low for hardmetal applications.

Substitution of Mo by Nb and Ta

Increases in β^m unit-cell dimensions are observed for nitrided mixes with starting compositions $Ni_8Mo_{11}Nb$ and $Ni_8Mo_{11}Ta$ indicating that these small concentrations of both Nb and Ta can be incorporated in the structure.

Substitution of N by C

Pelletised powder mixes containing up to one mole of carbon-black per mole of β^m-Ni_2Mo_3N were heat-treated under molecular nitrogen in sealed tubes at $1000^\circ C$ and showed a uniform increase of β^m cell dimensions up to a limit of 50% replacement of N by C;   at higher C concentrations, other phases appeared.   The β^m carbonitrides were prepared directly by heating mixed powders of C, Ni and Mo in N_2 at $1000^\circ C$.

CONCLUSIONS

β^m-Ni_2Mo_3N and its derivatives are readily prepared by nitriding mixed metal or oxide powders in $N_2:H_2$ and the hardmetal products can be pressureless sintered with Ni and other binders to give fully dense and oxidation-resistant compacts with acceptable hardness, fracture toughness and wear-resistance.   Limited substitution of Ni in the β^m-Ni_2Mo_3N can

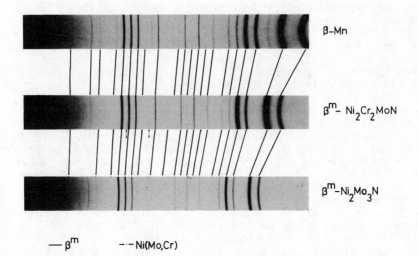

— β^m $--$ Ni(Mo,Cr)

Fig. 7   X-ray powder photographs of β^m-Ni$_2$Cr$_2$MoN and β^m-Ni$_2$Mo$_3$N compared with β-Mn

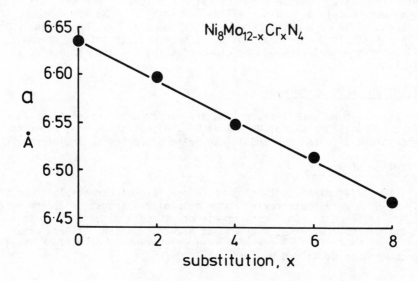

Fig. 8   Unit-cell dimensions of β^m-Ni$_8$Mo$_{12-x}$Cr$_x$N$_4$ with varying x values

be made by Fe and Co and of Mo by Nb and Ta. More remarkably, two-thirds of the Mo can be replaced by Cr and one-half of the N by C. These extensive replacements offer prospects of varying and improving the properties of the hard material.

ACKNOWLEDGMENTS

The work has been carried out with financial support from the Commission of the European Communities and the Wolfson Foundation. Presentation and publication of this paper has been assisted by the award of an Emeritus Fellowship to one of the authors (K.H.J.) by the Leverhulme Trust.

REFERENCES

Evans, D.A. and Jack, K.H. 1957, Acta Cryst. $\underline{10}$ 769
Goldschmidt, H.J. 1957, Acta Cryst. $\underline{10}$ 769
Grieveson P and Jack, K.H. Unpublished; see Grieveson P. 1960, M.Sc. Thesis, University of Durham.
Jack, K.H. 1967, British Iron and Steel Research Association Report MG/A/39/67
Jeitschko, W., Nowotny, H. and Benesovsky, F. 1964 Monatshefte für Chemie $\underline{95}$ 1212-1218
Kikuchi, M., Wakita, S. and Tanaka, R. 1973 Trans. I.S.I.J. $\underline{13}$ 226-228
Kuo, K. 1956 Trans. AIME, J. Metals $\underline{8}$ 97
Nutter, K.M., Grieveson, P. and Jack, K.H. Unpublished; see Nutter, K.M. 1969, Ph.D. Thesis, University of Newcastle upon Tyne

Inst. Phys. Conf. Ser. No. 75: Chapter 6
Paper presented at 2nd Int. Conf. Science Hard Mater., Rhodes

537

Effects of TiN on microstructure and properties of WC-based cemented carbides

T IGARASHI, M KOBAYASHI AND S TAKATSU

Toshiba Tungaloy Co Ltd, 8 Shinsugita, Isogo, Yokohama, Japan

ABSTRACT

The effect of TiN on the microstructure and properties has been
determined in (W,Ti,Ta)(C,N) solid solutions and in cemented carbides.
The cemented carbides exhibit an apparent **maximum** in transverse rupture
strength and a decrease in grain-size with increase in TiN content.
The fine grain-size is benefical in interrupted steel cutting. However
the substoichiometric carbonitride is harmful to performance in continuous
steel-cutting even though it improves the TRS.

INTRODUCTION

In WC-based cemented carbides for steel cutting tools, TiC, TaC and NbC
are added to improve the wear resistance and anti-adhesion. In general,
in spite of the good effects, the quantity of added carbides is kept
as low as possible because they cause a decrease in toughness of the
cemented carbides. The quantity of cubic carbides is changed according to
machining conditions, and the cemented carbides are classified into
several grades. In some applications, the conventional carbides have been
replaced by coated carbides and TiC-based carbides. The reason TiC-based
carbides are in wide use is that a fundamental improvement in toughness
has been achieved with the addition of nitrides. This study was performed
by Kieffer, Ettmayer and Freudhofmeier (1971) and followed by Rudy (1973).
Thus, TiC-based carbides containing nitrides exhibit good cutting
performance and are largely replacing P10 and P20 WC-based carbides.

Recently, Suzuki, Taniguchi and Hayashi (1982a) have studied properties
of WC-based carbides containing TiN and demonstrated a higher strength
and a better resistance against oxidation (Suzuki, Hayashi and Taniguchi,
1982b). However, the behaviour of TiN in carbides and its effect on the
cutting performance have been scarcely clarified.

Thus, it is interesting to examine the effects of TiN on properties of
both carbonitrides and the resultant WC-based cemented carbides. And it is
necessary to examine the change in cutting performance by added TiN in
order to make a further improvement in the reliability of WC-based
cemented carbides as is obtained in TiC-based cemented carbides.
Therefore, in this study, the effects of TiN on some properties and
cutting performance of WC-based cemented carbides were investigated by
changing TiN content.

EXPERIMENTAL PROCEDURE

Carbonitride Solid Solution

Solid solutions were prepared from WC, TiC, TaC and TiN in order to determine properties, and to add TiN successfully to WC-based carbides. The compositions were as indicated in Table 1. Namely, in pseudo-ternary solid solutions, the metal elements were fixed to W/Ti/Ta=0.189/0.619/ 0.192 in atomic ratio, while the non-metal elements were varied. Pseudo-binary solid solutions were composed of 70WC/30TiC double carbide and TiN.

The solid solutions were produced by hot pressing mixed powders at 2073 K at 20 MPa in vacuo. After hot pressing, the chemical analysis was carried out on the solid solutions and the results are indicated in Table 2.

TABLE 1 Nominal Compositions of Prepared Solid Solutions

Type	Code	N/(C+N)	Compositions in wt%
Ternary	A0	0	33.4WC-33.3TiC-33.3TaC
	A1	0.1	33.3WC-27.9TiC- 5.6TiN-33.2TaC
	A2	0.2	33.2WC-22.5TiC-11.1TiN-33.2TaC
	A3	0.3	33.2WC-17.1TiC-16.6TiN-33.2TaC
Binary	B1	0.17	63WC-27TiC-10TiN
	B2	0.32	56WC-24TiC-20TiN
	B3	0.44	49WC-21TiC-30TiN

TABLE 2 Chemical Compositions (wt%) of Solid Solutions

Code	W	Ti	Ta	Nb	Co	Fe	Ccomb	N	O
A0	31.9	26.4	30.6	0.02	0.04	0.02	10.54	0.02	0.15
A1	32.1	26.1	30.6	0.02	0.05	0.02	9.36	1.25	0.17
A2	31.9	26.2	30.4	0.02	0.05	0.02	8.57	2.23	0.17
A3	32.1	26.0	30.7	0.02	0.04	0.02	7.69	2.99	0.18
B1	58.8	29.5	---	---	0.05	0.02	9.06	2.01	0.16
B2	52.9	34.1	---	---	0.05	0.02	8.10	4.17	0.16
B3	46.0	41.4	---	---	0.05	0.02	7.05	5.05	0.19

Wettability by Co-WC eutectic, high temperature hardness and thermal conductivity were determined on hot pressed (W,Ti,Ta)(C,N) solid solutions in which the combined total of porosity and free graphite content was less than 2% by volume.

For wettability determinations, the sessile drop method was adopted and the contact angle was measured. In this case, the experiments were performed at 1623 K for 10 min in vacuo.

For high temperature hardness determination, the **Vickers micro-hardness** was measured at room temperature, 873 K, 1073 K and 1273 K with a load of 1.96 N.

For thermal conductivity determination, the laser beam flash method was adopted and the measurements were performed at temperatures of 293 K and 1273 K.

Cemented Carbides

WC-based carbides corresponding to ISO-P10, P20 and M10 were prepared using solid solutions.

The compositions of mixed powders were as follows:-
For pseudo-ternary solid solutions,
    P10; WC-38(TiC+TaC+TiN)-9Co,
    P20; WC-20(TiC+TaC+TiN)-9Co,
    M10; WC-10(TiC+TaC+TiN)-6Co, in wt%.
For pseudo-binary solid solutions,
    P10; WC-70(solid solution +TaC)-10Co,
    M10; WC-22(solid solution +TaC)- 7Co, in vol%.

Corresponding grades containing TiN were also made for comparison purposes. In this case, the mixing was carried out with 70WC/30TiC solid solution, pure TaC and TiN.

All were sintered at 1693 K for 50 min in vacuo and the carbon content was controlled to a medium carbon in the normal alloy. Hereafter, the code names used for the cemented carbides are the same as that of the solid solutions shown in Table 1, where the code names of pseudo-binary solid solutions and TaC content in wt% in the carbides are indicated together with a hyphen and "F" is used for carbides with pure TiN.

The metallographic analysis was performed mainly with a Scanning Electron Microscope by Backscattered Electron Image composition analysis for as-polished carbides and by Secondary Electron Image for etched carbides.

The mechanical properties of cemented carbides measured were:-
(1) The transverse rupture strength (TRS) was determined by the JIS method, ie, span/width/thickness=20/8/4mm and 3-point bending. The mean TRS value was determined by a Weibull analysis of 22 test bars.
(2) The chevron notched short bar fracture toughness (K_{IC}) was determined and 4 test bars were tested to obtain the average values.

Cutting tests, in which the inserts shaped SPG-422 (positive rake, whole ground type) were carried out on common steels and a cast iron. Details of the cutting conditions are given in the figures.

RESULTS AND DISCUSSION

Properties of Solid Solution

Fig.1 shows the variation in contact angle between the eutectic Co-WC and the pseudo-ternary solid solutions with nominal N/(C+N) ratio. The eutectic Co-WC droplet was used in this experiment because a liquid phase with near eutectic Co-WC composition should be generated in the actual sintering process. TiC is wetted far better than TiN by pure Ni (Ramqvist, 1965; Panasyuk, 1981) and the same result is expected for pure Co. Notwithstanding, the wettability of eutectic Co-WC increases with increase in N/(C+N) ratio of solid solutions as indicated in Fig.1. Indeed this result seems to be strange, but it is consistent with the results of Ramqvist (1965) and Kieffer, Ettmayer and Freudhofmeior (1971), where they found a better wettability for Co with decreasing atomic ratio of non-metal to metal and with carbonitrides of low N/(C+N) ratio. That is, the solid solutions show a deficiency of non-metal elements as can be estimated from Table 2, besides the denitriding of solid solutions occurs in larger quantity with increasing N/(C+N) ratio during heating as referred to later. So it should be considered that the present results are due to the substoichiometric effect of the solid solutions. However, it is still uncertain in respect of the influence of oxygen to wettability as pointed out by Tähtinen and Tikkanen (1979) because the present experiments were performed in vacuum of pressure range of 2 to 8 x 10^{-3} Pa.

Fig.2 shows the high temperature hardness of the pseudo-ternary solid solutions. Compared with the results on a single crystal of (W,Ti)C solid solution (Rowcliffe, 1981), the present results are considerably lower. A similar low hardness is found in hot pressed TiC (Miracle and Lipsitt, 1983). According to Lee (1983), the porosity of solid solutions used in present experiments could have reduced the hardness. Nevertheless, the present results show a similar tendency in temperature dependence of hardness as (W,Ti)C and show good accordance with the ratio of softer TiN in solid solutions at room temperature. Consequently it should be considered that the results are still useful for a qualitative comparison.

From the results, it is found that the solid solution with a lower N/(C+N) ratio exhibits a higher hardness at room temperature but a lower hardness at elevated temperature. Namely, TiN seems to improve the high temperature hardness of solid solutions by substituting TiC in spite of its low hardness at room temperature.

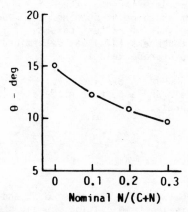

Fig. 1 Effect of nominal N/(C+N) ratio of (W,Ti,Ta)(C,N) on contact angle (θ) with eutectic Co–WC (Co–45WC) at 1623 K in vacuo.

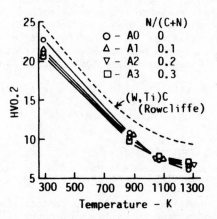

Fig. 2 Effect of nominal N/(C+N) ratio of (W,Ti,Ta)(C,N) on high temperature hardness of (W,Ti,Ta)(C,N).

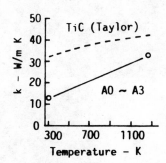

Fig. 3 High temperature thermal
conductivity (k) of (W,Ti,Ta)(C,N).

The thermal conductivity of pseudo-ternary solid solution was 13 ± 0.5
W/m K at 293 K and 33 ± 0.5 W/m K at 1273 K for each specimen. These
results are shown in Fig.3 with the results for pure TiC (Taylor, 1961).
According to Toth (1971), pure TiC, TaC and TiN exhibit almost the same
value of thermal conductivity. However, the present results are far lower
than that of pure TiC especially at room temperature. According to
Radosevich and Williams (1970), a strong point-defect scattering of phonon
has an important effect upon the thermal conductivity at high temperature.
In these pseudo-ternary solid solutions, no dependence of thermal
conductivity was observed on N/(C+N) ratio, so it is considered that the
extremely low and similar value for each solid solution was caused by
mutually dissolved W, Ti and Ta atoms.

Cemented Carbides

Some typical microstructures of cemented carbides corresponding to P10
are indicated in Fig.4 and 5. Since the SEM photograph represents the
composition image, the white, gray and black phases generally correspond
to WC, carbonitride and Co, respectively. However the cemented carbides
made from pure TiN also show black grains in the photograph, the black
grains are residual TiN as indicated by X-ray diffraction pattern. And
the residual TiN seems to form a Ti(C,N) because the peak position shifts
to the TiC side.

As indicated in Fig.4, when the total composition of (TiC+TaC+TiN) are
fixed by weight, the amount of free WC increases with incresing N/(C+N)
ratio because the miscibility limits of WC with carbonitride decreases
with increasing N/(C+N) ratio. The vol% WC determined metallographically
is shown in Fig.6.

Considering the grain size of the carbonitrides in Fig.4 and 5, it is
obvious that the carbonitride becomes finer with increasing N/(C+N) ratio
and with addition of TiN in the mixed powder. The measured grain size is
shown in Fig.7 for P10 grade carbides. In the present results, the effect
of TaC on grain size is far less than that of TiN.

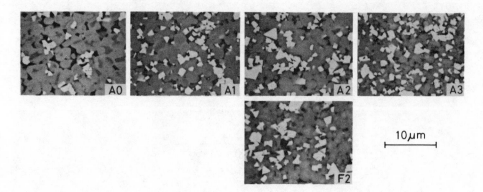

Fig. 4 Microstructures of P10 carbides made from (W,Ti,Ta)(C,N).

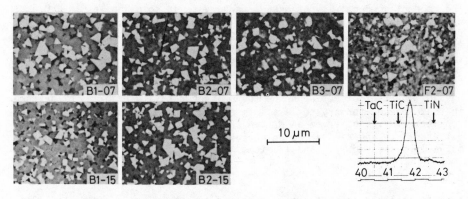

Fig. 5 Microstructures of P10 carbides made from (W,Ti)(C,N) and TaC.

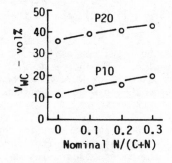

Fig. 6 Effect of nominal N/(C+N) ratio of (W,Ti,Ta)(C,N) on
volume fraction of WC (V_{WC}) in sintered carbides.

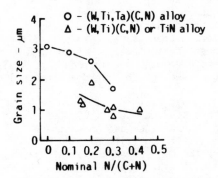

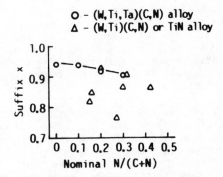

Fig. 7 Effect of nominal N/(C+N) ratio of cubic phase on carbonitride grain size in sintered carbides.

Fig. 8 Effect of nominal N/(C+N) ratio of cubic phase on estimated suffix x in Ti(C,N)x in sintered carbides.

Considerable denitriding occurs **during a vacuum** sintering process and this appears notably in carbides with a higher TiN content. Also the supply of carbon is not sufficient to compensate for the loss of nitrogen as shown in Fig.8. The suffix in Ti(C,N)x is estimated from C and N content assuming that W and Ta exist fully in forms of their stoichiometric carbides and Co does not dissolve neither C nor N.

Results for the mechanical properties, TRS and K_{IC}, of the carbides made from pseudo-ternary solid solutions are shown in Fig.9. In this figure, the TRS shows a maximum at a certain N/(C+N) ratio. The reason suggested is that carbide powder is sintered easily with addition of solid solution

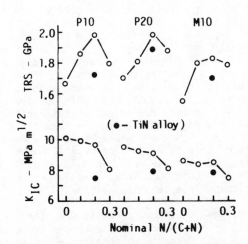

Fig. 9 Effects of nominal N/(C+N) ratio of (W,Ti,Ta)(C,N) on the transverse rupture strength (TRS) and fracture toughness (K_{IC}) of carbides.

with a higher N/(C+N) ratio because of good wettability between carbonitride solid solution and liquid phase as mentioned before, while micro-pores, which damage the TRS of cemented carbides, increase by denitriding of carbonitride in the sintering process: as a result of these opposite effects, the TRS shows an apparent maximum at a certain N/(C+N) ratio. However, when pure TiN is added to carbides, the increase in TRS is very small. And a similar tendency is observed for the pseudo-binary solid solutions.

Suzuki, Taniguchi and Hayashi (1982a) have studied the effect of (W,Ti)(C,N) on the TRS of WC-based cemented carbides and demonstrated a higher strength of the cemented carbides with (W,Ti)(C,N) than conventional cemented carbides. However, the high strength was attained after hot isostatic pressing, while a low strength was obtained in the same carbides in the as-vacuum-sintered state because of very large pores. And they concluded that this high strength was due to the fine carbonitride structure.

Though TRS, which is strongly affected by porosity, shows a maximum, K_{IC} decreases monotonically with increasing N/(C+N) ratio. This decrease is considered to be mainly due to the fineness in microstructure. The absolute influence of N/(C+N) ratio on K_{IC} is not known from the present results because K_{IC} is considered to be also sensitive to microstructure features such as Co mean free path and contiguity of the carbide phase.

Cutting Performance

In continuous steel cutting with P10 and P20 cemented carbides, the crater and flank wear resistance are plotted against the nominal N/(C+N) ratio of cubic phase in Fig.10a. A poorer resistance against crater wear is shown with increase in N/(C+N) ratio. However, the results are unexpected because TiN is considered to be a more effective addition against crater wear than TiC.

Crater wear is caused by the reaction between the tool and steel chip, and an increase in WC phase generally leads to greater crater wear because of a strong affinity between WC and steel. Therefore, a P10 grade is superior to a P20 grade in respect of the crater wear resistance. Similarly, it is considered that crater wear increases if the carbonitride phase has a high affinity for steel. As mentioned before, wettability increases with increasing N/(C+N) ratio, and the substoichiometric effect is considered to be a cause. Thus, the results are replotted against the suffix x in Ti(C,N)x as shown in Fig.10b. In this case, the crater wear depends on the suffix x very well and the **result can be interpreted as a decrease in suffix x representing a higher affinity.** The reason for the high affinity is considered to derive from the weakened bonding force by vacancies in the lattice due to deficiency of the non-metal elements and is paralleled in other properties such as a decrease in melting point and an increase in diffusivity in similar cubic compounds. This large decrease in the suffix x is not found in nitride-free cemented carbides.

In contrast to the above crater wear resistance, the flank wear resistance is almost the same for all the cemented carbides. However, the flank wear is considered to be mainly caused by abrasive wear, so the flank wear has been replotted with hardness in Fig.10c. This hardness dependence seems to support the abrasive wear mechanism, but it is not applicable to the high hardness cemented carbides which show poor crater wear resistance. In this case, therefore, it is considered that the above mechanism for crater wear contributes strongly to the flank wear, and the high hardness cemented carbides with poor crater wear resistance suffer a large flank wear.

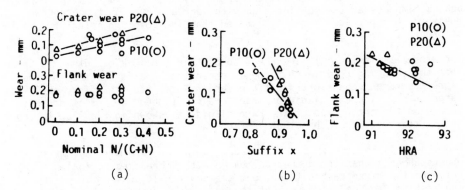

(a)   (b)   (c)

Fig. 10 Wear resistance in continuous dry turning.
    (a) Effects of nominal N/(C+N) ratio of cubic phase
       on crater wear depth and flank wear width.
    (b) Correlation between crater wear depth and estimated
       suffix x in Ti(C,N)x.
    (c) Correlation between flank wear and hardness.

Work material; Carbon steel JIS S48C (AISI 4047H), HB240
Tool; P11R-44 + SPG-422 (0.05 x 30°)
V=130m/min, d=1.5mm, f=0.3mm/rev, T=10min

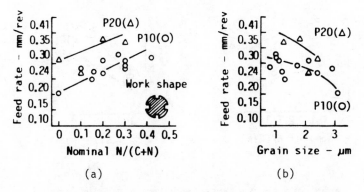

(a)   (b)

Fig. 11 Chipping resistance in interrupted dry turning.
    (a) Effect of nominal N/(C+N) ratio of cubic phase.
    (b) Correlation between chipping resistance and
      grain size.

Work material; Carbon steel JIS S48C (AISI 4047H), HB250
Tool; P11R-44 + SPG-422 (0.05 x 30°)
V=100m/min, d=1.5mm, f= varied from 0.1mm/rev
4000 impacts for each feed rate.

Fig.11a shows the chipping resistance in interrupted dry turning. In the present work, chipping resistance is improved with increasing nominal N/(C+N) ratio of the mixed carbonitride. And some of P10 grades exhibit the same resistance as that of a nitride-free P20 cemented carbide.

To understand the chipping resistance, it is necessary to know the strength against impacts at the cutting edge with wear and at high temperature. However, this problem is so complicated that the solution is not available at present.

Therefore, the effect of TiN is not completely explicable, but it is known that the chipping resistance strongly depends on characteristics of the microstructure such as phase ratio or grain size. And the effect of TiN is large on the fineness of carbonitride grains. Indeed the resistance does depend on the grain size of the carbonitride as shown in Fig.11b, though the scatter is larger than that in Fig.11a.

Fig.12a shows the number of thermal cracks in dry face milling. It is understandable that the number decreases with increasing nominal N/(C+N) ratio of the mixed carbonitride, but the scatter is considerably large. The results are replotted against the grain size of carbonitride in Fig.12b and show a better correspondence. In other words, the number of thermal cracks depends on the grain size of carbonitride. The reason is considered to be that the thermal stress caused by large carbonitride or largely agglomerated carbonitride grains is reduced when they dispersed finely and uniformly, and this uniformity is observed with a low suffix x or a high N/(C+N) ratio.

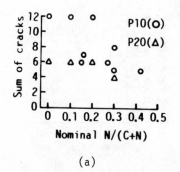
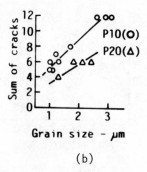

(a)   (b)

Fig. 12 Thermal crack resistance in dry face milling.
   (a) Effect of nominal N/(C+N) ratio of cubic phase.
   (b) Correlation between number of thermal cracks and
        carbonitride grain size.

Work material; Cr-Mo steel JIS SCM440 (AISI 4140H), HB280
                50W x 200L
Tool; PD1006R fly cutter + SPG-422 (0.1 x 30°), single insert
V=160m/min, d=2mm, f=0.2mm/tooth, 5 passes

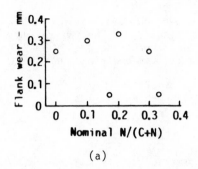
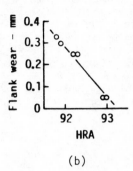

(a)   (b)

Fig. 13 Flank wear resistance in dry turning of gray cast iron.
(a) Effect of nominal N/(C+N) ratio of cubic phase.
(b) Correlation between flank wear width and hardness.

Work material; Gray cast iron JIS FC35
             (tensile strength 343 MN/m^2)
Tool; P11R-44 + SPG-422 (0.05 x 30°)
V=130m/min, d=1.5mm, f=0.3mm/rev, T=10min

For M10 grade carbides, the cutting test has been performed with dry turning of cast iron. The results are shown for the flank wear resistance in Fig. 13a. No correlation is observed. However, an obvious correlation appears when the results are replotted with the hardness as shown in Fig.13b. The effect of TiN on the flank wear resistance is uncertain because the hardness is dominated by the grain size of WC and the Co content in M10 grade.

CONCLUSIONS

In a determination of the properties of carbonitride solid solutions, the following important results were obtained:-
(1) The wettability of (W,Ti,Ta)(C,N) by eutectic Co-WC increases with increase in the N/(C+N) ratio in vacuo. In this case, substoichiometric effect seems to be important.
(2) At temperatures above 1100 K, (W,Ti,Ta)(C,N) increases in hardness with increase in the N/(C+N) ratio. The trend is reversed at lower temperatures. Thus the solid solution should be suitable for making a tool material. However this was not demonstrated convincingly in the cutting tests.

In the determination of the properties of the cemented carbides, the following effects of TiN (or (W,Ti,Ta)(C,N)) were observed:-
(1) The TRS shows an apparent maximum at a certain N/(C+N) ratio. This is considered to be due to the positive effect of good wettability and fine grained carbonitride and the negative effect of pores due to denitriding.
(2) K_{IC} decreases with increasing N/(C+N) ratio. However, this result was not related to cutting performance.

In the determination of the cemented carbides, the following effects of TiN (or (W,Ti,Ta)(C,N) were observed:-
(1) For steel **cutting** grades, TiN improves chipping and thermal crack resistance but reduces crater wear resistance. The improvements seem to be due to mainly the fineness of carbonitride grains. And the reduction seems to be due to the substoichiometric effect of carbonitride.
(2) For cast iron cutting grades, the effect of TiN was not significant because the amount of TiN was small in relation to WC.

REFERENCES

Kieffer, R., Ettmayer, P. and Freudhofmeier, M., 1971, Modern Developments in Powder Metallurgy. Plenum Press, New York, Vol 5, pp 201-214
Lee, M. 1983, Met. Trans. 14A 1625-1629
Miracle, D. B. and Lipsitt, H. A. 1983, J. Am. Ceram. Soc. 66 592-597
Panasyuk, A. D. 1981, Soviet Powder Met. Met. Ceram. 20 639-642
Radosevich, L. G. and Williams, W. S. 1970, J. Am Ceram. Soc. 53 30-33
Ramqvist, L. 1965, Int. J. Powder Met. 1 2-21
Rowcliffe, D. J. 1981, Proc. 1st Int. Conf. on Science of Hard Materials. Plenum Press, New York, pp 592-597
Rudy, E. J. 1973, J. Less-Common Met. 33 43-70
Suzuki, H., Taniguchi, Y. and Hayashi, K. 1982a, J. Japan Soc. Powder and Powder Met. 29 25-29
Suzuki, H., Hayashi, K. and Taniguchi, Y. 1982b, J. Japan Soc. Powder and Powder Met. 29 256-260
Tähtinen, K. and Tikkanen, M. H. 1979, Powder Met. Int. 11 80-82
Taylor, R. E. 1961, J. Am. Ceram. Soc. 44 525-526
Toth, L. E. 1971, Transition Metal Carbides and Nitrides. Academic Press, New York, pp 1-28

Inst. Phys. Conf. Ser. No. 75: Chapter 6
Paper presented at 2nd Int. Conf. Science Hard Mater., Rhodes

549

Binder phase strengthening through γ' precipitation in (Ti,W)C-base hardmetals

H DOI AND K NISHIGAKI

Central Research Institute, Mitsubishi Metal Corporation
Omiya, Saitama, Japan

ABSTRACT

The effect of γ' precipitation in the binder phase of $(Ti_{0.3}W_{0.7})C$-25.3%Ni-12.7%Co-0.57%Cr-(0~3.52)%Al hardmetals upon hardness, tranverse rupture strength, compressive strengths and total impact energy, and oxidation resistance was investigated. It was found that increased additions of Al considerably decrease the binder W content and contribute to coarsening of (Ti,W)C grains through an Ostwald ripening process. Addition of optimum amounts of Al (4-5% in binder) increases conspicuously both strength and toughness. Oxidation resistance at 1000°C is also improved due to formation of a thin protective (Al, Ti, Cr)-oxide layer.

INTRODUCTION

In previous work (Doi and Nishigaki, 1977; Nishigaki, Oosawa and Doi, 1979; Nishigaki, Yoshimura and Doi, 1980), the effect of Al additions to the binder phase of such cermets as TiC-15%Ni-8%Mo, $TiC_{0.7}N_{0.3}$-15%Ni-8%Mo on the mechanical properties was reported. The motivation for these studies was to improve mechanical characteristics such as strength, hardness, and wear resistance, etc. through γ' precipitation. The result confirmed the validity of such a concept; γ' precipitation produced significant increase in hardness, wear resistance, and in some cases the TRS. It was also shown that incorporation of potent solid-solution hardeners into the binder phase may be also effective in strengthening the cermet (Nishigaki and Doi, 1980; Yoshimura, Sugisawa, Nishigaki and Doi, 1981).

The present work is a continuation of the previous one on strengthening of cermets through precipitation. Specimens with chemical composition of $(Ti_{0.3}W_{0.7})C$-25.3%Ni-12.7%Co-0.57%Cr-(0~3.52)%Al were prepared and various mechanical and other properties were measured. A comparatively large volume fraction of the binder phase was chosen in order to distinctly demonstrate the effect of various phenomena relating to precipitation. Replacement of 1/3 of Ni with Co was intended for improving high temperature strength. A small amount of Cr was added to the binder phase to increase the oxidation resistance. The hard phase consists of a solid solution of TiC and WC. The maximum incorporation of WC into the lattice of TiC was intended to give various desirable properties for the wear-resistant tool applications. Chemical compositions are always expressed in wt% unless otherwise stated.

EXPERIMENTAL PROCEDURES

Predetermined amounts of $(Ti_{0.3}W_{0.7})C$, Ni, Co, AlN and Cr_2N powders were blended to give the desired compositions. Mechanical pressing of the powders, sintering and post heat-treatment were carried out as shown in Fig.1. Al and Cr were added in the form of chemically stable AlN and Cr_2N in order to avoid oxide-formation during processing. Nitrogen was found to diffuse away completely from the specimen in the course of sintering.

Microscopic observations were made on diamond-polished and electrolytically etched (10%KOH) specimens. Precipitation of γ particles was observed with as-sintered or variously heat-treated specimens which were subjected to etching in a solution containg 60%HCl, 20%HNO$_3$ and 20%glycerine as the etchant. Solution treatment of as-sintered specimens was carried out in air for 2h at 1150°C and fan-cooled. Aging was carried out in an evacuated ampule for 1~300h at 600, 700 and 800°C.

X-ray diffraction technique and SEM apparatus were utilized for identifying various phases of as-sintered and heat-treated specimens, measuring relative intensities of γ and γ' phases and obtaining concentration profiles of various elements in (Ti,W)C and binder phases. Mechanical properties such as TRS, compressive strengths, and hardness were carried out following the standard procedures. For the evaluation of the toughness of hardmetals, an instrumented Charpy impact machine with a standard capacity of 100N m was used. The shapes of the specimens for bending, compression and Charpy impact tests were 24x8x4mm, 4mm x10mm, and 4x3x40mm, respectively.

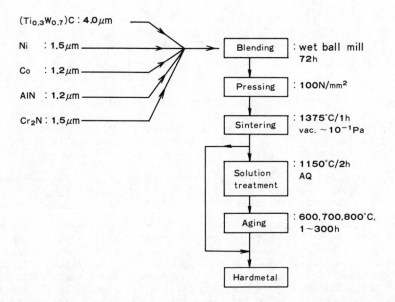

Fig.1 Manufacturing process of $(Ti_{0.3}W_{0.7})C$-Ni-Co-Cr-Al hardmetals.

RESULTS AND DISCUSSION

Table 1 shows the composition of the sintered hardmetals. 44~50% in volume of the hardmetal is occupied by the binder phase. Binder Al content varied up to 0~7.7% (0-22vol%). Considerable amounts of W contained in the original (Ti,W)C grains would have dissolved in the binder phase as will be discussed later, while the binder Ti contents were much smaller.

Fig.2 shows the microstructure of representative (Ti,W)C base hardmetals with various Al contents. Increasing the Al content leads to significant carbide grain coarsening.

X-ray diffraction profiles of as-sintered (Ti,W)C base hardmetals with various Al contents are shown in Fig.3. Precipitation of γ' particles ($Ni(Co)_3Al$) becomes noticeable with hardmetals containing more than 1.33%Al. It is to be noted that with an increase of Al additions small diffraction peaks corresponding to the existence of WC phase appear.

Variation of the microstructural features of γ' precipitation in the binder phase of as-sintered (Ti,W)C base hardmetals with the amount of Al addition are shown in Fig.4. γ' particles are uniformly small with the 1.73%Al containing hardmetal. However, when the amount of Al addition exceeds 2.13%, precipitation becomes characterised by the formation of a duplex structure consisting of fine and coarse particles; the proportion of coarse particles rapidly increases with increasing Al addition.

TABLE 1 Chemical Compositions of $(Ti_{0.3}W_{0.7})$C-25.3%Ni-12.7%Co-0.57%Cr-(0~3.52)%Al Hardmetals

Upper row : wt.%
Lower row : vol.%

Number of specimen	Carbide phase (Ti,W)C	Binder phase					Total binder content	Al content in binder
		Ni+Co	Cr	Al	Ti	W		
1	56.50	38.05	0.56	—	0.11	4.78	43.5	—
	56.25	40.40	0.74	—	0.23	2.34	43.71	—
2	56.10	38.13	0.58	0.47	0.11	4.61	43.9	1.07
	55.16	40.0	0.75	1.62	0.22	2.23	44.82	3.61
3	55.96	38.24	0.58	0.89	0.10	4.23	44.04	2.02
	54.36	39.65	0.74	3.04	0.20	2.02	45.65	6.66
4	55.62	38.35	0.57	1.33	0.09	4.04	44.38	3.00
	53.39	39.31	0.73	4.49	0.18	1.90	46.61	9.63
5	55.40	38.46	0.58	1.73	0.08	3.75	44.60	3.88
	52.61	38.99	0.72	5.77	0.16	1.75	47.39	12.18
6	55.03	38.58	0.58	2.13	0.08	3.60	44.97	4.74
	51.72	38.71	0.71	7.03	0.16	1.67	48.28	14.56
7	54.94	38.67	0.57	2.60	0.07	3.15	45.06	5.77
	50.96	38.30	0.70	8.47	0.14	1.43	49.04	17.27
8	54.74	38.73	0.55	3.17	0.05	2.76	45.26	7.00
	50.02	37.78	0.69	10.17	0.10	1.24	49.98	20.35
9	54.92	38.86	0.56	3.52	0.04	2.10	45.08	7.81
	49.64	37.49	0.69	11.17	0.08	0.93	50.36	22.18

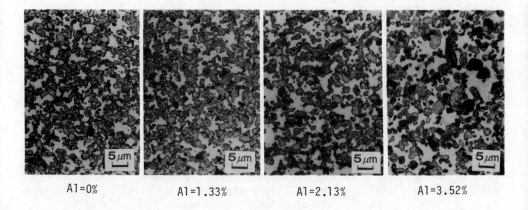

Al=0%   Al=1.33%   Al=2.13%   Al=3.52%

Fig.2 Photographs of $(Ti_{0.3}W_{0.7})C$-25.3%Ni-12.7%Co-0.57%Cr-
(0~3.52)%Al hardmetals.

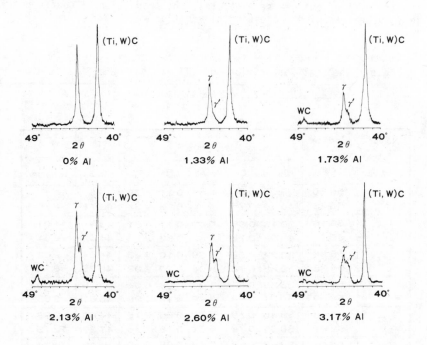

Fig.3 X-ray diffraction analyses of $(Ti_{0.3}W_{0.7})C$-25.3%Ni-12.7%Co-
0.57%Cr-(0~3.17)%Al hardmetals (as-sintered).

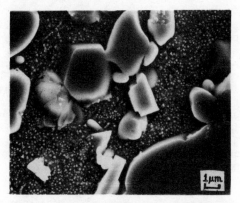

Al=1.73%

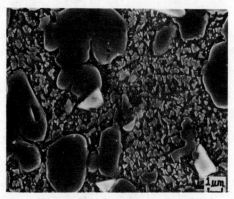

Al=2.13%

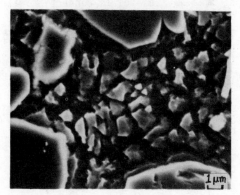

Al=3.52%

Fig.4 γ' precipitation in the binder phase of $(Ti_{0.3}W_{0.7})C-25.3\%Ni-12.7\%Co-0.57\%Cr-(1.73\sim3.52)\%Al$ hardmetals (as-sintered).

The effect of binder Al content of as-sintered hardmetals upon lattice parameters of the constituent phases, binder W and Ti contents and $\gamma'/(\gamma+\gamma')$ ratio is shown in Fig.5. It is apparent from the figure that binder W content considerably decreases with increasing Al content. A similar trend is also noticed with binder Ti content which is in general much lower than binder W content. The lattice parameter of γ phase first increases rather rapidly and then decreases with increasing Al content, whereas that of the carbide phase decreases only slightly. A comparison of the variation of the γ phase lattice parameter with variation of binder W content and $\gamma'/(\gamma+\gamma')$ ratio indicates that the initial increase of γ phase lattice parameter is caused by solution of Al, while its considerably rapid decrease above 2% in Al content can be explained by the reduction of both binder W content and binder Al content due to γ' precipitation.

Fig.6 shows the aging characteristics of $(Ti_{0.3}W_{0.7})C$-25.3%Ni-12.7%Co-0.57%Cr -2.13%Al. Aging at 600°C for 50h gives rise to the maximum hardness. On the other hand, aging for more than 50h at 800°C causes overaging.

Fig.7 shows precipitation of γ' particles in 600°C/50h, 700°C/10h, and 800°C/10h aged hardmetals. The precipitated γ' particles are in every cases uniformly small; γ' particles precipitated at 800°C are considerably larger than those precipitated at either 600°C or 700°C. Fig.8 shows that the size of particles in the overaged hardmetals is 0.1∿0.2μm. A comparison with Fig.4 indicates that precipitated γ' particles in as-sintered specimens are much coarser than those of solution-treated and aged specimens.

Fig.9 shows the effect of aging at 600, 700 and 800°C of $(Ti_{0.3}W_{0.7})C$-25.3%Ni-12.7%Co-0.57%Cr-2.13%Al hardmetal on the X-ray diffraction profile. It can be seen that the appearance of a γ' diffraction peak with aging time is rather slow; aging at 600°C for less than 10h **does not give rise to a distinct γ'** diffraction peak. After aging at the temperature for 300h, however, asymmetric line broadening of a γ diffraction peak which is indicative of the occurrence of γ' precipitation becomes noticeable. In connection with this, the trend of the age hardening curves shown in Fig.7 indicates that prolonged heating at this temperature is most effective for hardening. It appears that an initial stage of the precipitation of very fine γ' particles is effective for strengthening the binder phase.

Figs.10 and 11 show EPMA topographs, SEI and concentration profiles of various elements in carbides as well as binder phase of $(Ti_{0.3}W_{0.7})C$-25.3%Ni-12.7%Co-0.57%Cr-(0∿2.13)%Al hardmetals. It is apparent from these figures that (Ti,W)C phase has a kind of cored structure with higher Ti and lower W concentrations in the periphery of the double carbide grains in accord with the case of (Ti,W)C-Co hardmetals (Warren,1972). This indicates that dissolution and reprecipitation of the carbide on the surface of the existing carbide grains has taken place during sintering (Ostwald ripening) (May,1971). Taking into account the identification of the appearance of WC phase in the as-sintered $(Ti_{0.3}W_{0.7})C$-25.3%Ni-12.7%Co-0.57%Cr-(1.73∿3.52)%Al hardmetals through X-ray diffraction technique. It can be concluded that the following phase decomposition reaction (May,1971) has occured in the presence of Ni+Co binder during sintering

$$(Ti,W)C^{ov} \longrightarrow (Ti,W)C^{eq} + WC$$

(eq=equilibrium phase, ov=oversaturated phase). It is to be noted that increasi

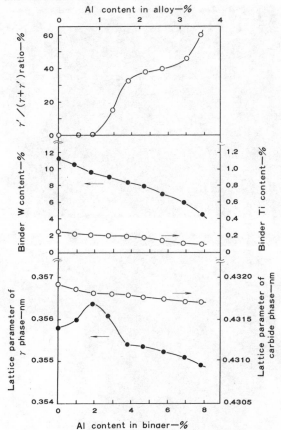

Fig.5 Effect of binder Al content upon lattice parameters of constituent phases, binder W and Ti contents and the amount of precipitation of $(Ti_{0.3}W_{0.7})C-25.3\%Ni-12.7\%Co-0.57\%Cr-(0\sim3.52)\%Al$ hardmetals.

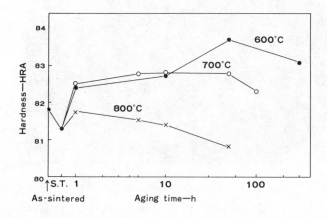

Fig.6 Aging characteristics of $(Ti_{0.3}W_{0.7})C-25.3\%Ni-12.7\%Co-0.57\%Cr$ -2.13%Al hardmetal.

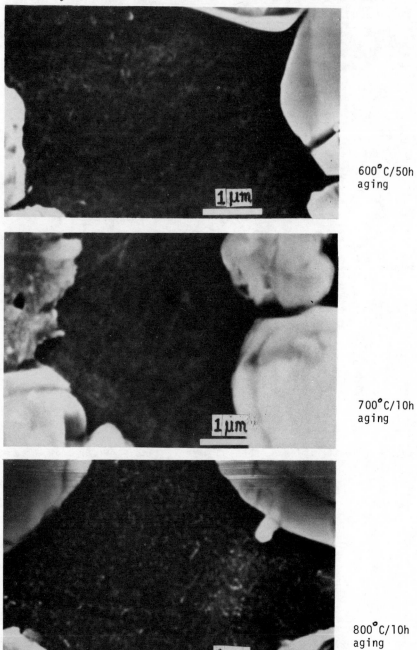

600°C/50h
aging

700°C/10h
aging

800°C/10h
aging

Fig.7 γ-precipitation in the binder phase of $(Ti_{0.3}W_{0.7})C-25.3\%Ni-12.7\%Co$
-0.57%Cr-2.13%Al hardmetal which was solution treated (1150°C/2h-
fan cooled) and aged under various conditions.

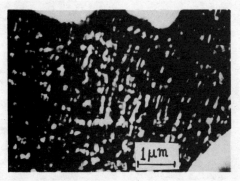

Fig.8 γ'-precipitation in the binder phase of $(Ti_{0.3}W_{0.7})C$-25.3%Ni-12.7%Co-0.57%Cr-2.13%Al hardmetal aged for 50h at 800°C after solution treatment.

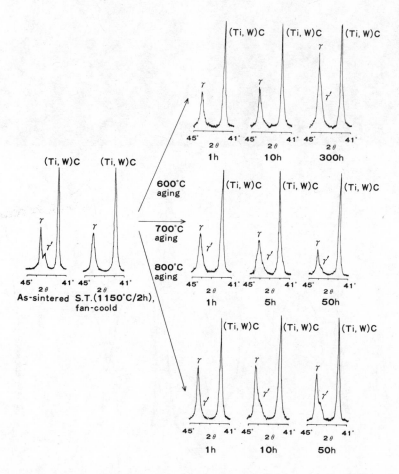

Fig.9 Effect of heat treatment on X-ray diffraction profile of $(Ti_{0.3}W_{0.7})C$-25.3%Ni-12.7%Co-0.57%Cr-2.13%Al hardmetal.

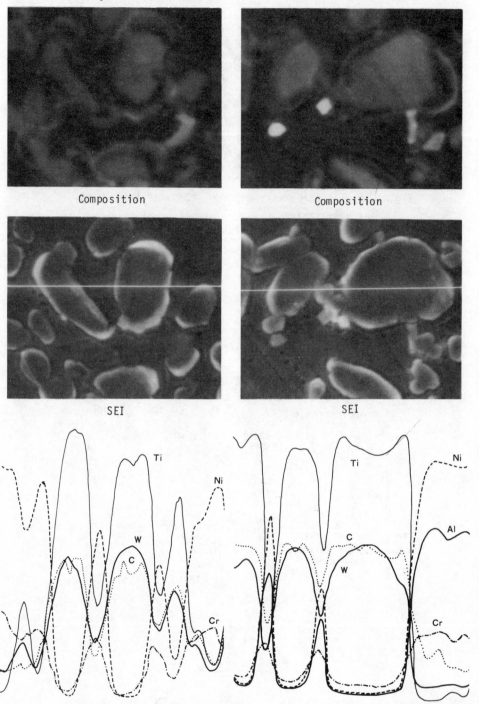

Fig.10 EPMA of $(Ti_{0.3}W_{0.7})C-25.3\%Ni-$
12.7%Co-0.57%Cr hardmetal.

Fig.11 EPMA of $(Ti_{0.3}W_{0.7})C-25.3\%Ni-$
12.7%Co-0.57%Cr-2.13%Al hardmetal

Al additions appear to enhance the above phase decomposition reaction. As shown in Fig.5, the amount of W in solution in the binder phase significantly decreases with increasing Al additions. This gives an explanation for an enhancement of precipitation of WC phase with increasing Al additions.

Fig.12 shows variations of the TRS and hardness with increasing binder Al content. It is apparent that γ' precipitation causes both of them to increase. The decrease of hardness for the hardmetal with increasing binder Al above 5% appears to indicate the effect of carbide coarsening. The effect of aging (700°C/10h) after solution treatment (1150°C/2h) is more clearly indicated with hardness than with TRS; the latter does not change significantly through heat treatment.

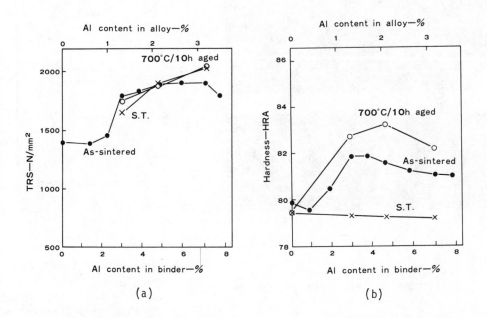

Fig.12 Effect of Al content upon TRS (a) and hardness (b) of $(Ti_{0.3}W_{0.7})C$-25.3%Ni-12.7%Co-0.57%Cr-(0~3.52)%Al hardmetals (as-sintered and after heat treated).

The effect of Al additions upon load-deflection curves in Charpy impact testing of (Ti,W)C base hardmetals is shown in Fig.13. Fig.14 shows the total absorption energy of (Ti,W)C base hardmetals plotted against binder Al content. An optimum amount of Al addition for giving the maximum absorption energy is about 2%, which is in accord with the result of TRS measurements (see Fig.12). An important point to note is that both toughness and hardness increase with an optimum amount of Al addition. Moreover, although toughness of the hardmetal somewhat increases with an adequate heat-treatment, the effect of precipitation during cooling from sintering temperature already gives rise to considerably large toughness and hardness.

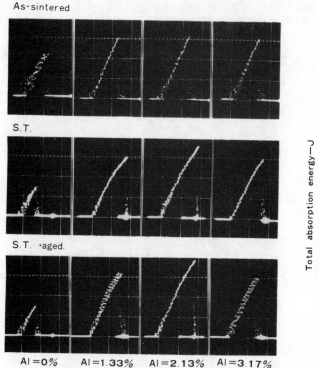

As-sintered

S.T.

S.T. aged.

Al=0%   Al=1.33%   Al=2.13%   Al=3.17%

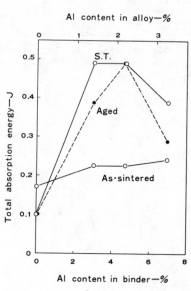

Al content in alloy—%

Total absorption energy—J

Al content in binder—%

Fig.13 Load-deflection curves obtained from
       instrumented Charpy impact tests of
       as-sintered, solution-treated and aged
       $(Ti_{0.3}W_{0.7})C-25.3\%Ni-12.7\%Co-0.57\%Cr-$
       $(0\sim3.17)\%Al$ hardmetals.

Fig.14 Effect of heat treatment of
       $(Ti_{0.3}W_{0.7})C-25.3\%Ni-12.7\%Co$
       $-0.57\%Cr-(0\sim3.17)\%Al$ hardmeta
       upon total absorption energy
       in Charpy impact tests.

The effect of binder Al content upon the compressive strength of the hardmetals
is shown in Fig.15. The compressive properties of the hardmetals were measured
after 700 C/10h aging. It can be seen that γ' precipitation leads to a marked
increase in compressive strength.

The effect of binder Al content upon oxidation resistance is shown in Fig.16.
Small amounts of Cr additions are believed to contribute considerably to
retard oxidation (Kieffer and Benesovsky,1965) although the relevant
experimental data are lacking. The addition of Al further improves the
oxidation resistance especially with increasing time of exposure. EPMA of
the surface layer of the oxidised hardmetals reveals that beneath the layer
of Ni and Co bearing oxides, Al-rich and (Ti,Cr) bearing oxide layer has been
formed, see Fig.17. It appears that this gives rise to an effective barrier
against oxidation.

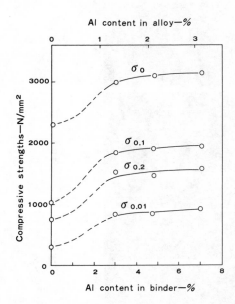

Fig.15 Effect of binder Al content upon compressive strength of $(Ti_{0.3}W_{0.7})C$ -25.3%Ni-12.7%Co-0.57%Cr-(0~3.17)%Al hardmetals (as-sintered and after heat-treated).

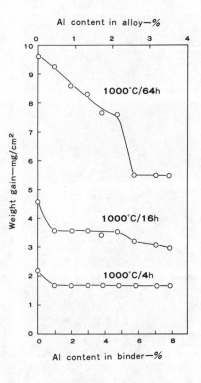

Fig.16 Effect of binder Al content upon oxidation resistivity of $(Ti_{0.3}W_{0.7})C$ -25.3%Ni-12.7%Co-0.57%Cr-(0~3.52)%Al hardmetals (as-sintered).

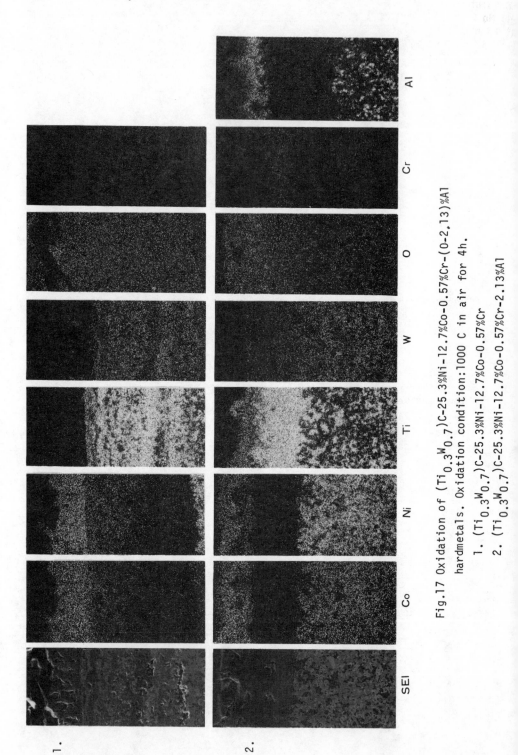

Fig.17 Oxidation of $(Ti_{0.3}W_{0.7})C-25.3\%Ni-12.7\%Co-0.57\%Cr-(0-2,13)\%Al$ hardmetals. Oxidation condition:1000 C in air for 4h.

1. $(Ti_{0.3}W_{0.7})C-25.3\%Ni-12.7\%Co-0.57\%Cr$
2. $(Ti_{0.3}W_{0.7})C-25.3\%Ni-12.7\%Co-0.57\%Cr-2.13\%Al$

SUMMARY

Strengthening of $(Ti_{0.3}W_{0.7})C$-25.3%Ni-12.7%Co-0.57%Cr-(0 3.52)%Al hardmetals due to γ' precipitation in the binder phase was investigated through measurements of hardness, TRS, compressive strength and total absorption energy in Charpy impact tests. The following results were obtained.

(1) γ' precipitation in the binder phase was observed in as-sintered hardmetals with binder Al contents of more than ~3%. The γ' precipitation was characterised by a duplex structure; coarse - grained γ' increased with increasing Al content.
(2) An increase in Al additions led to considerable coarsening of (Ti,W)C grains (Ostwald ripening).
(3) Dissolution of W into the binder phase from the periphery of (Ti,W)C grains was noted. Reprecipitation of W in the form of WC as the product of the phase decomposition reaction became enhanced with Al additions.
(4) Age-hardening treatment at 600°C for 50~100h gave rise to the highest hardness. A comparatively early stage of γ' precipitation appeared to be favourable for hardening.
(5) γ' precipitation due to Al additions increased TRS, compressive strength, and hardness of as-sintered (Ti,W)C base hardmetals. Heat treatment (1150 C/2h S.T. → Air cooling → 700°C/2h aging) further raised hardness, although the TRS remained almost the same.
(6) The toughness of as-sintered (Ti,W)C base hardmetals evaluated through Charpy impact tests increased considerably with Al additions; the optimum binder Al content was 5%.
(7) Addition of Al increased the 1000°C oxidation resistance of (Ti,W)C base hardmetals through the formation of a protective (Al, Cr and Ti)-oxide; the effect became more pronounced with 6% binder Al addition as the exposure time increased.

ACKNOWLEDGEMENT

The authors are indebted to Prof. Kobayashi of Toyohashi University of Technology for carrying out instrumented Charpy impact tests and for valuable discussions.

REFERENCES

Doi, H, and Nishigaki, K. 1977, Modern Development in P/M. 11 525-542
Kieffer, R, and Benevovsky, F. 1965, Hartmetalle, Springer, Wien,pp220-222
May, W. 1971, J. Mater. Sci. 6 1209-1213
Nishigaki, K., Oosawa, Y., and Doi, H. 1979, J. of Japan Soc. of Powder and Powder Metallurgy. 26 169-173
Nishigaki, K.,Yoshimura, H.,and Doi, H. 1980, J. of Japan Soc. of Powder and Powder Metallurgy. 27 50-55
Nishigaki, K, and Doi, H. 1980, J. of Japan Soc. of Powder and Powder Metallurgy. 27 130-136
Yoshimura, H., Sugisawa, T., Nishigaki, K., and Doi, H. 1981, Proc. of 10th Plansee Seminar. 2 HM16, 727-741
Warren, R. 1972, Planseeber. Pulvermetallurgie. 20 299-317

Inst. Phys. Conf. Ser. No. 75: Chapter 6
Paper presented at 2nd Int. Conf. Science Hard Mater., Rhodes

565

The influence of microstructure and composition on the mechanical behaviour of TiC based hardmetals

M KOMAK AND S NOVAK

J. Stefan Institute, E. Kardelj University, 61000 Ljubljana, Yugoslavia

ABSTRACT

Cemented titanium carbides with compositions $TiC-MoC_x-Ni$ and $TiC-NbC_x-Ni$ were studied with respect to their toughness and their behaviour under thermal shock conditions. By analysing the influence of particular compositional and microstructural parameters - composition, contiguity, mean grain size and grain structure of the carbide phase, the amount and mean free path of binder phase - similarities with WC materials could be established. Furthermore, it appears that improvement of TiC cemented carbides seems probable if by adjustment of composition and microstructure the interfacial cohesive forces, resistance to plastic deformation and thermal conductivity could be increased.

INTRODUCTION

Cemented titanium carbides are recognised as suitable tool materials for medium to high speed machining of steel. It has always been considered that the main obstacles to broader application especially in heavy machining are the disadvantages such as modest strength, low thermal conductivity as well as difficulties encountered in manufacturing a consistent product, which could result in variable and unpredictable performance (Lardner, 1970; Kalish, 1983).

TiC cemented carbides are less tough than straight WC-Co grades (Beger and co-workers, 1981). On the other hand, they exhibit equal or even higher fracture toughness than WC steel cutting grades (August and Kalish, 1983) but are lacking impact resistance and resistance to chipping. The efforts to overcome the deficiencies are substantial, however, it is our opinion that research and development work is to a considerable extent still empirical in nature. As compared to WC grades, the understanding of interrelationships between composition, microstructure, properties and performance of cemented titanium carbides is much less detailed, meaning that the basis for a scientific approach to material improvement is rather incomplete.

If high wear resistance of the tool material is the final goal, then it must be understood that this is not a material property, but rather a property of a complex tribological system. However, in order to simplify, tool wear can be generally classified into two main types, mechanical and thermochemical, and the main environmental factors are the temperature and the pressure at the tool - workpiece interface. These may be as high as 1500°C and 3500 MN m^{-2} (Perrot and Robinson, 1974). In discontinuous operations, the estimated thermal stresses may be in the range of 1500 MN m^{-2} (Loladze, 1975). Thus the magnitude of stresses may exceed the strength of the tool material, the failure being manifested as a fracture, either bulk or as a microfracture on worn surfaces.

Moreover, thermochemical mechanisms may result in the embrittlement of the binder, which promotes a decrease in fracture resistance of the material. Consequently, fracture resistance, irrespective of the origin of crack initiation, may become of major concern in the development of cemented carbide tools and of TiC grades especially. Thus a correspondence could be found between plots of tool life and transverse-rupture-strength (TRS) respectively vs. binder compositions in TiC-22.5 Ni-MoC$_x$ materials (Moskowitz and Humenik, 1978). Further, the resistance to flank wear of different TiC-MoC$_x$-Ni experimental compositions during cutting steel or gray cast iron^x was found to be proportional to the Palmqvist crack resistance (Komac and Kosmač, 1979).

Concerning interrupted cutting operations, tool lives of cemented carbides will be predominantly shorter than in continuous machining, the reason for earlier tool failure being an additional cumulative action of cycling mechanical and thermal impacts on the tool tip, which results in mechanical and thermal fatigue cracks (Bhatia and co-workers, 1979). Due to temperature fluctuations at the tool tip, the extent of wear will become, at least indirectly, related to the thermal properties of the tool material, e.g. thermal conductivity and thermal expansion. Consequently, better performance of WC grades as compared to TiC materials in intermittent cutting operations reflects the higher thermal conductivity and lower thermal expansion coefficient of the former. The increased thermal diffusivity of TiC cemented carbides which are alloyed with TaN is believed to enhance thermal fatigue resistance (Hara and co-workers, 1982). These tools are capable of performing intermittent cutting operations, although their resistance to thermal cracking is lower than that exhibited by tungsten carbide tools (Uehara and Kanda, 1981).

The present paper discusses the mechanical and thermal shock properties of two series of cemented carbides, TiC-MoC$_x$-Ni and TiC-NbC$_x$-Ni, in terms of their composition and microstructure. The composition of the samples was varied according to the following rationale:

- the stoichiometry of the carbide influences its physical properties as well as its solubility in the binder phase, which in turn determines the properties of the cemented carbide
- the substitution of Mo by Nb was shown to promote an increase in the TRS of the cemented carbide, accompanied by a slight drop in hardness (Samsonov and co-workers, 1976). The Palmqvist crack resistance, Charpy impact value and resistance to flank wear were reported to increase with additions of up to approximately 15 wt % NbC to TiC-Mo-Ni (Ishibashi and co-workers, 1973).

EXPERIMENTAL PROCEDURES

Cemented carbides were prepared by ball milling the appropriate powder mixtures, compacting, and subsequent liquid phase sintering at temperature of 1300 - 1400°C. The characterization involved chemical composition, microstructural parameters, Vickers hardness HV30, Young's modulus (ultrasonic method), K_{IC} (indentation method) (Niihara and co-workers, 1982), Palmqvist crack resistance, TRS in three-point bending (ASTM B406-76), thermal conductivity and linear thermal expansion. Fractured and polished surfaces were analyzed by scanning electron microscopy (SEM) and Auger electron spectroscopy (AES). Thermal shock measurements were performed by heating the samples in an argon atmosphere at temperatures 200 - 1000°C, with subsequent quenching in water at 20°C. Quenched samples were additionally characterized by the measurement of retained bend strength σ_a.

RESULTS AND DISCUSSION

Mechanical Behaviour

In order to evaluate the influence of the composition of the carbide phase on the mechanical behaviour of TiC-MoC$_x$-Ni samples, measurements of TRS, Palmqvist crack resistance and hardness were employed. If binder content was kept constant and the Ti/Mo ratio and carbon concentration were varied, no distinct differentiation between particular specimens was possible on the basis of TRS and hardness data. However, some trends were noticed: TRS generally increases with increasing Mo content of the carbide. Accordingly, hardness increases as Ti in the carbide increases. If the Palmqvist crack resistance W of TiC-MoC$_x$-Ni samples was plotted as a function of carbon content, the relationships presented in Fig. 1 were obtained.

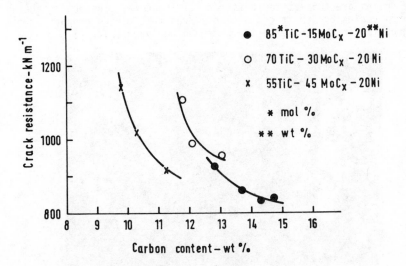

Fig. 1   Crack resistance of TiC-MoC$_x$-Ni cemented carbides as a function of total carbon content

Crack resistance increases with decreasing carbon content of the carbide at a constant Ti/Mo ratio. It is known that decrease of carbon content enhances the solubility of carbide metal components in the Ni binder, which is reflected in the increased TRS (Moskowitz and Humenik, 1978). Obviously, the same explanation applies for crack resistance as well, and was also used by Viswanadham and co-workers (1979) in interpretation of differences of crack resistance of different (Ti,V)C-(Ni, Mo) cermets. Additionally, two microstructural features - binder volume percent and binder mean free path - slightly increase, due to progressive dissolution of Ti and Mo in the binder, which could further increase the crack resistance. If carbon concentration is too low, $TiNi_3$ appears, which in turn diminishes the TRS and crack resistance. Moreover, crack resistance increases with increasing Mo content. This increase may be only apparent, since the volume fraction of binder phase increases with Mo content at constant Ni weight fraction, due to the increased density of the carbide. To exclude the possible influence of binder volume, the crack resistance values were normalised with respect to the binder volume fraction. A maximum appears in the range 20 - 30 wt % Mo in the cemented carbide compared to 10 wt % found by Moskowitz and Humenik (1978) for TRS.

The importance of carbon content may be further visualised by an inspection of its influence on grain growth during sintering. We could confirm earlier findings (Snell, 1974; Suzuki and co-workers, 1971) that the carbide grain size decreases with increased carbon content, whereas the growth of grains occurs mainly due to a formation of a $(Ti,Mo)C_{1-x}$ solid solution layer around the TiC core. A compromise is therefore needed in establishing the optimum carbon concentration in order to balance the opposite influences on crack resistance and grain growth. This becomes particularly important if one considers the reported data, which indicate that grain structure influences the rupture strength, the highest values being consistent with uniform, fine grained microstructure, and the grains possessing only a thin surrounding layer (Piljankevič and co-workers,1979).

Coarse grains are subject to transgranular fracture. Inspecting the fractured grains by SEM it could be observed that a high proportion of transgranular cracks traverse the surrounding layer (Fig. 2).

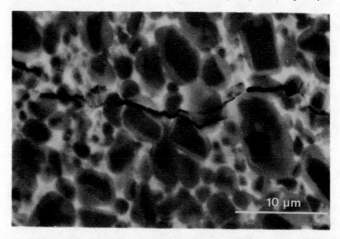

Fig. 2   Characteristic fracture path in $TiC-MoC_x-Ni$ sample.
The crack was introduced by Vickers indentation.

AES analysis of fractured surfaces in the regions where the density of
transgranular cracking was high (Fig. 3) suggests that transgranular
cracking through the surrounding layer occurs preferentially close to or
at the interface with the TiC core. At present no satisfactory interpretation can be given regarding the influence of cored grain structure
on energy dissipation during transgranular fracture.

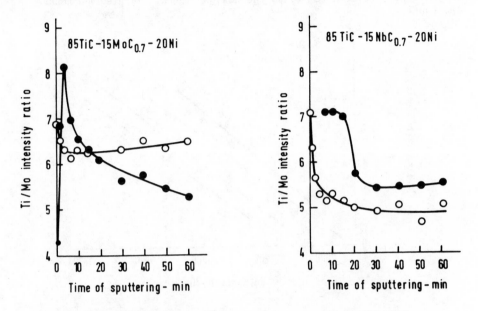

Fig. 3   Plot of Ti/Mo and Ti/Nb ratios vs. sputtering time of
polished and fractured surfaces of TiC cemented carbides.
O Polished surface;   ● Fractured surface.

The grains in the Nb series of samples are more uniform. No TiC core could
be observed; however, the gradients of Ti and Nb concentration were
established by electron probe microanalysis (EPMA), the concentration of Ti
increasing from grain boundary to grain centre and of Nb in the opposite
direction. Transgranular cracks preferentially traverse the centre of
grains and, with regard to the critical grain size for transgranular
fracture, no significant difference from TiC-MoC$_x$-Ni samples could be
found (2 - 3 μm).

In order to estimate the usefulness of the cemented carbides, a diagram of
1/W vs. H (Viswanadham and Venables, 1977) was constructed (Fig. 4) which
illustrates the position of the samples relative to WC-Co and some other
potential cemented carbides. Obviously, most of the specimens investigated fall well within the region of other experimental cermets. Nb containing samples are generally less favourable; however, compositions with
high Nb concentration seem to be very promising for good crack resistance,
but with some sacrifice of hardness.

The influence of microstructural parameters on fracture toughness was
studied on selected samples, characterized by constant carbide composition.
K_{IC} was related to binder volume, binder mean free path (approximated

as the ratio of the volume fractions of binder phase and carbide multiplied
by the carbide grain size) and carbide contiguity (Figs. 5 to 7). The
determination of fracture toughness was performed by the indentation method
and the formula used to calculate K_{IC} was that of Niihara and co-
workers (1982). Although the values obtained can hardly be considered as
the bulk fracture toughness of the as-sintered material, they are undoubtedly strongly related to it, so the designation K_{IC} is used throughout
the paper.

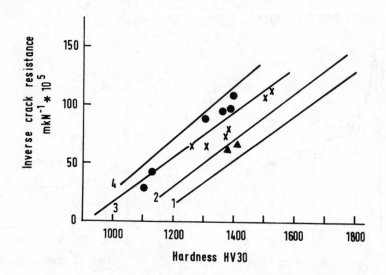

Fig. 4   Inverse crack resistance vs. hardness plots
          for several cemented carbides

1 - WC-Co
2 - (0.50 V, 0.50 Ti)C - (0.62 Ni, 0.38 Mo)   ⎫ (Viswanadham
3 - (0.66 V, 0.34 Ti)C - (0.71 Ni, 0.29 Mo)   ⎬ and
4 - TiC-Ni   ⎭ co-workers,
x - 85 TiC-15 MoC$_x$-Ni   1979)
● - 85 TiC-15 NbC$_x$-Ni
▲ - 50 TiC-50 NbC$_x$-Ni

No essential differences between the TiC-MoC$_x$-Ni and TiC-NbC$_x$-Ni
series were observed. In both cases the relationships closely resemble
those found in WC-Co alloys. Furthermore, different fracture modes which
were identified in WC-Co materials were found in TiC cemented carbides as
well. This indicates that in principle the interpretation of fracture
toughness could follow the philosophy utilised in establishing the models
proposed for WC-Co. However, certain peculiarities were noticed in TiC
materials. Slip lines could not be detected in carbide grains, even in the
case of highly deformed samples. Fracture surfaces are characterized by a
high proportion of interfacial decohesion. Deviation parameters Δ were
calculated accordingly to the method of Viswanadham and Sun (1977) from
peak-to-peak heights of Auger spectra of Ni and Ti on polished and fractured surfaces and are a measure of the amount of binder phase on the
fracture surface relative to polished surface. Hence, they define in a
qualitative manner the amount of fracture path associated with the binder

phase. The parameters Δ for TiC-MoC$_x$-Ni and TiC-NbC$_x$-Ni are lower compared to Δ of WC-Co with similar microstructural characteristics, indicating that the proportion of crack path associated with binder phase is higher in the case of WC-Co, the consequence being the higher toughness of the latter.

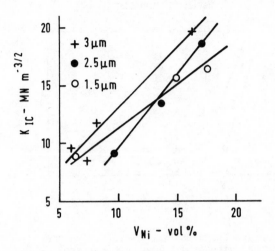

Fig. 5   Fracture toughness vs. binder content of TiC cemented carbide

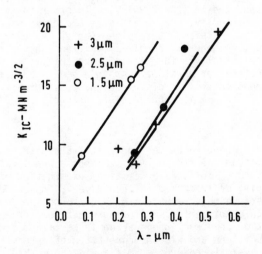

Fig. 6   Fracture toughness vs. binder mean free path
of TiC cemented carbide

Fig. 7 indicates the good correspondence between contiguity and deviation parameter Δ vs. K_{IC} relationship, respectively. This we believe is further evidence that the crack path is preferentially associated with the carbide phase, meaning that toughness will strongly depend on the properties of the carbide skeleton. Furthermore, the decohesions must be brittle in nature and once initiated propagate rather easily.

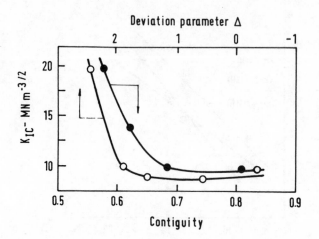

Fig. 7   Fracture toughness vs. deviation parameter and contiguity of TiC cemented carbide

On the other hand, the figure suggests that contiguity is very important in the high binder content region (20-25 vol %). In order to evaluate the relative importance of contiguity, mean grain size and grain size distribution with regard to K_{IC}, samples with a wide span of the above parameters were examined. A general trend was observed that at constant binder content K_{IC} is principally a function of contiguity. Accordingly, K_{IC} can be kept rather constant even in a highly unfavourable microstructure (broad grain size distribution, considerable fraction of grains in the range 4 - 7 μm) if contiguity is properly adjusted.

Thermal Shock Behaviour

The relative merit of different potential tool materials with regard to resistance to cycling mechanical and thermal stresses can be evaluated by conducting intermittent cuts, followed by critical examination of tool faces for cracks, chipping or fracture. Alternatively, impact resistance and thermal shock resistance measurements can be performed and materials graded by employing figures of merit. Following the theory of Hasselman (1969) of crack initiation and crack propagation in ceramic materials under thermal shock conditions and his interpretation of material behaviour in terms of physical properties, Mai and Atkins (1975) proposed two parameters as figures of merit. For crack initiation and crack propagation by thermal shock the appropriate parameters are $k.\sigma_f/E.\alpha$ and $(K_{IC}/\sigma_f)^2$ respectively, where k is the thermal conductivity, σ_f fracture strength in bending, E Young's modulus and α the coefficient of thermal expansion of the material under consideration. The larger the value of the two parameters, the lesser the chance that cracks will be initiated and will propagate. Mai and Atkins (1975) have already examined some commercial TiC cemented carbides, and in the present investigation thermal shock tests were performed on a series of TiC-MoC$_x$-Ni and TiC-NbC$_x$-Ni samples characterized by substantial differences in composition.

The mechanical and thermal properties of characteristic samples are given in Table 1.

TABLE 1 <u>Mechanical and Thermal Properties of Characteristic TiC Cemented Carbides</u>

Sample	E GNm^{-2}	σ_f MNm^{-2}	K_{IC} MNm$^{-3/2}$	Hardness HV30	α 10^{-6}K^{-1}	k Wm^{-1}K^{-1}
85 TiC-15 MoC-20 Ni	500	1245	12.6	1240	6.8	21
85 TiC-15 MoC$_{0.7}$-20 Ni	440	1280	12.7	1240	7.2	20
85 TiC-15 MoC$_{0.7}$-50 Ni	450	1620	25.3	990	7.6	30
85 TiC-15 NbC$_{0.7}$-50 Ni	460	1710	19.8	860	7.7	26
WC - 12 Co	570	2100	16.0	1300	5.5	67

Obviously, the differences in properties are mainly a function of changing the amount of binder phase. The thermal conductivity of Nb containing samples is believed to be lower due to the lower thermal conductivity of (Ti,Nb)C as compared to (Ti,Mo)C (Samsonov and co-workers, 1972).

Regarding the thermal shock behaviour, all compositions investigated (binder content in the region 10 - 50 wt %) followed the model of Hasselman (1969) which shows a distinct discontinuity in the tensile fracture strength vs. ΔT relationship, with considerable strength loss at a critical quenching temperature. Thus proportionality must exist between the critical temperature difference ΔT_c and the parameter k.σ_f/E.α and between the fractional fracture strength σ_a/σ_f of a sample which has undergone critical thermal shock and the crack propagation parameter $(K_{IC}/\sigma_f)^2$. Characteristic results are shown in Table 2.

TABLE 2   Thermal Shock Damage Parameters*

Sample	$\dfrac{k \cdot \sigma_f}{E \cdot \alpha}$ Wm^{-1}	ΔT_C $°C$	$\left(\dfrac{K_{IC}}{\sigma_f}\right)^2$ μm	$\dfrac{\sigma_a}{\sigma_f}$ $500°C$	$\dfrac{\sigma_a}{\sigma_f}$ $1000°C$
85 TiC-15 MoC-20 Ni	7.5	245	102	0.32	0.19
85 TiC-15 $MoC_{0.7}$-20 Ni	8.2	265	98	0.30	0.19
85 TiC-15 $MoC_{0.7}$-50 Ni	14.1	350	244	0.37	0.18
85 TiC-15 $NbC_{0.7}$-50 Ni	12.0	320	134	0.29	0.19
WC-Co	40-60	-	60-70		

* Room temperature values employed in the calculation.

The results suggest that resistance to thermal shock damage within the TiC series is dominated by the amount of binder phase, at least in the temperature range where K_{IC} and σ_f are temperature independent - up to approximately 600°C for TiC cemented carbide (Mai, 1976). At higher temperatures the fractional σ_a/σ_f displays a composition independent behaviour. We believe that for the compositions studied the amount of thermal strain accommodated is independent of the volume of the binder due to its plasticity in that temperature region.

An examination of the parameters $(K_{IC}/\sigma_f)^2$ in Table 2 suggests that TiC cemented carbides should exceed WC-Co grades with respect to resistance to crack propagation and the same figure is obtained if the high temperature data for K_{IC} and σ_f are employed in the calculation. Thus, there must exist some other reasons for the inferiority of simple TiC-MoC_x-Ni cemented carbides in interrupted cutting operations, since obviously, the interpretation in terms of the influence of transient thermal stresses alone is misleading.

It is believed that cracks which appear in carbide tools performing interrupted cutting can be attributed to tensile stresses which arise during the cooling period after part of the tool has undergone plastic deformation due to the combined effect of mechanical and thermal stresses (Braiden and Dugdale, 1970). Therefore, it was anticipated that improvement in the plastic deformation resistance of TiC tools should be reflected in improved performance in interrupted cutting and was virtually proven (Moskowitz and Humenik, 1981). The increase in deformation resistance could be accomplished by strengthening the binder and/or the carbide phase. The additions of TiN and TaN are of special benefit, since deformation resistance and thermal diffusivity can be enhanced simultaneously (Hara and co-workers, 1982).

Hence further improvement of TiC cemented carbides seems probable if by the adjustment of composition and microstructure, the interfacial cohesive forces, resistance to plastic deformation and thermal conductivity could be increased.

ACKNOWLEDGEMENT

This study was supported by Research Council of Slovenia and was performed within joint Yugoslav-German agreement on scientific co-operation
in the field of materials research and development.

REFERENCES

August, J.S. and Kalish, S. 1983, Int. J. Refractory and Hard Metals 2
    88-92
Beger, A., Fahrman, M., Gille, G., Kotsch, H. and Muller, K. 1981, Proc.
    7th Int. Conf. on Powder Metallurgy, ZFW, Dresden, pp 33-44
Bhatia, S.M., Pandey, P.C. and Shan, H.S. 1979, Precision Engineering
    148-152
Braiden, P.M. and Dugdale, D.S. 1970, Materials for Metal Cutting. The
    Iron and Steel Institute, pp 30-35
Hara, A., Yamamoto, T., Doi, Y., Sakanoue, H. and Takahashi, N. 1982,
    Int. J. Refractory and Hard Metals 1 32-36
Hasselman, D.P.H. 1969, J. Am. Ceram. Soc. 52 600-604
Ishibashi, O., Misumi, K., Nishimura, T. and Miyashita, H. 1973, Nippon
    Tungsten Rev. 6 74-79
Kalish, H.S. 1983, Metal Progress 21-27
Komac, M. and Kosmač, T. 1979, Contemporary Inorganic Materials 1979.
    Ed.I. Stamenković, Boris Kidrič Institute of Nuclear Sciences, Beograd,
    pp 174-186
Lardner, E. 1970, Materials for Metal Cutting. The Iron and Steel
    Institute, pp 122-132
Loladze, T.N. 1975, Ann. CIRP 24 13-16
Mai, Y.W. and Atkins, A.G. 1975, J. Mat. Sci. 10 1904-1919
Mai, Y.W. 1976, J. Am. Ceram. Soc. 59 491-494
Moskowitz, D. and Humenik, M. 1978, Int. J. Powd. Met. and Powd. Techn.
    14 39-45
Moskowitz, D. and Humenik, M. 1981, Modern Developments in Powder
    Metallurgy. Vol. 14, Ed. H.H. Hausner, MPIF, Princeton, pp 307-320
Niihara, K., Morena, R. and Hasselman, D.P.H. 1982, J. Mat. Sci. Letters,
    1 13-16
Perrot, C.M. and Robinson, P.M. 1974, J. Austr. Inst. Met. 19 241-253
Piljankevič, A.N., Šapoval, T.A., Dzodziev, G.T. and Paderno, V.N. 1979,
    Porosh. Met. No 10 73-79
Samsonov, G.V., Bogomol, I.V., Lvov, S.N. and Lesnaja, M.I. 1972, Porosh.
    Met. No 11 62-65
Samsonov, G.V., Voronkin, M.A., Linnikov, A.P. and Loktionov, V.A. 1976,
    Porosh. Met. No 12 37-41
Snell, P.O. 1974, Planseeber. Pulvermet. 22 91-106
Suzuki, H., Hayashi, K. and Terada, O. 1971, J. Jap. Inst. Metals 35
    936-942
Uehara, K. and Kanda, Y. 1981, Ann. CIRP 30 47-51
Viswanadham, R.K. and Venables, J.D. 1977, Met. Trans. 8A 187-191
Viswanadham, R.K. and Sun, T.S. 1977, Scripta Met. 13 767-770
Viswanadham, R.K., Sprissler, B., Precht, W. and Venables, J.D. 1979,
    Met. Trans. 10A 599-602

Inst. Phys. Conf. Ser. No. 75: Chapter 6
Paper presented at 2nd Int. Conf. Science Hard Mater., Rhodes

577

Residual stress relaxation in cemented carbide composites

A D KRAWITZ (1), R ROBERTS (1) AND J FABER (2)

(1) University of Missouri, Columbia, MO 65211 U.S.A.
(2) Argonne National Laboratory, Argonne, IL 60439 U.S.A.

ABSTRACT

The differential thermal residual stresses between the carbide and binder
phases in WC-Co and WC-(Co,Ni) cermets with 17 wt % binder have been
observed, using neutron diffraction, to relax upon compressive monotonic
and cyclic mechanical treatment. The relaxation is initially rapid with
increasing plastic strain and, for a given plastic strain, is greater for
cycled material. The strain response is consistent with a nominally
isotropic, hydrostatic stress state, although evidence of carbide anisotropy is present. Diffraction peak shape analysis indicates an initial
stress distribution in the cermets. Response is similar for all binders
although deformation mechanisms differ.

INTRODUCTION

The difference in thermal expansion coefficients of the carbide and binder
phases in cemented carbide composites (cermets) leads to a substantial
mutual residual stress state between the phases. The stress is tensile in
the binder and compressive in the carbide. Such a stress state is termed
pseudomacro because, while each phase is under a nominally constant stress
throughout the bulk, the stress through the material oscillates with a
frequency dictated by mean carbide and binder path lengths. Thus, over
distances large compared to the microstructure, the stress state averages
to zero.

In the present study (Roberts, 1984; Krawitz, Roberts and Faber, 1984),
relaxation of these residual stresses due to mechanical treatment is
observed via neutron diffraction in samples of WC-(Co,Ni) with Ni levels
of 0, 15 and 30 wt %. The binder deformation mechanisms were studied
previously (Krawitz and co-workers, 1983; Vasel and co-workers, 1985). As
Ni is added, the amount of HCP, created via the strain-induced FCC-to-HCP
martensitic transformation, is greatly decreased and dislocations and
twinning become the prevalent modes of deformation. The amount of HCP
produced as a function of plastic strain was determined using neutron
diffraction. In the course of this study, small but systematic shifts in
the binder peaks of neutron diffraction spectra were observed and reported
(Krawitz and co-workers, 1983). An entirely new set of measurements,

resulting in the present study, were made on the high resolution powder instrument at the Argonne Intense Pulsed Neutron Source.

The volumetric nature of the residual stress state makes neutron scattering an appropriate tool (Krawitz and co-workers, 1983). This is principally due to the increased penetrability of neutrons in cermets relative to X-rays. For the materials studied, a thickness of about 5mm is required to absorb 50% of an 0.13nm thermal neutron beam while the corresponding value for CuK X-rays is about 3μm. Thus virtually the entire volume of the cylindrical samples employed contributes to the spectra. In addition, the random variation of scattering cross-section with atomic number and its constancy with scattering angle make the binder phase more accessible to study and result in more usable peaks than is the case for X-rays.

EXPERIMENTAL

Samples

WC-(Co,Ni) cemented carbides with binder compositions of 0, 15 and 30 wt % Ni were prepared by milling of starting powders, pressing into cylinders, sintering at 1375°C and hot isostatic pressing at 1300°C under 100 MPa of argon. All samples contained 17 wt % binder and were in the form of 12.7mm diameter by 18.9mm high cylinders. A matrix of specimens comprised of as-produced material and two levels each of monotonic (under strain control) and cyclic (under stress control) treatments was generated. The treatments are summarized in Table 1, which also includes resultant plastic strains. Fatigue treatments were zero-compression-zero for 5×10^5 cycles. Relevant physical properties of the constituents are given in Table 2.

Neutron Diffraction

All measurements were made on the General Purpose Powder Diffractometer at the Intense Pulsed Neutron Source at Argonne National Laboratory. This is a high resolution time-of-flight instrument that provides, in principle, all allowable diffraction peaks from planes with interplanar spacings from about 0.05 to 0.50nm at fixed Bragg angles of, in this case, 90 and 150°2θ. Peak position and breadth (full width half-maximum) values and standard estimates of error were determined from non-linear regression fits of individual peaks. Further details regarding the facilities and data analysis procedures may be found in MacEwan, Faber and Turner, 1983.

RESULTS AND DISCUSSION

Residual Stress Relaxation

Diffraction measurements of stress are, in fact, measurements of strain which must then be converted to stress in an appropriate way. In the present study the strain was measured for crystallographic planes parallel to the cylinder axis of the specimens, i.e. in plane normal directions perpendicular to the cylinder axis. The strains were determined from shifts in interplanar spacings relative to the as-produced state. For the FCC binders, the 111, 200, 311 and 331 peaks were analyzed: for the hexagonal (wurtzite structure) carbide, the 101, 110, 002, 201, 112 and 211 peaks were employed. Results for selected planes are shown in Table 3.

TABLE 1 Mechanical Treatment Parameters
and Resultant Plastic Strain

Mechanical Treatment	Stress/Strain	Plastic strain %		
		0 Ni	15 Ni	30 Ni
As-produced	0	0	0	0
Low Fatigue	1.0GPa	-0.05	-0.05	-0.10
High Fatigue	2.1GPa	-0.12	-0.35	-0.61
Low Monotonic	0.75%	-0.27	-0.27	-0.30
High Monotonic	5.00%	-4.44	-4.33	-4.41

TABLE 2 Selected Physical Properties of the Constituents

Property	WC	Co	Ni
E Mpa	7×10^5	2×10^5	2×10^5
ν	0.2	0.32	0.32
K Mpa	3.89×10^5	1.85×10^5	1.85×10^5
$\alpha_\ell(°C^{-1})$	6.2×10^{-6}	13.8×10^{-6}	13.3×10^{-6}
ρ Mg/m^3	15.7	8.8	8.9

Considering for the moment the strain values, the data indicate that the tensile binder and the compressive carbide strain is relaxing, i.e. the interplanar spacings of the binder plane have decreased while those for the carbide have increased. Plots of the relaxation for the binder 311 and carbide 201 planes vs. plastic strain are shown in Fig. 1.

The relaxation is initially rapid and may generally be divided into monotonic and cyclic curves. This is because, for a given plastic strain, the relaxation is generally greater after cyclic treatment. This effect is generally true for the 0 and 15 Ni material and is particularly clear for the 30 Ni samples. Furthermore, the low monotonic treatment corresponds approximately to one cycle of the high fatigue treatment. Thus, although cyclic treatment produces greater relaxation per unit of final plastic strain, a considerable percentage of the ultimate relaxation appears to occur after only one cycle. The response seems to stabilize after only a few cycles, as is typical of hysteresis in cyclic stress-strain curves (Hertzberg, 1976).

TABLE 3   Stress (Strain) Relaxation for Selected Binder
and Carbide Planes

Cermet	Phase	hkℓ	Stress (MPa) (Strain x 10^5)			
			Low fatigue	High fatigue	Low monotonic	High monotonic
0 Ni	Binder	111	-128±83 (-23±15)	-799±83 (-144±15)	-566±83 (-102±15)	-1343±94 (-242±17)
		200	-183±122 (-33±22)	-638±122 (-115±22)	-622±122 (-112±11)	-977±166 (-176±30)
		311	-278±78 (-50±14)	-683±72 (-123±13)	-472±78 (-85±14)	-1750±89 (-317±16)
	Carbide	101	82±58 (7±5)	233±58 (20±5)	187±58 (16±5)	327±58 (28±5)
		002	198±140 (17±12)	362±128 (31±11)	467±140 (40±12)	992±140 (85±12)
		201	47±47 (4±4)	233±47 (20±4)	210±47 (18±4)	420±47 (36±4)
15 Ni	Binder	111	-422±67 (-76±12)	-888±61 (-160±11)	-827±67 (-149±12)	-1737±83 (-313±15)
		200	-422±83 (-76±15)	-960±83 (-173±15)	-749±83 (-135±15)	-1226±133 (-221±24)
		311	-422±67 (-76±12)	-1038±61 (-187±11)	-710±67 (-128±12)	-1504±72 (-271±13)
	Carbide	101	47±58 (4±5)	210±58 (18±5)	385±58 (33±5)	420±58 (36±5)
		002	152±152 (13±13)	525±140 (45±12)	420±140 (36±12)	1097±140 (94±12)
		201	152±70 (13±6)	327±58 (28±5)	292±58 (25±5)	548±70 (47±6)
30 Ni	Binder	111	-471±45 (85±8)	-1137±45 (-205±8)	-783±45 (-141±8)	-1683±54 (-295±10)
		200	-525±54 (-94±10)	-1203±51 (-205±8)	-765±54 (-138±10)	-1311±78 (-236±14)
		311	-489±45 (-88±8)	-1203±39 (-217±7)	-783±45 (-141±8)	-1548±51 (-279±9)
	Carbide	101	36±57 (3±4)	222±48 (19±4)	153±48 (13±4)	384±57 (33±5)
		002	117±153 (10±13)	537±141 (46±12)	327±141 (28±12)	933±153 (80±13)
		201	57±69 (5±6)	351±57 (30±5)	198±57 (17±5)	468±69 (40±6)

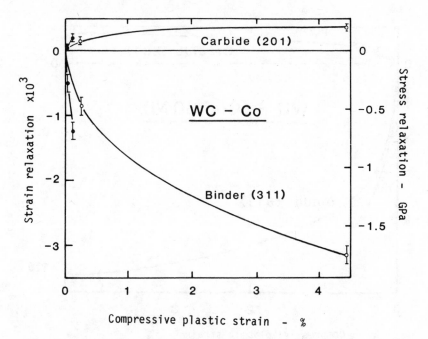

Fig.1a Strain relaxation versus compressive plastic strain for binder 311 and carbide 201 peaks in WC-Co

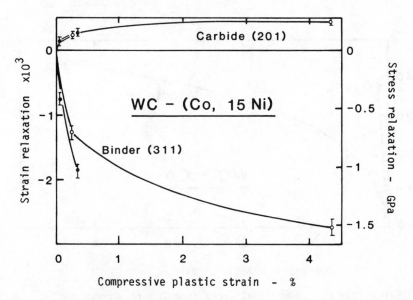

Fig.1b Strain relaxation versus compressive plastic strain for binder 311 and carbide 201 peaks in WC-(Co, 15Ni)

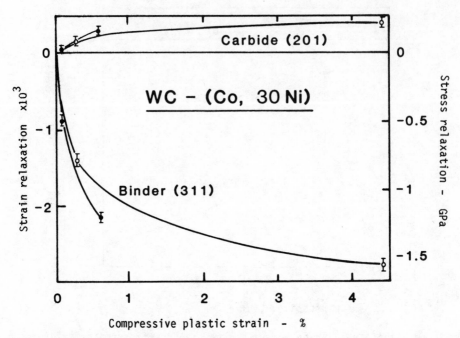

Fig.1c Strain relaxation versus compressive plastic strain for binder
311 and carbide 201 peaks in WC-(Co, 30Ni)

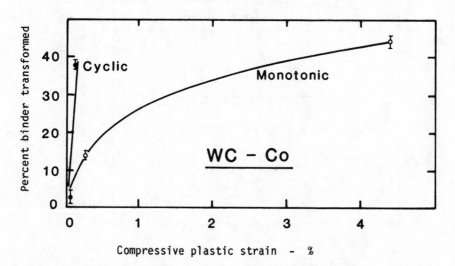

Fig.2 Percentage of binder transformed from FCC to HCP due to
mechanical treatment in WC-Co

The mechanism of relaxation appears to be plastic deformation in the binder. In Fig. 2 the percent of FCC binder transformed to HCP for the pure Co material is shown (Vasel, 1981). A striking similarity to the relaxation data is present. Thus it is possible to see directly in the pure Co material, in which binder deformation is via HCP formation, a parallel behaviour between stress relaxation and accumulation of plastic damage.

The binder strain response appears to be isotropic, within experimental error. The carbide strain response, however, shows consistently greater relaxation for the basal plane (002 peak) i.e. a crystallographic aniso-tropy is present. No systematic effects due to binder composition are discernible, i.e. the effect seems general with regard to binder com-position and, within experimental accuracy, insensitive even to binder deformation mechanism.

Values of stress were calculated assuming an isotropic, hydrostatic elastic stress state (Dieter, 1976):

$$\sigma_r = K\Delta_r \tag{1}$$

where K is bulk modulus ($E/3(1-2\nu)$), Δ_r is volume strain relaxation and σ_r is hydrostatic stress relaxation. The measured strain ε is the relaxa-tion in one direction so that the volume strain $\Delta = 3\varepsilon$. This model is an approximation, particularly in light of the carbide anisotropy. However, the measurement volume is very large, the carbide grain size is 2-3μm and these grains are equiaxed and randomly oriented (Krawitz and co-workers, 1983) so that a volume-averaged response may be expected to be nominally isotropic plastically even though a crystallographic anisotropy is present.

Equilibrium requires that the stress volume-average to zero:

$$V_f \,\bar{\sigma}_r\big|_B + V_f \,\bar{\sigma}_r\big|_C = 0 \tag{2}$$

where V_f is volume fraction (0.27 and 0.73 for binder and carbide, res-pectively), $\bar{\sigma}_r$ refers to average stress relaxation (calculated from the four binder and six carbide peaks cited above) and B, C are binder and carbide, respectively. The results are presented in Table 4. The low fatigue results are not included due to the large statistical errors in the data. These results indicate that, within error, the stresses volume-average to zero. This lends credence to the method used to calculate stress from strain. In this regard it is noted that a linear elastic conversion ($\sigma = E\varepsilon$) would not produce a balance.

The question of extent of relaxation cannot be addressed in an absolute sense with the present data. However, SEM observation of cross-sections of all samples show that the high monotonic material is extensively micro-cracked in a manner similar to that reported by Roebuck and Almond, 1979. This observation plus the observed decrease in rate of relaxation with increase in plastic strain suggests that complete relaxation at the high monotonic level is a reasonable approximation to reality. A calculation of the differential thermal residual stress assuming an isotropic, hydrostatic stress state with no interfacial shear or plastic flow can be made using (Kingery, Bowen and Uhlmann, 1975):

TABLE 4   Volume-Averaged Stress (MPa) for High Fatigue (HF), Low Monotonic (LM) and High Monotonic (HM) Treatments

Cermet	Treatment	Average Binder stress relaxation	Average Carbide stress relaxation	Volume average
0 Ni	HF	- 675	275	15
	LM	- 540	252	40
	HM	-1359	501	5
15 Ni	HF	- 921	324	-10
	LM	- 846	285	-20
	HM	-1473	588	40
30 Ni	HF	-1161	348	-60
	LM	- 774	204	-60
	HM	-1503	555	6

TABLE 5   Values of Corrected Peak Breadth ΔB (in µs) for Selected Binder and Carbide Peaks and Corresponding Silicon Standard Values

Cermet/ phase	hkℓ	As produced	Low fatigue	High fatigue	Low monotonic	High monotonic	Silicon standard
0 Ni/	101	43.4±2.0	43.0±1.3	41.7±0.7	43.1±1.2	64.1±0.8	24.2
Carbide	002	34.0±4.2	33.7±2.3	32.0±1.3	31.7±2.1	41.4±1.5	18.4
	201	28.6±1.2	26.9±0.7	26.5±0.4	26.7±0.6	40.1±0.5	15.1
15 Ni/	111	29.5±12.5	19.4±13.2	45.9±7.0	51.6±10.2	99.6±24.3	26.3
Binder	200	37.2±14.2	44.1±12.1	70.6±7.8	77.4±12.1	85.4±26.3	22.9
	311	26.8±3.7	23.2±2.4	35.5±1.7	35.6±2.5	86.4±5.1	14.1
15 Ni/	101	38.4±2.3	36.8±1.6	33.2±0.9	43.4±1.2	56.8±1.6	24.2
Carbide	002	34.6±4.5	31.8±2.7	31.7±1.5	25.7±2.3	35.1±2.4	18.4
	201	28.0±1.4	26.6±0.8	25.8±0.5	24.8±0.8	38.0±0.8	15.1
30 Ni/	111	38.7±10.5	30.3±5.5	54.1±2.9	49.3±4.4	116.0±15.9	26.3
Binder	200	45.6±12.8	51.9±6.2	60.0±3.1	55.8±4.4	194.2±27.1	22.9
	311	22.5±2.9	12.3±2.2	27.2±0.9	23.0±1.5	72.5±2.7	14.1
30 Ni/	101	35.9±2.6	39.7±1.5	30.9±0.7	35.6±1.1	54.5±1.7	24.2
Carbide	002	37.6±3.0	41.2±3.0	27.1±1.2	34.9±2.0	29.6±2.7	18.4
	201	29.1±17	28.6±0.9	24.6±0.4	26.1±0.7	39.4±0.9	15.1

$$\sigma_i = K_i(\bar{\alpha} - \alpha_i)\Delta T \qquad\qquad 3$$

where σ is hydrostatic stress, K is bulk modulus, α is volume coefficient of thermal expansion, $\bar{\alpha}$ is the average value for the composite (see Kingery, Bowen and Uhlmann, 1975 for a calculation of this term), ΔT is the temperature change from the stress-free state and i refers to the ith phase. Relevant properties are in Table 2. Note that α_ℓ, the linear coefficient of thermal expansion, is listed and that $\alpha = 3\alpha_\ell$. Using these values and $\Delta T = 700°C$, values of -975 and +2775MPa are obtained for the carbide and binder phases, respectively. These values are considerably larger than the maximum relaxations observed. This could mean that the values of K, α and ΔT are only imprecisely known, a particular possibility for α and ΔT, and/or that some plastic flow occurs in the binder during set-up of the stresses. In fact, the latter process does occur to some extent as is evidenced by the presence of faults in Co binders and dislocations in Ni binders (Drake and Krawitz, 1981). Thus, the observation of values of stress lower than those calculated is reasonable. The maximum relaxations measured are about half the calculated values.

Finally, it is noted that there would appear to be a considerable incentive for the system to relax. The stored elastic energy in each phase due to the residual stress state is given by (Dieter, 1976):

$$U_0 = \tfrac{1}{2}K\Delta^2 \qquad\qquad 4$$

Taking the levels of high monotonic relaxation to be the (negatives of the) equilibrium residual stresses in the as-produced material, then the carbide and binder are storing about 3×10^5 and 56×10^5 J/m³, respectively. These levels correspond to uniaxial elastic strains ($U_0 = E\varepsilon^2/2$) of 0.1 and 0.75%, respectively, for carbide and binder. It seems plausible, then, that the application of a uniaxial load would rather quickly provide, via an impressed deviatoric component, the means for the system to relax via shear processes in the binder.

Stress Distribution and Plastic Deformation

The results of the peak breadth analysis are shown in Table 5. The peak breadth parameter ΔB is given by:

$$\Delta B = (B_{sam}^2 - B_{stan}^2)^{\tfrac{1}{2}} \qquad\qquad 5$$

where B_{sam} is the FWHM breadth of the peak from a sample obtained from the fitting procedure and B_{stan} is the breadth that would be obtained from a well-annealed powder standard having a peak at the same position. The standard values are obtained from a plot of results for a silicon powder standard. The functional form of eqn. 5 assumes ideally Gaussian peaks. The corresponding standard peak breadths are included in the table.

It is seen that all peaks for both phases are broadened under all conditions. Considering first the as-produced state, some of the broadening for the binder peaks could be attributed to composition variation and/or plastic damage that accrued during set-up of the residual stresses during cooling. However, this cannot be the case for the carbide. The implication is that the primary source of broadening prior to mechanical treatment is a distribution of residual stress level. The high monotonic material, on the one hand, has experienced considerable, if not complete,

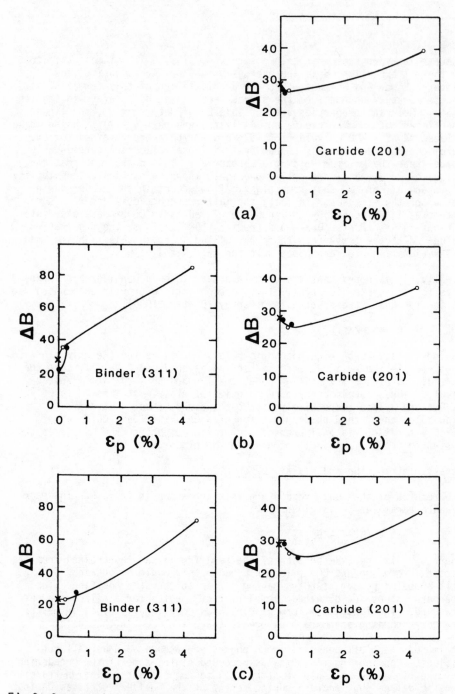

Fig.3   Corrected peak breadth versus plastic strain for binder 311 and
carbide 201 peaks for: a) WC-Co, b) WC-(Co, 15Ni) and c) WC-(Co, 30Ni).
The binder 311 is not included for WC-Co

relaxation of the residual stresses and, in addition, considerable plastic strain. Thus, the broadening in this case seems reasonably attributable to plastic deformation, i.e. microstrain. Consideration of intermediate treatments shows that the binder 111 and 311 (and others not included) actually sharpen as a result of the low fatigue treatment, then broaden as plastic strain increases. The effect is shown for the binder 311 in Fig. 3. A less pronounced trend occurs for the carbide; the 201 peak breadths are plotted vs. plastic strain in Fig. 3. In general, for the carbide peaks, the cyclic and monotonic treatments are not distinguishable. Rather, an approximately stable breadth or a shallow minimum occur at low plastic strain. Only the high monotonic treatment causes a clear increase in breadth providing, as for the binder, diffraction evidence of microstrain.

The following physical interpretation is suggested for the overall behaviour described above. The broadening in as-produced material is due principally to a distribution of residual stress. Since the relaxation commences rapidly, i.e. while plastic deformation is still at a low level, a sharpening of the peaks occurs so that broadening due to strain distribution decreases more rapidly than the increase in broadening due to plastic deformation. As discussed above, and shown in Fig. 1, the relaxation is more rapid for cycled material. For the binder, the strain levels are sufficiently high that their rapid decrease actually creates minima in peak breadth response in some crystallographic directions in cycled material. Ultimately plastic damage is very great in the binder so that profound broadening occurs after the high monotonic treatment. The same pattern occurs for the carbide, however, the strain levels are much lower, as is the degree of plastic damage, so that a general balancing results in the region of low plastic strain, and presumably, a gradual broadening until the high monotonic level. These responses are shown schematically in Fig. 4.

The binder 200 and the carbide 002 peaks exhibit behaviour seemingly at variance with the proposed model. The binder 200 response does not show a minimum due to its orientation to the $\{111\}$ slip planes in the FCC binder; plastic damage will most readily be reflected via $\{100\}$ interplanar spacings. In fact, there is a trend toward more pronounced minima as the $\{111\}$ are approached. The carbide 002 responses are somewhat erratic but generally exhibit a "flat" response which culminates in less broadening after the high monotonic treatment than any other peak. This cannot be attributed to basal plane slip because it has been established that slip occurs on prismatic planes in WC (Hibbs and Sinclair, 1981). The explanation appears to lie in the significantly greater stress relaxation observed normal to these planes; see Table 3. Presumably this is due to a greater residual stress in [001] caused by a higher coefficient of thermal expansion in this direction.

The presence of a distribution of residual stress sufficient to cause the observed broadening of as-produced binder and carbide peaks implies a wide range of local residual stresses in the cermets. This is certainly consistent with the heterogeneous character of microstructural changes observed using the TEM. In principle, it is possible to deconvolute instrumental contributions from the measured peaks and extract strain distributions. This would help bridge the gap between the volume-averaged information obtained thus far and the geometrically-induced local variations that are so characteristic of cermets. Intuitive inspection of the peaks suggest that the range of local variation will be found to be

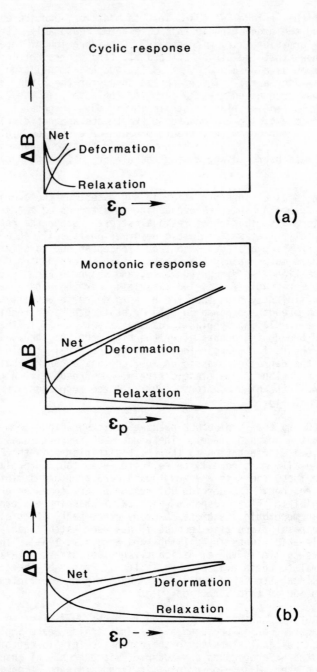

Fig.4   Schematic models for peak breadth behaviour versus plastic strain showing contributions of stress relaxation and plastic deformation for a) binder and b) carbide

substantial. It is also possible to perform a more standard Fourier analysis of the high monotonic profiles in order to obtain a separation and measure of particle size and micro-strain variance for the phases.

ACKNOWLEDGEMENTS

Reed Tool Co. produced and conditioned the samples, under the direction of Dr E.F. Drake. R.L. Hitterman, Argonne National Laboratory, provided technical assistance. Partial travel support was provided by Argonne Universities Association and IPNS.

REFERENCES

Dieter, G.E. 1976, Mechanical Metallurgy. McGraw-Hill, New York, pp 49-55.

Drake, E.F. and Krawitz, A.D. 1981, Met. Trans. 12A, 505-513.

Hertzberg, R.W. 1976, Deformation and Fracture Mechancis of Engineering Materials. Wiley, New York, pp 429-453.

Hibbs, M.K. and Sinclair, R. 1981, Acta Metall. 29 1645-1654

Kingery, W.D., Bowen, H.K. and Uhlmann, D.R. 1976, Introduction to Ceramics. Wiley, New York.

Krawitz, A.D., Drake, E.F., De Groot, R.L., Vasel, C.H. and Yelon, W.B. 1983, Science of Hard Materials. Eds. R.K. Viswanadham, D.J. Rowcliffe and J. Gurland. Plenum, New York, pp 973-989.

Krawitz, A.D., Roberts, R. and Faber, J. 1984, Adv. in X-ray Analysis, Vol. 27, Eds. J.B. Cohen, J.C. Russ, D.E. Leyden, C.S. Barrett and P.K. Predecki, Plenum Press, New York, pp 239-249.

MacEwen, S.R., Faber, J. and Turner, A.P.L. 1983, Acta Metall. 31, 656-676.

Roberts, R. 1984, Residual Stress Relaxation in Cemented Carbide Composites, M.S. Thesis, Univ. of Missouri, U.S.A.

Roebuck, B. and Almond, E.A. 1979, Conf. Recent Advances in Hardmetal Production, Loughborough Univ. Met. Powder Rep.

Vasel, C.H., Krawitz, A.D., Drake, E.F. and Kenik, E.A. 1985, Met. Trans. A, In press.

Inst. Phys. Conf. Ser. No. 75: Chapter 6
Paper presented at 2nd Int. Conf. Science Hard Mater., Rhodes

591

Effects of triaxial cyclic precompression on the length of Palmqvist cracks in WC-6% Co

G VEKINIS(1) AND S BARTOLUCCI LUYCKX(2)
(1) NIMR, CSIR, Pretoria, South Africa
(2) University of the Witwatersrand, Johannesburg
South Africa

ABSTRACT

This paper shows that the length of Palmqvist cracks in WC-6 wt % Co can be reduced up to $\sim$ 60% by means of triaxial cyclic precompression. It also provides a relationship between the reduction in crack length and the prestress and shows that shorter cracks are due to lower tensile residual stresses in the material.

INTRODUCTION

Some years ago it was observed that cyclic precompression increases the resistance of WC-Co to the propagation of Palmqvist cracks (Luyckx, 1981), i.e. increases its toughness (Luyckx, 1984). The effect, however, was then studied within a relatively narrow range of compressive stresses, the upper limit being set by the compressive strength of the material.

In the present work the upper limit of the compressive prestress was extended by encasing the WC-Co specimens in shrink-fitted steel rings, on the model of Pelepelin's system (Pelepelin, 1965). As a result, the WC-Co samples were subjected to non-uniform triaxial compression.

The previous investigation (Luyckx, 1981) produced only qualitative results, mainly on account of the specimens used, which were high pressure components of complex geometry where it was difficult to determine the stress distribution with precision. By contrast, the present work was carried out on specimens of simple geometry where the stress distribution was known, in an attempt to formulate the dependence of the Palmqvist crack length on the compressive prestress quantitatively.

The present work attempted also to establish if the increased resistance to crack propagation of precompressed material is due to changes in the residual stresses (Luyckx, 1981; Exner and Gurland, 1970).

EXPERIMENTAL

The specimens used in this investigation were cylinders of WC-6wt% Co ($\sim$ 2 μm average grain size) of 8.000 + 0.001 mm dia and 6.100 + 0.005 mm height. Steel rings (EN 30 B hardened to HRC51) of 5.850 $\pm$ 0.005 mm height and 50 mm outside diameter were shrink-fitted onto the WC-Co cylinders to a calculated interference of 0.1 $\pm$ 0.01%. This interference was achieved by observing very close tolerances ($\pm$ 1 μm) on the dimensions of both WC-Co cylinders and steel rings.

The specimens, i.e. the WC-Co cylinders encased in steel rings, were cyclically loaded in compression from zero to maximum compressive stresses ranging from 3 to 5.5 GPa. These compressive stresses will be referred to as "axial compressive stresses" as against the hoop compressive stresses exerted on the WC-Co cylinders by the steel rings. The hoop compressive stresses did not exceed 1.2 GPa at the highest axial compressive stress (5.5 GPa). The number of compressive cycles at each maximum axial compressive stress ranged from 1 to 100.

The axial compressive stress was applied at a constant loading rate of 65 MPa/s. The unloading rate was also kept constant, at 300 MPa/s.

After compression the cylinders were removed from the steel rings and spark-cut into halves along their axes. The cut surfaces were polished and indented along the central portion of the axis in order to nucleate Palmqvist cracks, as in a previous investigation (Luyckx, 1981). The indenting load was 500 N in all tests. The indentations were orientated so that two of the Palmqvist cracks nucleated by each indentation were parallel to the direction of the axial compressive stress (these will be called "parallel cracks") and two perpendicular to that direction ("perpendicular cracks").

By means of Finite Element Analysis under elastic-plastic conditions it was found that only a small central region along the axis of each WC-Co cylinder was under uniform stress when compressed triaxially as described above. Because of the small size of this region only two indentations were made and therefore only eight Palmqvist cracks were nucleated on each half-cylinder. As a result, each crack length value reported below represents an average over the lengths of 16 cracks.

The length of the Palmqvist cracks in precompressed samples was compared with the length of the Palmqvist cracks nucleated under identical conditions in identical uncompressed WC-Co half-cylinders. The average Palmqvist crack length, ℓp, was plotted against the maximum axial compressive stress, σ_a, at each number of compressive cycles N_c. The dependence of the Palmqvist crack length on the residual stresses in carbide and in Co was investigated by measuring the residual stresses in each phase before and after compression for 100 cycles. The stresses were measured on the cut surfaces of the half-cylinders of WC-Co before they were indented and exactly in the regions that were later indented. Great care had to be taken over the preparation of the specimen surfaces, as was demonstrated by French (1969). The measurements were carried out by means of the 2-tilt X-ray diffraction method (Cohen et al, 1979) on a compu-terized diffractometer. Details on the experimental conditions are given in Table 1. The radiation used for all measurements was CuKα (at 40 kV and 20 mA) with Ni filter, the divergence slit was 1°, the receiving slit 0,15 mm and the scatter slit 1°. Forty-one positions on both sides of each peak were sampled, to a total of approximately 4000 counts per position. The intensities were corrected for absorption and Lorentz polarization and the peak positions were determined by a 41 point Gaussian least-squares fit method.

TABLE 1 Experimental Conditions under which the X-ray Diffraction Measurements were carried out

	WC	Co(fcc)
Set of lattice planes	(103)	(111)
2θ	$\sim120°$	$\sim44°*$
Error in 2θ	$+0.002°$	$+0.005°$
ψ	$\overline{0}°$ and $25°$	$\overline{0}°$ and $5°$
Counter-to-specimen distance		
at $\psi = 0°$	185 mm	185 mm
" " " at $\psi = 25°$	107.6 mm	
" " " at $\psi = 5°$		118.5 mm

* The strongest peak was chosen, due to the very low intensity of the higher angle peaks.

In both phases the residual stresses were measured in the directions perpendicular and parallel to the axial compressive stress σ_a. The measured residual stresses will be indicated as $(\sigma_\phi)'_\eta$ and $(\sigma_\phi)''_\eta$ where:

$\phi = 0°$, when the measured residual stress is parallel to the direction of the axial compressive stress

$\phi = 90°$, when the residual stress is normal to σ_a.

η = WC, when the residual stress is measured in tungsten carbide

η = Co, when the residual stress is measured in cobalt

$(\sigma_\phi)'_\eta$ represents a residual stress in uncompressed material

$(\sigma_\phi)''_\eta$ represents a residual stress in precompressed material.

The results from the measurements of residual stresses will be reported as fractional changes in stresses, i.e.:

$$\frac{(\sigma_\phi)''_\eta - (\sigma_\phi)'_\eta}{(\sigma_\phi)'_\eta} = \frac{\dfrac{d''_{\phi,\psi} - d''_{\phi,\psi=0°}}{d''_{\phi,\psi=0°}} - \dfrac{d'_{\phi,\psi} - d'_{\phi,\psi=0°}}{d'_{\phi,\psi=0°}}}{\dfrac{d'_{\phi,\psi} - d'_{\phi,\psi=0°}}{d_{\phi,\psi=0°}}} \qquad 1$$

where:

$d_{\phi,\psi}$ = interplanar distance of a particular set of lattice planes of a specimen tilted by an angle of $\psi°$ with respect to the original normal to the sample surface

$d_{\phi,\psi=0°}$ = interplanar distance of the same lattice planes before specimen tilting

$d''_{\phi,\psi}$ and $d''_{\phi,\psi = 0°}$ = interplanar distances measured in precompressed specimens

$d'_{\phi,\psi}$ and $d'_{\phi,\psi = 0°}$ = interplanar distances measured in uncompressed specimens.

Since only fractional changes in stress were measured, it was not necessary to determine the X-ray values of the elastic modulus and the Poisson's ratio of either phase for any set of lattice planes (Cohen et al, 1979).

RESULTS

The dependence of the average Palmqvist crack length on the maximum axial compressive prestress σ_a is shown in Figs. 1 and 2 for various numbers of compressive cycles. Figs. 1 and 2 show respectively the dependence on σ_a of the length of the "perpendicular" Palmqvist cracks, ℓ_{pp}, and of the "parallel" cracks, $\ell_{p\ell}$. For the sake of clarity the error bars were omitted from these figures; however, their average magnitude is shown in Figs.3 and 4, after 100 compressive cycles.

The dependence of ℓ_{pp} on σ_a (Fig.1) may be described by an empirical relationship of the type

$$\ell_{pp} = \ell_0 - A\,\ell_0\,e^{-\left(k/\sigma_a\right)^3} \qquad\qquad 2$$

where: ℓ_0 = the average Palmqvist crack length in uncompressed material (97 $\pm$ 5 μm in the present WC-Co grade)

A = a dimensionless constant which depends on the number of compressive cycles and represents the maximum fractional decrease in the length of "perpendicular" Palmqvist cracks that can be obtained by means of the present precompression system

k = a constant equal to the axial compressive stress which yields a fractional decrease $\dfrac{\ell_{pp} - \ell_0}{\ell_0} = -\dfrac{A}{e} \sim -\dfrac{A}{2.7}$

Similarly the dependence of $\ell_{p\ell}$ on σ_a (Fig.2) may be expected to be of the type

$$\ell_{p\ell} = \ell_0 - B\ell_0\,e^{-\left(h/\sigma_a\right)^m} \qquad\qquad 3$$

where B and h are constants having the same physical significance as A and k in equation 2 and m >> 3.

Figs. 3 and 4 show the dependence on σ_a of the fractional change in the residual stresses in WC and in Co as well as the fractional change in crack length after 100 compressive cycles. The residual stresses in WC are indicated as $(\sigma_{0°})_{wc}$ and $(\sigma_{90°})_{wc}$ according to whether they are parallel or perpendicular to the direction of the axial compressive stress, and the residual stresses in Co are indicated as $(\sigma_{0°})_{Co}$ and $(\sigma_{90°})_{Co}$. The fractional changes in the stresses in Co are indicated as $(\Delta\sigma_{0°}/\sigma_{0°})_{Co}$ and $(\Delta\sigma_{90°}/\sigma'_{90°})_{Co}$ respectively and the fractional changes in WC are indicated as $(\Delta\sigma_{90°}/\sigma'_{90°})_{wc}$ and $(\Delta\sigma_{0°}/\sigma'_{0°})_{wc}$. According to these results, after compression WC is subjected to higher compressive

stresses and Co to lower tensile stresses than before compression. These
results do not agree entirely with those reported by Tumanov and co-workers
(1976) who used a magnetic polarity reversal method.

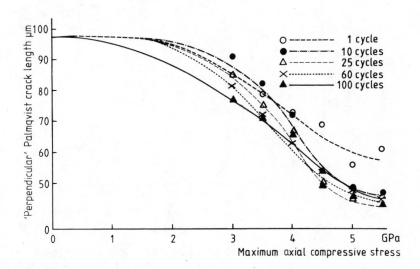

Fig.1. Summary of the results from the measurements of
"perpendicular" Palmqvist cracks at various maximum
axial prestresses and at various numbers of cycles.
The error bars have been omitted for the sake of
clarity.

In Fig.3 the change in residual stresses parallel to the direction of σ_a
is compared to the change in the length of cracks perpendicular to that
direction and in Fig.4 the change in the residual stresses perpendicular
to the direction of σ_a is compared to the change in the length of cracks
parallel to that direction.

Fig.3 shows that the changes in the length of "perpendicular" Palmqvist
cracks follow closely the changes in residual stresses. All the three
curves in Fig.3 exhibit inflection points at $\sigma_a \sim$ 3-4 GPa (which
corresponds to the yield stress of the material (Doi, 1974)) and all three
curves reach asymptotic values at $\sigma_a \simeq$ 5 GPa. The $\Delta\ell_{pp}/\ell_o$ curve appears
to be more closely related to the $(-\Delta\sigma_{0o}/\sigma'_{0o})_{wc}$ curve than to the
$(\Delta\sigma_{0o}/\sigma'_{0o})_{Co}$ one.

Fig. 4 shows that the scatter in the results from "parallel" cracks is too
large for any trend to be recognised. It is interesting, however, that,
in general, the change in the residual stresses in Co is larger than the
change in the residual stresses in WC, which is the opposite to what is

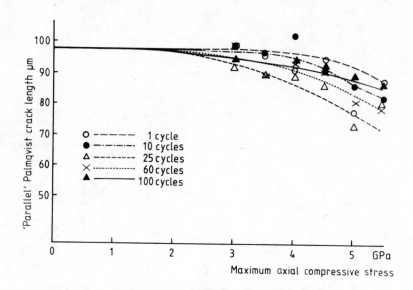

Fig.2.   Summary of the results from the measurements of
         "parallel" Palmqvist cracks at various maximum axial
         prestresses and at various numbers of compressive
         cycles.

observed in Fig.3. Again, the $\Delta\ell_{p\ell}/\ell_0$ curve appears to be more closely
related to the $(-\Delta\sigma_{90^\circ}/\sigma'_{90^\circ})_{wc}$ curve than to the $(\Delta\sigma_{90^\circ}/\sigma_{90^\circ})_{Co}$ one.

DISCUSSION

Fig. 3 shows a close similarity between the dependence of $\Delta\ell_{pp}/\ell_0$ and the
change in the residual stresses on σ_a. This suggests that the change in
the residual stresses controls the shortening of Palmqvist cracks after
compression.

The effect of residual stresses on the length of Palmqvist cracks can be
explained by considering separately the effect of the tensile stresses in
the cobalt and of the compressive stresses in the carbide.

First, it is reasonable to assume that the tensile hoop stress exerted on
the surrounding material by a Vickers indenter (which is used to nucleate
Palmqvist cracks (Luyckx, 1981)) is

$$\sigma_i \propto \frac{1}{\ell^2} \qquad\qquad 4$$

where ℓ represents the distance from the corners of the Vickers indentation. The dependence of σ_i on ℓ is schematically represented in Figs. 5
and 6 by curves AA.

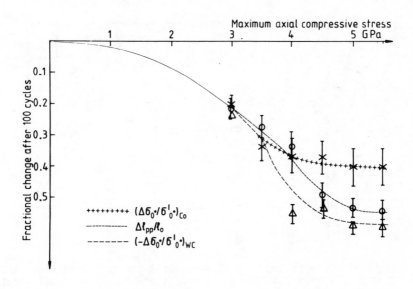

Fig.3.   Comparison between fractional changes in the length of
        "perpendicular" Palmqvist cracks and in the residual
        stresses normal to the crack path after 100 cycles.
        The symbols employed are defined in the text.

Let us assume next that the critical stress for propagation of a Palmqvist
crack in WC-Co is σ_c. The dependence of σ_c on ℓ is schematically
represented in Figs. 5 and 6 by lines BB. Let line C'C' in Fig.5
represent the average residual tensile stress in Co before compression,
σ'_{Co}, and C''C'' the average residual tensile stress in Co after
compression, σ''_{Co} ($\sigma''_{Co} < \sigma'_{Co}$). It follows that in the Co the resultant
stress acting on the tip of a Palmqvist crack in the direction normal to
the crack path is $\sigma_i + \sigma'_{Co}$ before compression, $\sigma_i + \sigma''_{Co}$ after compression.
The crack propagates only if $\sigma_i + \sigma'_{Co} \geq \sigma_c$ before compression (i.e.
within the distance ℓ_0 from the indentation, as shown in Fig.5) and if
$\sigma_i + \sigma''_{Co} \geq \sigma_c$ after compression (i.e. within the distance ℓ_p, shown in
Fig.5). Since $\sigma''_{Co} < \sigma'_{Co}$ it follows that $\ell_p < \ell_0$.

Similarly, let line D'D' in Fig.6 represent the average residual
compressive stress in WC, σ'_{WC}, before compression and D''D'' the average
residual compressive stress in WC, σ''_{WC}($|\sigma''_{WC}| > |\sigma'_{WC}|$), after compression.
It follows that in WC the resultant stress acting on the tip of a
Palmqvist crack in the direction normal to the crack path is $\sigma_i + \sigma'_{WC}$
before compression and $\sigma_i + \sigma''_{WC}$ after compression. The crack propagates
in WC only if $\sigma_i + \sigma'_{WC} \geq \sigma_c$ before compression (i.e. within the distance
ℓ_0, as shown in Fig.6) and if $\sigma_i + \sigma''_{WC} \geq \sigma_c$ after compression(i.e. within

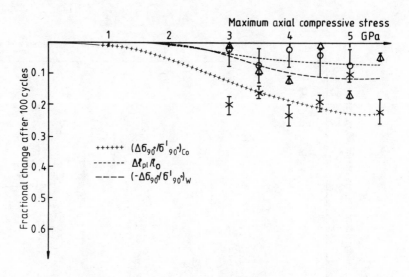

Fig.4. Comparison between fractional changes in the length of
"parallel" Palmqvist cracks and in the residual stresses
normal to the crack path after 100 cycles. The symbols
employed are defined in the text.

the distance ℓ_p shown in Fig.6). Since $|\sigma''_{WC}| > |\sigma'_{WC}|$ it follows that
$\ell_p < \ell_o$. Therefore, in both phases the measured change in residual
stresses is consistent with the observed shortening of Palmqvist cracks.

In WC-6wt% Co the changes in the length of Palmqvist cracks appear to be
more closely related to the changes in the residual stresses in WC than in
Co (Fig.3). This must be due to the fact that in low-Co material cracks
have a higher probability of propagating through WC grains or along WC-WC
boundaries than through binder-phase regions. It is expected, however,
that in higher Co-content alloys the role of the residual stresses in Co
would be more pronounced.

ACKNOWLEDGEMENTS

The authors wish to thank Professor F R N Nabarro for continued interest
in the project, the staff of the Boart Research Center, Krugersdorp, for
advice and assistance in the preparation of the samples, Dr S Hart, of
the CSIR, for making available the Rigaku diffractometer and Mr R Cooper
for making available specimen preparation and mechanical testing equipment.
They also gratefully acknowledge the financial support of Boart
International Ltd and of the Council for Scientific and Industrial
Research.

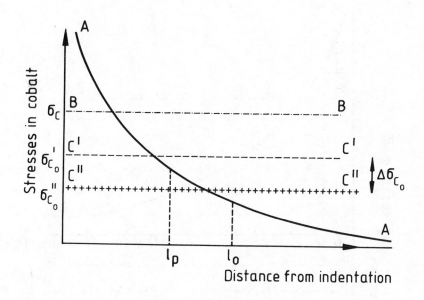

Fig.5.   Schematic representation of the stresses acting on the tip
of Palmqvist cracks in the direction normal to the crack
path in the binder phase. AA represents the tensile hoop
stress generated by the Vickers indenter, BB the critical
stress for crack propagation, C'C' the residual stress in
Co before compression and C''C'' the residual stress in Co
after compression. ℓ_0 is the length of the Palmqvist crack
in uncompressed material, ℓ_p the length in precompressed
material

REFERENCES

Cohen, J.B., Dölle, H., and James, M.R. 1979, Nat. Bureau of Std. Spec.
   Publ. 567. Proc. Symp. Acc. Powd. Diffr., Gaithersburg pp 453-477
Doi, H. 1974, Elastic and Plastic Properties of WC-Co. Freund Publ.
   House, Tel-Aviv p 66
Exner, H.E. and Gurland, J. 1970, J. of Mater. 5 75-85
French, D.N. 1969, J. Am. Ceram. Soc. 52 267-271
Luyckx, S.B. 1981, Proc. Intern. Conf. Science of Hard Materials.
   Wyoming, USA, pp 583-594
Luyckx, S.B. 1984, Proc. 6th Intern. Conf. on Fracture, New Delhi,
   India, pp 2665-2669
Pelepelin, V.M. 1965, Poroshk. Met. 35 76-80
Tumanov, V.I., Cheredinov, A.A., and Kuznetsova, K.F. 1976, Poroshk. Met.
   165 63-67

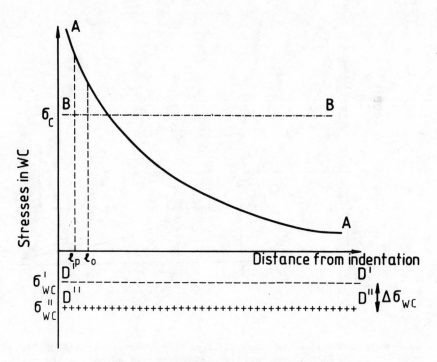

Fig.6.   Schematic representation of the stresses acting on the
tip of Palmqvist cracks in the direction normal to the
crack path in the carbide phase. AA represents the
tensile hoop stress generated by the Vickers indenter,
BB the critical stress for crack propagation, D'D' the
compressive residual stress in WC before compression and
D''D'' the compressive residual stress in WC after
compression. ℓ_0 is the length of the Palmqvist crack in
uncompressed material and ℓ_p the length in precompressed
material.

Discussion on Cermets and hardmetals — Part one

Rapporteur: J GURLAND (1)
Session Chairman: H E EXNER (2)

(1) Brown University, Providence, RI, USA
(2) Max Planck Institute, Stuttgart, FRG

The excellent properties of TiC-TiN-Mo-Ni cermets reported by Dr H Doi caused at least one member of the audience, Dr Santhanam, to wonder about the future of WC-Co-base hardmetals for metal cutting applications. In response, Dr H Doi predicted a kind of synergistic effect between the two alloy systems, whereby TiN additions would be incorporated into WC-Co base alloys and WC and TaC would be added to the TiC/TiN-base alloys, thereby, in effect, bringing the alloy compositions closer together. The addition of WC and TaC to TiC/TiN cermets is expected to improve the fracture toughness by increasing the Young's modulus. In answer to a question by Dr Shveikin, Dr H Doi discussed the influence of various carbides (WC, TaC, VC) on the properties of TiC-Mo-Ni hardmetals. The addition of WC and NbC hardens both carbide and binder phases, but TaC hardens only the carbide. WC and TaC additions were found to improve cutting properties, but Dr H Doi stressed that combined additions of TiN and suitable carbides such as WC, TaC or VC, are quite effective in improving mechanical and cutting properties.

Professor Jack introduced the interstitial multi-component complex nitrides, such as β^m-Ni_2Mo_3N, bonded with Ni, as a new class of hard-metals. This brought forth a number of questions about their properties relevant to tool performance (Drs Almond, Holleck and Schmid). In the discussion, Professor Jack made the following points:

1) <u>Hardness</u>. The hardness measurements given in the paper are for compacts with varying amounts of binder. By extrapolation, it can be inferred that the intrinsic hardness of β^m-Ni_2Mo_3N is lower than that of WC. However, no specimens of β^m phase are as yet available which would lend themselves to hardness testing, even on a microhardness scale.

2) <u>High-temperature stability</u>. The thermodynamic stability of β^m-Ni_2Mo_3N is remarkably high (decomposition temperature 1360°C for P_{N_2} = 1 atm) and compacts are processed at 1200°C. At 1000°C, oxidation resistance is much better than that of WC-Co. One could therefore expect the materials to have good wear resistance at temperatures up to 1000°C, although this is pure speculation and needs to be evaluated by quantitative measurements.

3) Crack resistance. Only very limited information exists. The only toughness value, based on crack propagation, was obtained on early hot pressed material with 15.5 wt % Ni (about 30 vol % binder) and gave a value of approximately 10 MPa m$^{1/2}$.

In elaboration of his talk, and in answer to a question by Dr Komac, Dr Igaraski discussed the solubility of WC in TiC/TiN, which decreases with increasing N/(C+N) ratio, because of lower miscibility limits. The solubility of the metal elements in the cobalt binder was not investigated in this study. The porosity observed in WC/TiN base materials is believed to be due mainly to free nitrogen gas evolved from the carbon-nitride grains during the sintering process. The porosity is present in spite of the improved wettability accompanying the use of the carbonitride.

Dr H Doi discussed and answered questions by Dr Almond, Professor Fisch-meister, Drs Leroux, Kny, Schmid and Viswanadham. Some of these were concerned with the high temperature properties of (Ti,W) C-Ni-Co-Cr-Aℓ cermets which Dr Doi believes to be superior to those of WC-TiC-TaC,Co hardmetals (Reference: K Nishigaki, Y Oosawa and H Doi, J. Japan Soc. of Powder and Powder Met., 26 (1979) 169). However, in the case of the γ' material, the operating temperature should not be higher than 700-800°C because of the aging characteristics. Therefore, perhaps the aged material is more suitable for wear applications than for cutting applications.

It was questioned by Dr Viswanadham where the γ' strengthening effects would be observed at binder volume fractions less than 0.2, since they were not seen at low binder content in the WC-(Ni,Aℓ) system (Reference: R K Viswanadham, P G Lindquist and J A Peck, in Science of Hard Materials, Eds. R K Viswanadham, et al, Plenum Press, New York, NY, 1983). Dr Doi referred to previously published work where γ' strengthening effects were observed in TiC-15% Ni-8% Mo and TiC$_{0.7}$ N$_{0.3}$-15% Ni-8% Mo cermets (K Nishigaki, Y Oosawa and H Doi, J. Japan Soc. of Powder and Powder Met., 26 (1979) 169; also K Nishigaki, H Yoshimura and H Doi, ibid, 27 (1980) 50).

In answer to questions by Drs Lazarev, Pope, Schmid and H Doi, Dr M Komac stated that the fracture resistance (in the work of M Komac and S Novak) was measured by the indentation crack method as well, without claiming necessarily an equivalent between the fracture toughness so measured and bulk fracture toughness. But, Drs Komac and Novak found a positive correlation between Palmqvist crack resistance and resistance to flank wear during turning of steel and cast iron.

Dr Komac also explained that in the case of NbC-containing TiC-Ni samples, the rim-core grain structure was not present. There is, however, a small Nb and Ti concentration gradient from grain centre to grain boundary. It was observed that in this case transgranular cracks traverse the centre of the grains.

A carbon-deficient NbC (e.g. NbC$_{0.7}$) was used to study the effects of carbon content variation in TiC-NbC$_x$-Ni, in analogy to similar studies of the TiC-MoC$_x$-Ni series. The variation of carbon content of TiC-NbC$_x$-Ni influences crack resistance in this system as well.

Professor Krawitz elaborated on his results by confirming to Professor Fischmeister that mechanical hysteresis was indeed observed during the cyclic loading experiments. Also, the continuous curvature observed in the microstrain (or low strain) region of cermet stress-strain curves, tends to become more linear on successive reloading. The relaxation of the differential thermal residual stresses may play a role in the curvature and its disappearance.

In answer to a question by Dr Schmid, Professor Krawitz said that the nature of the plastic deformation was studied in detail by HVEM. The results are reported in a paper submitted for publication to Metallurgical Transactions A.

Presentation of the last paper (Drs Vekinis and Luyckx) gave rise to an animated discussion relating to the effects and measurement of the different types of residual stresses. For instance, Dr Komac asked about the difference between the effects of the residual stresses introduced by surface grinding and those due to precompression. The answer is that the former are surface stresses which may toughen the surface, while the latter are bulk stresses which may affect the fracture toughness of the whole material.

As to the difference between the two types of residual stresses described by Professor Krawitz and Dr Vekinis, respectively, Dr Viswanadham asked why Drs Vekinis and Luyckx observed directional effects in the changes of Palmqvist crack length after compression, since Professor Krawitz showed that residual stresses are essentially hydrostatic. Drs Vekinis and Luyckx pointed out, in reply, that the changes in the residual stresses measured by X-ray diffraction are anisotropic because these stresses are measured near the surface, while Professor Krawitz measured bulk stresses by neutron diffraction. Obviously there is a surface effect, which is not fully understood, and on which the lengths of the Palmqvist cracks depend.

Regarding the inherent limitations of the X-ray method of surface stress determination, Professor Krawitz submitted the following discussion:

The X-ray measurement of stress in tungsten-bearing cermets only samples $\sim$3 µm below the surface. Now, a triaxial stress state must relax to a biaxial one at the surface. There is a surface-to-bulk transition region. The issue then becomes: what is the X-ray penetration depth relative to the depth required to get through the surface and transition region and therefore into the bulk? In my opinion, conventional X-ray methodology results in a considerable sampling of the surface and transition region in WC-base cermets. This means that X-ray studies (1) are sensitive to surface state and preparation, and (2) only indirectly yield information on the volumetric triaxial stress state even if no influence of preparation occurred.

With regard to (1), it seems unlikely that any surface polishing procedure will not influence the results obtained by X-ray analysis. The current practice of polishing until the measured shifts in peak position on psi-rotation have stabilized does not necessarily mean that surface preparation effects have been eliminated - it more likely means that they have stabilized. This is true even if linearity in d vs $\sin^2\psi$ plots is observed.

With regard to (2), even assuming no polishing effects, one cannot expect to directly observe the bulk triaxial stress state for such shallow beam penetrations. Indeed, if we accept the results of theory (and neutron data) that carbide particles are under hydrostatic stress, then the very observation of shift in peak position with sample rotation is evidence of surface relaxation because a hydrostatic stress would not lead to peak shifts upon rotation. Nevertheless, the sign should be correct even though the magnitude would be low. However, surface polishing effects could raise the magnitude, but for the wrong reason.

Turning more directly to the results presented, the remarks of Dr Luyckx are most appropriate - the surface region is being studied. The result that the carbide stress becomes more compressive and the binder stress less tensile means that the stress state has shifted in one direction (towards compression), a surface related behaviour. Nevertheless, the correlation of the stress pattern with the crack length anisotropy after compression is certainly real and significant. It is not, however, clear whether the anisotropic stress state is a surface and/or a volume effect. Our measurements were only for interplanar spacings normal to the uniaxial tensile axis. The magnitude of the relaxation and the consistency of the results with volumetric behaviour led us to presume that no substantial directionality was occurring in the bulk response of the system.

Inst. Phys. Conf. Ser. No. 75: Chapter 7
Paper presented at 2nd Int. Conf. Science Hard Mater., Rhodes

605

Some physical and metal-cutting properties of titanium carbonitride base materials

D MOSKOWITZ, L L TERNER AND M HUMENIK JR*

Ford Motor Co., Research Staff, 24500 Glendale Ave., Detroit, MI 48239
* Deceased

ABSTRACT

A series of unbonded Ti(C,N) solid solutions were prepared by hot-pressing, and lattice parameters and Knoop microhardness were measured on them. Two series of cemented Ti(C,N) alloys were prepared by bonding with Ni, 10 wt% Mo_2C and 10 wt% VC additions, and physical properties were measured in the TiN/(TiC+TiN) range from 0 to 0.6. Machining data were obtained for turning and milling gray cast iron, as well as for SAE 1045, 4140 and 4340 steels. Increased tool life with TiN addition was observed in all but one case, i.e., turning SAE 1045 steel.

INTRODUCTION

Titanium carbide (TiC) and titanium nitride (TiN) are isomorphous, and a continuous series of solid solutions can be prepared from them. Both materials can exist in compositions considerably below their stoichiometric values for carbon and nitrogen content, which affects their properties. TiO is also isomorphous with both TiC and TiN, so that oxygen, which is often present in these materials, appears on interstitial sites in Ti(C,N) solid solutions. Early work on these systems, which consisted essentially of X-ray studies, was published by Duwez and Odell (1950). More recently, work by Vil'k (1978), Vil'k and Danisina (1976) Grieveson (1967) and Yamamoto, Kieffer and Ettmayer (1974) investigated microhardness and lattice parameter variation in TiC-TiN solid solutions.

A number of studies involving cemented TiC base materials has shown that an improvement in physical properties can be obtained by TiN additions. Moskowitz and Humenik (1980) attributed an increase in resistance to plastic deformation to a grain refining effect observed when TiN was added to TiC-Ni-Mo materials. Tanaka (1981) noted improved tool life in turning 1045 carbon steel and turning and milling a Cr-Mo steel with TiN additions to TiC-5Mo$_2$C-10Ni compositions. This was correlated with observed increases in thermal conductivity and thermal shock resistance.

The present work consisted of two distinct investigations. In the initial portion, some properties of hot-pressed Ti(C,N) materials were measured and compared with recent literature values. The second section describes the results of studies involving cemented Ti(C,N) materials, with emphasis on their application in metal cutting.

EXPERIMENTAL PROCEDURE

Ti(C,N) materials were prepared as follows: discs were hot-pressed from approximately stoichiometric commercial TiC and TiN powders. Powders were first blended for 16 h, and then vacuum hot-pressed in graphite dies for five hours at 1800°C at a pressure of 27.6 MPa. Cemented Ti(C,N) speci-mens were prepared by ball milling these powders for four days with Ni, 10 wt% Mo_2C and 10 wt% VC additions, cold-pressing at 138 MPa, and vacuum sintering for 30 to 60 min at 1400° to 1550°C.

X-ray diffraction Debye-Scherrer photos were taken using CuK_α radiation and a 4 h exposure time. Three or four high-angle reflections were used, and lattice parameters were determined by extrapolation to 180° (2θ). Knoop microhardness measurements were made with 0.49N load.

SEM photomicrographs were taken. Fracture toughness was measured by the short rod method (Barker, 1977). Machining tests were run using lathe for turning, and a vertical milling machine for milling. 0.254 mm flank wear was used as the end-point on all machining tests. 12.5 mm square by 4.71 mm thick negative rake tool inserts were used, with 0.79 mm or 1.59 mm corner radii (styles SNG432 and SNG434, respectively).

RESULTS

Hot-Pressed Ti(C,N)

Table 1 lists the chemical analyses of three hot-pressed Ti(C,N) materials prepared at 25 wt% increments, together with those of the as-received powders.

TABLE 1 Chemical Analyses of Ti(C,N) Materials

Nominal TiC Content	Condition	Wt%			Interstitial Fraction			Total Interstitial Fraction
		C	N	O	C	N	O	
0	As Rec'd TiN	.065	23.4	.59	.003	1.035	.024	1.062
25	Hot-Pressed	4.92	17.3	.60	.246	.765	.024	1.035
50	Hot-Pressed	9.73	12.2	.56	.487	.540	.022	1.049
75	Hot-Pressed	14.51	6.0	.53	.726	.265	.021	1.012
100	As Rec'd TiC	19.23	.11	.52	.926	.005	.021	.952

Note that the total interstitial fraction of the elements carbon, nitrogen and oxygen is in all cases close to unity. The oxygen contents are in the range 0.56±0.04 wt%.

Fig. 1 shows the variation in lattice parameter as a function of the atomic fraction of carbon relative to the carbon plus nitrogen fraction C/(C+N). The relationship is linear, following Vegard's Law very closely. Vil'k and Danisina (1976) derived an equation that gives the lattice constant as a function of composition in the TiC_x-TiN_y quasi-binary system. This is also plotted in Fig. 1, using the appropriate "x" and "y" atomic fractions. As shown in the figure, there is excellent agreement between our experimental values and those calculated using Vil'K and Danisina's equation. Also shown in this figure are the ASTM values for the lattice constants of TiC and $TiN_{0.9}$ as measured by Christensen (1976).

Knoop microhardness measurements taken with a 0.49N load show a decrease in hardness from TiC to TiN (Fig. 2). The data follow a trend similar to that obtained by Smirnov (1978).

Cemented Ti (C,N) Properties

Two series of cemented Ti(C,N) composites were prepared by ball-milling TiC and TiN together with 10 wt% Mo+10 wt% VC+Ni powders. TiN/(TiC+TiN) contents covered the range from 0.10 to 0.60. Ni contents were chosen at levels of 12.5 and 22.5 wt%, corresponding to two grades of TiC-base cutting tool materials currently in commercial use.

TiN additions have a grain refining effect on cemented TiC base materials. Fukuhara and Mitani (1982) measured grain growth as a function of nitrogen content as well as sintering time in Ti(C,N)+20 wt% Ni+10 wt% Mo alloys. They found that the growth rate shows a minimum at N/(N+C) = 0.5, and that excessive growth is no longer observed where the ratio of nitrogen content, N/(N+C), exceeds a value of 0.2. Our observations confirm their findings.

Since increased sintering temperatures are required for materials containing larger amounts of TiN, relatively coarse carbonitride grain sizes result for compositions containing lesser amounts of TiN if all are sintered at the same elevated temperature. This is illustrated in Figs. 3A, 3B and 3C in the SEM photomicrographs of three compositions containing 0.13, 0.30 and 0.50 TiN/(TiC+TiN) ratios, all sintered at 1560°C for 30 min. Note the decrease in Ti(C,N) grain size with increasing TiN content.

In our investigation of the physical properties of these materials, we attempted to maintain a reasonably similar grain size for all the alloys. Accordingly, the following sintering conditions were chosen in order to minimize differences in Ti(C,N) grain size:

TiN (TiC+TiN)	Sintering Temperature °C	Sintering Time Min
0 - 0.15	1400	60
0.20 - 0.30	1475	60
0.40 - 0.60	1550	30

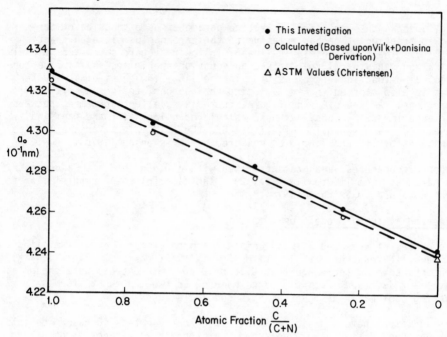

Fig. 1. Lattice parameters of Ti(C,N) materials vs. C/(C+N) ratio

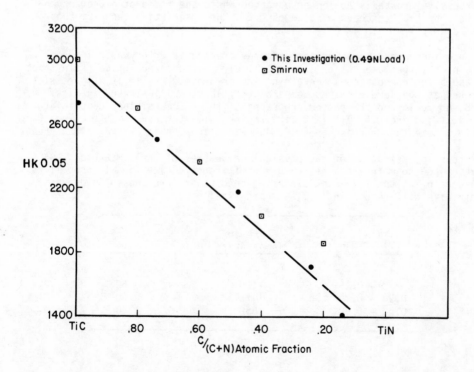

Fig. 2. Knoop microhardness of Ti(C,N) materials vs. C/(C+N) ratio

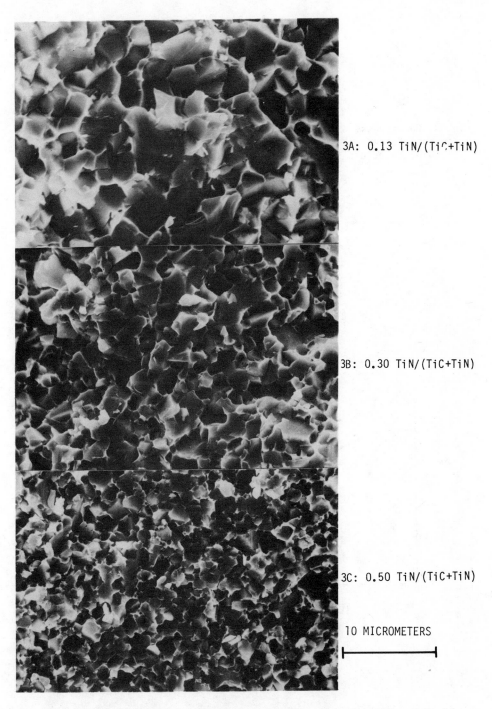

3A: 0.13 TiN/(TiC+TiN)

3B: 0.30 TiN/(TiC+TiN)

3C: 0.50 TiN/(TiC+TiN)

10 MICROMETERS

Fig. 3. Scanning electron micrographs (SEM) of Ti(C,N)+22.5 wt% Ni+10 wt% Mo+10 wt% VC sintered at 1560°C for 30 min

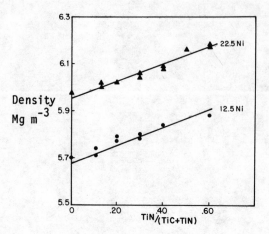

Fig. 4. Density vs. TiN
content of cemented
Ti(C,N) materials

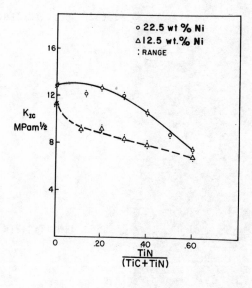

Fig. 5. Fracture toughness of
Ti(C,N)/Ni+10 wt% Mo+10 wt% VC
vs. TiN content
(short rod method)

This schedule yielded average Ti(C,N) grain sizes in the range from one to three micrometres, with the higher nitrogen content alloys tending toward the finer grain size.

Fig. 4 shows the variation in density as a function of TiN content with Ni content as parameter. Each series shows a moderate linear increase in density with TiN content, as expected.

Fracture toughness (K_{IC}) was measured as a function of TiN content at nickel levels of 12.5 and 22.5 wt%. These results are shown in Fig. 5. In each case there is a decrease in K_{IC} with increasing TiN/(TiC+TiN) ratio. This may be partially due to the decrease in grain size with TiN content, since fracture toughness of cemented carbides is known to drop with decreasing grain size. The tendency for K_{IC} to become less dependent on binder content at high TiN concentrations may be related to their very fine grain size, so that a sizable change in the amount of binder results in a relatively small change in its mean free path.

Fracture toughness was also measured at three different Ni levels for TiN/(TiC+TiN) fractions of 0 and 0.12±.01. These results are shown in Fig. 6, given as a function of the calculated vol%Ni as binder. There is an increase in K_{IC} with increasing binder content, regardless of TiN ratio. For such low TiN concentrations, in the range up to 0.12, K_{IC} appears to be independent of TiN content. Fracture toughness values are similar to those measured by August and Kalish (1983) using the same method for WC-Co base materials, when compared for similar volumes of binder.

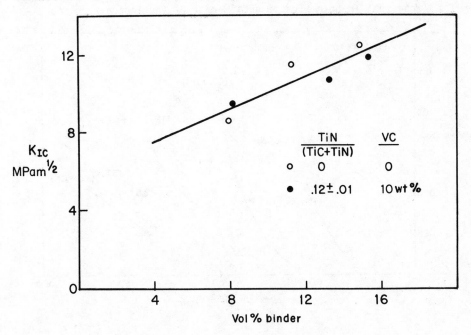

Fig. 6. Fracture toughness vs. binder content of Ti(C,N)-Ni-10Mo for TiN/(TiC+TiN) contents of 0 and 0.12

Machining Data

Earlier investigations in this system had shown that additions of TiN had a major effect on increasing the resistance to plastic deformation of the tool nose (Moskowitz and Humenik, 1980). Fig. 7 shows the nose deformation vs. cutting speed for materials containing 12.5 wt% Ni and 22.5 wt% Ni, after two minutes of cutting a 1.80 Ni, 0.80 Cr, 0.25 Mo, 0.40 C, 0.70 Mn steel (SAE 4340) of 300 HB. The positive effect of TiN/(TiC+TiN) fractions of 0.13 and 0.15, respectively, on deformation resistance was substantial. The allowable cutting speed for a similar amount of deformation was increased approximately 100 m/min.

If VC were added in addition to TiN, the effect on increasing deformation resistance appeared to be additive. Fig. 8 contrasts the tool nose deformation of a 12.5 wt% Ni material containing 0.25 TiN fraction and 10 wt% VC with one that had no additives. There was less deformation on the insert with additives after 32 min of machining SAE 4340 steel of 350 HB than there was on the one with no additive after only one minute of machining.

Turning tests were run on 12.5 wt% Ni materials at 183 m/min speed and 0.28 mm/rev feed on a pearlitic gray cast iron workpiece. As shown in Fig. 9, a substantial increase in tool life is obtained for increasing TiN/(TiC+TiN) ratios, reaching a maximum at a value of approximately 0.4. Fig. 10 shows the results of milling gray cast iron at 294 m/min and 0.15 mm/rev feed. An increase in tool life of approximately 50% over a material with no additive is observed for TiN ratios in the range of 0.1 to 0.3. Note that in both Figs. 9 and 10 most of the increase in tool

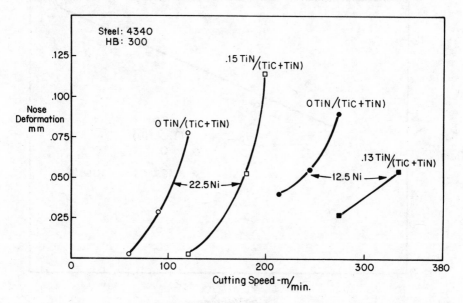

Fig. 7. Effect of TiN addition on the nose deformation of 12.5Ni and 22.5Ni materials

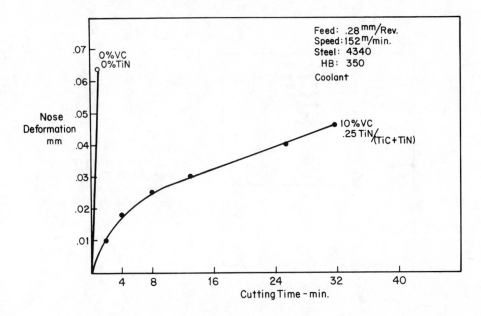

Fig. 8. Nose deformation of 12.5 wt% Ni-10 wt% Mo-TiC base materials
for TiN/(TiC+TiN) contents of 0 and 0.25 and VC contents of
0 and 10 wt%, respectively

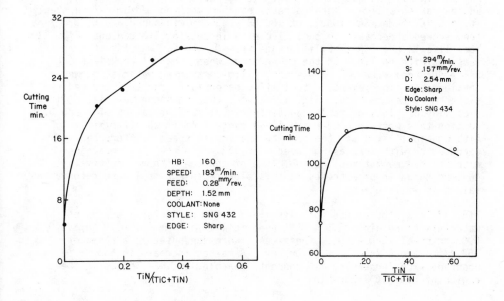

Fig. 9. Tool life of cemented
Ti(C,N) vs. TiN content turning
gray cast iron (12.5Ni grade)

Fig. 10. Tool life of Ti(C,N)+12.5
wt% Ni+10 wt% Mo+10VC vs.
TiN content milling gray
cast iron

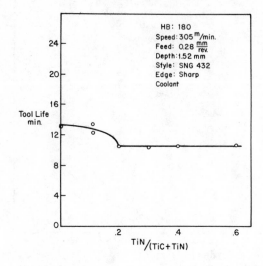

Fig. 11. Tool life of cemented Ti(C,N) vs. TiN content turning SAE 1045 steel (12.5 wt% Ni grade)

life occurs at a low TiN ratio. The reason for the appearance of a true maximum in milling cast iron (Fig. 9), but not in turning (Fig. 10) is unclear at this time. Turning tests on a normalized medium carbon steel (0.45 C, 0.75 Mn, SAE 1045), of 180 HB at 305 m/min speed and 0.28 mm/rev feed showed a decrease in tool life with increasing TiN content (Fig. 11). Tool life in the TiN/(TiC+TiN) range from 0.2 to 0.6 was 84% of that measured with no TiN. On the other hand, when a harder workpiece such as SAE 4340 steel with 300 HB was cut, a TiN-containing material showed a considerable improvement in tool life, essentially due to increased re- sistance to deformation. Fig. 12 shows a cutting speed versus tool life plot (Taylor plot) of tools containing TiC/(TiC+TiN) ratios of 0 and 0.25 cutting this workpiece over the speed range from 61 to 152 m/min. Although tool lives are the same at the lowest speed, differences are accentuated with increasing speed. At 152 m/min the TiN-containing material had many times the tool life of the one with no additive. Numbers shown in parentheses in this plot represent tool nose deformation values in millimeters.

Fig. 13 is a plot of the tool life versus TiN ratio for a series of materials containing 22.5 wt% Ni used to mill a 0.95 Cr, 0.20 Mo, 0.40 C, 0.85 Mn steel (SAE 4140). These tests were conducted at a speed of 158 m/min and a feed of 0.14 mm/rev. A marked increase in tool life is observed with increasing TiN content, reaching optimal values in the region of 0.4-0.6 TiN ratio.

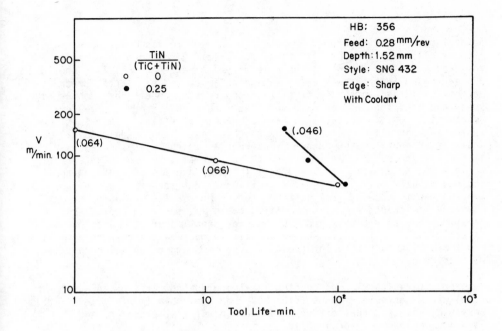

Fig. 12. Taylor plots of cemented TiC and TiC$_{.75}$N$_{.25}$ turning SAE 4340 steel (Ni content = 12.5 wt%)

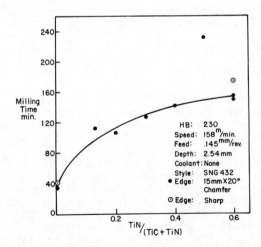

Fig. 13: Tool life of cemented Ti(C,N) vs. TiN content milling SAE 4140 steel (22.5 wt% Ni grade)

DISCUSSION

Several mechanisms have been proposed to account for increases in tool
life with TiN additions to TiC-base materials. Since nose deformation
resistance has been shown to increase with TiN content, tool life should
be lengthened in cases where the end-point is caused by plastic deformation. This is especially true if severe cutting conditions such as high
speed and high feed and/or hard workpiece are encountered. Increased
deformation resistance may also have a beneficial effect with respect to
inhibiting thermal cracks, as suggested by Braiden and Dugdale's (1970)
analysis. Tanaka's study showed that thermal conductivity increases with
TiN content, resulting in improvements in thermal shock resistance. This
would serve to explain increased tool life in intermittent cutting applications such as milling. However, this would not account for the improvements noted in turning gray cast iron or Cr-Mo steel. Variation in
abrasion resistance with TiN/(TiC+TiN) ratio may serve to explain this
behavior, but as yet this property has not been measured. Tanaka's
investigation convincingly showed a positive correlation between tool life
and thermal conductivity for these alloys, although no causal connection
was proved by him.

This investigation did not confirm the improved tool life on SAE 1045
steel with increasing TiN addition observed by Tanaka in his study. It is
possible, however, that differences between our test procedure and
Tanaka's may account for what appear to be conflicting observations. Our
data measured tool life after a constant value of flank wear, i.e.,
0.254 mm. Tanaka, on the other hand, plotted the flank wear after machining for a constant time of only three minutes. In our opinion, three
minutes is too short a duration of cutting time in which to compare values
of flank wear. In fact, the tool after 3 min may still be in the
"break-in" region of tool wear, which is far from linear. Our 0.254 mm
flank wear end-point was measured after from 11 min to 13 min of machining
time.

SUMMARY

Ti(C,N) solid solutions in the TiC-TiN quasi-binary system were prepared
by hot-pressing. Lattice constants and Knoop microhardness values were
measured as a function of C/(C+N) atomic fraction. These were compared
with recent studies by Vil'k in this system.

Two series of cemented Ti(C,N) alloys were prepared by bonding with Ni,
10 wt% Mo_2C and 10 wt% VC additions, and cold-pressing and sintering.
Physical properties were measured as a function of TiN/(TiC+TiN) ratio at
nickel levels of 12.5 and 22.5 wt%. Machining data were obtained for
turning and milling gray cast iron as well as for SAE 1045, SAE 4140 and
SAE 4340 steels. Increased tool life with TiN addition was obtained when
turning and milling gray cast iron as well as when machining SAE 4140 and
SAE 4340 steels. Only for the case of turning SAE 1045 steel at 305 m/min
was a decrease in tool life found with TiN additions.

Several mechanisms accounting for the observed increase in tool life were
discussed. Increased deformation resistance and improved thermal conductivity with TiN addition were presented as likely explanations.

ACKNOWLEDGMENTS

The authors wish to acknowledge the valuable assistance of C. R. Peters
for the X-ray work and R. L. Goss for the SEM photomicrographs in this
investigation. In addition, the valuable contributions of J. R. Dixon and
W. R. Felcher in obtaining the machining data in this study are gratefully
acknowledged. Thanks are given to T. J. Whalen and C. W. Phillips for
reviewing the manuscript.

REFERENCES

August, J. S. and Kalish, H. S., 1983, Int. Jour. Refr. and Hard Metals $\underline{2}$
    88-92
Barker, L. M., 1977, Engineering Fracture Mechanics, Pergamon Press, Great
    Britain, $\underline{9}$, 361-369
Braiden, P. M. and Dugdale, D. S., 1970, Materials for Metal Cutting, The
    Iron and Steel Institute, London, pp. 30-34
Christensen, A. N., 1976, J. Crystal Growth 33, pp. 99-104
Duwez, P., and Odell, F., 1950, J. Electrochem Soc., $\underline{97}$, 299-304
Fukuhara, M. and Mitani, H., 1982, Powder Metallurgy International, $\underline{14}$,
    No. 4, 196-200
Grieveson, P., 1967, Proc. British Cer. Soc., No. 8, 137-153
Moskowitz, D., and Humenik, Jr., M., 1980, Modern Developments in Powder
    Metallurgy, $\underline{14}$, 307-320, Metal Powder Industries Federation
Smirnov, G. V., 1978, Candidate's Dissertation, Leningrad (1971)
Tanaka, H., 1981, Cutting Tool Materials, pp. 349-361, American Society
    for Metals
Vil'k, Y. N. and Danisina, I. N., December 1976, Sov. Powder Met. and Met.
    Ceramics, $\underline{15}$, 932-936
Vil'k, Y. N., June, 1978, Sov. Powder Met. and Met. Ceramics, $\underline{17}$, 467-470
Yamamoto, T., Kieffer, R., and Ettmayer, P., 1974, 8th Plansee Seminar
    Proceedings, $\underline{4}$, 1-6

Inst. Phys. Conf. Ser. No. 75: Chapter 7
Paper presented at 2nd Int. Conf. Science Hard Mater., Rhodes

619

Estimation of the plastic zone size associated with cracks in cemented carbides

A-B LJUNGBERG(1), C CHATFIELD(1), M HEHENBERGER(1), B SUNDSTRÖM(2)

(1) AB Sandvik Hard Materials, Box 42056, S-126 12 Stockholm, Sweden
(2) Royal Institute of Technology, Division of Solid Mechanics,
    S-100 44 Stockholm, Sweden

ABSTRACT

It is shown by use of the finite element method that the maximum size of the plastic zone in front of a crack in cemented carbide is many times the dimension of the mean free path in the binder phase. Modelling grain boundaries between WC grains does not cause any change in plastic zone size.

INTRODUCTION

Recently a number of attempts have been made to develop equations that can predict the fracture toughness of cemented carbides (Murray, 1977; Pickens and Gurland, 1978; Chatfield, 1978). All these models assume that a critical parameter is the size of the plastic zone and that this is about the dimension of the mean free path in the binder phase. Another critical parameter is the plastic work required for crack propagation, γ_p. It is only by choosing extremely large values for this parameter (i.e. ten times larger than that obtained from a tensile test of the binder phase alone, Chatfield (1978)) that fracture toughness values in agreement with those measured are obtained.

The work to be presented was started with the aim of using the Finite Element Method (FEM) to follow the growth of the plastic zone in front of a stationary crack in a cemented carbide as a function of applied load to see what zone sizes are predicted at the stress intensity factor K_{1C} of the material.

COMPUTATIONAL DETAILS

Two cemented carbides grades with low and high cobalt contents were chosen for study and their microstructural parameters and mechanical properties are given in Table 1.

TABLE 1   Microstructural Parameters and Mechanical Properties of Grades studied, WC and Co.

Grade	Microstructural parameters				Mechanical properties			
	f_β	d_α /um	T_β /um	C	σ MPa	E GPa	ν	K_{1C} MPa m$^{1/2}$
	Idealized microstructure							
1	0.11	2.6	1.0	0.71	3300	626	0.21	
2	0.27	3.1	2.1	0.44	2000	470	0.22	-
WC					6000	707	0.18	-
Co					variable	176	0.31	-
	true microstructure							
1	0.11	2.8	0.9	0.57	3300	630	-	17.7
2	0.25	3.6	1.9	0.37	2100	540	-	23.1

Micrographs of the cemented carbides were used as a basis for the definition of the mesh used in the FEM calculations. An example of the FEM mesh used is shown in Fig. 1(a). This was chosen after testing a number of different mesh sizes and configurations. In trial calculations, assuming homogenous material, it gave values for K_1 expected from the linear elastic fracture mechanics (LEFM) equation:

$$K_1 = \sigma f(a,w)(\pi a)^{\frac{1}{2}}$$ 1

Note that the mesh size in the vicinity of the crack tip was deliberately reduced to increase the accuracy of the calculations in this region. The calculation of strains and stresses is performed at a number of gaussian integration points inside each 8-node element see Fig. 1(b). To apply the FEM mesh shown in Fig. 1(a) to a real microstructure is difficult thus the microstructures modelled were idealized somewhat before mesh definition. An example of a real and idealized microstructure is shown in Fig. 2. The microstructural parameters of the idealized microstructures were measured and are also given in Table 1. Values for E and ν for the WC-Co materials were estimated from the data of Doi, Fujiwara and Oosowa (1970). Values for their yield stress, σ , were estimated from the tensile fracture stresses reported by Doi (1974). This stress is somewhat higher than the true yield stress. Use of the fracture stress does, however, allow work hardening of the WC-Co composites to be taken into account.

To further simplify calculations the microstructure below the crack plane was made a mirror image of that above. To reduce computing times only the material in the vicinity of the crack was treated as two phase. The rest of the stressed area was comprised of material with the mean mechanical

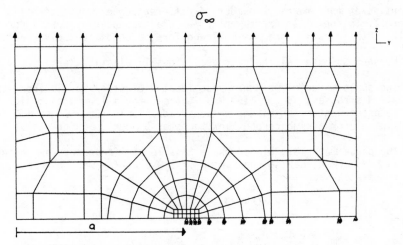

Fig.1a FEM mesh used to model microstructure.

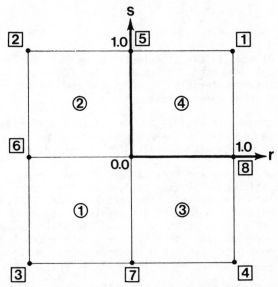

Fig.1b An 8-node element indicating the position of the 4 gauss
integration points at $r,s = \pm 0.57735$

Fig. 2a Real microstructure, grade 2.

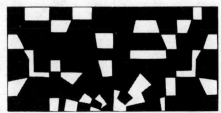

Fig. 2b Idealised version of microstructure.

properties of the grades. A number of different experiments were performed. In the first experiment WC/WC grain boundaries were ignored and growth of the plastic zone was followed for the conditions given below:

i)   homogenous material (both grades), ideally elastic/plastic.

ii) two phase material, binder phase ideally elastic/plastic and with yield stresses, $\sigma_{s\beta}$ = 1150 MPa, typical for a low carbon, high tungsten content in binder phase (grade 2 only) and $\sigma_{s\beta}$ = 870 MPa, typical for a high carbon, low tungsten binder (both grades).

iii) two phase material, binder phase capable of linear work hardening with a work hardening constant E_t = 2.9 GPa. The work hardening expression used is

$$\sigma = \sigma_{s\beta} + K\varepsilon_p \qquad\qquad 2$$

where

$$\frac{1}{K} = \frac{1}{E_t} - \frac{1}{E_o} \qquad\qquad 3$$

The assumptions of ideally elastic/plastic behaviour and linear work-hardening have been used previously in the FEM studies of deformation of two-phase composite materials (Fischmeister and co-workers, 1973; Karlsson and Sundström, 1974).

In (ii) above the yield stress, $\sigma_{s\beta}$, of the Co-based binder phase was estimated from the variation of hardness with tungsten and carbon determined by Roebuck and Almond (1981) for Co-base alloys.

In the second experiment the effect that WC/WC grain boundaries have on growth of the plastic zone was then studied for grade 2 using a network of "nodal-tie dislocation elements" (Bathe and Peterson, 1974) placed between the various WC "grains". These elements create a network of nodal points along both sides of a grain boundary so that the stiffness of the grain boundaries can be varied. The three stiffness parameters in the stiffness matrix D are d_{aa} and d_{cc} which are orthogonal to one another and lie in the boundary plane and d_{bb} which is normal to this plane.

An example of the "nodal-tie dislocation elements" is sketched in Fig. 3.

In Fig. 4 is illustrated the position of the maximum number of these elements used. This figure, when compared to Fig. 2, illustrates which grain boundaries between WC grains have been modelled.

Finally, the effect of the arrangement of the two phases directly in front of the crack upon plastic zone growth was also studied. The three cases examined are sketched in Fig. 5. Case 1, the crack lying in front of the binder phase, was the starting condition in experiments 1 and 2 above.

The microstructure in all the above cases was deformed in uniaxial tension by a load that increased in steps of 48 MPa. After each load increment the

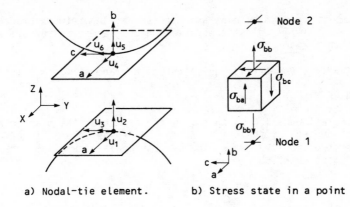

a) Nodal-tie element.   b) Stress state in a point

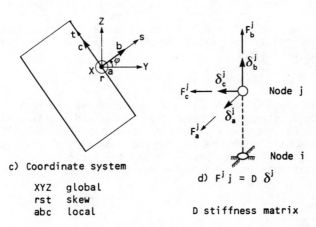

c) Coordinate system

XYZ   global
rst   skew
abc   local

d) $F^j{}_j = D\ \delta^j$

D stiffness matrix

Fig.3 A description of the nodal-tie dislocation element.

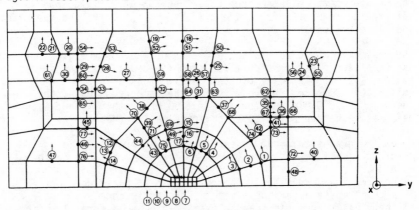

Fig.4 FEM mesh illustrating the position of the maximum number of
nodal-tie elements used. The elements a-direction is parallel
to the X-axis. The normal direction, b-direction, is indicated
for each element. The shear direction, c-direction, forms a
right-handed orthogonal system with a and b.

FEM programme calculated strains and stresses at the network of gaussian integration points. The plastic deformation of the microstructure was followed using the von Mises yield criterion.

The crack/loading geometry is equivalent to a plane strain K_{1C} test.

CALCULATION OF K_1

The FEM calculations allow us to follow the growth of the plastic zone, R_p and the "opening" of the crack. K_1 can be expressed as a function of r (r=distance from the crack tip) by rearranging the plane strain crack tip displacement equation

$$u = \frac{K_1}{\sigma}\left(\frac{r}{2\pi}\right)^{\frac{1}{2}}f(\theta,\nu) \qquad\qquad 4$$

Inserting the vertical displacements of the finite element nodal points along the crack surface into a fictitious stress intensity factor $K^*=K^*(r)$ given by the above mentioned equation and graphically extrapolating $K^*(r)$ to r=0 gives a K_1 value for each plastic zone size (Chan, Tuba and Wilson, 1970). However, R_p soon becomes larger than that allowed by LEFM namely:

$$a \geq 15\ R_p \qquad\qquad 5$$

where a is crack length and R_p plastic zone diameter in crack plane. In such a case the J-integral method should be used and compared to the J_{1C} value. This has not been done as time has not permitted the writing and testing of a computer programme for two phase materials. The diagrams to follow can thus contain K_1 values which are not strictly valid. However, this does not detract from the conclusions to be drawn.

RESULTS

Experiment 1

The growth of the plastic zone as a function of applied load is illustrated in Fig. 6 for grade 2 with $\sigma_{s\beta}$ = 870 MPa and the binder ideally elastic/plastic.

The maximum plastic zone size, R_p^{max}, is defined as the greatest extent of the plastic deformation from the crack tip. The stress intensity factor, K_1, calculated is plotted in Fig. 7 as a function of R_p^{max}. This figure compares the results for the homogenous case, for the case where $\sigma_{s\beta}$ = 870 MPa and for $\sigma_{s\beta}$ = 1150 MPa both with and without deformation hardening. From this figure we see that:

a) at the grade's critical stress intensity factor (K_{1C}=23.1) the value of R_p^{max} will be much greater than the mean free path in the binder.
b) deformation hardening of the cobalt binder phase does not significantly alter the size of the plastic zone.
c) the plastic zone size can be estimated quite accurately without having to resort to a two phase model of the material.
d) binder phase composition has a marked influence upon the plastic zone size.

WC/Co   Co/WC   WC/WC

Fig.5 Sketches of the various phase arrangements studied at the crack tip.

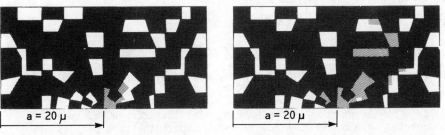

a = 20 µ   a = 20 µ

6a Applied stress, σ_∞ = 320 MPa, K_1 = 9.2 MPa m$^{1/2}$

6b Applied stress, σ_∞ = 464 MPa, K_1 = 17.4 MPa m$^{1/2}$

Fig. 6a,b Spread of plastic zone, idealized microstructure

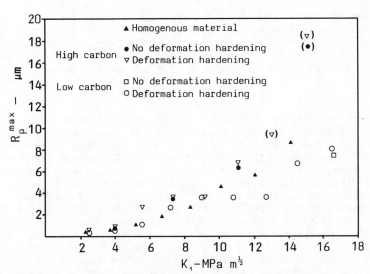

Fig.7 K_1 as a function of R_p^{max} for grade 2. Points in brackets are not valid.

Very similar results were obtained for grade 1.

It is also of interest to note that the plastic strain in the binder in the element directly ahead of the crack tip was 8 % at K_1=15 MPa m$^{\frac{1}{2}}$ for $\sigma_{s\beta}$ = 870 MPa.

Experiment 2

The effect of taking account of WC/WC grain boundaries is displayed in Fig. 8. Once again these results were obtained for grade 2. In Fig. 8 the plastic zone sizes and calculated K_1 values for the two phase material, mesh without nodal-tie elements, are compared to those obtained using the following nodal tie parameters:

i) $d_{bb} = d_{cc}$ = 0.5 E_{WC}, 49 nodal-tie elements

ii) $d_{bb} = 10^4 E_{WC}$, d_{cc} = 0.5 E_{WC}, 77 nodal-tie elements

In these three cases the binder phase was treated as elastic/plastic with $\sigma_{s\beta}$ = 870 MPa.

A large number of nodal-tie elements means that more WC/WC grain boundaries have been modelled. It would appear that the particular stiffness values chosen for d_{bb} and d_{cc} have very little effect upon the plastic zone size. Indeed it was found that only by setting d_{bb} and d_{cc} equal to zero, which is equivalent to introducing a large number of microcracks, could the plastic zone size be increased in size over that of the microcrack-free material.

Experiment 3

The local environment in front of a crack, i.e. whether there was binder phase or tungsten carbide or whether the crack stopped in a larger WC grain, had a significant effect initially on the morphology and extent of the plastic zone, Fig. 9(a). However, once the plastic zone had grown somewhat in size it became very similar for the three cases, Fig. 9(b).

DISCUSSION

In conclusion the results of these FEM calculations indicate that the plastic zone size in front of a stationary crack in cemented carbides is much larger than the dimension of a single mean free path in the binder phase as has been assumed previously (Murray, 1977; Pickens and Gurland, 1978; Chatfield, 1978). This conclusion is in agreement with Stjernberg's (1980) experimental evaluation of plastic zone size in cemented carbides. From his results he suggested that R_p^{max} was at least six times the carbide grains size. Our results also show that the plastic zone size can be further increased if failure of WC/WC boundaries occurs i.e. due to local strain concentrations formed in the plastic zone. An earlier FEM modelling of WC-Co microstructures by Sundström (1973) has shown that such strain concentrations occur. Heavy localized deformation of the binder phase resulting in dislocation pile-ups at WC grains has also been observed in electron-microscopy investigations of deformed WC-Co (Sarin and Johannesson, 1975; Roebuck and Almond, 1979).

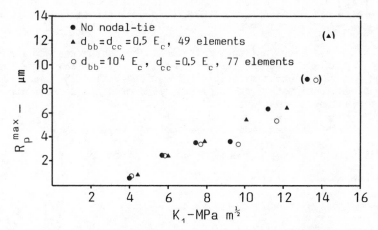

Fig. 8 Effect of modelling of WC/WC grain boundaries upon results in fig 7.

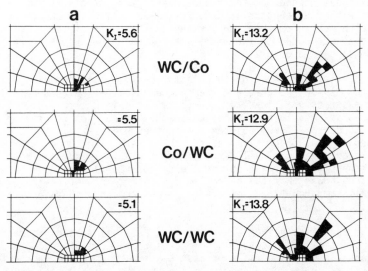

Fig. 9 Influence of phase arrangement at crack tip on plasticization at early, Fig. (a), and late, Fig. (b), stages of plastic zone growth.

Also, up to large R_p^{max} values little plasticity occurs in the WC. Generally only the particles in the immediate vicinity of the crack have yielded. This, however, will be partially a function of the yield stress chosen for the WC. This is function of particle size (Lee and Gurland, 1978) and also orientation (Lee, 1983). None of these factors have been taken into account in this work.

In this work plasticity in front of the crack has been treated as a two dimensional problem. Three dimensional modelling of the same type has been performed for single phase materials (Redmar and Dahl, 1980) and indicates that two dimensional modelling of plastic zone growth underestimates the extent of plastic deformation.

In these calculations two important assumptions have been made:

a) yield stress of binder phase is a function of composition only

b) stationary crack.

These assumptions will now be discussed.

Yield Stress of Binder Phase

It is well known (Roebuck and Almond, 1981; Rüdiger and co-workers, 1971) that the yield stress of Co-W-C alloys is very composition dependent. In cemented carbides it is generally assumed that in addition to this compositional dependence the binder phase yield stress is increased further either by constraint (Drucker, 1965; Almond, 1982) or by a Hall-Petch mechanism (Lee and Gurland, 1978). Both of these effects are assumed to occur due to the small size of the binder phase areas between the tungsten carbide grains.

In the earlier FEM study of the plastic behaviour of cemented carbide of Sundström (1973) reasonable agreement was obtained between calculated and measured stress-plastic strain curves if the yield stress of the binder phase was assumed to be that of unconstrained and coarse-grained binder phase. Sundström (1973) pointed out that even better agreement could be obtained if the binder yield stress could vary with composition. He did not rule out that the Hall-Petch effect could influence the binder phase yield stress to a certain extent. Sundström (1973) also noted that the FEM modelling showed that the hydrostatic stress component in the binder phase was small compared to the effective stress and thus that constraint is minimal.

The good agreement between Sundström's (1973) calculated and measured stress plastic-strain curves has lead us to the use of the composition dependent binder phase yield stress as a first approximation to the true yield stress in cemented carbides. The results of calculations using this yield stress and presented in Fig. 7 are felt to justify this choice. We see here that the plastic zone sizes calculated for the two phase microstructure are approximately of the same size at a given K_1 as that of the homogenous case.

Like Sundström (1973) we do not suggest that Hall-Petch effects are unimportant in WC-Co. We believe that they are smaller than suggested by

Lee and Gurland (1978). There is some experimental evidence in support of this suggestion. Lee and Gurland (1978) suggested the following equation relating **hardness and binder** mean-free path (1_β)

$$H_b = 304 + 12.7(1_\beta)^{\frac{1}{2}} \qquad\qquad 4$$

From a study of the deformation of a commercial Co-alloy Drapier and co-workers, (1970) have suggested the following relationship between composition independent yield stress and mean-free path between stacking faults or thin hexagonal lamellae

$$\sigma_{s\beta} = 640 + 7.83(1_b)^{\frac{1}{2}} \qquad\qquad 5$$

which, when converted to hardness (assuming $H=3\sigma$), yields:

$$H_b = 190 + 2.35(1_b)^{\frac{1}{2}} \qquad\qquad 6$$

Equation 6 has a much smaller dependence upon "grain size" than equation 4 and use of it would give much smaller increases in binder yield stress.

Stationary Cracks

Recent studies have shown that a surface crack in WC-Co advances by the formation of a number of microcracks at WC/WC interfaces in front of the main crack (Hong and Gurland, 1982; Almond, 1982; Sigl, Exner and Fischmeister, 1984). As the crack opens further these ligaments successively fail by ductile rupture.

The FEM calculations presented here are for a stationary crack as it is extremely difficult to perform such calculations for a moving crack in a two phase material. However, it is felt that the conclusions presented here are valid without restriction up to the point when the first WC/WC boundary ahead of the initial crack fails and the crack tip effectively advances. Our results also indicate that the occurrence of microcracks ahead of the original crack tip will increase the size of the plastically deformed binder phase zone. Indeed the process-zone observed by Almond (1982) can probably only be achieved by the occurrence of plastic deformation in the binder at large distance from the original crack. The strain concentrations occurring in this zone causing the weakest WC/WC boundaries to fail. As such his observations support the conclusion that the plastic zone in front of a crack in cemented carbides is much larger than one mean-free path in the binder phase.

ACKNOWLEDGEMENTS

The authors thank AB Sandvik Hard Materials for permission to publish this paper. The research reported was supported in part by a grant from the Swedish Board for Technical Development (STU).

REFERENCES

Almond, E. 1982, Speciality Steels and Hard Materials, Eds. N.R. Comins
    and J.B. Clark, Pergamon Press, Oxford (U.K), 353-360
Bathe, K.J. and Peterson, F.E. 1974, "Theoretical Basis for
    CEL/NONSAP, a non-linear structural analysis program", October 1974,
    CEL, Port Hueneme, California
Chan, S.K., Tuba, I.S. and Wilson, W.K. 1970, Eng. Fract. Mech., 2, 1-17
Chatfield, C. 1978, Proc. SEMP/5, P/M 78, Stockholm
Doi, H., Fujiwara, Y., Oosawa, Y. 1970, Met. Trans. 1, 1417-1425
Doi, H. 1974, Reviews on the Deformation Behaviour of Materials, Freund,
    Tel-Aviv, Israel, 1-108
Drapier, J.M., Viatour, P., Coutsouradis, D., Habraken, L. 1970,
    Cobalt,No. 49, 171-195
Drucker, D.C. 1965, High Strength Materials, Ed. V. F. Zackay, Wiley,
    New York, 795-830
Fischmeister, H.F., Hjälmered, J.O., Karlsson, B., Lindén, G.,
    Sundström, B. 1973, Proc. 3rd Int. Conf. Strength of Metals and Alloys,
    Cambridge (U.K), 621-625
Hong, J., Gurland, J. 1982, Science of Hard Materials,
    Eds. R.K. Viswanadham, D.J. Rowcliffe and J. Gurland, Plenum Press,
    New York, 649-666
Karlsson, B. and Sundström, B. 1974, Mat. Sci. Eng., 16, 161-168
Lee, H.C. and Gurland, J. 1978, Mat. Sci. Eng., 33,125-133
Lee, M. 1983, Met. Trans 14A,1625-1629
Murray, M.J. 1977, Proc. Roy. Soc. London A356, 483-508
Pickens, J.R. and Gurland, J. 1978, Mat. Sci. Eng., 33,135-142
Redmar, J. and Dahl, W. 1980, Proc. Int. Conf. Numerical methods
    in fracture mechanics, Swansea (U.K), 295-308
Roebuck, B. and Almond, E. 1979, Conf. Recent Advances in Hardmetal
    Production, Loughborough Univ. (U.K), Metal Powder Reports
Roebuck, B. and Almond, E. 1981, Proc. 10th Plansee Seminar, 493-508
Rüdiger, O., Hirschfeldt, D., Hoffman, A., Kolaska, J., Ostermann, G. and
    Willbrand, J. 1971, Tech. Mitt. Krupp Forsch. Ber., 29, 1-14
Sarin, V.K. and Johannesson, T. 1975, Metal Sci., 9, 472-476
Sigl, L.S., Exner, H.E., Fischmeister, H.F., 1984, this conference
Stjernberg, K.G. 1980, Metal Science, 14,189-192
Sundström, B. 1973, Mat. Sci. Eng., 12, 265-276

Inst. Phys. Conf. Ser. No. 75: Chapter 7
Paper presented at 2nd Int. Conf. Science Hard Mater., Rhodes

631

Characterization of fracture processes and fracture relevant parameters in WC-Co hardmetals

L S SIGL, H E EXNER, AND H F FISCHMEISTER

Max-Planck-Institut für Metallforschung, D-7000 Stuttgart 1, FRG

ABSTRACT

Experimental results demonstrating the important role of a multiligament zone at the crack tip and the limited width of deformation during crack propagation in WC-Co alloys are shown. The contributions of the four possible crack paths to the fracture energy are discussed. A quantitative model relating fracture toughness to the microstructural parameters is presented which is in agreement with the experimental facts and phenomenological observations.

INTRODUCTION

The fracture behaviour of cemented carbides has been the object of intense research work for more than three decades, but many aspects concerning crack propagation are still a matter of debate. Many attempts to design a physical model for the fracture toughness of WC-Co can be found in the literature (Johannesson 1975, Chermant and Osterstock 1976, Gille 1977, Murray 1977, Lindau 1977a, Chatfield 1978, Pickens and Gurland 1979, Nakamura and Gurland 1980, Nidikom and Davies 1980, Hong 1981, Viswanadham, Sun, Drake and Peck 1981, Almond 1982, Krstic and Komac 1985). All models which try to relate toughness parameters (critical strain energy release rate, critical stress intensity factor) to microstructural parameters (carbide grain size, binder phase intercept length) considering the mechanical properties of the individual phases suffer from more or less oversimplified, unjustified or unproven assumptions with respect to the details of crack propagation.

We studied the phenomena occurring during fracture of two-phase WC-Co alloys in detail in order to establish a sound model for predicting critical strain energy release rates as a function of microstructural geometry and individual phase properties.

EXPERIMENTAL

A series of WC-Co alloys (Table 1) were produced under carefully controlled conditions. The binder phase composition was adjusted to constant carbon and tungsten levels at 95 % of the maximum magnetic saturation. The binder phase content was varied between 10 and 20 vol % and the carbide grain size from 0.8 to 2.2 μm.

TABLE 1   Microstructural Parameters

Alloy	V_V^{Co} %	$\bar{D}_{WC}$* µm	$\bar{L}_\beta$* µm	ℓ_{50}^β µm	$\ln\sigma$	C
6 F	9.8	0.83+0.02	0.26+0.01	0.20	0.68	0.662
8 M	13.1	1.43+0.03	0.51+0.02	0.37	0.81	0.577
8 C	13.1	2.16+0.05	0.72+0.02	0.51	0.82	0.566
9 C	14.8	1.83+0.04	0.69+0.02	0.47	0.89	0.531
10 M	16.4	1.50+0.02	0.58+0.02	0.42	0.81	0.533
10 C	16.4	2.18+0.03	0.84+0.02	0.60	0.83	0.491
12 M	19.7	1.68+0.04	0.74+0.02	0.53	0.82	0.433

*95 % confidence intervals (mean $\pm$ 2 s where s = standard deviation)

SENB specimens (dimensions 35x6x2 mm) with a Chevron notch were used to determine G_{IC}. Precracking was carried out by stable crack growth in a displacement-rate controlled testing machine with a specially designed stiff loading cell (3 point bend test, roller spacing 28 mm). After marking the crack front by an oxidation treatment in air (300 °C, 1h) the SENB specimens were fractured under a loading rate of 20 Ns^{-1}. From crack length, a, which was measured in the light microscope (magnification 80x), the critical load, F_c, and the specimen dimensions, fracture toughness, K_{IC} was calculated. G_{IC} was obtained from K_{IC} using the elastic moduli measured by Doi and co-workers (1970).

The side planes of the specimens were polished and SEM micrographs (secondary electron mode) of the same area before and after passage of the crack were used in order to assess the details of deformation and microfracture during crack propagation in a more quantitative way. Semiautomatic linear image analysis was applied to SEM micrographs of randomly oriented polished cross sections for precise measurement of the linear intercept distributions in the binder phase.

CRACK PROPAGATION IN WC-Co ALLOYS

A large amount of fractographic work has been carried out during the past ten years which indicates that four types of fracture paths exist in WC-Co alloys (for a review see Fischmeister 1981).

B   =   Transgranular fracture through the binder phase

B/C =   Interface fracture along the carbide-binder phase boundary

C/C =   Intergranular fracture along the carbide grain boundaries, and

C   =   Transgranular fracture through the carbide crystals

Interface fracture B/C has been considered only by a few authors (Chermant and Osterstock 1976, Nakamura and Gurland 1980, Lea and Roebuck 1981).

It is generally agreed that the main contribution to fracture energy comes from path B. Nevertheless carbide fracture plays a decisive role since it precedes binder fracture and determines the direction and the type of path (B or B/C) in the binder as will be shown below. A prediction of the area fractions of the four types is difficult and, to our best knowledge, no reasonable model is yet available. Quantitative fractography has been used, but with disputable results (Fischmeister 1981). Even quantitative Auger spectroscopy (Viswanadham, Sun, Drake and Peck 1981, Lea and Roebuck 1981) is doubtful, since fractography using only one part of the fracture surface does not allow to differentiate clearly between interface (B/C) and binder fracture (B). Convincing evidence that macroscopically stable and unstable fracture proceed in an identical way has been obtained by comparing fracture energies and quantitative parameters of fracture surface geometry. Therefore, we believe that the area fractions can be determined from the line fractions measured on SEM micrographs of the polished surface containing a stably grown crack. This procedure seems to yield the most reliable data of all techniques used up to now since it allows a clear distinction between the four fracture types. An example is shown in Fig. 1 and results are given in Table 2.

TABLE 2   Area Fractions of Fracture Paths and Toughness Parameters

Alloy	A_A^B	$A_A^{B/C}$	$A_A^{C/C}$	A_A^C	$2\bar{r}_B$	R	$G_{IC}{}^*$
	%	%	%	%	μm	J/m²	J/m²
6 F	8.6	14.9	60.7	16.8	0.247	196	202±8
8 M	12.9	16.6	43.2	27.3	0.435	308	313±6
8 C	15.0	14.7	33.0	37.3	0.542	376	383±7
9 C	16.4	17.5	33.7	32.4	0.513	396	408±6
10 M	14.8	21.6	39.3	24.3	0.473	376	385±6
10 C	17.3	19.0	31.4	32.3	0.593	451	448±7
12 M	18.9	21.5	32.6	27.0	0.555	470	475±10

*95 % confidence intervals (mean $\pm$ 2 s where s = standard deviation)

Fig. 1
Side view of crack in
alloy 10 C

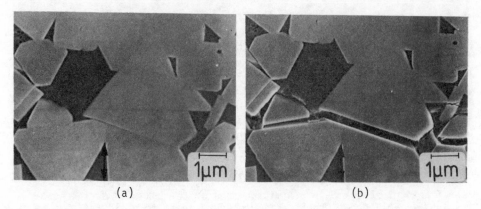

Fig. 2 Same field of microstructure before (a) and after (b) crack propagation in alloy 9 C

The Process Zone

Diverging views exist in the literature about the process zone around the crack tip. Gille (1977) and Almond (1982) assume an extended zone containing a multitude of carbide microcracks. In compressively deformed specimens Roebuck and Almond (1979) have demonstrated microcracking in the carbide and a process zone with microcracks has frequently been observed in ceramic materials (e.g. Al_2O_3, Al_2O_3-ZrO_2, Al_2O_3 + glassy phase). However, in the WC-Co alloys investigated in the study we did not detect any microcracks in the carbide which did not become part of the macrocrack nor did we observe crack branching.

Most authors assume that the size of the plastic zone is given by the mean linear intercept of the binder phase, $\bar{L}_\beta$. Johannesson (1975) as well as Chermant and Osterstock (1976) suggest an even larger extension, and work presented during this conference (Ljungberg and co-workers) seems to suggest a very large plastic zone associated with cracks.

Our observations clearly contradict these assumptions. Signs of plastic deformation, mostly in the form of a pronounced surface relief, are clearly visible only in binder phase regions through which the crack has passed (Fig. 1). Measurements of the dimensions of cobalt regions normal to the crack plane before (Fig. 2a) and after (Fig. 2b) passage of the crack (e.g. the chord marked in Fig. 2a,b) proved that the plastic strain in the cobalt regions not directly involved in crack propagation is zero at a statistical significance level of 90 %. Thus $\bar{L}_\beta$ is an upper limit for the mean extent of the plastic zone. A more detailed discussion on the actual extent of plasticity is given below.

Cracking of the Carbide: Formation of a Multiligament Zone

According to Evans, Heuer and Porter (1977) crack propagation in brittle materials with a small amount of ductile phase starts in the brittle phase. A crack in the brittle matrix circumvents the ductile regions leaving ligaments across the crack faces. Thus there is no sharply-defined crack tip: A "multiligament" zone is formed which consists of a continuous crack in the brittle phase interrupted by ligaments of ductile phase.

A multiligament zone was observed by Hong and Gurland (1982) in WC-Co alloys with 40 vol % binder phase during mode I fracture and by Almond (1982) at the tips of Palmqvist cracks for WC-Co alloys in the range of technical compositions. We have observed the multiligament zone at the crack tip of stably grown cracks in all alloys investigated. The length of this zone was found to be of the order of 5 $\bar{D}_{WC}$.

The typical features in the vicinity of the crack tip are shown in Fig. 3. At the right hand side of the figure, the microstructure is still fully coherent and both phases are deformed elastically. On the left hand side, both the carbide and the binder phase are fractured. In the multiligament zone between these two regions a three-dimensionally continuous crack path exists in the carbide skeleton while the binder phase is deformed plastically but is still coherent.

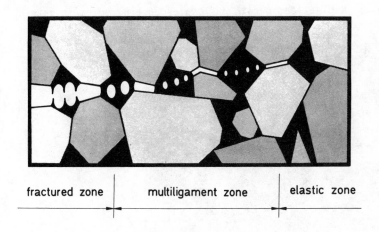

| fractured zone | multiligament zone | elastic zone |

Fig. 3 Crack tip region in a WC-Co alloy (schematic)

Deformation of the Binder Ligaments

A drastic change of the stress situation occurs when the multiligament zone begins to form: The parallel loading of both phases with the carbide carrying the major part of the load breaks down; the cobalt ligaments must now carry the full load, and they begin to deform plastically.

The plane strain state at the crack tip and the stiff carbide skeleton do not allow any contraction of the ligament when the crack opens. Instead, the elongation of the ligament must be compensated, to comply with the condition of constant volume, by formation of voids in the binder (Fig. 4).

From SEM micrographs of fracture surfaces (Fig. 5) the mean distance $\bar{s}_B$ between the dimples (corresponding to the mean distance of the pores) was found to be approximately 1.0 μm for all alloys studied here. Furthermore, the maximum depth of the dimples, h_B^{max}, was measured by means of an instrumented stereometer (Bauer and Haller 1981). As shown in Fig. 6, h_B^{max} is approximately 1.2 μm and does not vary systematically with alloy structure. This means that the geometry of the dimples (size, distance and shape) does not depend upon the microstructure in the range of $\bar{L}_\beta$ and V_V^β investigated here.

Fig. 4 Nucleation, growth and coalescence of voids in path B

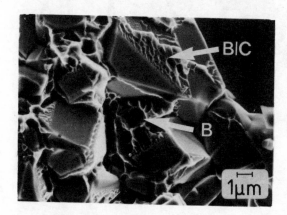

Fig. 5
Dimpled fracture path B
and B/C in alloy 10 C

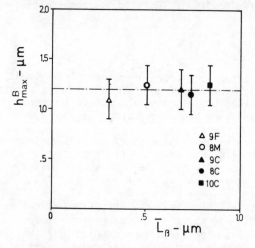

Fig. 6
Maximum dimple depth in path B, h_{max}^B, plotted against $\bar{L}_\beta$ for different alloys

The details of void nucleation are not clear yet. Under the cooling conditions and carbon content used, dispersed particles which usually give rise to heterogeneous nucleation are not present (Jonsson and Aronsson 1969, Schlump, Kolaska and Grewe 1982) and were not found in dimples. We assume that homogeneous void nucleation takes place which might be triggered by the stress induced cubic to hexagonal lattice transformation and assisted by the hydrostatic tensile component of the stress field.

Growth of the pores in the direction of normal stress is limited by impingement with a carbide/binder interface and by lateral coalescence as shown in Fig. 4. The binder is deformed to the maximum depth r_{max} where the coalescence of voids is the limiting factor for plastic deformation but to a smaller depth where the phase boundaries limit the ductile volume. The first type of event limits the maximum dimple depth while the second gives rise to a fraction of small dimples. The dimple depth determines the extent of plastic deformation in the ligament.

Due to the fact that the carbide crack in the multiligament zone serves as a precursor for the binder crack paths, the local geometry determines whether the crack proceeds through the binder or along the interface (Fig. 7). From scanning micrographs (Fig. 5) it becomes clear that in the latter case the crack does not propagate exactly along the interface. Rather, it proceeds parallel to it in the cobalt phase forming closely spaced shallow dimples. We suppose that the nucleation of these voids is heterogeneous and starts at the intersection of slip bands in WC with the phase boundary. Both the spacing and the height of dimples were found to be fairly uniform over the fracture surface. Stereometry yielded values close to 0.1 µm for both $\bar{s}_{B/C}$ and $h_{B/C}$.

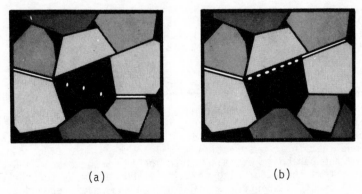

(a)   (b)

Fig. 7 Crack propagation through the binder (a) or along carbide/binder interface (b) determined by local position of carbide microcracks

CALCULATION OF THE CRITICAL STRAIN ENERGY RELEASE RATE

Obviously, the crack propagation resistance, R, (corresponding to G_{IC}), is the sum of energies dissipated in the four fracture paths B, B/C, C and C/C. This sum can be calculated if the specific fracture energies w_i (i.e. the energy consumed for creating unit area of fracture path i) and the area fractions A_A^i are known:

$$G_{IC} = \Sigma R_i = R^B + R^{B/C} + R^{C/C} + R^C \qquad \text{1a}$$

$$R_i = 2w_i A_A^{\ i} \qquad \text{1b}$$

The surface energies of carbide and binder are negligible compared to measured values of G_{IC}. Only the energies consumed by plastic deformation need to be taken into account and can be calculated or determined from independent measurements, as follows.

Crack Resistance of Crack Paths C and C/C

From indentation tests, Warren (1978) and Warren and Matzke (1982) estimated a critical fracture energy, G_{IC}^{WC}, of about 50 J/m². Extrapolation of fracture toughness values for a variety of WC-Co alloys to pure WC yields a similar value (Warren and Matzke 1982). Thus, the fracture energy is more than an order of magnitude higher than the surface energy of WC (2.8 J/m², Johannesson 1975). The fracture energy for C/C boundaries and for transgranular fracture is not known, but we will not go far wrong if we take

$$2w^C \simeq 2w^{C/C} \simeq G_{IC}^{WC} = 50 \text{ J/m}^2 \qquad 2$$

and

$$R^C + R^{C/C} = G_{IC}^{WC} (A_A^C + A_A^{C/C}). \qquad 3$$

Fracture paths C and C/C account for a large part of the fracture surface, and therefore their contribution to R is of the order of approximately 20 % for a WC-6wt% Co alloy. For a 12 wt% Co alloy it drops to 5 %.

Crack Resistance in Crack Paths B and B/C

The specific fracture energy of a dimple fracture can be estimated according to a model suggested by Stüwe (1980) which was modified for the present purpose. We assume that all dimples have identical shape, height and width and that the cross section normal to the dimple ridge can be described by the parabolic equation given in Fig. 8 where h, m and c are the parameters characterizing the dimple. Our estimate for the possible range of m is 1.5 to 3. Since the influence on the result proved to be minor, we used a value of 2 for the sake of mathematical tractability.

Local strain is a function of height within the dimple walls. The energy expended per unit volume that has been deformed to the true strain ϕ is then given by $\int_0^\phi \sigma(\phi)d\phi$. The total energy, W, needed to form the dimple structure from a flat slab of binder phase (Fig. 8) can be calculated, by integrating over all volume elements, as

$$W = \int_{V_0}\int^\phi \sigma(\phi)d\phi dV \qquad 4$$

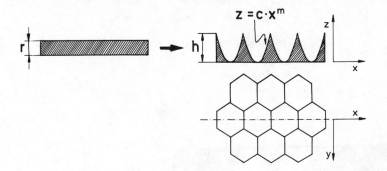

Fig. 8 A simplified model for a fracture surface with uniform dimples

The flow stress in the binder phase, $\sigma(\phi)$, is given by the Voce equation

$$\sigma(\phi) = \sigma_y + (\sigma_s - \sigma_y) \ (1 - \exp \ (-\phi/\phi_c)) \qquad\qquad 5$$

where σ_y is the yield stress, σ_s is the saturation stress, ϕ is the true strain and ϕ_c is given by

$$\phi_c = (\sigma_s - \sigma_y)/\Theta_o \qquad\qquad 6$$

Θ_o is the initial work hardening rate which, according to Almond and Roebuck (1981) is approximately $E_\beta/12$. Solving equation (4) for the fracture morphology in Fig. 8 and the Voce equation, the energy w for forming unit area of the dimple structure is obtained as

$$w = 1/4h \ (\sigma_y + (\sigma_s - \sigma_y)/(2\phi_c + 1)) \qquad\qquad 7$$

Volume conservation yields

$$h = 2r \qquad\qquad 8$$

and thus

$$w = 1/2r \ (\sigma_y + (\sigma_s - \sigma_y)/(2\phi_c + 1)) \qquad\qquad 9$$

where $\sigma_y + (\sigma_s - \sigma_y)/(2\phi_c + 1)$ is the mean flow stress in the binder. This result indicates that w is independent of dimple shape (i.e. c) and only depends on the mean flow stress and on the thickness of the undeformed layer r.

However, due to the impingement of pores with the carbide interfaces, r is not a constant for fracture path B. Fig. 9 shows a typical situation: The binder is deformed to the maximum depth r_{max} only when the coalescence of voids is the limitirา factor for plastic deformation, but the actual deformation depth is less than r_{max} where the phase boundaries limit the extension of plastification in the binder.

Taking this limitation into account, the average depth of the deformation, r_B, is given by

$$\bar{r}_B = \int_0^{r_{max}} r \cdot g(r)dr + r_{max} \int_{r_{max}}^{\infty} g(r)dr \qquad 10$$

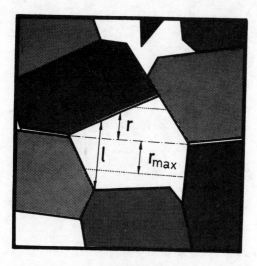

Fig. 9
The limits for the extension r of plastic deformation in path B: $0 < r < r_{max}$. ℓ indicates a random intercept in the binder

where $g(r)$ is the frequency of r. This is derived from the probability of dividing a binder intercept of length ℓ in such a way that a segment of length r is obtained, and from the intercept distribution $f(\ell)$. We assume that all partition ratios for dividing an intercept of length ℓ have the same probability. The binder intercept distributions $f(\ell)$ of all alloys studied here were found be log-normal and can be characterized by ℓ_{50} and σ, the median and the geometric standard deviation. After some algebraic manipulation we obtain

$$g(r) = \frac{\exp\left(1/2\,(\ln \sigma)^2\right)}{\ell_{50}} \int_{\ln r}^{\infty} \frac{1}{\sqrt{2\pi}\,\ln\sigma} \exp\left(-1/2\left(\frac{\ln \ell/\ell_{50}+\ln^2\sigma}{\ln \sigma}\right)\right)^2 d\,\ln\ell \qquad 11$$

From equation (8) we get r_{max} from the maximum dimple height h^B_{max}. Thus, $r_{max} = h^B_{max}/2$ and, with h^B_{max}, ℓ_{50} and $\ln\sigma$ measured experimentally, $\bar{r}_B$ can be calculated. The results are shown in Fig. 10. $\bar{r}_B$ approaches r_{max} for very coarse alloys and $\bar{L}_\beta$ for very fine alloys. Usually, neither of these limits is a correct estimate for $\bar{r}_B$.

For fracture path (B/C) the estimation of r in equation (9) is much simpler: Since all dimples in path B/C have nearly the same size and depth, we get again from equation (8)

$$r_{B/C} = 1/2 \; h_{B/C} \qquad\qquad 12$$

The Flow Stress of the Binder

The flow stress of the binder during fracture of WC-Co alloys has been estimated by Lindau (1977b). He, however, neglects the fact that the phase boundaries of the composite are much stronger obstacles to strain propagation than ordinary cobalt grain boundaries. Chou (1966) has suggested a modification of the Hall-Petch equation which is pertinent to this situation:

$$\sigma_y = \sigma_y^0 + k_y \; \bar{L}_\beta^{-1/2} \qquad\qquad 13$$

with

$$k_y = [G_{IC}^{WC} \cdot \frac{\mu_{Co}(2+(\pi-2))K}{2\pi(1-\nu_{Co})}]^{1/2} \qquad\qquad 14$$

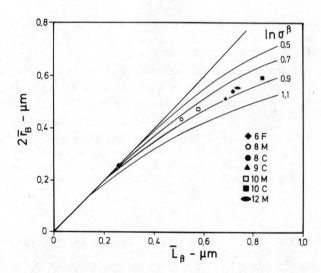

Fig. 10 Average width of deformed zone in the binder phase around the crack $2r_\beta$, as a function of mean free path in the binder $\bar{L}_\beta$, according to equation 10

where K is a function of the shear moduli μ and Poisson's ratios ν of binder and carbide. G_{IC}^{WC} is the critical strain energy release rate of the carbide, σ_y^0 is a friction stress depending on binder phase composition and $\bar{L}_\beta$ is the mean linear intercept of the binder phase. With $G_{IC}^{WC} = 50 \; J/m^2$ as above and K = 0.532, k_y is 1.55 $MN/m^{3/2}$. σ_y^0 was taken to be 480 MPa considering Almond and Roebuck's (1981) composition dependent data using the magnetic saturation (Tillwick and Joffe 1973) as an indicator for the composition.

The strain hardening of the Co phase was modelled by Voce's law (equation 5). The maximum strain hardening, $\sigma_s-\sigma_y$, is governed by the structure of the binder-phase at high strains (dislocations, twins, stacking faults), where an equilibrium between hardening and softening processes exists. A rough approximation for $\sigma_s-\sigma_y$ based on an analysis for the stress to pass a dislocation through a stacking-fault intersection (Rajan and Vandersande 1982) and on the assumption of a maximum dislocation density $\rho_{max}= 10^{16}m^{-2}$ yields a value of about 11000 MPa. Hence with equation (6) and E_β = 200 GPa we obtain $(\sigma_s-\sigma_y)/(2\phi_c+1)$ = 4740 MPa.

COMPARISON MODEL/EXPERIMENT

Summing up the energies spent in the 4 different crack paths the crack resistance of WC-Co alloys becomes

$$R = (\sigma_y^0+1550 \, \bar{L}_\beta^{-1/2}+4740) \, (\bar{r}_B \, A_A^B + r_{B/C} \, A_A^{B/C}) + 50(A_A^{C/C}+A_A^C) \quad J/m^2 \qquad 15$$

where σ_y^0 is taken from Almond and Roebuck (1982). $\bar{L}_\beta$ (in μm) is the mean linear intercept, $\bar{r}_B$ (in μm) is calculated from equation (10) using the experimental values of the parameters ℓ_{50}, $\ln\sigma$ of the binder intercept distribution, and the measured maximum dimple height, h_{max}^B. The value of $r_{B/C}$ (in μm) is given according to equation (12) from the measured dimple height $h_{B/C}$, and the area fractions for the four fracture paths, A_A^B, $A_A^{B/C}$, $A_A^{C/C}$ and A_A^A are determined experimentally. The values of R thus calculated for the seven alloys investigated are shown in Table 2 and plotted against the measured critical energy release rates G_{IC} in Fig. 11. The excellent agreement obtained without any adjustable parameters indicates that all essential factors are included in this model.

CONCLUSIONS

From the phenomenological fracture studies and the calculations of this work we can draw the following conclusions:
1.   Previous models describing the fracture toughness of WC-Co alloys as a function of their microstructural geometry are not in agreement with the processes actually occurring during crack propagation.
2.   Because of constraints by the rigid carbide skeleton plastic deformation of the binder phase is restricted to a very narrow zone adjacent of the crack. The mean width of the plastic zone in path B $2r_B$ is smaller that the mean free path in the binder $\bar{L}_\beta$.
3.   The plastic strain in the deformed binder is very high. The average true strain in the dimples is 0.5 corresponding to a technical strain of 65 %. This is in contrast to earlier models where plastic strains of the order of 10 % were assumed.

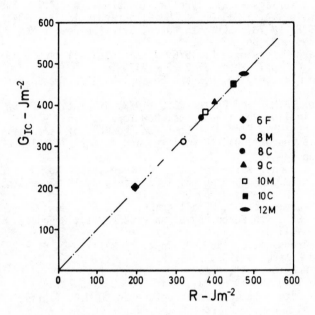

Fig. 11 A plot of measured G_{IC} against calculated fracture resistance R

4.   Since the crack to a large extent proceeds through the carbide phase, the contribution of this type of fracture must not be neglected in spite of its low specific fracture energy. In low cobalt alloys, this contribution amounts to 20 % of the total fracture energy, and it still amounts to 5 % in high cobalt alloys.

5.   The fracture resistance of the carbide exerts a strong influence on the flow stress of the binder via the constant k_y. This, combined with the strong work-hardening of the cobalt binder, is thought to be the reason why the fracture toughness of WC-Co alloys is superior to that of other carbide/binder combinations.

ACKNOWLEDGEMENT

The authors gratefully acknowledge assistance from Metallwerk Plansee AG Reutte, Austria, with specially produced alloy specimens and technical expertise, and from Kontron Bildanalyse GmbH, Munich, for the loan of an image analysis system IBAS.

REFERENCES

Almond, E.A. 1982, Speciality Steels and Hard Materials Ed. N.R. Comins
    and J.B. Clark, Pergamon Press, Oxford, pp. 353 - 360
Almond, E.A. and Roebuck, B. 1981, Proc. 10th Plansee Seminar Ed.
    H.M. Ortner, Verlagsanstalt Tyrolia, Innsbruck, pp. 493 - 508
Bauer, B. and Haller, A. 1981, Prakt. Metallographie 18 327 - 341
Chatfield, C. 1978, PM 78 - SEMP 5th Europ. Symp. on Powder Met.,
    Stockholm, Vol. 2, pp. 57 - 62
Chermant, J.L. and Osterstock, F. 1976, J. Mat. Sci. 11 1139 - 1951
Chou, Y.T. 1966, J. Appl. Phys. 37 2425 - 2429
Doi, H., **Fujiwara,** Y., Miyake, K., and Oosawa, Y. 1970, Met. Trans. 1,
    1417-1425
Evans, A.G. Heuer, A.H., and Porter, D.L. 1977, 4th Int. Conf. on Fract.
    Ed. D.M.R. Taplin, Pergamon Press, New York, pp. 529 - 556
Fischmeister, H.F. 1981, Proc. 1st Conf. on the Sci. of Hard Mat.
    Ed. R.K. Viswanadham, D.J. Rowcliffe, and J. Gurland, Plenum Press,
    New York, pp 1 - 42
Gille, G. 1977, VI. Int. Pulvermet. Tag. Dresden, Paper No. 6
Hoffmann, H., Bock H., and Wenzek, R. 1972, Wiss. Z. TH Magdeburg 16, 31-38
Hong, J. 1981, Ph. D. Thesis Brown University
Hong, J. and Gurland, J. 1981, Metallography 14 225 - 236
Johannesson, T. 1975, 4th Europ. Symp. on Powder Met., Grenoble,
    Paper No. 5 - 11
Jonsson, H. and Aronsson, B. 1969, J. Inst. of Metals 97 281 - 287
Krstic, V.D. and Komac, M. 1985, Phil. Mag. 51, 191-203
Lea, C. and Roebuck, B. 1981, Met. Sci. 15 262 - 266
Lee, H.C. and Gurland, J. 1978, Mat. Sci. Eng. 33 125 - 133
Lindau, L. 1977a, 4th Int. Conf. on Fracture Ed. D.M.R. Taplin,
    Pergamon Press, New York, pp. 215 - 223
Lindau, L. 1977b, Scand. J. Met. 6 90 - 91
Murray, M.J. 1977, Proc. R. Soc. Lond. A 356 483 - 508
Nakamura, M. and Gurland, J. 1980, Met. Trans. A 11A 141 - 146
Nidikom, B. and Davies, T.J. 1980, Planseeber. f. Pulvermet. 28 29 - 38
Pickens, J.R. and Gurland, J. 1978, Mat. Sci. Eng. 33 135 - 142
Rajan, K., and Vandersande, J.B. 1982, J. Mat. Sci. 17, 769 - 778
Roebuck, B. and Almond, E.A. 1979, Conf. on Recent Advances in Hardmetal
    Production, Loughborough Univ., Paper No. 28
Schlumpp, W., Kolaska, J., and Grewe, H. 1982, Metall 36 535 - 540
Stüwe, H.P. 1980, Eng. Fract. Mech. 13 231 - 236
Tillwick, D.L. and Joffe, I. 1973, Scripta Met. 7 479 - 484
Viswanadham, R.K., Sun, T.S., Drake, E.F. and Peck, J.A. 1981, J. Mat.
    Sci. 16 1029 - 1038
Warren, R. 1978, Acta Met. 26 1759 - 1769
Warren, R. and Matzke, H. 1981, Proc. 1st Int. Conf. on the Sci. of
    Hard Mat. Ed. R.K. Viswanadham, D.J. Rowcliffe, and J. Gurland,
    Plenum Press, New York, pp. 563 - 582.

Inst. Phys. Conf. Ser. No. 75: Chapter 7
Paper presented at 2nd Int. Conf. Science Hard Mater., Rhodes

645

Fracture toughness testing of WC-Co hardmetals precracked by indentation

LU YUANGMING

Zhuzhou Cemented Carbide Industry Company, Hunan, China

ABSTRACT

The transverse rupture test cannot be used to judge the intrinsic mechanical properties of hardmetals because it is sensitive to porosity and surface defects. Fracture toughness is a more relevant property but conventional testing methods require sophisticated equipment and specimen geometries. A test procedure has been developed which involves calculating the fracture toughness from measurements of the bend strength of specimens that have been precracked by indentation to provide surface cracks of known depth. The method is suitable for routine quality control of hardmetals.

INTRODUCTION

It is well known that the diagonal length of a Vickers hardness indentation increases with indentation load. For example, with a 294 N load for WC-Co hardmetals, the diagonal length of the indentation is 170-230 µm, its depth is 25-35 µm, and there is a long diagonal crack and a basal crack in the low Co hardmetals. This would be equivalent to a large surface flaw in a hardmetal and, assuming it can be used as the fracture origin, the procedure for precracking in fracture toughness testing will be simplified. An advantage of using an indentation is that the notch roots at the corners and bottom of the indentation are much smaller than those at the bottom of spark-cut or sawn notches, which are about 100-200 µm. Under the action of stress, the stress concentration at the corners of the indentation can be considered to be similar to the stress singularity at a crack tip. Therefore it is possible to deal with the fracture problems of indented specimens according to the basic principles of fracture mechanics.

EXPERIMENTAL

The work was done on seven WC-Co hardmetal grades: YG6X, YG6Y, YG8, YG105 YG11, YG11C, YG15, with different Co contents and different grain sizes, including fine (X), coarse (C) and medium. Table 1 lists their composition and relevant properties. The fracture toughness tests were performed in 3-point bend on single edge notch beam specimens, 10 x 10 x 55 mm, over a span of 40 mm, using a 5 mm deep precrack introduced by electro sparking with a Mo wire of 0.1 mm dia.

TABLE 1   Composition and Properties of Specimens

Grade	Co content wt %	Mean grain size μm	HRA	HV30	TRS kN mm^{-2}	K$_{IC}$ MN m$^{-3/2}$
YG6X	6	1.0-1.2	92.5	1700-1770	1.96	11.7
YG6	6	1.5-2.0	92.0	1650-1700	2.11	12.8
YG8	8	1.5-2.0	90.0	1500-1550	2.30	14.3
YG11	11	1.5-2.0	88.5	1300-1350	2.35	-
YG15	15	1.5-2.0	87.5	1150-1200	2.74	16.8
YG105	10	2.5-3.0	87.5	1220-1270	2.55	18.7
YG11C	11	2.5-3.0	86.5	1150-1200	2.55	19.5

The bend test specimens, 5 x 5 x 30 mm, were polished with diamond paste
in a metallographic polishing machine to remove the as-sintered outer
layer. A Vickers indentation was then produced at the centre of the
tensile stress face, and the length d of the indentation diagonal cracks
was determined. After this the specimens were oxidised at 400-500 °C in a
muffle furnace to indicate the depth h of the indentation crack. The
rupture load was measured and the strength σ_c was calculated.

RESULTS

Observation of the Fracture

Figs. 1a-e show the fracture of WC-Co hardmetal indented specimens that
had not been oxidised. The indented specimens' fracture surfaces are smoother than those of non-indented specimens; moreover, in almost all specimens
the fracture passed through the indentation diagonals, indicating
that the indentation acted as the fracture origin or as the site where the
fracture initiation processes started.

The fracture may be divided into two main types, one purely brittle
fracture, the other partly brittle fracture. The YG6X, YG6 and YG8 grades
exhibit the purely brittle type which is smooth and even. The partly
brittle fracture occurs mainly in high Co hardmetals, their fractures
involving three regions. The first region is the crack origin, it includes
the indentation cavity and a small area under the cavity which is smooth
and fine. The second region is large and rough, where radial markings
can be seen. The third is a lateral ripple section, it is also smooth and
fine but is not even. In this experiment the indentation is an artificial
flaw that corresponds to a "white spot" (Suzuki and Hayashi, 1974).

The above fracture characteristics describe the failure process of indented
crack specimens. When the specimen is subjected to external force, the
stress concentration will be established at the indentation base or at the
crack tip beneath it. When the concentrated stress reaches the ultimate
strength of the material, it will make a crack initiate at the tip of the
indentation bottom and make the crack propagate rapidly. The purely-
brittle smooth fracture will occur when the rapidly propagating crack is
not plastically controlled. In high Co hardmetals the crack propagation

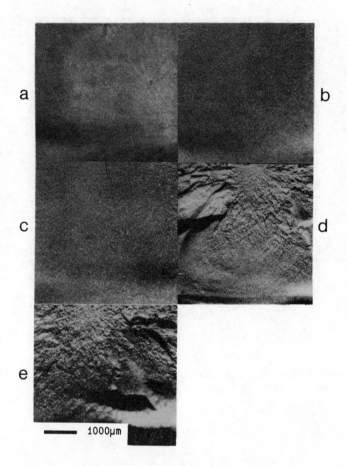

Fig. 1 a-e) Fracture surfaces of hardmetals indented with a Vickers
indentation load of 30 kg (294 N):

a) YG6X; b) YG6; c) YG8; d) YG11; e) YG11C

is plastically controlled and the speed of crack propagation decreases,
under alternate action of the propagation-control-propagation, the crack
propagation enters the stable stage, which produces many small plastic
deformation zones, thus making fracture of the second section become
coarse. When the length of the crack reaches a certain critical value,
propagation enters the unstable stage, and the crack propagates at an
extremely high speed, and produces a lateral ripple effect under the action
of compression, finally leading to catastrophic failure of the specimen.

Crack Shape and Stress Intensity Factor

After oxidation of the indented specimens, the morphology of a section
through the indentation could be seen on the fracture surfaces when the
specimen was subsequently tested as shown in Figs. 2a-f.

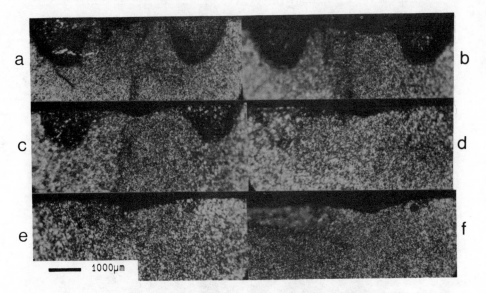

Fig 2. a-f) Fracture through heat-tinted indentation
cracks showing crack profiles

a) YG6X; b) YG6; c) YG8; d) YG11; e) YG105; f) YG15

The variation of the shape parameters (Fig. 3) of the indentation with the
indenting load is given in Table 2.

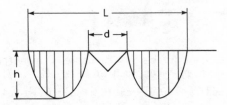

Fig. 3 Crack-shape parameters

For the low Co hardmetals such as YG6X, YG6, YG8, the crack shape is given
in Figs. 2a,b and c. The crack depth was much larger than that of the
indentation. The crack shape was determined by the propagation of a
crack from the bottom of the indentation to the surface where it connected
with the diagonal crack. Median vent cracks (Almond and Roebuck, 1979)
were not observed in this experiment. With increasing Co content the
possibility of forming this crack shape decreased because at lower
indenting loads it is difficult to produce a diagonal crack.

The crack shape can be characterised by the ratio h/L (or $2h/L$). For a
given hardmetal, the crack shape remained basically insensitive to indenting load, but for different hardmetals, different crack shapes developed.
The higher the Co content, the smaller the ratio h/L. For Co contents

above 10%, the ratio h/L approached 0.10-0.12. In fact, it was 0.1-0.125 of the depth of indentation.

TABLE 2   Indentation Load and Crack Shape

Grades	Load mass kg	Parameters of crack μm			Crack shape	
		h	C*	L	h/C	h/L
YG6X	5	35	115	160	0.26	0.21
	10	45	145	210	0.27	0.21
	20	65	240	320	0.25	0.20
	30	95	380	420	0.25	0.23
	50	140	550	610	0.25	0.23
	100	210	800	850	0.26	0.24
YG6	10	30	140	180	0.21	0.17
	20	50	245	300	0.20	0.17
	30	65	310	360	0.21	0.18
	50	100	465	515	0.22	0.19
	100	160	780	830	0.21	0.19
YG8	30	40	265	290	0.15	0.14
	50	70	395	470	0.17	0.15
	100	115	670	735	0.16	0.16
YG11	30	25	235	250	0.11	0.10
	50	40	350	370	0.11	0.11
	100	75	520	615	0.14	0.11

C* - length of the lateral diagonal crack including indentation diagonal.

It appears that there was no crack extension (change in precrack shape) prior to failure. For example, separate tests showed that the crack depth did not increase when a load was applied to the sample prior to fracture (Table 3).

TABLE 3   Parameters of Crack Shape in Rupturing Process of Specimens

Grades	Specimen No	Indentation mass kg	Prior bending load kN	Failure load kN	Parameter crack shape μm		
					h	L	h/L
YG6X	1	20	0.74	1.96	65	300	0.22
	2	20	1.13	1.76	70	330	0.22
	3	20	1.47	2.06	60	290	0.21
YG105	4	50	4.80	7.64	36	330	0.11
	5	50	3.63	7.15	40	350	0.12
	6	50	2.40	6.66	37	340	0.11

Thus, it appears that the crack shape was related only to the indenting load and not to the prior bending load. It is concluded from this that before failure, the vertical diagonal crack did not extend with increase in bending load to develop a crack across the width of the specimens. It appears that the indentation crack can be regarded as a semi-elliptical surface crack, in this case the long axis is L, and the short axis is approximately 2h.

If the indentation crack is regarded as a semi-elliptical surface crack, several expressions for the stress intensity factor are available. The Kies equation (Chu, 1979) can be used for a shallow surface crack in three-point bend. It is, however, simpler to use Irwin's analysis (1962) for the indentation crack which can be an extremely shallow surface non-through crack, and gives:

$$K_I = \frac{\sigma\sqrt{\pi a}}{\Phi}\left(\sin^2\phi + \left(\frac{a}{c}\right)^2 \cos^2\phi \right) \qquad 1$$

Where Φ is an elliptical integral of the second kind:

$$\Phi = \int_0^{\frac{\pi}{2}} \left(1 - \frac{c^2 - a^2}{c^2} \sin^2\phi \right)^{\frac{1}{2}} \qquad 2$$

The elliptical parameters are shown in Fig. 4 and Φ can be expressed as a series expansion:

$$\Phi = \frac{\pi}{2}\left\{ 1 - \frac{1}{4}\cdot\frac{c^2 - a^2}{c^2} - \frac{3}{64}\left(\frac{c^2 - a^2}{c^2}\right)^2 \cdots \right\} \qquad 3$$

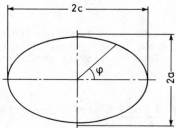

Fig. 4 Elliptical parameters used in calculating geometrical factor

Usually the terms after the third can be neglected, to give:

$$\Phi = \frac{3\pi}{8} + \frac{\pi a^2}{8c^2} \qquad 4$$

When K_I varies along the crack front, the stress intensity factors reaches a maximum at the end of the minor axis ($\phi = \frac{\pi}{2}$):

$$K_I(\phi = \frac{\pi}{2}) = \frac{\sigma\sqrt{\pi a}}{\phi} \qquad 5$$

For the critical condition, K_I is K_{IC}:

$$K_{IC} = \frac{\sigma_c\sqrt{\pi a}}{\phi} = Y\sigma_c\sqrt{a} \qquad \left[Y = \frac{\sqrt{\pi}}{\phi}\right] \qquad 6$$

where K_{IC} - fracture toughness
   σ_c - critical failure stress
   a - crack length
   Y - factor dependent on specimen and crack geometries.

For indentation cracks, there are different Y values (Table 4) owing to different ϕ values for various hardmetals, but generally speaking the difference is small. To a good approximation, Y is 1.5, which gives:

$$K_{IC} = 1.5\ \sigma_c\sqrt{a} \qquad \left[a = h\right] \qquad 7$$

TABLE 4   Crack Shape Factor for Various WC-Co Hardmetals

Grades	a/c (2h/L)	ϕ	Y
YG6X	0.44	1.25	1.42
YG6	0.36	1.23	1.44
YG8	0.30	1.21	1.46
YG11-15	0.22	1.20	1.48

Usually, K_I is corrected by 12% for the rear free surface. However, when the semi-elliptical crack propagates into the bulk of the material, the correction for the back free surface decreases from 1.12 to 1 (Broek, 1978). In addition, because hardmetals are basically brittle or semi-brittle materials, the size of the plastic zone at the crack tip is extremely small compared with the crack length and therefore the correction for the free surface and the plastic zone need not be considered.

Failure Strength and Fracture Toughness

The effect of the indentation load on the strength of the specimens is given in Fig. 5. Obviously, the bigger the indenting load, the deeper the indentation depth and the crack depth, and the lower is the failure

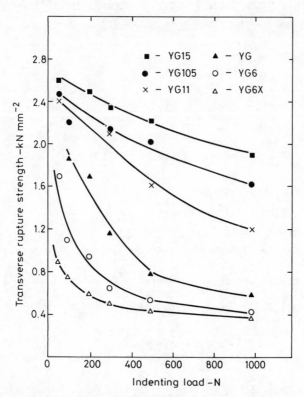

Fig. 5 The effect of indentation load on the transverse
rupture strength of the hardmetals

strength. The higher the rate at which the failure strength decreases with
the increase of the indenting load, the smaller is the ability of the
specimen to resist the crack propagation. In a sense, this gives an
indication of the fracture toughness of the material. If we choose the
failure strength at a certain indenting load for comparison, the results
obtained coincide with the order of the toughness of hardmetals determined
conventionally. For high Co hardmetals the failure strength can be chosen
for comparison only at high indentation loads, owing to the relatively
large scatter of the failure strength at low indentation loads.

When the transverse rupture strength was plotted as a function of the
inverse square root of the crack depth, $1/\sqrt{h}$, a good linear relationship
was obtained (Fig. 6). The fracture toughness can be determined from the
slope of the line. In fact, after measuring the crack depth of the specimens and failure strength, the fracture toughness of the material can be
calculated directly from equation (6) or (7).

For high Co hardmetals, because the crack depth is actually about one
eighth of the diagonal length of the indentation, K_{IC} can be obtained by
measuring the length of the diagonal and the transverse rupture strength
as shown in Table 5.

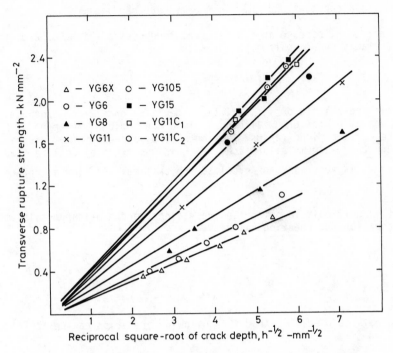

Fig. 6 Variation of transverse rupture strength with
reciprocal of square root of crack depth

TABLE 5   Fracture Toughness in $MNm^{-3/2}$ of WC-Co Hardmetals

Grades	YG6X	YG6	YG8	YG11	YG105	YG11C$_1$*	YG11C$_2$**	YG15
Calculated using h.	7.6	8.2	10.5	14.0	17.4	18.4	18.3	19.1
Calculated using d.	-	-	-	13.4	17.5	18.5	18.0	19.2

 * contains a little free carbon

** normal structure

It is concluded from this that the fracture toughness of high Co hardmetals
is inversely proportional to the fourth root of the hardness:

$$K_{IC} = \frac{A.\sigma_c}{\sqrt[4]{HV}}$$

8

where A is a constant and HV is the Vickers hardness.

Consequently, an increase in the fracture toughness of high Co hardmetals will inevitably decrease their hardness.

DISCUSSION

Hardmetals are brittle or semi-brittle materials; once crack propagation becomes unstable, failure will occur. In a conventional three-point bend test, the curve of load versus crack-opening-displacement (P-V diagram) is linear and the maximum failure load P_{max} is the critical failure load P_c. From this it is very easy to calculate the critical transverse rupture strength σ_c.

Since it is difficult to determine the yield strength of hardmetals, the normal equation for checking validity,

$$a \geq 0.5 \left(\frac{K_{IC}}{\sigma_s} \right)^2 \qquad\qquad 9$$

cannot always be used. Instead the following criterion can be applied:

$$a \geq 10\, r_y \left(= 0.5 \left(\frac{K_{IC}}{\sigma_s} \right)^2 \right) \qquad\qquad 10$$

where a - crack length

r_y - size of the plastic zone

σ_s - yield strength.

The above equation indicates that when $a \geq 10\, r_y$, the difference between the approximate solution $\sigma_y = K_I/\sqrt{2\pi r}$ and the accurate solution which is expressed by the K_I stress field is small ($\leq 7\%$) at the boundary of the plastic zone. That is, under the small area yield condition, the stress field at the crack tip is still controlled by K_I, so $K_I = K_{IC}$ can still be used as fracture criterion. For specimens in which a satisfies the above-mentioned condition, the K_I at the critical state is the fracture toughness K_{IC} of the material.

For hardmetals, the size of the critical plastic zone at the crack tip is probably a function of the average free path of the binder phase (Viswanadham and Venables, 1976), that is:

$$r_c = \alpha.\lambda \qquad\qquad 11$$

where r_c - radius of the critical plastic zone

α - constant (approximately $1/6\pi$ for plane strain)

λ - average free path.

From 10 and 11, it can be deduced that:

$$a \geq 10\,r_y = \frac{10\lambda}{6\pi} \qquad\qquad 12$$

The average free path of hardmetals is usually 1-5 µm. Obviously, in the present experiments the crack depth easily satisfies equation 12. Therefore, it is not necessary to carry out the validity check for the hardmetal indentation-bending strength method.

This method can be further simplified; for example, there is no need to grind and polish the specimens. Also, it has proved possible to calculate the fracture toughness from the results of two tests, instead of from the many measurements that were used to construct each curve in Fig. 6.

CONCLUSIONS

The method put forward in this paper for measuring the fracture toughness of WC-Co hardmetals is much simpler than the conventional three-point bend test and other methods.

1. Test specimens are of simple geometry.

2. Procedures for pre-cracking are simple.

3. The test procedure is simple and rapid.

4. The test results can be easily calculated.

5. The testing equipment required is simple and inexpensive.

Therefore, it is possible to use the method raised in this paper for testing fracture toughness for routine hardmetal quality control.

For high Co hardmetals it is more significant to calculate the fracture toughness directly using the indentation depth and the transverse rupture strength at certain indentation load. This is progress for industrial testing the fracture toughness of brittle or semi-brittle materials.

REFERENCES

Almond, E. A. and Roebuck, B. 1979, Recent Advances in Hardmetal
    Production, Loughborough U. and Metal Powder   Report, pp 31:1-31:30.
Broek, D. 1978, Elementary Engineering Fracture Mechanics. Sijthoff and
    Noordhoff, Netherlands, pp 77-83.
Chu, W. 1979, The Test for Fracture Toughness, Science Publishing House,
    Peking, pp 63-64.

Irwin, G. R. 1962, Trans. A.S.M.E., J. Appl. Mech. 29 651-654.
Suzuki, H. and Hayashi, K. 1974, J. Jap. I.M. 38 1013-1019.
Viswanadham, R. K. and Venables, J. D. 1976, Advances in Hard Material
    Tool Technology, Carnegie Press, Pittsburg, pp 245-259.

Inst. Phys. Conf. Ser. No. 75: Chapter 7
Paper presented at 2nd Int. Conf. Science Hard Mater., Rhodes

657

Micromechanics of fracture in WC-Co hardmetals

M ŠLESÁR, J DUSZA AND Ľ PARILÁK

Institute of Experimental Metallurgy, Slovak Academy
of Sciences, Košice, ČSSR

ABSTRACT

A study has been in WC-Co cemented carbides with grain sizes of 2.1-3.6 μm
and 13-32 vol % Co, of the relationship between the fracture toughness and
microstructural parameters and micromechanisms of fracture. Regression
analyses have been used to derive empirical relationships between fracture
toughness, and the binder spacing, the contiguity and the relative
proportions of fracture in the binder phase and between contiguous WC
grains.

INTRODUCTION

Owing to the excellent combination of useful properties of cemented
carbides WC-Co (hardness, strength, wear-resistance) they are amongst the
most frequently used tool materials. According to the purpose of application they are produced with different combination of grain size of the
carbide phase and volume fraction of Co-binder. Increasing their tool
life which is most often limited by brittle failure, necessitates a knowledge of the connection between the parameters of structure, the fracture
resistance and the individual fracture micromechanisms.

The WC-Co structure contains a hard and brittle carbide phase, and a
relatively soft and ductile binder, which is cobalt with dissolved carbon
and tungsten. The formation of the real WC-Co structure is defined by the
complex chemical and physical characteristics of the infiltrating phase
and the chosen technological regime of the production, which is reflected
in the final structure's quality and this determines the fracture deformation processes (size distribution of WC grains, distribution of Co-
binder, quality of grain boundaries, slip systems, dislocations, cracks
etc.). When studying fracture processes it becomes necessary to take all
the defects into account, which are a deficiency of the production
technology. In this connection we can include pores and their clusters,
structural heterogenities, inclusions, other phases and similar. It has
been proved that these defects exercise a decisive influence and introduce
scatter into the results of the three or four point bend strength. They
are the initiation sites for rupture and dependent on their size and
distribution in the volume they degrade strength (Almond and Roebuck,

1977; Dusza, 1982a). These defects overlap the structure's influence and its resistance against rupture. In order to, and with an aim to consider the structure's influence on fracture characteritics, several authors have applied LEFM and the evaluation of K_{IC} values (Inglestrom and Nordberg, 1974; Chermant and Osterstock, 1976; Pickens and Gurland, 1978).

The aim of this paper is to analyse the influence of the WC-Co system's structure on the K_{IC} values, to determine the basic micromechanisms of fractures, the conditions of their origin, the manifestation and its statistical representation on the fractures' surfaces, as well as their connection with fracture toughness.

BACKGROUND: FRACTURE TOUGHNESS, MICROMECHANISMS OF FRACTURE

In the case of brittle materials with limited ductility, including also cemented carbides of the WC-Co system, they to a great extent fulfil macroscopic conditions of stress - deformation characteristics of the linear elastic fracture mechanic. The determined values K_{IC} define the critical conditions of the material's resistance against crack initiation. For a complete understanding of its value in relation to the structure, it becomes necessary to decipher processes taking place in the process zone. From the point of view of experimental geometry there is the preparation of a prior artificial defect (notch, crack) of outstanding importance especially in relation to its radius. We can divide, as the most frequently used methods, the preparation of a notch into three groups:

a) the making of a sharp notch with a diamond wheel and by spark-erosion (Chermant and co-workers, 1974; Dusza 1982b)

b) the making of a sharp crack by means of an indenter (Almond and Roebuck, 1978; Warren and Johannesson, 1983)

c) the use of a chevron notch (Barker, 1980; Munz and co-workers, 1980).

In the first case it has been proved that as the notch radius decreases there is also a decrease in the value of K_{IC} (Fig. 1). This is connected with the different stress distributions ahead of the notch front for the blunt and sharp notch. When applying indenters the main problem rests with the reproducibility of the geometry of prior cracks, another problem is the elimination of residual stress. When making use of the chevron-specimen notch, there often occurs during the course of loading in three point or four point bend tests, formation of a natural defect - microcrack, the gradual growth of which is directly connected with the structure. For this reason this method has the most prospects. At present, the specimens SENB, short rod, short bar, are most frequently used. For the reason, that the value of K_{IC} in tests with the chevron notch expresses the resistance of the material to the fracture at a naturally formed structure crack and the attainment of its critical size, we consider it is important that in the WC-Co system the K_{IC} value is not dependent on the crack size and the crack growth resistance curve is flat (Munz, 1983).

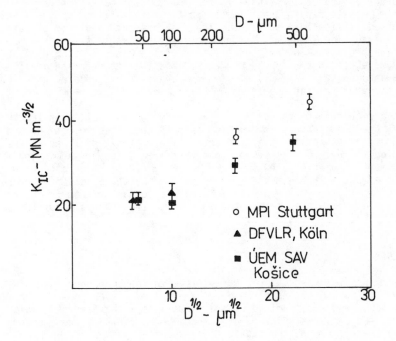

Fig. 1 Fracture toughness of WC-Co obtained with SENB specimens with notches of different width - D (Munz, 1983; Dusza, 1982a)

The aim of the K_{IC} value's determination for cemented carbides is to achieve information about the system's toughness, and thereby to define the critical defect size, which is permissible at a given level of stress. As already said, the deciphering of the structure's role depends on knowledge of microdeformations and microstress processes taking place in the process zone. It is necessary to recognise the extent of the influence of individual structure parameters, which makes it possible to suggest technological adjustments leading to structures of higher resistance against rupture.

The majority of authors attempt to relate K_{IC} values with structural parameters, mainly with D_{WC} - mean carbide grain size, L - mean free path in binder, and C_{WC} - contiguity. The evaluation of the relationship leads to the same conclusion: the K_{IC} values increase with the growth of D_{WC} and L_{Co}, and the C_{WC} has a degrading effect on K_{IC} (Chermant and Osterstock, 1976; Šlesár and co-workers, 1982).

On the other hand there exist only a few publications, which analyse fracture processes from a fractographic point of view, and K_{IC} values are related to individual fracture micromechanisms (Nidikom and Davies, 1980).

The fundamental methods of studying the micromechanisms of rupture are based on a metallographic analysis of the fracture path and on the fractographic study of fracture surfaces. The fracture path analysis presents us with information about metallographic manifestations of stress-deformation processes in front of the crack (slip-lines in Co and WC, microcracks etc.), but they do not make it possible to separate qualitatively and to quantify all types of micromechanisms of rupture, as we know it when studying fracture surfaces.

There have been efforts to use Auger spectroscopy, when studying fracture surfaces, but the obtained information is only quantitative, and does not make it possible to differentiate quantitatively single micromechanisms. In addition, it has to be stressed that this method is very demanding from an experimental point of view (Lea and Roebuck, 1981). According to our opinion the most suitable method is represented by the statistical stereofractographical observation of the related microareas on both fracture surfaces at a suitable enlargement according to the grain size of the system (Luyckz, 1977; Dusza, 1983).

EXPERIMENTS

The experiments were carried out on alloys with tungsten carbide grain size from 2.1 to 3.6 μm and with 13 to 32 vol % of binder phase. The fracture toughness was examined on specimens with a chevron notch which were fractured in three point bending mode. Dimensions of the chevron-notch specimens: the notch length at the specimen surface a_1 - 5 mm; the notch length to the tip of the chevron, a_0 = 1 mm, the width and the thickness of the specimen W = B = 5 mm. The width of the saw-cut and the span of the bend test rig was D = 100 μm and L = 15 mm, respectively. The load displacement curve departs from linearity due to stable crack extension before maximum load and unstable crack extension at maximum load. Specimen test load was applied at a constant test machine cross--head speed of 0.1 mm/min. With the assumption of a flat crack-growth resistance curve, maximum load for the chevron notch configuration can be used directly to determine the fracture toughness. The calculation of K_{IC} is based on previous work by the authors (Munz and co-workers, 1980). Metallographic examinations were made on a light microscope, the microstructural parameters were determined by statistical quantitative metallography. The characteristic features of the principal micro-mechanisms and their relative proportions on the fracture path or fracture surface were determined by the metallographic and stereofractographic methods on the scanning electron microscope.

RESULTS AND THEIR ANALYSIS

For the analyses, five structural stages have been used; firstly, the values L_{Co} and C_{WC} of which, as well as K_{IC} values are evident from Fig. 2. Thus the results are in full agreement with the literature on the influence of structural parameters on the K_{IC} values.

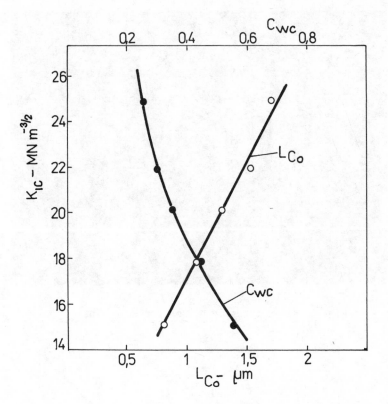

Fig. 2   Variation of K_{IC} with L_{Co} and C_{WC}

In the next stage we carried out metallographic analyses of fracture paths and their process zones. Within the process zone there are microdeformation processes with a high degree of heterogeneity taking place. They show themselves by the occurrence of slip lines in Co binders, but also in WC grains, Fig. 3. The exhaustion of the deformation ability of the Co-binder leads to the initiation of microscopic cracks, usually on the boundaries WC/Co by pile-up mechanism. For the relaxation - deformation processes, Co-binders are mainly responsible, which corresponds with the fact that with the increase of L_{Co}, the size of the process zone also increases (K_{IC} values are higher).

When applying the statistical evaluation of rupture mechanisms from the fracture paths it became difficult to differentiate between the fracture types WC/Co - interfacial rupture on the carbide-binder interface, and Co/Co - ductile rupture in the binder. Therefore we tried statistical stereo-fractographic analysis of fracture surfaces. The comparison of the results achieved by measurement with both methods is shown in Fig. 4. There exist fundamental differences in the quantification of rupture types Co/Co and WC/Co, therefore special attention has been given to those micromechanisms. From the analysis there are the following three basic

Fig. 3   Slip lines in binder and WC grains

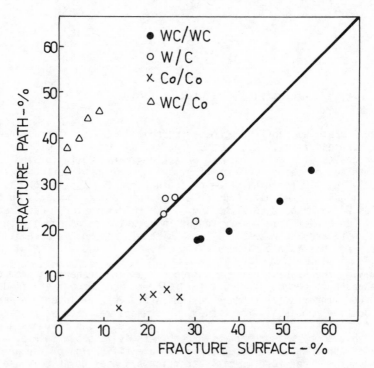

Fig. 4   Fracture micromechanisms on the
fracture path and on the fracture surface

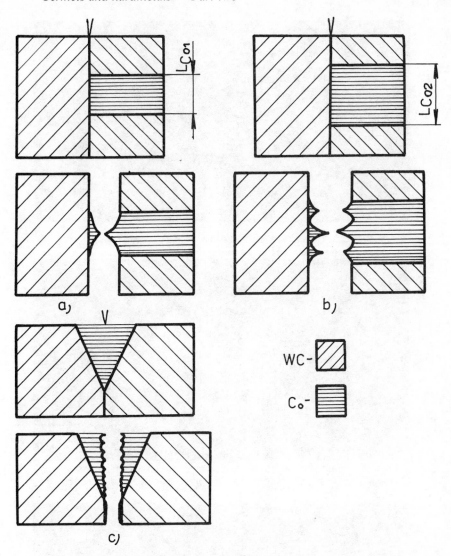

Fig. 5   Rupture modes in the binder phase

morphological manifestations of Co/Co failure, the model of which is shown
in Fig. 5 a,b,c and on fractographic pictures in the sequence Figs. 6 - 8.
The morphological manifestation of the ductile rupture is immediately
connected with the size L_{Co} and on the orientation of Co - ligaments
to the direction of major crack propagation and the operating stress. We
observed statistically at low L_{Co} values, ductile neck ruptures of the
binder, Fig. 9, with an increase of L_{Co} value, neckings are more
evident, Fig. 10 and the more spacious L_{Co} areas are ruptured by
dimple mechanisms, Fig. 11.

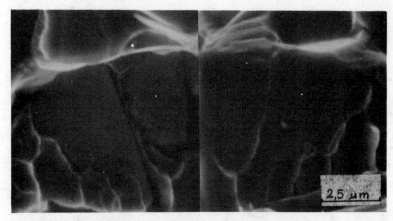

Fig. 6   Ductile rupture in binder (necking)
and decohesion between the WC grains

Fig. 7   Ductile rupture in binder, decohesion between
the WC grains and decohesion between the WC and binder

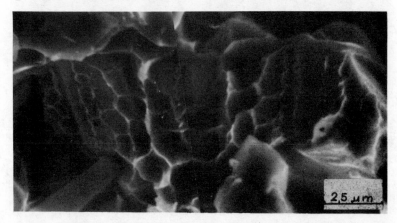

Fig. 8   Ductile rupture by dimple mechanisms in binder

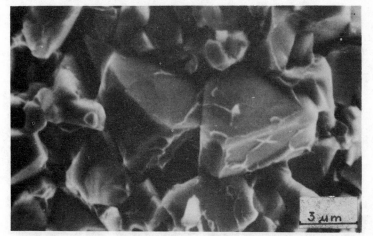

Fig. 9
L_{Co} = 0.75 μm

Fig. 10
L_{Co} = 1.2 μm

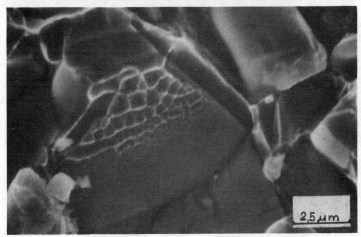

Fig. 11
L_{Co} = 1.75 μm

Figs. 9 - 11   Ductile rupture in binder with
different L_{Co} values

The microrelief of the ductile rupture Co/Co is so small (Fig. 9) that it cannot be identified on the fracture path. Therefore it is often confused with decohesion separation of WC/Co, which in fact occurs very rarely and on planes perpendicular to the direction of the propagation of the major crack as the consequence of the local triaxial stress (Fig. 7). A further significant rupture micromechanism is the cleavage of WC grains, W/C, Fig. 12 and the decohesion between WC grains, WC/WC, Fig. 13.

We show the mutual dependence of structural parameters and decisive micromechanisms in the process zone in Fig. 14. For the extent of the experimental program we have obtained very good correlations between the proportions of individual types of ruptures and the values controlling the structure of the system. In Fig. 15 we show the mutual relation between K_{IC} and the portion of individual rupture micromechanisms on the fracture surface.

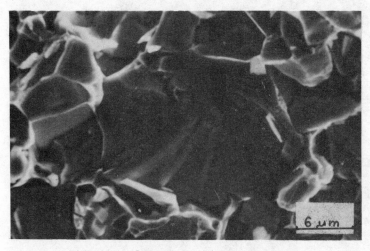

Fig. 12   Cleavage of the WC grains

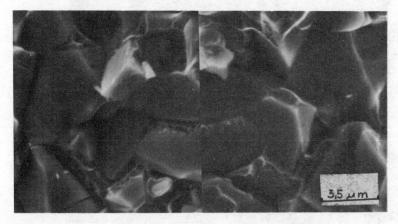

Fig. 13   Decohesion between the WC grains

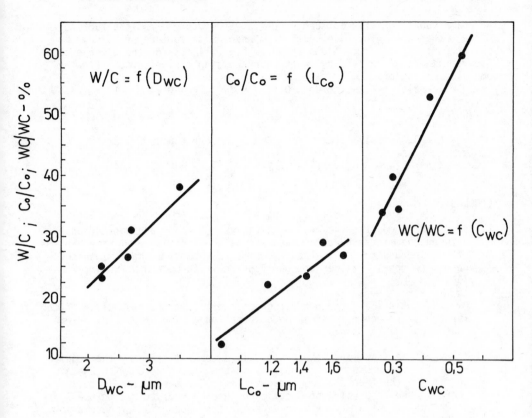

Fig. 14   Relations: W/C vs. D_{WC}, Co/Co vs. L_{Co}, WC/WC vs. C_{WC}

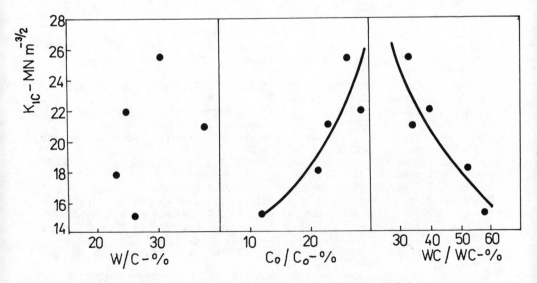

Fig. 15   Relations bewteen K_{IC} vs. W/C, Co/Co and WC/WC

An unambiguous positive influence on K_{IC} values comes from the rupture Co/Co; whereas degradation comes from decohesion rupture of WC/WC. By the linearization of relations shown in Figs. 14, 15 for the given extent of experiments we derived by means of regression analysis the following empirical relations:

$$K_{IC} = 21.1 + 0.3 . Co/Co - 0.17 . WC/WC \quad r = 0.94 \qquad 1$$

$$Co/Co = -2.3 + 18.7 . L_{Co} \quad r = 0.94 \qquad 2$$

$$WC/WC = 6.8 + 101.3 . C_{WC} \quad r = 0.98 \qquad 3$$

$$K_{IC} = 19.24 + 5.61 . L_{Co} - 17.22 . C_{WC} \quad r = 0.97 \qquad 4$$

CONCLUSIONS

1. There exist qualitative and quantitative relations and connections between the microstructural parameters, the individual micromechanisms of rupture and the K_{IC} values. The decisive micromechanisms which determine the K_{IC} values are the rupture of Co/Co and WC/WC interfaces.

2. By the statistical stereo-fractographic analysis the inspection of the rupture micromechanisms and the observation of the very fine rupture modes (Co/Co) is possible.

3. The acquired knowledge makes it possible to state exactly the requirements for the production technology in the system WC-Co with the aim to improve the fracture toughness values. The basic task is to increase the proportion of ductile rupture by redistributing the binder phase uniformly, and to decrease the proportion of decohesion of WC/WC boundaries by increasing the interface strength.

REFERENCES

Almond, E.A. and Roebuck, B. 1977, Met. Sci. 11 458-461
Almond, E.A. and Roebuck, B. 1978, Met. Technol. 5 92-101
Barker, L.M. 1980, Proc. ASM Conf. on Wear and Fract. Prev. Pretoria, Illinois
Chermant, J.L., Deschanvres, A. and Iost, A. 1974, Fract. Mech of Cer. Vol. 1, Plenum Pub. N.Y. 347-365
Chermant, J.L. and Osterstock, F. 1976, J. Mater. Sci. 11 1939-1951
Dusza, J. 1982a, Ph.D. Thesis. Slov. Acad. of Sci. Košice
Dusza, J., Parilák, L., Diblík, J. and Šlesár, M. 1982b, Proc 8th Congress on Mat. Testing. Budapest 2 527-536
Dusza, J., Parilák, L. and Šlesár, M. 1983, Proc. Fract.Conf.Czech.1 53-59
Lea, C. and Roebuck, B. 1981, Met. Sci. 15 263-271
Luyckx, S.B. 1977, Proc. ICF4 Waterloo Canada 2 170-176
Munz, D., Hohe, L. and Shannon J.L. 1980, Int. J. of Fract. 16 137-141
Munz, D. 1983 Proc. Fract. Mech. of Cer. Vol. 6 Plenum Pub. N.Y. 1-26
Nidikom, B. and Davies, T.J. 1980, Plans. für Pulv. Met. 28 29-38
Pickens, J.R. and Gurland J. 1978 Mater. Sci. Eng. 33 135-142
Šlesár, M., Dusza, J. and Parilák, L. 1982 Proc. Verformung und Bruch Magdeburg 1 123-129
Warren, R. and Johannesson, B. 1983 3rd Int. School on Sint. Mat., New Delhi

Inst. Phys. Conf. Ser. No. 75: Chapter 7
Paper presented at 2nd Int. Conf. Science Hard Mater., Rhodes

669

Tensile creep of WC-Co cemented carbides at 800–900°C

G WIRMARK(1), C CHATFIELD(2) AND G L DUNLOP(1)

(1) Chalmers University of Technology, Göteborg, Sweden
(2) AB Sandvik Hard Materials, Stockholm, Sweden

ABSTRACT

The tensile creep behaviour has been investigated in three alloys with varying binder phase contents (10-23 vol%) and WC-grain sizes (1-2μm). The tests were carried out at temperatures between 800°C and 900°C and in a stress range of 30-400 MPa. Evaluation of the steady state creep rates resulted in a stress exponent between 2 and 4. The creep rates were found to decrease with decreasing Co-content and increasing WC-grain size. Grain boundary sliding was observed at WC/WC boundaries. Transmission electron microscopy revealed the formation of subgrain boundaries in deformed WC.

INTRODUCTION

In many applications of cemented carbides, temperatures reach values where high temperature properties, such as creep strength are of vital importance. The aim of this investigation has been to improve the understanding of the creep behaviour of WC-Co cemented carbides and to relate tensile creep properties to microstructure.

Most investigations of the high temperature creep behaviour of WC-Co cemented carbides have been carried out with the specimen loaded in compression or in bending. However, these testing methods are not always simple to interpret. The former method usually suffers from frictional constraints on the loading surfaces and localised plastic deformation occurs in the latter. In order to avoid such difficulties the present investigation was carried out using tensile loading conditions.

Several investigations have demonstrated the influence of binder phase content and tungsten carbide grain size on high temperature creep rates. An increase in creep rate is usually noted when the volume fraction of the binder phase f_B is increased (Suzuki and co-workers, 1982; Moll and Wood, 1967). At high temperatures, (T=1150°C) Osterstock (1983) found an increase in the steady state creep rate, $\dot{\varepsilon}$, by a factor 3-5 when f_B was increased from 0.22 to 0.37.

The influence of tungsten carbide grain size D_{WC} on the creep rate is reported to vary with temperature and stress. Smith and Wood (1968) showed a decreasing creep resistance with increasing D_{WC} in compression creep under conditions of high stress or relatively low temperature (<900°C). The opposite behaviour was found at higher temperatures and lower stresses.

Similar results were also reported by Suzuki and co-workers (1982). The high temperature, low stress behaviour of increased resistance to creep deformation at larger, D_{WC} is confirmed by the work of Osterstock (1983) and Doi and co-workers (1984). These latter investigators did not however cover the range of temperatures and stresses where a decrease in creep strength with increasing D_{WC} might be expected to occur.

Some authors (Smith and Wood, 1968; Ueda and co-workers, 1975) have tried to explain their results in terms of models developed for dislocation creep in a dispersion hardened material (Ansell and Weertman, 1959). Such models assume that dislocation climb in the binder phase controls the rate of deformation. In later work (Ueda and co-workers, 1977; Doi and co-workers, 1984) grain boundary sliding GBS has been proposed as a possible mechanism of deformation. For small WC-grain sizes, sliding at WC/WC-boundaries is believed to be the rate-controlling factor (Doi and co-workers, 1984). Osterstock (1983) concluded for his work carried out at higher temperatures that the carbide phase was the rate-controlling constituent and he also suggested that GBS could be an important mechanism of deformation. Little experimental evidence for GBS in high temperature deformed WC-Co has however been presented.

Cavitation has been observed by several workers in crept specimens. The occurrenceof cavitation in the temperature range 800-1000°C appears to be dependent upon specimen geometry. Moll and Wood (1967) used short thick compressive creep specimens and found extensive cavitation in the vicinity of the 45° shear planes. Smith and Wood (1968) used long thin compressive creep specimens and did not detect any significant evidence for cavitation in the same temperature/stress regime. Ueda and co-workers (1975) noted small amounts of cavitation after tensile creep under similar temperature and stress conditions.

EXPERIMENTAL

Three alloys were investigated with **binder phase contents** and WC grain sizes as shown in Table 1. The flat-polished tensile creep specimens (Fig.1) had a rectangular cross-section of approximately 1x3 mm and a gauge length of 20 mm. A constant load was applied through Nimonic 105 adapters. These were connected to 3 mm holes in the grip section of the specimen by Astroloy rods.

The creep elongation was measured from ridges on the adapters using extensometry incorporating capacitive displacement transducers having a resolution of +0.1 μm. The distance between the outer edges of holes in the grip section was measured before and after the creep test. Differences in specimen extension of 2-3% were typically noted when comparing the elongation measured in these two ways. It was also found that deformation of the holes during creep was on average 15 % of the total elongation.

TABLE 1 Properties of the Alloys investigated

| Alloy | Co | WC | | Hardness | |
| | | linear intercept | RT | | 800°C |
	Vol%	μm	HV 3		HV 0.5
A	0.23	1.30	1180		420
B	0.11	1.00	1625		670
C	0.11	1.75	1375		490

The creep tests were carried out at temperatures between 800°C and 900°C in a vacuum furnace with a vacuum of approximately 1.5×10^{-3} Pa. The temperature along the gauge length was controlled to within +2°C during testing. Fiducial marker lines parallel to the tensile axis for observation of GBS were scratched onto the specimen surface prior to creep testing using 15 μm diamond paste.

After creep testing the specimens were examined by X-ray diffraction XRD , optical microscopy OM , scanning electron microscopy SEM and transmission electron microscopy TEM. The Vickers hardness was measured with a mass of 3kg (HV 3) at room temperature and 0.5 kg (HV 0.5) at 800°C. Specimens for TEM were prepared by grinding and polishing followed by ion beam thinning. They were examined in a JEOL 200 CX microscope.

RESULTS

The tensile creep tests were carried out under stresses of 30-400 MPa. The highest stress corresponds to about one quarter to the materials yield stress which has been estimated as 1/3 of the hardness at 800°C (Table 1). Some typical creep curves are shown in Fig. 2. The primary creep stage and steady state creep stage was observed in nearly all tests. In some cases a tendency for a reduction in creep rate was noted prior to fracture. The fracture was in all such cases located in the curved shoulder region which separate the gauge length from the broader ends of the specimens. The creep-rates obtained under such conditions are not included in the results. A tertiary creep stage, usually of very short duration prior to fracture, was observed in most of the creep tests.

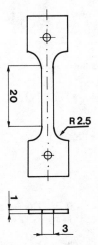

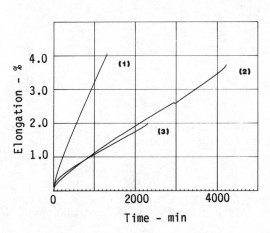

Fig. 1 Tensile creep specimen. Dimensions in mms.

Fig. 2. Representative tensile creep curves at 900°C. The alloys used and applied stresses are as follows:
(1) Alloy A, σ =170 MPa
(2) Alloy B, σ =150 MPa
(3) Alloy C, σ =200 MPa

The steady state creep rates, $\dot{\varepsilon}$, for tensile creep tests are shown in Figs. 3a and b. The decrease of creep rates with increasing D_{WC}, compare alloys B and C, is in agreement with Doi and co-workers (1984) and Osterstock (1983). The tensile stresses used in this investigation (<400 MPa) are also considerable lower than in the work of Smith and Wood (1968) (500-800 MPa), and our results do not therefore exclude the possibility of an opposite influence of D_{WC} on $\dot{\varepsilon}$ at higher stresses.

The influence of the Co-content on the creep rates is somewhat less simple to interpret as the Co-content could not be varied independently of D_{WC}. A comparison of the creep rates for alloy A with alloys B and C however clearly indicates a decrease in creep strength with increasing Co-content which is consistent with the findings of many previous investigations.

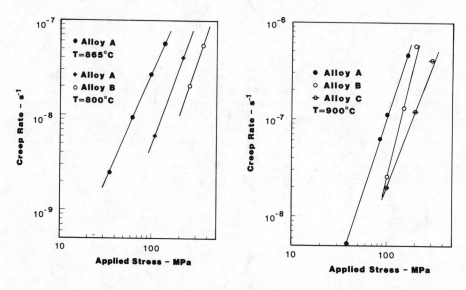

Fig. 3   Creep rates as a function of applied stress for various alloys.
   a) T=800°C and 865°C   b) T=900°C

The steady state creep rate, $\dot{\varepsilon}$, of many materials can be described by the power law (Dorn, 1954):

$$\dot{\varepsilon} = A\sigma^n \exp-(Q/RT) \qquad\qquad 1$$

where A is a material dependent constant, σ is the stress, Q is the activation energy, R is the gas constant and T is the absolute temperature.

Evaluation of the creep data in Figs. 3a and b results in the stress exponents, n, shown in Table 2. The exponents vary between 2 and 4 and no direct correlation to temperature or alloy can be made.

The apparent activation energy, Q, was estimated by changing the test temperature in small steps of 5°C for a given specimen and stress. The results obtained in this way were, however, considered to be somewhat unreliable since variations in Q between 60 and 90 kcal/mol were obtained depending on the direction of temperature change.

TABLE 2   Effect of Temperature on Stress Exponent (n)

Alloy	Temperature °C		
	800	865	900
A	2.9	2.3	3.0
B	2.9		4.1
C			2.6

The elongation at fracture was found to increase with increasing temperature, decreasing stress, decreasing D_{WC} and increasing Co-content. Elongations exceeding 10% were obtained for alloy A at 900°C.

The possibility of decarburisation during the high temperature deformation was checked by XRD and hardness measurements. No significant differences in hardness before and after creep deformation were found. XRD revealed no oxides in the specimen but the formation of η-phase was found in a few cases. The η-carbides were confined to a region within a few μm of the surface. It is also possible that decarburisation may have changed the degree of solution of tungsten in the binder phase.

Displacements of fiducial marker lines caused by grain rotation and grain boundary sliding can be observed in Figs. 4-6. The significant displacement visible at the WC/WC grain boundary in Fig. 4 indicates both rotation and inwards movement of the WC-grain in the center of the micrograph. The displacements at the grain boundaries shown in Figs. 5 and 6 of the order of 80-100 nm.

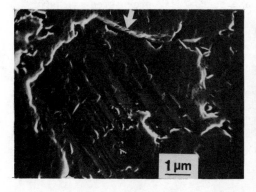

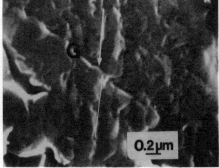

Fig. 4   SEM image of alloy C showing a rotation and inwards movement of a WC-grain. The arrow indicates WC/WC boundary having the largest displacement. (Elongation 1.3%, σ=300 MPa, T=900°C).

Fig. 5   Displacement (arrowed) of a fiducial marker line. The WC/WC boundary is marked G. (Alloy C, elongation 1.3%, σ=300 MPa, T=900°C).

Fig. 6 Displacement of a finducial marker line.
(Alloy B, elongation 5%, σ=200 MPa, T=900°C).

An estimate of the contribution to elongation from GBS, ε_b, can be made
using the following equation (Gates and Stevens, 1974):

$$\varepsilon_b = 2N(1+\varepsilon_{tot}) \ \left[\ \overline{\omega/\tan\theta} \ \right] \qquad\qquad 2$$

where N is the number of grain or phase boundaries intersected per unit
length of fiducial marker line, ε_{tot} is the total strain, ω is the offset
perpendicular to the stress axis and θ is the angle between the surface
trace of the boundary and the stress axis.

If GBS is assumed to be totally responsible for the deformation, average
offsets, $\overline{\omega/\tan\theta}$, of 10-20 nm should be expected for the creep conditions
in Figs. 5 and 6. The measured offsets thus indicate that GBS is an important mechanism of creep deformation under these conditions. The partial
evaporation of the binder phase at the surface during the test and the
limited resolution of the fiducial marker line technique make, however, a
quantitative estimation of GBS impossible.

Cavitation

Sections parallel to the stress axis were ground, polished and examined
for cavitation. No significant cavity formation was observed in the specimens (Fig. 7). In some cases cavitation was observed close to propagating cracks in the vicinity of the final fracture.

Transmission Electron Microscopy

Specimens of both as-sintered and crept materials were investigated by TEM. The binder phase in both types of specimen had a faulted internal structure with slip bands, twinning and stacking faults all being present (Fig. 8). Preferential etching of the binder phase during ion beam thinning made it difficult to investigate for the presence of microcavities in the binder phase.

Fig. 7 Section parallel to the tensile axis of a crept specimen. Alloy A. No cavities are visible between WC grains or at WC/Co interfaces.

Fig. 8 Binder phase in an undeformed specimen. Alloy C.

The WC-carbide grains contained a non-uniform dislocation density. Differences in dislocation density were usually larger between individual WC-grains of the same specimen than those between undeformed and deformed specimens (Fig. 9a and b).

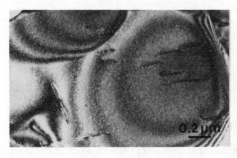
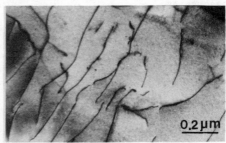

Fig. 9 WC-grains

a) As-sintered, alloy A.

b) Crept specimen. (Alloy B, elongation 2%, σ =250 MPa, T=800°C).

However, some features were more frequently observed in the deformed mate-rial. The formation of ordered dislocation networks and subgrains (Figs. 10-11) was common in crept material indicating that recovery occurred dur-ing high temperature deformation.

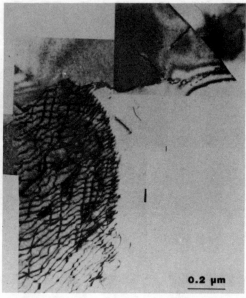

Fig. 10   Subgrain boundary partly
developed during creep.
(Alloy B, elongation 2%
σ =250 MPa, T=800°C).

Fig. 11   Subgrain boundary after
creep. (Alloy A, elongation
15%, σ=100 MPA, T=900°C).

DISCUSSION

The following general comments can be made concerning the results.

(i)   The absolute creep rates are in good agreement with previous data
for tensile creep of WC-Co cemented carbides in this temperature
range (Doi and co-workers, 1984). However, the measured elongations
to fracture are substantially larger in our experiments. It is possible that this difference could be associated with specimen volume.
The lower specimen volume used in this investigation, compared to ·
that used by Doi and co-workers (1984), should be advatageous if the
elongation to fracture is determined by the distribution of flaws in
the sintered material.

(ii) The model for dislocation creep in a dispersion hardened material (Ansell and Weertman, 1959) which has been proposed to be valid for high temperature deformation of cemented carbides (Ueda and co-workers, 1975), shows poor agreement with our experimental data. According to their model the rate controlling process is dislocation climb over second phase particles. Different expressions for the steady state creep rate are proposed for the low and high stress regimes. At low stresses $\dot{\varepsilon}$ is proportional to σ and to the inverse square of D_{WC}. At higher stresses $\dot{\varepsilon}$ is proportional to $\sigma^4 \lambda^2 / D_{WC}$ where λ is the inter-particle spacing. A constant volume fraction of the carbide phase thus gives, assuming the carbide contiguity to be constant, $\dot{\varepsilon}$ is proportional to $\sigma^4 \lambda$ or alternatively to $\sigma^4 D_{WC}$.

The measured dependence of $\dot{\varepsilon}$ on D_{WC} is opposite to that proposed for the high stress-regime. It should also be noted that the experimental stress exponents are inconsistent with the stress-dependence proposed for low stresses. Recent support for the skeleton model of WC-Co (Jayaram and Sinclair, 1983; Henjered and co-workers, 1984) indicates that a very limited amount of cobalt is present at WC/WC grain boundaries. These findings make dislocation movement confined to a thin Co-layer in these grain boundaries improbable. Deformation of a cemented carbide specimen must therefore include deformation of the skeleton itself. But, this does not rule out a possibility that segregation of Co to the WC/WC-interface may influence GBS at these grain boundaries.

(iii) Evidence for GBS has been presented and the magnitude of observed offsets at grain boundaries is consistent with GBS making a substantial contribution to the total strain. The measured stress exponents, n, are consistent with values of n usually noted for creep with a proportion of grain boundary sliding (Taplin and co-workers, 1982).

(iv) The absence of extensive cavitation indicates that this is not a major mechanism of accommodation in the temperature and stress range studied. Our results are consistent with those by Smith and Wood (1968) who used a similar specimen geometry. However, it would seem that cavitation is probably of vital importance in determining final fracture.

CONCLUSIONS

The observations of grain boundary sliding and grain rotation suggests that rearrangement of the WC grains in the polycrystalline composite provides most of the strain during creep of WC-Co under the conditions studied here. Local deformation of WC grains to accommodate the substantial shear at grain boundaries may well be rate controlling but, as yet this is not proven.

ACKNOWLEDGMENTS

This investigation was financed by the Swedish Board for Technical Development and AB Sandvik Hard Materials who also provided the materials for investigation.

REFERENCES

Ansell, G.S., and Weertman, J., 1959, Trans. Met. Soc. AIME, 215 838-843

Doi, H., Ueda, F., Fujiwara, Y., and Masatomi, H., 1984, Int. J. Refr. Hard Met., 3 146-148

Dorn, J.E., 1954, J. Mech. Phys. Solids, 3 85-116

Gates, R.S., and Stevens, R.N., 1974, Metall. Trans. A, 5 505-510

Henjered, A., Hellsing, M., Andrén, H-O., and Nordén, H., 1984 these proceedings

Jayaram, V., and Sinclair, R., 1983, J. Am. Ceram. Soc. 66 C137-C139

Moll, R.A., and Wood, J.D., 1967, Proceedings from seminar on "An understanding of the behaviour of WC-Co Cutting Tool materials", Jan 19, Lehigh University, Penn., USA

Murray, M.J., and Smith, D.C., 1973, J. Mater. Sci., 8 1706-1710

Osterstock, F., 1983, Science of Hard Materials Ed Viswanadham, R.K., Rowcliffe, D.J., and Gurland, J., Plenum Press, New York, 671-687

Smith, J.T., and Wood, J.D., 1968, Acta Metall., 16 1219-1226

Suzuki, H., Hayashi, K., Taniguchi, Y., and Matsubara, H., 1982, Trans. JIM, 23 77-84

Taplin, D.M.R., Dunlop, G.L., and Langdon, T.G., 1979, Ann. Rev. Mat. Sci., 9 151-189

Ueda, F., Doi, H., Fujiwara, F., Masatomi, H., and Oosawara, Y., 1975, Trans JIM, 16 591-600

Ueda, F., Doi, H., Fujiwara, F., and Masatomi, H., 1977, Trans. JIM, 18 247-256

Inst. Phys. Conf. Ser. No. 75: Chapter 7
Paper presented at 2nd Int. Conf. Science Hard Mater., Rhodes

679

Accelerated tests to study creep deformation in hard materials

A T SANTHANAM, K P MIZGALSKI AND W C McCOY

Philip M. McKenna Laboratory
Kennametal Inc., Greensburg, PA 15601, USA

ABSTRACT

Indentation and load-relaxation tests have been used to study creep deformation in two hardmetal compositions at 1000°C. The results are
compared with conventional compression creep data. It is found that the
best agreement with the compression creep data is obtained by the load-
relaxation technique. Unlike the indentation test, the load-relaxation
test yields strain rates which closely correspond to the steady-state
creep rates. The usefulness of the load-relaxation technique as a means
of rapid evaluation of creep resistance in hard materials is thus
indicated.

INTRODUCTION

As speeds increase in metal-cutting operations, thanks to the advent of
CVD coatings and CNC machinery, the cutting tool tip undergoes rapid
blunting which limits its useful life. It is estimated (Dearnley, 1983)
that under certain metal-cutting conditions, the tool tip temperature may
exceed 1000°C and localized stresses may reach the yield strength of the
tool material. Under these operating conditions, the tool tip blunting
is likely to occur by creep. A study of creep deformation in tool
materials as a function of their composition and microstructure is thus
essential for an understanding of tool tip deformation during service.
In this study, we have used three test techniques to evaluate creep
resistance in hard materials: conventional creep, load-relaxation and
indentation methods. The role of cubic carbide additions to WC-Co
alloys in creep deformation was also studied.

TEST TECHNIQUES

Conventional Creep Method

The conventional method of studying the creep phenomenon is to apply a
constant uniaxial load to a tensile or compression specimen and measure
its deformation or strain as a function of time. The creep rates are
obtained from the slopes of the strain-time curve. The principal item
of information obtained from the creep curve is the steady-state creep
rate, $\dot{\varepsilon}_s$. For a wide range of materials, $\dot{\varepsilon}_s$ can be related to applied
stress and temperature by the following empirical relation:

$$\dot{\varepsilon}_s = A\sigma^m \exp\left(-\frac{Q}{RT}\right) \qquad\qquad 1$$

where A and m are constants, σ is the applied stress, Q is the apparent activation energy for creep, R is the universal gas constant and T is the test temperature. A number of investigators have studied the creep phenomena in hard materials (Altmeyer and Jung, 1961; Dawhil and Altmeyer, 1963; Dawhil and Mal, 1965; Smith and Wood, 1968; Ueda and co-workers, 1975; Osterstock, 1981; Hirai and Niihara, 1979; Atkins and Tabor, 1966; Roebuck and Almond, 1982) and have reported a stress exponent m in the range 1 to 7. One of the difficulties with conventional creep tests is that, depending on the applied stress and temperature, they can be very time consuming. Development of test techniques that allow rapid scanning of materials for creep resistance is thus valuable in alloy development work.

Load-Relaxation Method

Another method by which the creep phenomenon may be studied is the load-relaxation technique. In this method, the specimen is deformed (in tension or compression) at test temperature past the yield load and then the crosshead of the test machine is stopped. The variation of load or stress with time is recorded as shown in Fig. 1. The applied load on the specimen relaxes due to the conversion of the elastic strain of the test system to plastic strain in the test specimen.

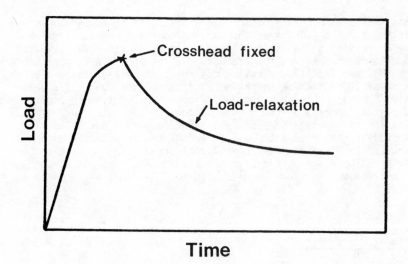

Fig. 1 Load-relaxation technique

It is observed that the relaxation process is a strong function of test temperature. At higher temperatures, the process occurs more rapidly due to the specimen creep deformation. The plastic strain rate or the creep rate of the specimen may then be calculated from the dynamics of the relaxation process.

Consider a specimen that is deformed to a total strain, ε_0 by a stress, σ. We wish to determine how σ must change to maintain the strain, ε_0. The total uniaxial strain, ε_0 can be separated into elastic and plastic components.

$$\varepsilon_0 = \varepsilon_{el} + \varepsilon_{pl} \qquad 2$$

At constant displacement, the strain rate satisfies the following equation.

$$\dot{\varepsilon}_0 = \dot{\varepsilon}_{el} + \dot{\varepsilon}_{pl} = 0 \qquad 3$$

$$\text{or } \dot{\varepsilon}_{pl} = - \dot{\varepsilon}_{el} = - \frac{1}{E'} \cdot \frac{d\sigma}{dt} \qquad 4$$

where E' is the elastic modulus of the test system, σ is the instantaneous stress on the sample at time t. The plastic strain rate in the specimen can thus be calculated from the rate of change of stress during the relaxation process. The results may be plotted as $\log_{10} \dot{\varepsilon}$ versus $\log_{10} \sigma$ to obtain the stress exponent m in the creep equation (Eq. 1). Murty and McDonald (1981) have successfully used load-relaxation tests to predict the steady-state creep rates in Zircaloy material. They found that best agreement between load-relaxation and tensile creep tests was obtained when the load-relaxation was carried out at the point of maximum load during the tensile test.

Indentation Method

Several investigators (Atkins and Tabor, 1966; Hirai and Niihara, 1979; Roebuck and Almond, 1982) have used hardness indentation tests to study the creep phenomenon in hard materials. Using a high temperature micro-hardness tester equipped with a capability to vary the dwell time of the indenter, they have measured the size of the hardness indentation as a function of indentation time. This information can be converted to a form equivalent to that given by Eq. 1 as shown below.

Following Geach (1974) and Roebuck and Almond (1982), the strain rate, $\dot{\varepsilon}$ can be defined as

$$\dot{\varepsilon} = \frac{d\varepsilon}{dt} = \frac{1}{D} \cdot \frac{dD}{dt} = \frac{d\log_e D}{dt} \qquad 5$$

where D is the indentation diagonal. The strain rates at various in-dentation diagonals (i.e. hardness levels) can be determined from the slopes of the plot $\log_e D$ versus time as shown in Fig. 2. Since in-dentation diagonals can be converted to hardness values, one obtains strain rate as a function of hardness. A plot of $\log_{10} \dot{\varepsilon}$ versus $\log_{10} HV$ should give information on the stress exponent m in the creep equation (Eq. 1).

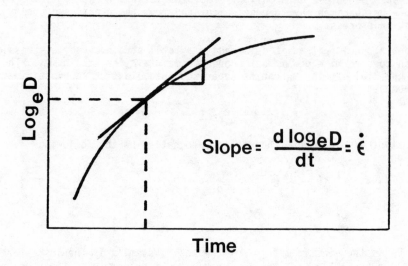

Fig. 2 Variation of indentation diagonal
with indentation time plotted as
$\log_e D$ versus t

APPLICATION OF THE TECHNIQUES

The above techniques were applied to two hardmetal compositions. Both are WC-Co base alloys with additions of cubic carbides (TiC and TaC). The cobalt content (10 vol %), grain size and magnetic saturation were kept constant, but the amount of cubic carbides was varied. Alloy A had a very small amount of cubic carbides (1.6 vol %) with a Ta/Ti ratio (wt %) of 2.75, whereas alloy B had 50 vol % of cubic carbides with the same Ta/Ti ratio. The constant Ta/Ti ratio enabled the composition of the solid solution carbides to be kept constant in the sintered alloy. The alloys were processed by conventional powder metallurgical techniques. Table I gives the chemical composition and physical and mechanical properties of the two alloys. Fig. 3 shows the microstructure of the two alloys. The mean linear intercept carbide grain size, measured from the SEM micrographs, was ∿0.7μm in both alloys A and B.

TABLE 1   Composition and Properties of Hardmetals

	Wt %			Vol %		
	Co	Ta	Ti	Co	Cubic carbides	WC
Alloy A	6.0	0.5	0.2	10.0	1.6	88.4
Alloy B	7.2	19.5	7.1	10.0	50.0	40.0

	Density $Mg\ m^{-3}$	Room temperature hardness HV30	Coercivity $kA\ m^{-1}$	Magnetic saturation $Tesla\ m^3\ kg^{-1}$	Carbide grain size μm
Alloy A	14.97	1639	17.7	112×10^{-7}	0.68
Alloy B	12.67	1646	14.5	128×10^{-7}	0.71

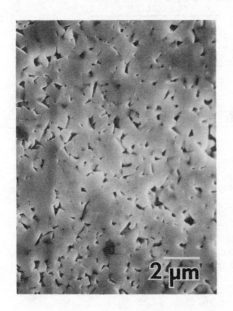 

Fig. 3   Scanning electron micrographs of alloys A and B

Conventional Creep Tests

Constant load compression creep experiments were performed on cylindrical specimens (10mm long and 5mm in diameter) of alloys A and B at 1000°C. The experiments were run using a test machine with a furnace mounted on the load frame. The furnace contains two semi-cylindrical tungsten-mesh heating elements and molybdenum heat shields. Loading of the specimen was accomplished using molybdenum alloy rams and Si_3N_4 platens. Temperature was measured by a Pt/Pt-10% Rh thermocouple fixed near the specimen. A controller capable of measuring temperature to ±1°C was used to monitor temperature. The specimens were compressed at 0.2mm/min. to the desired load which was then held constant to within ±1% by a load cycling control. Specimen displacement was measured with a transducer accurate to 1 mV (5×10^{-4} mm of displacement). The transducer was mounted on the upper ram between the crosshead and furnace. Displacement was converted to strain using the original length of the sample. The strain rate was determined from the slope of the steady-state portion of the strain-time curve. Three nominal stress values were chosen for each alloy. The associated stress exponents for Alloys A and B are shown in Table 2.

TABLE 2   Stress Exponents for Alloys A and B using
Conventional Creep and Load-relaxation Tests

	Conventional creep	Load-relaxation			
		∿0.5%	1.1%	2.6%	5.1%
Alloy A	3.3	3.4	3.2	2.9	-
Alloy B	7.0	6.8	-	-	4.9

Load Relaxation Tests

Cylindrical specimens of the same alloys (i.e. A and B) were used for the load relaxation tests at 1000°C using the test system described above. The tests were performed at two or three plastic strain levels in the range of 0.5 to ∿5.0%. The highest chosen strain level for each alloy corresponded to that point in the load time curve where the work hardening was negligible (i.e. flat part of the curve). The specimens were loaded at 0.2mm/min. to the different plastic strain levels and then the crosshead of the machine was stopped. The load-relaxation was monitored as a function of time. An example of the load-relaxation curves for both alloys A and B is shown in Fig. 4.

The load-time data were converted to stress-time from the initial cross-sectional area of the specimen, and the data were fitted with an exponential function. The plastic strain rates were calculated from Eq. 4. The $\log_{10}\dot{\varepsilon}$ - $\log_{10}\sigma$ data for the various plastic strain levels are given in Fig. 5 along with the data from the conventional creep tests. The calculated stress exponents are given in Table 2 for the load-relaxation tests performed at different levels of strain. Note that the best agreement with the conventional creep data is obtained when the load-relaxation was carried out at ∿0.5% plastic strain. With increasing pre-strain, the $\log_{10}\dot{\epsilon}$ - $\log_{10}\sigma$ curves are shifted upwards and away from the compressive creep data. No attempt was made to further optimize the prestrain level for an even closer agreement between the relaxation and constant load tests.

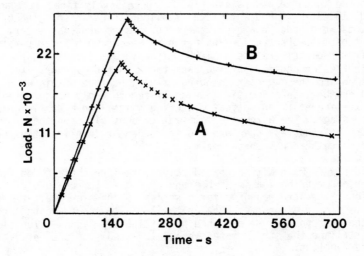

Fig. 4 Load-relaxation curves at ∿0.5% plastic
strain for alloys A and B at 1000°C

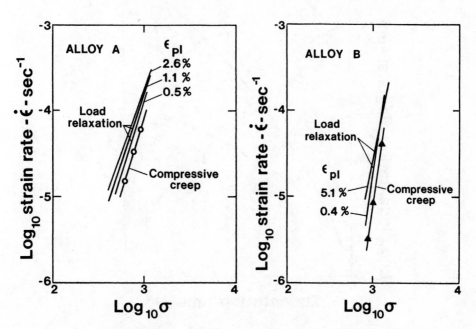

Fig. 5 Comparison of data obtained from load-relaxation
and compression creep tests at 1000°C

Indentation Creep Tests

The hardness indentation tests were performed on rectangular parallel-epiped specimens 10.2 mm long and 5.1 mm x 5.1 mm in cross-section. All the specimens were ground parallel with a diamond grinding wheel. The surface to be indented was polished with successively finer diamond paste (30, 15, 9, 6, 3 and 1μm) to a metallographic finish. Indentation measurements were made in a high temperature microhardness tester. The instrument permits observation and hardness measurement at temperatures ranging from room temperature to 1200°C in a vacuum or inert gas atmos-phere. The tester consists of two resistance heated furnaces (to heat the specimen and the indenter independently) contained within an evacuated chamber, an automatic loading system to apply preselected loads on the indenter, an optical measuring system and a precision micrometer stage. The automatic temperature controls keep the temperatures of the test specimen and the indenter identical thus en-suring accurate hardness measurements.

Indentations were made with a heated polycrystalline cubic boron nitride Vickers indenter using a mass of 1kg and at a test temperature of 1000°C. Indentation sizes were measured for a series of indenter dwell times (2 to 400 s). Fig. 6 gives the data for the two alloys. A smooth curve through the data points was obtained by fitting a polynomial function $\log_e D = f(t^{\frac{1}{2}})$ and the strain rates were calculated from the derivative of the fitted function.

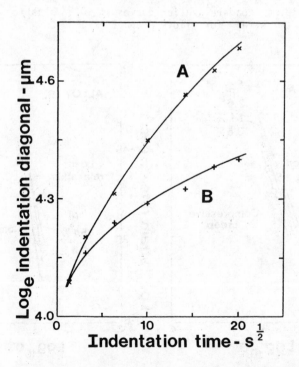

Fig. 6 Variation in indentation diagonal with indentation time for alloys A and B at 1000°C

Hardness values were calculated from the measured indentation diagonals.
The data are plotted as $\log_{10}\dot{\varepsilon}$ versus $\log_{10}$HV1.0 in Fig. 7. The
calculated slopes of the lines were 3.1 and 7.1 for alloy A and B,
respectively. Data from conventional creep tests and load-relaxation
at ∿0.5% plastic strain have also been included in Fig. 7.

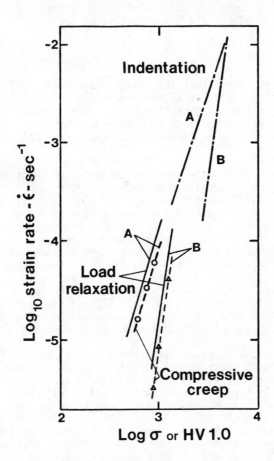

Fig. 7 Comparison of data obtained from indentation
creep, load-relaxation (∿0.5% plastic strain)
and compression creep tests at 1000°C

It can be seen in Fig. 7 that, for a given hardmetal composition, best
agreement with the compressive creep data is obtained by the load-
relaxation technique. Not only are the stress exponents similar in the
load relaxation and compressive creep experiments, but the data are very
close to each other. The indentation creep method also gives slopes
similar to the stress exponents seen in compressive creep, but the data
are obtained at levels of stress (hardness) which give high strain rates.

For alloy A (nearly straight WC-Co composition), this is not a problem, since an extrapolation of the indentation creep data to lower stresses nearly coincides with compressive creep data. On the other hand, for alloy B (high levels of cubic carbide), the indentation creep data are shifted to higher stresses. Similar shifts have been reported by Roebuck and Almond (1982) for a WC-10 wt % Co alloy at 600°C. Finally, the beneficial effect of cubic carbide additions in resisting the creep deformation is borne out by all three test techniques.

DISCUSSION

The present study has shown that the steady-state creep resistance of hard materials can be rapidly evaluated either by a load-relaxation technique or by a hardness indentation test. Under the test conditions employed, the load-relaxation test provides data that are in closer agreement with the results of conventional creep tests than the indentation method.

The effect of pre-strain on the load-relaxation response of alloys A and B would appear to contradict the results of Murty and McDonald (1981). Load-relaxation tests on Zircaloy material at 385°C (tensile specimens) at different levels of pre-strain indicated better agreement with conventional creep data as the strain level was increased. However, the effect of pre-strain on the load relaxation response of the Zircaloy material was fairly small. In the present study on hard materials, we have observed a pronounced effect of pre-strain on the stress exponent (see Table 2 and Fig. 5). The discrepancies between Murty and McDonald's results on Zircaloy material and our results on WC-Co base alloys are not clear at this time. However, the role of test temperature, differences between tension and compression tests and the role of alloy composition on load-relaxation response should be examined.

The results of indentation creep tests are indeed striking. Whereas for alloy A the indentation test gave a $\log_{10} \dot{\varepsilon}$ - $\log_{10} \sigma$ relationship which is merely an extension of the compressive creep data, a shift to the right is observed in the high cubic carbide alloy (B). While demonstrating equivalence between indentation creep tests and conventional creep tests for a straight WC-10 wt % Co alloy at 600°C, Roebuck and Almond (1982) found that the data from the indentation tests are shifted to the right, as was observed for alloy B. This shift between compressive creep and indentation data can be represented by the parameter, β in the equation

$$HV = \beta\sigma \qquad\qquad 6$$

where H is the hardness, σ is the stress and β is referred to as the constraint factor. Roebuck and Almond (1982) report that β is 1.5 in WC-10 wt % Co alloy at 600°C. In our study, for the WC-10 vol % Co - 50 vol % cubic carbide alloy, β is 1.8 at 1000°C. For the hardmetal that was almost free of cubic carbide, alloy A, β was 1.0. The parameter β may thus be dependent not only on the test temperature but on the material as well.

While both indentation and load-relaxation techniques are less time-consuming to perform than conventional compressive creep tests, the lack of precise knowledge of stress (associated with a given hardness) and the unpredictable shift in $\log_{10} \dot{\epsilon}$ - $\log_{10} \sigma$ (HV1.0) plots with composition and temperature would make the indentation test less dependable for obtaining information on the creep properties of hard materials. In contrast, the stresses are well-defined in the load relaxation tests and the measured strain rates closely correspond to the steady-state creep rates. It remains to be seen, however, whether this close correspondence between the load-relaxation test and the conventional creep test holds over a wide range of temperatures.

All three test techniques demonstrate the beneficial effects of cubic carbides in resisting creep deformation in WC-Co base alloys. Ueda and coworkers (1975) have previously shown the effect of TaC additions to WC-10 wt % Co alloy in decreasing the creep rate at 800°C. Also, it has long been known in the industry that cubic carbide additions to WC-Co base alloys increase the resistance to tool-tip deformation at cutting temperatures. However, additional studies are still needed to understand the mechanism responsible for the increased creep resistance of WC-Co base alloys containing cubic carbides. Use of accelerated tests such as the load-relaxation technique can be valuable in this regard.

SUMMARY

Three test techniques were used to study the creep resistance of two WC-Co base alloys. The cobalt content (10 vol %), grain size and magnetic saturation were kept constant, but the solid solution carbide content was 1.6 vol % for alloy A and 50 vol % for alloy B. Conventional compressive creep, load-relaxation and hardness indentation tests were performed on the two alloys at 1000°C. The data from each type of test were then compared. All three test techniques yielded similar stress exponents (m=3.3 for alloy A; m=7.0 for alloy B); however, for both hardmetal compositions, best agreement with the compressive creep data was obtained by the load relaxation technique if the relaxation was carried out at ~0.5% plastic strain. Use of the load-relaxation technique can thus provide a rapid means of evaluating the creep resistance of hard materials.

ACKNOWLEDGEMENTS

The authors wish to thank G. P. Grab, A. P. Pantano and J. A. Hillebrecht for their efforts in providing the specimens used in this study, and W. C. Eversole, Jr. for conducting the tests. Stimulating discussions with Dr. K. L. Murty are gratefully acknowledged.

REFERENCES

Altmeyer, G. and Jung, O. 1961, Z. Metallkunde 52 576-583.
Atkins, A. G. and Tabor, D. 1966, Proc. Roy. Soc. A 292 441-459.
Dawhil, W. and Altmeyer, G. 1963, Z. Metallkunde 54 645-650.
Dawhil, W. and Mal, K. 1965, Cobalt 26 25-35.
Dearnley, P. A. 1983, Metals Tech. 10 205-214.
Geach, G. A. 1974, Int. Metall. Rev. 19 255-267.
Hirai, T. and Niihara, K. 1979, Mtls. Sci. 14 2253-2255.
Murty, K. L. and McDonald, S. G. 1981, Proc. 6th Int. Conf. on Struc.
  Mech. in Reactor Technology, Eds. J. Rastoin and B. A. Boley. North
  Holland, Paris, Paper C2/3.
Osterstock, F. 1981, Proc. 1st Int. Conf. on Sci. of Hard Materials,
  Eds. R. K. Viswanadham, D. J. Rowcliffe and J. Gurland. Plenum Press,
  New York, pp. 671-687.
Roebuck, B. and Almond, E. A. 1982, J. Mat. Sci. Letters 1 519-521.
Smith, J. T. and Wood, J. D. 1968, Acta Met. 16 1219-1226.
Ueda, F., Doi, H., Fujiwara, F., Masatomi, H. and Oosawa, Y. 1975, Trans.
  Japan Inst. of Metals 16 591-600.

Inst. Phys. Conf. Ser. No. 75: Chapter 7
Paper presented at 2nd Int. Conf. Science Hard Mater., Rhodes

691

Plastic deformation of TiC/Ni-Mo composites at high temperatures

C F WOLFE AND W S WILLIAMS

Department of Ceramic Engineering
University of Illinois at Urbana-Champaign, Urbana, IL 61801

ABSTRACT

Specimens of TiC/Ni-Mo cemented carbides were subjected to uniaxial
compression at temperatures from 1100°C to 1350°C. The yield stress for
macroscopic plastic deformation at a strain rate of 5 x 10^{-4}/s was higher
than that of single crystal TiC at temperatures below 1200°C and lower
above this temperature. Arrhenius behavior of the yield stress was
observed, with two linear regions joined at 1250°C. Grain boundary
sliding or diffusion along the binder phase is suggested as the mechanism
for high temperature yielding, while Hall-Petch strengthening may account
for the lower temperature behavior.

INTRODUCTION

The binder phase in a cemented carbide acts as a crack inhibitor to
complement the hard but brittle carbide and reduce the probability of
fracture. The role of the binder in macroscopic plastic deformation is
less clear. As the cemented carbides which have been studied suffer a
serious loss of hardness with increasing temperature (Baldoni and
Williams, 1978; Shenck, Gottschall and Williams, 1978), and as it has been
suggested that the binder phase may be partly responsible (Gottschall,
Williams and Ward, 1980), it is important to explore this possibility
further. However, neither the binder phase nor the carbide phase in
WC/Co-based materials has been completely characterized as to high temperature yield behavior, so it is difficult to isolate the contributions
of each phase to the observed yield of the composite.

To bypass this deficiency, in the present study the yield behavior of
cemented titanium carbide (TiC/Ni-Mo) was examined, and reference was made
to the detailed information generated earlier on plastic deformation of
single crystal TiC at high temperatures. The role of the binder phase was
expected then to be apparent by comparison.

The plastic yield behavior of TiC has been characterized. The
{111} <1$\bar{1}$0> slip system dominates at high temperatures. Williams (1964)
and Hollox and Smallman (1966) measured the high-temperature critical
resolved shear strength of single-crystal TiC, and Williams (1967) studied

the deformation of large-grained, polycrystalline TiC. Both groups reported that TiC has a high yield stress at lower temperatures but noted that it decreases exponentially with increasing temperature. The yield behavior of large-grained, polycrystalline TiC is similar.

Less is known of the mechanical properties of cemented TiC composites although some data have been reported. Moskowitz and Humenik (1978, 1981), using transverse rupture tests at room temperature, showed that the binder phase in TiC/Ni-Mo composites is strengthened by increasing its Ti content. A maximum is reached at 6 wt% Ti (7 at%). Above this point, there is formation of Ni_3Ti which probably lowers the strength of the material. The increase in fracture strength with Ti addition is presumably by solid solution strengthening. In TiC/Ni-Mo, the binder is composed of Ni alloyed with Ti and small amounts of Mo; the Ti level is inversely proportional to the C content at a constant Mo content.

Evidence for an intergranular film of binder phase at grain boundaries in samples of TiC/Ni-Mo has been found (Ruchlewicz, 1982). This film should influence the high temperature properties of the material, perhaps allowing grain boundary sliding to occur at temperatures approaching the liquidus of the binder, as Sharma, Frazer, and Williams (1980) suggested for WC/Co composites, where they found an analogous film.

The present study is the first to generate high temperature yield stress data for the TiC/Ni-Mo system, and, by comparison with single crystal data, to provide a basis for evaluating the role of the binder phase in the plastic deformation of cemented carbides.

EXPERIMENTAL PROCEDURE

Equipment and Procedures for Compression Testing

The yield stress measurements on TiC/Ni-Mo were made in a vacuum furnace with a graphite load train, previously described (Williams, 1964), which rested on a 100 kg load cell of a testing machine. Tests were performed under argon atmosphere to minimize oxidation of the sample and apparatus; control tests in vacuum showed no difference in yield stress from those run with argon.

Load cell output was recorded on a strip chart recorder. The rate of cross-head travel produced a strain rate of about 5×10^{-4}/s. The time versus force curve was linear except at the initial loading and at the start of plastic deformation. Yield force was defined, as in our previous work, to be the force at the onset of plastic deformation determined as the point of intersection of extrapolations of the elastic and plastic portions of the force-time curve. Control tests run on a nondeforming material (TiB_2) established that the time-force curve of the apparatus was linear, and, therefore, that the response of the apparatus to the force could be neglected for determination of yield stress.

Temperature was measured with a 0.13mm Pt-Pt/13%Rh thermocouple held in a alumina thermocouple shield, and was read directly from a digital thermometer (± 0.5°C). The temperature was confirmed with a micro-optical pyrometer by sighting on a small hole in the graphite specimen chamber.

Specimens were measured with a micrometer before and after testing to confirm plastic deformation. The yield force divided by the original cross sectional area was defined as the yield stress.

Sample Preparation

The composition of the cemented carbides was TiC/12.5Ni-11Mo. Samples were cut to 1 mm x 1 mm x 2 mm either with a diamond saw or with an electric discharge machine. There was no difference in the yield stresses of the samples prepared by the two methods. Nonuniform samples were discarded. The surfaces of uniform samples were finished by mechanical polishing. One TiC/Ni-Mo sample was etched with Murakami's reagent for 15 s and the sample was photographed under Nomarski interference conditions to illustrate the microstructure of the composite.

RESULTS

Microstructure of the TiC/Ni-Mo

The TiC grains in the cemented TiC/Ni-Mo composite are slightly more rounded than WC grains in WC/Co composites. The grains are small and of a uniform size, about 1-2 μm in diameter. This composite appears similar to the TiC/Ni-Mo composite characterized by Moskowitz and Humenik (1966, 1978, 1981).

Compression Tests

The high temperature plastic deformation of the TiC/Ni-Mo cemented carbide resembles the deformation behavior of other cemented carbides which we have studied earlier (Baldoni and Williams, 1978; Schenck and co-workers, 1978). The yield stress shows a sharp drop with increasing temperature (Fig. 1), and the Arrhenius curve is broken into two sections with

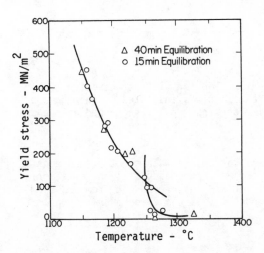

Fig. 1 Temperature dependence of the yield stress for macroscopic plastic deformation in uniaxial compression for TiC/12.5Ni-11Mo cemented carbide. Note rapid loss of hardness above 1250°C.

different slopes, suggesting two mechanisms (Fig. 2). This feature also appears as a distinct cusp in the temperature-yield stress data at about 1230°C (Fig. 1). This cusp has not been shown in similar stress-temperature curves before, although it is implied by the breaks in the Arrhenius plots. The close spacing of the experimental points from the TiC/Ni-Mo material probably accounts for the appearance of the cusp in this series of experiments.

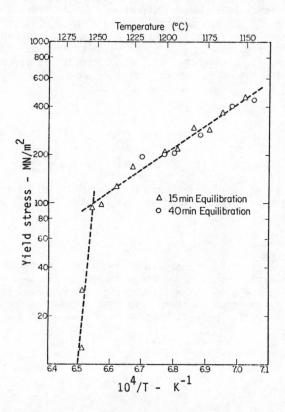

Fig. 2   Arrhenius plot for yield stress data of Fig. 1.
Two different mechanisms indicated. Grain boundary
sliding on film of binder phase suggested as mechanism
for higher temperature regime.

TiC/Ni-Mo samples tended to rupture when the test was continued longer than the minimum time needed to detect plastic deformation. Some of the samples broke into several smaller pieces when removed from the sample chamber. Rupture during the test was evident as a drop in the time-force curve.

DISCUSSION

Ni-bonded TiC showed higher yield stress values than had single crystal TiC (Williams, 1964) or large grained polycrystalline TiC (Williams, 1967) at temperatures below 1175°C (Fig. 3). The measurements were made in the same apparatus. (An effective Schmid factor of 0.33 was used in the comparison of the cemented material to the single crystal to take into account the random orientation of the grains in the cemented carbide.) The yield stress dropped much more sharply with increasing temperature for the cemented TiC than for either of the two types of single-phase TiC samples; as a result, the curves cross, and the yield stress of the cemented carbide falls below the yield stresses of the single phase carbides at temperatures around 1200°C.

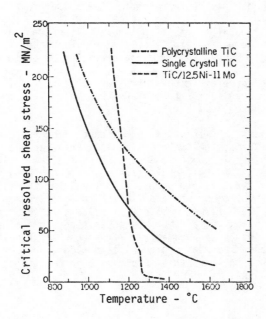

Fig. 3 Comparison of yield stress versus temperature
curve for TiC/Ni-Mo composite (Fig. 1) with similar
curves for TiC single crystal (Williams, 1964) and
polycrystal (Williams, 1967). The cemented carbide
is harder than the single crystal at temperatures
below about 1250°C, but softer at higher temperature
due probably to the presence of the grain boundary
(binder) phase.

The high temperature yield behavior of TiC/Ni-Mo is similar to that of the WC-Co-based cemented carbides studied previously (Baldoni and Williams, 1978; Schenck, Gottschall and Williams, 1978), although the yield stress drops more precipitously for TiC/Ni-Mo and falls below that of the other carbides at 1250°C. This behavior is expected in view of the greater temperature dependence and lower values of the yield stress for single-phase TiC than for single-phase WC.

In view of the lower values of the yield stress for the TiC/Ni-Mo composite than for pure TiC crystals at high temperatures, we are led to attribute an important role in macroscopic plastic deformation to the former's binder phase at these temperatures. The argument has some force in the present case, as it is based on a comparison between the pure carbide and the cemented form. Grain boundary sliding has been suggested for WC/Co deformation at higher temperatures, and the same mechanism is likely to be effective for TiC/Ni-Mo deformation.

Another mechanism involving the binder phase which should not be overlooked is rapid diffusion of mass through this interconnected network in response to applied stress at high temperatures.

The break in the Arrhenius curve representing the temperature dependence of the yield stress for TiC/Ni-Mo (see Fig. 2) occurs at a temperature (1230°C) approximately equal to the temperature of the cross-over point (1250°C) on the graph comparing single crystal TiC with the TiC/Ni-Mo composite (see Fig. 3). Above this temperature, the composite yields more readily than the crystal, whereas below this value, the reverse is true. This approximate congruence of temperatures further supports identifying the higher temperature regime for the composite with a yield mechanism not found in the single-phase TiC.

If grain boundary sliding or mass diffusion accounts for the deformation at the higher temperatures, the increase in yield resistance at lower temperatures might be attributable to a version of the Hall-Petch effect for the carbide phase. The small grain size (1-2 micrometers) would be expected to lead to a larger yield stress than found for the single crystal, though it is not clear why a different temperature dependence would be exhibited. The Hall-Petch effect might be particularly effective in raising the yield stress of a composite containing a grain boundary phase, as slip might not be propagated readily across the boundary - providing the grain-boundary phase (binder) itself does not contribute to the deformation. Hence this mechanism would be expected to be active only at the lower temperatures.

ACKNOWLEDGEMENTS

This work was supported by the Ceramics Program, Division of Materials Research, National Science Foundation, under Grant NSF DMR-80-18695. The TiC/Ni-Mo specimens were obtained from D. Moskowitz of the Ford Motor Company.

REFERENCES

Baldoni, J.G. and Williams, W.S. 1978, Am. Cer. Soc. Bull. 57 1100-1102
Gottschall, R.J., Williams, W.S. and Ward, I.D. 1980, Phil. Mag 41 1-7
Hollox, G.E. and Smallman, R.E. 1966, J. Appl. Phys. 37 818
Moskowitz, D. and Humenik, M., Jr. 1966, Mod. Devel. Powd. Metall. 3 83
Moskowitz, D. and Humenik, M., Jr. 1978, Int. Jour. Powd. Metall, 14 39
Moskowitz, D. and Humenik, M., Jr. 1981, Mod. Devel. Powd. Metall. 14 307
Ruchlewicz, J.P. 1982, M.S. Thesis, U. of Illinois
Schenck, S.R., Gottschall, R.J. and Williams, W.S. 1978, Mater. Sci. Eng.
   32 229-239

Sharma, N.K., Ward, I.D., Fraser, H.L. and Williams, W.S. 1980, J. Am. Cer. Soc. 63 194-196
Williams, W.S. 1964, J. Appl. Phys. 35 1329-1338
Williams, W.S. 1967, Propriétés Thermodynamiques Physiques et Structurales des dérives semi-metalliques (Editions du Centre National de la Recherche Scientifique, Paris) 181-189

Inst. Phys. Conf. Ser. No. 75: Chapter 7
Paper presented at 2nd Int. Conf. Science Hard Mater., Rhodes

699

High temperature mechanical behaviour of WC-6 wt % Co cemented carbide

G FANTOZZI (1), H SI MOHAND (2) and G ORANGE (1)

(1) G.E.M.P.P.M. (U.A. 341), C.R.R.A.C.S., I.N.S.A.,
69621 Villeurbanne, France.
(2) Commissariat aux Energies Nouvelles, Alger, Algérie.

ABSTRACT

The mechanical and fracture behaviour of a cemented carbide WC-6 wt % Co has been studied from room temperature up to 1000°C. Elastic modulus, fracture strength and fracture toughness have been measured in inert atmosphere by four point bend tests. Between room temperature and 800°C, there is only limited plasticity and linear elastic fracture mechanics can be applied. Above 800°C, linear fracture mechanics theory is no longer valid on account of significant plasticity. So the material toughness has been characterized with the J integral. Several methods have been used and discussed.

INTRODUCTION

Cemented carbides consist of relatively equiaxed grains of transition metal carbides dispersed in a metallic binder phase. These composites exhibit both high hardness and good fracture toughness and are widely used in industry.

The mechanical behaviour at room temperature has been studied extensively, particularly in the cemented carbides WC-Co. In general, these materials have a brittle behaviour and the linear fracture mechanics concepts can be used. Some correlations have been established between the microstructural parameters, the cobalt content and the flexural fracture strength and the fracture toughness (critical stress intensity factor). (Gurland and Bardzil, 1955; Doi, Fuji Wara and Mikaye, 1968; Ingelstrom and Nordberg, 1974; Pickens and Gurland, 1978; Nikidom and Davies, 1980; Luyckx, 1981; Almond, 1983; Osterstock, 1983; Warren and Johannesson, 1984).

In general, WC-Co cemented carbides operate at elevated temperatures (the tip of a cutting tool can reach temperatures of the order of 800°C-1000°C). Consequently, particular attention must be paid to strength and toughness measurements at elevated temperatures. Hitherto, some studies have been made of the high temperature mechanical behaviour using hardness, bend and creep tests (Si Mohand 1983). These studies allow the deformation or rupture mechanisms to be specified and their evolution as a function of temperature to be followed.

Concerning the high temperature fracture toughness, very few studies have been made (Warren and Johannesson, 1984). They show no significant change in fracture toughness up to about 600°C and an increase in K_{IC} above.

In the present work, we present the evolution of the flexural strength and the fracture toughness of a WC-6 wt % Co cemented carbide up to 1000°C. From 800°C, significant deviation from linearity of the load – displacement curve occurs, indicating the onset of a large scale plasticity. Consequently, linear fracture mechanics theory is inapplicable and an energy concept like the J integral is applied to characterize the high temperature rupture behaviour.

MATERIALS AND EXPERIMENTAL PROCEDURES

Materials

The materials used were liquid-phase sintered WC-Co 6 wt % cemented carbide. They were manufactured by pre-sintering at 850°C and then by sintering at about 1400°C for 40 min in a hydrogen atmosphere. Their specific gravity was 14.8 Mg.m^{-3}, the Vickers hardness was 1550 HV and the mean diameter of the tungsten carbide crystals was about 2 μm.

The samples used were parallelepipeds with dimensions 30 x 4 x 2 mm for the measurement of the flexural rupture strength and with dimensions 30 x 6 x 4 mm for the toughness measurements.

Before testing, all specimens were polished mechanically using diamond paste (6 and 1 μm).

Mechanical Tests

The mechanical tests were carried out in a testing machine with a loading capacity of 100 kN. The fracture strength σ_F, the yield stress σ_y, the elastic modulus E and the fracture toughness have been measured up to 1000°C, in controlled atmosphere (95 % N_2 + 5 %H_2) by four-points bend test (Orange and co-workers, 1980). The specimens were loaded by means of rollers (upper span : l' = 8 mm, lower span : l = 24 mm). The jig is made of molybdenum alloy which has a good mechanical behaviour at elevated temperatures.

Values of Young's modulus were deduced from the applied load and displacement of specimens measured with a transducer.

The fracture toughness measurements were made on single edge notched beam specimens, with straight through notches.

The critical stress intensity factor (K_{IC}) was calculated following (Brown and Srawley, 1966) :

$$K_{IC} = \frac{3\ P(l-l')}{2B\ W^2} Y \sqrt{a}$$

1

where P is the rupture load, B and W are the specimen thickness and width respectively, l and l' the upper and lower spans, a is the notch depth and Y is a geometric factor given by the Brown and Srawley relation :

$$Y = 1.99 - 2.47(a/W) + 12.97(a/W)^2 - 23.17(a/W)^3 + 24.80(a/W)^4 \qquad 2$$

The critical stress intensity factor is related to the critical energy release rate G_{IC}; for plane strain conditions, the relation takes the form :

$$G_{IC} = \frac{K_{IC}^2}{E}(1-\nu^2) \qquad 3$$

Where ν is the Poisson's ratio.

For K_{IC} measurements of cemented carbide, the nature of the artificial pre-crack is very important. The use of machined notches with a diamond saw, very often satisfactory for ceramics, leads to overestimates of K_{IC} values in cemented carbides (Osterstock, 1980).

Alternative methods proposed for producing sharp "natural" cracks have been reviewed by Warren and Johannesson (1983). We have chosen spark erosion which allows us to obtain very sharp cracks with a controlled geometry. Firstly the specimens were notched with a 0.3 mm width diamond saw; then a crack is introduced by spark erosion under an electrical potential of 40-50V, using a tungsten blades (of 20 μm thickness) as a tool. Which this method, correct values of K_{IC} were obtained (Si Mohand, 1983).

For all the K_{IC} measurements, the relative notch depth a/W varies between 0.5 and 0.6 and the notch tip radius is about 20 μm.

The methods used for the measurement of the J integral will be described later.

All the mechanical tests were performed with a cross-head speed of 0.1 mm/min.

RESULTS

Elastic Modulus, Yield Stress and Fracture Strength.

The effect of temperature on the shape of the load-displacement curves is shown in Fig. 1. Plastic deformation appears only above 600°C and becomes significant at 1000°C. From these curves, we can deduce the variation of the elastic modulus E, the yield stress σ_y ($\varepsilon_p = 0.02$ %) and the transverse rupture strength σ_F with temperature : the results are shown in Figs. 2 and 3 respectively.
At low temperature, the yield stress σ_y, is higher than the fracture strength σ_F, and cannot be measured. Figs. 2 and 3 show that E, σ_Y and σ_F decrease slightly up to 600°C and more quickly above 600°C. Finally, the elastic modulus and the strength drop rapidly above 900°C.
All the results are listed in Table 1 (we indicate also the central deflection F_{max} and the strain ε_{max} at the maximum load).

TABLE 1 <u>Mechanical Properties of WC-6wt%Co Hardmetal vs Temperature</u>

T °C	E GPa	σ_F MPa	σ_y MPa	F_{max} μm	ε_{max} x10³
20	602	1650		114	2
600	474	1090	520	116	2.1
700	412	1020	420	163	3
800	361	890	330	217	4
900	310	710	110	889	16
1000	193	340	60	1800	33

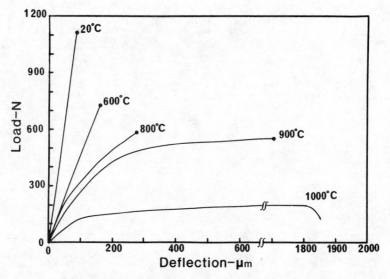

Fig. 1 - Load-deflection curves at different temperatures.

Critical Stress Intensity Factor.

The values of the critical stress intensity factor K_{IC} obtained by using the linear elastic fracture mechanics (eq. 1) are given in Table 2. The critical energy release rate G_{IC} can be deduced from eq. 3.

For materials which exhibit plastic behaviour, there is a zone ahead of the crack tip where plastic deformation occurs. This crack tip plastic zone may be considered as a cylinder of radius r_y. In plane strain conditions, this radius is given by (Broek, 1978):

$$r_y = (1/6\pi) \ (K_{IC}/\sigma_y)^2 \qquad\qquad 4$$

values of r_y are indicated in Table 2.

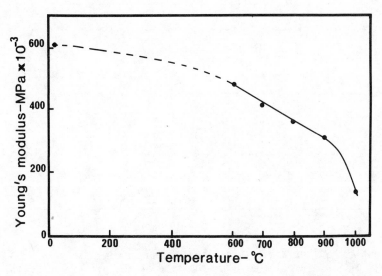

Fig. 2 - Variation of Young's modulus E with temperature.

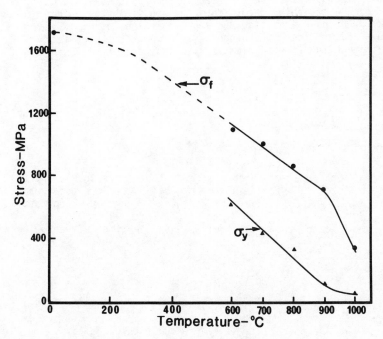

Fig. 3 - The influence of temperature on the yield stress σ_Y and the fracture stress σ_F.

TABLE 2   Fracture Toughness Critical Size and Size
of the Plastic Zone as a Function of Temperature

T	K_{IC}	G_{IC}	$2.5(K_{IC}/\sigma_y)^2$	a_c	r_y
°C	MPa m$^{1/2}$	J/m^2	m	μm	μm
20	10.5	180	$< 10^{-3}$	10	
600	10.7	230	10^{-3}	25	22
700	10.9	260	1.5×10^{-3}	30	35
800	13.7	500	4×10^{-3}	63	90
900	14.1*	610*	45×10^{-3}		
1000	9.5*	450*	60×10^{-3}		

* Denotes nominal value

Consistent K_{IC} values are only otained if the specimen thickness B
and crack size a are larger than the size of the plastic zone. The
ASTM requirement indicates that B and a must be higher than
$2.5(K_{IC}/\sigma_y)^2$ (Broek, 1978). The values of this parameter are given
in Table 2. Up to 800°C, the ASTM requirement is satisfied. Above
800°C, the size of the crack tip plastic zone becomes so large that
linear fracture mechanics theory does not apply. Although the results
are invalid, we have indicated the apparent values of K_{IC}. Thus, at
high temperatures, there is an appreciable crack tip plasticity and
the requirement for limited plasticity can be dropped by using the J
integral.

Critical Value J_{IC}

The J integral, as proposed by Rice, (1968) which characterizes the crack
tip stress and strain field is defined by :

$$J = \int_{\Gamma} (w\ dy - \vec{T} \frac{\partial \vec{u}}{\partial x} ds) \qquad 5$$

where Γ is a contour which surrounds crack tip, ds is an element of Γ, $\vec{T}$ is
the tension vector on Γ according to an outside vector $\vec{n}$ perpendicular to Γ
$(T_i = \sigma_{ij} n_j)$, $\vec{u}$ is the displacement and w is the strain energy density.

For the elastic case, the J integral is equivalent to the energy release
rate G.

Rice also pointed out that :

$$J = -\frac{1}{B}\frac{dU}{da} = -\frac{dU'}{da} \qquad 6$$

where U (U') is the potential energy (for unit thickness).

So, J is a generalized relation for the energy release rate due to crack propagation, which is valid with a large crack tip plasticity.

Several test methods can be applied for calculating J values. Some of them require the use of several specimens whereas others allow to calculate J from a single specimen. We have applied the two types of methods, but we shall present only the results obtained from a single load-displacement curve. Three J integral methods are applied here.

Firstly, we can use the method of Rice, Paris and Merkle (1973). The load-displacement curves of an uncracked specimen and a cracked specimen are compared. The J integral value is given by :

$$J_R = \frac{2(A-A')}{B(W-a)} \qquad\qquad 7$$

where A is the area under the load-displacement curve of the cracked specimen and A' the corresponding elastic area for the uncracked specimen. In the case of bend loading, A' may be appreciable.

Secondly, the method of Sumpter and Turner (1976) takes into account the subcritical crack extension. We have :

$$J_S = G + J_p \qquad\qquad 8$$

where $G = K_I^2 (1-\nu^2)/E$, K_I being calculated with final load P_f and $J_p = 2\, U_{pl}\, (W-a_f)/B\, (W-a_i)^2$. U_{pl} is the energy expended on slow crack growth and plastic deformation, a_i and a_f are the initial and final crack depths respectively.

For both methods, the cracks must be deep enough that plasticity is confined to the uncracked ligament region ahead of the crack (the ratio a/W must be $\geqslant 0.5$, Broek, 1978). Furthermore, difficulties arise for the determination of the load corresponding to the onset of slow crack growth. Very often, the maximum load is chosen but we shall see that, this method is not convenient for cemented carbides at high temperatures.

Thirdly, a method for determining the onset of crack propagation and the corresponding critical value J_{IC} is the J resistance curve calculation. The J resistance curve was determined by the iterative relation of Garwood, Robinson and Turner (1975) :

$$J_n = J_{n-1} \frac{W-a_n}{W-a_{n-1}} + \frac{2\, A_n}{B(W-a_{n-1})} \qquad\qquad 9$$

where the initial value of J is calculated from relation 7 (A' is neglected) and A_n is the increase of the area under the load-displacement curve when there is a small increment of the crack size from a_{n-1} to a_n. This increment can be measured by using the variation of the specimen compliance C (Clarke and co-workers, 1976).

$$a_n = a_{n-1} + \frac{W-a_{n-1}}{2} \left(\frac{C_n-C_{n-1}}{C_n} \right) \qquad\qquad 10$$

The J values calculated by the Garwood's method are denoted by J_G.

The critical value J_{IC} at which crack extension occurs is given by the intercept of the J-resistance $(J - \Delta a)$ curve with the crack blunting line (Landes and Begley, 1974). For ceramic materials, the blunting line can be considered as the ordinate axis (Kromp and Pabst, 1983).

We have applied the method of Rice and co-workers from room temperature up to 1000°C. Fig. 4 shows the load-displacement curves obtained at 1000°C for an uncracked specimen and a cracked specimen.

At high temperatures, we have equally used the method of Sumpter and Turner (eq. 8).

The J value obtained by both methods are given in Table 3. For the two methods, we have chosen the maximum load as the point of onset of slow crack extension. With the Rice method, we have also calculated J by using for the load of onset of crack growth the point P_1 where the load-displacement curve begins to be nonlinear. This value is denoted by J_{R1} whereas the value measured at the maximum load is denoted by J_{Rm}.

TABLE 3 G_{IC} and $J(J/m^2)$ Values obtained
by Different Methods as a Function of Temperature

T °C	20	600	700	800	900	1000
K_{IC}	180	230	260	500	610 [*]	450 [*]
J_{Rm}	200	245	300	530	790	2700
J_{R1}	200	245	300	530	500	370
J_S					120	3000

[*] Denotes nominal value.

At high temperatures, we have also calculated the J-resistance curve by the Garwood's method, the crack extension being determined from the variation in compliance.

In the case of steels, the compliances are measured by periodically unloading and reloading. The J values are not affected, even by 30 % unloading. At high temperatures, the cemented carbides show load-displacement curves which present a significant hysteresis effect, which should be studied in more detail.

This cycling effect is all the more important as the crack length increases. For alumina, Kromp and Pabst (1983) have performed the compliance measurement with total unloading, the hysteresis effect being very important. So, in our experiments, total unloading was also performed and Fig. 5 shows an example of a load-displacement curve at 1000°C.

To calculate the compliance, the tangent at the point of minimal curvature of the loading curve is used. After the last cycle, the load was removed and specimen was cooled down to room temperature and was broken. Then the crack propagation was measured optically. The optically measured crack sizes were 15 %-20 % larger than the crack lengths calculated by compliance.

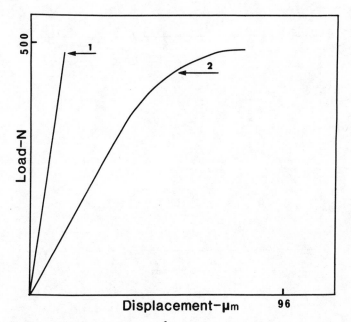

Fig. 4 - Load (P) displacement (δ) curves at 1000°C : 1, for uncracked specimen; 2, for specimen (a = 3.2 mm)

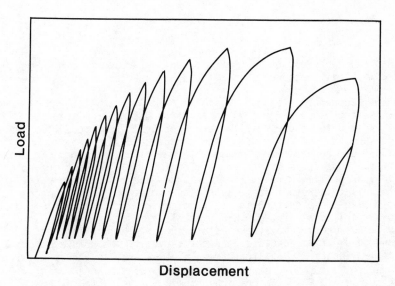

Fig. 5 - Example of a load-displacement curve with total unloading (at 1000°C).

The J-resistance curve is calculated by using the iterative formula (eq 9), the initial value of J being calculated at the load from which the compliance varies. Fig. 6 shows a J-resistance curve at 1000°C, the intersection with the ordinate axis ($\Delta a=o$) being equal to the critical value J_{IC}.

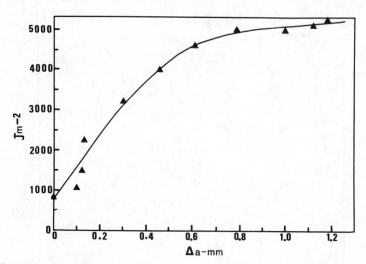

Fig. 6 - J-resistance curve calculated by the Garwood's method (at 1000°C).

By the J-resistance curve analysis, we obtained for J_{IC} 550 J/m^2 at 900°C and 830 J/m^2 at 1000°C.

DISCUSSION

The mechanical behaviour of WC-6 % Co cemented carbide is controlled by different mechanisms and we can distinguish three temperature ranges.

Between room temperature and 600°C, the material is brittle and the fracture is controlled by unstable crack propagation from critical defects. These defects are mainly microstructural inhomogeneities due to sintering such as pores, inclusions or exaggerated-size grains. Observations of the fracture surface with scanning electron microscope show that the rupture is intergranular for the small carbide grains and transgranular for large grains (Chermant and Osterstock, 1976; Osterstock 1980). In this temperature range, the linear elastic fracture mechanics theory is valid and therefore the critical energy release rate G_{IC} must be equal to the critical value J_{IC}. This equality is clearly shown in Table 3. The J values measured by the Rice's method at the maximum load correspond to J_{IC} (the load-displacement curves of the notched specimens exhibit only linearity up to fracture). The fracture toughness K_{IC} is constant up to 600°C, as observed by others authors (Warren and Johannesson, 1984). From the critical stress intensity factor K_{IC} (Table 2) and the rupture strength σ_F, we can deduce the value of the size of the critical fracture-initiating defect a_c by using eq. 1. The evolution of a_c as a function of temperature is shown in Table 2; the critical defect size increases slightly.

From 600 to 800°C, load displacement curves obtained with unnotched specimens exhibit some plasticity, as shown by Fig. 1. However, the notched specimens behaviour is linear elastic up to fracture. So, in this temperature range, the crack tip plasticity is low and the linear elastic fracture mechanics can be applied, the ASTM requirement is satisfied as indicated by Table 2. This limited plasticity can be correlated to an increase of the energy expended in the plastic work around the crack tip and so to an increase of the critical strain release rate G_{IC} and of the critical stress intensity factor. Others mechanisms have been proposed by Warren and Johannesson (1984) for explaining this K_{IC} increase : increased plasticity of the carbide, blunting of the crack tip due to creep during loading and crack healing by sintering. Other experiments are necessary to clarify this point.

The size of the critical defect increases in this temperature range (Table 2). This increase can be due to an onset of subcritical crack growth which leads to a coalescence of the defects. Indeed, microfractographs of the surface show that some microcracks have several interconnected pores to form a more important defect.

Above 800°C, the mechanical behaviour of the WC-6 % Co cemented carbide changes. The load-displacement curves of the notched specimens exhibit a strong non-linearity as shown by Fig. 4 and the linear fracture mechanics theory cannot be applied because of large scale plasticity. As observed by Ueda and co-workers (1977), the macrofractographs of the surface show an uneven region, of which the size increases strongly at 1000°C. Electron microfractographs show that the fracture is totally intergranular. Ueda and co-workers (1977) have observed that, in this temperature range, microstructure presents essentially microvoids which are generated near the triple point of WC/WC/Co boundaries or the corner of WC particles.

We have characterized the fracture toughness with the J integral. Several methods have been used to calculate the critical value J_{IC} which corresponds to the beginning of the crack growth. We can note (Table 4) that the J values obtained by the Rice method with the maximum load (J_{Rm}) and by the Sumpter and Turner method (J_S) are larger than the values given by the Garwood method J_G. The J_S values are slightly higher than the J_{Rm}. The Sumpter and Turner method takes into account the slow crack propagation up to the maximum load, whereas this effect is not considered with the Rice method. Thus the J_{Rm} value is underestimated.

The Garwood method gives the J value at the load of onset of slow crack growth. This load is not the maximum load at high temperatures and thus J_{Rm} and J_S are larger than J_G. The J_{R1} value obtained by the Rice method and calculated by using the load where the load-displacement curve begins to be non linear is much lower than J_{Rm} and J_S, particularly at 1000°C. This J_{R1} value is near the J_G value at 900°C but is lower at 1000°C.

The difference between all the J values arise from the choice of the point of onset of slow crack growth.

By compliance measurement, the critical load P_c where the crack growth begins can be determined and thus the corresponding J integral, denoted by J_{RC}. At 900°C, this load is equal to the load P_1 at which there is a beginning of non-linearity of the load-displacement curve and $J_{RC} \simeq J_{R1} < J_{Rm}$ and J_S. At 1000°C, the load P_C is situated between P_1 and P_{max}, the plasticity taking place before the crack growth and J_{R1} is much lower than $J_{RC} < J_{Rm}$. Naturally, the J_{RC} values are nearer the J_G values, the two values corresponding to the initiation of the crack propagation.

TABLE 4   J Values (J/m^2) obtained by Different Methods

T °C	J_G	J_{R1}	J_{Rm}	J_S	J_{RC}
900	550	500	790	1200	500
1000	830	370	2700	3000	850

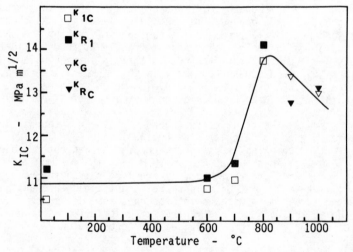

Fig. 7 - Effective critical stress intensity factor K_{IC} as a function of temperature.

Thus, the Garwood and Rice method give good results provided that the J value is calculated at the critical load for initiation of crack growth. The compliance measurements allow us to determine precisely this load.

The critical J values increase between 800°C and 1000°C. The rupture behaviour in the 800°C-1000°C range is probably controlled by the plasticity of the binder phase. Several models have been proposed to explain the dependence of fracture toughness on microstructural parameters

(Almond, 1983). In all cases, the toughness is linked to the ductile behaviour of the binder phase. The models are not enough quantitatively developed for making a precise analysis of our results which needs further developments. Furthermore, in the same temperature range, microvoids are nucleated near triple points of boundaries or the corners of WC grains. These microvoids coalesce by a diffusional flow of vacancies in the Co phase (Ueda and co-workers, 1977). Interconnection between these voids leads to a formation of microcracks and to a slow crack growth process.

From the critical J values, the effective critical stress intensity factor values can be deduced by the relation $J_{IC}=K_{IC}^2(1-\nu^2)/E$. Fig. 7 shows the K_{IC} variation as a function of temperature. Fracture toughness exhibits a maximum around 800°C. The K_{IC} decrease is due to the large reduction of the Young's modulus E. The presence of this maximum must be confirmed.

ACKNOWLEDGEMENTS

This work has been supported by the Agence Francaise pour la Maitrise de l'Energie under contract Number EMP. OD. n° 596. Procurement of the materials were made by the Society Tykram-Pedersen. The authors would like to thank Dr. P. Goeuriot and F. Thevenot for valuable discussions.

REFERENCES

Almond, E.A. 1983, Sciences of Hard Materials, Eds R.K. Viswanadham, D.J. Rowcliffe and J. Gurland, Plenum Press, New York, pp 517-562.
Broek D. 1978 in Elementary engineering fracture mechanics, Sijthoff and Noordhoff, Netherlands.
Brown, W.F. and Srawley, J.E. 1966, Plane Strain Crack Toughness Testing of High Strength Metallic Materials, ASTM STP 410 pp 1-145.
Chermant, J.L. and Osterstock, F. 1976, J. Mat. Sci. 11 1939-1951.
Clarke, G.A. Andrews, W.R., Paris, P.C. and Schmidt, D.W. 1976, Mechanics of Crack growth, ASTM STP 590 27-42
Doi, H., Fujiwara, Y. and Mikaye, K. 1968, Trans. J. Inst. Met. 9 616-622
Garwood, S.J., Robinson, J.N. and Turner, C.E. 1975, Int. J. Fract. 11 528-530
Gurland, J. and Bardzil, P. 1955, J. Metals, Trans. AIME, 203 311-315
Ingelstrom, N. and Norberg, H. 1974, Eng. Fract. Mech. 6 597-607
Kromp, K. and Pabst, R.F. 1983, J.Am. Ceram. Soc., 66 106-110
Landes, J.D. and Begley, J.A. 1974, Fracture Analysis, ASTM STP 560 pp 170-186
Luyckx, S.B. 1982, Fracture 81, Proc. V Internat. Conf. on Fracture Ed. D. Francois, Pergamon, Oxford, pp 1075-1081
Nikidom, B. and Davies, T.L. 1980, Planseeber. Pulvermet 28 pp 29-38
Orange, G., Dubois, J., Fantozzi G. and Gobin, P.F. 1980, Mem. Sci. Rev. Met. 2 pp 131-143
Osterstock, F. 1980, Thesis, University of Caen, France
Osterstock, F., Fracture Mechanics of Ceramics 6, Eds R.C. Bradt and co-workers, Plenum Press, New York, pp 243-254
Pickens, J.R. and Gurland, J. 1978, Met. Sci. Eng. 33 135-142
Rice, J.R. 1968, J. Appl. Mech. 35 pp 379-386
Rice, J.R., Paris, P.C. and Merkle, J.G. 1973 in Progress in Flaw Growth

and Fracture Tonghness Testing, ASTM STP 536 pp 231-245

Si Mohand, H. 1983, Thesis INSA Lyon

Sumpter, J.D.G. and Turner, C.E. 1976, in Cracks and Fracture, ASTM STP 601 pp 3-18

Ueda, F., Doi, H., Fujiwara, F. and Masatomi, H. 1977, Trans Jap. Inst. Met. 18 pp 247-256

Warren, R. and Johannesson, B. 1984, Proc. 3rd Int. School on Sintered Materials New-Dehli

Inst. Phys. Conf. Ser. No. 75: Chapter 7
Paper presented at 2nd Int. Conf. Science Hard Mater., Rhodes

713

Fracture of hardmetals up to 1000°C

B JOHANNESSON and R WARREN

Department of Engineering Metals, Chalmers University of Technology
S-412 96 Göteborg, Sweden

ABSTRACT

The fracture toughness of three hardmetals has been studied between 20 and
1000°C using a four-point bend test. Between 400 and 500°C for a WC-15%Co
alloy and between 700 and 800°C for a WC-6%Co alloy, the two alloys
exhibited a transition in fracture behaviour from brittle/elastic to sub-
critical crack growth in association with significant plastic deformation
at the crack tip. A mixed carbide alloy (WC-20%TiC-15%(Ta,Nb)C-9%Co) remained brittle up to 1000°C. This study, together with those of other
investigators, indicate that hardmetals generally exhibit such a transition but that the transition temperature varies considerably depending on
the composition of the alloy.

INTRODUCTION

It has long been appreciated that, in many service situations, the high-
temperature mechanical properties of hardmetals are more relevant than
their room-temperature properties. It is only recently that attempts have
been made to measure the fracture toughness (e.g. K_{IC}) of these materials
at elevated temperatures. Earlier measurements of transverse rupture
strength at temperatures up to 1000°C, made by e.g. Ueda and co-workers
(1977) have shown that for common hardmetals the TRS falls and fracture
strain increases markedly at about 600°C. Novikov and co-workers (1982)
studied the notch fracture up to 1000°C of three WC-Co alloys (6, 9 and 15
wt % Co), reporting the results in terms of K_{IC}. Mohand, Orange and Fantozzi (1984) investigated the notch fracture of a WC-6%Co alloy to the
same temperature. Loshak and co-workers (1983) report on a similar investigation of TiC- and Ti(C,N)-Ni-Mo-alloys. Mohand and co-workers found in
their study of the 6% Co alloy that above about 750°C the fracture was
associated with a large amount of plasticity and sub-critical crack growth
and consequently that the toughness could not be decribed in terms of a
valid K_{IC} value. In their studies, Novikov and co-workers and Loshak and
co-workers do not state explicitly that they observed plasticity.

In the present work a study has been made of the fracture toughness and
fracture behaviour of three commercial hardmetals (two WC-Co alloys and
one mixed carbide alloy) also up to 1000°C. Particular attention was paid
to the transition from elastic to plastic-dominated fracture.

EXPERIMENTAL

The main characteristics of the alloys studied in the present work are given in Table 1. The microstructural properties were determined using standard quantitative metallographic techniques on SEM-micrographs.

TABLE 1 <u>The Three Alloys Investigated</u>.

Nominal compositions wt %	WC-6%Co +(0.2%NbC-0.2%TaC)	WC-15%Co	WC-20%TiC-15%(Ta,Nb)C-9%Co
Wt % C (analysis)	5.81	5.21	--
Wt % Co (- " -)	5.95	14.6	--
Vol % Co	10	23.9	10.5
Vol % binder phase*	7.5	24.5	9
" WC phase*	92.5	85.5	6.5
" γ-carbide*	--	--	84.5
Carbide grain size (mean			1.6 (γ)
linear intercept; μm)	0.5	2.9	1.1 (WC)
Hardness (HV 30)	1750	1050	1500
K_{IC} (room temp; $MNm^{-3/2}$)	9±0.5	17±1	8.5±0.5

* Measured by point counting on SEM micrographs of unetched sections; the binder fractions tend to be underestimated.

The fracture of the alloys was studied by means of a single edge notch four-point bend test. The specimens were rectangular bars, 6 mm high, 3 mm in width and 46 mm long, lapped and polished down to a finish of 1 μm diamond paste. Pre-cracks were created by bridge indentation. This method, which avoids many of the difficulties of pre-cracking, was first described by Sadahiro and Takatsu (1981) and then studied in more detail by Warren and Johannesson (1984) and Johannesson (1985). The depth of the crack is dependent on the bridge punch spacing and so is conveniently predetermined. Fig. 1 shows the top and side surface of a sample precracked by this method. Pre-cracks with depths between 1.5 and 4 mm were created. The crack depths were measured on the sides of the specimen using an optical microscope. Observation of fracture surfaces after fracture indicated that the crack fronts were straight

The 4-point bend test, using rolls spaced at 19 and 38 mm, was performed in a screw-driven 50 kN testing machine. The deflection of the sample was recorded during the test by measuring with a linear displacement transducer the relative movement of three vertical rods touching the underside of the sample, one at the midpoint and two reference rods under the inner rolls. This permitted observation of the load-deflection behaviour. Furthermore, sub-critical crack growth could be estimated in situ via relative changes in the specimen compliance (Okamura, Watanabe and Takano, 1972). For testing at high temperatures, the bending fixture was placed in an electric resistance furnace in a vacuum chamber, tests being carried out in a vacuum of about 0.01 Pa. Except where stated otherwise, tests were performed at a cross-head speed of 60 μm/min.

K_{IC} values were calculated from the fracture load and pre-crack length using the conventional linear elastic fracture mechanics formula for the four-point SENB test. The interpretation of tests in which the load-deflection behaviour deviated from linear-elastic are discussed in the next section.

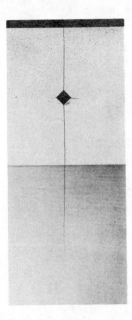

Fig. 1 Top and side surface of indented sample.

In order to study elevated temperature fracture in greater detail, a
number of supplementary experiments were carried out. These included tests
interrupted before fracture, fracture testing of specimens at lower temperatures after pre-loading at high temperatures and fracture tests carried
out at different strain rates (cross-head speeds between 6 and 600 μm/min)
Details of these experiments are given in the results section.

After testing, specimens were examined using optical and scanning electron
microscopy. Both fracture profiles and fracture surfaces were investigated

RESULTS

General Fracture Behaviour

Fig. 2 shows typical load deflection curves for the 6% and 15% Co alloys.
At lower temperatures the fracture can be termed brittle; the load-deflection curves were very nearly linear up to the load at which unstable
fracture occurs. Nevertheless, a small deviation from linearity was often
observed just before fracture. This was interpreted as being the result of
a small amount of sub-critical crack extension, Δa. The magnitude of Δa
was estimated from the corresponding change in sample compliance and added
to the initial pre-crack size (measured by optical microscope) when calculating K_{IC}. The effect of this phenomenon on the K_{IC} value was relatively
small.

Between 400 and 500°C for the 15%-alloy and between 700 and 800°C for the
6% alloy, the load-deflection curves began to exhibit significant non-
elastic behaviour. Examination of fractured samples and samples from
interrupted tests revealed that this was associated with both subcritical

crack growth and significant plastic deformation. This means that the fracture behaviour above these temperatures cannot be expressed in terms of a valid K_{IC} value. Hitherto in this work, tests that permit a reliable estimate of fracture energy parameters such as J_{IC} have not been performed. As a convenient alternative means of expressing the fracture results, use has been made of two nominal fracture parameters, K_C^* and K_C^{**} (see Table 2). K* is a nominal stress intensity based on the prevailing load and the initial crack depth, a, while K** is a stress intensity based on the load and the true crack length, i.e. including any sub-critical crack growth, Δa, but ignoring any crack-tip blunting that may have taken place.

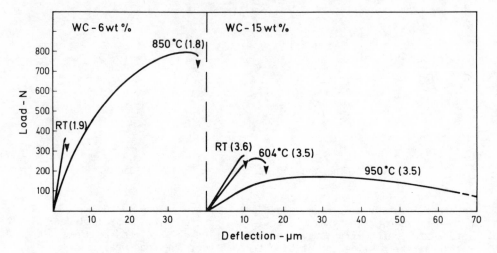

Fig. 2   Load vs. midpoint deflection curves for WC alloy samples.
Numbers in brackets give pre-crack depth.

TABLE 2 Definition of the Three Toughness Parameters Used.

Load	Crack-length		Fracture toughness
F(max)	a_0	(original pre-crack length)	K_C^*
F(max)	$a_0 + \Delta a$	(sub-critical crack growth estimated from the change in compliance)	K_{IC}
F(critical)	$a_0 + \Delta a$	(Δa is subcritical crack growth above the transition temperature, estimated from the fracture surface)	K_C^{**}

As will be shown below, such crack growth is usually identifiable from the fracture profile or fracture surface. K_C^{**} is defined as the value of K** at the onset of unstable fracture while K_C^* is based on the initial crack length and the maximum load attained during loading to fracture. These parameters depend on test geometry and have little value as general material constants. However, they do give an indication of the relative resistance to fracture in the materials of this study. The difference between K_C^{**} and K_C^* for a given test indicates the sub-critical crack growth that has taken place. Furthermore, K_C^* provides a means of comparison with other investigations in which values of K_{IC} are quoted on the basis only of maximum load and initial crack length regardless of any non linear-elastic behaviour that might have occurred.

The mixed-carbide alloy exhibited brittle behaviour up to 1000°C.

Fig. 3a and b show micrographs of the crack and crack-tip region in a specimen of the WC-6%Co alloy that was loaded at 850°C to a K* value giving sub-critical crack growth but not complete failure (K* = 14.8 MNm$^{-3/2}$). The result is typical for all specimens of both this alloy and the 15%Co alloy tested above their respective brittle/elastic transition temperatures. The transition from the straight pre-crack, formed by bridge indentation and the sub-critical crack is easily discerned. The sub-critical cracks were always more serpentine than the pre-cracks and unstable cracks and often exhibited crack branching. This difference was also apparent on the fracture surface observed after final failure (Fig. 4). Although it may not be clearly seen in the micrographs of Fig. 3, regions of deformation could be seen extending from the crack, up to 250 μm in some samples.

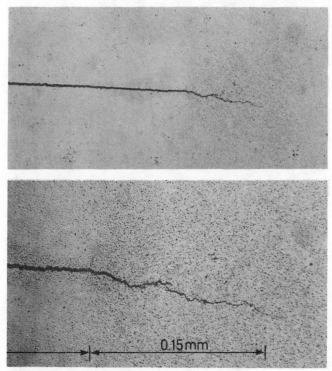

Bridge indented | Sub-critical crack growth
crack at RT. | at 850°C.

Fig. 3a and b Micrographs (same crack, two magnifications) demonstrating sub-critical crack-growth (and plastic deformation) in WC-6%Co at 850°C.

50 µm

Fig. 4 Fracture surface of sub-critical crack region; note fissures and
partially-detached protuberences (SEM).

By interrupting tests at various loads, the sub-critical crack growth and
plastic deformation (measured as the ratio of plastic deflection to total
deflection) could be measured as a function of increasing K**. The results
for the WC-6%Co alloy at 850°C are shown in Fig. 5.

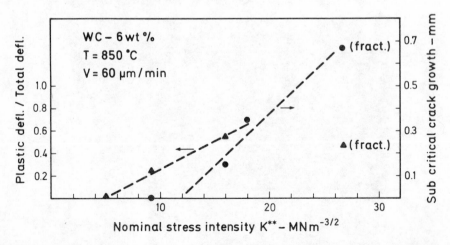

Fig. 5 Sub-critical crack growth and plastic deformation as a function
of applied stress intensity for WC-6%Co, at a temperature of 850°C.

The sub-critical crack growth does not begin until a threshold value, corresponding roughly to the RT K_{IC}, is reached and then shows a roughly linear increase with K**. The increase of K** is the result of increases in both load and crack length. The ratio of plastic deflection to total deflection vs. K**, calculated from the load-deflection curves, also exhibits linear behaviour initially. Here, the threshold value of K** marking the onset of plastic deformation was obtained from the point where the load-deflection curve deviates from linearity. The two curves illustrate the fact that plastic deformation precedes crack extension.

Fracture Toughness Results

Fig. 6 shows the effect of temperature on fracture toughness of the three alloys. Above the transition temperatures, the fracture is described in terms of K_C^* and K_C^{**}, as discussed above. The WC-6% alloy exhibits a sharp increase in apparent toughness above 700°C. This occurs in association with the onset of the large scale plastic deformation described above. The rise in K_C^* values reflects a marked increase in the fracture load for a given crack depth.

In the WC-15%Co alloy significant plastic deformation and subcritical cracking begins between 400 and 500°C. Unlike the corresponding processes in the 6%Co alloy these continue at nominal stress intensities around the room-temperature critical value and consequently cause an apparent lowering of toughness with temperature when measured in terms of K_C^*. Even expressed in terms of K_C^{**} only a small increase in apparent toughness is observed.

The fracture toughness of the γ-carbide alloy is relatively unaffected by temperature up to 1000°C, where the first signs of an increase occur.

To better understand the reason for the sharp increase in toughness of the WC-6%Co alloy above 700°C, tests were carried out on pre-loaded, pre-heated samples. These experiments are summarised in Table 3. Samples heated to above the transition temperature and loaded to just below the fracture load, when subsequently tested to failure at lower temperatures, retained the high apparent K_{IC}. Similar samples heated to the same temperature, but only lightly loaded, retained a low K_{IC} when subsequently tested at the lower temperature. These results indicate that the apparent increase in toughness can be attributed to crack blunting rather than to an increase in the intrinsic toughness of the microstructure or to crack healing.

Table 3 Summary of Tests with Pre-heated WC-6%Co Specimens

pre-conditions		test-conditions	
Temp. ^{o}C	K* MNm$^{-3/2}$	Temp. ^{o}C	K_{IC} MNm$^{-3/2}$
800	2-5	200	10.0
870	"	400	9.9
800	"	650	12.2
871	9.6	221	16.3
800	15.3	RT	>15.3 (no fract.)

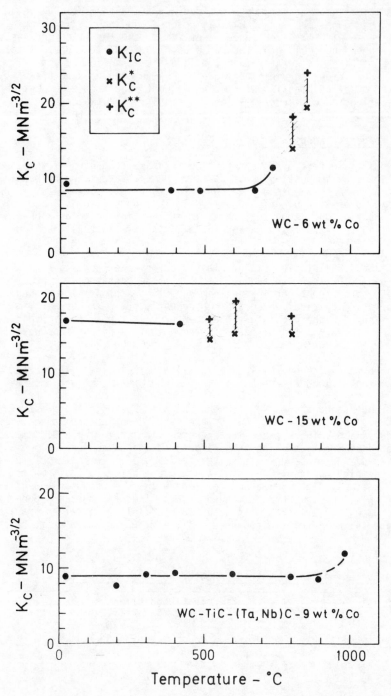

Fig. 6 Effect of temperature on the fracture toughness of the three alloys.

Fig. 7 shows the effect of testing rate on K_C^* for WC-6%Co at 750°C. The K_C^* value changed from 8 to 15 $MNm^{-3/2}$ when the crosshead rate was changed between 600 and 6 μm/min. The $\dot{K}^*$ scale is approximate only, being derived from the crosshead speed and an assumed value of a=2.5 mm. At 750°C the onset of plastic deformation is suppressed at $\dot{K}^*$ values above 8 $MNm^{-3/2}$ s^{-1} (The ASTM standard E399-72 recommends stress intensity rates between 0.55-2.75 $MNm^{-3/2}$ s^{-1} for room temperature tests. At 850°C plastic behaviour was observed even at the highest rate tested viz. 600 μm/min ($\dot{K}_C^*$ = 24 $MNm^{-3/2}$).) At room temperature, the K_{IC} value was found to be insensitive to strain rate.

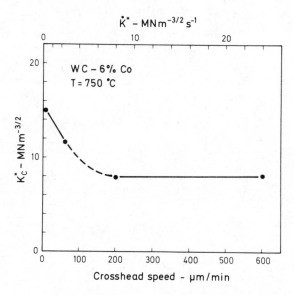

Fig. 7   Effect of cross-head speed on K_C^* of the WC-6%Co alloy at 750°C.

DISCUSSION

The main results of this work are summarised in Table 4 together with the results of high-temperature studies of other investigators. Although neither Novikov and co-workers (1982) nor Loshak and co-workers (1983) mention plasticity or sub-critical crack growth explicitly, it has been assumed that these must have occurred in their experiments and therefore their toughness values are interpreted as K_C^* in the table. The different investigations of WC-6%Co and WC-15%Co alloys are in good agreement.

It is apparent that, in common with most other materials, hardmetals are characterised by a brittle/ductile transition temperature. It is also clear that this temperature varies over a large range, depending on the alloy in question. For the WC-15%Co alloy studied here it is between 400 and 500°C while for the mixed-carbide alloy it is above 1000°C. It is proposed here that the transition in the WC-15%Co alloy is determined by the onset of yield and subsequent ductile fracture occurring largely by dislocation and transformation mechanisms in the cobalt binder phase. It is suggested that in the WC-6%Co alloy these processes are suppressed by the existence of a much stronger carbide skeleton. Deformation first becomes possible when this skeleton can deform by creep processes, in this

case probably carbide grain-boundary sliding and cavitation. This suggestion is consistent with the observations of Ueda and co-workers (1977)
in their transverse rupture experiments. They observed a transition to a
creep-like deformation and failure in WC-Co alloys occurring between 600
and 800°C. The marked strain-rate dependence of the fracture above 700°C
observed in the WC-6%Co alloy in the present study is also an indication
of a time-dependent creep process.

Table 4   Summary of Elevated Temperature Fracture Toughness Studies

Alloy wt %	(ref)*	Grain size	$K_{IC}(RT)$ MNm$^{-3/2}$	Temperature behaviour
WC-6%Co	(J)	<1μm	9	Significant plastic deformation and sub-critical crack growth begin at approx. 750°C; K_C rises sharply.
	(M)	2	10.5	-"-
	(N)	2	11	K_C rises moderately, from about 500°C.
WC-9%Co	(N)	3	15	Slow rise in K_C from about 700°C.
WC-15%Co	(J)	3	17	Plastic defn and sub-critical cracking begins between 400 and 500°C. K_C changes little with temp. up to 900°C.
	(N)	5	20	From RT, K_C decreases moderately.
WC-TiC-(Nb, Ta)C-9%Co	(J)	1.6(γ)	8.5	K_{IC} relatively unaffected by temp. up to 1000°C.
TiC-Ni-Mo (12vol% binder)	(L)	1-4	8	K_{IC} unaffected by the temp. up to 650°C above which a rapid increase in K_C.
Ti(C,N)-Ni-Mo (binder % not given)	(L)	"	11.5	-"-

* J: Johannesson and Warren (1984); M: Mohand, Orange and Fantozzi (1984);
  N: Novikov and co-workers (1982); L: Loshak and co-workers (1983).

The absence of a transition in the mixed carbide alloy suggests that the
carbide forms a strong and creep-resistant skeleton. This can be attributed partly to the inherently greater creep-resistance of the cubic carbide
which has been demonstrated by creep measurements (e.g. see Santhanam,
Mizgalski and McCoy (1984)) and partly to the larger grain-size in this
alloy which would reduce the tendency to grain-boundary creep processes.

The transition at around 650°C in TiC-Ni-Mo and Ti(C,N)-Ni-Mo alloys observed by Loshak and co-workers (1983) is interesting since it indicates
that there is a significant difference between the microstructure of this
family of alloys and the mixed WC-γ-carbide-Co alloy examined in this
work. Considerably more information is required to obtain a clearer understanding of the mechanisms of deformation in these alloys.

CONCLUSIONS

- The notch fracture behaviour of hardmetals is characterised by a transition temperature below which the material behaves in a brittle manner and above which final failure is preceded by significant plastic deformation at the notch tip and sub-critical crack growth.

- Different hardmetals can have significantly different transition temperature.

- For WC-6%Co alloys the transition occurs between 700 and 800°C. The plastic deformation and sub-critical crack growth leads to blunting of the pre-crack tip and a consequent marked increase in the ultimate fracture load for a given pre-crack length.

- For a WC-15%Co alloy with a 3 µm carbide grain size the transition occurred between 400 and 500°C. The plastic deformation and sub-critical crack growth did not cause a significant increase in fracture load.

- For a WC-20%TiC-15%(Ta,Nb)C-9%Co alloy no transition was observed up to 1000°C.

ACKNOWLEDGEMENTS

Specimens of the WC-Co alloys were kindly supplied by Sandvik AB, Stockholm, and the γ-carbide alloy be SECO Tools, Fagersta. We are also indebted to the Swedish Silicate Research Institute, Göteborg, for use of their testing machine. The work was financed by the Swedish Board of Technical Development (STU).

REFERENCES

Johannesson, B. 1985, J. Matls. Sci. Letters, 4 299-301 .
Loshak, M., Konovalenko, N., Shapoval, T. and Dudkin, M. 1983, in Carlsson I. and Ohlson, N.G. (Eds.), Proc. 4th Internat. Conf. on Mech. Behaviour of Matls., Pergamon, Oxford, pp 865-871.
Mohand, H.Si., Orange, G. and Fantozzi, G. 1984, Proc. 2nd International Conference on the Science of Hard Materials (Rhodes)(see this volume)
Novikov, N.V., Majstrenko, A.L., Konovalenko, N.K. and Uljanenko, A.P. 1982, P/M-82 in Europe, Proc. Internat. Powder Met. Conf., Florence, Associazione Italiana di Metallurgia, Florence, pp 685-591.
Okamura, H., Watanabe, K. and Tahano, T. 1972, ASTM publication STP 536, ASTM, Philadelphia, pp 423-438.
Sadahiro, T. and Tahatsu, S. 1981, in "Modern Developments in Powder Metallurgy", (ed. H.H. Hausner, H.W. Antes and G.D. Smith), Vol. 14, New York, Plenum, pp 561-572.
Santhanam, A.T., Mizgalski, K.P. and McCoy, W.C. 1984, Proc. 2nd International Conference on the Science of Hard Materials, Rhodes (see this volume).
Ueda, F., Doi, H., Fujiwara, F. and Masatoni, H. 1977, Trans. Jap. Int. Metals, 18 (3) 247-256.
Warren, R. and Johannesson, B. 1984, Powder Metallurgy, 27 25-29.

Discussion on Cermets and hardmetals — Part two

Rapporteur: R K VISWANADHAM (1)
Session Chairman: E A ALMOND (2)

(1) Reed Tool Company, Houston, Texas, USA
(2) National Physical Laboratory, Teddington, Middlesex, UK

The paper by Dr Moscowitz on Ti(CN) based hardmetals attracted a
question from Dr Chatfield on possible effects of TiN in reducing
edge-chipping of cutting tools. No correlation was observed although
the author expected that changes in resistance to edge-chipping should
correlate with fracture toughness. In response to a request from Dr
Exner, Dr Moscowitz listed the properties that he rated as important
in developing a brittle material for metal cutting applications:
    -minimal tool/work-piece reactivity (chemical inertness)
    -maximum tool wear resistance with respect to the particular
    workpiece (cutting-edge wear resistance)
    -maximum fracture toughness
    -maximum resistance to tool deformation
    -maximum thermal shock resistance (good thermal conductivity and
    low thermal expansion coefficient)
The model microstructure would consist of a hard brittle phase, no
more than several micrometres in grain size, uniformly dispersed by a
thin sub-micrometre film of a refractory ductile binder phase which
fully wets the hard phase.

To a question from Dr Komac regarding an influence on properties from
altering the processing route, Dr Moscowitz said he had no results on
using pre-alloyed carbonitride instead of a blend of carbide and
nitride powders. However this would be the subject of future research.

Dr Landingam thought that the improved resistance to nose deformation
attributed to the higher thermal conductivity obtained with increasing
TiN, would be accompanied by a similar correlation between TiN content
and the flank wear of tools used to cut hard and soft metals. However,
Dr Moscowitz believed that the improved resistance to nose deformation
could be related to grain refinement. The results on thermal
conductivity had been obtained by Dr Tanaka, not him. Flank wear
resistance was not affected by increasing TiN content except in a
turning test on a 1045 steel of hardness 180HB at moderate speed and
feed.

To questions from Dr Pastor on the preparation of TiN-rich
carbonitrides and chemical analysis, Dr Moscowitz agreed that 5h at

1800 C should be suitable; he stated that vacuum fusion was used to determine the nitrogen and oxygen contents.

Dr Santhanam sought further information on fracture toughness and hardness. The author observed a tendency for bulk hardness to increase with increasing TiN content for cemented Ti(C,N) materials, which he attributed to a grain-refinement effect since unbonded Ti(C,N) materials exhibited a drop in microhardness with increasing TiN fraction. Therefore the fracture toughness data were obtained on specimens whose hardness was slightly higher at higher TiN fractions.

Dr Moscowitz informed Mr B Williams that no improvement was obtained in strength after hot-isostatically pressing except where there had been significant residual porosity or other flaws.

In reponse to a series of questions from Dr Komac on his paper on the high temperature deformation of Ni-TiC hardmetals, Professor W S Williams made the following comments. The data for a crossover of the CRSS curves for polycrystalline and cemented forms of the carbide had so far only been obtained for TiC materials, but it was planned to extend the research to the WC-systems. He speculated that the ideal microstructure for high deformation resistance would be fine-grained and have a low binder content. The most advantageous carbon-to-metal ratio should be close to the stoichiometric value, but this idea had not been tested yet in the cemented TiC-(Ni,Mo); only one composition had been studied.

Dr Schmid remarked that the yield stress versus temperature curves, were similar to those obtained for the thermally activated motion of dislocations in bcc metals, and asked about the mechanism and activation energy. Professor Williams replied that a yield stress that decreased rapidly with increase in temperature had been observed in all the carbide single crystals and cemented composites he had studied. Thermally activated dislocation glide over a high Peierls barrier was one possible mechanism in the range of temperature where deformation was not controlled by the binder phase. For TiC-(Ni,Mo) he had not reported an apparent activation energy, though the data fitted an Arrhenius equation. It was often difficult to attribute such an energy to a specific mechanism in a complex system.

The paper on estimation of the plastic zone size associated with cracks in cemented carbide was presented by Dr Chatfield. Professor Fischmeister criticised the use of a coarse FEM mesh because it prevented the model developing high stress concentrations at places where the microcracks in the carbide enter the cobalt phase. The refinement of the net at the crack tip did not help because the elements were assigned carbide and binder properties alternately. To model the behaviour observed by Sigl, Exner and Fischmeister it would be necessary to subdivide a large area of binder phase into many small elements. Dr Chatfield believed that his FEM mesh was sufficiently fine to give good accuracy. The first Gaussian integration point was only a few nanometres from the crack. Even there, no signs of the high stresses proposed by Sigl and co-workers were found. Dr Laugier argued that if the plastic zone at the crack tip could be modelled by a homogeneous model, it followed that toughness should be independent of microstructure, and he asked whether this was found experimentally. Dr Chatfield claimed that since toughness was generally found to be a linear function of hardness, WC-Co grades with the same hardness but different microstructures should have the same toughness. However he agreed with a comment from Dr Warren that it was difficult to understand the FEM prediction of plastic strain occurring in binder phase regions surrounded by carbide that only deforms elastically.

The paper on characterisation of fracture processes and fracture relevant parameters in WC-Co hardmetals was presented by Mr Sigl. Professor Larsen-Basse asked whether the model took into account the stress concentrations which exist at carbide grain corners in the binder phase region. This should give rise to plastic deformation of the binder phase even at some distance from the crack. According to Mr Sigl, the FEM calculations indicate that such events do occur. However, when a crack forms the stresses are redistributed so as to concentrate deformation at the points where the microcracks enter the binder phase. This happens before general yielding of the binder phase, but it is important to note that the flow stress of the binder phase is very high since the effective 'Hall-Petch' constant is determined not by Co/Co boundaries but by the more rigid WC/Co boundaries.

Dr Schmid asked Dr Fantozzi about the testing machine used in the investigation of high temperature mechanical behaviour of 6%Co-WC cemented carbide. Dr Fantozzi replied that it had a capacity of 100kN and the tests had been performed at a cross-head speed of 0.1mm/min. A second question concerned the possible existence of an adhesion zone behind the crack tip which could account for the J/a effect. However, this possibility had not been investigated, and Dr Fantozzi thought that the explanation for the J-curve was linked to the plastic zone at the crack tip, but this would need to be confirmed by electron microscopy.

Mr Knox concluded the session by drawing attention to the round robin exercise performed by the ASTM for the purpose of standardising the measurement of fracture toughness with chevron-notched short-rod specimens.

Inst. Phys. Conf. Ser. No. 75: Chapter 8
Paper presented at 2nd Int. Conf. Science Hard Mater., Rhodes

729

CVD of Al_2O_3 for coated cemented carbides

B LUX(1), C COLOMBIER(1), H ALTENA(1) AND K STJERNBERG(2)

(1) Institute for Chemical Technologie of Inorganic Materials, Technical
University Vienna, Getreidemarkt 9, A-1060 Vienna, Austria
(2) AB Sandvik Coromant, Box 42056, S-126 12 Stockholm, Sweden

ABSTRACT

A short survey of nucleation, growth and orientation relationships
during α-Al_2O_3 deposition is given. It is shown that the equilibrium
shape of α-Al_2O_3 is strongly influenced by impurities, such as Cr, Ni,
Fe, Co and Ti, which could also originate from the reaction between the
construction materials or the substrates and the gases used. These
impurities often cause undesirable crystal defects, holes, whisker
growth and dendritical branching, combined with a change of the Al_2O_3
growth rate.

INTRODUCTION

Since the introduction of cemented carbides, turning and milling
operations have become much more efficient. Even today they are still
used extensively for steel cutting. Around 1970 another revolution was
brought about by the application of thin coatings to the cemented
carbides. This permitted further reduction of cutting time and an
increase in the lifetime of the turning tools by a factor of 4 to 10
(Ekemar, 1977; Lux and Schachner, 1977) (Fig.1).

The impact of coatings on the cutting tool industries can be deduced
from Fig. 2. While the chip volume showed a steady increase each year,
both the annual carbide consumption per chip volume and the annual
consumption costs for carbide tools were reduced.

Fig. 3 shows a global view of the market situation for cutting tools as
well as the optimal cutting speeds for some specific grades (Kalish,
1983). It can be seen that currently about half of all cemented carbide
tools are coated. Furthermore, it becomes apparent that the largest gain
for new coatings can be expected in the lower cutting speed range. The
diagram also reflects the excellent behaviour of today's coated tools
based mainly on coatings containing Al_2O_3 layers and designed for use at
high cutting speeds.

The typical microstructure of a coating on a cemented carbide consists
of an inert layer of about 5 μm of TiC, with an outer layer of Al_2O_3.
Today, Al_2O_3 is generally recognized as being one of the best coatings

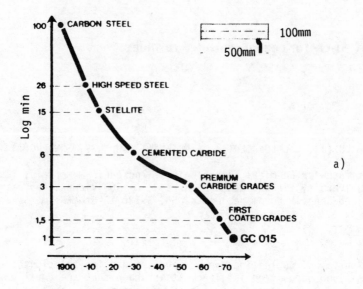

a)

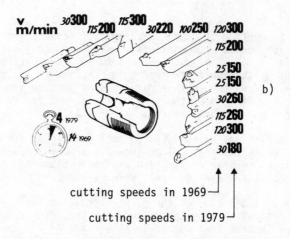

b)

cutting speeds in 1969⌐

cutting speeds in 1979⌐

Fig.1 New tool materials and machining times of steel.
a) The cutting time for a work piece could be drastically reduced by
using cemented carbides. After S.Ekemar, 1977.
b) Comparison of real cutting time/speed between 1969 and 1979.
13 different cutting tools must be used to manufacture this piece.
14 mn work were necessary in 1969 and only 4 mn in 1979.
After E.Lundgren.

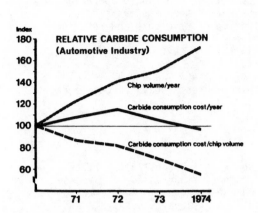

Fig.2 Effect of new cutting
tools on productivity in a
Swedish automotive plant.
After S.Ekemar, 1977.

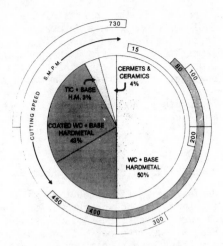

Fig.3 Percentage use of various
metal cutting materials with
their optimal cutting speed
After H. Kalish, 1983.

available as far as increased performance for turning operations at high
cutting speeds is concerned. It is found in many commercial coatings as
one of the layers applied to the tool. In the following only the
formation and problems associated with CVD-applied α-Al_2O_3 (corundum)
layers will be discussed.

PREPARATION OF ALUMINA COATINGS BY CVD

Today CO_2 is used as an oxygen donor for industrial production, reacting
with hydrogen to form water, which reacts with the Al donor ($AlCl_3$) to
form Al_2O_3 (Lindström and Schachner, 1980). Thus, the chemical equations
involved in the Al_2O_3 CVD deposition are :

1) formation of the O-donor (H_2O) in the gas phase:

$$CO_2 + H_2 \longrightarrow H_2O + CO \qquad (1)$$

2) reaction between the adsorbed Al-donor and the O-donor at the
   solid surfaces:

$$2AlCl_3 + 3H_2O \longrightarrow Al_2O_3 + 6HCl \qquad (2)$$

The water formation taking place in the gas phase is the most important
rate-limiting factor of this CVD process (Lindström and Schachner, 1980;
Lindström and Stjernberg, 1985).
Basic requirements for the α-Al_2O_3 CVD layer are:

    1) uniform thickness around the specimen, including
       edges,
    2) a certain minimum thickness,
    3) a specific microstructure and
    4) regular surface morphology.

Moreover, all the inserts of one reactor batch, consisting of about 5000 to 10000 pieces, must be coated identically in one single operation. The reactor design, gas flow (Fig. 4), the reaction parameters such as temperature, pressure, time, concentration of reactants and reaction products, as well as the condition of the insert surface prior to coating, are of greatest importance in this context.

Temperatures above 1200 K are needed since the $\mathcal{H}$- and α-Al$_2$O$_3$ modifications do not form at lower temperatures. Today, reduced and normal pressure conditions are used in industrial reactors.

Nucleation and Growth of Al$_2$O$_3$ Layers

Nucleation starts on certain single spots at the substrate surface (Kornmann and co-workers, 1975). The number of nuclei for a given set of reaction parameters is dependent on substrate, time and temperature. Fig. 5 shows the early stages of growth on a sapphire substrate and how these single spots grow together to form a layer. On the perfect α-Al$_2$O$_3$ substrate, epitaxial growth occurs and nucleation of new crystal orientations is prevented (Altena, Colombier and Lux, 1983a). On pure single- or polycrystalline alumina, epitaxial growth can be disturbed by surface imperfections. Dislocation networks created by

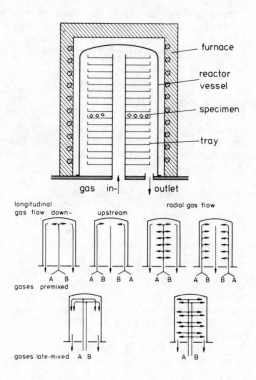

Fig.4 Different gas flow patterns in CVD reactors
for coating of cemented carbides.

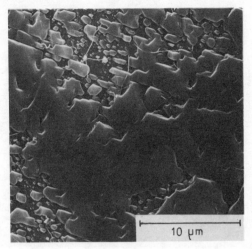

Fig.5 Early stages of α-Al$_2$O$_3$ growth on the (1.0.-1.0) plane
of an α-Al$_2$O$_3$ single crystalline substrate.

Fig.6 Influence of dislocation network (created by scratches) and of
substrate orientations on nucleation and growth.
a) α-Al$_2$O$_3$ polycrystalline, b) sapphire (1.0.-1.0)
and c) sapphire (0.0.0.1)

scratches from cutting and polishing can favour nucleation or stop
crystal growth, depending on the orientation of the grain (Fig.6). The
Al$_2$O$_3$ crystals nucleated within the scratched areas can still have clear
orientational relationships with the substrate. Deep etching removes
such defects and favours the epitaxial growth mode (Manasevit and
Morritz, 1967; Korec, 1979). Nucleation certainly also depends on the
chemical nature of the substrate (Kornmann and co-workers, 1975;
Schmitt, Altena and Lux, 1983). Orientational relationships may also
exist between Al$_2$O$_3$ and substrates different in chemical nature from
Al$_2$O$_3$ (Chatfield, 1984).

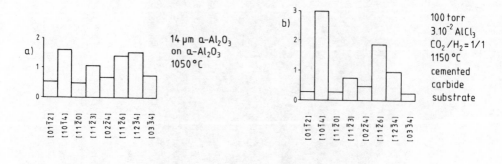

Fig.7 Preferred orientation in Al₂O₃-CVD layers.
a) after Colombier, 1984 and b) after Park, Kim and Chun, 1983.

Influence of Substrate Orientation on Growth

The grains of a polycrystalline sapphire substrate exhibit different
growth rates depending on their respective orientation (Altena,
Colombier and Lux, 1983a). Generally the low-index faces grow much
slower than the high-index faces; (0001) has the slowest growth rate. A
coating consists of many individual single crystals grown together. Each
single crystal grows at a different rate characteristic for its
orientation. This leads to naturally preferred orientations in the
layers (Fig.7) (Colombier, 1984; Park, Kim and Chun, 1983) It can be
questioned whether this "natural" orientation represents the optimum for
later use as a wear resistant and tough layer.

Crystal Habitus and Equilibrium Shapes in $AlX_3/H_2/CO_2$ Environments

Crystals grown slowly in the gas phase develop equilibrium facets which
depend strongly on the gas environment in which they were grown.
Typical equilibrium shapes where fivefold pyramids shown in Fig. 8. In
some cases 7-fold pyramids have also been observed. These fivefold
pyramids are bounded by (0.0.0.1), (1.1.-2.1) and (3.-4.1.15) crystal
faces (Colombier and co-workers, 1986). It was surprising to discover
the high-index plane as an equilibrium facet in the $AlX_3/H_2/CO_2$ systems.
Its existence can be most likely explained by the presence and strong
adsorption of HX on this crystal facet (Lux and co-workers, 1985). As
shown in Fig.8 only those CVD systems containing hydrogen showed 5- (or
7-) fold pyramids while systems free of hydrogen, such as $AlX_3/CO_2/Ar$,
developed quite different equilibrium shapes.

In the $AlX_3/H_2/CO_2$ systems other equilibrium shapes are also found quite
frequently, bounded mainly by (0.0.0.1) and (1.0.-1.0) faces
(Altena, Colombier and Lux, 1983a; Lux and Co-workers, 1985). Changes in
the relative growth rates of these two faces lead to pyramidal or coin-
like crystals. Particularly pronounced are C-type crystals or whiskers
and so-called "coin like" shapes formed in the presence of certain
impurities (Altena, Colombier and Lux, 1983b; Lux and Co-workers, 1985).

AlCl₃/CO₂/H₂, 1323 K, 65 mbar

⊢—⊣ 1 μm

AlCl₃/CO₂, 1523 K, 65 mbar

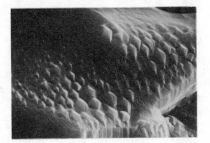

⊢—⊣ 1 μm

AlBr₃/CO₂/H₂, 1323 K, 65 mbar

⊢————⊣ 1 μm

AlBr₃/CO₂/Ar, 1523 K, 65 mbar

⊢—⊣ 1 μm

AlI₃/CO₂/H₂, 1323 K, 65 mbar

⊢—⊣ 1 μm

AlI₃/CO₂/Ar, 1323 K, 65 mbar

⊢—⊣ 1 μm

a)   b)

Fig.8 Comparison of the equilibrium crystal facets in the AlX₃/CO₂/H₂ and AlX₃/CO₂/Ar System. a) AlX₃/CO₂/H₂ system, formation of fivefold pyramids and b) AlX₃/CO₂/Ar system, no formation of fivefold pyramids.

INFLUENCE OF IMPURITIES ON Al$_2$O$_3$ LAYER FORMATION

Impurities can have a detrimental influence on the Al$_2$O$_3$ coating formation (Altena, Colombier and Lux, 1983b). A relationship between the thickness of the underlaying TiC and the critical thickness of the Al$_2$O$_3$ before layer breakdown occurred is illustrated in Fig.9. This suggests that an element diffusing from the underlaying WC-Co substrate to the surface disturbs the "planar" Al$_2$O$_3$ growth front. Another example supporting this is shown in Fig. 10, where a layer was submitted to a second coating operation after cooling down to room temperature. It is well known that alumina layers develop cracks when cooled. After reheating these cracks apparently permitted more rapid diffusion from the underlaying substrate to the outer surface, influencing the second Al$_2$O$_3$-deposition along these cracks (Fig.10). In both cases cobalt most likely influenced the Al$_2$O$_3$ coating deposition (Altena and Co-workers, 1983).

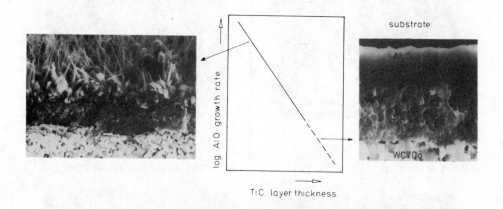

Fig.9 Layer breakdown due to impurities. Influence of TiC thickness on whisker formation and Al$_2$O$_3$ growth rates.

Fig.10 Diffusion of impurities from the substrate through a cracked layer.

Testing Impurity Interactions During Al$_2$O$_3$ Growth

Some simple tests applied to screen the many possible impurity
influences on the alumina growth are briefly described below.

1) **Rub-on test**

A substrate - either pure alumina or cemented carbide - is rubbed with
the impurity, as for example with nickel, before coating. The Al$_2$O$_3$
layer formed is disturbed where nickel was deposited (Fig. 11).
Mechanical deformation of the substrate surfaces could also have a
disturbing influence, but it can be assumed that the impurity has the
dominant effect (Altena, 1982).

2) Wire test

A wire is put across the substrate surface and reacts with the gas
environment to form chlorides. This set-up, where the impurity
concentration decreases with the distance from the wire, also allows
the observation of the dependence on concentration (Fig.12) (Peng, 1986;
Colombier and Co-workers, 1986).

3) Sputtering impurities onto the substrate surface

Prior to CVD the impurity can be deposited on a substrate by vacuum
evaporation or sputtering, whereby the thickness of the deposited layer
can be varied (Altena, Colombier and Lux, 1983b; Gass, Mantle and
Hintermann, 1975). Again the influence can be clearly seen by comparing
the Al$_2$O$_3$ grown on areas with and without impurities.

These tests are valid for elements which are solid up to 1400 K and
which react slightly with the gas phase. For liquid or gaseous
impurities and for quantitative measurements, it is necessary to
introduce the impurity in a carrier gas stream as vapor or volatile
compound in appropriate amounts (Altena, 1982). This is of course more
difficult and requires a thorough study for each individual element.

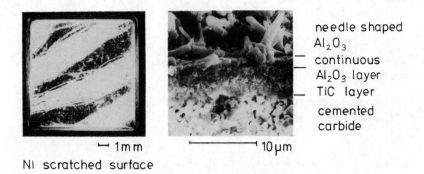

needle shaped
Al$_2$O$_3$
— continuous
Al$_2$O$_3$ layer
— TiC layer
cemented
carbide

⊢ 1mm   ⊢――――――⊣ 10μm

Ni scratched surface

Fig.11 Rub-on impurity test: influence of Ni.

Typical Growth Changes due to Impurities

Formation of holes

In the presence of impurities, Al_2O_3 crystals develop holes during the early growth stages (Fig.13). Such crystals have been observed with many different impurities if present in high concentration e.g. underneath and close to the wire in the "wire test". The presence of crystal defects such as dislocations (Vuorinen, 1984) - normally helpful for the Al_2O_3 growth - are also preferred sites for the adsorption of the impurity, thus blocking the growth sites. A hole remains and a "negative growth spiral" develops.

Layer breakdown, needlelike crystals and whisker formation

Most of the above screening tests showed that impurities can favour formation of needlelike crystal growth or whiskers. This early breakdown of the layer formation can be caused by traces of Co, Cr or Si (Altena and Co-workers,1983). These elements are adsorbed preferentially at certain spots on the surface - most likely where lattice defects occurred - leading to the growth of individual single crystals (Fig.14). In a later stage whiskers develop due to the influence of the impurity, the concentration and the temperature field (Altena and Co-workers, 1983). Sometimes dendritic branching occurs on top of the whiskers, leading to a club-like appearance of the whiskers (Fig.14). The layer breakdown can also be affected by wet reaction gases or a reaction chamber leakage, both of which lead to a change in the gas environment.

Another type, so-called zigzag whiskers, has been known for many years. According to Amelinckx they are associated with impurities or dislocation movements (Amelinckx, 1958). Fig. 15 shows typical zigzag whiskers due to the presence of Fe, Co or Au during Al_2O_3 CVD. For the industrial coating of cemented carbides all these growth modes are highly undesirable.

CONCLUSIONS

Nucleation growth and preferred orientation during $\alpha-Al_2O_3$ deposition by means of the $AlCl_3/CO_2/H_2$ CVD reaction have been briefly discussed. Process parameters and impurities influence the equilibrium shapes of the crystals. Impurities such as Cr,Ni,Fe,Co or Ti are also present during industrial production. The raw materials, the construction material of the CVD unit and the substrates themselves are sources of such impurities.

Impurities can considerably influence growth rate and crystal morphology, leading to crystal defects, needlelike formations and whisker growth as well as dendritically branched and cauliflower-like structures.

Various examples of such undesirable artifacts observed during $\alpha-Al_2O_3$ CVD on cemented carbides have been given. It is however also conceivable that such foreign elements ,if present in small concentrations, may accelerate heterogeneous nucleation and growth of Al_2O_3 on the substrate surface, or modify the Al_2O_3 microstructure in a beneficial manner.

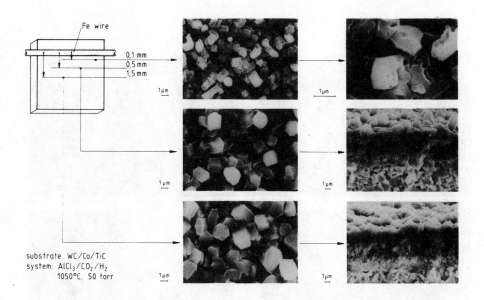

Fig.12 Impurity test with Fe wire.

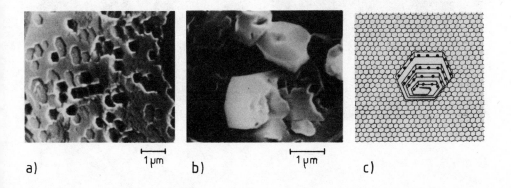

Fig.13 Formation of holes in Al_2O_3 CVD crystals due to impurities.
a) the formation of etch pits shows the existence of dislocations in normally grown Al_2O_3 CVD crystals.
b) the adsorption of an impurity (Fe) on the growth steps emerging from a screw dislocation leads to the formation of holes during Al_2O_3 growth.
c) schematic presentation of a "negative growth spiral" as created by strong adsorption of impurities on growth steps.

example 1   example 2

Fig.14 Sequence of a layer breakdown favoured by
presence of impurities.

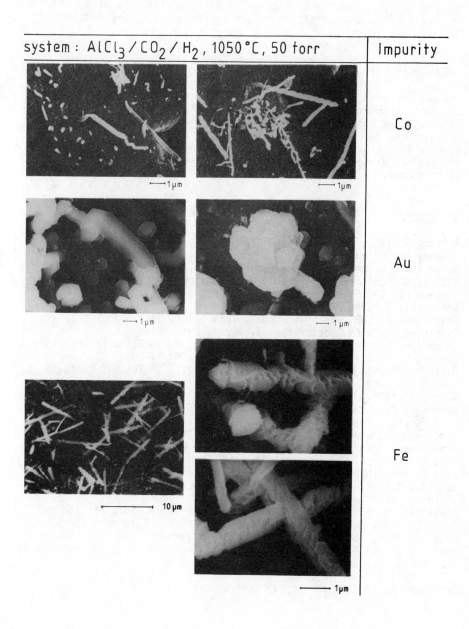

system : $AlCl_3 / CO_2 / H_2$, 1050°C, 50 torr | Impurity

Co

Au

Fe

Fig.15 Zigzag whiskers occurring during Al_2O_3 CVD with high impurity concentration level.

Nevertheless, production of high-grade Al_2O_3 coatings by many industrial producers today demonstrate that such problems can be overcome during industrial production.

AKNOWLEDGEMENT:

The authors wish to thank the Sandvik AB, Hard Materials Division for financial support; Director Prof. Dr. Bertil Aronsson for his interest, his continuous personal encouragement and for permission to publish; and Dr. Jan Lindström, Dr. Mats Söjstrand and many other co-workers of the Sandvik research team for many helpful and open discussions as well as their practical assistance during the performance of this work.

REFERENCES

Altena, H. 1982, doctor thesis, Techn. Univ. Vienna
Altena, H., Colombier, C. and Lux, B. 1983a, Euro CVD IV, Eindhoven, 435-443
Altena, H., Colombier, C. and Lux, B. 1983b, Euro CVD IV, Eindhoven, 451-458
Altena, H., Colombier, C., Lebl, A., Lindström, J. and Lux, B. 1983,
    Euro CVD IV, Eindhoven, 428-434
Amelinckx, S. 1985, Phil.Mag. 38 425
Chatfield, C. 1984, 6 th Int. Conf. Thin Films, Stockholm
Colombier, C. 1984, doctor thesis, Techn. Univ. Vienna
Colombier, C., Peng, J., Altena, H. and Lux, B. 1986, to be published
Ekemar, S. 1977, Soc.Manufacturing Engineers MR 77 201
Gass, H., Mantle, H. and Hintermann, H. 1975, 5 th Int.CVD Conf. 99-110
Kalish, H. 1983, Euro CVD IV, Eindhoven 58-69
Korec, J. 1979, J. Crystal Growth 46 362-370
Kornmann, M., Schachner, H., Funk, R. and Lux, B. 1975,
    J. Crystal Growth 28 259-262
Lindström, J. and Schachner, H. 1980, Euro CVD III 208-217
Lindström, J. and Stjernberg, K. 1985, Euro CVD V, Uppsala 169-182
Lundgren, E., priv. communication
Lux, B. and Schachner, H. 1977, 9. Planseeseminar, Reutte 34
Lux, B., Colombier, C., Altena, H. and Stjernberg, K. 1985,
    to be published in Thin Solid Films
Manasevit, H. and Morritz, F. 1967, J. Electrochem. Soc. 114 204-207
Park, C., Kim, J. and Chun, J. 1983, Euro CVD IV, Eindhoven 410-420
Peng, J. 1986, doctor thesis, Techn. Univ. Vienna
Schmitt, T., Altena, H. and Lux, B. 1983, Euro CVD IV, Eindhoven 421-427
Stjernberg, K., Gass, H. and Hintermann, H. 1977,
    Thin Solid films 40 81-88
Vuorinen, S. 1984, 9 th Int.CVD Conf., USA

Inst. Phys. Conf. Ser. No. 75: Chapter 8
Paper presented at 2nd Int. Conf. Science Hard Mater., Rhodes

743

Preparation of TiC, TiN, Al₂O₃ and diamond-like carbon by a glow discharge technique

A DOI, N FUJIMORI AND T YOSHIOKA

Itami R&D Department, R&D Group, Sumitomo Electric Industries Ltd,
1-1-1, Koya-kita, Itami, 664, Japan

ABSTRACT

TiC, TiN and Al_2O_3 were deposited by means of modified plasma CVD (named TP-CVD by the present authors), and diamond-like carbon was deposited by cracking hydrocarbon gas in a glow discharge atmosphere. Structural and compositional analyses were done for these thin films by SEM, AES, ESCA, X-ray, optical microscope and other techniques. It was found that the TiC, TiN or Al_2O_3 film had almost the same characteristics as those of the corresponding bulk materials. Diamond-like carbon film turned to be electrically conductive when doped with boron during the deposition process.

INTRODUCTION

A glow discharge plasma reaction is known to reduce the vapour deposition temperature for ceramics (Kato and Koga, 1971; Archer, 1981) and is applied for the production of amorphous Si solar cell and for the basic research on the synthesis of diamond-like carbon from vapour phase, as well. A new plasma CVD method, TP-CVD (Three Dimensional Plasma CVD) was developed and has been adopted for TiC coating on molybdenum parts on a production scale (Abe and co-workers, 1984). These parts are the constituents of a first wall for the nuclear fusion testing device, JT60, now under construction in Japan. Conventional induction coupled plasma decomposition is recognized as one of relevant methods to obtain diamond-like carbon films (Doi, Yoshioka and Fujimori, 1983); possible impurity doping into such films could be of value to develop a unique semiconductor device.

The microstructure of these films is normally fine grained or amorphous, characteristics of thin film ceramics synthesized by a glow discharge plasma. In the present work, a study has been made of the basic properties of TiC, TiN, Al_2O_3 and diamond-like carbon films synthesized or deposited by this technique.

TP-CVD METHOD AND TiC, TiN AND Al_2O_3 DEPOSITION

The conventional plasma CVD was modified in such a manner that all surfaces of a substrate could be covered, in this case by layers of ceramic materials. The deposition conditions for TiC, TiN or Al_2O_3 are shown in Table 1, and the substrate was molybdenum.

TABLE 1 <u>Deposition Conditions for TiC, TiN or Al_2O_3 by TP-CVD Method</u>

Substance	Substrate Temp. (°C)	Source Gas	Deposition Rate (m/hr)
TiC	900	$H_2, CH_4, TiCl_4$	2.5
TiN	800	$H_2, N_2, TiCl_4$	3.0
Al_2O_3	600	$H_2, CO_2, AlCl_3$	2.0

SEM photographs of each substance are shown in Fig. 1. AES and ESCA in analysis were performed on the TiC and Al_2O_3. AES indicated no appreciable amount of impurities (such as oxygen or chlorine) present in these ceramic films. The binding energy for Ti(2S) electrons was found to coincide with that for corresponding electrons in a sintered TiC. A similar result was obtained for Al_2O_3 (Al(2S) electrons in this case).

TiC   TiN Al_2O_3

$\longleftarrow$

2μm

Fig. 1 SEM photographs showing surface of TiC, TiN and Al_2O_3 deposited by TP-CVD.

X-ray diffraction showed that TiC was of crystalline structure having a lattice parameter of 4.3280 Å, a value very close to that for a single crystal of TiC. Al_2O_3 had amorphous structure, however. The Vickers microhardness with 0.98 N load was measured on a sectional surface of the coating layer of the specimen with about 20 μm thick TiC or TiN film. The values of 3100 HV and 1800 HV were obtained for TiC and for TiN, respectively.

For the Al_2O_3 sample, the relationship between current and voltage was studied. The order of $10^{11}\Omega$ cm was obtained as a specific electrical resistance and the break-through voltage was 5×10^6 V/cm.

Scatter in the thickness of TiC layer was examined and one of the results is shown in Fig. 2. The schematic description of the TP-CVD assembly can be found in the literature (Fujimori and co-workers, 1984).

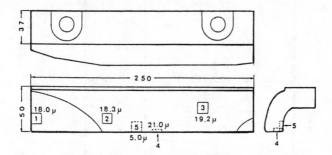

Fig. 2   Thickness distribution measured over molybdenum first wall parts specimen coated with TiC.

HIGH FREQUENCY WAVE PLASMA DECOMPOSITION METHOD AND DIAMOND-LIKE CARBON

It is well known that visually transparent and either amorphous, microcrystalline or crystalline carbon having properties similar to those of diamond can be deposited by cracking hydrocarbon gas as a source material (Doi, Yoshioka and Fujimori, 1983; Holland and Ojika, 1976; Setake and co-workers, 1981). Radio frequency plasma or microwave plasma is the most common means to crack hydrocarbon gases like methane and these two methods were chosen for the deposition of such carbon films. Under the conditions listed in Table 2, a crystalline structure was observed in the film by condition 1. In contrast, the film obtained by condition 2 turned out to be amorphous according to either X-ray or reflective electron diffraction analysis. The film thickness was about a few microns ($1{\sim}5 \times 10^{-4}$ cm).

TABLE 2   Deposition Conditions for Carbon by Cracking Methane Diluted with Hydrogen

Condition		1	2
Cracking condition	frequency	2.45 GHz	13.56 MHz
	power density	4.5 w/cm^3	4.0 w/cm^3
Gas composition	CH_4	1	1
	H_2	100	100
Gas velocity		50 mm/sec	52 mm/sec
Total pressure		1×10^3 Pa	1×10^3 Pa

The carbon film synthesized by either method exhibited a specific electrical resistance of 10^9 to 10^{10} Ω cm and reflective index of about 2.40. Emission spectra from the gas plasma were obtained during the deposition and the spectra corresponding to each deposition condition are shown in Fig. 3. Substrates used for the deposition were Si, SiO$_2$ and molybdenum. The film characteristics were found to be independent of the substrate.

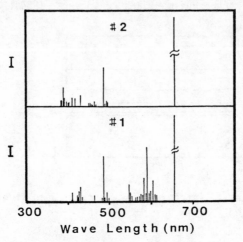

Fig. 3 Emission spectra of gas plasma.

Doping of these films with boron adding 1 vol % B$_2$H$_6$ to the methane resulted in a decrease of specific electrical resistivity down to the order of 10^{-1} Ω cm. The V-I curve obtained for the boron doped diamond-like carbon film deposited under condition 1 is shown in Fig. 4 as reference. Such film could be of value when used as thin film thermistor for high temperature use.

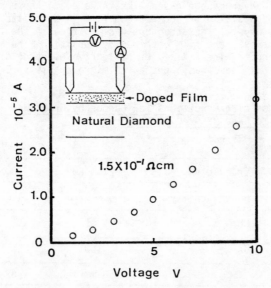

Fig. 4 V-I curve of boron doped diamond-like carbon film.

CONCLUSIONS

Use of a glow discharge plasma atmosphere was found to permit a reduction of the deposition temperature for TiC, TiN or Al_2O_3 when they were synthesized by CVD.

TiC coating using TP-CVD process was applied for the protective coating on the first wall parts of the nuclear fusion testing device, JT60. The impurity level and the thickness uniformity of the TiC layer was found satisfactory.

Diamond-like carbon was successfully deposited on Si, SiO_2 or molybdenum.

Boron doping of the film resulted in a reduction of the electrical resistivity.

REFERENCES

Abe, T., Murakami, Y., Itou, Y., Mizoguchi, T., Kajiura, S., Sagawa, J., Doi, A., Fujimori, N., Yoshioka, T. and Yashiki, T. 1984, J. of Vac. Soc. of Jap. Vol 27, No. 5, 398
Archer, N.J. 1981, Thin Solid Films, 80 221
Doi, A., Yoshioka, T. and Fujimori, N. 1983, Proceedings of the 30th Conference of Japan Soc. of Applied Physics, 214
Doi, A., Fujimori, N., Yoshioka, T. and Doi, Y. 1983, Proceedings of International Ion Engineering Congress, 1137
Fujimori, N., Yoshioka, T., Doi, A., Abe, T., Mizoguchi, Y. and Itou, Y. 1984, Thin Solid Films, 118 5-14
Holland, L. and Ojiha, S.M. 1976, Thin Solid Films, 38 L17
Kato, H. and Koga, Y. 1971, J. Electrochem Soc. 118 1619
Setaka, N., Matsumoto, S., Kamo, M. and Sato, Y. 1981, Proceedings of the Annual Meeting of the Carbon Soc. of Japan, 70

Inst. Phys. Conf. Ser. No. 75: Chapter 8
Paper presented at 2nd Int. Conf. Science Hard Mater., Rhodes

Effects of substrate material on the growth and hardness of TiN films prepared by reactive sputtering

J E SUNDGREN, M-K HIBBS, B-O JOHANSSON AND U HELMERSSON

Dept. of Physics and Measurement Technology,
Linkoping University, S-581 83 Linkoping, Sweden

ABSTRACT

TiN films have been reactively sputtered on high speed steel and stainless steel substrates in the temperature range from 200°C to 650°C. In
the case of the high speed steel, it is thought that epitaxial growth may
occur due to good lattice match between the TiN film and MC particles in
this particular steel. This gives rise to a bimodular grain size distribution where the shape of discrete and enlarged grains is sensitive to the
deposition temperature. Consequently, it is observed that the hardness of
the TiN film is dependent on its structure. These phenomena do not occur
with the stainless steel substrate.

INTRODUCTION

A large number of papers reporting different physical properties of TiN
thin films have been written during the last ten years (Aronson, Chen and
Class, 1980; Hummer and Perry, 1983; Igasaki and co-workers, 1978; Igasaki
and Mitsuhashi, 1980; Jacobson, Nimmagadda and Bunshah, 1979; Mumtaz and
Class, 1982; Münz and Hessberger, 1981; Nakamura and co-workers, 1977;
Poitevin, Lemperiere and Tardy, 1982; Schutz, 1983; Sundgren and co-
workers, 1983; Ting, 1982; Wittmer, Studer and Melchior, 1981). The majority of these papers concerns film deposited by various physical vapour
deposition methods. However, for almost all of the reported properties
there is a large scatter in the data presented. For example, Vickers
hardness values between 340 and 3000 HV are found (Hummer and Perry, 1983;
Jacobson, Nimmagadda and Bunshah, 1979; Münz and Hessberger, 1981;
Nakamura and co-workers, 1977). This scatter is mainly due to the different process conditions used, resulting in different nucleation and growth
phenomena and thus variations in composition and structure of the films.

Preparation of stoichiometric TiN films using reactive sputtering or ion
plating processes requires a careful control of the process parameters,
especially the flows of the reactive and inert gases (Aronson, Chen and
Class, 1980; Mumtaz and Class, 1982; Nakamura and co-workers, 1977;
Sundgren, Johansson and Karlsson, 1983a; Sundgren, Johansson and Karlsson,
1983b; Wittmer, Studer and Melchior, 1981). Also substrate temperature
(Jacobson, Nimmagadda and Bunshah, 1979; Münz and Hessberger, 1981),

deposition rate (Jacobson, Nimmagadda and Bunshah, 1979), and substrate material and geometry (Johansson and co-workers, 1984) are of importance. Small changes in any of these parameters can cause large differences in the properties of the TiN films.

The variation in the properties of the films will also affect their performance in different areas of application. For instance, the use of TiN as a wear resistant coating on tools shows a very promising development, and on the average results in a significant increase in the lifetime of tools. However, the range in lifetime between the best and the worst tool is very large (Ting, 1982). This spread is very unsatisfactory since TiN coated tools are used, to a large extent, in machines in continuous production systems requiring a minimum of stops. A better understanding of the relationship between the microstructure and the processing parameters is accordingly of essential importance.

In this paper we report on the variation of the microstructure with substrate material and substrate temperature for TiN films prepared by reactive d.c. magnetron sputtering. Also considered are the effects of substrate bias. The different microstructures observed are also correlated with the hardness of the films, and also to the adhesion of the films.

EXPERIMENTAL PROCEDURES

The films were prepared by reactive d.c. planar magnetron sputtering in a diffusion pumped vacuum system with an ultimate pressure of 5×10^{-7} Torr. Titanium with a purity of better than 99.95% was sputtered from 6" target in a mixture of argon and nitrogen. The flow rates of both nitrogen (99.9992% pure) and argon (99.9997% pure) were kept constant, at 0.18 Torr x $1s^{-1}$ and 1.08 Torr x $1s^{-1}$ respectively, resulting in a total pressure of 5 m Torr. The power applied to the discharge was 1.9 kV (480V and 4A) which results in a deposition rate of 200 A/min at a target to substrate distance of 6 cm. Films were deposited to a thickness of approximately 4 μm. For some depositions a substrate bias of $_{\sim}300V$ was applied, resulting in an ion current density of 1.6 mA/cm^2.

Two different kinds of steels were used as substrate materials, one powder metallurgical (ASP23) high speed steel (HSS) and one stainless steel (SIS 2304). The substrates were in the form of long strips that were ground and polished using 1 μm diamond paste for the final polish. Prior to deposition they were ultrasonically cleaned in trichlorethylene, acetone and ethanol. The substrates were also sputter cleaned immediately before the deposition. During the deposition one end of the strip was heated while the other was cooled. A temperature gradient ranging from 200°C to 700°C was obtained as measured with thermocouples point welded onto the strip.

The hardness of the films was measured in an optical microscope with a Vickers indenter. The presented results are mean values of 20 indentations made at a load of 0.1 N, where no effect from the substrate material is found. The adhesion of the films to the substrate was measured with the scratch-test technique where a loaded diamond tip makes a scratch on the film surface. At a critical load the film close to the scratch will be removed from the substrate. This critical load will be a measure of adhesion.

The samples for the electron microscope studies, using 120 kV, were prepared by first grinding the substrates to a thickness of 100 μm and then
punching out a disc 3 mm in diameter. These discs were thinned to electron transparency using an ion milling apparatus. An argon beam operated
at 6 kV was directed towards the substrate side of the sample at an angle
of 30° from the sample surface. During the final 0.5 - 1 hour, milling
was performed from both sides of the sample in order to clean the film
surface. It is estimated that the areas observed in the microscope were
located 1 to 2 μm below the original surface of the film.

RESULTS AND DISCUSSION

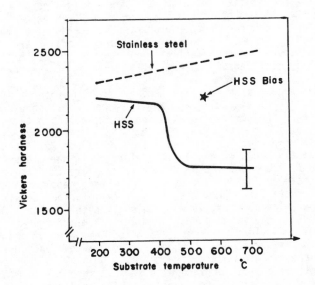

Fig. 1 Vickers hardness as a function of substrate
temperature for films deposited on both stainless
steel (-----) and high speed steel, HSS (———).
The error, indicated for one hardness value, is
approximately the same for all measured hardness values.

The variations in film hardness as shown in Fig. 1 can be correlated
with film structure. For films deposited on high speed (HSS) the microstructure changes drastically at the temperature (775K) where the drop
in film hardness occurs. At all temperatures the grain size is inhomogeneous, ranging from 0.05 μm to 0.5 μm. For T_s < 775K the grain orientation is random and some voids are observed both at grain boundaries and
within the grains. However, at T_s = 775K large (0.5 μm) square- and
triangular-shaped grains appear having the <112> direction perpendicular
to the substrate. The void content of the films increases drastically,
with the grains. Fig. 2 shows TEM micrographs of films deposited on HSS
at two different temperatures 400°C and 550°C that is before and after the
drop in film hardness.

The mean spacing between the large grains found in films deposted on HSS substrates is approximately the same as that of the MC carbides in the substrates. The size of the large grains is also approximately the same as the carbide size. The MC carbides, which mainly consist of VC, have the same crystal structure as TiN with a mismatch in the lattice parameters of only 1.4%. This good lattice match stimulates a localized expitaxial growth on these carbide particles. This results in a microstructure consisting of large grains surrounded by small grains. This has been verified by Hultman and co-workers (1984), who used TEM to study thin TiN films grown on TEM foils of HSS. In that study the substrate temperature was varied between 310-920 K and air exposed substrates were deposited with and without sputter etching. Examination of the as-deposited films showed fine equiaxed grains of TiN on the substrate that was not sputter etched. On the sputter etched substrate, TiN grew epitaxially on the MC-carbides and with fine equiaxed grains on the surrounding steel matrix. The size and number of TiN grains per unit area are the same for the non-sputter etched substrates and for the steel matrix in the sputter etched substrates. The carbide grains are randomly oriented and distributed in the steel matrix. Epitaxial growth of TiN was observed on the six most densely packed lattice planes of the MC-carbides.

The shape of the large grains is influenced by the temperature. Some of the large grains take on a peculiar triangular- and square-shaped form when the temperature is increased. This can be explained in terms of minimization of surface energy (Winterbottom, 1967). The sides of the square are almost perpendicular to the <311> direction. Even though the angles do not exactly match with the (311) plane it is the closest plane that simultaneously is perpendicular to the (211) plane (the plane parallel to the substrate surface). However, it has been reported in the literature that crystals can grow with a low energy surface consisting of a number of steps (Winterbottom, 1967). This gives the envelope of the surface a different angle with respect to a certain plane than the low energy surface plane would give. The reason for this kind of growth is that the steps in the surface can act as preferential nucleation sites. This means that the surface must not necessarily be a (311) surface, it could just as well be a (111) surface with steps. In order to minimize the energy and for these grains to develop, a high atom mobility is needed since the atoms have to move long distances. At lower temperatures the surface diffusion is not sufficient to form the triangular- and square-shaped grains. After the formation of the first triangle the second triangle starts to grow on the first one. This process must then take place in the bulk of the film during and/or after deposition. However, at this temperature the process is not sufficient to create fully developed grains, the grains are penetrated by long cracks or voids. Also, a lot of voids are visible along their periphery. At T_s = 550°C only about 20% of the large grains have this square or triangular shape. Since the VC particles in the substrate have no preferred orientation the square or triangular shaped grains are probably nucleated on the (211) VC surface, but possibly also on some other surfaces (e.g. (111)). On the other surfaces present large grains without any particular shape will develop during growth. The cracks and voids formed during the development of these grains lower the hardness by about 20% as can be seen in Fig. 1. For many applications this unsatisfactory result can be avoided by applying a substrate bias which gives rise to continuous renucleation during the film growth and the development of large grains will be suppressed and hence a dense fine grained structure develops. Such a structure is shown in Fig. 4. This also increases the hardness value, see Fig. 1.

Fig. 2 Microstructure of films deposited on HSS, at
two temperatures, a) T_s = 400°C, b) T_s = 500°C.

In the case of films deposited on stainless steel substrates the microstructure of the TiN films changes slowly with temperature. The grain
size is rather small (0.05 μm) and uniform, increasing slightly with
temperature. A micrograph from a film deposited at 550°C is given in
Fig. 3.

Fig. 3 Microstructure of films deposited
at T_s = 550°C on stainless steel substrate.

Fig. 4 Microstructure of a film deposited at 550°C on a HSS substrate. A substrate bias of -300V and 1.6 mA/cm² was used during growth.

The presence of MC-carbides in the HSS was also found to influence the adhesion between TiN and HSS. In Fig. 5 the normalized critical load values as a function of the mean distance between MC-carbides in the HSS surface are shown. It is found that a higher density of MC-carbides particles gives a better adhesion.

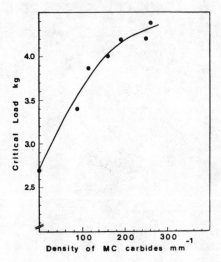

Fig. 5 Critical load values as a function of surface density of MC-carbides in the high speed steel ASP 23. T_s = 450°C.

SUMMARY

The presence of MC-carbides in HSS has been found to have a large
influence on the structure and properties of TiN films. A localized
epitaxial growth occurs on the MC-carbides which gives rise to a bimodal
grain size distribution. If the substrate temperature exceeds 500°C the
large grains take on a peculiar shape which also causes a lot of voids and
cracks to occur in the films. These voids/cracks lower the hardness of
the film. A film simultaneously deposited on stainless steel shows a
homogeneous grain size distribution and also a higher hardness value.
The problem with low hardness values for films on HSS can be avoided by
applying a substrate bias during deposition. The ion bombardment suppresses the development of large grains due to a continuous renucleation
process. An increased amount of MC-carbides in the substrate has also
been found to increase the adhesion.

REFERENCES

Aronson, A.J., Chen, D. and Class, W.H. 1980, Thin Solid Films 72 535
Helmersson, U., Johansson, B.O., Sundgren, J.-E., Hentzell, H.T.G. and
    Billgren, P. 1984, Submitted to 6th Int. Conf. on Thin Films, Stockholm
Hultman, L., Hentzell, H.T.G., Sundgren, J.-E., Johansson, B.O. and
    Helmersson, U. 1984, Submitted to 6th Int. Conf. on Thin Films,
    Stockholm
Hummer, E. and Perry, A.J. 1983, Thin Solid Films 101 243
Igasaki, Y., Mitsuhashi, H., Azuma, K. and Muto, T. 1978, Jpn. J. Appl.
    Phys. 17 85
Igasaki, Y. and Mitsuhashi, H. 1980, Thin Solid Films 70 17
Jacobson, B.E., Nimmagadda, R. and Bunshah, R.F. 1979, Thin Solid Films
    63 333
Johansson, B.O., Sundgren, J.-E., Hentzell, H.T.G. and Karlsson, S.-E.
    1984, Thin Solid Films 111 313
Mumtaz, A. and Class, W.H. 1982, J. Vac. Sci. Technol. 20 345
Münz, W.D. and Hessberger, G. 1981, Vak. Tech. 30 78
Nakamura, K. Inagawa, K., Tsurvoka, K. and Komiya, S. 1977, Thin Solid
    Films 40 155
Poitevin, J.M., Lemperiere, G. and Tardy, J. 1982, Thin Solid Films 97
    69
Schutz, R.J. 1983, Thin Solid Films 104 89
Sundgren, J.-E., Johansson, B.O. and Karlsson, S.-E. 1983a, Thin Solid
    Films 105 353
Sundgren, J.-E., Johansson, B.O., Karlsson, S.-E. and Hentzell, H.T.G.
    1983, Thin Solid Films 105 367
Sundgren, J.-E., Johansson, B.O. and Karlsson, S.-E. 1983b, Surface
    Science 128 265
Ting, C.Y. 1982, J. Vac. Sci. Technol. 21 14
Winterbottom, W.L. 1967, Surfaces and Interfaces I. Eds. J.J. Burke,
    N.L. Reed and V. Weiss, Suracuse University Press, Suracuse, New York
Wittmer, M., Studer, B. and Melchior, H. 1981, J. Appl. Phys 52 5722

Inst. Phys. Conf. Ser. No. 75: Chapter 8
Paper presented at 2nd Int. Conf. Science Hard Mater., Rhodes

757

On carbide matrix reactions in wear resistant TiC-Ni-Cr-B-Si coatings

O KNOTEK AND P LOHAGE

Institut für Werkstoffkunde, Lehrstuhl B, RWTH Aachen
D-5100 Aachen, F.R.G.

ABSTRACT

Matrix carbide reactions in heterogeneous TiC-Ni-Cr-B-Si alloys during
coating and their influence on hardness and wear resistance are investigated. (Ti,W)C containing alloys give interesting results in wear tests.

INTRODUCTION

Corrosion- or wear-resistant coatings of heterogeneous or homogeneous
nickel hard alloys are presently produced by thermal spraying with simultaneous or subsequent melting onto substrates. Due to their self-fluxing
properties Ni-Cr-B-Si alloys can be also directly melted for surface
coatings in vacuum furnaces or inert gas atmospheres after simple application of the powders or pastes (Knotek, Lugscheider and Eschnauer, 1975).
This procedure ensures a close control of the melting temperature and
holding time as well as the heating and cooling rates.

To increase the wear resistance homogeneous Ni-Cr-B-Si alloy powders were
mixed with additional hard compounds. These heterogeneous - mechanically
alloyed - powders behave more or less like the homogeneous ones except
that carbide-matrix reactions occur.

The properties of nickel hard alloys have been determined by the ductility
of the nickel solid solution matrix with the dissolved chromium, as well
as by the kind, amount and the distribution of the hard phases; in the
homogeneous alloys these are mostly CrB and some Cr_7C_3, in heterogeneous
alloys other carbides are also present. In addition the dissolved chromium
forms a close grained and strongly adherent oxide coating on the overlay.

For the formation of heterogeneous (pseudo) alloys a matrix alloy of low
internal hardphase volume and good corrosion resistance has been selected.
The following composition was chosen because of the best melting properties for titanium carbide:

Cr	Si	B	Fe	C	Ni	T_S	T_L
10.0	3.0	2.25	4.25	0.45	bal.	965°C	1180°C

The metalloids Si and B lower the melting point of nickel to the range 960°C to 1200°C mainly by forming the Ni_3B-Ni eutectic and cause the self-fluxing properties of the alloys. Oxides on the surface of the substrates are reduced by silicon and boron forming boronsilicates. By these surface reactions a metallurgical bond of the coating layer to the substrate with excellent adhesion is achieved and diffusion of alloying elements from the substrate into the overlay, and vice versa, can be observed.

Industrially produced Ni-Cr-B-Si powders always contain iron and carbon. Therefore the selected matrix alloy contained the hard phases Ni_3B, CrB, M_7C_3 and other complex compounds (Fig. 1).

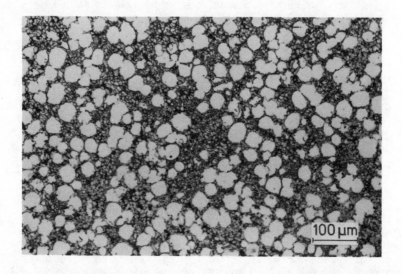

Fig. 1 Micrograph of NiCrBSi matrix alloy used for all experiments.

Changes by diffusion of alloying elements from and into substrates have been reported (Knotek, Lugscheider and Reimann, 1975).

EXPERIMENTAL

For the experiments powder mixtures of the hard alloy with 25 vol.-% and 50 vol.-% TiC, $(Ti_{0.5}, W_{0.5})C$ and WC have been prepared. A plain carbon steel (St 37), an austenitic corrosion resistant steel (X10CrNiTi 18 9 = AISI 321) and a tool steel (C105W1) were selected as substrates.

The powder mixtures were mounted on round substrates (Ø 50 mm x 10 mm) and melted by means of a laboratory vacuum furnace applying a vacuum of at least 10^{-4} mbar. Due to the different wetting behaviour of the carbides it was not possible to apply identical melting temperatures for all mixtures.

To observe the reactions dependent on time and temperature the melting tests were executed at TE: 1100°C-1350°C over dwell periods of t_E: 5 min 15 min. The structures and properties of the coatings were determined by metallographic and electron optical investigations, by the X-ray diffraction method, as well as by further measurements of the bulk hardness and the wear-resistance.

The properties concerning wear have been measured on a pin-disc apparatus using chromium containing tool steel X210Cr12 and silicon carbide grinding paper as counter material. The laboratory results cannot be considered as those to be expected in practical use. Nevertheless the method yields an approximation of the wear behaviour and its dependence on alloy compositions and processing conditions. In the case of the steel counter material, wear caused by tribo-oxidation and abrasion was observed. In the case of silicon carbide counter material only abrasion was observed.

OBSERVATIONS ON RESULTANT PHASES AND MICROSTRUCTURES

During the coating procedure by melting the applied powders, boron diffuses into the steel substrate. This lowers the melting temperature of the base material in the diffusion range and a considerable amount of iron is dissolved in the coating alloy dependent on the holding time. Both the lowered concentration of melting point decreasing elements, caused by the diffusion into the substrate, as well as the absorption of iron and other alloying metals from the base material lead to crystallisation of a nickel solid solution zone without hard phase precipitation in a width of up to 50 μm at the interface substrate-coating alloy.

Fig. 2   Micrograph of a NiCrBSi - 50 vol.% WC alloy
t_E - 5 min; T_E - 1100°C.

In the case of heterogeneous alloys containing WC (Fig. 2) the attainment of thermodynamical equilibrium means the decomposition of the tungsten carbide and a formation of embrittling η-carbides Ni_2W_4C and - because of the iron absorption from the steel substrate - Fe_3W_3C (Fiedler and Stadelmaier, 1975; Pollock and Stadelmaier, 1970).

The dissolution and/or decomposition of TiC could not be observed under the applied coating conditions but can be expected under much higher energy input. There is a great difference in formation enthalpies of WC (H_{298} = -35.1 KJ/mol) and TiC (H_{298} = -183 KJ/mol) (American Society for Metals, 1973).

The wide stoichiometric range of homogeneity of TiC makes a loss of carbon feasible: above 15 min/1200°C the formation of Cr_7C_3 could be observed. All chromium carbide crystals grew from the TiC grains. The carbon content of these Cr_7C_3 crystals originates from the titanium carbide while the source of the chromium metal was the matrix alloy as well as the stainless steel substrate when the conditions permitted diffusion. The diffused iron of the substrate remains in the matrix first and then forms M_7C_3 above 1250°C/15 min. No diffusion of chromium into TiC crystals could be observed (Fig. 3).

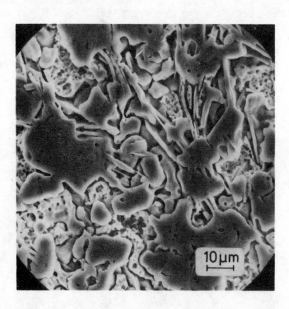

Fig. 3   Scanning electron micrograph of a NiCrBSi - 50 vol.% TiC alloy (Deep etching). t_E - 15 min; T_E - 1300°C.

At these temperatures a segregation of TiC can be observed in 25 vol.% TiC alloys due to the lower viscosity of the matrix hard alloy. The nickel solid solution can dissolve approx 15 at.% titanium. We could observe a maximum content of 4 at.% under the applied conditions.

At lower energy input (1150°C/5 min) coalescence of TiC grains could be observed (Fig. 4).

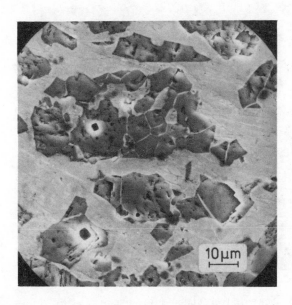

Fig. 4   Scanning electron micrograph of a NiCrBSi - 50 vol.% TiC
t_E - 15 min; T_E - 1150°C.

In alloys containing 50 vol.% TiC segregation could not be observed even
at temperatures of 1300°C.

INFLUENCE OF REACTIONS ON HARDNESS AND WEAR

Up to 1200°C/5 min in heterogeneous alloys containing $(Ti_{0.5}, W_{0.5})C$ a
carbide skeleton is formed even partially in 25 vol.% carbide mixtures. At
1250°C/15 min and at higher temperatures the (Ti,W)C solid solution
decomposes via an η-phase and the matrix dissolves titanium and tungsten.

Due to the small carbon reactions, the small segregations up to 1200°C and
the low exchange carbide-matrix (Fig. 5) linked with a slow boron diffusion into the substrate (no iron diffusion into the coating), these
coatings promise a good wear resistance.

In heterogeneous tungsten carbide alloys under high energy input, gravi-
tational segregation can be observed at 25 vol.%.

In 50 vol.% WC layers tungsten carbide segregation was observed above
1250°C/15 min. In all experiments η-carbides of $Ni_2W_4C_2$ type were formed.
Increasing holding times at melting temperatures permit a considerable
high diffusion of metalloids (C,B) into the substrate and iron diffusion
into the coating alloy matrix. Consequently the formation of η-carbides
of Fe_3W_3C type was observed and an almost complete WC-decomposition.
Therefore after a 1300°C/15 min cycle hardly any WC could be registered.
At 1100°C/5 min however the wear resistance of these heterogeneous alloys
is excellent (Knotek, Lohage and Reimann, 1983).

Fig. 5 Scanning electron micrograph of a NiCrBSi -
25 vol.% $(W_{0.5}, Ti_{0.5})$ C alloy
t_E - 5 min; T_E-1100°C.

The hardness of the pseudo-alloys is considerably higher than that of the
matrix alloy with HRC 40. The hardness of the layer falls, however, with
increasing energy input as a result of the carbide decomposition.

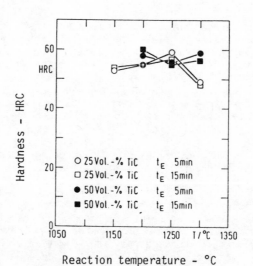

Fig. 6 Hardness of NiCrBSi - TiC alloys.

In the case of TiC and (Ti$_{0.5}$, W$_{0.5}$)C (in the following figures called (W,Ti)C-50/50) containing alloys the decrease of hardness is slightl due to the carbide stability and the formation of chromium carbide (Figs. 6 and 7).

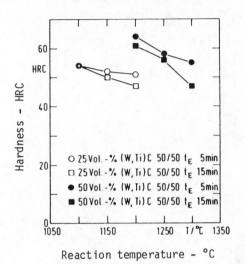

Fig. 7 Hardness of NiCrBSi - (W,Ti)C 50/50 alloys.

Wear-tests on the pin and disc apparatus with different counter materials show the following results (Figs. 8-11).

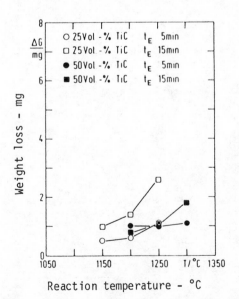

Fig. 8 Wear of NiCrBSi - TiC alloys (wearing part X210 Cr12, distance 40 km; load 10N; 200 rev min^{-1}).

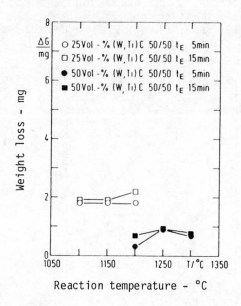

Fig. 9   Wear of NiCrBSi - (W,Ti)C 50/50 alloys (wearing part X120 Cr12, distance 40 km;   load 10N;   200 rev min^{-1}).

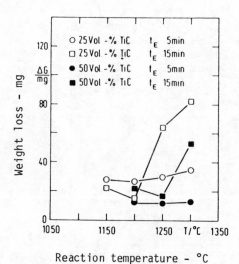

Fig. 10   Wear of NiCrBSi - TiC alloy (wearing part SiC; wear distance 0.8 km;   load 10N;   200 rev min^{-1}).

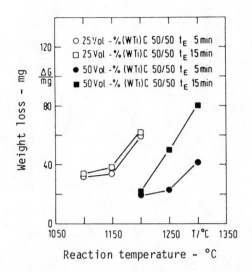

Fig. 11 Wear of NiCrBSi - (W,Ti)C 50/50 alloys (wearing part SiC;
         wear distance 0.8 km; load 10N; 200 rev min^{-1}).

The wear figures underline again the relations between carbide and carbide
matrix reactions and wear resistance. The extremely small loss of weight
of TiC- and (W,Ti)C-containing heterogeneous alloys is the reason for
further coating experiments.

REFERENCES

Fiedler, M.L. and Stadelmaier, H.H. 1975, Z. Metallk. 66 pp 402
Knotek, O., Lohage, P. and Reimann, H. 1983, Thin Solid Films 108, pp 449
Knotek, O., Lugscheider, E. and Eschnauer, H. 1975,. Hartlegierungen zum
    Verschleißschutz, Stahleisen, Düsseldorf
Knotek, O., Lugscheider, E. and Reimann, H. 1975, J. Vac. Sci. Technol. 4
Pollock, C.B. and Stadelmaier, H.H. 1970, Metallurg. Trans. 1, pp 767
American Society for Metals, Selected Values of the Thermodynamic
    Properties of Binary Alloys, 1973, ASM, Metals Park, Ohio

Inst. Phys. Conf. Ser. No. 75: Chapter 8
Paper presented at 2nd Int. Conf. Science Hard Mater., Rhodes

767

Microanalysis of the interface region between titanium carbide and substrate in CVD coated cemented carbides

J SKOGSMO(1), A HENJERED(1), H NORDEN(1) AND K-G STJERNBERG(2)

(1) Dept. of Phys., Chalmers Univ. of Tech., S-412 96 Göteborg, Sweden
 (2) AB Sandvik Hard Materials, S-126 12 Stockholm, Sweden

ABSTRACT

Needle shaped specimens of cemented carbides have been coated with thin layers of TiC using the CVD method. The coatings have been analysed, in particular by TEM and FIM-AP, but also by SEM-EDS, SIMS and X-ray diffraction. The very thin coatings examined can be seen as representative of the coating-substrate interface region for thicker coatings.

INTRODUCTION

A cutting tool must resist high strains, posses a high resistance to wear and have a high thoughness. To resist, wear the tool, or at least its surface, must have a high hardness, a good chemical stability and resistance to oxidation and a smooth surface to reduce friction.
The difficulty of combining these characteristics with a high thoughness has made it necessary to produce a tool composed of two zones.

The bulk material of the tool must be as tough as possible to take up the load and hard enough to give satisfactory support to the outer zone, which is essential for a high wear resistance.

Due to the high temperatures during machining, chemical wear is dominant over physical wear. The wear surface must therefore posses high chemical stability, thus creating a barrier for diffusion between the tool and machined metal. Moreover the layer must prevent crater wear and resist abrasive wear through its hardness.

The most common coating materials today are TiC, TiN, Ti(C,N) and Al_2O_3 either used as single layers or combined to form multiple layer coatings. The vast majority of coated cemented carbide cutting tool inserts have TiC as the innermost or the only layer. A 4-8 μm thick layer of TiC grown by chemical vapour deposition (CVD) will generally improve tool performance by a factor of four.

The TiC coating must not only be hard, inert and stable, but must also have very good adhesion to the cemented carbide substrate. The adhesion depends critically on the microstructure and composition in the interface region between the coating and the substrate. Very few microanalyses of this interface region have been carried out (Sarin, 1983: Sharma, Williams and Gottschall, 1977: Vourinen and Horsewell, 1983), mainly due to the difficulty in preparing thin foil specimens (Sarin, 1979) for analytical electron microscopy (AEM). Electropolishing of these multiple phase materials is very difficult, and ion etching is very time consuming.

To study the interface region, cemented carbide specimens in the form of very sharp needles (tip radius less than 100 nm) were coated with a thin layer (less than 200 nm) of TiC. Such specimens can be studied both with AEM and atom-probe field-ion microscopy (AP-FIM). Electron microscopy gives information about thickness, structure and morphology of the coating. Atom-probe analysis offers elemental analysis with equal detectability for all elements. The spatial resolution is of the order of 1 nm, which allows accurate analysis at or close to interfaces.

EXPERIMENTAL TECHNIQUE

The substrate material used in this study was cemented carbide of the WC-Co type with cubic carbide additions. The composition (in weight %) of the cemented carbides and the carbide grain size are shown in Table 1.

TABLE 1   Composition and Grain Size of the Cemented Carbide Substrate

Composition	WC	(Ta,Nb)C	Co	Grain size
	92	2	6	1-2 μm

The material was in the form of cylindrical rods with a diameter of 0.8 mm. Specimens useful for AP-FIM analysis were prepared by electropolishing these rods, to the shape of sharp needles, in a solution of 8 % sulphuric acid in methanol at 12-14 Volts DC (Henjered and Nordén, 1983).

The specimens were inspected in TEM, and those with a tip radius less than 100 nm were selected to be coated with TiC.

The coating process was performed in a hot wall vertical type CVD apparatus of 1/20 size of the production apparatus.

The total reaction for chemical vapour deposition of TiC is:

$$TiCl_4(g) + CH_4(g) \xrightarrow{H_2} TiC(s) + 4\ HCl\ (g) \qquad (1)$$

The deposition parameters are given in Table 2.

TABLE 2   Coating Conditions

Temperature	$1000^{\circ}C$
Total pressure	50 torr
$TiCl_4$ concentration	4 vol %
CH_4 concentration	4 vol %
H_2 concentration	balance

The reaction chamber was filled with hydrogen during heating to the deposition temperature and during cooling to room temperature after deposition.

Specimens for AP-FIM analysis have to be very sharp. In order to produce specimens with coatings thin enough to be useful for AEM and AP-FIM analysis, very short coating times of 30 s., 3 min. and 10 min. were used. Analytical electron microscopy was performed, using a JEOL 200 CX with a Link systems 860 Analyser.

The AP-FIM used in this investigation is basically a field-ion microscope equipped with a time of flight mass spectrometer. The area to be analysed, typically a few nanometers in diameter, is selected with an aperture. The surface atoms are then analysed one by one, and the analysed volume is a narrow cylinder parallel to the specimen axis. The instrument and the technique have been described in detail elsewhere (Andrén and Nordén, 1979; Henjered and co-workers, 1983). The very high spatial resolution and the equal detectability for all elements means that atom-probe analysis of coatings (Henjered and co-workers, 1981; Hellsing and co-workers, 1983) can give information about composition, stoichiometry and very localised concentration variations such as segregations to interfaces.

In addition to the FIM specimens, cutting tool inserts with the same composition were coated at the same time. These inserts were used for analysis by other methods such as scanning electron microscopy (SEM), scanning auger microscopy (SAM), secondary ion mass spectroscopy (SIMS), and X-ray diffractometry.

RESULTS

TEM

The short coating times, 30 s., 3 min. and 10 min. produced very thin coatings of TiC. The appearence of the FIM specimens as electropolished and after different deposition times are shown in the TEM micrographs in Fig. 1. The coating thicknesses were determined from the micrographs and the results are shown in Fig. 2. The thickness measurements were performed on several specimens for each coating time, and also on specimens coated on different occasions.

The spread in thickness for different coatings with the same nominal deposition time is to a large extent due to the difficulty of reproducing exactly the same deposition parameters. In addition to this variation, the coating thickness in the initial stage also seems to vary with the distance to the closest binder phase area (Fig.3).

The needle shaped specimens had a WC grain at the tip, and lamellas of Co rich binder phase were found at some distance from the tip apex. This distance was determined for each specimen using TEM. The coating at the apex tended to be thicker the closer to the apex the binder phase lamella was situated, but this dependence seemed to be weak. To reduce the influence of the deposition parameters on this effect, these measurements were made on specimens coated on the same occasion.

The grain size of these TiC coatings was extremely small (Table 3).

The diffraction patterns from these polycrystalline coatings on the FIM
specimens represented a f.c.c. structure with a lattice parameter of
0.43 nm, which is in good agreement with the lattice parameter of TiC.
The thinnest coatings consisted to a large extent of a single layer of
TiC grains on the substrate. Diffraction patterns from these coatings
indicated that TiC grains in direct contact with the substrate had grown
with some specific orientation relationship to the substrate grains. This
observation was supported by the occurence of Moiré fringes, as can be
seen in Fig. 1b.

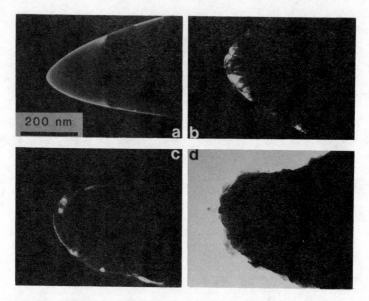

Fig. 1   TEM micrographs of needle shaped specimens before
(a) and after coating with TiC with an exposure time of
30 s (b), 3 min. (c) and 10 min. (d).

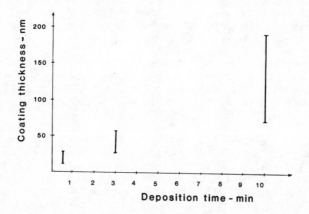

Fig. 2   Coating thickness as a function of deposition time.

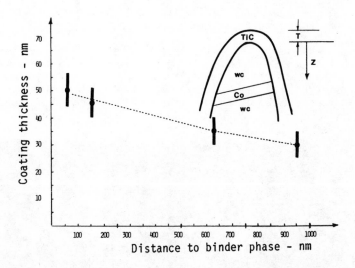

Fig. 3 Coating thickness at the tip apex as a function of the distance to the closest binder phase area.

TABLE 3 TiC Grain Size for Different Coating Times

Exposure time	30 s	3 min	10 min
Grain size (nm)	5-15	5-35	5-50

X-Ray Diffraction

The cutting tool inserts, that were coated at the same time as the FIM specimens, were examined with X-ray diffraction. The diffractograms showed mainly WC and the eta phase Co_6W_6C. Small peaks of what could be interpreted as the eta-phase Co_3W_3C were also observed. There was no significant difference in the amount of eta-phase observed after different coating times. The coatings were very thin, so the peaks from the TiC layer were only observed after the longer deposition times, and even then they were very weak. There is also some peak overlap with WC and the eta-phases, so no accurate determination of the lattice parameter of the coatings could be achieved.

SEM-EDS

The cutting tool inserts were also analysed with SEM-EDS. Fig. 4 shows
SEM micrographs of a specimen before deposition and after an exposure
time of 3 min. A commercial TiC coating, which is approximately 6 μm thick,
is shown for comparison.

Chemical analysis using EDS showed that besides Ti, W and Co, small amounts
of Fe and Cr also were present. The spatial resolution of this method,
however, is not sufficient to determine the location of the elements. It
was also observed that the concentration of Co near the substrate surface
had increased after the deposition process for all coating times.

Fig. 4. SEM micrographs of a) uncoated cemented carbide insert
        b) TiC coating after an exposure time of 3 min.
        c) a commercial, 6 μm thick, TiC coating.

SAM

The scanning auger microscope (SAM) is an instrument by which the outermost atomic layers of a specimen can be analysed. Depth profiles can
also be obtained by sputtering through the coating (Fig. 5). The depth
calibration was estimated from the sputtering yields. The outermost
atomic layers of a coating contained rather large amounts of oxygen,
and less carbon than was present deeper within the coating. This indicates that the outermost layers were oxidized. In this oxidized layer
the concentration of W was very low. It was also observed that Co
appeared before W in depth profiles. Unfortunately the depth resolution
is gradually destroyed during sputtering so elements from both the coating
and the substrate appear at the same time. Small amounts of Fe and Cr were
detected as in the EDS analysis. Because of the limited spatial resolution
it is not possible to draw any conclusions about the composition of the
interfaces, TiC substrate or TiC TiC, from the SAM analyses.

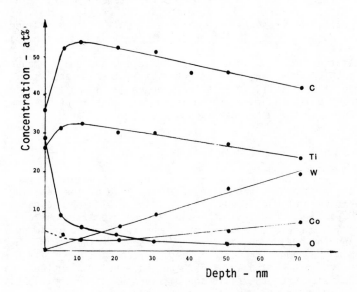

Fig. 5 SAM depth profile through a TiC coating.
The depth scale was estimated from the sputtering yields.

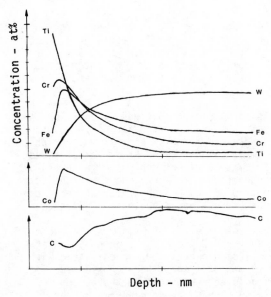

Fig. 6 SIMS depth profile through a TiC coating.

SIMS

Secondary ion mass spectrometry (SIMS) has a higher detectability for trace elements than has SAM. Depth profiles were taken by sputtering with oxygen ions so the presence of this element in the coatings could not be detected directly. The presence of oxygen could only be inferred indirectly from the lower carbon signal in the outermost atomic layers (Fig. 6).

Very low concentrations of W in the oxidized layer, and small amounts of Fe, Cr and Co in the coating, were observed as in the SAM analysis. Quantification is difficult, so note that the vertical scales in Fig. 6 are in arbitrary units, and that there is no relationship between the relative heights for the different curves.

FIM-AP

The only method with sufficient spatial resolution to determine the location of the elements in these very thin and fine grained coatings, is atom-probe field-ion microscopy. Fig. 7 shows field-ion micrographs which were obtained using Ne as the image gas. The specimen was cooled to liquid nitrogen temperature. Fig. 7a shows the specimen at an image voltage of 13.7 kV, and the analysing aperture can be seen at the TiC grain in the center of the micrograph (arrowed). In Fig. 7b the specimen voltage has been increased to 16.0 kV and about 350 ions have been collected in an atom-probe analysis. From the field-ion micrographs it can be clearly seen that the analysing aperture was situated on the same TiC grain during the analysis. The atom-probe analyses were performed at liquid nitrogen temperature to reduce surface diffusion, and with a background pressure less than $5 \cdot 10^{-8}$ Pa. A typical mass spectrum from a coating is shown in Fig. 8.

In the initial part of an analysis, TiO^{2+} ions were often detected. This oxidized outer layer only contained small amounts of W. This observation agrees with the results from the SAM and SIMS analyses. The oxidized layer was usually only a few atomic layers thick. The interior of the coatings had, in addition to the major coating elements Ti and C, a relatively high concentration of W.

Atom-probe analysis was performed on coatings of different thicknesses. Fig. 9 shows the W concentration as a function of the distance to the substrate, which was determined from TEM and FIM micrographs. The diagram is composed of data from specimens coated at different times, so the spread in W content can to some extent depend on the difficulty of reproducing the same deposition parameters. However, the trend was that grains closer than about 30 nm to the substrate, showed W contents of 9-20 at %, whereas grains further out had W contents of less than 8 at %.

Low concentrations (less than 2 at %) of Fe, Cr and Co were also observed, uniformly distributed in the TiC grains. The composition of a typical grain was very close to stoichiometry and varied little with depth, when only the main coating elements Ti and C were considered (Fig. 10). When W, Fe, Cr and Co were included, the stoichimetry became $MC_{0.6}$ to $MC_{0.8}$. Fig. 11 shows a calculated ternary phase diagram for the Ti-W-C system at 1100°C (Uhrenius, 1984), in which the compositions obtained by atom-probe analysis have been marked with black dots.

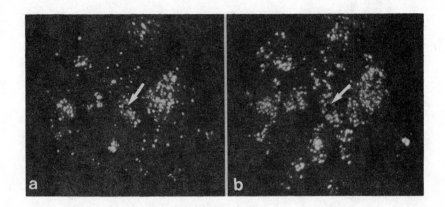

Fig. 7 Field-ion micrographs of a TiC coating before (a)
and after (b) atom-probe analysis.

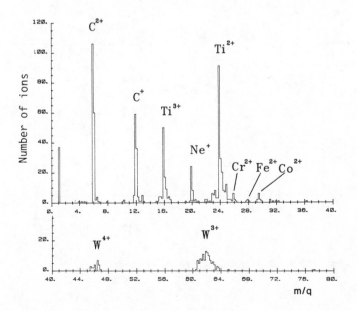

Fig. 8 Atom-probe mass spectrum from a TiC coating.

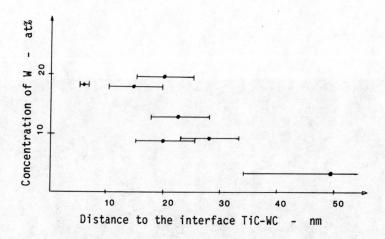

Fig. 9 The concentration of W in TiC grains as a
function of the distance to the substrate.

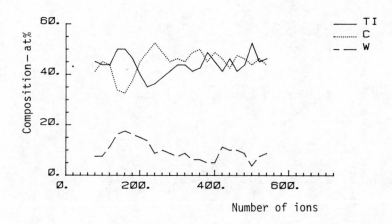

Fig. 10 Atom-probe depth profile from a TiC grain.

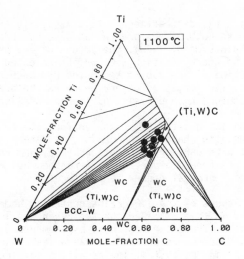

Fig. 11 Calculated ternary phase diagram for the Ti-W-C system at 1100°C (Uhrenius, 1984), including compositions obtained by atom-probe analyses.

DISCUSSION

The initial stages of the growth of TiC on cemented carbides have been studied and the composition and morphology of the extremely thin coatings examined here are considered to be representative of the coating-substrate interface region in thicker coatings.

An almost linear growth rate was measured in these early stages, which should be compared with the observation that the growth rate for thicker coatings decreased with time (Stjernberg, Gass and Hintermann, 1977). This gradual decrease in growth rate depends probably on the carbon supply. Initially most of the carbon is taken from the substrate (Jonsson and Hintermann, 1980) but this becomes more difficult as the coating gets thicker and more and more of the carbon has to be supplied by the gas phase.

The carbon is probably transported from the substrate to the coating by the Co from the binder phase. Previous work has shown that Co is present at the substrate-coating interface (Henjered and co-workers, 1981) and that the initial growth rate depends strongly on the Co content of the substrate material (Lee, Nam and Chun, 1981). This agree also with the present observation that, under otherwise identical deposition conditions, the coating formed on the tip apex was thicker the closer it was to a Co rich binder phase lamella.

If the substrate is decarburized, new phases will form. Deposition at temperatures less than 1100°C gives mainly the eta phase Co_6W_6C (Lee and Chun, 1981, Sharma and co-workers, 1977, Vuorinen and Horswell, 1982), which was observed by X-ray diffraction in this study. In addition small amounts of the eta phase Co_3W_2C were present as also has been observed earlier (Jonsson and Hintermann, 1980). The amount of eta phase did not vary significantly with the deposition time which indicated that it was formed during the heating of the substrate in the hydrogen atmosphere before the deposition. The formation of eta phase can be prevented by carburizing the surface prior to deposition. (Sarin and Lindström, 1979b).

The nucleation and initial growth of TiC was uniform and produced a very fine grained structure. The shape of the grains, as determined by TEM, was equiaxed. This agrees with earlier observations (Lee, Nam and Chun, 1981) that equiaxed grains are formed at growth temperatures below 1050°C whereas elongated grains grow at higher temperatures.

The outermost atomic layers of the TiC grains contained some oxygen, and the concentration of W in this oxidized surface layer was lower than deeper within the grains. The oxide was found by atom-probe analysis to be TiO. One deposition was performed when there was a leak, giving a higher oxygen partial pressure in the system. The resulting coating was richer in oxygen and the atom-probe analysis showed the presence of both TiO and TiO_2. This is in agreement with other observations (Chatterjee and Lipsitt, 1982, Horvath and Klemme, 1980) that in the first stage of oxidation of TiC a dense film of TiC-TiO in solid solution is formed. As oxidation continues the TiO is likely to be converted to higher oxides, TiO_2 and Ti_2O_3.

The Fe and Cr found in the coatings are probably impurities that have been dissolved from metal parts in the reaction chamber by the gaseous HCl formed during the reaction (Zeman and Kulmberg, 1980). X-ray analysis of parts from a reaction vessel showed that the Cr and Ni content of the surface layer had decreased considerable after extended use.

CONCLUSIONS

Needle shaped specimens of cemented carbides coated with thin layers of TiC have been analysed with TEM and FIM-AP.

Cutting tool inserts coated simultanously with the needle shaped specimens were analysed with SEM-EDS, SAM, SIMS and X-ray diffraction.

Considerable amounts of W were dissolved in the TiC grains together with small amounts of Fe, Cr and Co.

The outermost atomic layers of the coatings contained oxygen and the amount of W in the oxidized layer was much lower than in the interior of the TiC grains.

The TiC grains were equiaxed and the grains in direct contact with the substrate showed specific orientation relations to the substrate.

Eta phase, mainly Co_6W_6C but also some Co_3W_3C, was formed.

ACKNOWLEDGEMENT

This work was financially supported by the National Swedish Board for Technical Development (STU).

REFERENCES

Andrén, H.-O., Nordén, H. 1979, Scand. J. Metall. 8 147-152

Chatterjee, D. K. and Lipsitt, H. A. 1982, Metall. Trans. A, 13A 1837-1841

Hellsing, M., Henjered, A., Nordén, H. and Andrén, H.-O. 1983, Sci. of Hard Mater. 931-945 ed Viswanadham, R. K., Rowcliffe, D. J., and Curland, J. Plenum Press, New York

Henjered, A., Kjellsson, L., Andrén, H.-O. and Nordén, H. 1981, Scr. Met. 15 1023-1027

Henjered, A., Nordén, H. 1983a, J. Phys. E: Sci. Instrum. 16 617-619

Henjered, A., Nordén, H., Kjellsson, L., and Skogsmo, J. 1983b, Proc. Euro CVD IV. ed Bloem, J., Verspui, G. and Wolff, L. R. Eindhoven, The Netherlands

Horvath, E. and Klemme, K. E. 1980, Proc. Euro CVD III. ed Hintermann, H. E. Neuchatel, Switzerland, pp 218-229

Jonsson, H. and Hintermann, H. E. 1980, Proc, Euro CVD III. 200-207 ed Hintermann, H. E. Neuchatel, Switzerland

Lee, C. W., Nam, S. W., and Chun, J. S. 1981a, Thin Solid Films 86 63-71

Lee, C. W. and Chun, J. S. 1981b, Thin Solid Films 86 73-78

Ruppert, W. 1972, CVD III. 340-352 Glaski, F. A., Amer. Nucl. Soc. Hinsdale, Illinois

Sarin, V.K. 1979a, Proc. Int. CVD VII. 79-3 476-487 Sedgwich, T. O. and co-workers Princeton, N. J. USA

Sarin, V. K. and Lindström, J. N. 1979b, J. Electrochem. Soc. 1281-1287

Sarin, V. K. 1983, Sci. of Hard Mater. 395-411 ed Viswanadham, R. K., Rowcliffe, D. J. and Gurland, J.

Sharma, N. K., Williams, W.S., and Gottschall, R. J. 1977, Thin Solid Films, 45 265-273

Stjernberg, K. G., Gass, H., and Hintermann, H. E. 1977, Thin Solid Films 40 81-88

Uhrenius, B. 1984, Calphad 8 101-119

Vuorinen, S. and Horsewell, A. 1983, Sci. of Hard Mater. 433-449 ed Viswanadham, R. K., Rowcliffe, D. J., and Gurland, J. Plenum Press, New York

Zeman, F. and Kulmberg, A. 1980, Proc. of Euro CVD III. ed Hintermann, H. E. Neuchatel, Switzerland, pp 191-199

Inst. Phys. Conf. Ser. No. 75: Chapter 8
Paper presented at 2nd Int. Conf. Science Hard Mater., Rhodes

781

Transverse rupture strength of cemented carbides multi-layer coated by PVD

T SUGISAWA and A NISHIYAMA

Tokyo Metal Plant, Mitsubishi Metal Corp.
Nishishinagawa 1-27-20, Shinagawa-ku, Tokyo, Japan

ABSTRACT

A multi-layer coating on a cemented carbide was investigated, the multi-layer consisting of repeated layers of TiN and Ti which were from 0.05 to 0.2 μm. Multi-layer coated carbide deposited at 800°C had higher TRS than TiN coated material deposited at the same temperature. Both the toughness in cutting and wear resistance of multi-layer coated carbides were evaluated by interrupted cutting tests as well as continuous cutting tests.

INTRODUCTION

Much work has been carried out on PVD and CVD coatings on cemented carbides to improve the cutting tool life (Bunshah, 1978; Jacobson, Nimmagadda and Bunshah 1979). For PVD coatings until quite recently, transverse rupture strength of the coated carbide was believed to be almost the same as that of the uncoated one. In 1983, we reported that the TRS of a carbide which has been TiN coated by PVD depends very much on the coating temperature and the layer thickness (Itaba, Nishiyama and Sugisawa, 1983).

In our previous work, the following results were obtained:
1) The TRS of coated carbide at 450°C was almost equal to that of the substrate. The TRS of material coated at 800°C decreased with increasing thickness of the layer.
2) The inserts coated at 800°C had higher wear resistance than those coated at 450°C.
3) These two properties was probably influenced by the morphology of the layers. Low temperature coating gave a fine crystaline structure; on the other hand higher temperature coating such as at 800°C gave a columnar structure.
4) Regarding the fracture mechanism of PVD coated carbide, no apparent prior cracking was observed before the fracture. The fracture occurred when the first crack developed in the layer or in the substrate. In the case of CVD coating, prior cracks were always observed in the layer under relatively low stress and the fracture occurred when this pre-crack began to propagate into the substrate.

EXPERIMENTAL PROCEDURE

For TRS measurement a WC-Co alloy containing 12%Co was used, because if a cutting grade which has relatively low TRS were used as substrate, the rupture would be initiated from internal fracture sources. Specimens were prepared by following procedure.

1. powder mixing
2. pressing
3. sintering
4. HIP'ing
5. grinding
6. lapping
7. PVD coating

Ten pieces of specimens were prepared for each TRS measurement. HIP'ing is necessary to avoid residual pores. The HIP'ed specimens were ground with a diamond wheel and polished with fine diamond powder to remove scratches on the surface and to avoid the effect of residual compressive stress introduced by grinding.

Multi-layer coating was carried out using an ion plating apparatus; a combination of continuous deposition of Ti metal and intermittent introduction of N_2 gas gave the multi-layer coating. The time interval of N_2 gas introduction was intentionally varied to form coating with various thickness ratio; Ti/Ti+TiN. Substrates were heated during deposition at 800°C. A TiN single-layer was deposited on the same substrate in the same apparatus at 450°C and 800°C.

Continuous as well as interrupted cutting tests were carried out to evaluated both the wear resistance and the toughness of the multi-layer coated carbide. A P30 cutting grade was selected as the substrate of the cutting test specimens. Comparison was made with TiN single-layer coated samples.

RESULT AND DISCUSSION

A slant cross-section of a typical multi-layer is shown in Fig. 1. White colored zones correspond to Ti rich layers, while grey colored zones to TiN layers. This photograph is of a sample prepared intentionally to reveal the multi-layer structure. Multi-layer coatings investigated in this study have finer repeated layers as shown in Fig. 6-9. Since it was very difficult to measure the thickness of the sublayers directly, this is explained indirectly in terms of a deposition time ratio e.g. the Ti deposition time over the deposition time for Ti and TiN together.

The dependence of TRS of the multi-layer coated carbide (WC-12%Co) upon the thickness of the layer is shown in Fig. 2. The data of single layer coated carbides deposited at 450°C and 800°C are also shown in Fig. 2 for comparison. The coating time ratio of the multi-layer samples was 0.25. The TRS of the TiN single layer coated carbide deposited at 450°C are almost the same as TRS of the substrate for up to a 3 μm thick layer. The TRS of the single layer coated carbide prepared at 800°C decreases with increasing layer thickness. The TRS of the multi-layer coated carbide deposited at 800°C shows a similar tendency.

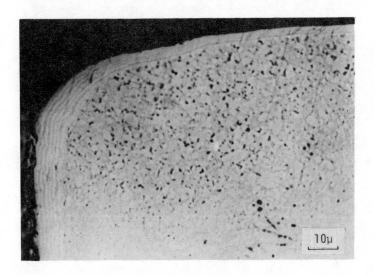

Fig. 1 Slant cross section of multilayer.

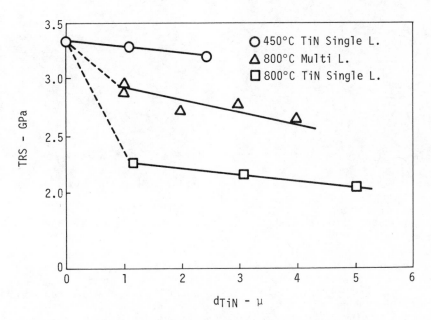

$d_{TiN} - \mu$

Fig. 2   The effect of layer thickness on
TRS of WC - 12%Co with various coating layers.

However it is clear that the TRS of the multi-layer coated carbide higher than those of the TiN single layer coated carbide when compared at same coating temperature.

The fracture surface of the TiN single layer sample coated at 450°C is shown in Fig. 3. The surface is smooth and no large crystal can be see on it. In Fig. 4, the fracture surface of the single layer sample coated at 800°C is shown. The layer exhibits a typical columnar structure.

The effect of the coating time ratio on the TRS of the multi-layer coated carbide is shown in Fig. 5. In this test the layer thickness was about 2 μm, and the coating temperature was 800°C. The TRS increases from the value of the single layer coated material with increasing fraction of Ti in the layer. The curve has a broad peak at the coating time ratio of 0.2 to 0.4.

The structure of the multi-layer coating for a deposition time ratio of 0.1, is shown in Fig. 6. There is one thick Ti layer which was produced by missing N_2 gas introduction, but the normal, very thin Ti rich layers are hardly distinguished. X-ray diffraction analysis on the multi-layer indicated that the layer in fact consisted of TiN and Ti_2N.

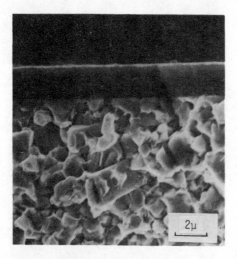

Fig. 3   Fracture surface of TiN single layer coated at 450°C

Fig. 4   Fracture surface of TiN single layer coated at 800°C

The structure of the multi-layer in the case of a coating time ratio of 0.33 is shown in Fig. 7. Each of the fourteen Ti rich and TiN sublayers are clearly observed on the slant cross section. The rupture surface shown in Fig. 8 revealed a fine structure but not a columnar structure like the TiN single layer. In X-ray analysis of this case, TiN, Ti_2N and Ti metal were identified.

Fig. 6   Slant cross section of
the multi-layer.
Coating time ratio : 0.1

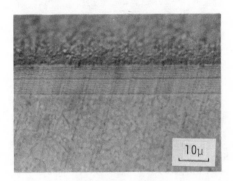

Fig. 7   Slant cross section of
multi-layer.
Coating time ratio : 0.33

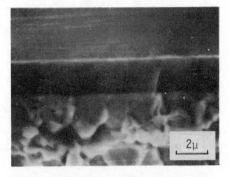

Fig. 8   Fracture surface of
multi-layer.
Coating time ratio : 0.33

Fig. 9   Slant cross section of
multi-layer.
Coating time ratio : 0.55

Fig. 10   Fracture surface of
multi-layer.
Coating time ratio : 0.55

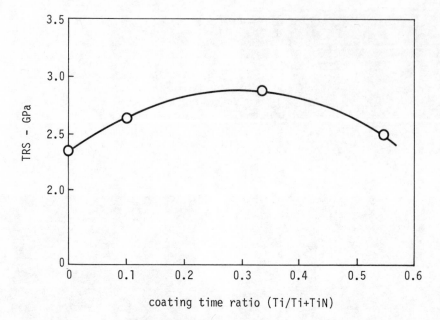

Fig. 5   The effect of the coating time
ratio; Ti/Ti + TiN on TRS of WC - 12%Co
with multi-layer coating.
Total thickness : 2 μm

The structure of the multi-layer for a deposition time ratio of 0.55 is
shown in Fig. 9. Many fine sublayers can be seen on the slant cross
section. The rupture surface also presents a fine structure as shown in
Fig. 10. TiN, Ti_2N and Ti metal were identified from X-ray analysis.
These result indicate that Ti and TiN layers reacted together yielding
Ti_2N when multi-layer coatings were made at 800°C.

The dependence of flank wear on cutting time in continuous steel cutting
using inserts coated by PVD is shown in Fig. 11. Cutting conditions are
written under the figure. P30 grade inserts (shape : SNGN120408) were
used as substrate. The thickness of the layers were about 2 μm. If the
tool life is considered to be the time for a flank wear 0.3 mm, the TiN
single layer coated insert had about twice the tool life of the non-coated
insert. The multi-layer coated insert had about three times the tool life
of non-coated one.

The interrupted cutting test was performed on a turning lathe with a steel
bar having rectanglar cross section. The results of test are shown in
Fig. 12. The test conditions; cutting speed, depth of cut and cutting
time are written under the figure. Cutting edges were tested to cut for
two minutes. In cases where edge chipping occured in less than two
minutes, the time is marked - X in the figure. Mark - o means that the
edge cut through 2 minutes without chipping. Accordingly, this figure
indicates the safety limit of feed rate for each specimen. The toughness
in cutting can be compared from the limit of feed rate. Non-coated P30
inserts had superior toughness to both coated inserts.

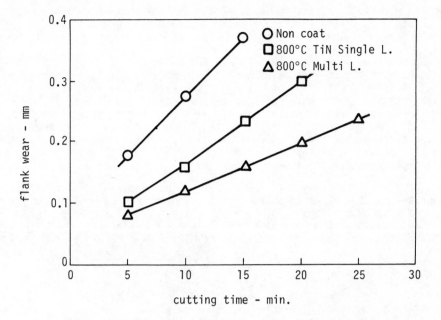

Fig. 11   Progress of flank wear of PVD coated P30
in continuous cutting of steel - Work : AISI 4340 (H_B270)
Cutting Speed : 125 m/min.
Depth of Cut : 1.5 mm
Feed rate   : 0.32 mm/rev.
Dry Cut

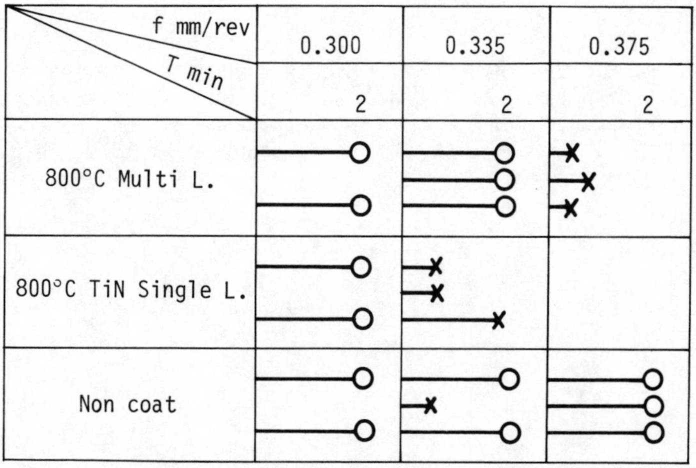

f mm/rev / T min	0.300	0.335	0.375
	2	2	2
800°C Multi L.	──○ ──○ ──○	──○ ──○ ──○	─✗ ──✗ ─✗
800°C TiN Single L.	──○ ──○	─✗ ─✗ ──✗	
Non coat	──○ ──○	──○ ─✗ ──○	──○ ──○ ──○

Fig. 12   Interrupted cutting test of PVD coated P30.
Work : AISI 4340 (H_b300)
Cutting Speed : v = 100 m/min.
Depth of Cut : d = 2 mm
Cutting Time : T = 2 min. max.

Comparing coated inserts, the multi-layer ones have clearly higher toughness than single layer TiN coated ones. These results indicate good corespondence between toughness of cutting and TRS of the coated carbides.

In our previous work (Itaba, Nishiyama and Sugisawa, 1983) we suggested that the high TRS of PVD coated carbide can be attributed to the high deformability of this type of coating. The fact that the multi-layer coated material has higher TRS than the carbide coated with a single TiN layer can also be explained by the higher deformability of the multi-layer coating. The morphology of this coating has not be clearly revealed but on the basis of the fracture appearance it is assumed that it consists of a mosaic of short columnar crystals of TiN, their growth obstructed by the Ti interlayers.

The successful development of a PVD coating technique which gives large toughness as well as wear resistance to cemented carbide would be an important advance not only for cutting tool but also for wear resistant tool application. The new multi-layer coating technique is hopefully anticipated to provide a way of achieving this.

SUMMARY

Multi-layer coatings on cemented carbides have been prepared by a PVD technique. The coating consist of alternate layers of crystalline TiN, Ti_2N and Ti. The carbides coated with such multi-layer coatings have a higher TRS than similar material coated with single TiN layers.

The cutting performance of multi-layer coated inserts is superior, both with regard to toughness of cutting and wear resistance, to inserts coated with a single TiN layer at the same deposition temperature of 800°C.

REFERENCES

Bunshah, R. F. 1978, High Temp - High Pressure 10 187-196
Itaba, T., Hishiyama, A. and Sugisawa, T. 1983, Proc. Advances in Hard
    Metal Production, MPR Publishing Services Ltd., UK, pp 20-1 - 20-15
Jacobson, B. E., Nimmagadda, R. and Bunshah, R. F. 1979, Thin Solid
    Films 63 357-364

Inst. Phys. Conf. Ser. No. 75: Chapter 8
Paper presented at 2nd Int. Conf. Science Hard Mater., Rhodes

789

The effects of ion-implantation on the surface mechanical properties of some engineering hard materials

P J BURNETT AND T F PAGE

Department of Metallurgy and Materials Science,
University of Cambridge, Pembroke Street, Cambridge CB2 3QZ, U.K.

ABSTRACT

Ion-implantation into ceramics is known to produce changes in surface mechanical properties. This paper describes the changes in hardness and indentation fracture that occur in reaction-bonded silicon carbide, sialon, partially stabilized zirconia and WC. These modifications will be correlated to the structural changes brought about by the implantation process.

INTRODUCTION

Ion-implantation is a surface treatment process that enables the structure, microchemistry and resultant surface properties to be controllably changed. The implantation affected layer is typically <0.5µm thick. Beneficial effects upon the wear and corrosion behaviour of metals and cermets such as WC/Co have been extensively reported (e.g. Dearnaley, 1982; Hartley, 1979; Potter, Ahmed and Lamond, 1983). By contrast, the properties and possible uses of ion-implanted ceramic surfaces have been relatively neglected. However, previous work, principally on single crystal behaviour, has identified three major areas of modification to the hardness and indentation fracture behaviour of ceramic materials. Firstly, ion-implantation generally results in the production of a large number of point defects, the consequence of which is to hinder dislocation motion in the surface; i.e. radiation hardening occurs. Other contributions to this hardening may arise from solid-solution effects, which are dependent upon the chemical nature (e.g. size and valency) of the ion species. Secondly, above a critical dose (ϕ_{crit}), the cumulative effect of radiation damage is to amorphise the ceramic surface. The precise value of ϕ_{crit} is dependent upon the ion species/energy used and the nature of the substrate - ionically bonded (e.g. MgO) and electronically conducting (e.g. WC, TiB_2) ceramics amorphise at much higher doses than covalently bonded materials (e.g. SiC). This amorphisation (in all known cases) produces a layer of softer material, which initially forms at the peak of the Gaussion damage profile and thus lies beneath the surface (see Burnett and Page, 1984a,b). On increasing the dose this layer then thickens until it extends to the surface. The production and growth of this layer leads to a progressive loss of hardness (at small indentation

depths) and may eventually result in an absolute softening. Finally, the
point defects and implanted ions themselves produce a volume change in the
surface layer. Since expansion is constrained by the underlying material,
sizeable surface compressive stresses may be generated. These stresses
(which may be of the order of GPa) have been shown to markedly affect the
indentation fracture behaviour (Burnett and Page, 1984b,c).

This paper investigates the extent to which these previously characterised property modifications may occur in commercially available ceramics
and hard metals. The hardness and indentation fracture behaviour of four
important classes of materials have been investigated. These are reaction
bonded α-silicon carbide (RB-SiC), β'-sialon, MgO-partially stabilized
zirconia (Mg-PSZ) and a WC/Co hard metal. Particular attention has been
paid to the role of the microstructure upon the implantation induced
effects produced.

EXPERIMENTAL

Samples of reaction-bonded silicon carbide, β'-sialon, MgO-partially
stabilized zirconia and WC/9%Co were sectioned using a high speed annular
diamond saw, to produce slices 1-2mm thick. These were then mounted and
polished on a succession of diamond-impregnated laps and cloths to a 1/4μm
finish. Implantations were performed with a variety of ions; nitrogen
(as N_2^+ 90keV or N^+ 100keV) was implanted into RB-SiC and WC/Co to
allow comparison with previous work, whilst Ti^+ at 300 and 400keV was
implanted into sialon, Mg-PSZ and WC/Co. This ion was chosen as one
likely to produce an intermediate carbide in WC and one that creates a
relatively large degree of damage per ion whilst producing a reasonably
thick implanted layer. N_2^+ implantations were performed using the
PIMENTO prototype implantation machine at AERE Harwell. All other
implantations were performed using the Cockcroft-Walton accelerator at
AERE. Table 1 summarises the implantation conditions used.

TABLE 1   Implantation Conditions

| Target material | Ion | Energy keV | Ion range parameters μm | | Dose range ions cm-2 |
			mean	std.deviation	
RB-Sic	N_2^+	90	0.0847	0.0267	$5.10^{14}-10^{18}$
sialon	Ti^+	300	0.153	0.0434	$10^{15}-5.10^{17}$
Mg-PSZ	Ti^+	400	0.190	0.0587	$10^{15}-5.10^{17}$
WC	N^+	100	0.100	0.0508	$10^{16}-10^{18}$
	Ti^+	400	0.103	0.456	$10^{16}-10^{18}$

Using a Leitz Miniload (under ambient laboratory conditions), low-load
Knoop microhardness tests (loads in the range 98-491mN) with penetration
depths similar to the implanted layer thicknesses (see Table 1) were used
to determine the changes in hardness behaviour. Vickers hardness tests
were used to induce indentation fracture in all but the WC/Co specimens,

for which it was found that even loads up to 491N produced no cracking.
For the remaining tests, loads of between 4.9N and 73.6N (tests at >9.8N
were performed using a standard Vickers hardness tester), the radial crack
trace lengths were measured using the optics of the Leitz Miniload. The
crack length data was analysed using the relations derived by Lawn and co-
workers (1979, 1980) linking crack lengths (through hardness, elastic
modulus and load) to toughness (K_c).

RESULTS AND DISCUSSION

Reaction-Bonded Silicon Carbide

The RB-SiC used in this study is 100% dense consisting mainly of $\sim$10μm
equiaxed grains, principally of the alpha (hexagonal) polytypes.
Approximately 10% residual silicon is present after the reaction bonding
process. The microstructure of REFEL is described in detail by Sawyer and
Page (1978). The variations of both low-load Knoop hardness and indentation fracture toughness with dose for N_2^+ implanted REFEL are shown in
Fig. 1(a),(b). At the lowest dose considered, a distinct hardening is
observed, this being followed by a re-softening until a dose of 3-4 x 10^{17}
ions cm^{-2} is reached. At this dose a dramatic decrease in hardness is
observed. Fig. 1(c) shows the predicted variation of amorphous layer
thickness with dose (from the model of Burnett and Page (1984a) using the
amorphisation criterion of Williams, McHargue and Appleton (1983)), and it
can be seen that the decreasing hardness in the dose range 10^{15}-10^{17}
ions cm^{-2} may be clearly correlated with the increasing thickness of the
amorphous layer. Note that the datum point at 10^{16} on Fig. 1(a) corresponds to a 60keV implant which is expected to give a thinner layer, hence
a higher hardness. The large surface-softenings observed above
4.10^{17} N_2^+ cm^{-2} (the critical dose of Roberts and Page (1982)) have been
attributed to an unpredicted thickening of the amorphous layer in silicon
(Burnett and Page, 1984a) and in silicon carbide (Burnett, 1983). This
thickening is allied to the occurrence of blistering in the surface (e.g.
Fig. 2), this blistered surface being markedly weaker than the unblistered
amorphous material. The hardening observed on the lowest dose specimen
(less than ϕ_{crit}) may be attributed to radiation hardening of the surface.
This overall pattern of hardness variation with dose is analogous to that
observed for single crystal MgO and $A\ell_2O_3$ (see Burnett and Page, 1984d,e).

Fig. 1(b) shows that all the implanted specimens showed a greater K_c than
the unimplanted control (i.e. the radial crack lengths decreased). This
reflects the presence of an implantation-induced surface stress inhibiting
the propagation of the crack through the surface (as described by Burnett
and Page, 1984b). Using the relationship derived by Lawn and Fuller
(1984), the magnitude of the stress required to give this change is
$\sim$500MPa for the lowest dose specimen. However, we have shown (for
sapphire) that this relation underestimates the actual stress for very
thin layers such as those considered here (Burnett and Page, 1984c).
Consequently, a more accurate estimate may be $\sim$3GPa. The subsequent
decrease in K_c from its peak value at higher doses to a relatively
constant value reflects the variation of the implantation-induced stress
with dose and, in particular, highlights the stress relieving effects of
amorphisation. Another manifestation of the surface stress is the
suppression of the lateral mode of indentation fracture first noted by
Roberts and Page (1982), described further by Burnett and Page (1984b) and
shown in Fig. 2.

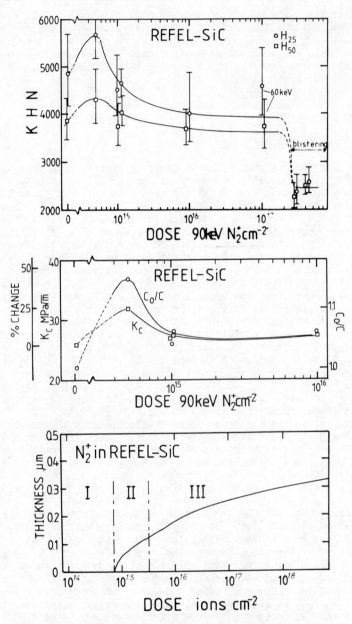

Fig. 1   Showing for REFEL-SiC the variation with dose (of N_2^+) of
(a) low-load Knoop microhardness, (b) indentation
fracture toughness (K_c) and ratio of radial crack trace
length to crack length on an unimplanted control (C_0/C),
(c) predicted amorphous layer thickness. The regions
marked I, II and III correspond to the presence of a
damaged but crystalline layer, a sub-surface amorphous
layer and a surface amorphous layer respectively.

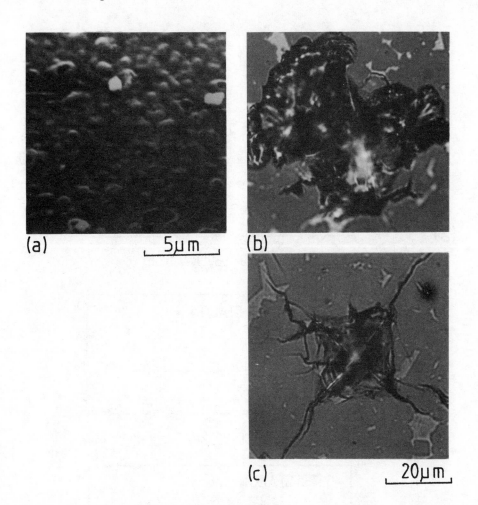

Fig. 2 (a) Blistering of the surface apparent at high dose
        $(8 \times 10^{17} N_2^+ cm^{-2})$ implantations.
        (b) Fracture around 4.9N Vickers indentation in unimplanted RB-Sic.
        (c) Fracture around 4.9N indentation in RB-SiC implanted with
        $10^{17} N_2 cm^{-2}$. Note less fracture than in (b) is present.

Sialon

Sialons, generically, are silicon nitrides which have been modified by
substitution of aluminium oxide within the nitride crystals. In addition,
large amounts of sintering additives ($\sim$6%) such as yttria may be added
and may also take part in this substitution. It is by controlling these
additives that the tailoring of the properties of subsequently produced
grain boundary phases may be effected. Being basically covalently-bonded,
the sialon is expected to undergo implantation induced structural modifications in a similar manner to the covalently-bonded silicon carbides.
Fig. 3(a) shows the variation of low load hardness with dose, the hardness
falling off to a fairly constant value as the dose increases. This hard-

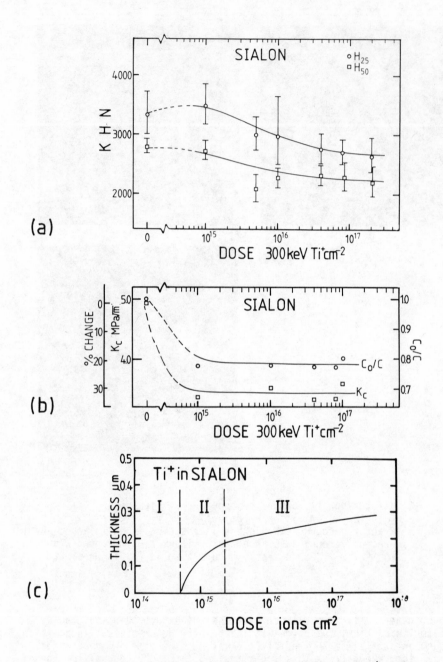

Fig. 3   Showing for SIALON the variation with dose (of Ti$^+$) of
(a) low-load Knoop microhardness, (b) indentation fracture
toughness and C$_o$/C (see Fig. 1 for definition),
(c) predicted amorphous layer thickness.

ness is less than that of the unimplanted control.   This softening is undoubtedly due to the formation of an amorphous layer. However, at the lowest dose considered here there is an indication that some (but not substantial) hardening may be occurring; i.e. this dose may be close to ϕ_{crit}. Estimating ϕ_{crit} as $10^{15} Ti^+$ cm^{-2} yields the predicted variation of amorphous layer thickness with dose shown in Fig. 3(c). As before, it can be seen that there is a close correlation between the amorphous layer thickness and the hardness measured.

However, unlike RB-SiC, the fracture toughness appears to drop markedly after implantation at all doses considered here (Fig. 3(b)). Allied with this is a change in radial crack morphology (Fig. 4). The cracking in the unimplanted control appears to be intergranular, the crack trajectories appearing "wiggly"; after implantation the crack trajectories appear to be much straighter. It is tempting to attribute this change to the fracture mode changing from intergranular to transgranular; however, it is more likely that the surface is now sufficiently homogeneous (due to the

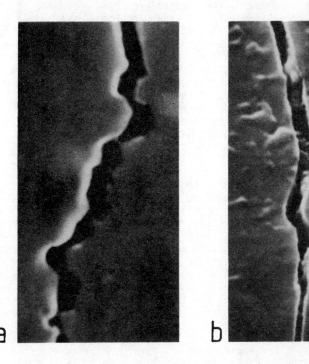

Fig. 4   (a) Part of a "wiggly" intergranular radial crack around an
          indentation in unimplanted SIALON.
          (b) Part of a crack around an indentation in SIALON
          implanted with $8.10^{16} Ti^+$ cm^{-2}.
          Note that this crack appears straighter near the surface
          than in (a).

formation of an amorphous layer) that the grain boundaries which acted as obstacles to crack propagation are no longer present. Hence the crack propagates more easily, resulting in a longer crack trace, despite the presence of an implantation-induced compressive stress field. Subsequent removal of the amorphous surface by polishing reveals once again the wiggly crack morphology beneath the straight crack traces on the surface. Thus, in summary, implantation results in surface softening and decreased fracture resistance due to the formation of an amorphous layer. It may be possible to both harden and increase fracture resistance of sialon by implantations to doses less than those required to form amorphous material. This will be investigated in future work.

Mg-Partially Stabilized Zirconia

Zirconia is largely ionically-bonded and thus, by comparison with the implantation-induced structural changes observed in other ionically-bonded materials, it is expected that amorphisation of zirconia will occur at doses higher than either of the two materials considered so far. The hardness dose data in Fig. 5(a) reflects this; an increased hardness is apparent for all doses considered, the hardness reaching a maximum at $\sim 4.10^{16} Ti^+$ cm^{-2} and then subsequently decreasing at higher doses. If we assume that hardness starts to decrease at the onset of amorphisation (as demonstrated by Burnett and Page, 1984d,e) then we can estimate ϕ_{crit} as $\sim 4.10^{16} Ti^+$ cm^{-2}. Thus, the predicted variation of amorphous layer thickness with dose may be evaluated as before (Fig. 5(c)). It can be seen that, even at the highest dose considered here, only a sub-surface amorphous layer is expected. Hence, at the doses above ϕ_{crit} the hardness is composed of contributions from both the softer amorphous material and the hardened crystalline material. This results in the material still retaining some degree of hardening. This behaviour is in good agreement with that of ionic single crystals (e.g. Burnett and Page, 1984d,e).

The variation of K_c with dose is shown in Fig.5(b). Prior to amorphisation, the toughness decreases with dose until after amorphisation, a fairly constant value is attained. This is contrary to the observations of Cochran, Legg and Solnick-Legg (1984), who found increases in toughness in fully stabilized single crystal zirconia after implantation. However, microscopical studies have now suggested an explanation for this discrepancy. Fig. 6(a),(b) shows the appearance of indentations in both the unimplanted specimens; increased fracture is apparent around the indentation made in the implanted specimens. Using oblique illumination it can also be seen (in Fig. 6(c),(d)) that the surface topography characteristic of the regions in which the tetragonal -> monoclinic transformation toughening mechanism has occurred is present in the unimplanted case (e.g. as the four lobes characterised by Swain and Hannink, 1983; Fig. 6(c)) but this is much decreased in the implanted case. This indicates that the transformation toughening mechanism is being suppressed by implantation, this being clearly linked with the decrease in toughness observed. This effect can be attributed to either the action of the compressive stress upon the stability of the tetragonal precipitates (upon transformation a compressive stress is set up, hence superposition of another compressive stress may be expected to inhibit transformation) or the titanium itself acting as a stabilizing agent for the tetragonal phase (as yet there is no information available on this topic).

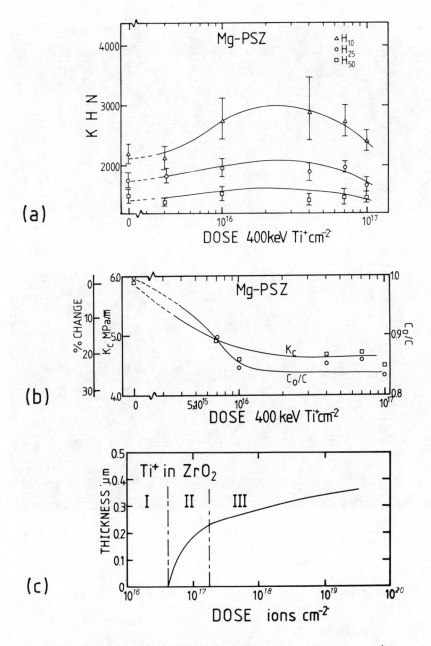

Fig. 5   Showing for Mg-PSZ the variation with dose (of Ti^+) of
(a) low-load Knoop hardness, (b) indentation fracture
toughness (K_c) and C_0/C (see Fig. 1 for definition),
(c) predicted amorphous layer thickness.

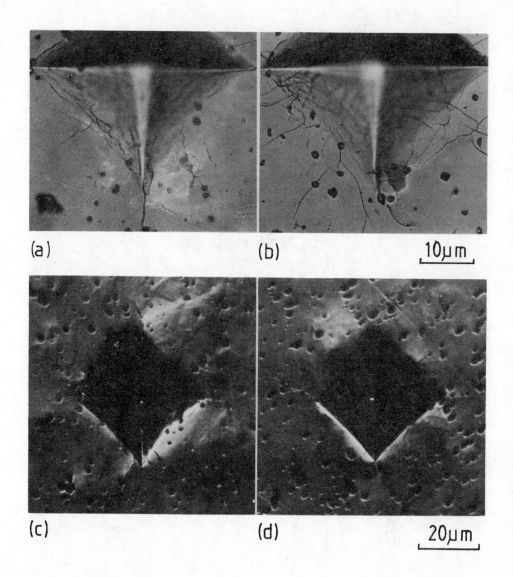

Fig. 6   (a) Reflected light micrograph of cracking around 73.6N
Vickers indentation in unimplanted Mg-PSZ.
(b) as for (a) only in a specimen implanted to $4 \times 10^{16} Ti^+ cm^{-2}$
Note more extensive cracking than in (a).
(c) Reflected light micrograph (oblique illumination) of a 73.6N
indentation in unimplanted Mg-PSZ showing the surface topography
generated by the action of the shear transformation occurring
around the indentation, (d) as for (c) but after implantation
with $4 \times 10^{16} Ti^+ cm^{-2}$. Note that near the transformation associated surface topography is not as apparent.

Thus, in summary, although the usual hardness-dose behaviour is observed for implantation of titanium into Mg-PSZ, the toughness is found to decrease as a result of the suppression of the transformation-toughening mechanism.

WC/Co Hard Metal

Tungsten-carbon bonding in the cubic monocarbide is principally covalent in character with some hybridization present as a consequence of the tetravalent carbon occupying octahedral sites in a tungsten matrix. Allied to this is a retention of a substantial proportion of the original tungsten metallic bonding; hence WC is a good metallic conductor. This is of great importance when assessing the susceptibility of the material to amorphisation by ion-bombardment. On a bonding criterion argument, (see Naguib and Kelly, 1975), it is expected that WC will be difficult to amorphise. The hardness-dose behaviour confirms this, surface hardening reaching a peak at $\sim 4.10^{16}$ Ti^+ cm^{-2} and $\sim 3.10^{17} N^+$ cm^{-2} (Fig. 7). Considering these doses as critical energy densities for amorphisation (see Burnett and Page, 1974b, for details) yields similar values for both species $(4-6.10^{23} keV$ $cm^{-2})$. For both ion species similar degrees of hardening for similar energy deposition (dose x energy deposited into atomic displacements/ion) are observed, thus inferring that the observed hardening is indeed radiation damage dependent. This observation is in disagreement with the data of Kolisch and Richter (1983) who found nitrogen a more potent hardener than either the heavier phosphorous or argon. These authors claimed that a possible martensitic reaction within the cobalt binder phase may be responsible for this anomalous result. However, Roberts (1982), upon implanting pure cobalt with nitrogen, found a surface softening rather than a hardening. A more likely reason for this discrepancy is that, unlike the present case where implantation parameters have been closely controlled to give very similar ion-range parameters (i.e. the implanted layers are of similar thickness and hence hardness results are comparable; see Table 1), their implantations were carried out at a single energy resulting in layers of varying thickness (hence hardness results are not directly comparable), those due to heavier ion species being the thinner. The result of this is to cause an underestimate of the hardness of this layer.

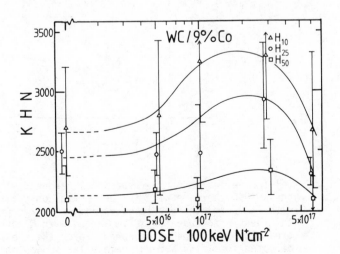

Fig 7(a)

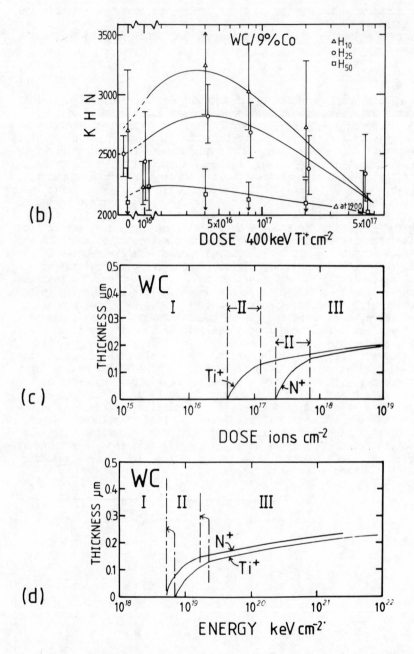

Fig. 7 (a) The variation of low load Knoop hardness with dose of N⁺
for WC/9%Co hard metal. (b) as for (a) only after implantation
with Ti⁺. (c) the predicted variation of amorphous layer
thickness with dose for WC implanted with N⁺ of Ti⁺. (d) the
predicted variation of amorphous layer thickness with energy
depicted into displacement damage for WC implanted with N⁺ or Ti⁺.

Using the values of ϕ_{crit} given previously, the predicted amorphous layer thicknesses can be determined as before for both variation of dose and energy deposition (Fig. 7 (c),(d)). As for the Mg-PSZ, the doses considered here give rise only to a sub-surface amorphous layer. This is reflected by the hardness changes, no absolute softening (characteristic of a surface amorphous layer) being seen.

Thus, in summary, WC/Co may be hardened by ion-implantation, this hardening being principally due to radiation damage. However, it is unclear whether the hardening occurs in the WC, the binder or both.   No indentation fracture could be induced in the specimens used here, so no toughness measurements could be made.

CONCLUSIONS AND SUMMARY

The wide range of materials studied have shown a number of effects in addition to those previously reported in studies of single crystals (e.g. Burnett and Page, 1984d,e). The results of the previous sections may be summarized as follows:

In general, all the materials studied showed softening upon amorphisation. The dose at which amorphisation first occurs being determined both by the ion species/energy and by the bonding characteristics of the target. The more directional ionic and metallic bond types requiring higher energies to amorphise than covalently bonded materials. Amorphisation may be controlled to produce either a thin sub-surface layer or a thicker surface amorphous layer.

Implantation-induced hardening was observed to occur in all materials; however, the extent of the hardening was much greater in the zirconia and tungsten carbide/cobalt materials. This may have been due solely to lack of data at low doses for the SiC and sialon materials and this will be investigated further.

The indentation fracture toughness of RB-SiC was found to increase after implantation. This has been attributed to the presence of a surface compressive stress of ∿3GPa.

The indentation fracture toughness of sialon was found to decrease with an accompanying change in crack morphology after implantation. Despite the undoubted presence of a compressive surface stress, the radial crack trace length was observed to increase. This indicates that the amorphous surface layer formed during implantation is less tough (despite being softer) than the unimplanted crystalline material.

The indentation fracture toughness of Mg-PSZ was found to decrease with implantation, a concomitant suppression of the transformation toughening mechanism is believed responsible.

A dependence of the surface hardening of WC on radiation damage has been observed; the ion species chemical nature apparently not contributing to this hardening.

In conclusion, the materials studied here have provided graphic examples of both the role of crystal chemistry upon the susceptibility of materials to radiation damage and the relationship between this damage and the surface mechanical properties observed. The observations made previously on single crystals have been reaffirmed whilst new observations concerning the role of microstructure upon implantation-induced mechanical effects have been made. Of these, the suppression of the transformation toughening mechanism in zirconia is probably the most important. A more detailed study of this will be published later.

ACKNOWLEDGEMENTS

We wish to thank Prof. R.W.K. Honeycombe, F.R.S. for provision of laboratory facilities, P.J.B. wishes to acknowledge S.E.R.C. and A.E.R.E. for the provision of a CASE award. We also wish to thank Dr. G. Dearnaley at A.E.R.E. for provision of implantation facilities. Experimental materials were supplied courtesy of J.N. Ness/U.K.A.E.A. ("REFEL" reaction-bonded silicon carbide), LUCAS Industries (sialon), L.J. Schioler, AMMRC, Mass (Mg-PSZ: Coors origin) and Sandvik U.K. (WC/Co hard metal).

REFERENCES

Burnett, P.J. 1983, unpublished work
Burnett, P.J. and Page, T.F. 1984a, J. Mater. Sci. 19 845-860
Burnett, P.J. and Page, T.F. 1984b, J. Mater. Sci. 19 3524-3545
Burnett, P.J. and Page, T.F. 1984c, J. Mater. Sci. in press
Burnett, P.J. and Page, T.F. 1984d, Proc. Mat. Res. Soc. 27 401-406
Burnett, P.J. and Page, T.F. 1984e, Plastic Deformation of Ceramic
    Materials, Eds. R.C. Bradt and R.E. Tressler, Plenum Press, New York,
    669-680
Cochran, J.K., Legg, K.O., Solnick-Legg, H.F. 1984, Proc. Mat. Res. Soc.
    24 173-178
Dearnaley, G. 1982, J. of Metals, 35 18-28
Hartley, N.E.W. 1979, Thin Solid Films, 64, 177-190
Kolitsch, A. and Richter, E. 1983, Cryst. Res. and Tech. 18 K5-K7
Lawn, B.R. and Marshall, D.B. 1979, J. Am. Ceram. Soc. 62 347-350
Lawn, B.R., Evans, A.G., Marshall, D.B. 1980, J. Am. Ceram. Soc. 63
    574-581
Lawn, B.R. and Fuller, E.R. 1984, J. Mater. Sci., in press
Naguib, H.M. and Kelly, R. 1975, Radiat. Eff. 25 1-12
Potter, D.I., Ahmed, M. and Lamond, S., J. of Metals 36 17-22
Roberts, S.G. 1982, Ph.D. Thesis, University of Cambridge
Roberts, S.G. and Page, T.F. 1982, Ion-Implantation into Metals
    Eds. V. Ashworth, W.A. Grant and R.P.M. Proctor, Pergamon, Oxford
    135-146
Sawyer, G.R. and Page, T.F. 1978, J. Mater. Sci. 14 885-904
Swain, M.V. and Hannink, R.H.J. 1981, J. Mater Sci. Letts. 16 1428-1431
Williams, J.M., McHargue, C.J. and Appleton, B.R., 1983, Nucl. Instrum.
    Methods 209/210 317-323

Inst. Phys. Conf. Ser. No. 75: Chapter 8
Paper presented at 2nd Int. Conf. Science Hard Mater., Rhodes

803

The surface mechanical properties and wear behaviour of ion implanted TiB$_2$

C J MCHARGUE, C S YUST, P ANGELINI, P S SKLAD, AND M B LEWIS

Oak Ridge National Laboratory, Oak Ridge, Tennessee   37831

ABSTRACT

Implantation of 1×10^{17} Ni·cm^{-2} (1 MeV) into polycrystalline TiB$_2$ produced an altered microstructure to a depth of 750 nm. The implantation damage was in the form of coarse dislocation tangles near the surface and fine damage structure at greater depths. The surface mechanical properties, as measured by Knoop micro-indentation hardness, indentation fracture tough-ness and scratch-wear resistance, were increased by 50 to 110% by implan-tation. During pin-on-disk tests, entire grains were removed by the process of grain boundary cracking.

INTRODUCTION

The compound TiB$_2$ has a high hardness (2550–3050 HK), a high melting point (3253 K or 2980°C), high thermal and electrical conductivities, and nonreactivity with many liquid metals. Such properties are of interest for technological applications such as cutting tools and dies, wear-resistant components in fossil energy conversion systems, and Hall cell electrodes.

At the Oak Ridge National Laboratory, we are studying the effect of ion beam modification of the near-surface structure of materials and the effect of the modified structure on the surface mechanical properties. Since the properties of ceramics as a class of materials are extremely sensitive to surface flaws and the surface stress state, a number of these materials have been included in this study. Some initial observations on the response of Al$_2$O$_3$, SiC, and TiB$_2$ were reported in the proceedings of this conference held in 1981 (McHargue and co-workers, 1983a). This paper sum-marizes recent results on TiB$_2$.

Ion implantation was developed by the semiconductor industry as a technique for introducing controlled levels of dopants. Ions from a suitable source are accelerated through a potential of tens to thousands of volts and allowed to strike a target surface. When such an ion impinges onto a target, atoms are displaced from their normal lattice sites by atomic collisions, point defects are produced, sputtering of the outermost layers may occur, and the injected ion comes to rest as an impurity or alloying element. Since the alloying species is forced into the substrate, the resulting composition and structure may differ significantly from those resulting from more conventional processing methods wherein thermodynamic and chemical kinetic factors are controlling.

Studies at this Laboratory (McHargue and co-workers 1982, 1983a, 1983b; McHargue and Williams 1982; Naramoto and co-workers 1983; Yust and McHargue 1984) and by Burnett and Page (1984) have shown that ion implantation of ceramics may produce highly defective but crystalline structures with increases in hardness or amorphous phases that are softer than the unimplanted material. Our earlier results on TiB_2 showed an increase in hardness of 69% due to implantation of 1×10^{17} Ni·cm^{-2} (1 MeV) (McHargue and co-workers 1983a). Padmanabhan and Sorensen (1981) reported that implantation of sputter-deposited TiB_2 films with 4×10^{16} Kr·cm^{-2} (300 keV) caused a hardness increase of about 50%. This paper extends the range of properties measured and reports observations on the damage structure responsible for the property changes.

EXPERIMENTAL PROCEDURE

Two sources of polycrystalline TiB_2 were used. One set of specimens were prepared in our laboratory from powder that initially contained approximately 1 wt % 0 as the major impurity. Samples having 98.4% theoretical density were produced by vacuum hot pressing at 2025°C under 25 MPa uniaxially applied pressure for 4 h. The grain size was in the range of 50 to 100 μm (Tennery and co-workers 1983). The other material was obtained as a sintered plate. The TiB_2 powder was produced in an arc plasma-reactor from vapor reactants, blended with paraffin wax as a binder, cold-pressed and fired under vacuum for 1 h in the temperature range of 1900 to 2275°C (Baumgartner and Steiger 1984).

Samples were cut from these plates by electric discharge machining (EDM). The surfaces were prepared to a metallographic finish by mechanically polishing with diamond paste.

The ORNL 5-MV Van de Graaff facility was used to implant Ni ions at 1 MeV. A fluence of 1×10^{17} Ni·cm^{-2} was used, giving a Ni to Ti ratio of 0.12 at the position of the maximum Ni concentration. A portion of the sample surface was protected from the ion beam by a metallic mask to provide a reference unimplanted region.

Samples for transmission electron microscopy (TEM) were prepared in a manner that allowed the cross section of the implanted region to be viewed. The implanted samples were cut into pieces approximately $1 \times 2 \times 5$ mm and pairs were glued together with the implanted surfaces facing each other. Slices about 250 μm thick were then cut, mechanically polished to about 75 μm, mounted on Cu washers, and Ar ion milled. The TEM specimens were then examined in a Philips EM400T/FEG equipped with 6585 STEM and EDAX 9100 x-ray energy dispersive spectroscopy (EDS) system. In addition, a Gatan double-tilt liquid-nitrogen-cooled specimen holder with a Be cup was used during EDS analysis. The Ni/Ti ratios were determined from standardless analysis routines developed at ORNL.

Knoop micro-indentation hardness was determined using 0.147 N (15 g) loads. The long diagonals of the impressions were in the range of 5 to 8 μm and the corresponding depth of the impressions were in the range 150 to 240 nm, or less than one half the thickness of the implanted zone. Because TiB_2 is hexagonal, the hardness is expected to vary as the crystallographic orientation of the grains varies. To permit direct hardness comparisons, where possible, measurements were made in individual grains where the mask had protected part of the grain from the ion beam. In instances where the

grain size was too fine to permit this practice, twenty impressions were
made and the average value and standard deviation determined. Since the
hardness values may reflect a contribution from the underlying unmodified
lattice, these values are reported only as relative values, expressed as
the ratio of the hardness of the implanted area to that of an unimplanted
region on the same specimen, indicating the minimum hardness changes rather
than absolute hardness values.

Indentation fracture toughness values were determined from Vickers hardness indentations. Vickers indents produced by 0.49 and 0.98 N (50 and
100 g) loads generated cracks at the corners of the indenter impressions.
The crack lengths and indentation diagonals were measured and used to
calculate an apparent fracture toughness value by the method of Marion
(1979).

An evaluation of the response of the surface to mechanical abrasion was
made by means of a scratch test. In this test, a stylus was slowly
translated across the surface under normal loads of 0.098 to 0.49 N (10 to
50 g) while the tangential force on the stylus was continuously measured.
The scratch origin was positioned to cause the stylus to cross the
unimplanted/implanted interface during the traverse of the surface. The
stylus was a Vickers diamond indenter, positioned to move over the sample
surface with one of the pyramid faces normal to the direction of motion.
The stylus moved at a velocity of 28 μm/s. The cross-sectional area of the
scratch was determined from profilometer traces and used with the recorded
tangential force to calculate a work of material removal. The tracks were
examined in optical and scanning electron microscopes for information on
the deformation behavior.

Pin-on-disk wear tests were conducted at room temperature in a dry nitrogen
atmosphere. Normal forces of 0.49 to 4.9 N (50 to 500 g) were used with
pins of diamond and cubic boron nitride (CBN). The pins had spherical tips
with radii of 200 μm. The tangential force was recorded as the disk
rotated at a speed of 250 rpm. Test periods ranged from 1 to 72 h. The
coefficient of friction was given by the ratio of the tangential to normal
force.

ELECTRON MICROSCOPY

Figure 1 is a TEM micrograph of a cross section of the as-implanted
material. The micrograph shows the change in the nature of the damage as a
function of distance from the implanted surface. There is a uniform,
moderate density of tangled dislocations in the region extending from the
surface to approximately 550 nm. The areas between the dislocations appear
to be free of defects. Between 550 nm and approximately 750 nm the
microstructure is noticeably different. This region contains a high density of small defects 5 to 10 nm in size. Because of the contrast from the
defects, it is impossible to determine whether any tangled dislocations of
the type seen nearer the surface are present. At approximately 750 nm the
observable damage ends abruptly. The presence of dislocations and other
defects in the microstructure demonstrates that the implanted layer was
still crystalline.

Measurements of the Ni content of the implanted layer were made using x-ray
energy dispersive spectroscopy (EDS). The measurements were made in the
STEM mode with a beam diameter of approximately 2 nm. Precautions were

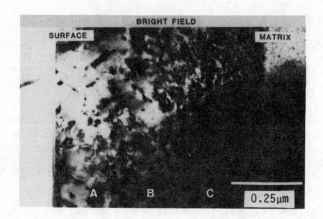

Fig. 1 Bright-field TEM photograph showing the implanted layer in cross section. Note the dislocations near the surface and the fine damage structure nearer the substrate TiB_2.

taken to ensure that each measurement was made under the same conditions. These results are plotted in Fig. 2 as the ratio of Ni to Ti versus distance from the implanted surface. They show a maximum in the Ni/Ti ratio at approximately 450 nm. Only a small amount of Ni was detected in the region deeper than 650 nm, despite the obvious presence of a modified microstructure. Although Ni was present in fairly high levels, there is no evidence of Ni precipitation either in the bright-field images or the diffraction patterns from the implanted layer. The regions of the two types of damage seen in the photographs of Fig. 1 are marked on Fig. 2.

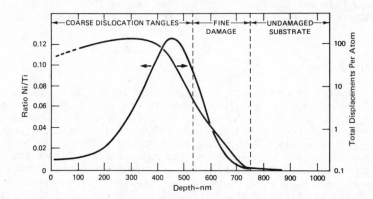

Fig. 2 The measured nickel concentration and calculated defect production as functions of distance from surface. The damage microstructure is indicated.

The damage production profile (or deposited energy profile) calculated by Lewis (1984) is also shown in Fig. 2. In this calculation, an accounting is made of the energy dissipated by the recoil atoms; thus the damage due to both the primary Ni ions and the recoiling Ti and B ions is included. An earlier calculation which ignored the recoil damage indicated that the damage should end at about 500 nm.

The microstructure observed in Ni-implanted TiB_2 is complicated. The dislocation structure near the implanted surface is similar to the microstructure that develops in metallic alloys during ion irradiation. There is no indication that these dislocations are the result of the build-up of stress due to the presence of extra cations in the lattice. The dislocations are the result of the accumulation of the point defects (in this case undoubtedly interstitials) into loops that eventually grew and interacted to form dislocation tangles and segments. In the case of ion-irradiated metallic alloys, the dislocation density is a function of distance from the surface and usually reaches a maximum near the depth of peak ion deposition. The maximum in the Ni concentration occurred at approximately 450 nm while the peak in the deposited energy occurred at 200 nm. It is evident from Fig. 1 that the microstructure in this specimen does not exhibit a peak in dislocation density and is affected to a much greater depth. Moreover, a change in the character of the microstructure, from dislocation tangles to a high density of small defects, occurred at about 550 nm.

Attempts to determine experimentally the nature of the fine defects observed in the deeper regions have been unsuccessful to date. The apparent high number density of these defects makes imaging difficult due to overlapping images. Calculations indicate that the B atoms which receive the maximum energy transfer from the 1 MeV Ni ions during a primary knock-on (PKO) event near the surface will have a range of about 750 nm. Such high energy knock-ons can, of course, cause further displacements. This calculated end of the range of these displaced B knock-on atoms is in agreement with the observed end of the modified layer.

The coarse damage extending to about 550 nm may be due to defects produced by the Ni bombarding ions and by the displaced Ti ions. The width of the damage region is consistent with the ranges of these ions. The underlying region of fine damage may then have its origin in the boron PKOs and subsequent B displacements.

MECHANICAL PROPERTIES

A summary of the Knoop micro-indentation hardness, indentation fracture toughness, and scratch-wear results is given in Table 1. All results are presented as relative values (implanted value/unimplanted value) because of uncertainties associated with testing thin surface layers.

Significant hardening of TiB_2 resulted from the implantation of 1×10^{17} Ni·cm^{-2} at 1 MeV. Relative values were in the range of 1.7 to 2.1 for both materials (increases in hardness of 70 to 110%). The maximum indicated hardness was 6405 HK15.

The indentation fracture toughness was increased by 50 to 80%. Because of the relatively low loads used, the cracks and indentation diagonals were in the range of 8 to 20 μm. Measuring difficulties thus caused considerable scatter in the data and large uncertainties in the true value of K_{IC}. Nevertheless, the trend shows an increase in toughness due to implantation.

TABLE 1   Surface Mechanical Properties

Material	Specimen Number	Relative Knoop hardness	Relative fracture toughness	Relative work of material removal
Hot-pressed	1	1.69	1.81	1.5
	2	2.05	--	2.4
	3	2.10	--	2.4
Sintered	4	1.77	1.96	2.0

The scratch-wear test gives an indication of the wear resistance under gouging or abrasive conditions, and is equivalent to a single pass pin-on-disk test. There was no or little change in the measured tangential force as the diamond stylus moved from the unimplanted region to the implanted area, but the profiles of the grooves were markedly different. The ratio of the tangential force to the normal force was in the range of 0.1.

The cross-sectional areas of the scratches were measured with a profilometer. The work to remove a unit volume of material was calculated from the tangential forces and areas and are summarized in Table 1. Implantation caused increases in the work to remove a unit volume of material of 50 to 140%. The SEM of Fig. 3 illustrates a scratch traversing both implanted (to the right) and unimplanted (to the left) areas for a normal force of 0.196 N. The scratch has scarcely disturbed the implanted region but has caused cracking and pull-outs of entire grains in the unimplanted region. The mode of material removal appears to be by grain boundary cracking and removal of entire grains. At higher normal forces, cracks form both within grains and along grain boundaries in the unimplanted region but only along boundaries in the implanted zone.

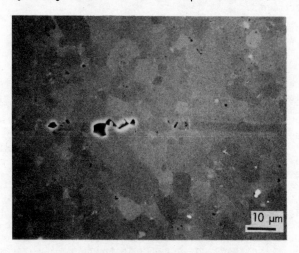

Fig. 3   SEM photographs of single pass wear track.   The implanted region is to the right.

The results from pin-on-disk tests provide some insight into the deformation of both implanted and unimplanted TiB_2. The high hardness of the material has caused excessive wear of the diamond and CBN pins and has prevented the determination of quantitative wear rates. The area of the pin contacting the disk increased by a factor of 2 to 3 during the tests. From measurements of the initial and final areas, an average maximum Hertz stress was calculated for each test. These fell in the range of 60 to 360 MPa, compared to a fracture stress of about 750 MPa for this material.

The friction coefficients (F_t/F_n) were extremely low, 0.02 to 0.04 for diamond on TiB_2. Because the value was so low, no differences between implanted and unimplanted areas were detected.

The wear tracks were characterized by a deposit that had the appearance of a smeared film and heavy deposit of fine particles. Figure 4a is a bright-field optical photomicrograph showing these features for a track made in implanted material with an average maximum Hertz stress of 86 MPa (F_n = 1.96 N, 24 h). Dark-field photomicrographs (Fig. 4b) show the grain structure and indicates that some of the deposited layer is transparent.

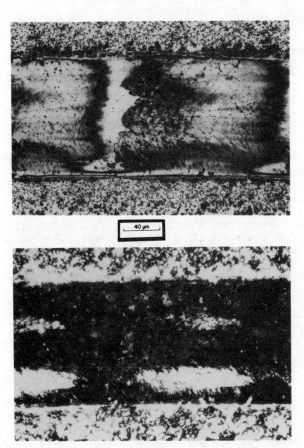

Fig. 4 Photomicrographs of wear track made by diamond stylus in 24 h at an average maximum Hertz stress of 86 MPa: (a) bright field; (b) dark field.

Profilometer traces showed the presence of the deposit but very little removal of the TiB_2. Figure 5a gives such a trace for the specimen of Fig 4. The profile for the track made in the unimplanted region (Fig. 5b) also indicates the deposit and may show an indication of a small amount of material removal.

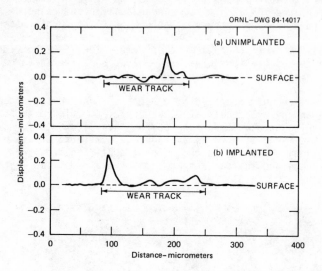

Fig. 5   Profilometer trace of wear track shown in Fig. 4: (a) implanted region and (b) unimplanted region.

As in the case of the single-pass scratch test, the main mode of material removal is by removal of entire grains due to grain boundary cracking. Figure 6 is a SEM photograph from an unimplanted area (Fn = 1.96 N, 24 h) that clearly shows this characteristic.

Dark-field optical micrographs showed that about 2% of the grains in the as-received material contained transgranular cracks. The frequency of such cracks was two to four times higher in grains within the wear tracks, showing that deformation had indeed occurred. The frequency of such cracks was about the same in implanted and unimplanted regions. However, the wear tracks in unimplanted regions contained numerous intergranular cracks. The arrows in Fig. 7 denote two regions in which grain boundary cracks run for the length of several grains. In many instances, cracks were found along three of the four of five boundaries shown in this plane of polish. This observation again suggests that the entire grains are removed and the general abrasion of the surface does not occur under the condition of these tests.

CONCLUSIONS

Implantation of TiB_2 with a fluence of 1×10^{17} Ni·cm^{-2} (1 MeV) produced a complex, two-part near-surface damage structure. The affected region extended to a depth of approximately 750 nm, significantly deeper than the measured range of the Ni ions.

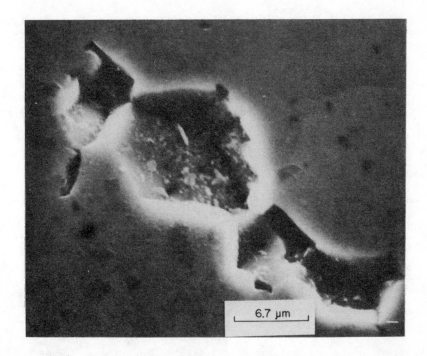

Fig. 6 SEM photograph of wear track showing removal of entire grains.

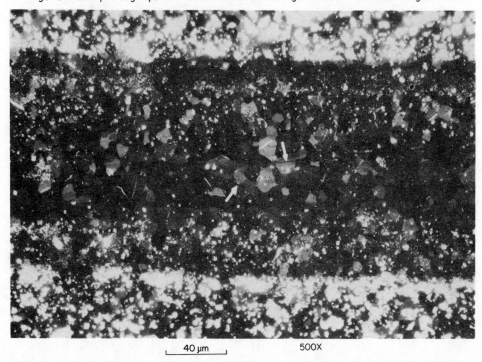

40 µm   500X

Fig. 7 Dark-field photomicrograph showing grain boundary cracking
in wear track on unimplanted TiB$_2$.

The surface mechanical properties as measured by Knoop micro-indentation hardness, indentation fracture toughness and scratch-wear test were significantly enhanced by this ion beam treatment. Increase of 50 to 110% in these properties were measured.

Material removal during scratch-wear or pin-on-disk wear tests was primarily due to intergranular cracking that resulted in the removal of entire grains. Ion implantation suppressed this form of cracking. Pin-on-disk tests were characterized by very low coefficients of friction, large wear rates of the diamond or CBN pins, and formation of much fine debris.

ACKNOWLEDGMENT

The authors acknowledge the invaluable help of C. B. Finch in preparing one of the TiB_2 compacts, B. C. Leslie for metallographic assistance, J. T. Houston for the very difficult task of preparing electron microscopy specimens, and J. E. Pawel, J. S. Pullium, and C. E. DeVore for determination of the surface mechanical properties. The research was sponsored by the Division of Materials Sciences and Energy Conversion and Utilization Technologies Program of the U.S. Department of Energy, under contract DE-AC05-840R21400 with the Martin Marietta Energy Systems, Inc.

REFERENCES

Baumgartner, H. R. and Steiger, R. A. 1984, J. Amer. Ceram. Soc. <u>67</u>(3) 207–212

Burnett, P. J. and Page, T. F. 1984, J. Mater. Sci. (in press)

Lewis, M. B. 1984, to be published in Nuclear Instruments and Methods B

Marion, R. H. 1979, Fracture Mechanics Applied to Brittle Materials Ed. S. W. Frieman. American Society for Testing and Materials, Philadelphia, pp 103–111

McHargue, C. J., Naramoto, H., Appleton, B. R., White, C. W., and Williams, J. M. 1982, Metastable Materials Formation by Ion Implantation. Eds. S. T. Picraux and W. J. Choyke. Elsevier Science Publishing Company, New York, pp 147–153

McHargue, C. J. and Williams, J. M. 1982, ibid. pp 303–309

McHargue, C. J., Lewis, M. B., Appleton, B. R., Naramoto, H., White, C. W., and Williams, J. M. 1983a, The Science of Hard Materials. Eds. D. J. Rowcliffe, R. K. Viswanadham, and J. Gurland. Plenum Publishing Company, New York, pp 451–465

Naramoto, H., White, C. W., Williams, J. M., McHargue, C. J., Holland, O. W., Abraham, M. M., and Appleton, B. R. 1983b, J. Appl. Phys. <u>54</u> 683

Padmanabhan, K. and Sorensen, G. 1981, Thin Solid Films <u>81</u> 13–19

Tennery, V. J.., Finch, C. B., Yust, C. S., and Clark, G. W. 1983, The Science of Hard materials. Eds. R. K. Viswanadham, D. J. Rowcliffe, and J. Gurland. Plenum Publishing Company, New York, pp 891–909

Yust, C. S. and McHargue, C. J. 1984, Emergent Process Methods for High Technology Ceramics. Eds. R. F. Davis, H. Palmour III, and R. L. Porter. Plenum Publishing Company, New York, pp 533–547

Inst. Phys. Conf. Ser. No. 75: Chapter 8
Paper presented at 2nd Int. Conf. Science Hard Mater., Rhodes

Cavitation erosion of thermally sprayed coatings

A ADAMSKI AND R McPHERSON

Department of Materials Engineering,
Monash University, Clayton,
Victoria, Australia, 3168

ABSTRACT

The process involved in the cavitational erosion of several types of
metallic and ceramic thermally sprayed coatings have been studied using
the ultrasonic probe technique. The weight loss occurring during three
minute intervals of exposure to cavitational erosion was determined, the
surface of the specimens was examined by scanning electron microscopy (SEM)
and the wear debris were examined by SEM and "ferrography". All specimens
exhibited similar erosion rate curve types. The results suggest that an
improvement in coating performance could be achieved by controlling the
effective contact between the lamellae of which the coating is made up.

INTRODUCTION

Cavitational Erosion - the State of the Art

It is well known that many problems associated with fluid systems are
caused by cavitational erosion of the materials used in construction. In
normal liquids, when the local pressure decreases below a specific
critical value, cavitational bubbles are generated. In liquids subjected
to an oscillating pressure field (vibratory cavitation) bubbles are
formed within its volume, near the vibrating surface, and the local
pressure fluctuation leads to repeated bubble growth and collapse [Mørch,
(1979); Hansson and Mørch, (1980)]. The resultant stresses on the front
side of the specimen resulting from bubble collapse are responsible for
the cavitational erosion of materials.

For a long time cavitational erosion was discussed on the basis of a
single bubble collapse theory. The erosion mechanism was described as a
result of the impact jet formed by asymmetrical bubble collapse [Hansson
and Mørch, (1980); Vyas and Preece, (1976)], and the shock wave
originating from spherical bubble collapse [Hansson and Mørch, (1980)].
Recently, this process has been explained as an effect of concerted
collapse of cavity clusters. Two models have been given [Preece, (1979)].
The first one assumes the superposition of the shock waves from individual
bubbles collapsing and formation of a single high-intensity damaging
shock wave.

In the second model collective collapse of the first cavities gives an
energy which is transferred to those not yet collapsed. The liquid jet
is formed by asymmetric collapse of the bubbles near the surface of the
solid.

The first model can be used to explain an appearance of large scale
deformation (craters, undulations) in the surface of tested materials.
Concentration of the erosion in the central part of the specimen can
result from the concerted collapse of the cluster, as the central cavities
give the strongest contribution to the damaging action, due to local
increase of hydrostatic pressure.

Erosion Mechanism of Metals

Large numbers of both ductile and brittle metals have been tested by
various methods.   Vyas and Preece (1977), Preece (1979) and Preece,
Vaidya and Dakshinamoorthy (1979) made a detailed study of the mechanism
of deformation and erosion in FCC, BCC and HCP bulk metals using the
vibratory technique.   SEM observations of tested surfaces showed that
ductile, isotropic metals are deformed by plastic flow during the first
few seconds of test. Small pits and depressions appear on their surfaces
and grow with increased exposure time. The surface of the metal becomes
undulated (Fig. 1) and deep, large craters ($\sim$ 150 μm) develop from these
undulations.   After a sufficient time they become the only form of
surface deformation (Fig. 2). Material loss then occurs by necking and
breaking of the crater rims.

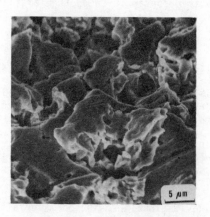

Fig. 1   Pitting and surface undu-
         lations. Exposure time
         45 sec. Brass

Fig. 2   Crater formation and crack-
         ing. Exposure time
         6 min. Brass

Vyas and Preece (1977) showed that work hardening of FCC metals occurs at
the early stages of the test.

The mode of failure of BCC metals was reported to be very sensitive to
the strain rate and hence, dependent on the test conditions (Preece,
Vaidya and Dakshinamoorthy, 1979). Low strain rate at high temperature
will favour plastic mechanisms, whilst material tested under high strain
rate at low temperature fails by brittle fracture.

Wade and Preece (1978) in experiments on zone refined iron showed that the first material removal occurs by a ductile mechanism. This was followed by development of large craters by cleavage which resulted in an increased erosion rate.

In general, the mechanism of material removal in HCP metals shows a brittle nature. However, a high density of slip and twinning has been observed in favourably oriented grains of zinc [Preece, Vaidya and Dakshinamoorthy (1979); Vaidya and Preece (1977)].

Erosion of zinc resulted from the growth of cracks at the grain boundary triple points and grain boundary twin intersections and their propagation across the grains. In monocrystalline zinc the erosion mechanism was found to be more dependent on the crystal orientation than in FCC and BCC metals.

The erosion was reported to be time dependent for all tested metals. A graph of the material removal rate versus exposure time consists of four regimes: incubation regime, maximum rate regime, decreased rate regime and steady regime. During the incubation period plastic deformation occurs without any detectable weight loss.

Erosion Behaviour of Sprayed Coatings

In comparison with bulk materials, little work has been done on coatings, at least plasma sprayed coatings. Akhtar (1982) has investigated resistance to cavitational erosion of plasma sprayed alumina, alumina-titania and chromium oxide coatings. All tested coatings showed an absence of an incubation period, in which no weight loss is observed. He reported a linear relationship between the erosion rate and exposure time for alumina and chromium oxide coatings, which became parabolic in shape for alumina-titania. SEM observations revealed that in alumina and chromium oxide coatings nucleation of interlamellar microcracks leads to formation of triple points which act as sites for crater formation.

Intercrystalline microcracks developed on the surface of $Al_2O_3 - TiO_2$ coatings, coalesced and formed nodal points. These nodes then developed into craters. With time, the diameter of these craters increased to 0.4mm and remained constant. Further exposure resulted in development of new craters. Thus the resistance to the nucleation and propagation of the intercrystalline microcracks can be a major factor in determining the cavitational erosion of plasma sprayed oxides.

Response to the cavitational erosion of flame sprayed Al, Zn and Zn - 15% wt Al coatings on steel substrates was analysed by Dorfman, Vargas, Clayton and Herman (1979). Porosity and work hardening was postulated as having the greatest effect on the resistance of tested coatings to cavitational erosion. Zinc coatings appeared to be least porous, but due to the work hardening effect, showed the greatest overall weight loss. By rolling, the authors obtained a reduction in the initial rate of erosion of the Al-coating (the sealing of pores), a small increase in erosion of Zn-coating (work hardening) and no significant change for Zn - Al coating.

Frees (1983) in experiments on TiC coatings deposited by the CVD technique on various substrates, showed a correlation between the erosion rate and the initial state of the tested coatings. Defects such as holes in the coating, pits, unevenness of the surface, microcracks and poor adhesion determined their resistance to cavitational attack.

All the papers cited above show that the quality of the coatings, resulting from the deposition processes and post-spraying treatment, have a
major effect on their resistance to cavitational erosion

The present paper reports a study of the cavitational erosion behaviour of
several sprayed coatings on steel.

EXPERIMENTAL DETAILS

The cavitational erosion of materials can be determined by various testing methods. Flow channels, disks and vibratory devices are widely used
for studying cavitational effects. The fastest, cheapest and easiest to
use is vibratory cavitation equipment.

The vibratory apparatus used in this work was a Vibrason 150, solid state
ultrasonic generator with direct reading, amplitude control and lead
zirconate titanate high efficiency transducer assembly, with an aluminium
bronze velocity transformer (Fig. 3). Its nominal frequency was 20 kHz

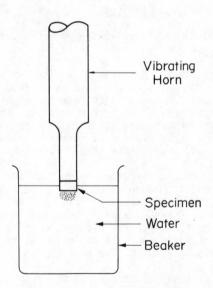

Fig. 3 Schematic diagram of cavitational erosion vibratory test.

and maximum strain, in terms of tip-to-tip amplitude, between 28 and 30 µm.
Acoustic stepped couplers were used to increase the transducer amplitude
to a level of 50 µm. The amplitude transformation was given by a ratio of
the areas of the ends. The specimens were screwed into a threaded hole in
the end of the horn and were immersed into a water bath to the depth of
3.5 mm. The water temperature was maintained constant (45°C) for all
tested materials. The exposure time intervals used were 180 sec;

specimens were weighted after each test period and SEM micrographs were taken of the deformed and/or eroded surfaces. Collected debris were studied by Ferrography and SEM. Ferrographic analysis of wear particles is a technique which allows the qualitative and quantitative character- istics of wear particles to be evaluated (Adamski, Stachurski and Stecki, 1984; Scott, Seifert, Westcott, 1975).

SPECIMEN MATERIALS

Tested materials were selected on the basis of the manufacturer's recom- mendation and literature data. The characteristics and some selected properties of the powders and deposited coatings are listed in Table 1. Surfaces of the substrate samples were sand blasted with alumina grit. Prior to spraying, specimens of cast α-brass (Fig. 1,2) and aluminium bronze were also tested for comparison purposes.

TABLE 1   Selected Properties of Powders and Coatings

Specimen Type	Composition %	Size Powder Range	Melting Point °C	Porosity	Hardness
T15E	Cr-17,Fe-4, Si-4,B-3,5 C-1,Ni-Bal.	45 ± 15 μm	1025	Essentially non-porous	Rc62
P105	Al_2O_3-98, SiO_2-0.5, Other oxides 1.5	-	2035	Low	Rc60
P130	TiO_2-13, Al_2O_3-Bal.	53 ± 15 μm	1840	Negligible	Rc63
P106	98%-Cr_2O_3	90 ± 15 μm	2435	Less than 1%	Rc70
P451	Cr-9.5, Si-2.5,B-1.5, Al-0.5,Ni-Bal.	160 ± 45 μm	1010	-	Rc33
Aluminium Bronze *	Al-10, Fe-5,Ni-5, Cu-Bal.	-	1050-1080	-	Rc16

T - thermo-sprayed and fused
P - plasma sprayed
* - as-cast

EXPERIMENTAL RESULTS

Analysis of Powders and Coatings

X-ray diffraction was used to determine the phase constitution of powders and coatings deposited from them. Results are presented in Tables 2 and 3.

TABLE 2   The Phase Constitution of Powders

Powder	Phases Present
105	α - alumina
130	α - alumina
106	α - Cr_2O_3
15E	$\gamma + \alpha$
451	γ + residual α

TABLE 3   The Phase Constitution of Coatings

Coating	Phases Present
105	γ - alumina, traces of α and δ - alumina
130	γ - alumina, traces of α and δ - alumina, ρ - Al_2TiO_5
106	α - Cr_2O_3
15E	γ + traces of α
451	γ

In general, the microstructures of thermally sprayed coatings are very complex. X-ray diffraction study of the Al_2O_3 - 13 wt% TiO_2 coating revealed that it consisted mostly of γ - Al_2O_3 with a small fraction of α, δ and β-Al_2TiO_5 phases.

105 coatings containing 98 wt% Al_2O_3 consisted of γ-alumina with some fraction of α and δ phases. The presence of some residual α-alumina in 105 and 130 coatings indicates that a small fraction of the particles was not completely melted (McPherson, 1980). The presence of δ-phase indicates slower cooling in the latter stage of deposition. A small fraction of a Cr-rich phase (α) was found in both metallic powders and in the 15E coating which consisted predominantly of the γ-phase.

Erosion Rates

Both metallic and ceramic coatings exhibited similar erosion rate versus exposure time curve type. The absence of an incubation region appears to be characteristic for them.

During the first few minutes of the exposure the erosion increased for all tested coatings but decreased after approximately 11-12 minutes and then remained constant with further exposure times (Fig. 4).

The thermo-sprayed and fused Ni-Cr alloy appeared to be the most resistant among all the tested coatings in the initial rapid erosion regime. However, the lowest erosion in the steady regime occurred for both alumina based coatings (Fig. 4).

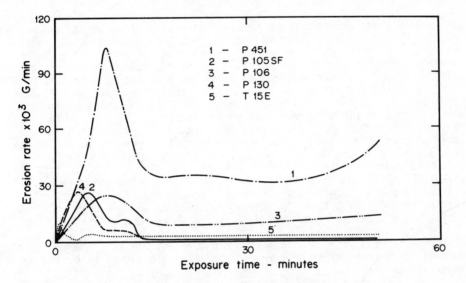

Fig. 4 Erosion rates versus exposure
time curves for tested coatings.

The only bulk material tested, aluminium bronze, exhibited a long incubation period with negligible weight loss (Fig. 5). However, the erosion rate for the steady regime was higher than for most of the tested materials.

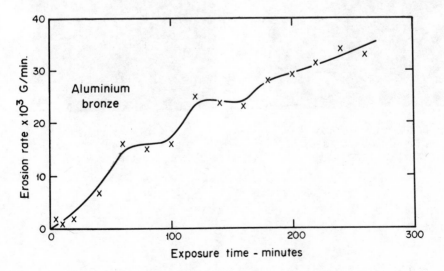

Fig. 5 Erosion rates versus exposure time
curves for tested aluminium bronze.

SEM Observations

All specimens, except the thermosprayed and fused coatings and aluminium bronze, showed very similar erosion mechanism with material removal occurring due to pitting and crater formation.

Small microcracks present in the material propagate along the inter-lamellar boundaries and form nodal points which, in time, develop craters (Fig. 6). The presence of pores and not fully melted particles also leads to crater formation (Fig. 7). Material can be removed by cracking and breaking of parts of the lamellae of which the coatings are made up, from the rims of the craters (Fig. 8). Small spheres of unmelted particles (or formed during the impact of the melted droplets on to the surface) were stripped from the surface leaving small craters (Fig. 9). Large smooth areas inside the craters were produced by the removal of complete lamellae or a large part of them, and pores present between the deposited particles were opened up (Figs. 9,10). Initial microcracks readily nucleated and propagated across the coating (Fig. 11). Comparison of the plasma deposited and thermosprayed fused Ni-Cr alloys showed that 12 minutes exposure to cavitational attack produced only a few groups of craters randomly distributed on the fused specimen surface (Fig. 12), whilst the plasma deposited coating appeared to be severely eroded (Fig.13).

Aluminium bronze, after 42 minutes of exposure, showed a high degree of surface distortion which, in turn, led to crater formation in the latter stage of the test.

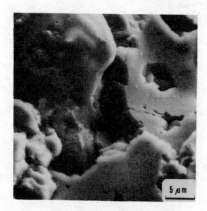

Fig. 6 SEM micrograph of 451
plasma sprayed coating.
Exposure time 12 min.

Fig. 7 SEM micrograph of 451
plasma sprayed coating.
Exposure time 6 min.

Fig. 8 SEM micrograph of Cr_2O_3 plasma sprayed coating. Exposure time 6 min.

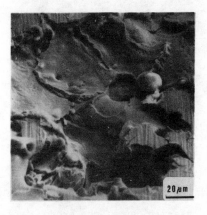

Fig. 9 SEM micrograph of Cr_2O_3 coating after 12 min. of the test.

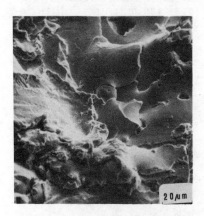

Fig. 10 SEM micrograph of plasma sprayed alumina coating. Exposure time 12 min.

Fig. 11 SEM micrograph of alumina-titania coating. Exposure time 12 min.

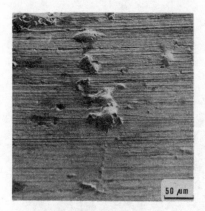

Fig. 12 SEM micrograph of thermo-
sprayed and fused nickel-
chromium alloy after
12 minutes exposure time.

Fig. 13 SEM micrograph of plasma
sprayed nickel-chromium alloy.
Exposure time 12 min.

Wear Debris Analysis

SEM observations of the eroded surfaces have been followed by wear debris
analysis. Collected debris (particle size 15-300 μm) were analysed by
Ferrograph, bichromatic microscope and SEM. The first sample was taken
after 3 minutes exposure time and the second sample after 6 minutes.
Observations revealed that the first removed material contains mainly
single spherical particles, or clusters of particles embedded in the
fractured, splashed lamellae (Fig. 14). Most of them were removed from
the surface layer and originate from the deposition processes. With increase in the exposure time (6 minutes) eroded particles appeared to be
large, flattened lamellae, or their fragments (Fig. 15). In general, the
observed debris had flat smooth surfaces with sharp edges resulting from
fracture (Fig. 15). Some deformation present is associated with areas of
good interlamellar contact and are produced by a cohesive failure.

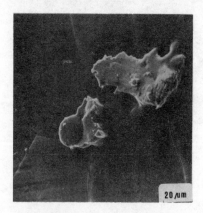

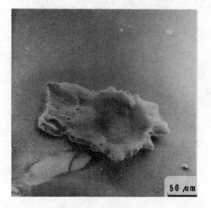

Fig. 14 SEM micrograph of wear
debris collected after 3
minutes of exposure time,
plasma sprayed nickel-
chromium alloy.

Fig. 15 SEM micrograph of particle
collected after 6 minutes
of exposure time, plasma
sprayed nickel-chromium
alloy.

DISCUSSION

It is a well established that plasma sprayed coatings are formed by the build up of successive layers of material. Melted particles are flattened due to the impact and solidify to give a lamellar structure. The particles have to be completely melted and their velocity has to be sufficient to obtain good bonding with the lamellae which have already solidified. However, the liquid particles must not be broken into small droplets during impact. The liquid particles can deform and partially adopt the shape of substrate irregularities, nevertheless absorbed and entrapped gas, oxide films (in metallic coatings) and other contamination leads to relatively high porosity of sprayed coatings. Thus, coatings consist of regions of good interlamellar contact and regions of poor contact with gaps of 0.01-0.1 μm. It has been suggested that approximately one-fifth of the area between individual lamellae is in true contact [McPherson (1982, 1984)]. Residual stresses reulting from the high cooling rates and other effects, together with high impact energy of molten particles can generate crack formation and interlamellar contact failure.

The presence of these defects within the lamellae appears to be a major factor which determines the resistance of coatings to cavitational attack. The SEM observations indicate that the presence of surface porosity and surface microcracks is very significant in the early stages of the test (Fig. 4). Pores which develop into craters contribute to the material loss. Crack propagation results in cohesive failure within the coating as well as adhesive failure on the coating-substrate interface.

The present study shows that a short initial period of enhanced weight loss occurs during the cavitational erosion of thermally sprayed coatings, whereas bulk metallic materials exhibit an incubation period, during which the rate of material loss is small, which corresponds to the initiation of craters by a plastic deformation process. The evidence suggests that the high initial wear rate of the coatings is associated with the removal of some lamellae and small droplets of splashed material not well bonded to the underlying deposit. The erosion rate after this initial period is however relatively low for the ceramic coatings but high for the Ni-Cr coating tested, unless the latter is "fused".

The microstructure of plasma sprayed ceramic coatings consists of fine grain size polycrystalline lamellae, only a few μm thick, which are attached to underlying lamellae by strongly bonded regions, equivalent to grain boundaries, over relatively small regions of true contacts [McPherson and Shafer, 1982]. As a result of this structure the elastic modulus of the coating is only about one-fifth of bulk value [Boch et al. (1981)]. Thus it is possible that the plasma sprayed ceramic coatings deform elastically under the influence of cavitation and the local stresses do not reach values sufficient to cause rapid damage to the inherently high strength lamellae. The situation is rather different for plasma sprayed metallic coatings because of the presence of oxide at the interlamellar interfaces, so that inherently weaker interlamellar bonding would be expected compared with oxide ceramics. This may explain the poor performance of the plasma sprayed Ni-Cr alloy and the considerable improvement achieved using a similar alloy as a fused coating.

CONCLUSIONS

The resistance to cavitational erosion of thermally sprayed coatings is apparently related to the defects between the lamellae of which the coating is made up. An initial high rate of erosion is a result of the removal of poorly bonded lamellae and spherical particles. The relatively good performance of plasma sprayed ceramic coatings after the initial period may be related to the low elastic modulus associated with the lamellar microstructure. Control of the microstructural features of coatings, for example by spraying under low pressure conditions, heat treatment or surface melting could lead to improved resistance to cavitational damage.

ACKNOWLEDGEMENTS

All tested coatings were prepared and supplied by Brenco Services, Melbourne.

REFERENCES

Adamski, A., Stachurski, Z. and Stecki, J. 1984, Wear 97 129-137
Akhtar, A. 1982, Materials Performance. Aug. 15-18
Boch, P., Fargeot, D., Gault, G. and Platon, F.P. 1981, Rev. Int. Hautes Temp. Refract. 18 85-87
Dorfman, M., Vargas, J., Clayton, C. R. and Herman, H. 1979, Thin Solid Films. 64 351-357
Frees, N. 1983, Wear. 88 57-66
Hansson, I. and Mørch, K. A. 1980, J.Appl.Phys. 51 4651-4658
McPherson, R. 1980, J.Mater.Sci. 15 3141-3149
McPherson, R. and Shafer, B. V. 1982, Thin Solid Films. 97 201-204
McPherson, R. 1984, Thin Solid Films. 112 89-95
Mørch, K. A. 1979, Treatise on Materials Science and Technology. 16 309-355
Preece, C. M., Vaidya, S. and Dakshinamoorthy, S. 1979, Erosion: Prevention and Useful Applications. ASTM 409-433
Preece, C.M. 1979, Treatise on Materials Science and Technology. 16 249-308
Scott, D., Seifert, W. W. and Westcott, V. C. 1975, Wear. 34 251-260
Vaidya, S. and Preece, C. M. 1977, Metallurgical Transactions A. 9A 299-307
Vaidya, S. and Preece, C. M. 1977, Scripta Metallurgica. 11 1143-1146
Vyas, B. and Preece, C. M. 1976, J.Appl.Phys. 47 5133-5138
Vyas, B. and Preece, C. M. 1977, Metallurgical Transactions A. 8A 915-923
Wade, E. H. R. and Preece, C. M. 1978, Metallurgical Transactions A. 9A 1299-1310

Inst. Phys. Conf. Ser. No. 75: Chapter 9
Paper presented at 2nd Int. Conf. Science Hard Mater., Rhodes

Fundamentals of the wear of hard materials

D H BUCKLEY(1) AND E RABINOWICZ(2)

(1) NASA Lewis Research Center, Cleveland, Ohio 44135
(2) Massachusetts Institute of Technology, Cambridge, Massachusetts

ABSTRACT

The wear of hard materials is best understood by initially identifying the operable mechanism of wear relative to a particular surface. Properties shown to influence the wear of materials include the presence or absence of crystallinity, crystal orientation, anisotropy in mechanical properties as well as surface chemistry and changes therein. Segregation of species from the bulk to the surface and bond dissociation in the surficial layers influences wear. Molecular-orbital energy and density of state studies for ceramic oxides and diamond respectively when in contact with metals provides insight into observed adhesion, friction and wear. A reduction in wear of hard materials can be accomplished even with adsorbed gases such as with H on diamond. Some solid lubricants because of their anisotropic hardness characteristics can under certain conditions act as abrasives, even wearing hard steels.

INTRODUCTION

Whenever two solids are in contact, under load and with relative motion, wear to one or both solids can and usually does occur - even under well lubricated conditions. The nature and quantity of wear will depend upon the materials in contact, the composition of their surfaces and the environment in which they are found.

Wear occurs to the hardest of materials. Diamond, for example, can wear just as other materials (Bowden and Tabor, 1964). Hardness is important in the calculation of the wear of materials (Rabinowicz, 1972). At asperity contacts even extremely hard materials will undergo plastic deformation and wear similar to less hard materials (Buckley, 1967). The chemistry at the surface of very hard materials such as ceramics is important just as for softer metals (Duwell and Butzke, 1964; Brown, Eiss and McAdams, 1964).

Fundamental studies on the wear of hard materials (hardness greater than 10 GPa) have been conducted by a number of investigators. Studies have been conducted on diamond (Bowden and Tabor, 1964), aluminum oxide (Duwell and Butzke, 1964; Steijn, 1961), rutile (Duwell, 1969; Steijn, 1969), silicon carbide (Miyoshi and Buckley, 1982), SiN_4 (Rabinowicz, 1976), metal carbides (Tipnis, 1980), and other solids (Buckley, 1972) by way of example.

The objective of this paper is to review the fundamental nature of the wear of hard materials. Representative materials to be discussed in order of increasing hardness will include metallic glasses, SiC, TiO_2, Al_2O_3 and diamond. These hard materials will be examined in contact with themselves, harder materials and with metals. The basic interfacial bonding mechanisms, anisotropy and surface chemical effects will be discussed as they relate to wear behavior.

WEAR MECHANISMS

There are certain well recognized forms of wear. These include adhesive, abrasive, corrosive, erosion, cavitation and fatigue (Rabinowicz, 1965). Adhesive wear happens when adhesion occurs at the interface between two solids and the bonding is stronger at the interface than in the cohesive bonds in the cohesively weaker of the two materials. Abrasion, as a wear process, takes place when one material is harder than the other and the asperities from its surface shear those of the softer surface. This is referred to as two body abrasion as distinguished from three body abrasion where a hard third body (i.e. dirt particle) is trapped at an interface (Rabinowicz, Dunn and Russell, 1961). In three body abrasion the third body is harder than either or at least one of the two mating solids.

Corrosive wear occurs where surfaces react excessively with constituents of the environment or the lubricant reacts excessively with the solid surfaces accelerating material loss due to simple mechanical surface interactions. Some forms of lubrication are really controlled corrosion, for example, extreme pressure lubrication where lubricant additives are intended to chemically interact with the surfaces.

Erosion of solids will take place when solid particles are caused to impinge on the surface. The energy associated with the incoming particles is of a sufficient quantum to remove material. Material hardness has been thought to affect wear rates with this form of wear.

Cavitation loss of material occurs when gases trapped in fluids generate into bubbles with mechanical agitation and these bubbles collapse against the surface of solids. The energy of collapse is transferred to the solid and is of sufficient magnitude so as to result in material removal.

Wear by the process of fatigue happens when repeated cycles or traversals are made across a surface as experienced, for example, in a rolling element. Repeated stress cycles cause, after some period of time depending upon the material, cracks to develop in the material - either surface or subsurface. These cracks grow and eventually cause the loss of particles from the solid.

While hardness per se does not control the occurrence of adhesive wear it does affect wear particle size (Sasada, Norose and Mishina, 1979). Abrasive wear does definitely relate to hardness - with wear resistance increasing with hardness (Kruschov and Rabichev, 1960). Erosion (Rao and Buckley, 1984) and cavitation (Rheingans, 1959) both involve hardness and harder materials experience less wear via these mechanisms.

RESULTS AND DISCUSSION

Amorphous Alloys

In 1960 it was shown that by very rapidly quenching certain alloy compositions from the melt, amorphous solid phases could be formed. This
discovery launched a new field of research activity. Well over 200 alloy
systems now have been identified as being capable of quenching into the
amorphous state. These alloys are referred to as amorphous alloys or
metallic glasses.

Amorphous metals constitute, in effect, a totally new class of materials
with physical properties that are distinctly different from those of the
corresponding crystalline alloys. Amorphous alloys are currently being
used in the aerospace industry and offer an outstanding combination of
high-temperature strength, fatigue properties, and oxidation and corrosion
resistance. Amorphous alloys are also becoming important as magnetic
materials for highly developed magnetic recording devices (e.g. video tape
recorders and computer memory devices). Amorphous alloys can also be used
in foil bearings.

While mechanical strength, ferromagnetic effects, and corrosion behavior
offer the most promise at this time, others, namely the tribological
properties of amorphous metals are also of interest; they are hard
($\sim$10 GPa), yet, unlike silicate glasses they possess substantial plasticity. They are among the strongest known engineering materials, and
resist the propagation of cracks. Very little research has been done,
however, on the tribological properties of amorphous alloys (Miyoshi and
Buckley, 1982).

The surface of ferrous-based amorphous alloys were analyzed by XPS.
The surfaces of amorphous alloys contain, in addition to the alloying
elements, oxides and an adsorbed film of O and C. To investigate the role
of the oxide film on wear, sliding experiments were conducted with three
amorphous alloys in contact with a 6.4 mm dia Al_2O_3 spherical rider in an
Ar atmosphere. Comparisons were made with 304 stainless steel, which is a
very common and practical material.

When the amorphous alloy and 304 stainless steel slid against the Al_2O_3
rider at a load of 2.5 N for 30 min, very little difference in the
friction behavior resulted between the two alloys. The wear results were,
however, markedly different. No detectable wear on the surface of the
amorphous alloy was found by optical and scanning electron microscopic
examinations. However, considerable wear to the 304 stainless steel
surface was observed as indicated in the scanning electron micrograph of
Fig. 1. Considerable plastic flow occurred, a copious amount of wear
debris was generated on the 304 stainless steel, and lumps of metal
appeared in the wear track. The surface oxide layers present on the
amorphous alloy clearly provide a protective film against wear. The
amorphous alloys have, therefore, superior wear resistance to the
crystalline stainless steel alloy.

Since no visible wear was observed on the amorphous alloy, sliding
friction experiments were conducted with a smaller spherical rider (3.2 mm
dia), with sliding times extending to 150 h to provide a high contact
pressure and more severe experimental conditions. At a sliding period of
150 h, considerable plastic flow occurred, and much oxide debris was

Fig. 1 Photomicrograph of 304 stainless steel wear surface.
Rider, Al_2O_3; load, 2.5 N; dry sliding in an Ar atmosphere;
sliding velocity, 0.3 mm/s.

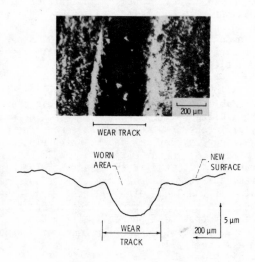

Fig. 2 Scanning electron micrograph and surface profile of wear track
on $Fe_{67}Co_{18}B_{14}Si_1$ amorphous alloy after a sliding period of 150 h.
Rider, 3.2 mm dia Al_2O_3 sphere; sliding velocity, 1.5 mm/s;
sliding distance, 810 m; room temperature, laboratory air atmosphere.

generated on the amorphous alloy (Fig. 2). The wear rate, which is defined as the quantity of amorphous alloy removed under a unit load and with a unit distance of sliding, was 5×10^{-9} mm^3/N mm or less. The transmission electron microscopy and diffraction studies indicated that the worn surface of the amorphous alloy can be crystallized during the sliding process. Crystallization of the worn surface causes high friction.

Fig. 3 presents the friction data obtained for the amorphous alloy both with a new (as-received) surface and a worn surface with 150 h of sliding together with the relevant electron diffraction patterns. The diffraction pattern of the new surface indicates that the amorphous alloy was not completely amorphous but contained extremely small grains of a few nanometers in size. The electron diffraction patterns of the worn surface clearly indicate that crystallization results from sliding friction on the amorphous alloy surface. Crystallites with sizes ranging from 10 to 150 nm are produced on the worn surface of the amorphous alloy during sliding. The difference in friction between the new and worn surface is primarily based on crystallization characteristics. The coefficient of friction of the amorphous alloy increased with crystallization of the alloy. To avoid the crystallization of the worn surface and to prevent the high friction, designs such as optimizing contact pressures and temperatures can be used to achieve the desired performance level of the wear resistance with the amorphous alloys.

Amorphous alloys are expected to be used for high-temperature tribological applications. To quantify the tribological properties of amorphous alloys at high temperature, sliding friction experiments were conducted in an ultrahigh vacuum system with the amorphous alloys in contact with an Al_2O_3 rider.

Table 1 summarizes the composition of the surficial layer of the amorphous alloy analyzed by XPS. Generally, the XPS results indicated that the surface which was cleaned by Ar ion-sputtering and heated to 623 and 1023 K, consisted of Fe, Co, B, Si, C, and O. The relative concentrations of the various constituents were very different at the surface from the nominal bulk compositions as indicated in Table 1. Contaminants can come from the bulk of the material to the surface upon heating and impart the presence of B_2O_3 and SiO_2 at 623 K, and BN above 773 K.

Fig. 4 presents the coefficients of friction as a function of the sliding temperature of the amorphous alloy specimen. Typical XPS spectra of boron obtained from the amorphous alloy at 623 and 1023 K are shown in Fig. 5. In Fig. 4 the coefficients of friction for clean amorphous alloy in contact with clean Al_2O_3 oxide are in the range of 1.0 to 1.5. The coefficients of friction increased with increasing temperature from room to 623 K. The increase in friction is primarily due to an increase in the adhesion, which results from segregation of the oxides (Fig. 5), to the amorphous alloy surface. Above 773 K the coefficient of friction decreased dramatically. The rapid decrease in friction above 773 K correlated with the segregation of BN onto the surface of the amorphous alloy, as clearly indicated by the spectrum of Fig. 5.

Thus, the segregation of compounds such as B_2O_3 and BN to the metallic surface is responsible for the observed friction behavior. BN is a solid lubricant, and its presence on the amorphous alloy surface at high temperature provides lubrication; it has, therefore, an important function in the wear reduction.

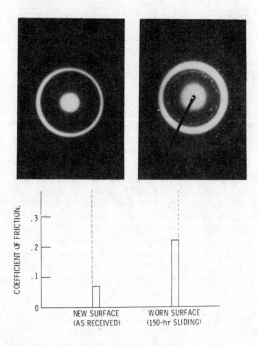

Fig. 3 Coefficients of friction and electron diffraction patterns for
new and wear surface (sliding period of 150 h) of $Fe_{67}Co_{18}B_{14}Si_1$
amorphous alloy. Rider, 3.2 mm dia Al_2O_3 sphere;
sliding velocity, 1.5 mm/s room temperature.

TABLE 1   Composition of Metallic Glass Surficial Layer
(Manufacturer's Analysis)

Nominal bulk composition, at %: $Fe_{42}Co_{11}B_{46}S_1$

Treatment	Composition on surface, at %
Ar ion sputtering	$Fe_{49}Co_{14}B_{17}Si_6C_9O_5$
Heating to 623 K	$Fe_{52}Co_{11}B_{18}Si_6C_8O_5$
Heating to 1023 K	$Fe_{54}Co_9B_{19}Si_5C_8O_5$

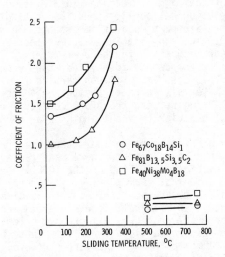

Fig. 4 Coefficient friction as a function of temperature for Al_2O_3 sliding on $Fe_{67}Co_{18}B_{14}Si_1$, $Fe_{81}B_{13.5}Si_{3.5}C_2$, and $Fe_{40}Ni_{30}Mo_4B_{18}$ in vacuum. Normal load 0.2 N; sliding velocity, 0.05 mm/s; vacuum, 10 nPa.

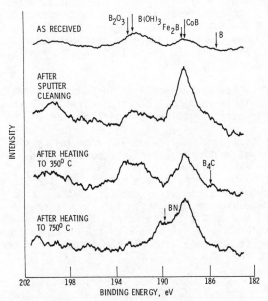

Fig. 5 Representative, B_{1s}, XPS peaks on $Fe_{67}Co_{18}B_{14}Si_1$ surface.

Carbides

Since fracture is one of the main limitations to a wider use of hard
ceramic materials in tribological applications, understanding and new
developments in this area are particularly important. To reduce fracture
wear in practical tribological applications, it is important to understand
the various fracture wear mechanisms for hard materials. If the mechanism
of wear in a particular system can be identified, then some significant
reduction in wear is often feasible. Silicon carbide (SiC) is one of the
principal hard ceramic materials currently being considered for wear
resistant applications. Studies were therefore conducted on single
crystal SiC tribological behavior (Miyoshi and Buckley, 1982).

Coefficient of friction and widths of the permanent wear grooves in plastic flow accompanied with surface cracking were measured as a function of
the crystallographic direction of sliding on the (0001), (10$\bar{1}$0), and
(11$\bar{2}$0) lanes of SiC for a conical diamond rider in mineral oil. The
apical angle of the conical rider was 117±1° and the radius of curvature
at the apex was less than 5 μm. Mineral oil was used to minimize adhesion.
Fig. 6 indicates the correlation of crystallographic orientation with the
coefficient of friction and groove width (wear).   Thus, the [11$\bar{2}$0]
direction on the basal (0001) plane has the larger groove, primarily as a
result of wear by plastic flow and is the direction of high friction for
this plane. The <0001> directions on the (10$\bar{1}$0) and (11$\bar{2}$0) planes have
the greater groove width (wear) and likewise are the directions of high
friction when compared with the <11$\bar{2}$0> on the (10$\bar{1}$0) plane and the <10$\bar{1}$0>
on the (11$\bar{2}$0) plane.

The contact pressure calculated from the data of the groove width is also
represented in Fig. 6, together with the Knoop hardness obtained by
Shaffer (1964). The anisotropies of this contact pressure and Knoop
hardness clearly correlate with each other. The anisotropies of friction,
wear, (groove width), contact pressure, and Knoop hardness on the (0001),
(10$\bar{1}$0), and (11$\bar{2}$0) planes of SiC are primarily controlled by the slip
system (10$\bar{1}$0) [11$\bar{2}$0] and are explained by a resolved shear stress calculation (Amelinckx, Straumane and Webb, 1960; French and Thomas, 1965;
Brookes, O'Neill and Redfern, 1971).

An increase in surface temperature of a solid tends to promote surface
chemical reactions. These chemical reactions cause products to appear on
the surface which can alter adhesion, friction and wear.

Fig. 7 presents the surface chemistry of SiC analyzed by XPS. The
as-received crystal was heated at various temperatures in a vacuum.

The Si_{2p} photoelectron peak energies are associated with SiC at various
temperatures. The photoelectron lines for C_{1s} of the SiC surface are split
asymmetrically into double energy peaks. The results show a significant
influence of temperature on the SiC surface. The double energy peaks are
due to distinguishable kinds of carbon: (1) a C contamination peak and a
carbide peak at room temperature, and (2) the graphite and the carbide
peaks at temperature from 673 to 1773 K.

At room temperature and 523 K, the primary peaks were the adsorbed amorphous C contamination and carbide. For specimens heated to 673 K the C
contamination peak disappears from the spectrum. Above 673 K both the
peak heights of graphite and the carbide are increased with an increase of

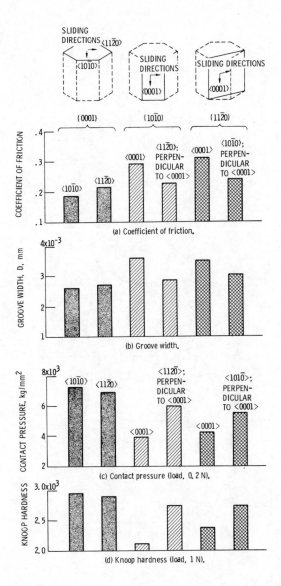

Fig. 6 Anisotropies on {0001}, {10$\bar{1}$0}, and {11$\bar{2}$0} surfaces of SiC.

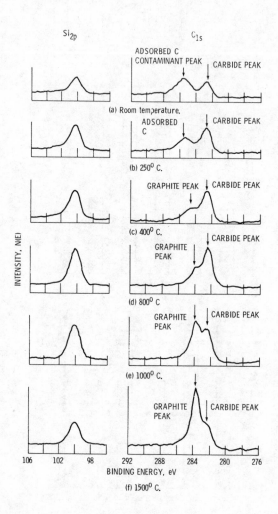

Fig. 7   Representatives Si_{2p} and C_{1s} XPS peaks on SiC {0001} surface
preheated at various temperatures to 1500°C.

heating temperature. A large carbide peak was distinguished at 1073 K. At 1173 K the carbide peak height was smaller than that at 1073 K, but the graphite peak height was larger. At 1273 K the height of the carbide peak decreased and became smaller than that of the graphite. At 1773 K the height of carbide peak became very small, but a very large graphite peak was observed. The results indicate that temperature affects significantly the surface chemistry as the surface of SiC graphitizes.

Three questions arise as a result of the foregoing observations:
(1) How thick is the graphite layer formed on the SiC?
(2) What is the graphitization mechanism for SiC?
(3) Does the graphite affect the friction and wear properties of SiC?

The thickness of the outermost surficial graphite layer on the SiC heated to 1773 K is 1.5 to 2.4 nm. This result suggests that the collapse of the C is the most probable mechanism for the graphitization of the SiC surface (Fig. 8).

A complete elemental depth profile for the SiC surface heated to 1773 K is presented as a function of sputtering time in Fig. 9. The graphite peak decreases rapidly in the first 30 min of sputtering and thereafter decreases gradually to 18 h. The depth at the sputtering time of 18 h is about 100 nm. From these results the depth of the mixture of graphite and SiC is of the order of 100 nm (Fig. 10).

To answer the third question, sliding friction and wear experiments were conducted with SiC in contact with Fe in vacuum. Fig. 11 indicates the coefficients of friction of the SiC surface in contact with Fe as a function of sliding temperatures. The Fe rider was sputter-cleaned with Ar ions. The SiC was in the as-received state. After the specimen was heated to the sliding temperature, the coefficient of friction increased slightly with increasing temperature at temperatures below 673 K. The coefficient of friction decreased with increasing temperature in the range 673 to 873 K. The general decrease in friction at these temperatures was due to the removal of the contaminants C and O from the surface. The coefficient of friction increased with increasing temperature in the range 873 to 1073 K. The increase in friction at these temperatures was associated with increased adhesion and plastic flow in the area of contact. Above 1073 K the coefficient of friction decreased rapidly with an increase of temperature. The rapid decrease in friction above 1073 K correlated with the graphitization of the SiC surface, as already discussed.

Once the graphitized layer is formed on the SiC surface, one further question arises: Will the graphite layer produced on the SiC surface above 1073 K influence the friction behavior at all temperatures? To answer this question the SiC surface was heated to 1773 K and then cooled to room temperature. The surface was graphitized. The SiC specimen was then reheated to the sliding temperature before the friction experiment was initiated. Friction experiments were conducted on the reheated surface of the SiC. The Fe rider was Ar ion sputter-cleaned.

The friction and wear properties of SiC (0001) surfaces in contact with Fe were measured at temperatures to 1473 K. The coefficients of friction are generally lower as is the wear after the surface has been graphitized.

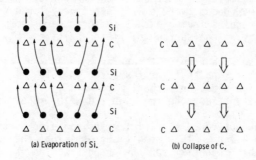

(a) Evaporation of Si.   (b) Collapse of C.

Fig. 8 Graphitization of SiC surface.

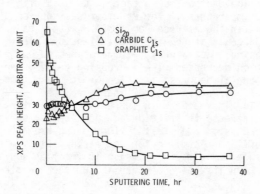

Fig. 9 Elemental depth profile of SiC {0001} surface
preheated at a temperature of 1500°C for 1 h.

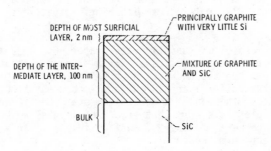

Fig. 10 Surface condition of SiC after heating at
temperatures above 1200°C.

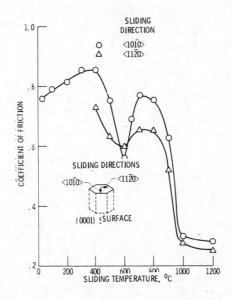

Fig. 11 Effect of temperature on coefficient of friction
for SiC (0001) surface sliding against an iron rider.
The Fe rider was Ar ion sputter cleaned before experiments.
Normal load, 0.2 N; vacuum, 10 nPa.

The low friction and wear at the high temperatures appears to correlate with the graphitization of the SiC surface. The coefficients of friction on this surface at the high temperatures are nearly the same as those on pyrolytic graphite in sliding contact with single-crystal Fe in the vacuum of 10nPa.

The removal of adsorbed films (usually water vapor, Co, CO_2, and oxide layers) from the surfaces of ceramics and metals results in very strong interfacial adhesion when two such solids are brought into contact. For example, when an atomically clean SiC surface is brought into contact with a clean metal surface, the adhesive bonds formed at the SiC-to-metal interface are sufficiently strong that fracture of cohesive bonds in the metal and transfer of metal to the SiC surface results. This is observed in the scanning electron microscope.

Fig. 12 presents scanning electron photomicrographs of the wear tracks generated by ten passes of Rh and Ti riders on the SiC (0001) surface along the $[10\bar{1}0]$ direction. Metal transfer is evident in the sliding contact. The sliding of a metal on a SiC surface also results in local cracks along cleavage planes. The cracks, which are observed in the wear tracks, primarily propagate along cleavage planes of the $\{10\bar{1}0\}$ orient-ation. In Fig. 12(a), the hexagonal light area is the beginning of a wear track, and there is a large crack where cracks primarily along the $\{10\bar{1}0\}$ planes were generated, propagated and then intersected during loading and sliding of the Rh rider on the SiC surface. It is postulated from Fig. 12(a) that subsurface cleavage cracking of the (0001) planes, which are parallel to the sliding surface, also occurs. Fig. 12(b) reveals a hexagonal pit and a copious amount of thin Ti film around the pit. The hexagonal fracturing is primarily due to cleavage cracking along $\{10\bar{1}0\}$ planes and subsurface cleavage cracking along the (0001) plane and, accordingly, the generation of wear debris is observed.

Oxide Ceramics

In addition to carbides such as SiC, oxides like TiO_2 (rutile), Al_2O_3 and MgO have been considered for wear resistant applications. With these hard materials, both structure and surface chemistry are also extremely important to wear behavior.

The anisotropic wear behavior for TiO_2 has been measured (Duwell, 1969). The change in wear rate with crystallographic orientation is presented in Fig. 13. A seven-fold difference in wear rate is observed with a change in orientation. The rate of wear was influenced not only by crystallo-graphic plane but direction as well. For example, when sliding on the (001) surface the least amount of wear occurs in the [100] direction (Duwell, 1969). Thus, wear for TiO_2 just as for SiC is sensitive to both plane and direction of rubbing or sliding.

When hard oxide ceramics are in solid state contact with softer materials such as metals the marked difference in elastic and plastic deformation of the two materials can result in plowing of the softer material. This, then can contribute to the wear of one or both materials. In Fig. 14 a rider (hemisphere) of sapphire slid on a single crystal flat of Cu. The specimen materials were then reversed so that a single crystal Cu rider slid on a sapphire flat. The coefficient of friction for the sapphire sliding on Cu was 1.5. With Cu sliding on sapphire it was 0.2. In both instances, adhesion of Cu to sapphire occurred. The difference in

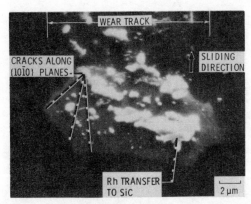

(a) Hexagonal cracking.

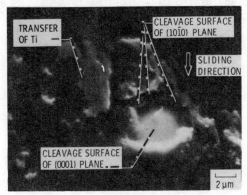

(b) Hexagonal pit.

Fig. 12 Scanning electron photomicrographs of wear tracks
on the {0001} surface of single-crystal SiC in contact with Rh
and Ti as a result of ten passes of a rider in vacuum.
Sliding direction, <101̄0>; sliding velocity, 3 mm/min;
load, 0.3 N; room temperature; pressure, 30 nPa;
metal pin rider, 0.79 mm radius.

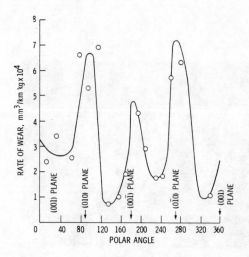

Fig. 13 Rate of wear of a rutile single-crystal sphere on a great
circle in the plane of the a- and c-axes is normal to a plane of
sliding at 0 and 180°. Slide direction in plane of the great circle
(Powell, 1969).

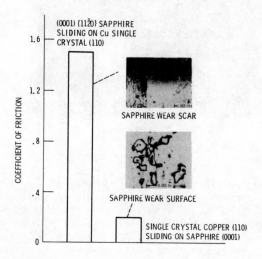

Fig. 14 Friction for Cu in sliding contact with sapphire
on vacuum 30 nPa; load, 0.98 N; sliding velocity, 0.13 mm/s.

friction coefficient for the two experiments is due to the effects of plowing (Buckley, 1967).

In both experiments the sapphire underwent wear as indicated in the photo-micrographs of Fig. 14. The wear on the sapphire flat was occasioned by fracture along (0001) planes and subsurface and parallel to the sliding interface. When metals contact hard oxide ceramics, surface chemistry plays a very important role in the observed friction and wear behavior. Various metals were slid on a flat of sapphire with the basal orientation in the sapphire parallel to the sliding interface. With the metals which form stable oxides such as Cu, Ni, Re, Co and Be, adhesion of the metal occurred to the O ions in the outermost atomic layer of the sapphire. The manner of bonding is shown in Fig. 15.

When the same sliding experiments are conducted with metals that do not form stable oxides in vacuum (Au and Ag) strong chemical bonding does not occur at the interface, adhesion is weak and wear to the hard oxide ceramic is absent. This is demonstrated in the data of Fig. 16 for Au and Ag sliding on the basal orientation of sapphire (Buckley, 1967).

The photomicrographs of Fig. 16 indicate an absence of any wear to the sapphire such as observed in Fig. 14 with Cu. The only surface markings in Fig. 16 were polishing scratches. The coefficients of friction were also one-half in Fig. 16 of that observed with Cu sliding on sapphire in Fig. 14. In Fig. 14 fracture occurs in the sapphire because in the interfacial metal to the sapphire bond strength is greater than the cleavage or fracture strength in the sapphire and accordingly sapphire wear occurs. With Au and Ag in Fig. 16, the weakest region is the interface, and simple shear takes place there with bond strength no stronger than that found with lubricant molecules - as evidenced by the friction coefficient of 0.1 comparable to that experienced with effective boundary lubrication.

Experiments subsequent to those of Buckley (1967) confirmed a chemical bond between metals and the O ions indicated in Fig. 15 (Pepper, 1976). The shear strength of the metal to sapphire contact were correlated with the free energy of formation of the metal oxide (Pepper, 1976).

Further attempts have been made to explain in a more fundamental manner the metal to sapphire bonding. This could assist in understanding the wear of hard ceramic oxides in general. Molecular-orbital energies have been examined for clusters in bulk sapphire and the metal-sapphire interface (Johnson and Pepper, 1982). Figs. 17 and 18 represent an attempt to explain the bonding mechanism in greater detail (Johnson and Pepper, 1982).

The primary interaction at the metal-sapphire interface, as revealed by these studies, occurs between the metal d orbitals and otherwise nonbonding p orbitals of the O ions at the surface of the sapphire crystal, i.e. the nonbonding p orbitals at the top of the valence band. It will be noticed in Fig. 17 that the d-orbital energies of the isolated metal atoms, as determined by this method, are in close proximity to the sapphire valence band, and the position of the atomic level relative to the top of the valence band changes systematically through the series Fe, Ni, Cu and Ag. The metal-sapphire contact interaction produces at the interface manifolds of spatially localized occupied metal (d)-O(p) bonding molecular orbitals of energies near the bottom of the sapphire valence band and metal (d)-O(p)

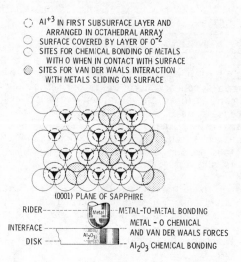

Fig. 15   Nature of surface interaction and bonding of metal to Al_2O_3.

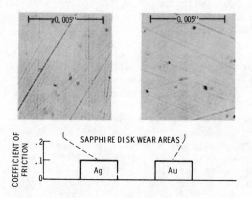

Fig. 16   Coefficient of friction for Au and Ag riders sliding on
sapphire in vacuum (30 nPa).   Sliding velocity, 0.13 mm/s;
ambient temperature, 25°C; duration, 1 h.

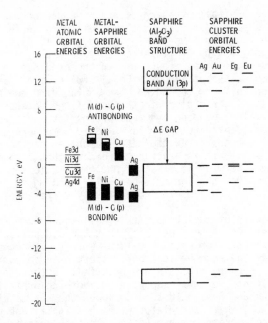

Fig. 17 Molecular-orbital energies, as determined by the self-consistent-field X-alpha scattered wave method, for clusters representing bulk sapphire and metal-sapphire interfaces (ref. 28).

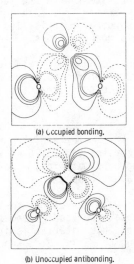

(a) Occupied bonding.

(b) Unoccupied antibonding.

Fig. 18 Fe(d) - O(p) molecular-orbital wave-function contour maps for a Fe atom supported on sapphire, plotted in the plane of the Fe atom and two surface O atoms. The solid and dashed contours represent the positive and negative phases of the wave function (Johnson and Pepper, 1982).

antibonding molecular orbitals of energies near the top of the sapphire valence band - as exemplified for Fe by the bonding and antibonding orbital wave-function contour maps in Figs. 18(a) and 18(b), respectively. For Fe and Ni, the antibonding orbitals are only partially occupied and are located well above the valence band within the band gap, as shown in Fig. 17.   For example, the antibonding Fe(d)-O(p) orbital mapped in Fig. 18(b) is unoccupied. For Cu and Ag, the antibonding orbitals are fully occupied and are located close to the top of the valence band. In addition to these covalent bonding and antibonding interactions between the metal and sapphire substrate there is an ionic component associated with metal-to-O charge transfer at the metal-sapphire interface. These results on covalent and ionic interactions indicate that a chemical bond is in fact established between metal atoms and the O anions on the sapphire surface, and that the ionic component of bonding is proportional to the metal(d) and O(p) orbital electronegativity difference.

Relative to the strength of this bond, it is well known from the most elementary principles of quantum chemistry that the occupation of anti-bonding molecular orbitals tends to cancel the effects of occupied bonding orbitals, and reduces the net chemical bond strength in comparison to the situation where only bonding orbitals are occupied.   Therefore, the increase in occupancy of the metal-sapphire antibonding orbitals through the series Fe, Ni, Cu and Ag should tend to lower the net metal-sapphire chemical bond strength, which correlates with the significant reduction in metal-sapphire contact strength measured in adhesion, friction and wear experiments.   The decrease of the friction coefficient from 0.2 for Cu to 0.1 for Au or Ag is explained qualitatively by the combined effects of increasing antibonding orbital occupancy and decreasing metal-to-O charge transfer.   The relatively small friction coefficients and shear strengths of Au (0.1) and Ag (0.1) are consistent with the fact that the fully occupied antibonding orbitals shown in Fig. 17 essentially cancel the covalent contributions of the bonding orbitals, leaving only small residual ionic and van der Waals contributions.

Similar studies should be conducted on other oxide ceramics to determine if a general relationship between interfacial bonding and fracture can be developed. This could lead to a better understanding of wear.

Diamond

In a discussion of the wear of hard materials, the hardest of materials, diamond, (Mohs, 1825) must be considered.   If an arbitrary number of 10 GPa is selected for defining hard materials, then diamond exceeds that hardness by a factor of ten (Tabor, 1956).

Diamond because of its superior wear resistance to many other materials has been heavily used in the metal fabrication industry as a tool. One of the difficulties with the use of this material is its reaction with carbide forming elements leading to rapid wear (Schey, 1983).

The wear of diamond when in contact with other solids has not only been affected by interaction with contacting solids which form carbides but also can be affected by thermal oxidation, mechanical abrasion and the transformation to another form of C, either amorphous or graphite (Bowden and Tabor, 1964).

Environment strongly influences the wear behavior of diamond. The data of Table 2 taken from Bowden and Tabor (1964) indicates that the wear rate for diamond is least in O and greatest in its absence (N). This occurs despite the possible oxidative degradation of the diamond surface at the high load and sliding speed indicated in Table 2 which would lead to high surface temperatures.

In Table 2 the diamond is sliding on glass and during the rubbing process diamond wear fragments were observed to transfer to the glass surface. The glass eventually becomes charged with diamond and diamond is rubbing against diamond.

Studies have been conducted with metals contacting diamond just as those already reported herein for SiC and Al_2O_3. It has been found that strong bonding between metals and the diamond surface can damage diamond. Even H will seriously retard bonding (Pepper, 1982).

A detailed study of the bonding of the metal, Cu to the diamond surface has been conducted which can be related to the friction behavior of metal-diamond contacts. A correlation exists between bonding and unoccupied surface states in the band gap of the diamond surface.

The correlation of high interfacial strength with the presence of unoccupied surface states in the band gap leads to consideration of a possible interfacial chemical bond based on the energy level diagram of Fig. 19 for Cu in contact with diamond. First, from elementary molecular-orbital theory of chemical bond formation, a bond is formed by partially occupied orbitals of similar energy. The orbitals that constitute the valence band of diamond are fully occupied, while the orbitals that con-stitute the conduction band are empty. Thus, for the diamond surface, without gap states, the Fermi level electrons - those electrons with the highest energy in the metal - must interact with the (empty) conduction band orbitals in the diamond to form a chemical bond. It may be that the energy difference between the bottom of the conduction band and the Fermi level is too large to allow a bond to be formed and this may account for the low friction and little material transfer. On the other hand, the unoccupied states in the band gap of the annealed surface lie much closer in energy to the Fermi level and this smaller energy difference may allow a bond to be formed by the Fermi level metal electrons and the empty gap states. Such a bond would increase interfacial strength and lead to higher friction and greater material transfer. Since electron bombardment of diamond generates unoccupied gap states, interfacial bonds should be possible after this treatment as well, leading to the observed higher friction and material transfer. Note that this chemical bonding explan-ation requires the energy of the unoccupied surface states to be in the band gap of the crystalline ground state.

Lubrication and Wear

The presence of a lubricant on a solid surface generally retards wear. Both liquid and solids have been applied to the surface of solids to reduce wear. It must be remembered, however, that adsorbed gases from the environment can act as lubricating films and even H, with its effect on diamond adhesion to metal (Pepper, 1982), in a strict sense is a lubricant.

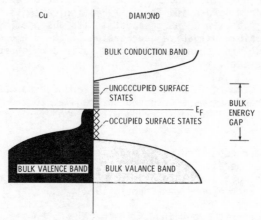

Fig. 19 Schematic representation of the density of slates for the Cu
and diamond surface near the energy gap of diamond. The diamond has
been depicted with both occupied and unoccupied surface states in
the gap, which are absent on the as-polished surface.

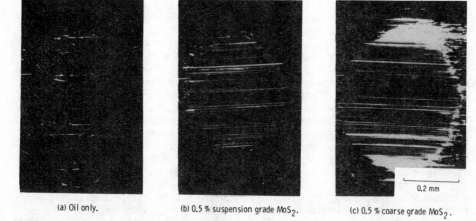

(a) Oil only.   (b) 0.5 % suspension grade MoS_2.   (c) 0.5 % coarse grade MoS_2.

Fig. 20 Effect of MoS_2 in 78% cS oil on abrasion of polished
steel ball during sliding contact.

TABLE 2 Influence of Atmosphere on the Wear of Diamond
(Bowden and Tabor, 1964)

[Load = 4.9N; sliding speed = 840 cm/s; {100}/{101}; extra dense flint glass in dry conditions]	Atmosphere (dry)	Wear rate (10^{-6} cm^3/h)
	0	464±31
	Air	476±54
	N	756±21

Both graphite and MoS_2 are hexagonal layer lattice solids, with generally easy shear along basal planes providing for good lubrication. Because of their structures, however, these materials have anisotropic mechanical properties. For example, hardness of 7-8 Mohs have been reported for MoS_2 (Lancaster, 1966) presumably this hardness is along the prismatic orient- ation of the crystallites.

While graphite, MoS_2 and liquids separately provide resistance to wear, mixing solid and liquid lubricants can in fact promote wear. This is demonstrated in the photomicrographs of Fig. 20.

A hardened bearing steel ball was brought into contact with a glass plate. The contact region between the ball and glass plate was viewed, with an ordinary optical microscope, through the glass. The glass plate was caused to rotate and mineral oil as well as a mixture of mineral oil and MoS_2 were used to lubricate the contact.

With oil lubrication of the glass-steel contact very little surface damage occurred with sliding (Fig. 20(a)). The addition of MoS_2 to the oil produced serious abrasive wear to the steel surface (Figs. 20(b) and (c)). These abrasion studies by what are normally considered solid lubricants have been reported for MoS_2 and for graphite (Cusano and Sliney, 1981).

The use of the layer lamellar solids in the dry state allow for the basal planes to be parallel to the sliding interface with easy shear along these planes. In the fluid, however, these crystallites can align such that the crystallite edges cut into the hardening bearing steel because of the increased mobility provided by the fluid.

CONCLUDING REMARKS

The wear behavior of hard materials like that of other solids is highly dependent upon a number of fundamental considerations. Understanding the particular wear mechanism involved can elucidate material properties important to the wear process.

The presence or absence of crystallinity, and the structure in crystalline solids influences wear. With metallic glasses transformation from the amorphous to the crystalline state results in an increase in wear and friction. Wear in the amorphous state is extremely low.

With hard materials in the crystalline state, wear behavior is aniso-tropic. The data for SiC and TiO_2 demonstrate the dependency of wear on orientation. Wear is generally lower on the highest atomic density crystallographic planes when sliding in the most closely packed directions.

Surface chemistry plays an important role in wear behavior of hard materials. With the metallic glasses, the surface segregation of elemental species and subsequent combination to form inorganic compounds, such as BN, reduce both the wear and friction. With materials such as SiC the Si to C bond scission in the surface layers, with the subsequent graphite formation, reduces friction but consumes surface layers of the SiC in a form of wear.

Studies with hard oxide ceramics, such as Al_2O_3, indicate the importance of surface atomic bonding mechanisms - both to the wear of the ceramic and the ceramic in contact with a metal. With metals contacting the oxide a consideration of molecular-orbital energies and the resulting chemical bonds provide insight into adhesion and adhesive transfer. When the diamond surface is considered with contacting metals the density of state near the energy gap of diamond lends insight into the nature of interfacial bonding which ultimately influences adhesion, friction and wear behavior.

Lubrication of hard materials generally reduces wear. In a strict sense even adsorbed H monolayers are lubricants for such surfaces as diamond. Some lubricants, for example MoS_2 when suspended in oil will, because of anisotropic hardness characteristics, act as abrasives and abrasively wear hardened bearing steels.

REFERENCES

Amelinckx, S., Straumane, G. and Webb, W.W. 1960, J. Appl. Phys. 31 8 1359-1370

Bowden, F.P. and Tabor, D. 1964, The Friction and Lubrication of Solids, Part II, Oxford at the Clarendon Press, 180-184

Brookes, C.A., O'Neill, J.B. and Redfern, B.A.W. 1971, Proc. R. Soc. A322 73-88

Brown, W.R., Eiss, N.S. Jr. and McAdams, H.T. 1964, Am. Ceramic Soc. J. 47 157-162

Buckley, D.H. 1967, ASLE Trans. 10 [2] 134-145

Buckley, D.H. 1972, Am. Ceramic Soc. Bull. 51 (12) 884-890

Cusano, C. and Sliney, H.E. 1981, ASLE Trans. 25 (2) 183-189

Cusano, C. and Sliney, H.E. 1981, NASA TM-81709, ASLE Trans. 25 (2) 190-197

Duwell, E.J. and Butzke, H.C. 1964, ASLE Trans. 7 101-108

Duwell, E.J. and Butzke, H.E. 1964, J. Appl. Phys. 35 3385-3390

Duwell, E.J. 1969, ASLE Trans. 12 34-35

French, D.N. and Thomas, D.A. 1965, Trans. Metall. Soc., AIME 233 950-952

Johnson, K.H. and Pepper, S.V. 1982, J. Appl. Phys. 53 (10) 6634-6637

Kruschov, M.M. and Babichev, M.A. 1960, (In Russian) U.S.S.R. Academy of Sciences, Moscow

Lancaster, J.K. 1966, Wear 9 169-188

Miyoshi, K. and Buckley, D.H. 1982, ASLE Trans. 26 1 53-53

Miyoshi, K. and Buckley, D.H. 1982 Wear 75 253-268

Miyoshi, K. and Buckley, D.H. 1982, NASA TM-82973

Miyoshi, K. and Buckley, D.H. 1983, NASA TP-2140

Mohs, F. 1825, English Trans. by W. Haldinger, Treatise on Mineralogy, Edinburgh, Constable and Co Ltd

Pepper, S.V. 1976, J. Appl. Phys. 47 3 801-808

Pepper, S.V. 1982, J. Vac. Sci. Technol. 20 (3) 643-646

Rabinowicz, E., Dunn, L.A. and Russell, P.G. 1961, Wear 4 345-355

Rabinowicz, E. 1965, John Wiley and Sons, Inc.

Rabinowicz, E. 1972, Wear 21 401-402

Rabinowicz, E. 1976, Wear 39 101-107

Rao, P.V. and Buckley, D.H. 1984, Journ. of Pipelines 4 193-205

Rheingans, W.J. (ed) 1959, Cavitation Damage, American Society of Mechanical Engineers, New York

Sasada, T., Norose, S. and Mishina, H. 1979, ASME (ed) K.C. Ludema, W.A. Glaeser and S.K. Rhee, New York, 72-80

Schey, J.A. 1983, Tribology in Metal Working, ASM, Metals Park, Ohio, 660

Shaffer, P.T.B. 1964, J. Am. Ceramic Soc. 47, 9, 466

Steijn, R.P. 1961, J. Appl. Phys. <u>32</u> (10) 1951-58
Steijn, R.P. 1969, ASLE Trans. <u>12</u> 21-33
Tabor, D. 1956, Brit. J. of Appl. Phys. <u>7</u> 159-166
Tipnis, V.A. 1980, Wear Control Handbook, Ed. M.B. Peterson and
    W.O. Winer, ASME, New York, 891-930

Inst. Phys. Conf. Ser. No. 75: Chapter 9
Paper presented at 2nd Int. Conf. Science Hard Mater., Rhodes

A comparative study of the wear resistance of hard materials

A BALL

Department of Materials Engineering
University of Cape Town

ABSTRACT

A number of different ceramic materials and WC cermets have been tested for their resistance to abrasion and erosion. Five different experimental methods are described and the performances of the materials have been determined for each test and ranked against that of a standard material. The extensive use of scanning electron microscopy has provided insight into the different modes of wear and discussion is focused on the microstructural factors which provide resistance to material loss.

INTRODUCTION

The loss of material by abrasive and erosive wear of critical components such as valves, dies, pump impellors and cutting and drilling tools is a factor which causes a significant reduction in the efficiency and the useful lifetime of machinery. Extremely small changes in dimensions or geometry can result in the failure of a component and downtime of a system. This problem is particularly relevant in hydraulic power systems where control valves can suffer from both cavitation erosion and flow erosion by entrained particles. Clearly the development of erosion resistant materials in conjunction with research into design configurations is a matter requiring urgent attention.

In the past most researchers have concentrated on one type of erosive or abrasive wear. Our current approach is a comparative study of the performance of the same wide range of materials when subjected to several different types of wear tests. These have included abrasive wear produced by a moving belt of abrasive particles, cavitation erosion of a specimen held in water beneath the tip of an ultrasonic drill, erosion produced by silica particles entrained in water moving at a high velocity ($250 \, m \, s^{-1}$), erosion produced by silicon carbide particles carried in an air stream moving at low velocity ($40 \, m \, s^{-1}$) and erosion produced by coal ash carried in an air stream moving at sonic velocities.

Related experiments have been undertaken on ceramic materials and cermets by previous workers. For example, Blombery, Perrott and Robinson (1974) studied the abrasion wear of WC-Co when used in drill and drag bits. Heathcock, Ball, Yamey and Protheroe (1981) assessed the resistance to cavitation erosion of a variety of Stellite alloys and cemented carbides while Conrad, Shin and Sargent (1983) examined the influence of grain size and cobalt content of WC-Co alloys when subjected to particles carried in a low velocity air stream. The erosion of hard materials by particles carried in a liquid stream has received little attention apart from the works of Shetty, Wright and Clauer (1983), Wright, Shetty and Clauer (1983) and Forse and Ball (1983). Limited success has been achieved in relating wear resistance to individual macroscopic parameters such as hardness and fracture toughness. Clearly the prediction of a material's performance is difficult and factors which must be taken into account include grain size, the presence of second phases, impurities and voids, the thermal properties of a material in relation to frictional heating and adiabatic shear processes and phase stability under impact and local heating.

This paper gathers some experimental evidence on the mode of material loss of hard materials including alumina, silicon carbide, Sialon and in particular WC-Co and WC-Ni cermets.

EXPERIMENTAL

The abrasive wear tests were carried out on a modified belt sander (for details see Allen, Ball and Protheroe (1981-82)) where the ends of 11.5 mm cylindrical samples were held with a load of 30N against a belt containing 80 grit alumina moving at $0.1 m s^{-1}$. The weight loss of a previously 'worn-in' specimen was determined after a path length of 14.64 m. The relative abrasion resistance is defined as the ratio of the volume loss of a standard mild steel (BS970 070M20) specimen to the volume loss of the hard material.

The cavitation erosion tests were carried out under distilled water at 25°C on 12 mm diameter x 3 mm thick discs using an ultrasonic signal generator (20 kHz, 75 μm amplitude) in the manner described by Vyas and Preece (1976). The weight losses of previously polished (1 μm diamond) specimens were measured after time intervals up to 10 hours and converted into volume losses.

Specimens, identical to the ones used for the cavitation tests were subjected to particle erosion. The polished specimens were held at an angle of 45 degrees at the end of a flight tube which emitted a stream of compressed air containing 100 μm diameter SiC particles. The particle velocity was determined to be $40 m s^{-1}$ and mass losses were determined after successive erosions produced by 5 g. of the erodent.

The remaining two tests were designed to simulate flow through valves. The apparatus used for liquid flow erosion has been described by Forse and Ball (1983). A pump delivers water, contaminated with a known amount of quartz dust, at a maximum pressure of 35 MPa to a test cell consisting of a conical valve and seat. (82 degree cone with a 23 mm dia.base). The velocity of the liquid and entrained quartz is controlled by adjusting the valve gap. Typical test conditions utilised water moving at a velocity of $250 m s^{-1}$ and containing a concentration of 100 ppm by weight of silica particles with diameters less than 38 μm and a mean of 8 μm. The volume loss of the valve was determined after successive time intervals up to 3 hours and in some cases up to 8 hours.

The erosion by very high velocity coal ash particles was achieved by allowing a cylinder containing compressed air and a quantity of ash to decompress via a flight tube and then a small gap between a specimen lying at 45 degrees to the stream and a 90 degree steel block. Full details are described in the thesis of Willmott (1984). On account of the geometry of the test cell the velocity of the air and entrained ash is near sonic (approximately $340 \, \mathrm{m \, s^{-1}}$). The ash, which consists mainly of silica and alumina, has a particle size range of $0 - 600 \, \mu\mathrm{m}$ and a mean of $150 \, \mu\mathrm{m}$. The weight loss is measured after an erosion period of 40 secs and converted to volume loss. The relative erosion resistance is calculated as the ratio of the volume loss of a BS970 080M40 steel to the volume loss of the specimen.

With the exception of the abrasion specimens, all samples were polished to a $10 \, \mu\mathrm{m}$ diamond paste finish prior to testing. After testing the specimens were preserved in a dessicator and subsequently coated and examined in the scanning electron microscope. The gradation of damage from the as-polished condition to the areas of severe damage provided some insight into the initiation or incubation periods of erosion. In some cases it was possible to monitor the modes of erosion by examining the specimens at intervals during testing. Optical metallography and hardness tests were undertaken on a polished cross-section of the specimens after the completion of the scanning electron microscopy. In the case of some specimens subjected to cavitation erosion an X-ray diffraction trace was obtained prior to and after erosion.

RESULTS

Abrasion

The values obtained for the relative abrasion resistance (RAR) and bulk hardness of the five compositions at three grain sizes are shown in Fig. 1. The wear resistance increases sharply for carbide contents of greater than 90 per cent and the influence of grain size becomes significant at these higher carbide contents. These two facts indicate that either the contiguity or the intergranular separation of the WC grains or both of these parameters have an important influence on the abrasion resistance of these alloys. This conclusion is supported by the analysis of wear tracks in the scanning electron microscope. At low WC contents distinct tracks with considerable associated plastic deformation or ploughing can be observed, whereas the harder alloys with high WC contents (> 90 per cent) show indistinct wear tracks and an obvious resistance to indentation and plastic deformation.

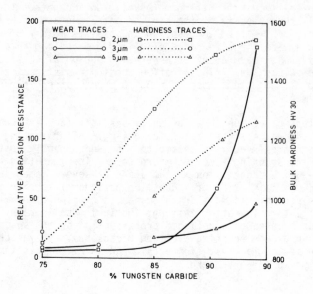

Fig. 1 The relative abrasion resistance and hardness of WC-Co alloys as a
function of composition and grain size.

Cavitation Erosion

Over one hundred different metallic, ceramic and polymeric materials have
been tested. Confining our attention to hard materials the order of
performance, based on a 5 hour cumulative volume loss, is given in Table 1
for the best ten materials.

TABLE 1   Cavitation Erosion Performance of Hard Materials

Rank	Material	Grain Size (μm)	5h Cum. Vol. loss (mm^3)	Hardness HV30
1	WC-5Ni	2	0.01	1478
2	Stellite$^{(R)}$3*	-	0.015	610
3	Stellite$^{(R)}$6*	-	0.02	399
4	WC-9Ni	2	0.03	1246
5	Carburized Steel		0.03	900
6	Tool Steel A2		0.03	650
7	Syndite**		0.08	-
8	Amborite**		0.10	-
9	WC-15Co	2	0.21	1048
10	WC-15Co	5	0.27	933

* Stellite$^{(R)}$ is a registered Trademark of the Cabot Corporation
** Registered Tradenames of De Beers (Pty) Ltd

The performance of ten grades of WC cermets in terms of their 10 hr
cumulative volume losses is shown in Fig. 2 as a function of binder
content and grain size. Although the Ni alloys appear to show a linear
relationship between erosion loss and binder content, the Co alloys show a
minimum loss (i.e. maximum resistance) at about 15 percent Co. The
evidence also suggests that a fine grain size is beneficial to resistance
to cavitation erosion.

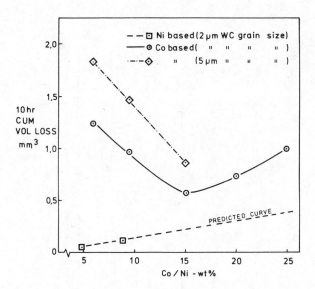

Fig. 2   The cavitation erosion loss of WC cermets as a function of
composition and grain size.

Although the hardness of the WC cermets show a smooth decrease with
increasing binder content, the hardness of a material is not a good
indication of erosion resistance. This is exemplified by the fact that
the alloy WC-6Co (2 μm) with a hardness of 1450 HV30 shows a volume loss
of almost twice that of the alloy WC-15Co (2 μm) with a hardness of 1048
HV30. The transformation of the cobalt matrix from the face-centred cubic
structure to the close-packed hexagonal structure during cavitation was
detected in both the Stellite[R]3 and WC-15Co (5 μm) alloys. Fig. 3
shows the X-ray diffraction traces taken from the WC-15Co (5 μm) alloy
before and after the erosion. Clearly the ratio of the relative X-ray
diffraction peaks indicates that the relative amounts of the hexagonal
phase has increased during the cavitation.

The scanning electron microscopy studies have been previously reported for
the Stellite alloys and WC cermets (Heathcock, Ball, Yamey and Protheroe,
1981). Briefly, the carbides in Stellites are eroded selectively in a

brittle manner with the carbide-matrix interface acting as an initiating
site. The cobalt-rich matrix shows transformation and twin markings prior
to a more ductile mode of erosion. In the case of the WC cermets, the
matrix is lost initially, leaving the carbide particles unsupported. The
eroded surface therefore consists of denuded angular carbide particles
with an appearance similar to that shown in Fig. 11 for the case of liquid
flow erosion. Specimens of the Syndite and Amborite compacts show a
similar intergranular mode of erosion.

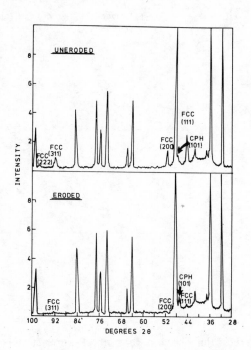

Fig. 3 The change in the X-ray diffraction traces taken from a WC-15Co
        alloy before and after cavitation erosion.

Low Velocity Particle Erosion

Ten WC cermets were tested for their resistance to erosion by $100\,\mu$m SiC
particles travelling at $40\,\mathrm{m\,s^{-1}}$ and an impact angle of 45 degrees. A
typical mass loss curve as a function of the mass of erodent delivered is
shown in Fig. 4. The slope of the straight line portion is taken as the
steady state erosion rate $(10^{-4}\mathrm{g\ g^{-1}})$ and its intercept is defined as the
incubation. Fig. 5 illustrates the dependence of the erosion rate on the
binder content. The plots of incubation and cumulative volume loss for
10 g of SiC show identical shapes with a maxima at 10 percent binder and a

minima at about 20 percent binder. Although the Vickers hardness values
for the alloys decrease smoothly with increasing binder content, the
resilience numbers as measured by a Shore Schleroscope, show similar
maxima and minima.

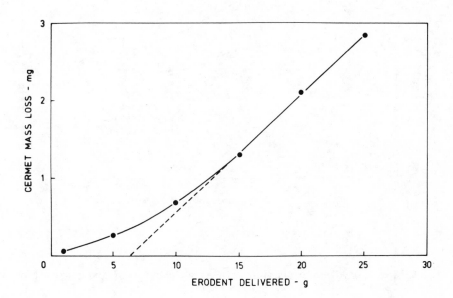

Fig. 4   Mass loss due to low velocity particle erosion as a function of
mass in grams of erodent delivered.

Scanning electron microscopy has been carried out on previously polished
specimens of all the grades tested after the erosion by 1g and 25g of
erodent. This was undertaken in order to obtain information on both the
incubation and steady state periods of erosion. After the short period of
erosion, the craters produced by individual strikes of the SiC particles
could be clearly observed (Fig. 6), whereas the specimens in the steady
state period show a multitude of overlapping strikes (Fig. 7). No
differences in the modes of deformation or fracture could be discerned in
specimens of different binder type, binder content or hardness. The
plastic deformation and fracture processes appear to be insensitive to the
detailed microstructure and massive movement of the two phase alloy occurs
by a combination of plastic flow and fracture of both the hard carbide
particles and the matrix. Isolated slip lines, twins or transformation
markings have been observed on individual carbide particles but ragged
multifracturing is a much more common observation.

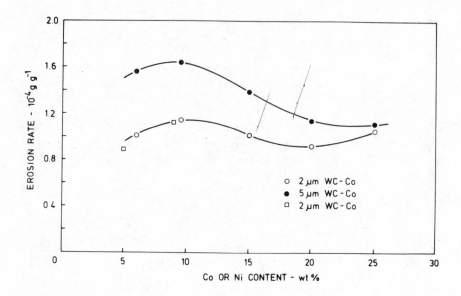

Fig. 5   The low velocity erosion rate as a function of the composition of
WC cermets.

Fig. 6 An erosion impact site of
Wc-20Co after a short erosion period

Fig. 7 Overlapping impacts on a WC-
20Co after an extended
erosion period

The features on the steady state samples clearly indicate that repeated cross-strikes are required before material is finally detached from the surface. The microtoughness of the material must be an important factor controlling the magnitude of the steady state erosion rate.

High Velocity Particle Erosion

Table 2 lists the hard materials tested according to their performance. An extremely fine grained WC cermet containing 5.7 percent Co performs intermediately between two commercially available sialon ceramics.

In general, the performances of the WC-Co cermets, as measured by volume loss over a fixed test period, show a nearly linear relationship with grain size. Alloys with constant grain size show an increase in volume loss with an increase in the volume of binder phase. The volume loss decreases with increasing hardness. Deviations from these relationships occur with materials obtained from different manufacturers, and process variables probably account for these discrepancies.

Examination of the eroded surfaces by scanning electron microscopy revealed evidence for individual particle strikes in lightly eroded regions away from the centre of erosion. In the case of WC-Co cermets the region of strikes and the heavily eroded surface have a scarified or glazed appearance (Fig. 8). This glazed appearance, together with teardrop shapes strongly suggest that the high velocity erosion has caused localised melting of the cermet. In the cause of the sialon specimens, the heavily eroded areas had a hammered appearance and the spallation of flakes was a frequent observation (Fig. 9).

Fig. 8 'Glazed' appearance of WC-10Co alloy after high velocity particle erosion

Fig. 9 The surface of Kyon 2000 after high velocity particle erosion

TABLE 2   The Performance of Hard Materials to High
Velocity Particle Erosion

Rank	Material	Grain Size (μm)	Relative erosion resistance	Hardness HV30
1	Kyon 2000*		9999	2026
2	WC–5.7Co	1	6296	1971
3	Syalon**		3253	1448
4	WC–6Co	1.26	1661	2081
5	WC–9.5Co	2.5	1525	1434
6	WC–6Co	1.88	1373	1591
7	WC–7Co	1.86	1123	1493
8	WC–5Co	1.87	1120	1608
9	WC–0.5Co–6Ni	1.98	1026	1634
10	WC–7.8Co	2.75	1023	1463
11	WC–10Co	1.87	999	1557
12	WC–10Co	2.82	609	1363
13	WC–6Co	2.87	596	1429
14	WC–8Co	3.20	550	1391
15	WC–10Co	3.10	511	1333
16	WC–12Co	2.98	473	1251
17	WC–15Co	3.14	421	1192
18	WC–9.5Co	5.7	259	1149
19	WC–5.7Co	5.57	230	1304
20	WC–12.2Co	1.94	225	1324

*Kyon 2000 is a registered tradename of the Kennametal company.
** Syalon is a registered tradename of Lucas (Pty) Ltd.

The Erosion by Particles in a Liquid Stream

The relative performance of nine hard materials is given in Table 3 and
Fig. 10 shows the plots of the cumulative volume losses as a function of

time for a Syalon specimen and four WC cermets. It will be noted that the sialon specimen has a long incubation period followed by a rapid rate of erosion and a subsequent reduced rate of erosion. This pattern repeats itself and that would suggest that the material requires a reincubation strain prior to periods of erosion. The rate of erosion of the WC-Ni hard metals decreases with time of erosion while the WC-Co specimens have a relatively rapid and constant rate of erosion. Over an extended period of time the WC-Ni alloys out-perform the Syalon specimen.

Scanning electron micrographs of eroded regions of WC-11Co, alumina and sialon specimens are shown in Figs. 11, 12 and 13. The cermet has lost its binder phase leaving the carbides denuded and unsupported. This mode of erosion is identical to that observed for cavitation erosion of cermets but distinctly different from the modes observed for both low and high velocity particle erosion in a gaseous carrier stream. Both the reaction bonded SiC and hot pressed alumina specimens eroded by an intergranular separation mechanism (Fig. 12) while the Syalon lost material by intermittent flaking or spallation (Fig. 13).

TABLE 3   The Performance of Hard Materials to Particle Erosion in a Liquid Carrier

Rank	Material	Grain Size (μm)	Erosion Rate mm^3 min^{-1} x 10^3	Hardness HV30
1	WC-5Ni	2	1.1	1550
2	WC-9Ni	2	1.1	1375
3	Syalon	-	1.4	2000
4	WC-6Co	2	6.2	1290
5	Alumina (99.5 percent density)	-	8.4	1530
6	Wc-11Co	2	12.0	1100
7	Silicon carbide (Reaction bonded)	-	13.3	2100
8	Alumina (90 percent density)	-	19.7	1200
9	Stellite**6	-	102.0	400

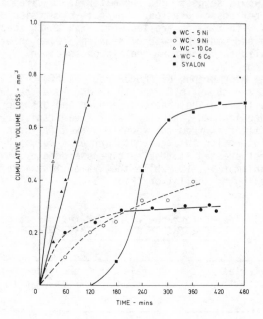

Fig. 10   The erosive loss of hard materials as function of time by
particles in a liquid stream.

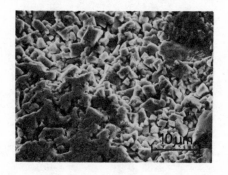

Fig. 11(a)   The eroded surface of
WC-11Co after erosion by particles
in a liquid stream.

Fig. 11(b)   Higher magnification
of Fig. 11(a).

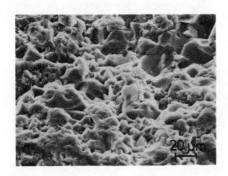

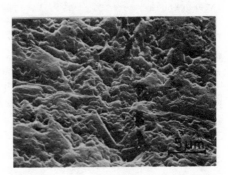

Fig. 12 The eroded surface of
   alumina after erosion by
particles in a liquid stream

Fig. 13 The eroded surface of
R.B. silicon carbide after erosion
by particles in a liquid stream

DISCUSSION

This work, using different test facilities, has demonstrated that a given
material can show significant changes in relative performance and
different mechanisms of erosion under different wear conditions.
Irrespective of this observation, the surface of a specimen will only lose
material when two conditions have been satisfied. The strain imposed must
reach a critical value at which microfractures are initiated. These
microfractures must then propagate in order that the fragments are
released. An ideal wear resistant material will therefore have an ability
to absorb strain prior to the initiation of microfractures and, in
addition, have a toughness which resists the propagation of the
microfractures. In the case of brittle ceramic materials or coatings the
first stage of the erosion process will be absent if any structural
defects or flaws exist on the surface or in the near-surface regions.
These defects will propagate under the imposed deformation.

The abrasive or erosion loss via the above mechanisms may occur after a
single impact or strike. Alternatively, an incubation period may be
required during which strain and microfractures are accumulated in the
surface. In this case a virgin surface will show a low rate of wear
followed by steady-state conditions. The extent of the incubation period
and the rate at which steady-state wear occurs will be a reflection of
both the severity of the imposed erosion, and the microstructure and
mechanical properties of the material. The incubation period and rate of
the steady-state erosion are reflected in the cumulative volume loss of
material for a given time of wear or in the relative wear rate as defined
previously.

The results from the five test facilities, although different in detail, clearly indicate that fine grained WC cermets with low volume fractions of binder phase possess a good combination of hardness and microtoughness which is required for erosion resistance. The sialon materials also have microstructures which confer outstanding erosion resistance. The sintered aluminas and reaction bonded silicon carbide appear to contain intergranular defects, which initiate failure along interfaces, and hard coatings can also be susceptible to the propagation of cracks from inherent defects.

These initial studies have demonstrated the need for detailed studies of the mechanisms of erosion. For example, the evidence for localised melting during the high velocity erosion of WC cermets and the preferential leaching of the binder phase in a liquid environment are significant and further studies will give attention to aspects such as these.

ACKNOWLEDGEMENTS

The author acknowledges the experimental contributions of Dr A.W. Paterson, Mr D. Graham, Mr C. Forse, Mr G. Fogel, Dr J. Heathcock, Mr S. Willmott and Miss A. Resente.

REFERENCES

Allen, C., Ball, A. and Protheroe, B.E. 1981-82, Wear, 74, 287-305
Blombery, R.I., Perrott, C.M. and Robinson, P.M. 1974, Wear 27, 287-305
Condrad, H., Shin, G.A. and Sargent, G.A. 1983, Speciality Steels and Hard Metals, ed. N.R. Comins and J.B. Clark, Pergamon Press, New York, pp. 423-429
Forse, C. and Ball, A. 1983, Proc. 6th Int. Cong. on Erosion, by Liquid and Solid Impact, Cambridge, pp. 61.1-61.7
Heathcock, C.J., Ball, A., Yamey, D. and Protheroe, B.E. 11981-82, Wear 74, 11-22
Shetty, D.K., Wright, I.G. and Clauer, A.H. 1983, Proc. 6th Int. Conf. on Erosion by Liquid and Solid Impact, Cambridge, pp. 62.1-62.9
Vyas, B. and Preece, C.M. 1976, J. Appl. Phys. 47 5133
Willmott, S. 1983, M.Sc. Thesis, University of Cape Town
Wright, I.G., Shetty, D.K., and Clauer, A.H. 1983, Proc. 6th Int. Conf. on Erosion by Liquid and Solid Impact, Cambridge, pp. 63.1-63.8

Inst. Phys. Conf. Ser. No. 75: Chapter 9
Paper presented at 2nd Int. Conf. Science Hard Mater., Rhodes

865

Fine scale abrasive wear of ceramics by a plastic cutting process

S B TOH AND R McPHERSON

Monash University, Clayton. Victoria. 3168

ABSTRACT

Abrasive wear of ceramics results in plastically deformed debris if the scale of deformation is sufficiently small. The microscopic debris produced is similar in appearance to large metallic machining chips and is composed of very fine, equiaxed crystallites. There is some evidence for metastable phase formation in several ceramics. It is suggested that cracking is inhibited along scratches in ceramics as the scale is reduced so that below a critical load the debris is formed by a plastic cutting process similar to that observed in metals.

INTRODUCTION

X-ray diffraction studies have shown that the structure of debris produced during the abrasive wear of several ceramics is comparable with that of heavily cold worked metals, and recovery and recrystallization processes occur on heating (Cutter and McPherson, 1969, 1973). Scanning electron microscopy (SEM) demonstrated that small segmented chips, similar in appearance to chips formed during the conventional machining of metals, were formed during the fine scale abrasive wear of rutile (Aghan and McPherson, 1973). Transmission electron microscopy (TEM) of these chips showed that they had an extremely fine polycrystalline structure with crystallite size <10 nm. The present paper reports a further study of the fine scale abrasive wear mechanism of some ceramics and metals.

EXPERIMENTAL PROCEDURE AND RESULTS

Specimens of polycrystalline Al_2O_3, TiO_2, Y_2O_3, Sm_2O_3 partially and fully stabilized ZrO_2, and the metals Cu, C-Mn steel, Nb, Mo and ferritic stainless steel were polished using conventional techniques and abraded with SiC paper, diamond impregnated laps, or scratched with a single diamond tool. The single point experiments were conducted at loads of between 0.05 and 1.0 N, and at a constant velocity of 1 mm/min. The diamond tool used was of triangular pyramid shape with included angle of 87° (leading face). The angle of attack was 54°. The debris produced were collected on a carbon film for TEM examination after the surfaces had been examined by SEM.

The fine grooves, Figs. 1(a) and 1(b), produced by scratching Cu and
3 wt% Y_2O_3-ZrO_2 with the single point diamond, illustrates the similarity
in behaviour of these two materials during fine scale abrasion.   In both
cases the grooves appear to be completely plastic in nature with no
apparent cracking, and fine debris particles were produced by detachment
of extruded fins at the groove edge.   Similar observations were made on
the other metals and ceramics.

Cracking along grooves in ceramics was observed as the load was increased
beyond a critical level, which depended upon the material, but this effect
was not observed in metals.   Figure 1(c), for example, shows cracks along
a plastic groove in 14wt%.MgO-ZrO_2 at a load of 1.0 N.   As the load was
further increased cracking became more extensive and the debris was
produced predominately by fracture processes, as illustrated in Fig. 1(d)
for soda glass at a load of 0.9 N.   The ceramic debris produced at low
loads by abrasion with SiC abrasive papers or diamond impregnated metal
laps was in the form of serrated chips or detached fins as previously
reported (Aghan and McPherson, 1973).

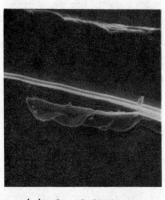
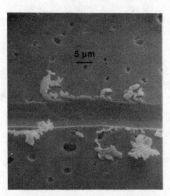

(a) Cu, 0.2N load.   (b) 3 wt.% Y_2O_3-ZrO_2
                                  0.5 N load.

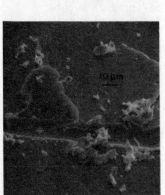
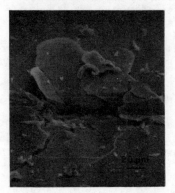

(c) 14 wt.%MgO-ZrO_2 (d) Soda glass, 0.9 N load.
     1.0 N load.

Fig. 1 Grooves in various materials produced
by single point diamond at velocity of 1 mm min^{-1}.

A conventional two stage plastic/carbon extraction technique was used to collect the debris from the surfaces of abraded specimens for TEM examination. The ceramic debris produced at low loads were usually sufficiently thin for direct examination whereas this was only possible for a small proportion of the metallic debris particles without ion-beam thinning.

Serrated chips less than ~1 μm wide were observed for all the ceramics which had been abraded at low loads. Very similar chips were also observed from metals however it was difficult to control the experimental conditions so that chips as fine as those observed from ceramics could be prepared. The marked similarity in morphology observed for all the fine scale debris is illustrated in Figs. 2(a), (b) and (c) for Cu, 8 wt.%. MgO-ZrO_2 (PSZ) and sintered Y_2O_3 respectively.

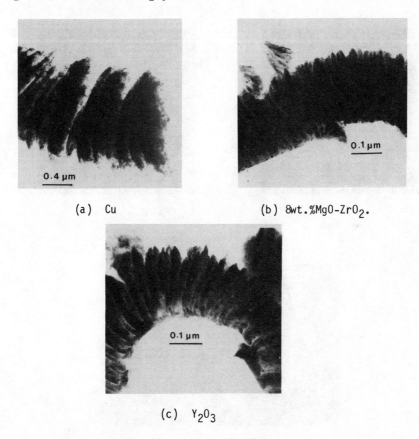

(a) Cu   (b) 8wt.%MgO-ZrO_2.

(c) Y_2O_3

Fig. 2 TEM debris produced by abrasion on SiC paper.

Selected area diffraction patterns (SAD) showed that both the metallic and ceramic chips had a very fine polycrystalline structure; Figs. 3(a) and (b). The substructure, seen more easily by dark field examination (Fig. 4), appeared remarkably similar for all of the ceramics examined, with a

mean crystallite size of ≈7.5 nm.   The crystallite size was somewhat
larger in the metallic debris.

SAD patterns of rutile debris showed additional rings, which disappeared
on heat treatment at 700°C for 1 hour.   These additional rings were
consistent with the presence of the metastable brookite form of TiO_2,
however this interpretation is limited because it is based on a few
additional diffraction rings only which could not be measured with high
accuracy.

The debris from fully stabilized (cubic) ZrO_2 had the cubic structure, but
that from 3 wt% Y_2O_3-ZrO_2 (initially tetragonal), and 8 wt% MgO-ZrO_2
(initially cubic plus tetragonal) gave SAD patterns consistent with the
presence of monoclinic ZrO_2.

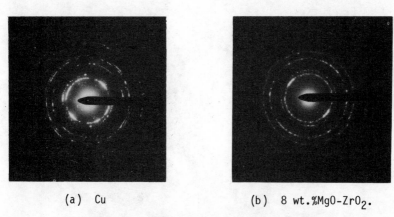

(a) Cu   (b) 8 wt.%MgO-ZrO_2.

Fig. 3 Selected area diffraction patterns of chip-like debris.

Fig. 4 TEM, dark field image, TiO_2 abrasion debris.

DISCUSSION

The very similar morphologies and substructures observed for fine scale abrasive wear debris of a wide range of ceramics and metals suggests that the process of debris formation is largely independent of the nature of the material. The major difference observed between the metals and "brittle" ceramics is that plastic chips are only observed if the abrasion of ceramics is carried out on a fine scale (≈ 1 μm) whereas they are observed to a macroscopic scale in the case of metals. The observations of debris formation are consistent with the model for removal of material during plastic groove cutting in metals by the formation of chips at the working face of the abrasive particle, if the particle geometry is appropriate, and by the detachment of fins at the edge of the groove (Aghan and Samuels, 1970).

The plastic deformation of normally brittle materials by indentation is a well-known phenomenon and is usually accompanied by cracks which nucleate below the indentation (Lawn and Wilshaw, 1975). As the load on the indentor is decreased, however, a stage is reached at which cracks are no longer formed; for the case of soda glass the critical indentation size is ≈ 1 μm (Lawn, Jensen and Arora, 1976). Swain (1979) reported that cracks were not observed beneath scratches in soda glass below a critical load of ~ 0.05 N but at loads in the range 0.1 to 1 N cracks often formed beneath the plastic groove although they did not intersect the surface. Lomdahl and McPherson (1981) also observed that cracks did not form along plastic scratches less than 1-2 μm wide in soda glass.

Kendall (1978) has quoted a number of reports of the size dependence of the fracture strength of brittle materials in compression and has suggested that, as the size is reduced, the fracture strength increases to the level at which even the most brittle materials will deform plastically rather than fracture. He estimated that the critical size (d) below which a particle will not fracture is:

$$d = 10 \ ER/Y^2$$

where E is Young's modulus, R is the fracture energy and Y the yield stress. This is similar in form to the relationship proposed by Lawn, Jensen and Arora (1976) for the critical indentation size (a*) for cracking:

$$a^* = \xi \ ER/H^2$$

where H is the hardness (and H $\sim 3Y$).

These conditions therefore suggest that, below a critical dimension of the abrasion event, material is removed from ceramics by a plastic cutting process because fracture is inhibited.

The formation of serrated chips during metal cutting has been extensively studied and the formation of such structures by catastrophic shear is well established (von Turkovich and Black, 1970). Thus, a lamellar structure is formed, with individual lamellae separated by very thin regions of highly concentrated shear. It has been proposed that catastrophic shear will occur within a plastically deforming material when the slope of the true stress-strain function becomes zero.   That is, when the local rate of change of temperature decreases the strength at a rate equal to, or greater than, the rate of strain hardening.

The unusual polycrystalline structures produced between the shear bands in the fine scale debris of both metals and ceramics may result from a form of dynamic recrystallization.   Such an effect could arise from the temperature increase produced in the chips because of the large energy dissipation, resulting from plastic deformation which occurs at strain rates of the order 10^4 sec^{-1} (Toh and McPherson, 1984).   The fine scale of the process together with the large strain and strain rate occurring at unknown levels of stress, are outside the range of previous studies of the plastic deformation of materials and the nature of the process must therefore be largely speculative.

The apparent observation of brookite in rutile chips is surprising because it is metastable with respect to rutile at all temperatures and pressures (Dachelle, Simons and Roy, 1968) and it would not therefore be thermodynamically possible for brookite to form directly from rutile. Brookite can crystallise from amorphous TiO_2 (Yamaguchi and co-workers, 1975) suggesting that it may have formed from amorphous regions, possibly within the shear bands.

There is evidence in the literature that surface melting can occur during the wear (Hornbogen, Klein and Schmidt, 1982) or high velocity impact (Lawn Hockey and Weiderhorn, 1980) of refractory materials.   Metastable δ-Al_2O_3 has also been observed in the finely recrystallized, abraded surfaces of alumina, (Hines, Bradt and Biggers, 1979) an observation which is consistent with rapid quenching of a thin molten layer.   The formation of thin regions of austenite within the shear bands of a martensitic high-Ni steel provides evidence that higher temperatures are reached within the shear band than in the bulk of a metal cutting chip (Jovane, Carro-Cao and Pomey, 1969).   The formation of a transient liquid phase within shear bands of ceramic microchips formed at extremely high stresses and strain rates is thus not unreasonable but attempts to locate the brookite in the present study, and indeed, confirm its presence, have not been successful.

The presence of monoclinic ZrO_2 in the chips produced by abrasion of PSZ, indicates that transformation of metastable tetragonal ZrO_2 to the monoclinic form has occurred as expected. This result also shows that the chip temperature could not have risen much above the equilibrium temperature for the monoclinic to tetragonal transformation ($\approx750°C$ for 3 wt.%Y_2O_3), since tetragonal ZrO_2 would be retained to room temperature on cooling at the extremely fine grain size observed in the debris (Lange, 1982).

The present study suggests that plastic mechanisms of material removal dominate during the fine scale abrasive wear of normally brittle materials such as ceramics.   This has considerable significance in relation to the abrasion of "wear-resistant" hard materials since, by definition, the rate of material removal is low during service.   It is therefore probable that, under these circumstances, wear occurs by plastic deformation processes similar to those described in the present paper.

CONCLUSIONS

The abrasive wear of ceramics occurs by plastic cutting processes when the scratch width is less than 1-10 μm.   Plastic deformation occurs in normally brittle ceramics because cracking is inhibited by a size effect.   The deformation process appears to be similar for ceramics and metals during fine scale abrasion and chips may be formed which have an ultrafine polycrystalline substructure between catastrophic shear bands.   The polycrystalline substructure probably forms by dynamic recrystallization because of the transient temperature increase produced during high strain, high strain rate deformation.   The plastic wear process could be a significant mechanism in the industrial wear of hard, wear-resistant materials.

REFERENCES

Aghan, R.L. and Samuels, L.E. 1970, Wear 16 292-301.
Aghan, R.L. and McPherson, R. 1973, J.Amer.Ceram.Soc., 56 46-47.
Cutter, I.A. and McPherson, R. 1969, Phil.Mag.20 489-494.
Cutter, I.A. and McPherson, R. 1973, J.Amer.Ceram.Soc. 56 266-269.
Dachille, F., Simons, P.Y. and Roy, R. 1968, Amer.Mineral 53 1929-1939.
Hines, J.E., Bradt, R.C. and Biggers, J.V. 1979, Proc.Int.Conf.Wear
    Materials ed. K.C. Ludema, W.A. Glasser and J.K. Rhee, A.S.M.E., New
    York, pp.540-550.
Hornbogen, E., Klein, K.H. and Schmidt, I. 1982, J.Mater.Sci.Letters 1 94-
98.
Jovane, F., Carro-Cao, J. and Pomey, J. 1971, C.R. Acad.Sc. C272 629-631.
Kendall, K. 1978, Proc.Roy.Soc., 361A 245-263.
Lange, F.F. 1982, J.Mater.Sci. 17 225-234.
Lawn, B.R. and Wilshaw, R. 1975, J.Mater.Sci. 10 1049-1081.
Lawn, B.R., Jensen, J. and Arora, A. 1976, J.Mater.Sci. 11 573-575.
Lawn, B.R.Hockey, B.J. and Wiederhorn, S.M. 1980, J.Amer.Ceram.Soc. 63
356-358.
Lomdahl, G.S. and McPherson, R. 1981, Wear 73 205-208.
Swain, M.V. 1979, Proc.Roy.Soc. A366 575-597.
Toh, S.B. and McPherson, R. 1984, "Deformation of Ceramics II" ed.
    Tressler, R.F. and Bradt, R.C., Plenum New York, pp.723-732.
von Turkovich, B.F. and Black, J.T. 1970, Trans.A.S.M.E. B92, 130-134.
Yamaguchi, O., Omaki, H., Takeoka, K. and Shimuzu, K. 1975,
    J.Jap.Soc.Powder and Powder Met. 22 173-175.

Inst. Phys. Conf. Ser. No. 75: Chapter 9
Paper presented at 2nd Int. Conf. Science Hard Mater., Rhodes

873

Design and wear resistance of silicon nitride based composites

S T BULJAN AND V K SARIN

GTE Laboratories, Inc., Waltham, MA 02254

ABSTRACT

Because of its high fracture toughness, strength, and thermal stability, silicon nitride provides an excellent base for the development of composites for wear applications (Buljan and Sarin, 1983). The complexity and variety of wear phenomena require that material design for a particular application consider the specific factors controlling wear in each circumstance. This study focuses on a silicon nitride-based composite design for cutting tool applications and explores the effect of mechanical and chemical parameters on the wear of a broad range of composites in the Si_3N_4-Y_2O_3-Al_2O_3-TiC system.

INTRODUCTION

The complexity and variety of wear processes and conditions require that factors controlling wear be examined for each specific application in order to design and optimize the materials.

Attrition wear, defined as the removal of microscopic or slightly larger amounts of material from the surface, occurs as a result of physical or mechanical and chemical factors and their interreaction. In that sense, two types of wear are usually recognized: abrasive and adhesive.

Abrasive wear occurs when two surfaces of considerably different hardness are brought into contact. Asperities (protrusions) or free particles of the harder material are imposed at the interface. These can cut into and remove the softer material. The presence and location of subsurface defects such as dislocations, vacancies, impurities, microstructural inhomogeneities and microcracks in the superficial layers of the material in contact will generally dictate the zones from which wear particles are generated. For wear to occur, the fracture strength of one of the two materials in contact has to be less than that of the interfacial junction. It follows then that analyses of materials' abrasive wear resistance are inseparable from the considerations of their mechanical properties such as fracture toughness, hardness, and strength.

Observations of wear of ceramic materials in contact with metals have also
shown that a rough correlation exists between attritious wear and the
chemical reactivity of particular ceramic metal systems (Trent, 1978).
Attritious wear tends to be high in the systems with lower heats of
reaction. In a more general sense, wear due primarily to adhesional
friction is caused by chemical bonding across the interface. Friction
results in material removal by shearing and initiates wear processes
activated by heat. The heat generated by friction can cause thermal
spalling, can weaken or melt surfaces, or result in increased reaction
between the constituents of the wear couple, leading to the formation of
solid solutions, compounds, or liquid eutectic phases, thus contributing
to attritious wear. It should be recognized, for example, that only a few
surface atoms have to dissolve in the molten interface for solution wear
to occur. Calculations for abrasive grains show that a solution layer 100
atoms thick, with each pass in normal grinding ($\approx$300 m/min), could easily
produce an attrition rate of about 0.01 cm/min. Several types of solid-
state reactions which can occur at lower temperature through inter-
diffusion may cause liquid formation at temperatures substantially lower
than the melting temperature of either wear couple constituent. Finally,
at the elevated temperatures produced by friction, reaction with the liquid
and gaseous environment is also possible.

The degree to which any of these processes dominates the wear mechanism
depends on the composition and physical mechanical properties of both wear
couple constituents as well as on the load, environment, etc.

With these considerations, it is apparent that in order to design and
optimize a material for a specific application, considerable information
regarding the wear mechanism is a necessary prerequisite. Tools for metal
cutting provide such an example.

RESULTS AND DISCUSSION

In machining operations, the cutting tool is subjected to stress and
elevated temperatures resulting from friction and metal shear. Under
these conditions, the cutting edge undergoes continuous change due to
deformation and mechanical and chemical wear. Several concurrent wear
processes and their severity determine the tool life. The dominant wear
mode of the tool varies depending on the environment, which in a broader
sense includes all conditions of use (workpiece material, stress, atmosphere, etc.). Abrasion (mechanical) wear is observed at low cutting
speeds due to the lower temperatures generated. At higher temperatures
such as those observed in high speed machining chemical wear is expected to
become more dominant. If high temperatures are not generated or if the
tool is chemically inert, abrasion will remain the dominant mode of wear.

In practice, based on extensive testing of Al_2O_3, Al_2O_3-TiC, Si_3N_4, and
Si_3N_4-TiC cutting tools, it has been found that ceramic tools in gray cast-
iron machining wear primarily by abrasion (Buljan and Sarin, 1981; Sarin
and Buljan, 1983).

Material removal by fracture that occurs in abrasion can be assumed to
take place when lateral cracks of adjacent indentations that are caused by
the penetration of sharp surface protrusions (or abrasive particles as in
pin-and-disc) of the opposing surface intersect. Considering the dependence of the size of the indentation and the length of cracks emanating

from such angular indentations on the hardness (H) and fracture toughness (K_{IC}), respectively, the following expression for maximum volume removed (V) by the system of indentors in a grinding operation was derived by Evans and Wilshaw (1976).

$$\hat{V} \propto \frac{1}{K_{IC}^{3/4} H^{1/2}} \sum_{i=1}^{i=N} P_i^{5/4} \ell_i \equiv \frac{1}{K_{IC}^{3/4} H^{1/2}} N \bar{P}^{5/4} \bar{\ell} \qquad 1$$

where N is the number of abrasive particles, P is a vertical force on the particle and ℓ is a length of travel. From the experimental wear data obtained using a pin-on-disc method under constant load (Buljan and Sarin, 1981; Sarin and Buljan, 1983), the abrasion resistance for the series of Al_2O_3 - and Si_3N_4-based ceramic cutting tool materials was found to be directly proportional to $K_{IC}^{3/4} H^{1/2}$. The abrasive wear resistance parameter (Fig. 1), measured as the inverse of volume removal per unit of travel, provides a first approximation to a relative ranking of materials.

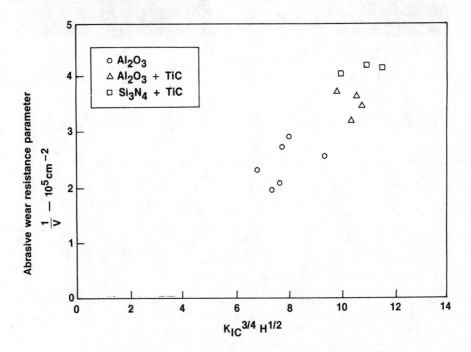

Fig. 1 Abrasive wear resistance of ceramic cutting tools as a function of their hardness and fracture toughness.

Since the mechanical properties pertinent to abrasive wear resistance (hardness and fracture toughness) in the ceramic materials investigated maintain their relative rankings at elevated temperatures, it was also possible to correlate properties determined at room temperature with the abrasive wear resistance of tools in machining gray cast-iron.

While silicon nitride-based cutting tool materials and Si_3N_4-TiC composites in particular show outstanding wear resistance in cast-iron machining, their application to steel machining represents a much more complex problem. A series of tests using silicon nitride-based tools in steel machining have shown their relatively low wear resistance and have indicated that the predominant mode of tool wear is chemical in character. As shown in Fig. 2, tool wear in cast-iron machining is observed only on the nose and flank of the tool, while in steel machining, crater formation on the rake face becomes a prominent wear feature, indicating considerable contributions of chemical interaction between the tool and the workpiece.

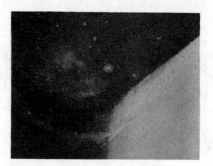 

Gray cast iron — BHN 180
Cutting speed: 426 m/min
Cutting time: 12 min

4340 Steel — BHN 300
Cutting speed 213 m/min
Cutting time: 1 min

Fig. 2 Appearance of tool nose after machining of gray cast-iron (a) and 4340 steel (b) Feed: 0.254 mm/rev. Depth of cut: 1.27 mm.

Having established that a chemical process dominates the wear of Si_3N_4 materials in steel machining, we considered it prudent to modify the chemical character of the composite and explore more closely the chemical wear. From the broadly explored silicon nitride composite systems shown in Fig. 3(a), a number of composites representing the indicated composition plane were investigated in order to examine and define factors controlling wear. A series of composites in the system Si_3N_4-Y_2O_3-Al_2O_3-TiC and Si_3N_4-Y_2O_3-HfC are used as an example. The examples represent Si_3N_4 (2 wt. % SiO_2) + 6 wt. % Y_2O_3 base material, with additions of up to 50 vol % of Al_2O_3 and up to 50 vol % refractory carbide (TiC or HfC) having an average particle diameter (APD) of 2 μm. The consolidated composites consist of a two-phase - Si_3N_4 and glass binder phase - matrix throughout which a particulate refractory carbide dispersoid is evenly distributed. The glass binder phase is composed of SiO_2, Y_2O_3, and Al_2O_3.

As can be seen from Fig. 3(b,c,d), additions of TiC dispersoid prominently increases the hardness (KH1) of the composite and result in a relatively small decrease of fracture toughness measured by indentation. The room temperature modulus of rupture is only moderately affected. Additions of Al_2O_3 result in a decrease of the values for all three properties.

In order to examine their wear resistance in machining operations, a series of cutting tools were fabricated from the prepared materials and tested in gray cast-iron and AISI 4340 steel machining. Tool wear resistance was evaluated using SNG432 inserts with equivalent edge preparations

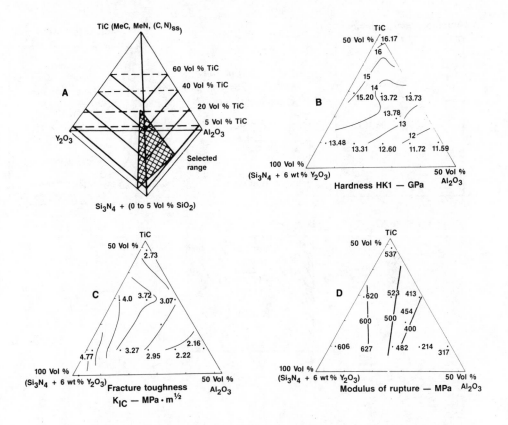

Fig. 3 Composition and mechanical properties of selected
silicon nitride-based composites (b,c,d) in the
Si_3N_4-Y_2O_3-Al_2O_3-MeC system (a)

on a 40 HP N/C lathe equipped with a direct tool wear monitoring system.
The system consists of a device placed adjacent to the tool which measures
and registers workpiece radius changes (ΔR) due only to the tool wear and
deformation (Baldoni and co-workers, 1983).

The wear of silicon nitride-based tools in machining gray cast-iron was
found to be predominantly mechanical in nature and therefore proportional
to a wear resistance parameter $K_{IC}^{3/4} H^{1/2}$. The abrasive wear resistance
of a series of selected composites is given in Fig. 4(a).

Tool wear after 10 minutes of cutting of 4340 steel is given in Fig. 4(b)
and indicates that wear is reduced both with additions of Al_2O_3 and refractory carbide dispersoid. HfC dispersoid shows a more pronounced effect on
wear reduction than TiC.

Taking into account the known abrasive wear resistance of these materials,
it becomes possible to construct a behavior model for the wear of these
materials in steel machining. The model considers both contributions to
wear: abrasive and chemical. While the absolute magnitude cannot be

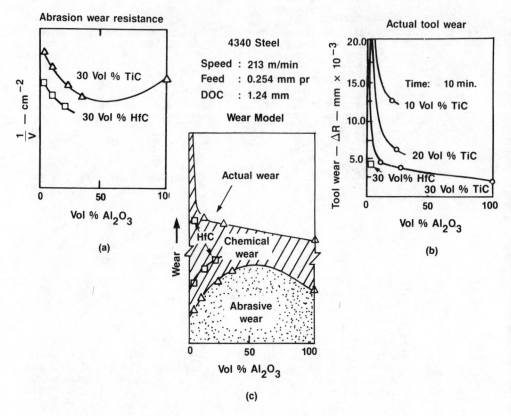

Fig. 4 Cutting tool wear: (a) abrasive wear resistance;
(b) actual tool wear in AISI 4340 steel; (c) tool wear model.

determined, it suffices to observe that an overwhelming contribution of chemical wear obliterates any improvements in abrasive wear resistance. Under these circumstances, it is apparent that the outcome of efforts expended on further mechanical property optimization (such as micro-structural tailoring for fracture toughness improvements) for increased wear resistance would be futile or limited at best.

Based on the degree of adherence and material transfer between the tool and the workpiece material when the two materials in contact are heated at 1000-1200°C, it has been reported that Si_3N_4 and TiC show greater reactivity with type 4340 steel than does Al_2O_3 (Whitney, 1974). These are diffusion tests and are indicative of the degree of possible chemical interaction at high machining speeds. It has also been suggested (Trent, 1952a; Trent, 1952b; Trent, 1978; Kramer and Suh, 1980) that free energy of formation, diffusion and solubility of tool in workpiece materials can be used to predict tool wear.

Estimates of the solubility limit of single phase coating materials in α-iron have been used to predict performances of the coating in steel machining (Kramer and Suh, 1980).

Taking into account the complex makeup of our composite tool materials, their solubility (C_i) in α-iron at 1400°K was estimated by considering fractional contributions (Vf) of individual components.

$$C_{tool} \atop 1400°K = \sum_{i=1}^{i=n} Vf_i C_i \qquad\qquad 2$$

For the sake of simplicity, this estimate considers only the solubility of precursor materials and disregards their interaction on the densified composite.

While these thermodynamic estimates do not consider reaction kinetics and therefore may not be exactly and directly transferable to the complex and dynamic situation of tool wear, they at least reflect the potential for chemical reaction. The resulting estimate of relative tool resistance to chemical wear (solubility limit) is shown in Fig. 5. As can be seen, this simplified approach yields an approximation of actual tool performance in 4340 steel (Fig. 4b), although it apparently fails to account for the strong effect of small additions of Al_2O_3 or TiC. More importantly, this model could imply that similar wear behavior may be expected in the machining of other ferrous alloys, including, for example, gray cast-iron. Observations of wear behavior of silicon nitride cutting tools in high-speed machining of cast-iron (up to 1825 m/min) fail to reveal a substantial chemical wear component. In addition, diffusion experiments indicate virtually no reaction between the Si_3N_4 composite and cast-iron, but a considerable reaction with 4340 steel (see Fig. 6). Therefore, it may have been fortuitous that solubility limit-based estimates of coating performance in steel machining (Kramer and Suh, 1980) were only tested on

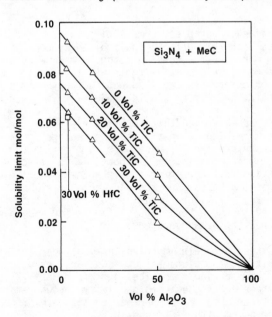

Fig. 5 An estimate of composite tool solubility in α-iron at 1400°K, (Si_3N_4 + 6 w/o Y_2O_3)-Al_2O_3-TiC system

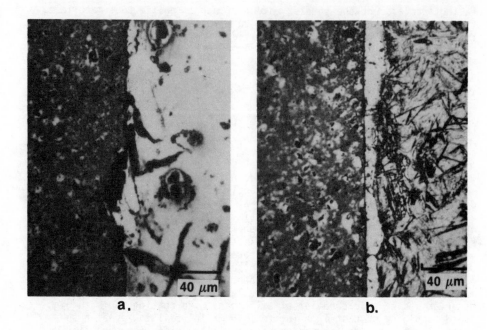

Fig. 6 Photomicrograph of workpiece and Si_3N_4-TiC
composite tool material diffusion couple hot-pressed
in argon at 1375°K, 15 MPa for 30 min.
(a) gray cast-iron; (b) AISI 4340 steel.

4340 steel. The change in workpiece composition (concentration of carbon
or alloying elements) strongly affects the tool-workpiece chemical interaction and requires more careful consideration in estimates of wear
resistance. Furthermore, in a complex tool material consisting of two or
more phases, interaction effects may also considerably change the cumulative chemical potential and render simple projections inaccurate.

Although silicon nitride-based cutting tool materials have proven their
superiority for the machining of cast-iron, their usage in steel machining
has been severely restricted by the excessive chemical wear. Utilizing
the approach of chemical tailoring, it is possible to effect substantial
improvements in cutting tool performance. Improvements of Si_3N_4-Y_2O_3-TiC
cutting tool wear resistance through the additions of Al_2O_3 are illustrated in Fig. 7. The tool of life of alumina-containing composites is
considerably improved to the extent that their use for machining of steel
is not only viable but can be advantageous.

Tailoring of composites to obtain improved chemical wear resistance can
utilize either matrix modifications (Al_2O_3 additions) or dispersoid substitution. HfC-bearing composites containing only 2 vol % of
Al_2O_3 had, due to their lower chemical reactivity, shown remarkable
chemical wear resistance (Fig. 7) comparable to that exhibited by 20 vol %
Al_2O_3 containing composite with TiC dispersoid.

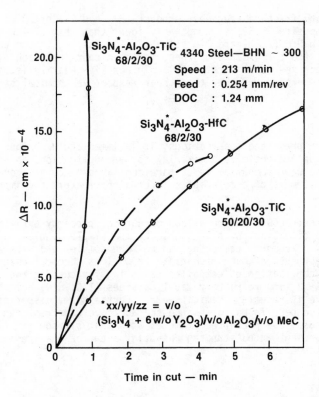

Fig. 7   Wear (ΔR) of silicon nitride-based composite cutting tools in AISI 4340 steel machining.

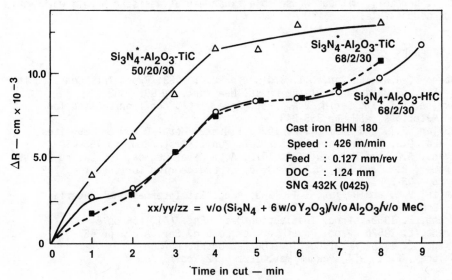

Fig. 8   Wear (ΔR) of silicon nitride-based composite cutting tools in gray cast-iron machining

The hardness and fracture toughness product ($K_{IC}^{3/4}H^{1/2}$) of 30 vol % Al_2O_3 containing composite (8.59) is comparably lower than that of low alumina composites with HfC (9.81) and TiC (11.91). Expectedly, in the absence of chemical wear, the performance of these tools in cast-iron machining (Fig. 8) is observed to be related to their abrasive wear resistance (mechanical properties), rather than their respective chemical stability.

CONCLUSIONS

Considering the presented observations, it is immediately obvious that in order to develop and optimize a material for an application, an understanding of the wear mechanism is of paramount importance and governs the choice of adjustments of a material's internal parameters for improved performance.

Composite microstructure design allows mechanical property optimization, sometimes in a direction not addressable by the simple single-phase or binary systems. Property tailoring can be more complex and extensive, and it allows adjustments of both mechanical as well as chemical character. In a specific application of composite tailoring for cutting tools, additions of Al_2O_3 and refractory metal carbides (TiC, HfC) have been found effective in reducing chemical wear. Furthermore, dispersoids such as TiC have also been found to enhance the abrasive wear resistance of these composites. The presented examples show that silicon nitride-based composites represent an outstanding potential area for wear resistant material development.

ACKNOWLEDGEMENT

The authors acknowledge the contributions of Mr E. Geary and the late Mr R. Wentzell, without whose able experimental assistance this work would not be possible.

REFERENCES

Baldoni, J.G., Buljan, S.T. and Sarin, V.K. 1983, Proc. Int. Conf. on the Sci. of Hard Mat., Plenum Press, New York, pp 991-1002
Buljan, S.T. and Sarin, V.K. 1981, Proc. Intl. Conf. on Cutting Tool Materials, ASM, pp 335-348
Buljan, S.T. and Sarin, V.K. 1984, Sintered Metal-Ceramic Composites, Ed. G.S. Upadhyaya, Elsevier Sci. Publ., Amsterdam, pp 455-468
Evans, A.G. and Wilshaw, T.R. 1976, Acta Metallurgica, No. 24, pp 939-956
Kramer, B.M. and Suh, N.P. 1980, J. of Engineering for Industry, 102 pp 303-309
Sarin, V.K. and Buljan, S.T. 1983, Proc. 1st International Material Removal Conf. Paper MR83-189
Trent, E. 1952a, Proc. Institution Mech. Eng. (A), 166 pp 84
Trent, E. 1952b, Proc. Royal Soc. of London, Ser. A 212 pp 467
Trent, E. 1978, Metal Cutting, Butterworths, London
Whitney, E.D. 1974, Powder Met. Int. 6, 2, pp 73-76

Inst. Phys. Conf. Ser. No. 75: Chapter 9
Paper presented at 2nd Int. Conf. Science Hard Mater., Rhodes

Binder extrusion as a controlling mechanism in abrasion of WC-Co cemented carbides

J. LARSEN-BASSE AND N DEVNANI
University of Hawaii
Honolulu, Hawaii 96822, U.S.A.

ABSTRACT

Abrasive wear of cemented WC-Co and WC-FeNi compositions was simulated by single and multiple pass sliding of a conical diamond indenter over polished specimen surfaces. The two types of material behave quite differently under the different loading conditions with the WC-Co alloys showing better resistance to single passes and poorer resistance to multiple passes. The results are explained on the basis of differences in wear mechanism: cobalt binder extrusion being dominant for the WC-Co alloys, microfracture for the WC-FeNi materials.

INTRODUCTION

Cemented carbides generally have excellent resistance to wear by abrasion and that property is, indeed, a major reason for the use of these alloys in many industrial applications. Abrasion resistance is a rather poorly defined compound property which has a general relationship with hardness, so long as the abrasive is considerably harder than the material being abraded. When the hardness of the abrasive falls below a certain value, the abrasive is no longer able to indent and scratch the surface and there is a marked drop in wear rate (Nathan and Jones, 1966) and apparently a complete change in wear mechanism (Larsen-Basse and Premaratne, 1983). The transition hardness is generally taken to be 20% greater than the metal hardness, following Tabor's (1956) comparison of the Vickers and Moh's hardness scales which showed that in order for one material to indent and scratch another it must be at least 20% harder. Data for metals (Nathan and Jones, 1966; Larsen-Basse and Premaratne, 1983) have confirmed this value. For cemented carbides in the WC-CO family, Larsen-Basse and Koyanagi (1979) have shown that a similar transition in wear rate and wear mechanism takes place but at a slightly lower transition hardness which is only a few percent above the bulk hardness of the material.

For most engineering materials the commonly occurring abrasives can be classified as relatively hard abrasives because their hardness is well above the transition value -- compare, for example, quartz at 1000-1200 HV and a hardened tool steel at up to 850 HV. This factor is probably a major reason that the mechanism of abrasion by relatively soft abrasives has received considerably less attention than abrasion by relatively hard abrasives.

Cemented carbides in the WC-Co family are used in a number of applications, such as rock drilling, ore crushing, etc., where the main abrasive is quartz. The hardness of the alloys generally ranges between 1000 and 1400 HV. Thus, many of them could be expected to wear by a soft-abrasive type of mechanism. The abrasion behavior of these alloys under laboratory conditions has recently been reviewed (Larsen-Basse, 1984). As mentioned above they do show a hard-soft abrasive transition with a substantial change in wear rate and a complete change in wear mechanism.

When abrasion is by relatively hard abrasives (e.g. SiC at 2400 HV) the abrasives form grooves and/or craters in the specimen surface and thereby remove the material lost in wear. Surface deformation is primarily by plastic flow assisted, occasionally, by microfracture in the binder phase. The scale of each removal event is one or more orders of magnitude greater than the WC grain size and bulk hardness is the property which best correlates with resistance to abrasion in this region of abrasion by relatively hard abrasives.

Soft abrasives, such as SiO_2 and $CaCO_3$ remove material from the surface of WC-Co specimens on a much smaller scale. Each wear event is one or more orders of magnitude smaller than the WC grain size and the wear mechanisms are closely related to the specific microstructure and composition of the specimen material. The subsurface material is exposed to multiple contact stress loadings before it reaches the exposed surface; for example, in abrasion by 120 μm SiO_2 abrasives it was estimated that close to 10,000 abrasive grains must pass over a given surface point in order to remove material to the depth of one WC grain diameter (about 1.5 μm) (Larsen-Basse and Koyanagi, 1979). Due to the multiple contact loading there is a gradual removal of Co binder metal and fragmentation of the WC grains; these two mechanisms are responsible for the wear.

Preferential removal of the Co binder was mentioned already by Doeg (1960) who studied worn rock drilling tool bits. He did not propose a mechanism for this process but suggested that it could be rate controlling for the wear. The present author initially proposed (Larsen-Basse, 1973) that fine fragments of crushed SiO_2 abrasives might preferentially abrade the softer binder phase away since SiO_2 is a hard abrasive for the Co phase while it is a soft abrasive for the WC grains and, generally, for the WC-Co material as a whole. After SEM observation of many abraded surfaces he later proposed (Larsen-Basse and Koyanagi, 1979; Larsen-Basse, 1980; Larsen-Basse, 1984) that the preferential binder removal is an extrusion process which is caused by movement of individual WC grains in the surface due to frictional and contact stresses. Another suggestion which has been made is that high local temperatures in wear contacts can result in formation of Co silicides. The binder removal would then occur by alloying with the SiO_2 abrasives (Blomberry, Perrott and Robinson, 1974).

Since the wear performance of hard materials against relatively soft abrasives is of great technological importance and may be directly correlated with binder removal at least for the WC-Co family of alloys (Doeg, 1960; Blomberry, Perrott and Robinson, 1974), the present effort was initiated to study this process in some detail.

By their nature abrasives are uneven in shape and size. Normally, it is impossible to directly control the loading and the deformation pattern when they abrade a surface. This makes it difficult to obtain a detailed understanding of the wear process. In the present work a microtool, in the form of a diamond indenter, was used to minimize some of these experimental problems.

EXPERIMENTAL

A diamond stylus tip of 5 μm tip radius and an included angle of 100° was used as the model abrasive. It was loaded against the polished surface of the sample and slid over the surface in a reciprocating motion. The length of the stroke was 6 mm and the applied loads ranged up to 3.4 N. To simulate abrasion both single pass and multiple pass testing was used; the latter was done at a frequency of 20-25 cycles per minute.

The materials used were WC-Co and WC-FeNi alloys. The WC-Co materials had a nominal range of Co content of 8-15%. The WC grain size, as determined from microstructural analysis, fell in the range 1.3-2.2 μm and the mean free path in the range 0.40-0.62 μm. The WC-FeNi materials were used for comparison. They all contained 25% binder but the Ni content in the binder varied between 10 and 40%. The WC grain size of these alloys was a uniform 0.55 μm and the mean free path 0.22-0.32 μm.

After abrasion the groove width was measured optically and the grooves were observed in the scanning electron microscope. Groove depth was measured optically for the larger grooves, in the SEM on a cross-section for the smaller grooves. Groove volume was calculated from the measured width and depth values.

RESULTS AND DISCUSSION

Single Stroke Tests

The track width is proportional to the square root of the applied load as illustrated in Fig. 1. This shows a general similarity with hardness, except that the grooves are wider than expected from the diamond pyramid indentation test and narrower than expected from calculated scratch hardness values. For example, at 1.25 N load the diagonal of the hardness indentation is 7.5-9.2 μm for a hardness range of 800 to 1300 HV. For the same scratch hardness values, where the applied load is carried only by the leading surface of the indenter, the expected groove widths for the present indenter are 21.5 and 27 μm, respectively. The actual grooves have widths of 12 and 22 μm, respectively. The slopes, S, of the lines are plotted versus the inverse square root of bulk hardness in Fig. 2. This value is used to put the two data in the same unit basis. The similarity between the static hardness indentation and the groove is confirmed by the linear relationship between S and HV. The two alloy families show slightly different behavior -- data for the WC-FeNi materials all fall on one line while the hardest, most brittle WC-Co materials have higher S-values than the extrapolated WC-FeNi values and the more ductile WC-Co compositions have lower S-values. The data in the various ranges all appear to lie on parallel lines. These results show that, while bulk hardness gives a good indication of groove width, some other material properties such as ductility and work hardening also play a role.

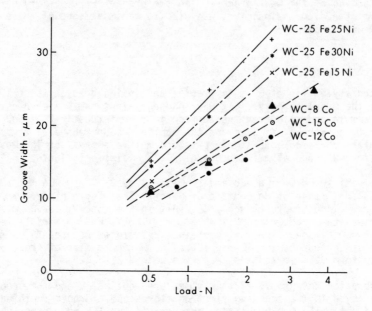

Fig. 1 Width of single stroke grooves vs. applied load
for various compositions

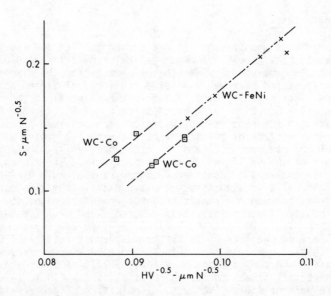

Fig. 2 Slope of lines in Fig. 1 vs. inverse square root
of bulk hardness

No permanent groove depth develops below a threshold load which is 0.4 N
for the WC-Co materials and 0.05 N for the WC-FeNi materials. The difference between the two alloy families is probably due to the difference
in WC grain size (1.3-2.2 μm for the Co binder materials vs. 0.55 μm for
the FeNi binder materials). Above this threshold load the groove depth
and the cross-sectional area, which is proportional to volume removed and
therefore to wear rate, increase linearly with load. This is as expected
from conventional scratch hardness results.

The effect of hardness on wear volume is shown in Fig. 3. There is a
tendency for a general decrease in wear rate with increasing hardness, as
usually found for abrasive wear. An exception is found for the WC-Co
alloys abraded under the highest load used here, 2.5 N. In this case
wear increases with bulk hardness. This is most probably due to brittle
fracture along the edges of the groove. These were occasionally observed
and it is known that the amount of cracking is dependent on indentation
size -- there is a size below which no cracks may form (Moore, 1980). It
has been proposed (zum Gahr, 1979) that for materials which wear by a
combination of ductile groove formation and brittle microcracking a compound property would best describe abrasion resistance. The property he
proposed was the product of hardness and fracture toughness. The local
fracture toughness of the surface layers has been used here. It was
evaluated from the length of the cracks which form at the corners of
Vickers hardness indentations using the method of Evans and Wilshaw
(1976). The results are shown in Fig. 4. For the high load the wear
volume is, indeed, closely related to the product of hardness and fracture toughness. For the lower load the data show an increase in wear
with increasing value of this product. This does not appear reasonable
in view of the removal processes which are operational -- plastic deformation and cracking -- and indicates that cracking plays little role at
this lower load, while plastic deformation dominates and hardness alone
gives the best measure of wear volume.

SEM observation of the grooves supported these general concusions. For
the low loads the groove is shallow and has cracked WC grains in its
bottom and along the sides. These cracks are probably caused by the tensile friction forces which appear immediately behind a passing indenter.
The Co binder appears to have sufficient local ductility to resist this
type of cracking. Wear fragments along the groove edges consist of
minute (0.01-0.1 μm) angular WC particles removed from the WC grains by
brittle fracture, and small Co particles. The same characteristics are
seen for the grooves formed under the higher load. In addition, along
the sides of these grooves there are whole WC grains which have been
tilted in the binder due to the passage of the indenter. These grains
will readily be removed by any additional loading in the groove. The
main crack which removes a grain runs in the binder phase, as do the
cracks which form at the corners of a hardness indentation and which were
used to calculate the local fracture toughness. Thus, it is reasonable
to expect that a combination of hardness and crack resistance will give a
good correlation with wear volume under these conditions, as shown for
the higher load in Fig. 4. The hardness value determines the size of the
groove and the amount removed by plastic deformation while the local
fracture toughness is a measure of how much material is lost along the
sides of the groove by brittle type of fracture in the binder phase.
Additional work will be necessary to quantify the respective roles of the
two properties.

The data of Fig. 3 also show that the WC-Co materials have inherently greater resistance to this type of wear by hard abrasives than do the WC-FeNi materials. The reason for the 200-250% difference in wear volume is not quite clear at this stage. Some of it may be due to inherent differences in the performance of the two binder metals. Some of it may be due to the greater fraction of WC grains in the WC-Co alloys and to their greater contiguity and larger size; this could result in differences in overall work hardening and ductility at the very high strains and strain rates which are responsible for ductile material removal in abrasive wear.

Only a very qualitative discussion of these possible factors can be given at this point since they are not well understood or documented. It may be envisioned that the large fraction of very hard WC grains ($\sim$2400 HV), which is distributed in a cemented carbide material of considerably lower overall hardness ($\sim$1200 HV), can have an effect on the local contacts between a hard abrasive and the surface which is not fully described by bulk parameters or properties such as indentation hardness. And one might expect that the WC-Co materials used here, with their larger fraction of WC grains, would show a greater resistance to formation of small grooves than the WC-FeNi materials of the same hardness, because in the latter case a larger fraction of the bulk hardness is due to contributions from the martensitic binder. Comparing the two materials at adjoining ends of the two alloy families, a WC-15.4 weight-% Co alloy of 1085 HV and a WC-25 weight-% Fe10 Ni alloy of 1075 HV one finds a ratio between wear volumes (FeNi/Co) of 2.0 at 2.5 N load and 1.6 at 1.25 N load. The ratio between carbide phase areas (or volumes) is around 1.1 and between binder phase areas (or volumes) around 1.6. When considering that some of the wear of the WC-Co material at 2.5 N load is due to cracking, as discussed above, it is seen that binder area (and/or volume) gives a fairly good indication of the difference in wear between these two families of materials for the low load region where plastic deformation dominates.

The hard carbide grains do have an effect at very low loads. The larger fraction of WC grains combined with their larger individual size in the WC-Co materials -- 1.3-2.2 μm vs. 0.05 μm in the FeNi materials -- is clearly responsible for the larger load required to form a groove of permanent depth ($\sim$0.4 N vs. $\sim$0.05 N). At low load the indenter primarily interacts with the WC grains of high hardness rather than with the composite as a whole with a considerably lower hardness. Not until the load reaches a certain value where a substantial volume of material is stressed -- a volume which depends on the WC grain size -- does significant plastic deformation take place. It is interesting to note that at the critical load for groove formation the groove width (or the width of the track of contact) corresponds to 3-5 grains plus their associated binder widths in both material families. The behavior is in agreement with earlier results on abrasion of these materials by SiC abrasive papers (Larsen-Basse, Shishido and Salem, 1976).

Some contributions to the difference in wear rate between the two types of material may also come from difference in behavior under the conditions of high strain and strain rate which control material removal in groove formation by plastic flow. Cobalt binder material is known to be particularly ductile and able to sustain very large strains under conditions of contact loading (Brainard and Buckley, 1976), while the martensitic binder of the WC-FeNi materials is expected to be less able to

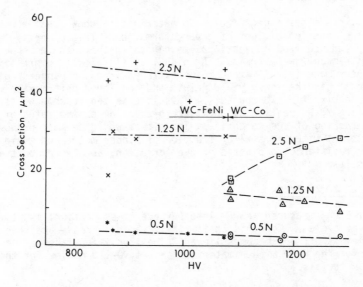

Fig. 3   Wear volume in single pass abrasion, expressed
as groove cross-sectional area, vs. bulk hardness

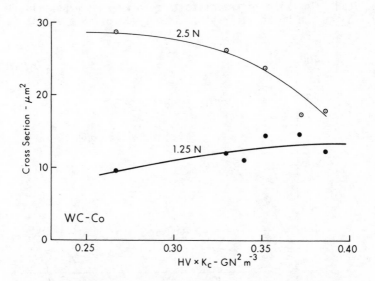

Fig. 4   Wear volume in single pass abrasion for WC-Co alloys vs.
the product of bulk hardness and fracture toughness.

sustain such strains. The latter materials would thus be expected to show greater wear, as demonstrated by the results.

Finally, one might expect the WC-Co materials to show greater work hardening under conditions of groove formation. This property is not readily deduced from conventional mechanical property data because these deal with much lower strains. At high levels of local compressive strain the binder material will tend to be extruded from the binder regions, because it has the lower yield point, and continued deformation will result in direct WC-WC contacts. This is expected to show up as increasing WC-WC contiguity and therefore in increasing rates of overall work hardening of the composite. The effect would be more pronounced for the material with the smaller fraction of binder metal and would give it greater resistance to plastic groove formation, as also demonstrated by the results.

Multiple Stroke Tests

In Fig. 5 the measured wear volume (groove cross section) is plotted vs. number of cycles. There is an initially rapid increase and then the curve levels off and becomes linear. The transition takes place at 1-10 cycles for the FeNi binder materials and at 90-110 cycles for the Co binder materials.

The volume removed due to the cyclic part of the load (V_N-V_1) is proportional to the applied load for all materials tested. The amount of material removed per stroke is then independent of the width of the pre-existing groove; this is because apparent contact area and apparent contact stress have almost no effect on the wear which is controlled by applied load. The lines extrapolate to zero wear at a positive load of 0.25-0.4 N for the WC-Co materials. The origin of this effect is the

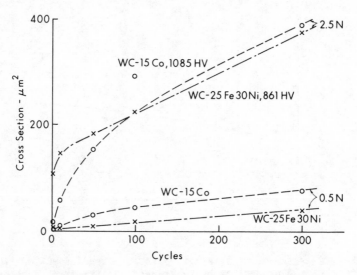

Fig. 5   Wear volume vs. number of cycles in multiple
         pass abrasion for a WC-Co and a WC-FeNi material
         under two different loads.

same as discussed for the single stroke tests: it is due to the requirement that a minimum volume be plastically deformed before these materials
behave according to their bulk properties. The size of this volume
depends on WC grain size and it dictates the minimum load for significant
groove formation and wear.

The amount of material removed per unit of applied load during the cyclic
loading to any given number of cycles has here been named S_N, where

$$S_N = (V_N - V_1)/P \qquad\qquad 1$$

where V_N and V_1 are the groove volumes after N and 1 cycles, respectively, and P is the applied load.

The value of S_N depends on N, as shown in Fig. 6. It is seen that the
two families of materials behave quite differently. The FeNi binder
materials attain a steady state immediately, i.e., the amount of material
removed per cycle and unit of applied load is constant after the first
few cycles. This again points to a typical wear condition. Also, the
slope is lower for the harder materials, as expected in abrasive wear
situations.

The Co binder materials show a contrasting behavior: the curves do not
reach steady state until well after 100 cycles and the harder material
shows the greater wear. It is also seen that under cyclic loading the
WC-Co materials show much higher wear than WC-FeNi materials of the same
hardness -- opposite to the behavior in single stroke groove formation.

It is felt that this behavior can be explained, qualitatively, on a combination of three factors -- brittle fracture around the grooves, extrusion of cobalt binder due to repeated loading, and the lubricating
effects of the extruded cobalt.

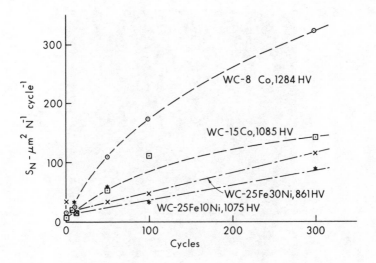

Fig. 6 Cyclic wear volume per unit of applied load,
S_N, vs. number of cycles.

The brittle microfracture of the WC-Co alloys around the grooves was discussed above. Its contribution to wear will increase with the hardness of the material and (disproportionately) with the size of the groove. It is probably a contributing factor responsible for giving the harder materials a greater wear rate. Its exact contribution cannot be determined on the basis of these experiments.

Extrusion of the binder in WC-Co materials is demonstrated in the micrograph shown in Fig. 7. It is seen that even the relatively mild straining of the surface material some distance from a groove results in preferential extrusion of the binder. This extrusion changes the surface material gradually, decreasing the amount of binder and also the residual compressive stresses in the WC grains which oppose their fracture. Cobalt on the surface can also have lubricating properties. Lagerquist (1975) noticed this effect of the low shear strength of cobalt which became smeared over the surface of specimens during testing for thermal fatigue; and Brainard and Buckley (1976) noticed the effect in sliding friction tests.

The SEM observation of the grooves in the WC-Co materials showed clear evidence of smearing of binder material and there was some cracking of WC grains and a few regions where binder had been removed. Immediately outside the groove there was some tilting and cracking of WC grains and particles of very fine debris were found. They consisted of WC fragments and cobalt particles. For specimens exposed to many cycles the smeared layer became so thick that the underlying structure could not be resolved. The WC-FeNi materials showed some, very minor smearing of a layer over the groove surface.

It is thought that the extrusion of the cobalt in combination with its lubricating properties is responsible for the behavior of the WC-Co

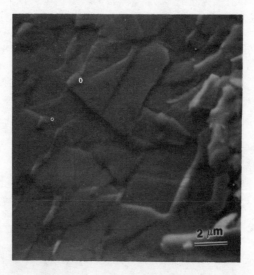

Fig. 7   Deformation pattern outside the groove formed on a WC-10.4% Co specimen during one pass of an R_c indenter loaded at 50 N.

alloys in the cyclic tests. Since the cobalt is extruded only gradually it will take a substantial number of cycles before a steady state groove surface is established. Because of its lubricating effects, materials with the greater Co content will show the lesser wear, as the data show. Also, since it is possible to gradually damage these materials by extrusion of the Co binder in the cyclic loading, their wear resistance under these conditions is considerably less than that of the WC-FeNi alloys for which binder extrusion has not been observed and is not expected to occur. A similar effect was reported by Montgomery (1969) who noticed that various metallic coating on WC-Co had no effect on friction but reduced wear due to contact fatigue. He attributed this to an effect of the diffusing coating metal on the residual tensile stresses of the binder and subsequent improved fatigue life.

The slope of the curves in Fig. 6 is the increase in groove cross sectional area per unit load and per cycle, and can be considered as a cyclic wear coefficient. Attempts were made to correlate this parameter with bulk hardness and microstructural parameters. Only quite general trends could be observed, the data showed substantial scatter.

In general the cyclic wear coefficient, K, decreases with increasing fraction of Co and with increasing binder mean free path in accordance with the extrusion-lubrication explanation offered above. K increases with increased WC grain size. It is not clear if this is due to increases in brittle fragmentation of the WC grains as their average size increases or to the interrelation between binder volume, binder mean free path, and WC grain size.

In correlating with bulk hardness it was found that the WC-FeNi materials show the type of behavior one might expect -- decreasing wear coefficient with increasing hardness -- while the WC-Co materials show the opposite trend, i.e., their wear behavior is not determined primarily by hardness.

It should be noted that the wear test used in this study is somewhat unique for abrasion in that the wear debris remains in the track. For situations where the cobalt is continually removed from the wear scar one would expect that the hardness effects would be quite different, as usually found (e.g., Larsen-Basse and Koyanagi, 1979).

CONCLUSIONS

The simulated abrasion tests which were used in this study show that there is a great difference in the relative performance of cemented tungsten carbide materials with different binder metals, depending on the type of simulated abrasive wear. In simulated abrasion by hard abrasives by single stroke groove formation the following observations were made:

- no permanent grooves form below a minimum load which corresponds to a groove width of 3-4 WC grains,

- groove size is directly related to indentation or scratch hardness for most conditions,

- under high load the more brittle materials show an increased groove size with hardness and the wear correlates well with the product of hardness and local fracture toughness,

- the mechanism of groove formation is plastic deformation aided by brittle fracture along the groove sides for the larger grooves in the more brittle materials, and

- the WC-Co materials have substantially better resistance to this type of wear than do the WC-FeNi materials of the same hardness; this is attributed to the greater ability of cobalt to sustain large strains without fracture and to the expected greater work hardening rate of the WC-Co materials due to their greater content of the hard WC phase.

For cyclic stroke tests in the same track it was found that:

- the WC-FeNi materials show a constant wear loss per cycle and unit load (cyclic wear coefficient) which decreases with hardness; this corresponds to a conventional wear situation,

- the WC-Co materials show an initially large cyclic wear coefficient which gradually stabilizes at a lower value after about 100 cycles,

- the wear of a WC-Co material is greater than the wear of a WC-FeNi material of the same hardness,

- the wear for the WC-Co materials decreases with increasing binder content; this is attributed to the lubricating effect of extruded cobalt on the surface, and

- the mechanism of wear of the WC-Co alloys appears to be a gradual extrusion of binder which may weaken the groove surface material and make it susceptible to removal by subsequent passes of the indenter; the extruded Co also acts as a lubricant which gradually reduces the wear loss per cycle.

ACKNOWLEDGMENTS

The experimental part of this work was supported by the National Science Foundation under Grant DMR 76-17158. The work is based, in part, on a thesis submitted by Nirmal Devnani to the University of Hawaii in partial fulfillment of the requirements for the M.S. degree in Mechanical Engineering. The WC-FeNi materials were kindly provided by Dr. D. Moskowitz of Ford Motor Company and the WC-Co specimens by Dr. Colin Perrott of CSIRO, Australia.

REFERENCES

Brainard, W.A. and Buckley, D.H. 1976, ASLE Trans. 194, 309-318
Doeg, H.H. 1960, J. South African Inst. Mining Met. 60, 663
Evans, A.G. and Wilshaw, T.R. 1976, Acta Met. 24, 939-955, 977
Lagerquist, M. 1975, Powder Met. 18, 35
Larsen-Basse, J. 1973, Powder Metallurgy, 16, 1-32
Larsen-Basse, J. 1980, J. Lub. Technol. 102, 560-565
Larsen-Basse, J. 1983, J. Metals, 35, (11) 35-42
Larsen-Basse, J. and Koyanagi, E.E. 1979, J. Lub. Technol. 101, 208-211

Larsen-Basse, J. and Premaratne, B. 1983, Wear of Materials 1983,
    Ed. K.C. Ludema, ASME, New York, pp 161-166
Larsen-Basse, J. Shishido, C.M. and Salem, L.K. 1976, Proc. 1976 Int.
    Conf. Hard Material Tool Technology, Ed. R. Komanduri, Carnegie-
    Mellon University, Pittsburgh, pp 231-243
Montgomery, R.S. 1969, Trans. AIME 244, 153-157
Moore, M.A. 1980, Fundamentals of Friction and Wear of Materials,
    Ed. D.A. Rigney, ASM, Metals Park, OH, pp 73-118
Nathan, G.K. and Jones, W.J.D. 1966, Proc. Inst. Engrs. 181, 215-226
Tabor, D. 1956, Brit. J. Appl. Phys. 7, 159-166
zum Gahr, K.-H. 1979, Metal Progress, September 45-52

Inst. Phys. Conf. Ser. No. 75: Chapter 9
Paper presented at 2nd Int. Conf. Science Hard Mater., Rhodes

897

Frictional behaviour of diamond and cubic boron nitride abrasives on hard materials

T MATSUO(1), S TOYOURA(2) AND H KITA(2)

(1) Kumamoto University, Kurokami 2, Kumamotoshi, Japan
(2) Yatsushiro National College of Technology, Yatsushiro, Japan

ABSTRACT

Single point abrasive friction tests have been performed by the pin-on-disc sliding method for abrasive materials (including diamond and cubic BN) on various hard disc materials - such as hardened alloy steel, cemented carbide and sintered alumina. The volumetric wear of the abrasive was also measured. In addition, an over-cut fly-milling test was conducted on the same work materials to evaluate the mechanical wear of the abrasives.

INTRODUCTION

Diamond and cubic BN are widely used as tool materials for the machining of hard materials. In these applications, sliding friction and wear are both important but there is little precise information and the frictional behaviour is not well understood (Casey and Wilks, 1973; Enomoto and Tabor, 1973; Tanaka, 1968). In particular, there is scarcity of data on cubic BN. An understanding of the relevant physical properties could lead to further improvements in these materials and their use as abrasives in engineering.

EXPERIMENTAL PROCEDURE

Pin-on-disc-Sliding

The friction measurement, as well as the abrasive grain wear test, was carried out with the geometry as shown in Fig. 1. The friction force was measured by the strain gauge mounted on the thin part of the steel holder. Thus, both the variation of the friction force with sliding distance and the coefficient of friction under steady state conditions were measured. The abrasive grain was either diamond ('D'), cubic BN ('B'), Al_2O_3 ('A') or black SiC ('C'). The grit size was 2.0mm with an exception of cubic BN which was 1.6mm. The disc (100 to 200mm dia.) materials are shown in Table 1. The vertical load was either 2.5 or 4.5N and the sliding speed could be varied between 5.0 and 15 m s^{-1}. Scanning electron microscopy was used to examine the worn surface of the abrasive material. A given abrasive grain slid in its particular track during the measurement of

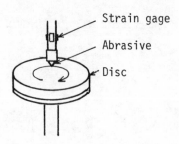

Fig. 1   Measurement of frictional force

TABLE 1   <u>Disc Materials and Hardness</u>

Material type	Compositions, wt %	Hardness	Remark
Hardened steel	1.0%C, 1.5%Cr, Rem. Fe	HRC63	High-C Cr steel
Co-alloy	3.5%C, 17%Cr, 27%W, Rem.Co	HV30 960	Stellite No.6
Cemented carbide	76%WC, 14%(TiC+TaC), 10%Co	HRA90	Type P20
Sintered alumina	95%Al$_2$O$_3$, 4%SiO$_2$	HRA86	s.g.$\mathcal{P}$ =3.6
Sintered magnesia	97%MgO, 1%SiO$_2$	HRA85	s.g.$\mathcal{P}$ =3.5

friction but each separate grain had its own track. Scatter in the co-efficient of friction was small and, therefore, was determined as the average for three individual grains. The wear volume of the abrasive was determined from its loss of height and the worn flat area as detailed in an earlier paper (Matsuo, 1981). Wear tests on the abrasive were carried out at the same speeds and normal loads as in the friction tests, where the sliding distance was almost 4000m, but the volumetric wear was based on the average of about six experiments due to the increased scatter of the data. The groove volume of wear for the disc was also measured - using a surface profilometer.

<u>Fly-milling</u>

The overcut fly-milling tests on the abrasive grains were done with a surface grinder (1.1KW) and with the workpiece slightly inclined - as shown in Fig. 2. The method was similar to that of Stanislao (1969) and Brecker (1973). When making a measurement, the metal wheel was lowered until it just touched the workpiece surface at its lowest end. As the work was fed past the wheel the depth of the groove was automatically increased, due to the slope of the work surface, but it may have cycled up and down if the abrasive grain chips. Each subsequent cut then increased in depth, until the grain fractured, whilst attritious wear of the tip occurred. The measurement of attritious wear was made by tracing across the grooves with a stylus instrument to determine the decrease in depth of cut between successive grooves. The conditions of the fly-milling tests are summarised in Table 2 and it should be noted that no cutting fluids were used.

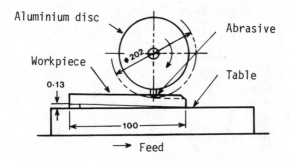

Fig. 2 Arrangement for overcut fly-milling with abrasive specimen

TABLE 2 <u>Milling Test Conditions</u>

Abrasive	Diamond (natural)	'D'	HK	8000
	Cubic boron bitride	'B'	HK	4700
	Regular alumina	'A'	HK	2100
Work material	Hardened steel		HRC	63
	Cemented carbide		HRA	90
	Sintered alumina		HRA	86
Cutting condition	Cutting speed	Vg=38 m s^{-1}		
	Work speed	Vw=4.0 m/min (f=1.1mm/rev.)		

RESULTS AND DISCUSSION

Coefficient of Friction

Fig. 3 shows the variation in the coefficient of friction with sliding distance, as three different abrasive grains rub against the hardened steel, with a normal load of 2.5N and at three sliding speeds. The co-efficient of friction for the 'D' abrasive decreased initially with increasing sliding distance and then settled down to a very low constant value of 0.06 to 0.11, whilst the 'A' and the 'B' abrasives developed a constant and relatively high value of 0.4 to 0.5. It is also apparent that increasing the sliding speed reduced the coefficient of friction for the 'A' grain but had little effect on the behaviour of the 'D' grain. This is consistent with a friction coefficient of 0.1 for diamond sliding on Cu as reported by Casey and Wilks (1973). According to an observation of the worn grain surface (Fig. 4a and b), there is no substantial differ-ence in appearance between the two abrasives. But, the 'D' abrasive dis-closes a little more flat surface than the 'A' abrasive.

Fig. 5 shows the measured coefficient of friction for three different abrasives on the Co-alloy disc using normal loads of 2.5 and 4.5N. It is evident that the steady state and low value for abrasive 'D' is independ-ent of sliding speed and load. Furthermore, the friction is directly

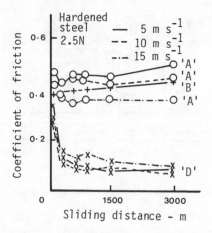

Fig. 3 Friction of abrasives against hardened steel

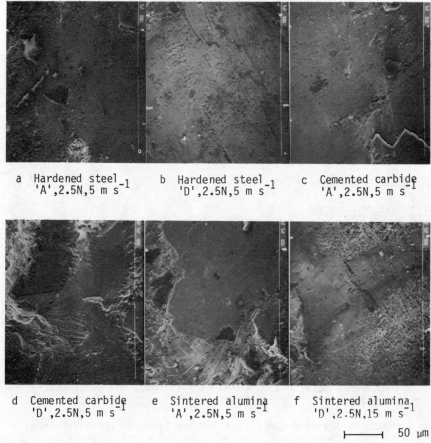

a Hardened steel
   'A',2.5N,5 m s⁻¹

b Hardened steel
   'D',2.5N,5 m s⁻¹

c Cemented carbide
   'A',2.5N,5 m s⁻¹

d Cemented carbide
   'D',2.5N,5 m s⁻¹

e Sintered alumina
   'A',2.5N,5 m s⁻¹

f Sintered alumina
   'D',2.5N,15 m s⁻¹

⊢────┤ 50 μm

Fig. 4 SEM micrographs of worn abrasives surfaces

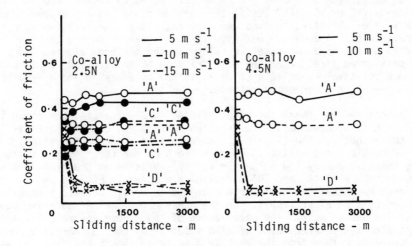

Fig. 5 Friction of abrasives against Co-alloy

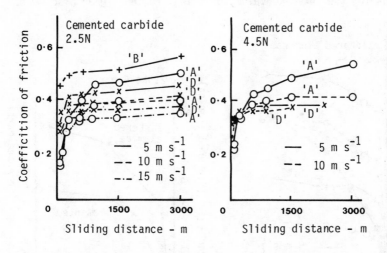

Fig. 6 Friction of abrasives against cemented carbide

similar, at a given load and speed, for this abrasive when it slides on both the hardened steel and the Co-alloy.

In Fig. 6 the effect of sliding speed and normal load is shown for three abrasives against a cemented carbide disc and, here, the abrasive 'D' has a similarly high level of friction as 'A' and 'B'. This fact can be considered an important frictional characteristic of the 'D' grain against cemented carbide. In Fig. 4c and d, the worn surface of abrasive 'A' and 'D' are shown. It is noted that, while the 'A' abrasive reveals a relatively smooth surface, the 'D' abrasive shows a picked off surface.

This rugged surface with the 'D' can be considered to result from a marked abrasion by TiC or WC particles in the carbide disc. In a previous study on the wear of abrasives (Matsuo and Toyoura, 1982), the relative wear of 'D' abrasive to 'A' abrasive was significantly greater on the carbide disc than on the hardened steel, Co-alloy or the sintered alumina discs. It is considered likely that the high friction of abrasive 'D' on carbide, and hence relatively high rate of wear, is due to adhesion and a dynamic friction peculiar to cemented carbide.

Results obtained for these abrasives sliding on alumina disc are shown in Fig. 7. It is apparent that the friction coefficient on this material is about twice that for the 'A' and 'C' abrasives on the hardened steel, Co-alloy or carbide disc whilst it is much the same for the 'B' on all discs. Again, there is a very low value for the 'D' abrasive on the alumina disc. The worn surface of 'A' and 'D' abrasives are shown in Fig. 4e and f. It is evident here that the worn surface of the 'D' grain reveals a peculiar flat surface like a sandy soil. On the other hand, the 'A' abrasive shows a typical adhesive wear. The measurement of the groove volume (Fig. 8) demonstrates a greater removal of disc material by the 'D' abrasive from the alumina disc than from the carbide disc, even though the friction was significantly lower. This surely simply reflects the relative toughness of alumina and cemented carbide. Namely, the fracture toughness, K_c, is about 4 and 6MNm$^{-3/2}$ for the alumina and carbide respectively. The alumina disc, having the lower Kc value, would lead to the higher disc wear.

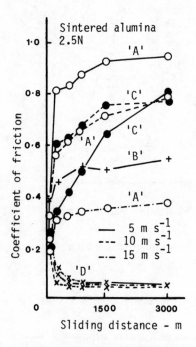

Fig. 7 Friction of abrasives against sintered alumina

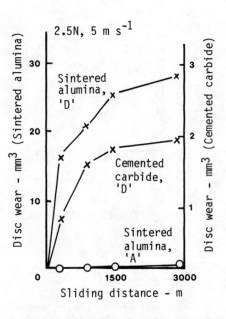

Fig. 8 Measurement of disc wear

Further measurement of the friction of abrasives 'A' and 'D' on sintered
magnesia is shown in Fig. 9. In general, the behaviour is similar to the
results for the other discs.

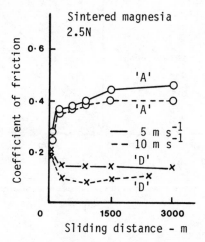

Fig. 9   Friction of abrasives against sintered magnesia

Wear of the Abrasives

It is a matter of concern to know the wear of abrasive tip in these
pin-on-disc friction tests. Examples of the measurement of the tip wear
of the abrasives is represented in Figs. 10 and 11. It is evident that
wear of 'D' abrasive is always very small but a distinct wear flat was
invariably developed. Similarly, small wear was obtained for the 'D'
grain on the Co-alloy and sintered magnesia discs, whilst the wear of the
'D abrasive on cemented carbide was relatively large, as mentioned above.
It can be considered from the results on the abrasive wear that the small
wear of the 'D' grain is related to its low friction. However, a conclusion cannot be drawn due to an insufficient number of samples for the
relation between friction and wear.

Fly-milling

In these experiments tip wear of the abrasive and its micro-fracture
properties were evaluated by measuring the variation of groove depth. The
depth of the groove can be theoretically calculated when there is no wear
of the abrasive but, in practice, the actual depth deviated from this
'ideal' value (i.e. was usually less than the 'ideal' value). The extent
of the deviation will be much dependent upon abrasive and workmaterial.
Attritious wear as a function of the actual depth of cut is shown in Figs.
12 and 13. The wear parameter Δr_a was obtained by dividing the total wear
height by the number of cuts and the line shown is a regression plot with
a confidence level of 95%. It is clear that the wear Δr_a decreases
slightly with depth of cut in all cases. The relatively high rate of wear
of abrasive 'D' on the hardened steel is somewhat surprising.   It is
considered, as a reason for the large wear of 'D' grain, that there exists
a chemical wear mechanism based on the solution of carbon in the steel.
Similar experiments using abrasives 'A' and 'B' on cemented carbide and
sintered alumina were fruitless since these abrasives fractured before
cutting the workpiece.

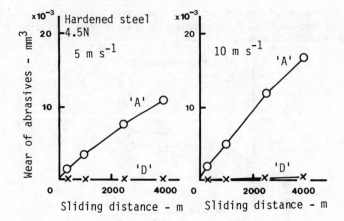

Fig. 10 Wear of abrasives vs. sliding distance

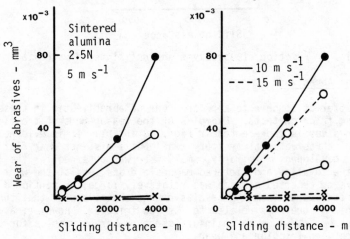

Fig. 11 Wear of abrasives vs. sliding distance

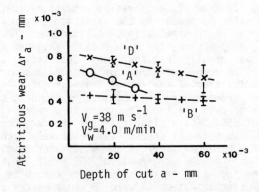

Fig. 12 Variation of attritious wear for all abrasives
with depth of cut on hardened steel

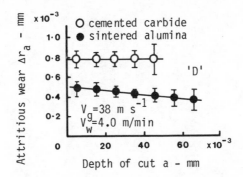

Fig. 13 Variation of attritious wear for 'D' abrasive with
depth of cut on cemented carbide and sintered alumina

SEM examinations of the worn surface on the abrasives usually revealed a
very smooth worn flat on the ultrahard materials together with a large
number of cracks.   However, the worn surface of abrasive 'C' showed
evidence of marked plastic flow. In addition, X-ray analysis confirmed the
presence of Fe as a metallic film on the surface of all the abrasives as
used on the hardened steel workpiece. It was thought likely that this
film would tend to blunt the cutting edge and increase the temperature of
the abrasive workpiece interface. If so, then the relatively high rate of
wear of abrasive 'D' could have been due to solution of carbon in the Fe
film.

It is clear from this work that there is no simple relationship between
the friction measured on the pin-on-disc apparatus and the wear both under
those conditions and in the fly-milling experiments. This suggests that
there is no universal mechanism of wear for these combinations of
materials and therefore, at this time, no obvious basis for a predictive
model.

ACKNOWLEDGEMENTS

The experimental work was carried out at Kumamoto University and the
authors wish to thank E. Oshima, A. Ogata and Y. Shibata (Department of
Production Engineering) for their technical assistance. Also, they thank
N. Kawabata (Noritake Diamond Industry) for supplying the abrasive;
Y. Uehara (Toshiba Tungaloy Co.) and T. Hagio (Nippon Tungsten Co.) for
their preparation of disc and workpiece materials.

REFERENCES

Brecker,J.N., Komanduri,R.K. and Shaw, M.C. 1973, Annals of CIRP,22 219-225
Casey, M. and Wilks, J. 1973, J. Phys. Appl. Phys, 6 1772
Enomoto, Y. and Tabor, D. 1973, Proc. Roy. Soc. A-373 405-412
Matsuo, T. 1981, Annals of CIRP, 30 233-236
Matsuo, T. and Toyoura, S. 1982, Yogyo-Kyokai-Shi, 90 25-30
Stanislao, J. and Gielisse, P.J. 1969, Tech. Report, AD-704611,
    University of Rhode Island
Tanaka, T. 1968, Thesis (Ph.D) Osaka University

Inst. Phys. Conf. Ser. No. 75: Chapter 9
Paper presented at 2nd Int. Conf. Science Hard Mater., Rhodes

907

Microstructure and wear of cubic boron nitride aggregate tools

R M HOOPER AND C A BROOKES

Dept. of Engineering Science, Univ. of Exeter, Exeter EX4 4QF, U.K.

ABSTRACT

The microstructure of some metal cutting tools based on aggregates of cubic boron nitride has been examined using etching techniques. The aggregates were found to consist of strongly bonded particles of boron nitride and a 'filler' phase consisting of aluminium nitride and aluminium diboride. Investigation of typical worn surfaces, produced on cutting tool inserts of this material, showed that the wear mechanism had been influenced by the development of the intrinsic defect structure of synthesised cubic BN; the interaction of the tool with its environment; and the formation of protective surface layers.

INTRODUCTION

Cutting tool inserts based on polycrystalline aggregates of cubic boron nitride (cubic BN) allow the use of significantly increased cutting rates on relatively hard (i.e. 800HV30) workpiece materials. The inserts possess properties of high hardness and wear resistance which are retained at temperatures in the region of 1000°C. However, wear scars are ultimately observed on these tools and, in common with other ultra-hard materials, it is likely that the chemical interaction between the tool and its environment - including the workpiece - will have an important role in the wear mechanism. By analogy with etching, these chemical wear processes will also be influenced by the intrinsic and developed defect structure of the tool material. In this paper, the microstructure of cubic BN aggregates is outlined and a mechanism of wear proposed in terms of their chemical interaction with the environment.

MICROSTRUCTURE

An analysis of the mechanical properties of aggregate materials requires a knowledge of the properties of their components, their volume fractions and information on the contiguity or amount of contact between the phases. Such a model has been applied by Lee and Gurland (1978) to the hardness data for carbide based cermets. Attempts to rationalise the hardness data of cubic BN based aggregates and single crystals, using this model, have generally been unsuccessful. Consequently, chemical etching techniques have been used to investigate their microstructure further.

X-ray diffraction analysis of tool inserts showed them to contain boron nitride in the cubic phase alone, together with the borides and nitrides of aluminium. In the case of some inserts heated to temperatures in excess of 1500°C, for extended periods (~ 1 h) in a vacuum ($<10^{-5}$ mbar), there was some evidence of transformation to the hexagonal form together with the loss of the aluminium boride. In addition Poduraev (1979) has reported evidence of transformation to the hexagonal phase in used tool tips. It was therefore possible to reveal the hard phase structure by chemical leaching of the aluminium compounds using boiling sodium hyrdoxide solutions.   The microstructure of the cubic BN particles themselves was etched by oxidation using the technique described in a companion paper presented at this Conference. Thus, at temperatures above a critical level (i.e. approximately 900°C for cubic BN) the highly localised defects introduced by point loading at lower temperatures, or during synthesis, are developed into much larger dislocated volumes which may be subsequently etched by atmospheric oxidation.   Fig. 1 shows a typical microstructure produced by the application of these two techniques and the following salient features have been revealed:

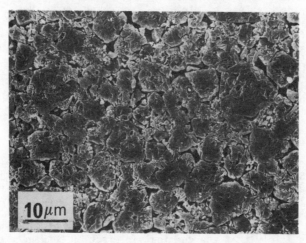

Fig. 1   Etched microstructure of cubic BN aggregate tool.

(a) The cubic BN particle size was variable but typically of the order of ten micrometres. The dimensions are comparable with those of the analogous diamond based aggregates but are about an order of magnitude larger than the hard phase particles in a conventional cermet material.

(b) The particles are well bonded together by direct sintering and intergrowth - although it must be noted that the oxidation etching procedure is likely to reduce the apparent density of the aggregate from the intrinsic value.

(c) The individual particles often contained inclusions which appeared to serve as three dimensional indenters and during heat treatment extensive defect structures were developed around these inclusions which were sub-

sequently removed by oxidation. In parallel experiments on the effects of
thermal shock on cubic BN particles, it was noted that the stresses
produced by differential thermal expansion of these inclusions sometimes
led to fracture of the particle.

Fig. 2 Etched fracture face of a cubic BN aggregate tool.

The complete aggregate therefore can be considered as consisting of a continuous skeleton of cubic BN formed during its synthesis. The response of
the aggregate to applied stresses will therefore principally involve the
deformation of this skeletal structure. This is supported by the
appearance of the fracture faces (see Fig. 2, an etched fracture face)
where the fracture is seen to have propagated through the intergrown
structure as a whole. The aluminium boride and nitride may therefore be
more appropriately described as a filler within the cubic BN inter-
particulate spaces. Thus, since information on the grain size; contiguity;
and intrinsic hardness of these phases is not available from the microstructures, it has been impossible to quantify their influence on the
mechanical properties of the aggregate as a whole.

WEAR

Cubic boron nitride tool inserts are used with conventional cutting
geometry, the work is rotated about an axis and the tool forced into its
surface removing metal as a chip from the cut. The chip moves over the
upper or rake face of the tool and experiments with conventional materials
(e.g. Smart and Trent, 1975) have shown that the load on the tool is high
and that temperatures of the order of 900°C may be generated. In order to
exploit their high hardness at elevated temperatures, these aggregates are
often used with a negative rake angle which increases the shear strain in
the chip and hence the temperature at the cutting interface. This reduces
the yield stress of the workpiece and facilitates the cutting process.

The conditions at the tool/workpiece interface for cubic BN aggregates cutting grey cast iron at various speeds and associated temperatures have been investigated by examination of the tools at the end of a cutting sequence, and of the chips themselves. In addition, instantaneous information has been obtained using the so-called quick stop technique (Wright and Trent, 1974) which, by explosively detaching the tool from the work during cutting, leaves the chip adhering to the work or tool surface and hence preserves the information otherwise lost if the tool is slowly withdrawn from the cut.

Since earlier observations of aggregates and single crystal cubic BN indicated that etching by atmospheric oxygen was an important process (Brookes and co-workers, 1983), the influence of the environment was also studied by application of jets of oxygen or nitrogen to the rake face of the tool during cutting.

CHIP MORPHOLOGY

The conditions at the cutting interface, and common features of the resultant chips, are illustrated by the typical quick stop section shown in Fig. 3. This particular micrograph shows a polished and etched section of a mild steel chip, cut by a cubic BN tool, which has been rapidly removed in the quick stop test. According to the extensive work of Trent (1979)

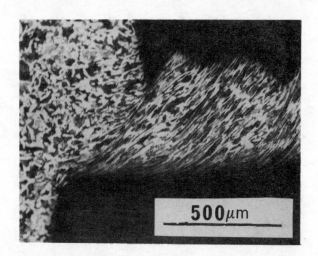

Fig. 3 Quick stop section of mild steel cut by cubic BN aggregate tool. (Cutting speed 13 m s^{-1}).

the conditions at the cutting interface - high compressive stresses on clean material at high temperatures - can produce a friction weld of the chip to the tool. In extreme cases this can lead to the formation of a built-up edge of seized material which protects the tool. The material close to the tool is subjected to very high strains - and rates of strain - and may be considered to behave in a fluid-like fashion. A flow zone is then formed with a steep velocity gradient away from a zero value at the rake surface. This zone is clearly seen in the etched section together with the shear zone produced by the cutting edge of the tool. In the case of the grey cast iron, the graphite flakes (Fig. 4) do not support these large shear stresses and lead to a segmented structure, with the consequent production of small chips.

Metallographic sections of the actual chips, cut under various conditions, revealed a layered structure at the base of these chips formed at lower cutting speeds (e.g. 1.25 m s^{-1}, Fig. 5). The incidence of this structure

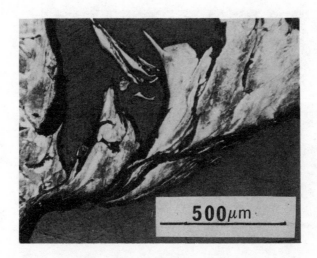

Fig. 4 Quick stop section of grey cast iron cut by cubic BN aggregate tool at 1.25 m s^{-1}.

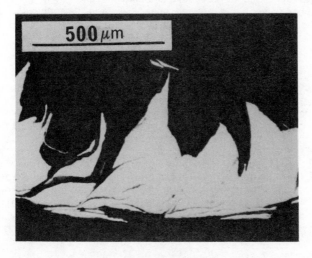

Fig. 5 Polished metallographic section of similar grey cast iron chip.

was markedly reduced on increasing the cutting speed and became almost
negligible at 14 m s^{-1} (Fig. 6). A similar but less pronounced effect was
obtained by enriching the environment of the tool with oxygen. Examination of these chips, using scanning electron microscopy, confirmed that
the base of chips produced at high speed or in an enriched atmosphere were
smooth; and that in contrast, those formed at lower speeds or in the
presence of nitrogen showed evidence of fragmentation into a layer-like
structure.

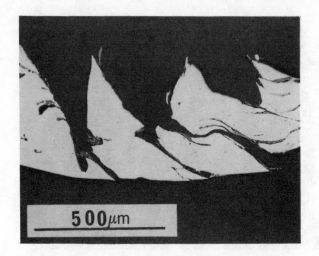

Fig. 6 Polished metallographic section of
grey cast iron chip cut at 14 m s^{-1}.

Extreme examples of these effects are shown in Figs. 7 (a) (low speed,
nitrogen atmosphere) and 7 (b) (high speed, oxygen atmosphere).

TOOL INSERTS

Examination of the tool inserts themselves revealed a number of pertinent
features. In the case of the higher speed tests, a shallow crater had
formed on the rake face extending almost to the cutting edge - although
this edge had remained sharp. The surface of the tool in the region of
the crater was coated with a metallic layer of similar composition to the
workpiece, (A in Fig. 8). This layer proved extremely tenacious and was
removed only after extensive chemical attack. The particles below were
then found to be polished with some grooving in the direction of flow of
the chip. Application of the two stage etching sequence described earlier
had little effect on this area - other than to emphasise the grooving
present (Fig. 9) and in cases of more protracted etching, revealed the
etch features shown in Fig. 10. These features are essentially similar to
those observed on single crystals of cubic BN reported in the companion
paper and we believe that they are evidence of the plastic deformation
produced during the initial fabrication of the aggregate. This suggestion
might be verified by further observation on unused tool inserts. The
etching process had a dramatic effect on the morphology of those particles

Fig. 7 Base of chip cut (a) at 1.25 m s^{-1} in nitrogen
atmosphere and (b) 14 m s^{-1} in oxygen atmosphere.

Fig. 8   Used cubic BN aggregate tool showing
cratering and metal coating.

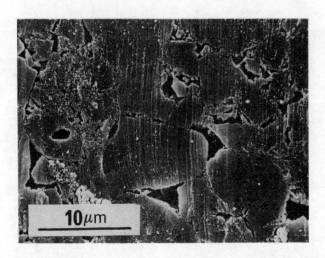

Fig. 9 Polished grains in crater area of cubic BN tool.

Fig. 10   Etch features developed on grains
in crater area of cubic BN tool.

in the adjacent zone (B) which became severely etched as shown in Fig. 11.
Particles in an area further removed from the cutting edge (C) were unaffected and yielded the same etched structure as an unused insert. In
the lower speed quick stop tests the tools invariably fragmented and in
most cases delamination fracture occurred, i.e. fracture parallel to the
rake face, suggesting the formation of a strong bond between the layer and
the tool. Fig. 12 shows the edge of one such fracture with a small frag-

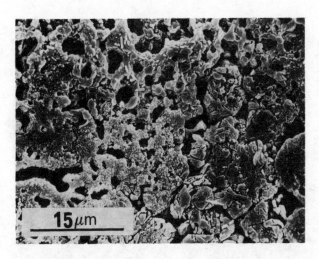

Fig. 11 Developed etch structure on grains
adjacent to the crater on a cubic BN tool.

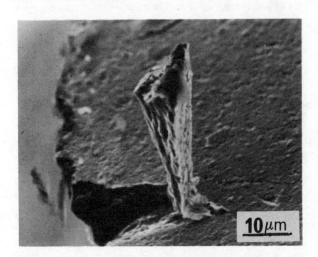

Fig. 12 Cubic BN fragment attached to fractured
tool by metallic layer.

ment of cubic BN firmly attached to the metallic layer. Poduraev (1979)
found similarly that low speed cutting leads to more frequent fracture of
cubic BN inserts than higher cutting speeds.

Large, smooth grooves were noted in the wear scar on the clearance face of
the tool when used in air and also in an oxygen enriched atmosphere. This
region was reduced in area and showed more evidence of fracture when a
nitrogen atmosphere was employed (Fig. 13 (a) and (b)).

a   b

Fig. 13 Clearance face wear scars produced (a) in
oxygen atmosphere and (b) nitrogen atmosphere.

In summary, three zones may be identified on the cutting tool:

(a)   The cratered and grooved area near the cutting interface, on both
the rake and clearance faces of the tool, which has a polished appearance
and may be coated with a thin metallic layer.

(b)   A second zone which shows little evidence of polishing but which,
on subsequent heat-treatment, develops an extensive etched structure.

(c)   The remainder of the tool which is unaffected by the cutting process.

DISCUSSION AND PRELIMINARY CONCLUSIONS

Throughout this work there has been no evidence to suggest that fracture
processes in cubic BN play a significant role in the wear of cutting tools.
Our observations of the wear of a number of cubic BN tools used to cut
grey cast iron have been interpreted in terms of the interaction between
the tool and its environment (including the chemical composition of the
workpiece). On the clearance face of the tool, wear proceeds by the
continuous production and development of defective areas and their subsequent removal by oxidation in a process referred to as etching wear. This
process may be regarded as akin to chemical-mechanical polishing and
produces a smooth surface and slight cratering of the tool. The intrinsic
defect structure of the individual particles may then be revealed by
subsequent oxidation etching of these polished surfaces.

The material close to this zone may be deformed at lower temperatures but does not develop the extensive defect structure of the first zone. However, this can be achieved by subsequent heat treatments. The creep of cubic BN under conditions of point loading at relatively low homologous temperatures, reported in the companion paper, would serve to exacerbate this process.

It is postulated that a metallic protective layer, whose exact composition is yet to be determined, is formed on the rake face. At lower temperatures, or cutting speeds, the cohesion between the chip and the layer is comparable with that between the layer and the tool. Thus, the layer is intermittently removed by a process of sub-surface fracture of the tool due to the high adhesive bonds developed by the chip. The wear of the rake face takes place by this attrition and by its periodic exposure to the environmental oxygen and etching wear. At higher temperatures, the cohesion between the chip and the layer is reduced and the chip flows smoothly over its surface. These effects are also influenced by the oxygen partial pressure in the environment. Hence, both the favourable physical and chemical properties of the cubic BN aggregates are responsible for their superior performance at high cutting rates.

ACKNOWLEDGEMENTS

The authors are indebted to Dr M. Wise and Mr E. Smart, of the University of Birmingham, for the use of their quick stop apparatus and to De Beers Industrial Diamond Division Ltd for a grant to the laboratory.

REFERENCES

Brookes, C.A., Hooper, R.M., Lambert, W.A. and Ross, J.D.J. 1983, Ultrahard Materials Application Technology, Vol. 2. Ed. P. Daniel. De Beers Industrial Diamond Division, Ascot, U.K., pp 72-80.
Lee, H.C. and Gurland, J. 1978, Mater. Sci. Eng. 33 125-133.
Poduraev, V.N., Yumatov, V.A., Karavlov, A.K. and Zhbanov, P.D. Vestn. Mashinestr. 1979, 58 pp 47-49.
Smart, E.F. and Trent, E.M. 1975, Int. J. Prod. Res. 13 265-290.
Trent, E.M. 1979, Treatise on Materials Science and Technology - Vol. 13 - Wear. Ed. D Scott. Academic Press, London, pp 443-489.
Wright, P.K. and Trent, E.M. 1974, Met. Techol. 1 13-23.

Inst. Phys. Conf. Ser. No. 75: Chapter 9
Paper presented at 2nd Int. Conf. Science Hard Mater., Rhodes

919

Comparison of sliding and abrasive wear mechanisms in ceramics and cemented carbides

E A ALMOND, L A LAY AND M G GEE
National Physical Laboratory, Teddington, UK

ABSTRACT

The wear of various ceramics and WC/Co cemented carbides has been examined for a range of conditions using abrasive slurries, a sliding indenter and a vertical pin-on-disc test. The cemented carbides performed better than the ceramics when abrasive-wear predominated, probably because of differences in the surface cracking characteristics of the two materials when indented by abrasive particles and wear debris.

INTRODUCTION

The implicit assumption in wear testing is that the results will in some way be relevant to service performance and that data can be uniquely defined in terms of specific parameters of the test system. The degree of fulfillment of such expectations ranges from seldom to sometimes, and only qualitative conclusions can be safely drawn from the results and preferably these should be no more than descriptions of characteristics of the worn surfaces.

The situation for hard materials requires further qualification since the majority of conventional wear test geometries were devised to accentuate and identify wear failure mechanisms that predominate in metals rather than those peculiar to hard brittle materials. Also, unlike in metals, there have been no major investigations into test reproducibility for hard materials, except for recent ASTM-B611 studies where certain deficiencies have been revealed.

The present paper describes an initial pragmatic stage of investigating the feasibility of obtaining relevant reproducible data from laboratory wear tests on specimens of simple geometry of hard materials. For this purpose, a study was made of the wear characteristics of ceramics and cermets using empirical wear tests originally developed for metals, coatings and cemented carbides. Four types of test were performed on a range of ceramics and cemented carbides in order to study microstructural wear damage and identify differences in the behaviour of the materials under a range of conditions. The tests were:

a) pin-on-disc;
b) sliding-indenter;
c) abrasive-slurry test to ASTM-B611;
d) abrasive-slurry test on a modified polishing wheel.

At the time of testing, no major modifications had been made to alter the test geometries nor loading conditions to make them more suitable for measuring the properties of ceramics and cermets.

MATERIALS

The test materials are given with some microstructural details in Tables 1 and 2, together with details of the tests performed on each material.

The cemented carbides ranged in hardness from 1890 HV30 for a 4wt%Co binder phase content to a hardness of 1060 HV30 for a 20wt%Co binder phase content. Radial cracks were formed at the corners of the hardness impressions and the total lengths of the four cracks as measured from the indenter corners to the crack tips are recorded in Table 2. Although radial cracks were also formed in the ceramics, they were not reproducible nor easily distinguished from the general surface damage around the indentation.

Referring to the tests by the letters used above, the combination of tests used on each material was as follows. Various aluminas were tested in all four tests. A range of Co/WC cemented carbides containing 3-15wt%Co was tested in (b) and (c). Fine- and coarse-grained 6wt%Co/WC cemented carbides, and hot-pressed and reaction-bonded forms of Si_3N_4 were tested in (a) and (d). Reaction-bonded SiC and a fully stabilized ZrO_2 were tested in (a),(b) and (d). A boro-silicate glass was tested in (a), a sialon and partly-stabilized zirconias in (b), and a cubic BN in (d).

TEST METHODS

a) Pin-on-Disc Tests

The disc material in all these tests was a 95wt%-purity sintered alumina. The test material was the pin. The procedure (Fig. 1a) consisted of applying a load of 37.5N to a 4mm dia. cylindrical pin which pressed against the top of the rim of a 70mm dia. disc rotating in a vertical plane at 100 rev/min to give a sliding speed of 0.37 m/s at the pin/disc interface. In some tests, the lower part of the wheel was immersed in water to a depth of 20mm so that the wear interface was kept moist during rotation. Friction was measured from the flexure of a strain-gauged cantilever bar which restrained the sideways motion of the specimen holder. The relative wear rate (displacement at the pin/disc interface) was measured with a displacement transducer.

b) Sliding Indenter Tests

In this test (Fig. 1b) the specimen was mounted on a sliding table which was moved in a horizontal plane against the apex of a vertically mounted

TABLE 1   Ceramic Materials

Material	Composition wt%	Minor additions	Grain size μm	Hardness HV0.2	Hardness HV2.5	Hardness HV30	Tests
A11	99.5%Al_2O_3	CaO,SiO_2,MgO	15	1640	1480	1270	b,c
A12	99.5%Al_2O_3	CaO,SiO_2,MgO	12	1825	*	*	a,c,d
A13	95% Al_2O_3	CaO,SiO_2	5	1740	1200	1080	a,b,c
A14	94% Al_2O_3	CaO,SiO_2		1490	880	900	b,c
ZR1	95% ZrO_2	CaO	30	1250	850	*	a,b,d
ZR2	96.6%ZrO_2	MgO	40-60	1170	970	960	b,d
ZR3	96.6%ZrO_2	MgO	40-60	1140	950	890	b,d
HPSN	Si_3N_4	MgO	1	1590			a,d
RBSN	Si_3N_4		0.2	1200	560	450	a,d
Sialon	Si-Al-O-N	Y,Mg	1	1440	1370	1310	b,d
RBSC	SiC	Si	5-50	2470	1810		a,b,d
B	Borosilicate glass	Na_2O trace	N.A.				a,d
CBN	Cubic boron nitride			4310			d
Agate	SiO_2		N.A	770			d

* Hardness value not measurable due to fragmentation around indentation

TABLE 2   Properties of Cemented Carbides

Material	Co wt%	Other additions vol%	Grain size* μm	Hardness HV2.5	Hardness HV30	Crack length** μm	Tests
HA	4	0.5%Cr_3C_2	<1	1890	1890	72	b,c
HB	8	1%Cr_3C_2	0.8+0.5	1740	1750	190	b,c
HD	6	0.1%TaC	<$\overline{1}$	1715	1660	360	c
HE	5	0.5%TaC	0.6+0.3	1595	1550	146	b,c
HF	5.8	None	$\overline{0}$.96	-	1540	326	a,d
HG	14	0.5%Cr_3C_2	0.7+0.3	1375	1375	164	b,c
HH	6	None	3.$\overline{2}$2	-	1230	150	a,d
HI	11.5	None	2.5+1.1	1175	1195	38	b,c
HJ	10	None	$\overline{4}$	1290	1190	18	c
HK	13	None	1.6+0.9	1220	1150	100	b,c
HL	20	None	3.$\overline{5}$	1065	1060	0	c

* Mean linear intercept measurement   ** At 294N load
a Pin-on-disc   b Sliding indentation   c ASTM slurry test
d Dish slurry test

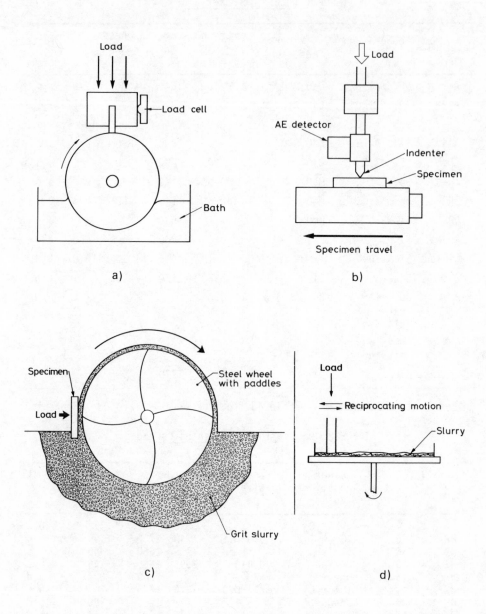

Fig. 1 Geometry of: a) sliding wear test, b) sliding indenter test, c) ASTM abrasive slurry test, d) abrasive slurry test

Rockwell-C diamond indenter of 0.2mm radius. The instrument was modified to enable measurements to be made of the horizontal force exerted on the specimen by the indenter. Also, the existing acoustic emission device supplied with the instrument was replaced by a more sensitive version. All tests were performed on polished cross sections of the specimens at a loading rate of 4N/s at table speeds of 0.5mm/s except for a few tests on small specimens where the speed was reduced to 0.2mm/s. The load was increased linearly to a maximum of 100N except in tests on the alumina, AL1. Multiple-pass tests involving up to 20 passes down the same track were also performed.

c) ASTM-B611 Abrasive-Slurry Tests

This is a standard test used for measuring the abrasion resistance of mining grades of cemented carbides. A 169mm dia. steel disc rotating in a vertical plane is fitted with curved vanes on both faces and is immersed to a depth of 60mm in a slurry of 30 mesh (630um ave. dia) alumina grit in water in the proportion of $4g/cm^3$ (Fig. 1c). In the present tests, in accordance with the standard, the 40 x 20mm face of a 4mm thick cemented-carbide specimen was pushed with a horizontal load of 1962N against the rim of the disc. When rotated at 100 rev/min the rim became coated with alumina which was fed into the specimen/rim interface for a period of 10 min. In addition to using the test on the cemented carbides, it was tried on the aluminas, AL1, AL2, and AL3, but the severity of the damage was so great under standard conditions that it was necessary to reduce the load and times to 981N and 2.5-5 min respectively.

d) Abrasive Slurry Tests on Polishing Wheel

For this test the abrasive slurry was contained in a Petri dish mounted on a polishing wheel which rotated at a speed of 100 rev/min. The 5 x 5mm face of a vertically mounted specimen was pressed into the slurry under a load of 10N and the specimen holder was moved in an arc across the wheel for the duration of a 10 min test. The calculated relative speed between the specimen and disc had a mean value of 0.35m/s, with a minimum of 0.31m/s and a maximum of 0.46m/s.

The grits used were Al_2O_3, SiO_2, SiC and B_4C with a nominal particle size of 29μm, and some smaller grits to study the effect of grit size on abrasive wear.

RESULTS

Pin on Disc Tests

Examination of the wear surfaces produced in the dry tests on the majority of the ceramic specimens showed linear striations which ran parallel to the direction of rotation of the disc (Fig. 2a). These consisted of alternate areas of rough and smooth surfaces representing areas which had recently broken away from the specimen and areas where the surface layers were deformed and fractured and were in the preparatory stages of detachment from the specimen (Figs. 2b and c). A small amount of chipping of the trailing edge was visible in some pin

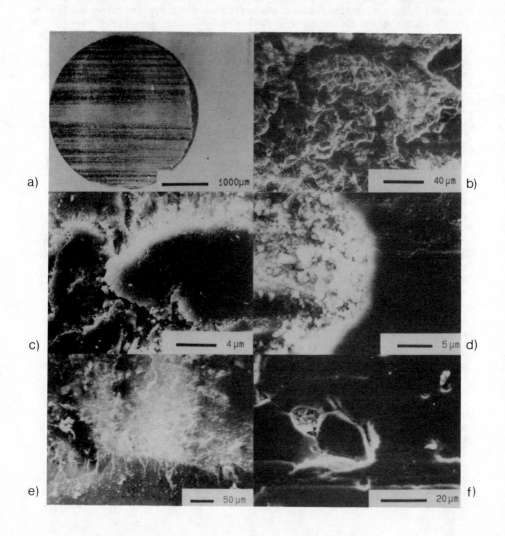

Fig. 2 Wear surfaces from dry sliding wear tests:
a) RBSC pin, b) ZR1 pin, c) ZR1 pin,
d) AL3 pin, e) borosilicate pin, f) AL2 pin

specimens. Wear debris accumulated in the cavities of the rough area and there were microcracks at the edges of the cavities (Fig. 2d).

There were exceptions to this type of wear surface. The surface of the glass pin B was rough and worn, and there were cracks leading down into the glass from the surface. There was a high density of cracks near to the surface (Fig. 2e).

Specimens which gave low wear rates, characteristically gave a smooth wear surface. The wear surface of the AL2 pin was an example (Fig. 2f), but it was also grooved with some cracking from pores and along the grain boundaries. There was also anisotropic behaviour, with pitting of some grains. This has been noted before in alumina (Wallbridge, Dowson and Roberts, 1983) and in magnesia partly stabilized zirconia (Hannink, Murray, and Marmach, 1983).

The wear surfaces of the ceramic pins in the wet tests were normally smooth and polished, revealing porosity in the material (Fig. 3a). However, in both of the wet tests on alumina pins, ring cracks were formed. These cracks were intergranular and often involved cracking between the two phases in the alumina (Fig. 3b).

The wear debris was very similar in appearance in most of the tests despite the variation in materials. Very fine particles (Fig. 4a) constituted the greater part of the debris. These had often agglomerated into large particles which were a few tens of micrometres in diameter, and were often very rounded in appearance (Fig. 4b). Fragments of material from the wear surfaces (Fig. 4c) were also produced. Particle size measurement using SEM micrographs indicated that the fine wear debris particles collected during the wear of pins AL3 and RBSC had a size range of 0.1 to 0.3μm, with a minimum size 0.05μm.

The wear surfaces in the dry tests on the cemented carbide specimens contained numerous WC grains which had fragmented (Fig. 5a). In the wet test, the fragments were absent and had presumably been dislodged from the surface of the material, but in the dry test they remained in place (Fig. 5b).

In the dry tests the relative wear rates between pin and disc were higher in all of the ceramics (except pin AL2) than in the cemented carbides (Table 3). However, in the wet tests the ceramic wear rates were significantly decreased and the cemented carbide wear rates increased so that the cemented carbides wore more than the ceramics. The explanation appeared to lie in the action of the water in decreasing the amount of "self-abrasion" in the ceramics by washing away the accumulated wear debris from the surfaces. Conversely, the survival rate of fragmented WC grains in the wear surfaces of the cemented carbides appeared to decrease when the surface was wet.

Friction coefficients for the ceramics were approximately unity in the dry tests, and 0.6 in the wet tests. The exception was the RBSC pin which gave a low value of 0.1 in the wet test. The cemented carbides gave a friction coefficient of 0.5-0.8 in the dry tests and 0.2-0.3 in the wet tests.

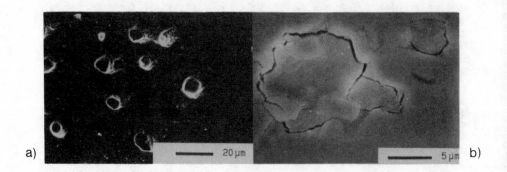

Fig. 3 Wear surfaces from wet sliding wear tests:
a) ZR1 pin, b) AL3 pin

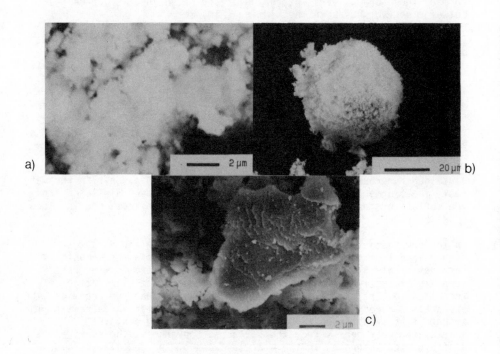

Fig. 4 Three different types of wear debris from dry sliding wear test
with AL3 pin

A feature observed in some of the wet tests on ceramics was a periodicity in the friction and wear results (Fig. 6). The peak in the friction coincided with an increase in noise, and often with a sharp increase in the wear. It is possible that they were caused by the sudden breaking away of raised areas on the wear surface followed by a period of steady wear and the formation of further raised areas.

Sliding Indenter Test

The tracks in the surface of the cemented carbide specimens were relatively smooth and were bordered by a piled-up ridge of material displaced from the track. Considerable plastic deformation was evident just outside the ridge (Fig. 7a) and the edges of the track were ragged in some of the cemented carbides (Fig. 7b) where material had broken away. The latter behaviour was more common in the ceramic specimens (Fig. 8a) and became more pronounced as the number of indenter traverses increased. Cracks perpendicular to the direction of motion of the indenter were found at the base of the tracks in both the ceramic and cemented carbide specimens (Figs. 7c, 8b and c), and in the latter, the density of cracking increased with decrease in Co content. Some of the cracking in the tracks of the ceramic specimens had probably initiated on planes parallel to the base of the track and was manifested by the presence of loosely held flakes of material surrounded by rough areas (Fig. 8d). Examination of the wear track in the AL1 specimen indicated that the cracks followed a mixed intergranular/transgranular path.

The wear tracks in the cemented carbide specimens contained fragmented WC grains (Fig. 7d) which had not been dislodged from the surface. The track base was coated with a layer which became more distinct as the number of traverses by the indenter was increased (Figs. 7e and f). X-ray analysis of the layer revealed the presence of a high Co content. When the diamond indenter was examined after a single pass test on cemented carbide HL, a coating rich in Co was also found on the surface of the indenter (Fig. 9). These observations indicate that an adhesion/re-deposition process had taken place between the layer and the diamond.

Examination of a section through the wear tracks in cemented carbide HI showed that the depth of significant damage to the material microstructure was about 3μm, which is similar to the 2.5μm grain size of the specimen. The depth of damage beneath the wear track surface did not increase with increase in number of traverses, but there was an indication of an increase in the density of cracking within the damaged layer. Also there was a decrease in the Co content of the damaged zone, suggesting that the Co from this zone has been removed to form the Co rich surface layer. This layer was too thin to be observed in cross-section. It was estimated to be of the order of 50nm thick from a knowledge of secondary electron sampling depth in the scanning electron microscope and from observations of rate at which the layer built up as the number of passes increased.

Examination of the wear debris showed that there were many small fragments of material, and a large proportion was in the form of flakes in both the ceramic and cemented carbide specimens (Figs. 10a and b). Many flakes showed linear sliding marks on one face, indicating that

TABLE 3   <u>Wear Rates and Friction Coefficients in Pin-on-Disc Tests</u>

Material	Wear rate µm/min		Friction coefficient		Duration min	
	dry	wet	dry	wet	dry	wet
AL2	0.1	0.02	1.0-1.1	0.3-0.6	1085	523
AL3	2.3	<0.01	-	0.7-0.8	170	385
ZR1	125	0.4	1.0-1.1	0.5-0.6	50	235
HPSN	1.2	0.2	0.8-1.1	0.5-0.9	65	650
RBSN	23.1	0.5	0.9-1	0.5-0.6	95	402
RBSC	0.9	0.03	0.7-0.9	0-0.1	2125	1488
B	95	-	0.7-0.9	-	5	-
HH	0.02	0.1	0.5-0.8	0.2-0.3	1200	255
HF	0.3	0.9	0.5-0.8	0.2-0.3	1058	436

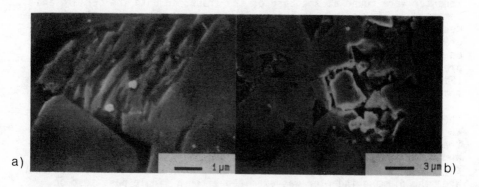

Fig. 5 Wear surface of HH cemented carbide pin in sliding wear:
a) dry test, b) wet test

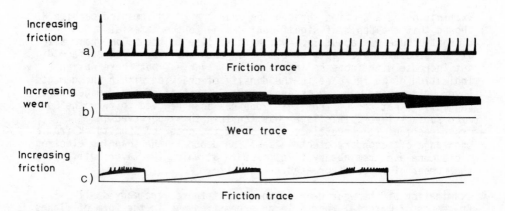

Fig. 6 Intermittent wear behaviour:
a) shows the friction trace from HPSN, and b) and c) the wear and
friction traces for RBSC

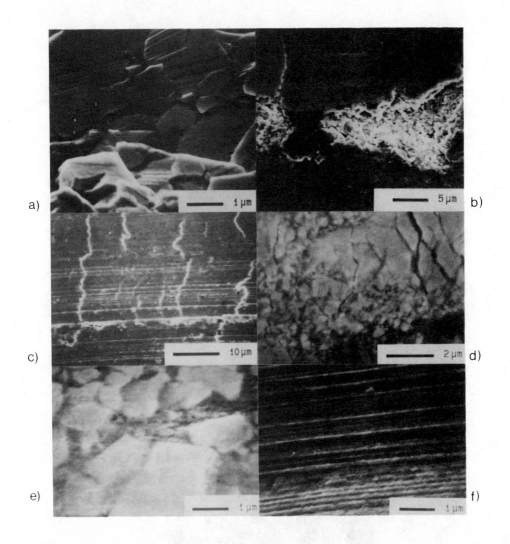

Fig. 7 Features of sliding indenter tests on cemented carbides:
a) pile up at edge of single traverse of HK, b) flaking at edge of
five-traverse track on HB, c) traverse cracking at base of
single-traverse track on HB, d) WC grain cracking at base of
single-traverse track on HI, e) base of single-traverse track on
HK, f) base of twenty-traverse track on HK.
All micrographs were taken at a position corresponding to the
maximum load of 100N

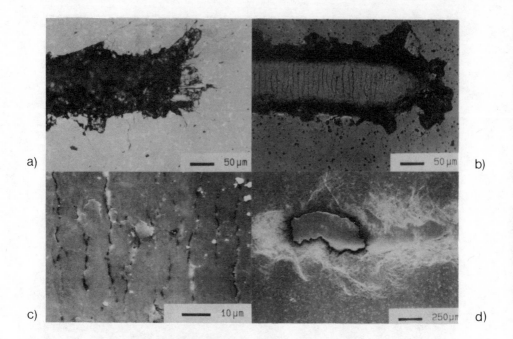

Fig. 8 Features of sliding indenter tests on ceramics:
    a) RBSC single-traverse track, b) ZR3 five-traverse track,
    c) base of AL3 single-traverse track, d) ZR1 five-traverse track.
    All micrographs were taken at the maximum load position

Fig. 9 Co-rich coating on tip of diamond indenter after single-traverse
    of HK cemented carbide

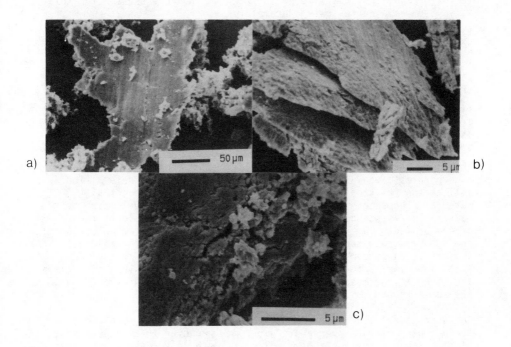

Fig. 10 Debris from sliding indenter tests:
a) from ten-traverse tests on RBSC, b) and c) from five-traverse
HG cemented-carbide test

this face had been the base of the wear track in a previous traverse of the indenter. Closer examination of flakes from the cemented carbide test revealed characteristic networks of microcracks which are often observed when these materials are subjected to high compressive stresses.

The main results of the sliding indenter tests are presented in Table 4, in terms of the following parameters:

i) The nominal coefficient of friction is defined as the ratio of horizontal force F to applied load, F/P.

ii) The Meyer-O'Neill (O'Neill,1934) scratch hardness H_s is calculated from $H_s = 8P/D^2$ where P is the applied load and D the track width obtained from optical microscopy measurements.

iii) The mean flow pressure p is F/A, where A is the cross-sectional area of the wear track calculated from D on the assumption that the track retains the shape of the indenter after a traverse has occurred (Brookes and co-workers, 1972).

The first parameter is not a true friction coefficient since it incorporates a measure of the difficulty of subsurface deformation as well as the friction at the indenter's interface with the test material. The second and third parameters provide an indication of the material's resistance to penetration by the indenter load to form a groove of width D, and the material's resistance to the extension of the groove across the surface. Since both parameters are empirical and arbitrary, in common with hardness, they are dependent on applied load, indenter geometry and machine stiffness. Their values in Table 4 are given at the maximum load which was 100N, except for the tests on AL1 where it was 60N.

The average friction coefficient remained approximately constant throughout a given traverse of a specimen. However in all the tests, fluctuations in the horizontal force F were observed which increased in amplitude as the load increased, and also as the number of traverses increased (Fig. 11a). The average friction coefficient at 100N (60N for AL1) varied between 0.18 for the sialon to 0.37 for the HI cemented carbide. Although for most specimens the average friction coefficient remained constant throughout all the multipass tests, this was not true for some of the ceramics. For example, in the RBSC tests, the average friction coefficient increased to a value of 0.54 on the twentieth traverse from the initial value of 0.27. For the AL4 tests the average friction coefficient dropped to a value of 0.11 on the twentieth traverse from the initial value of 0.28. There was no correlation of friction coefficient with scratch hardness.

The scratch hardness at 100N varied from 3.5 kNmm^{-2} for the fully stabilised zirconia ZR1, to 28.82 kNmm^{-2} for cemented carbide HA. There was a general increase in scratch hardness with hardness, with H_s approximately equal to 1.5H, where H is the hardness (for H_s and H both expressed in Nmm^{-2}), which agrees with the observations of Brookes and co-workers, 1972. The ceramic scratch hardness values lie mostly below the linear regression line expressing this relationship, with the cemented carbide results above (Fig. 11b).

TABLE 4   Sliding Indenter Test at 100N and 0.5mm/s

Material	Friction coef.	Approximate threshold -load for AE N	Average AE rate counts/s 1st scratch	Hardness HV2.5	Scratch hardness kNmm^{-2}	Mean flow pressure kNmm^{-2}	$\dfrac{H_{s20}}{H_{s1}}$	$\dfrac{P_{20}}{P_1}$	Friction behaviour with increased traverses
A11 +	0.31	10	3.2×10^4	1480	7.28	10.52	0.53	1.23	decreased
A13 x	0.26	50	2760	1200	7.69	9.49	0.82	0.56*	decreased
A13	0.31	30	52000	1200	9.13	14.91	2.11	1.21	decreased
A14	0.28	30	1.24×10^5	880	4.49	5.75	1.15	0.88	decreased
ZR1	0.34	none	2.3×10^5	850	3.5	5.67	1.31	0.94	constant
ZR2 x	0.19	25	1.6×10^5	970	17.63	24.99	1.95	0.99*	constant
ZR3 x	0.20	25	1.6×10^5	950	14.05	18.64	1.16	0.67*	constant
RBSN	0.28	20	2×10^5	560	6.00	5.14	0.59	0.78	decreased
RBSC	0.27	none	1.92×10^4	1810	20.83	47.91	0.70	1.33	increased
Sialon	0.18	none	2.2×10^5	1370	20.83	31.00	0.26	0.13	constant
HA	0.34	0	120	1890	28.82	96.60	3.70	2.31	constant
HB	0.34	0	72	1740	15.70	38.37	4.39	3.65	constant
HC	0.26	0	51	1690	26.52	64.84	2.77	2.57	constant
HE	0.32	0	20	1600	21.41	56.02	3.32	2.48	constant
HG	0.33	0	16	1370	17.68	44.97	4.47	2.61	constant
HI	0.37	0	4	1180	16.30	44.10	3.97	3.28	constant
HK	0.26	0	40	1220	16.30	30.99	8.3	7.57	constant

x - specimen speed 0.2 mm/s

+ - maximum load 60N

* - H_{s10}/H_{s1} ; P_{10}/P_1

The mean flow pressure p, which is proportional to the energy needed
to remove unit volume of material in a given time varies from the
highest value of 96.6 kNmm^{-2} for HA cemented carbide to 5.14 kNmm^{-2} for
the RBSN specimen.

It is of interest to consider how materials retain their resistance to
the passage of the indenter as the number of traverses increases. One
measure of this is H_{s20} /H_{s1} which is the ratio of the effective
scratch hardness calculated from the increment in track area dA during
the twentieth traverse to the value of scratch hardness obtained after
one traverse. (Values for dA_{20} were calculated from values of dD
obtained by drawing a tangent to curves of D versus the number of
traverses at the twentieth traverse.) This ratio and the corresponding
ratio for the flow pressure, p_{20} /p_1 , are given in Table 4. The values
are typically lower for the ceramics than for the cemented carbides,
with the lowest value of 0.26 and 0.13 for the hardness and flow
pressure parameters respectively for the sialon. This indicates that the
ceramics did not stand up as well to the repeated passage of the
indenter as the cemented carbides did.

Little acoustic emission was detected below a threshold load in many of
the multipass tests on ceramic specimens. The existence of a threshold
load, F_T in Fig. 11c, became more evident as the number of traverses
increased (Figs. 11a and c). Typical values were 25N for the ZR2 and
ZR3. Steady acoustic emission output occurred after these values were
exceeded. The output was several orders of magnitude greater for the
ceramic specimens than for the cemented carbide specimens. However, even
though the output from the cemented carbide specimens was very low,
there was evidence of a dependence of output rate on hardness.

ASTM Abrasive Slurry Test

Examination of the surfaces of the cemented carbide specimens revealed
fine wear tracks parallel to the direction of wheel rotation (Fig. 12a).
Considerable fragmentation of WC grains had occurred and there was
evidence of a Co rich layer on the surface of the specimens (Fig. 12b).
On the surface of cemented carbide HK, ridges had been created during
the movement of the grit particles over the specimen. In Fig. 12c, the
passage of one grit particle is shown at the top of the micrograph, and
another at the base of the micrograph leaving two ridges nearly touching
in the centre. Closer examination of one of the ridges showed that the
heavily deformed and fragmented material had been pushed to one side by
the passage of the grit particle but had not been completely detached.

The surfaces of the alumina specimens were dull and had not developed an
obvious wear track pattern. However, when the AL3 wear surface was
examined, a few wear tracks of heavily deformed material were observed
(Fig. 12e). The major part of the wear surfaces showed intergranular and
transgranular fractures (Fig. 12f).

It should be noted that the nominal diameter of 620μm of the grit
particles was large compared with the grain-size of the test materials.
When the size and shape of the particles were measured in cross-section
using an image analysis system, the average equivalent circle diameter
of the particles was 590 + 181μm. When the same parameter was measured
on grit after it had passed between the specimen and the wheel, the

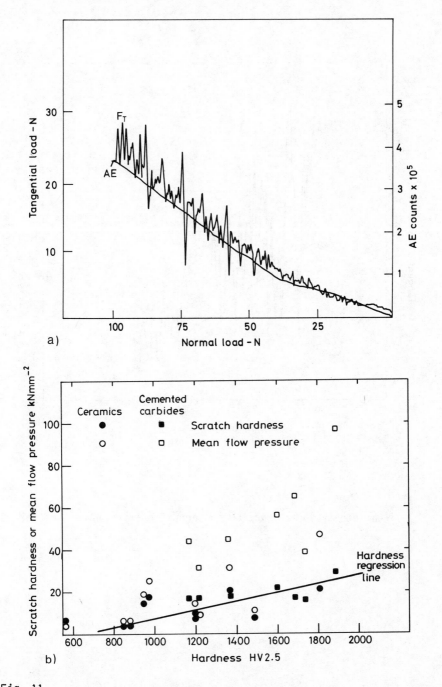

Fig. 11
a) First-traverse friction and acoustic emission results for RBSN.
b) Relationship of scratch hardness Hs to Vickers hardness, HV2.5.

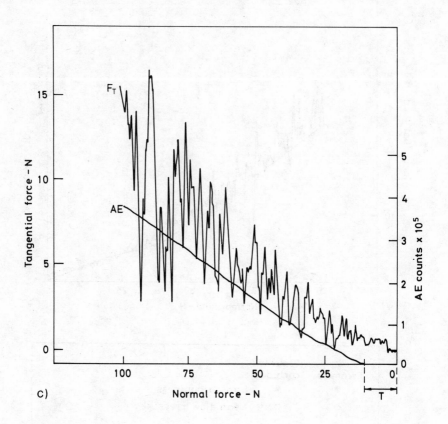

c)

Fig. 11

    c) Twentieth-traverse friction and acoustic emission results for RBSN

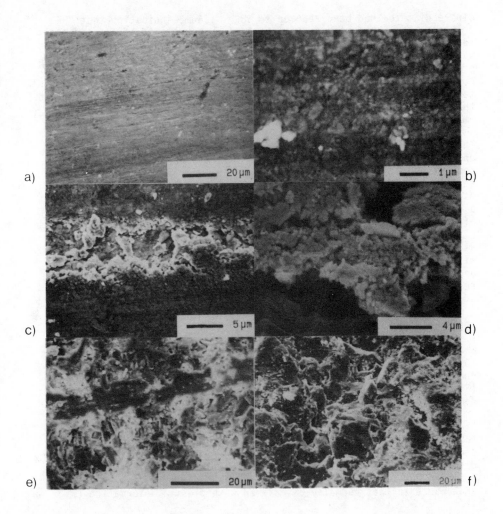

Fig. 12 ASTM test specimens:
   a) Wear surface of HB cemented carbide showing wear tracks
   b) Wear surface of HK cemented carbide showing Co-rich layer
   c) Wear surface of HK cemented carbide showing wear tracks
   d) Wear debris from cemented carbide
   e) Wear surface of AL3 alumina showing wear tracks
   f) Wear surface of AL3 alumina showing intergranular and
      transgranular fracture

circle diameter had been reduced to 380 + 136μm, indicating that the
grit had been comminuted during the test. Measurements of shape factors
for the grit indicated that the grit initially had smoooth surfaces but
was angular after the test. Examination of the used grit confirmed the
presence of freshly fractured grit particles, and also showed heavily
deformed fragments of debris from the cemented carbide (Fig. 12d). By
taking flash photographs of the wheel rim at the exit side it was found
that the grit was in a single layer at a density of 0.21 particles per
mm².

The results of the tests are given in Table 5. The ceramics were much
more severely worn than were the cemented carbides, although a
comparison is difficult to make because the wear area increases as the
test proceeds, so that the average volume loss per revolution is not
directly comparable for the two types of material. It is suggested that
a lower resistance to fracture was responsible for the greater severity
of wear in the ceramics with respect to the cemented carbides. For
example, cracks of length 5mm, parallel to the surface, were revealed
at a depth of few millimetres by sectioning an AL2 specimen. However,
within the range of cemented carbides tested, the property which
appeared to have the greatest influence in determining good wear
resistance was high hardness.

Abrasive Slurry Tests

Examination of the cemented carbides, HF and HH, after abrasion by the
160 mesh (80μm dia. ave.) SiO_2 revealed WC grains standing proud from a
roughened surface (Fig. 13a). Some fractured WC grains were visible
(Fig. 13b), but there was no sign of any directionality in the wear
damage.

The surface of the fully stabilised ZR1 specimen after abrasion by the
160 mesh (80μm dia. ave.) SiO_2 showed regions where the porosity had
been revealed in regions of fracture. There was little evidence of heavy
deformation of the specimen surface (Fig. 13c). In contrast, the RBSC
specimen abraded under the same conditions had a surface of similar
appearance to that normally associated with plastic deformation (Fig.
13d).

The cubic BN specimen abraded by SiO_2 grit had a surface that suggested
fracture had occurred at or near to grain boundary interfaces, leaving
some grains standing proud, and holes where other grains had been pulled
out of place (Fig. 13e).

There was a general trend for abrasion resistance to increase with
hardness, but the cemented carbides had better abrasion resistance than
all the ceramics except for the cubic BN specimen (Table 6). The 29μm
B_4C grit gave abrasion results approximately equal to the results for
the SiC, but the tests with the alumina grit gave volume losses
approximately three times lower than those for the SiC and B_4C. The
larger SiO_2 grit nevertheless gave even smaller volume losses. The
greatest amount of wear observed was for the ZR1 specimen with a slurry
of B_4C, and the least wear with the cubic BN, where wear was not
detectable by the weight loss method.

TABLE 5   <u>ASTM Slurry Tests</u>

Material	Applied load N	Number of revs	Total volume loss mm^3	Average volume loss per rev mm^3	Hardness HV30
AL1	98.1	500	410	0.82	1270
AL1	196.2	250	460	1.84	1270
AL2	98.1	500	500	1	-
AL2	196.2	250	570	2.28	-
AL3	98.1	500	840	1.68	1080
AL3	196.2	250	1000	4	1080
AL4	98.1	500	300	0.6	890
AL4	196.2	250	330	1.32	900
HA	196.2	1000	13	0.013	1890
HB	196.2	1000	11	0.011	1750
HD	196.2	1000	21	0.021	1660
HE	196.2	1000	59	0.059	1550
HG	196.2	1000	167	0.167	1375
HI	196.2	1000	182	0.182	1195
HJ	196.2	1000	185	0.185	1190
HK	196.2	1000	323	0.323	1150
HL	196.2	1000	435	0.435	1060

TABLE 6   <u>Abrasive Slurry Tests with Different Grits</u>

Material	Volume loss - mm^3			
	82µm SiO$_2$	29µm Al$_2$O$_3$	29µm SiC	29µm B$_4$C
CBN	<0.3	<0.3	<0.3	<0.3
HF	0.1	0.9	3.7	4.6
HH	0.1	0.7	4.4	4.8
Sialon	0.3	1.9	6.0	6.9
RBSC	0.3	1.3	6.1	10.0
ZR3	0.7	2.2	8.1	9.5
HPSN	0.5	2.8	8.8	9.1
ZR2	0.6	2.2	9.2	10.8
AL2	0.4	2.6	18.4	20.8
Agate	0.46	13.5	34.1	38.6
RBSN	0.64	25.5	62.1	60.1
ZR1	0.64	27.4	74.6	83.8

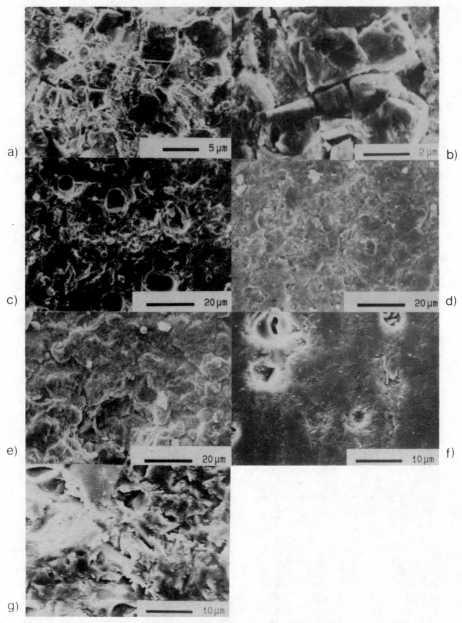

Fig. 13 Wear surfaces from abrasive slurry test:
   a) and b) HH abraded with 160 mesh SiO_2
   c) ZR1 abraded with 160 mesh SiO_2
   d) RBSC abraded with 160 mesh SiO_2
   e) CBN abraded with 160 mesh SiO_2
   f) AL1 abraded with 3μm B_4C grit
   g) AL1 abraded with 20μm SiC grit

Measurements of the wear of AL2 specimens showed that as the particle size of the abrasive increased, there was an increase in the wear (Table 7). There was also an increase in the roughness of the wear surface, and in the proportion of intergranular fracture on the wear surface (Figs. 13f and 13g).

It is interesting to note that the order of abrasion resistance found here is almost the same as that found by Moore and King (1979) in abrasive tests on abrasive paper.

The alumina grit size was measured before and after one of the slurry abrasion tests. No significant change was found.

DISCUSSION

The overall trend in all four tests was for the cemented carbides to have better wear resistance than the ceramics, and for the wear resistance to increase with increase in hardness. From a practical point of view, some inferiority in wear performance may be tolerable in some of the ceramics because of their lower cost and other attractive properties such as lightness, chemical inertness and high temperature strength. Nevertheless, greater benefits can be obtained if the mechanistic sources for low wear resistance can be identified and modified, and if effective test procedures can be developed for this purpose.

Consider first the conditions in the present tests and the materials' responses. The conditions prevailing in the tests involving wear by an abrasive-slurry favour free particle abrasion, a process whereby each abrasive particle is free to rotate and make succesive indentations in the surface. The absence of wear tracks in the slurry tests, indicates that this process probably predominates except in the ASTM tests on the cemented carbides where a well defined pattern of striations was produced parallel to the direction of relative motion of the opposing wear surfaces.

The conditions in the sliding indenter test are meant to simulate abrasion by a single fixed abrasive particle. It is evident that the present test geometry is unrepresentative of the other tests since the 0.4mm dia. conical-indenter is not as sharp as the fragmented abrasive particles of 380μm dia. in the ASTM test, and the indenter samples a greater volume of the test surface than is sampled by 30μm dia. grit in the disc tests. At the other extreme, the pin-on-disc test is meant to involve two 'smooth' surfaces sliding across each other without any source of abrasion. However the wear process generated an abrasive debris 0.1-0.3μm dia. which produced wear tracks similar to those obtained by fixed particle abrasion.

With this information on test conditions it is possible to offer a tentative qualitative explanation about the differences in the behaviour of ceramics and cemented carbides, looking initially at the general properties that govern the wear of all materials and then looking at specific characteristics of the harder materials.

For example, it is generally found that the macroscopic hardness

TABLE 7   <u>Slurry Tests on AL2 with Different Slurries</u>
<u>of Different Grit Size</u>

Grit	Grit size μm	Volume loss mm^3
B$_4$C	3	0.8
SiC	12	5.4
SiC	20	13.9
SiC	29	17.6
SiC	43	36.9

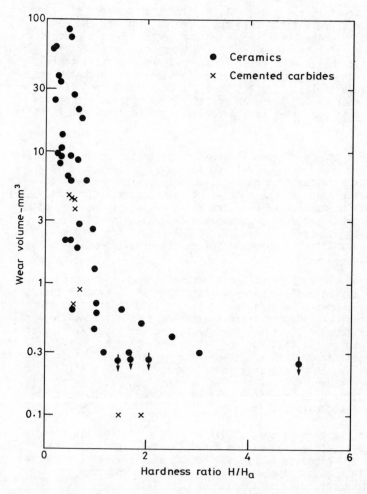

Fig. 14 Variation in abrasive wear with hardness ratio   H/H$_a$

provides a good indication of potential abrasive wear resistance. The basis for such a relationship is that the mechanism and extent of material displacement by a hardness indenter are large scale representations of material removal by indentation and by grooving by abrasive particles. This dependence was observed in the abrasive slurry test, as can be seen in Fig. 14 where the wear volume is plotted against the normalised hardness H/H_a and the divisor is the hardness of the abrasive in its bulk form. It can also be seen that there was a decrease in wear when the hardnesses of the sample and the abrasive were almost equal, and this agrees with results obtained elsewhere for abrasion of ceramics against abrasive papers (Moore and King, 1979) and in a three component system (Misra and Finnie, 1983).

However, when examined more closely the results from the ASTM and disc slurry tests show that cemented carbides had superior wear resistance to the sialon, the SiC and the aluminas of similar hardness, and this trend was present to a lesser extent in the pin-on-disc tests. Consequently high hardness was not the sole criterion for abrasion resistance for the materials examined; for a more complete explanation, other surface damage mechanisms need to be examined, and the most obvious candidate is the characteristic crack patterns that are associated with the indentation of hard materials.

The geometry of indentation cracks is dependent on indenter shape and the material. For a Vickers indenter, the radial cracks that emanate from each corner of the indentation are either the traces of four separate semi-elliptical cracks if they occur in a cemented carbide or, if they occur in a ceramic they are the traces of two intersecting semi-circular cracks that have propagated from sub-surface median cracks during unloading of the indenter. An additional crack geometry observed in ceramics, is lateral cracking which occurs between the radial cracks on planes at a shallow angle to the surface. Lateral and median cracking is rarely observed in WC/Co cemented carbides except in very hard grades or in corrosive environments (Almond and Roebuck, 1976 and 1983).

The extent of abrasive-wear damage in ceramics has been related to lateral cracking by Evans and Wilshaw (1978), who proposed that the tendency to lateral cracking was dependent on H/K^2, where H and K are hardness and fracture toughness parameters. If macroscopic values for K and H are substituted in this ratio, the theory would predict that ceramics should be about ten times more likely to exhibit lateral cracking than cemented carbides with similar hardnesses. An extension of the theory proposes that the rate of material removal by lateral cracking during abrasion is proportional to $F\,H^{-m}\,K^{-n}$, where m is 0.75 and n is 0.5, and F is the load per abrasive-particle. There are obviously other conditions involving combinations of minimum loads and particle-sizes below which co-ordinated fracture does not occur, and the plastic deformation and microfracture processes that make surface-lapping possible predominate.

Since the macroscopic values of the fracture toughness K-parameter are an order of magnitude greater for cemented carbides than for ceramics of similar hardness, at least a qualitative explanation can be given for the low abrasion resistance of ceramics. The contribution from lateral cracking in weakening the structure neighbouring an indentation can also be invoked to explain why ceramics have lower resistance than that of cemented carbides to material removal by repeated passes of a sliding indenter along the groove. The results are expressed in Table 4 in terms of the rate at which the flow pressure increases with increase in number of passes: it is interesting that the trend for the incremental change in the Meyer-O'Neill scratch hardness is similar to that for flow pressure. Since the former quantity is related to the depth of damage, a possible explanation why the ceramics suffered a greater reduction in the Meyer-O'Neill scratch hardness than was observed in the cemented carbides is that it was a manifestation of a greater tendency in the ceramics to form subsurface cracks of cone and median vent geometry. The major contribution from microcracking to the damage incurred by the ceramics during the sliding indentation tests was also evident in the results of the acoustic emission monitoring.

The above explanations indicate that it is probably important to take into account the fracture resistance of a hard material when judging its suitability for an abrasive wear application. In adhesive wear also, the contribution of toughness to wear and friction mechanisms appears to be a major factor. Thus it is now recognized that a criterion based on a strain energy release parameter, similar to the G or R value of fracture mechanics, is more important in determining the difficulty of separating an adhesive bond at a wear interface than is a thermodynamic comparison of the bond strengths in the wear materials and at their interface (Johnson, 1981).

Although the introduction of fracture mechanics concepts represents a significant advance in understanding wear phenomena, the treatment of problems remains qualitative because of uncertainties in appropriate parameters to use. Thus macroscopic values of fracture toughness and hardness are probably inapplicable in microscopic wear mechanisms since hardness is volume dependent and fracture toughness often has a crack-length dependence. Also the properties that are important are not the hardness and toughness of the bulk material but those of a worn surface layer that will probably be plastically deformed, cracked and subjected to a high loading rate, an elevated temperature and a reactive lubricant. An additional complication is that the fracture toughness parameters of interest will not be those for the simple Mode I geometry but will be those for a mixed Mode I and II configuration.

Specific Wear Processes

Another limitation on explaining wear of hard materials is the relatively small amount of experimental data available for theoretical analysis. There is also a need for more characterization of wear processes at a microstructural level before substantive theoretical models can be formulated. Of special interest in the present work was: i) the reversal that occurred in the rating of ceramic and cermets when a water lubricant was used in the pin-on-disc test and, ii) the various deformation processes that were associated with the sliding indenter test and the effects of grit size and geometry.

Considering first the pin-on-disc test, the best resistance to wear was obtained in the coarse-grained cemented carbide HH which had wider binder-phase regions than HF; this matrix would have been expected to be a source of weakness in abrasive wear conditions because of its relatively low hardness. However, in the dry tests it was found that the binder-phase had been impregnated with wear-debris from the alumina wheel and from detached WC fragments which reinforced the matrix and enhanced its wear resistance. By inhibiting preferential wear of the binder-phase this mechanism appears to have also increased the capacity of the binder-phase for supporting and retaining WC grains that had been fragmented by wear debris but were not dislodged from the wear surface. The elevated temperature at the wear interface would have assisted the incorporation of wear particles into the binder phase by increasing its ductility. In contrast, the wear debris abraded the ceramic surfaces and caused grooving and fracture. A mechanism of wedging by wear-debris that accumulated in pits and in holes in the wear surfaces, appears to have made a major contribution to cracking and breakaway of further material (Fig. 15).

The effect of water lubrication was to remove the wear-debris before it could re-enter the rubbing surfaces. As a result the binder-phase of the cemented carbides was not reinforced by wear debris and suffered preferential wear. A secondary effect was that the protruding WC grains became easily detached from the wear surface when they were damaged. Conversely the ceramics benefitted from the removal of wear debris. This explanation is compatible with the observations in the other tests and work elsewhere that ceramics have better resistance to sliding wear than cemented carbides except when abrasives are present. However water lubrication may not always be beneficial since in the long term it can promote stress corrosion of some ceramics (Wallbridge, Dowson and Roberts, 1983).

Other Microfracture Effects

The sliding indenter provided a useful large scale demonstration of the types of cracks that could occur when a hard material is scratched by a fixed hard abrasive particle (Fig. 16). As the diamond indenter moved over the surface, four sets of cracks formed. Their location was consistent with the stress fields predicted by the analyses by Hamilton and Goodman (1966) for a Hertzian stress field, and by Hills and Ashelby (1982) for a stress field that has been redistributed as a result of plastic deformation introduced during the first traverse of the indenter. For example there was microcracking caused by the high tensile stresses in front of the indenter. Associated with this was a piling-up of material and debris in front of the indenter and fracture of material as the indenter moved through the surface. The compressive stresses generated behind the indenter produced cracks perpendicular to the direction of motion at the base of the track. In addition, the shear stress generated by the motion of the indenter was at a maximum a short distance beneath the surface of the material, and produced shear cracks which caused delamination on planes parallel to the surface.

The exact form of the cracking developed would have been different in each type of test. This is partly due to differences in the geometry of the abrading element, which was a surface asperity or wear-debris particle in the pin-on-disc test, the hemispherical indenter in the sliding indenter test, or angular grit particles in the two slurry tests. The shape of indenters has been shown to have an effect on the friction and wear in sliding indenter tests (Van Groenou, Maan and Veldcamp, 1975). The sharpness of the tip of the indenter will obviously

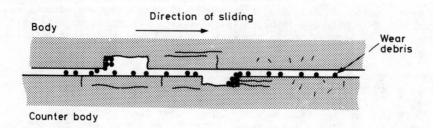

Fig. 15 Schematic diagram of wear processes in sliding wear

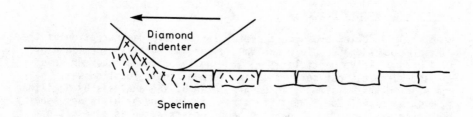

Fig. 16 Schematic diagram of wear processes in sliding indentation test

have an important influence on the stress field produced in the material. For example, Moore and Swanson (1983) found that there was an increase in abrasive wear of steel for angular crushed quartz abrasive as compared with rounded grit. The grit used in these abrasive slurry tests was also angular. Fracture of the abrasive grit took place in the ASTM slurry test. In the classification scheme of Misra and Finnie (1979), this type of abrasion is classified as high stress abrasive wear. By comparison the grit in the abrasive-slurry disc-test remained unbroken after the test, which is classified as low stress abrasive wear. The fracture of the grit particles under high stress conditions creates new sharp edges which maintain a high wear rate during the test. In this respect high loads are doubly damaging in producing abrasive damage in ceramics since they increase the severity of surface cracking not only through the strong load-dependence of cracking but by also producing sharp grit which promotes the lateral and median geometry cracks.

CONCLUSIONS

The overall trend in all four tests was for the cemented carbides to have better abrasive wear resistance than the ceramics, and for the wear resistance to increase with increase in hardness.

A mechanistic explanation for the lower wear resistance of ceramics is that they are more susceptible than cemented carbides to indentation fracture by abrasive particles.

For ambient temperature applications the best wear resistance will be obtained from ceramics by using them at low load conditions and preferably where abrasives are not present or are prevented from entering the wear interface.

REFERENCES

Almond, E.A. and Roebuck, B. 1976, J. Mater. Sci. 11 565-568
Almond, E.A. and Roebuck, B. 1983,Science of Hard Materials, Plenum Press, New York, pp 597-614
ASTM B611, Abrasive Slurry Wear Test, American Society for Testing of Materials (1976)
Brookes, C.A., Green, P., Harrison, P.H. and Moxley, B. 1972, J. Appl. Phys. 5 1284-1295
Evans, A.G. and Wilshaw, T.R. 1978, Acta Metall. 24 939-956
Hamilton, G.M. and Goodman, L.E. 1966, Trans. ASME J. Appl. Mech. 371-376
Hannink, R.H.J.,Murray, M.J. and Marmach, M. 1983, Proc. 4th Int. Conf. Wear of Materials, ASME, New York pp 181-186
Hills, D.A. and Ashelby, D.W. 1982, Wear 75 221-240
Johnson, K.L. 1981, Friction and Traction. IPC STP Ltd. Guildford, UK pp 3-12
Misra, A. and Finnie, I. 1979, Proc. 2nd Int. Conf on Wear of Materials ASME, New York pp 313-318
Misra, A. and Finnie, I. 1983, Wear 85 57-68

Moore, M.A. and King, F.S. 1979, Wear <u>60</u> 123-140

Moore, M.A. and Swanson, P.A. 1983 Proc. 4th Int. Conf on Wear of
    Materials, ASME, New York pp 1-11

O'Neill, H. 1934, The Hardness of Metals and Its Measurement. Chapman
    and Hall, London pp 144-149

Van Groenou, A.B., Maan, N. and Veldkamp, J.D.B. 1975, Phillips Res.
    Rep. 30 320-359

Wallbridge, N., Dowson, D. and Roberts, E.W. 1983 Proc. Int. Conf on
    Wear of Materials, ASME, New York pp 202-211

Inst. Phys. Conf. Ser. No. 75: Chapter 9
Paper presented at 2nd Int. Conf. Science Hard Mater., Rhodes

949

Erosion of multiphase materials

K ANAND, C MORRISON, R O SCATTERGOOD, H CONRAD (1),
J L ROUTBORT (2) AND R WARREN (3)

(1) Dept. of Materials Engr., Box 7907, N.C.S.U., Raleigh, NC 27695
(2) Argonne National Lab., Materials Science Div., Argonne, IL 60439
(3) Dept. of Engr. Mat., Chalmers Univ. of Tech., S40220, Goteburg, Sweden

ABSTRACT

Erosion rate measurements were made for a series of WC-Co alloys with
varying WC-Co grain size and for a fully aligned composite consisting of
Al_2O_3 rods in a stainless steel matrix. The erosion rate vs. angle-of-
incidence curves change from a brittle to a ductile signature, or vice
versa, under appropriate changes in microstructure or erodent particle
size and velocity. This change is due to a microstructural effect, and
appears to be directly linked to the relative size of the impact damage
event and the pertinent microstructural size scales.

INTRODUCTION

Solid particle erosion occurs when a flux of erodent (abrasive) particles
impact against a target surface. The impact of a particle against the
surface produces a local damage event, which ultimately leads to the removal of material by a number of different mechanisms. Since many technological systems involve process streams with entrained particulate
matter, erosion has important engineering aspects as well as offering a
number of interesting and challenging materials science problems.

Although the basic phenomenology of erosion has been established for a
wide range of materials, there still exists some uncertainty and contro-
versy over the underlying mechanisms. Theoretical models of erosion
(Ruff and Wiederhorn, 1979) in single-phase ductile materials have met
with partial success. One class of models assumes a micromachining or
ploughing mechanism that correctly accounts for a maximum in the erosion
rate at low angles of incidence of the impacting particle (Finnie and
McFadden, 1978). However, the models predict no erosion at normal incidence, whereas experiments show significant erosion for this case. The
latter is generally attributed to a damage accumulation or extrusion and
forging mechanism that becomes effective under normal incidence conditions
(Bellman and Levy, 1981). In single-phase brittle materials, which behave
in a somewhat simpler fashion, the particle impact event generates a localized plastic zone, and residual stresses cause a subsurface lateral
crack to propagate outward and subsequently remove a chip of material.
The mechanism is similar to the material chip removed by static hardness
indentation tests at high loads on brittle samples. Several models have

been developed for this case (Evans et. al., 1978; Wiederhorn and Lawn, 1979).

The above models have been applied in numerous investigations and, while quantitative predictions are rather difficult to develop, the general features of erosion are correctly predicted. The role of microstructure has, however, been mostly ignored in both the development of models and in measurements of erosion rates. Very few systematic studies are available which intentionally introduce microstructure variations along with changes in the operational variables for the erosion test. The purpose of the work reported here was to measure erosion rates on samples with different microstructures and then attempt to evaluate the microstructural effects in terms of the erosion properties of the individual components.

The first system used for this investigation was a series of WC-Co alloys with varying WC grain size and fixed Co content. These alloys provide an interesting microstructure, which combines a brittle, hard phase dispersed in a ductile matrix. The alloys are also quite erosion resistant and, hence, have potential application in technological situations. The second system used is a rather unique composite consisting of fully aligned Al_2O_3 rods in a 304 stainless steel matrix. The microstructure here also combines a brittle phase in a ductile matrix, but the microstructural size scale is very large relative to the WC-Co alloys. The fully aligned continuous microstructure also eliminates any effects that might arise from random, discrete particle geometry. The rationale for selecting these systems was that, if microstructural effects exist, then the scale of the microstructure relative to the scale of the impact damage event should be important. The WC-Co alloys offer the possibility of making large changes in the relative scales. On the other hand, the composite has a very large microstructural scale, and this offers the possibility of examining the erosion process without large changes in the relative scales, and without possible geometric effects caused by the distribution of the individual components.

The erosion measurements were made over a range of conditions for these systems. The principal results presented in this paper involve the measurement of the erosion rate as a function of the angle of incidence of the impacting particle. These curves give a characteristic signature which is useful in assessing the changes in mechanisms or averaging processes imposed by the microstructure.

EXPERIMENTAL PROCEDURE

The measurement of erosion rates can be done using various types of test apparatus, but the basic measurement techniques are similar (Ruff and Wiederhorn, 1979). A flux of abrasive particles of average diameter D with velocity v impinges on a target sample surface at an incident angle θ ($\theta = 90^\circ$ is normal incidence). Material will be removed from the sample surface as a result of the particle impacts. The material loss is measured, for a given mass of abrasive particles impacted, by weighing. A series of intermittent weight-loss measurements are made until the rate of loss from the target sample plotted as a function of the mass of particles impacted reaches a steady-state value. This is the steady-state erosion rate W, which is given here as the mass loss of the target sample per unit mass of particles impacted. The operational variables in the erosion test are the abrasive particle size D, the particle velocity v and the angle-of-incidence θ. The results presented in this paper will focus primarily upon the W vs. θ curves for different values of D and v, and for different microstructural states of the target samples.

Two different types of erosion apparatus were used for the work reported here. The first type was an air-blast apparatus in which the abrasive particles are accelerated in an air-carrier stream to obtain a specified value of particle velocity v. Details of this apparatus and the methods used for calibration have been discussed elsewhere (Hovis, et. al., 1985). The second type was a slinger apparatus in which the particles are accelerated in vacuum by a rotating arm, the speed of which determines the particle velocity. Details of this apparatus are also available elsewhere (Kosel, et. al., 1979). The air-blast apparatus was used for the WC-Co alloys while the slinger apparatus located at the Argonne National Laboratory was used for the composite.

Erosion samples were tested by fixing the samples in the abrasive particle beam at a specified angle θ. In the slinger apparatus, several samples can be run simultaneously at different θ values. The samples are impacted with a fixed mass of abrasive particles, then removed from the apparatus and carefully cleaned, dried and weighed. The accumulated mass-loss values are plotted against the accumulated mass of particles impacted until the curve becomes linear. A least-squares fit is made through the linear portion and this value is reported as the steady-state erosion rate W for fixed D, v and θ conditions.

The abrasive particles used as the erodent were standard Al_2O_3 abrasives obtained from the Morton Co. The abrasives were obtained in grit sizes from 280 grit to 80 grit. The mean particle diameters D were in the range from about 40 micrometers to 270 micrometers, and in all cases the values of D reported here are the mean sizes quoted by the supplier for the particular grit size used. The WC-Co alloys were kindly supplied by the Kennametal Co. and these were specially prepared alloys. Three alloys each containing 6 wt % Co were used for the results reported here. The alloys are designated F, M and C and had maximum carbide grain sizes of about 2 micrometers (Fine), 6 micrometers (Medium) and 16 micrometers (Coarse), respectively. These values are the values quoted by the supplier and were confirmed by preliminary metallographic investigations. A full metallographic examination (grain size distributions, mean-free path lengths, etc.) will be made as part of the continuation of the present research program. The stainless steel - Al_2O_3 composite (SSAC) was a fully aligned composite with continuous rods of Al_2O_3 in a 304 stainless steel matrix. The rod diameter was 500 micrometers and the average center-to-center spacing of the rods was 625 micrometers, thus giving an area fraction of about 0.5 Al_2O_3 on a plane section. The SSAC was fabricated by hot-pressing a bundled sample made by inserting the Al_2O_3 rods into stainless steel tubes. The composite fabrication technique has been discussed elsewhere (Warren, 1979). All of the erosion results reported here for the SSAC samples were obtained using specimens with the Al_2O_3 rods oriented normal to the target sample surface.

RESULTS

Typical results for the W vs. θ curves for a single-phase ductile material and a single-phase brittle material are shown in Figs. 1 and 2. A ductile material like the 304 stainless steel shown in Fig. 1 has a characteristic

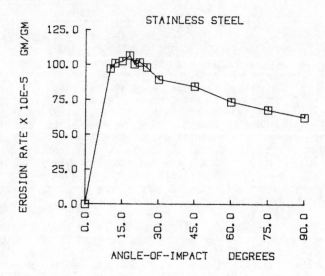

Fig. 1   W vs. θ for 304 stainless steel.   v = 100 m/s and
D = 130 micrometers.

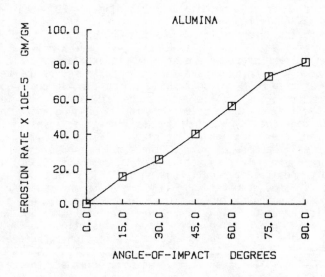

Fig. 2   W vs. θ for GTE Al$_2$O$_3$.   v = 59 m/s and D = 270 micrometers.

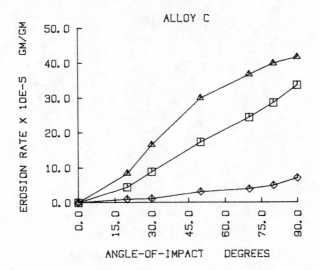

Fig. 3 W vs. θ for WC-Co alloy C. v = 50 m/s and the values
of D in micrometers are: 60 (diamonds); 140 (squares);
270 (triangles).

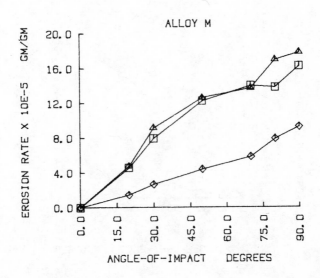

Fig. 4 W vs. θ for WC-Co alloy M. v = 50 m/s and the values
of D in micrometers are: 60 (diamonds); 140 (squares);
270 (triangles).

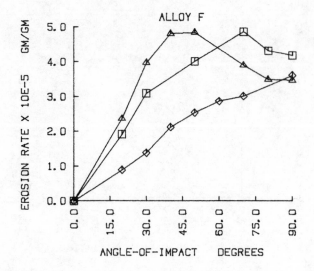

Fig. 5   W vs. θ for WC-Co alloy F.   v = 50 m/s and the values
of D in micrometers are: 60 (diamonds); 140 (squares);
270 (triangles).

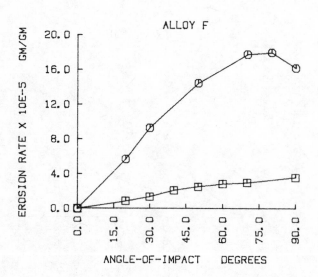

Fig. 6   W vs. θ for WC-Co alloy F.   D = 60 micrometers and
the values of v in m/s are: 50 (squares); 93
(circles).

"signature" in the W vs. θ relation such that a peak occurs at about 20 degrees. As was stated earlier, the currently accepted interpretation of this behavior is that a micromachining mechanism is dominant during the impact events at low θ angles, and this causes the peak to occur there. However, there is still an appreciable erosion rate at 90 degrees in a ductile material and, while the mechanism is not completely understood, it is no doubt due to a damage accumulation process that causes ductile rupture after repeated impact cycles. In contrast, a brittle material like Al_2O_3 shown in Fig. 2 has a characteristic W vs. θ curve that peaks at 90 degrees. A single fracture mechanism is believed to operate for erosion in brittle materials and, as noted earlier, this occurs by the formation of a subsurface lateral crack that removes a chip of material during the impact event. The process is most effective at 90 degrees because the normal component of the impact force generates the lateral cracks.

Figs. 3 to 5 show a series of W vs. θ curves for the WC-Co alloy samples C to F, respectively. Each figure shows three curves obtained with abrasive particle sizes of 60, 140 and 270 micrometers. The particle velocity was fixed at 50 m/s in all cases. There is an increase in the erosion rate as the particle size increases, as is indicated by the upward shift in the curves in each figure. The magnitude of the shift depends on the particular alloy and, furthermore, the shape of the W vs. θ curve changes between samples C to F tested under otherwise similar conditions. Noting that the carbide grain size decreases in going from sample C to F, Figs. 3 to 5 show the following important features:

1. For the smallest WC grain size, which is alloy F in Fig. 5, the erosion rate curves at larger abrasive particle sizes display a peak at θ less than 90 degrees and the peak shifts to lower θ values as the particle size increases. No peak at θ less than 90 degrees is observed for alloys C and M.

2. As the WC grain size decreases with all else constant, the erosion rate decreases. Compare the rates under identical conditions for alloys C, M and F in Figs. 3 to 5, respectively.

Figs. 6 and 7 show W vs. θ curves for alloy F using 60 and 140 micrometer abrasive particle sizes, respectively. Each figure shows two curves for particle velocities of 50 m/s and 93 m/s. The erosion rates increase with velocity as is indicated by the upward shift of the curves, but there is also a noticeable change in shape of the curves. Figs. 6 and 7 show the following features:

1. The ersoion rate increases and the curves display a peak at θ less than 90 degrees as the abrasive particle velocity is increased at fixed abrasive particle size. The peak at 93 m/s is more pronounced and shifts to lower θ values as the particle size increases from 60 to 140 micrometers in Figs. 6 and 7, respectively.

Figs. 8 and 9 show a series of W vs θ curves for the SSAC using abrasive particle sizes varying from 37 to 270 micrometers. The microstructure of the SSAC target samples is the same for both figures and the particle velocities were 59 m/s and 100 m/s for Figs. 8 and 9, respectively. The erosion rate increases as the abrasive particle size increases, as is

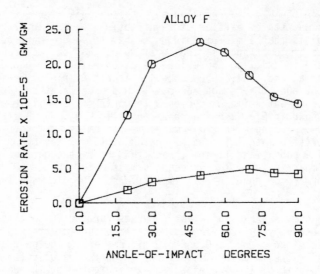

Fig. 7   W vs. θ for WC-Co alloy F.   D = 140 micrometers and
         the values of v in m/s are: 50 (squares); 93
         (circles).

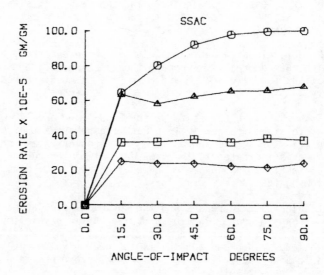

Fig. 8   W vs. θ for SSAC.   v = 59 m/s and the values of D in
         micrometers are: 37 (diamonds); 63 (squares); 130
         (triangles); 270 (circles).

indicated by the upward shift in the curves, and the rate also increases with particle velocity as a comparison of Figs. 8 and 9 will disclose. It is clear from these results that, for a fixed SSAC microstructure, the shape of the W vs. θ curve can change significantly as erosion conditions are varied, as was the case for the WC-Co alloys. The following important features are shown by Figs. 8 and 9:

1. As the abrasive particle size increases at fixed particle velocity, the peak at low θ values gradually disappears and is replaced by a peak at θ = 90 degrees.

DISCUSSION

The results presented in Figs. 3 to 9 provide a systematic set of erosion rate data for a range of different abrasive particle sizes, velocities and angle-of-incidence values for two rather different, but well-defined target sample microstructures. The microstructures each contain a hard, brittle component and a softer, more ductile component. The aim of the discussion given here is an attempt to rationalize the variation in the W vs. θ "signatures" for the various conditions and microstructures used. The microstructure will enforce an averaging of the intrinsic behaviors of the single-phase ductile and brittle components shown in typical form in Figs. 1 and 2. While the discussion and interpretations are qualitative at this point, the effects of microstructure can be significant and the challenge for future work will be further documentation of the microstructural effects along with the development of more quantitative analyses and models.

According to currently accepted models, which have been verified by many investigations, the erosion rate W in either ductile or brittle single-phase materials obeys a power-law relationship for both the abrasive particle-size dependence and the particle velocity dependence. The investigations show that the average volume loss V per particle impact event has the form $V = fD^m v^n$ where m and n are the size and velocity exponents, respectively. Note that the erosion rate W is usually given as mass loss per unit mass of particles impacted, so that W will follow a similar relation to V, but with m replaced by m - 3 because a D^3 factor is normalized out. f is a function of θ and the target sample material properties, and is determined by the superposition of micromachining and damage accumulation mechanisms for ductile phases. For brittle phases there is a single lateral crack fracture mechanism, and the θ dependence can usually be included by replacing θ with vsinθ, keeping a constant value for f. The values of n and m typically fall in the range of n = 2 to 4 and m = 3 to 4. For purposes of the discussion to follow, the precise values of the exponents are not critical, but one should note that brittle materials normally have m = 3.6 to 3.8 while ductile materials have m = 3. Thus, in terms of W, which includes the above mentioned D^3 factor, there is no size dependence for ductile phases except for possible threshold effects at small particle sizes (Ruff and Wiederhorn, 1979), but there is an intrinsic size dependence for brittle phases.

In brittle phases, the erosion mechanism is usually viewed in the following manner. A subsurface lateral crack is formed by the abrasive particle impact, and this crack causes the removal of a chip of material. Thus, there is a "damage event" associated with a single impact that causes material loss. As a first approximation, the size or scale of the damage

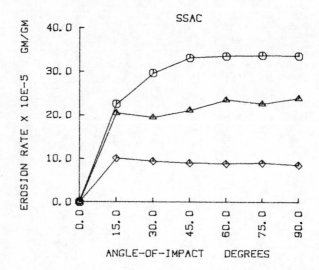

Fig. 9   W vs. θ for SSAC. v = 100 m/s and the values of D
          in micrometers are: 37 (diamonds); 130 (triangles);
          270 (circles).

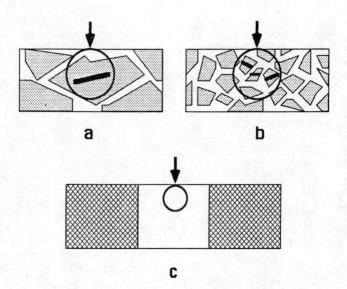

Fig. 10   Schematic of the impact event size vs. microstructural
           scales, as discussed in the text. The dark circles
           denote the size of impact events. The dark lines in
           a and b denote cracks in the WC-Co phase (shaded).

event varies with D and v according to the variation in V, i.e., as $D^m v^n$. For the range of D and v values used in this investigation, which are typical for erosion rate measurements, the scale of the damage event is a small fraction of the impacting abrasive particle size D. In the case of ductile phases, it is more difficult to associate the damage event that causes material loss with a single particle impact; this is especially true for the damage accumulation mechanism operative at higher θ values where repeated impacts are needed to remove a chip of material.

The two principle effects that are thought to underlie the behavior of the W vs. θ curves shown in Figs. 3 to 9 are the relative differences in the erosion rates of the brittle vs. the ductile phases as the abrasive particle velocity or size are increased, and the scale of the damage event relative to the scale of the microstructure.

First consider the situation for the WC-Co alloys. The coarse-grained alloy C shown in Fig. 3 displays W vs. θ curves that have a distinctly brittle signature, with the peak in W occurring at θ = 90 degrees and a power-law particle size dependence such that W varies as D^m with m about .6 (the latter is obtained from a cross plot of lnW vs. lnD). The curves in Fig. 4 for alloy M show similar trends except that the dependence of W on particle size tends to saturate at larger particle sizes. In contrast, Fig. 5 for alloy F shows a peak in the curves at θ less than 90 degrees, and this peak becomes more prominant and shifts to lower θ values as the particle size increases. Thus, according to the W vs. θ signature, there is a change from brittle to ductile response in the WC-Co alloys as the erosion conditions are varied. This change can be attributed to an effect related to the scale of the impact damage event relative to the scale of the microstructure, which in this case is the WC grain size. As shown schematically in Fig. 10a, when the damage event (dark circle) is small compared to the WC grain size, lateral cracks reach their full extension with single grains and there is no apparent constraint on the brittle-fracture erosion mechanism. Since WC is the major component in these alloys, the erosion rate of the WC phase dominates and so a brittle W vs. θ signature is observed. In contrast, when the damage event (circle) is large compared to the WC grain size as shown in Fig. 10b, the lateral cracks in the small WC grains are constrained by the Co binder phase (the cracks would normally propagate over distances large compared to the WC grain size). Evidently, the constraint on the brittle-fracture erosion mechanism and the presence of the ductile Co binder phase now present within the spatial extent of each damage event is sufficient to produce a transition to more ductile-like erosion mechanisms. Insofar as the absolute erosion rates are concerned, the values of W decrease as the WC grain size decreases with all else constant, as a comparison of Figs. 3 to 5 shows, and this reflects the decreased effectiveness of the erosion mechanism in the brittle WC phase due to the microstructural constraint. Furthermore, logarithmic plots of W vs. D at fixed v and θ will, in light of the results shown in Figs. 3 to 5, show very different behavior depending on the signature characteristic of the W vs. θ curves. Variable and seemingly anaomolous behavior for the abrasive particle-size dependence of the erosion rate in multiphase alloys is clearly a manifestation of effects of this kind.

Changes in the abrasive particle velocity should also produce changes in the W vs. θ curves because, as was noted earlier, the damage event changes in scale according to the relation $D^m v^n$. This effect is clearly shown for alloy F in Figs. 6 and 7 where increasing the velocity at fixed particle size produces a transition to a ductile signature in the W vs. θ curves due to the increased scale of the damage event. With a larger particle size (Fig. 7), the peak is at lower θ values, consistent with the larger-scale damage event for this case. Very recent work not reported here shows that similar trends can be observed in the M alloy. A quantitative SEM investigation of damage events, size scales, etc., for single impact events in the WC-Co alloys is currently in progress.

The erosion behavior of the SSAC also shows a noticeable change in the W vs. θ signature as abrasive particle size is increased at fixed velocity. Figs. 8 and 9 show that the ductile peak present at lower particle size diminishes as particle size is increased, and the W vs. θ curve has a brittle signature at the largest particle size. This is opposite from the behavior observed in the WC-Co alloys (compare Figs. 5 and 8) and it is a result of a distinctly different microstructural effect. The microstructural size scale is very large in the SSAC samples, and damage events will always be small relative to this size scale (for the range of conditions used here). As shown schematically in Fig. 10c, damage events will always occur in an unconstrained fashion in either the brittle Al_2O_3 phase or the ductile stainless steel. The change in the W vs. θ signature must then be a manifestation of the change in relative erosion rates for the two components; the microstructure imposes no obvious constraint as the abrasive particle size is increased in this case, and the microstructural "averaging" remains unchanged. As the abrasive particle size increases at fixed velocity, the erosion rate of the brittle phase should increase relative to that of the ductile component. As was mentioned earlier, the erosion rate W for brittle materials has an intrinsic particle size dependence whereas W for ductile materials is normally independent of size (tests on 304 stainless steel showed that W is essentially size independent for particle sizes in excess of about 60 microns). Thus, in a first approximation, as the abrasive particle size is increased, the Al_2O_3 erosion rate increases relative to stainless steel at fixed velocity and the W vs. θ signature manifests this change in Figs. 8 and 9. The velocity dependence of Al_2O_3 and stainless steel appear to be rather similar, and an effect due to velocity change at fixed abrasive particle size is not evident in these figures.

Finally, the interpretation of the change in W vs. θ signature is SSAC is supported by preliminary SEM examinations of the erosion surfaces, where for small abrasive particle sizes the Al_2O_3 rods protrude out of the eroded surface while for larger particle sizes they are undercut below the surface, consistent with the expected relative changes in W. There are more subtle aspects of the behavior of the SSAC, which relate to the nature of the "steady-state" erosion surface for two components that have different rates, and which have a continuous microstructure. No true steady-state erosion surface need exist, and work currently in progress is aimed at a detailed, quantitative examination of these aspects of the erosion behavior and also models for the averaging laws for SSAC.

CONCLUSIONS

The results of the work reported here support the following conclusions:

1. If the scale of the erosion damage event is varied such that it is large or small relative to the microstructural scale, then the erosion response of the material measured by the W

vs. θ signature can be changed significantly. In WC-Co alloys with 6 wt % Co, a brittle signature results if the WC grain size is large relative to the scale of the damage event because the dominant mechanism is the (unconstrained) brittle-fracture erosion process which occurs in the WC phase. If the WC grain size is small relative to the scale of the damage event, then a ductile signature develops, reflecting the constraints imposed on the brittle-fracture erosion process in WC grains and also the presence of the ductile Co within the spatial extent of the average damage event.

2. The W vs. θ signature can be changed by changes in the relative erosion rates of the components present even if the microstructural scale imposes no apparent constraint. In SSAC, where the microstructural scale is always large relative to the scale of the damage event for the test conditions used, the W vs. θ signature changes from ductile to brittle because of the increased relative erosion rate of the Al_2O_3 with increased abrasive particle size.

ACKNOWLEDGEMENTS

The authors wish to acknowledge the U. S. Department of Energy for support of this work under Grant No. DE-FG05-84ER45115. The authors also acknowledge support from the U. S. Department of Energy for work performed at the Argonne National Laboratory under the Thesis Parts Program.

REFERENCES

Ruff, A. W. and Wiederhorn, S. M. 1979, Treatise on Mat. Sci. and Tech., 16, p. 69.
Finnie, I. and McFadden, D. H. 1978, Wear, 48, p. 181.
Bellman, R. and Levy, A. V. 1981, Wear, 70, p. 1.
Evans, A. G., Gulden, M. E. and Rosenblatt, M. 1978, Proc. Roy. Soc., London, A361, p. 343.
Wiederhorn, S. M. and Lawn, R. B. 1979, J. Amer. Cer. Soc., 62, p. 66.
Hovis, S. K., Anand, K., Conrad, H. and Scattergood, R. O. 1985, Wear, 101, p. 69.
Kosel, T. H., Scattergood, R. O. and Turner, A. P. L. 1979, Proc. Intl. Conf. on Wear of Mat., A.S.M.E., p. 192.
Warren, R. 1979, Composites, p. 126.

Inst. Phys. Conf. Ser. No. 75: Chapter 10
Paper presented at 2nd Int. Conf. Science Hard Mater., Rhodes

963

Adhesive wear of boride and nitride layers on steel

K-H HABIG

Federal Institute for Materials Research and Testing,
D-1000 Berlin 45, Federal Republic of Germany

ABSTRACT

The adhesive wear behaviour of nitride and boride layers and of uncoated
steel was tested under vacuum conditions with the pin and disc system.
Pin and disc were equally heat treated so that self-mated materials were
examined. The results reveal that only the nitrided steel couple is able
to avoid adhesive wear after removal of the outer adsorption and reaction
layers while the other couples show strong adhesive wear. The different
adhesive wear behaviour of the couples is discussed with regard to the
different plastic deformability and type of atomic bonding of the
materials.

INTRODUCTION

In many tribotechnical systems, for instance gears or guidances, steel
couples are used. If such couples are poorly lubricated or overloaded
they can fail by adhesive wear. To avoid adhesive failure tribo-elements
are protected by different types of surface layers. Protective surface
layers on steel can be formed by metalloids which are found in the
periodic table of elements in the vicinity of oxygen (Fig. 1).

Two groups of metalloids can be distinguished:

    Group I: Boron, carbon, nitrogen, sulfur forming reaction
               layers by thermochemical treatments.

    Group II: Phosphorus, sulfur, chlorine, oxygen forming reaction
               layers as components of additives in lubricants.

$\boxed{B}$ FeB Fe$_2$B	$\boxed{C}$ Fe$_3$C	$\boxed{N}$ ε-Fe$_x$N γ'-Fe$_4$N	$\boxed{O}$ Fe$_2$O$_3$ Fe$_3$O$_4$ FeO	F
Al	Si	$\boxed{P}$ Fe$_x$P$_y$	$\boxed{S}$ FeS Fe$_2$S$_3$ FeS$_2$	$\boxed{Cl}$ FeCl$_2$ FeCl$_3$
Ga	Ge	As	Se	Br
In	Sn	Sb	Te	J
Tl	Pb	Bi	Po	At

Fig. 1   Iron compounds
to reduce wear of steels

In the following only the metalloids boron, nitrogen and oxygen of the
first group will be considered. Metallic materials are normally covered
with natural oxide layers. The thickness of the oxide layers can be
increased by a treatment in steam, but their wear resistance is limited.
If the oxide layers are worn away, direct contact of virgin metallic
surface regions occurs so that adhesive interactions become effective.

By boriding and nitriding, thicker compound layers are formed which are
also covered with natural oxide layers. The aim of this paper is to show
whether these compound layers can prevent adhesive wear of steel couples
after removal of the outer oxide layers.

TEST PROCEDURE

The wear tests were carried out with the pin and disc system under vacuum
conditions. The test conditions are presented in Fig. 2. Pin and disc
were equally heat-treated. Vacuum was chosen to reduce the velocity of
rebuilding of the worn oxide layers so that direct contact of virgin
surfaces could occur. The heat treatment parameters and the properties of
the samples are compiled in Table 1. During nitriding in the salt bath
not only nitrogen but also carbon diffuses into the surface regions of
steel, but the structure of the compound layer is that of iron nitride
ε -Fe$_x$N. By boriding a thick layer of iron boride, Fe$_2$B was formed
with small portion of FeB at the outer surface region.[2]

Vacuum 10^{-5} mbar

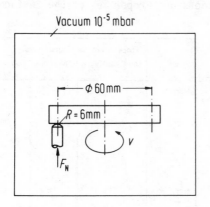

Normal load F_N : 10 N
Sliding speed v : 0,1 m/s
Sliding distance s : 1 km

Fig. 2   Pin and disc test system
and test conditions

TEST RESULTS

The friction-test duration curves are presented in Fig. 3. The friction
coefficients (f) of the hardened and the borided steel couples start at
nearly 0.1. The friction coefficient of the nitrided steel couple is
visibly higher at the beginning. With increasing test duration the
friction coefficients of all the couples tested rise to values between
0.45 for the nitrided couple and 0.65 for the borided couple. More
important than these differences are the friction peaks observed for the
hardened and borided couples. These peaks are typical for adhesive interactions. It is striking that the nitrided steel couple does not show any
friction peak.

The wear-test duration-curves in Fig. 4 show a very low wear of the
nitrided steel couple. The wear of the borided steel couple is higher
than that of the hardened steel couple. This may be caused by the greater
brittleness of the boride layer in comparison with the martensite
structure of the hardened steel. The morphology of the worn surfaces was
observed by means of a scanning electron microscope (Fig. 5). The
hardened and borided surfaces were strongly roughened while the nitrided
surface was smooth.

The result of the friction and wear measurements, including the morphologies of the worn surfaces, reveal that only the nitride layer can
avoid adhesive wear after disruption of the outer oxide layer - while the
boride layer and the hardened steel are not able to withstand adhesive
wear without an additional oxide layer. The high resistance of nitrided
steel couples is in agreement with other publications for instance of
Rogalski and Senatorski (1967) and Williamson (1980). Golego and co-
workers (1973) and Eyre (1975) confirmed the low adhesive wear resistance
of borided steel couples under vacuum conditions.

TABLE 1　　Heat Treatment Conditions and Properties of the Surface Layers

Steel DIN	AISI	Heat treatment	Phase	Hardness HV 0.2	Thickness μm	Roughness R_z μm
C 45	1045	**Hardening** 830°C → NaOH (10 %)	Martensite	760	—	~ 1
		Annealing 220°C 30' → H_2O				
42 CrMo 4	4140	Hardening 830°C → oil 50°C				
		Annealing 600°C 1 h → H_2O				
		Nitriding in TF1-bath *) 580°C 2 h → H_2O	ϵ-Fe_xN ($2 \leq x \leq 3$)	720	12	~ 1
42 CrMo 4	4140	**Boriding** in Ekabor 2 **) 900°C 4 h → oil 50°C	Fe_2B (FeB)	1560	100	~ 1
		Annealing 570°C 2 h				

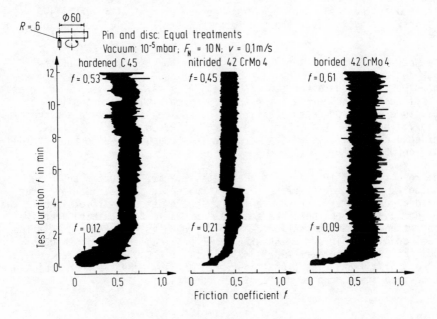

Fig. 3　　Friction curves

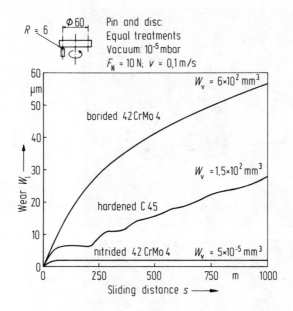

Fig. 4   Wear curves

DISCUSSION OF THE RESULTS

To find an explanation for the different adhesive wear behaviour of the steel couples tested a look into the contact area of the tribo-elements is necessary. According to Bowden and Tabor (1950, 1964) two bodies contact only in microcontact areas which form the so called real contact area. During tangential motion adhesive interactions can become effective in the microcontact areas depending mainly on two points:

I.   Hardness and ductility of the surface regions of the contacting materials

II. Probability and strength of adhesive bonding by interactions of electrons of the contacting materials.

The size of the microcontact areas can be kept low by high hardness and low ductility of the materials. Adhesive bonding should be restricted by a low density of free electrons (Czichos, 1972).

If the properties of martensite, iron nitride and iron boride are compared, martensite has the strongest metallic character. In the martensite of steel C 45 only 2 atomic percent carbon are dissolved, while iron nitride contains 25 to 33 atomic percent nitrogen and iron boride 33 to 50% boron. Because of its metallic character martensite has the highest ductility and the highest density of free electrons for adhesive bonding. Therefore its strong adhesive wear is understandable.

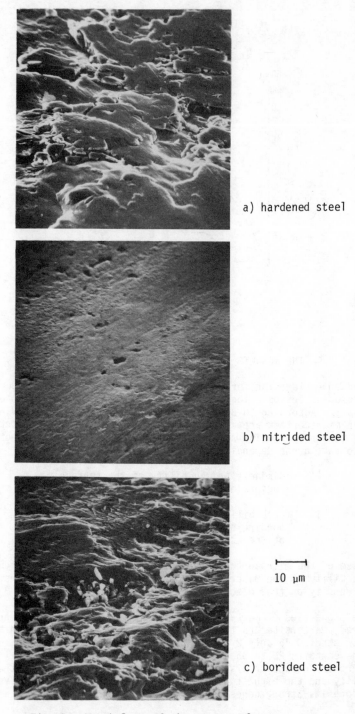

a) hardened steel

b) nitrided steel

10 μm

c) borided steel

Fig. 5   Morphology of the worn surfaces

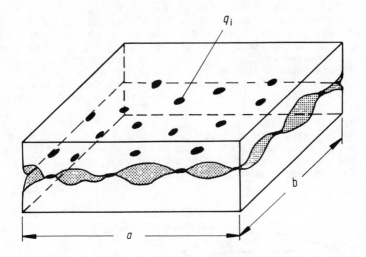

Geometrical contact area $A_{geom} = a \cdot b$

Real contact area $A_r = \sum\limits_{i=1}^{n} 9i$

Fig. 6   Geometrical and real contact area

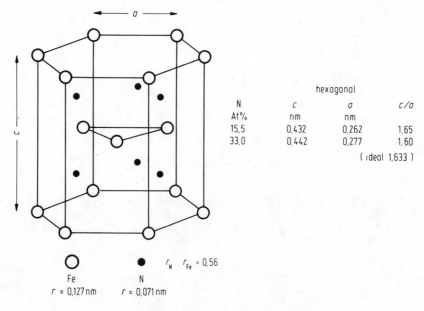

N At%	c nm	a nm	c/a
15,5	0,432	0,262	1,65
33,0	0,442	0,277	1,60

hexagonal

(ideal 1,633)

r_N r_{Fe} = 0,56

Fe
r = 0,127 nm

N
r = 0,071 nm

Fig. 7   Lattice of iron nitride of ε-Fe_xN

Though iron nitride is not much harder than martensite it has a lower ductility as it consists of hexagonal ε-Fe_xN (Fig. 7). In this phase the iron atoms form the hexagonal cell, the nitrogen atoms occupy the octahedral holes. The c/a-ratio, which depends on the nitrogen concentration, is close to the ideal ratio of 1.633 so that plastic deformation can mainly be achieved by sliding processes of the three basal slip systems. Therefore the growth of the microcontact areas by plastic deformation is limited. Free electrons of the iron lattice are covalently bonded by nitrogen atoms. Therefore the formation of metallic adhesion bonds is restricted.

The iron boride layers were composed of tetragonal Fe_2B (Fig. 8) with small regions of rhombohedral FeB. Both borides are relatively hard and brittle. The covalent type of bonding predominates, so - at first sight - the high adhesive wear of the borided steel couple is surprising.

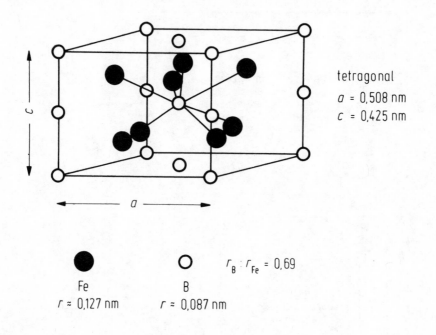

tetragonal
$a = 0.508$ nm
$c = 0.425$ nm

Fe
$r \approx 0.127$ nm

B
$r \approx 0.087$ nm

$r_B : r_{Fe} = 0.69$

Fig. 8   Lattice of iron boride Fe_2B

But according to Pauling (1964) borides have a special type of bonding. As the distance of the boron atoms in the iron borides is small (Table 2), there are not only bonds of Fe-B and Fe-Fe but also of B-B, while in iron nitride only Fe-Fe- and Fe-N-bonds are formed.

TABLE 2   Shortest Distances of the Atoms in Iron Borides (Kiesling, 1959)

Compound	Structure	Shortest distance in nm		
		Fe - Fe	Fe - B	B - B
Fe_2B	tetr.	0.241	0.218	0.220
FeB	rhomb.	0.263	0.215	0.180

It can be assumed that the greater variety of forming bonds of iron boride can also be effective in the microcontact areas so that the covalent adhesive bonds cause severe adhesive wear. Borided steel couples have only a resistance against adhesive wear if the boride layers are covered by an oxide layer preventing a direct contact of the boride layers.

REFERENCES

Bowden, F.P., Tabor, D. 1950, The Friction and Lubrication of Solids.
    Clarendon Press, Oxford.
Bowden, F.P., Tabor, D. 1964, The Friction and Lubrication of Solids.
    Part II. Clarendon Press, Oxford.
Czichos, H. 1972, J. Phys. D. Appl. Phys. 5 1890-1897
Eyre, T.S. 1975, Wear 34 385-397
Golego, N.N., Epik, A.P., Derkach, V.D., Labunets, V.F. 1973,
    Protective Coatings on Metals 257-260
Kiesling, R. 1959, Acta Chemica Scandinavia 4 209-227
Pauling, L. 1964, Die Natur der chemischen Bindung, Verlag Chemie,
    Weinheim
Rogalski, Z., Senatorski, J. 1967, IfL-Mitt. 6 444-452
Williamson, P.K. 1980, Tribology international 13 51-59

Inst. Phys. Conf. Ser. No. 75: Chapter 10
Paper presented at 2nd Int. Conf. Science Hard Mater., Rhodes

Bonding dense Si$_3$N$_4$ components to steel for elevated-temperature cycling applications

R LANDINGHAM AND T SHELL

University of California, Lawrence Livermore National Laboratory
P.O. Box 808, L-369, Livermore, California 94550 USA

ABSTRACT

A ceramic-to-metal bond has been developed for dense Si$_3$N$_4$ and steel. The large thermal expansion mismatch was taken into account in the design of such bonds to prevent failure upon thermal cycling. Selection of bonding materials was restricted by the bonding temperature that would allow diffusion of such materials but would not overheat the steel. Previous bond developments had minimized such restrictions as expansion mismatch and bonding temperature by selecting a refractory metal (Mo) to be bonded directly to reaction-bonded Si$_3$N$_4$. Here, these temperature/expansion restrictions were minimized by going to a two-stage bonding technique.

INTRODUCTION

Previous bonding studies at LLNL provided the incentive for attempting the more complex ceramic-to-metal bond described in this report. A bond of a metal shaft to a ceramic turbine rotor was needed. In the course of developing a unique ceramic turbine system on the Ceramic Helical Expander Program (CHEP) (see Landingham and Taylor, 1979; Wells and co-workers, 1967; Myers and co-workers, 1975; Myers, Deis and Shell; Taylor, 1977; Taylor and Shell, 1978; and Mohr, 1977) at the Lawrence Livermore National Laboratory (LLNL). These ceramic helical rotors were made of reaction-bonded silicon nitride (RBSN), because of cost considerations and the reduced thermal and mechanical stress levels imposed by this turbine design in comparison with thin ceramic blade turbine designs. TZM (0.5 wt % Ti - 0.08 wt % Zr - bal. Mo) was selected as the metal shaft because of low expansion and high temperature bonding and strength properties. Simple pressure bonding (1573 K at 69 MPa) of the TZM to RBSN was not successful until an intermediate bonding layer (10 μm thick) of MoSi$_2$ was slurry-coated at the interface (see Huffsmith and Landingham, 1978; U.S. Patent Office, 1981).

The RBSN produced in the 1970s was still under rapid development and had non-uniform porosity with high SiO$_2$ and Si contamination. Such RBSN provided easier bonding conditions as a result of porosity and lower melting contaminates. The silicon nitride (Si$_3$N$_4$) used in the present study was a high-strength nearly full-dense sintered Si$_3$N$_4$ (doped with Y$_2$O$_3$ and Al$_2$O$_3$)

to satisfy the higher stress levels required by the thin blade designs. The use of a steel shaft was also selected for its high-temperature properties but obviously not for its thermal expansion match to Si_3N_4. The objective of this study was restricted by funding and time to the question of feasibility and potential promise of developing a bond from the previous work performed at LLNL. The final parameters need an optimum ceramic-to-metal bond for these materials and will be determined in future studies.

EXPERIMENTAL PROCEDURES

Evaluation of Materials

Two centreless ground solid cylinders (3 cm dia by 6 cm long) of Si_3N_4 were provided for this study. The typical properties and composition as given by the suppliers are shown in Table 1. Wafers ($\sim$0.3 cm thick) were

TABLE 1   Properties of Sintered Si_3N_4 (doped with Y_2O_3 and Al_2O_3)

● Effect of Stress Level on Failure:

Probability of failure	Stress MN/m^2	Probability of failure	Stress MN/m^2
0.1	444.80	10^{-4}	267.65
0.01	374.68	10^{-5}	226.20
0.001	316.75	10^{-6}	191.24

● Elastic Modulus and Thermal Expansion:

Temperature K	Elastic Modulus GN/m^2	Thermal Expansion Coefficient 10^6 mm/mm K
300	294	2.48
1003	286	4.09
1173	284	4.48

● Typical Properties of Si_3N_4:

	Hot Pressed	Sintered
Bulk Density (kg/m^3)	3260	3200
Thermal Conductivity (W/m K)	29.3	14.6
Specific Heat (J/kg K)	.712	.712
Flexual Strength (GN/m^2)	1.18	85
Compressive Strength (GN/m^2)	441	392
Thermal Expansion Coefficient ($\times 10^{-6}$/K)	3.2	2.8
Young's Modulus (GN/m^2)	313.6	274.4
Max Usable Temp (K)	1873	1473
Normal Usable Temp (K)	1473	1273

● Composition:

(X-ray fluorescence – no standards available)	$Y_2O_3 \leq 5$ wt %	$TiO_2 \leq 0.24$ wt %
	$Al_2O_3 \leq 8$ wt %	$Si \stackrel{\sim}{\sim} 46$ wt %
	$Fe_2O_3 \leq 0.008\%$	Balance N

cut off one end of these cylinders for preliminary bonding studies. The cut surfaces were polished with 1 μm diamond before bonding. One long bar (3.5 cm dia by 25 cm long) of steel was also provided for this study. The composition and typical properties of this steel are given in Table 2. Wafers (∿0.6 cm thick) were cut from this bar for preliminary bonding studies. These cut surfaces were also polished with 1 μm diamond before bonding to Si_3N_4.

Bonding Conditions for Stainless Cr-Ni-W Steel

Initial investigations were directed to exploring the bonding conditions that would not adversely affect the Cr-Ni-W steel.

Pressure bonding (6.9 MPa) of two steel wafers together with a thin (∿20 μm thick) slurry layer of $MoSi_2$ powder was demonstrated at 1273 K. Although a good bond was observed, the microstructure of the steel had been significantly altered during the 1/2 h pressure bonding treatment (see Figs. 1(a) and 1(b)). Microhardness across the bonded region did not indicate any significant difference in hardness or the development of a brittle interface at the bonded regions (see Fig. 2). The general hardness of the steel had increased (312 to 521 HV 0.5) as expected due to the high temperature used to achieve this bond. Pressure bonds with the $MoSi_2$ layer attempted below 1073 K were not successful. The lowest temperature capable of achieving a pressure bond was 1133 K (see Fig. 3). The microstructure of the steel was not changed and the microhardness only changed slightly (312 to 296 HV 0.5).

TABLE 2   Composition and Properties of the Cr-Ni-W Steel

| Element | Composition (Weight %) | |
	Typical	Actual (Heat No. 1G3046)
Carbon	0.15	0.20
Manganese	0.40	0.37
Silicon	0.30	0.35
Chromium	13.00	12.44
Nickel	2.00	1.96
Tungsten	3.00	2.66
Sulphur	-	0.01
Phosphorus	-	0.02
Iron	Rem.	Rem.

● Elastic Modulus and Thermal Expansion:

Temperature K	Elastic Modulus GN/m^2	Thermal Expansion Coefficient 10^6 mm/mm K
300	196.5	10.62
473.2	193.0	10.80
873.2	179.1	11.02
1473.2	158.6	12.38

● Hardness: 40 R_c or 260/285 BHN (Spec AMS-5616F--2023 K min oil)

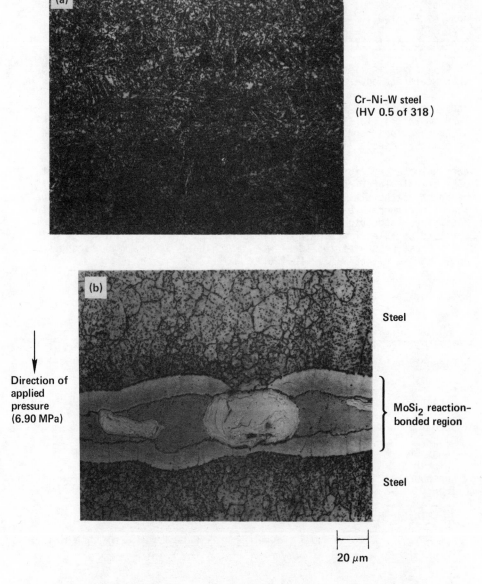

Cr-Ni-W steel
(HV 0.5 of 318)

Steel

Direction of
applied
pressure
(6.90 MPa)

MoSi$_2$ reaction-
bonded region

Steel

20 μm

Fig. 1 Microstructure of Cr-Ni-W steel (a) as received and
(b) after pressure bonding with MoSi$_2$ at 1273 K for 1/2 h.
(Etched with 5% HCl-methanol)

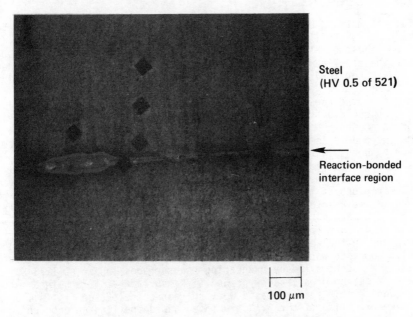

Steel
(HV 0.5 of 521)

← Reaction-bonded
interface region

├─────┤
100 μm

Fig. 2 Pressure-bonded (1273 K) regions of MoSi$_2$ between two plates
of steel with microhardness to show no embrittlement at the
interface. (Etched with 5% HCl-methanol)

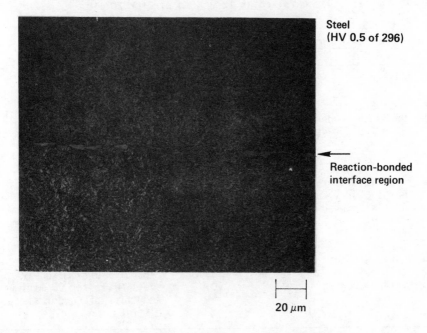

Steel
(HV 0.5 of 296)

← Reaction-bonded
interface region

├─────┤
20 μm

Fig. 3 Pressure-bonded (1133 K) region of MoSi$_2$ between two plates
of steel. (Etched with 5% HCl-methanol).

Bonding Conditions for Si_3N_4

Initial attempts to bond Si_3N_4 to the Cr-Ni-W steel with an intermediate slurry layer of $MoSi_2$ powder were made at 1133 K with no success. As expected, the thermal expansion mismatch sheared the relatively weak bond between the steel and $MoSi_2$. The $MoSi_2$ layer had been previously bonded to the Si_3N_4 at 1673 K in vacuum.

To reduce this expansion mismatch and develop a metallic layer on the Si_3N_4 that could be vacuum-brazed to the steel, a $Ni/MoSi_2$ slurry coating was prepared and bonded to the Si_3N_4 at 1673 K in vacuum.[2] Slurries of Ni and $MoSi_2$ powders as well as Ni, Mo, and Si powders were evaluated (see Table 3). Both slurry coatings produced enough wetting of the Si_3N_4 surface to get a metallic layer for subsequent vacuum brazing to steel. While the bond at the interface was excellent, microcracks were observed in thick regions of the metallic layer. The Ni to $MoSi_2$ weight ratios of 1:2 and 1:1 were investigated during this study. Slurries prepared early in the study were mixtures of $Ni/MoSi_2$ or Ni/Mo-2Si of the 1:2 ratio, which usually resulted in poor wetting of the Si_3N_4 surfaces. Significant improvement in the wetting was observed by changing this ratio to 1:1 by weight. Complete wetting was not achieved until surface preparation of the Si_3N_4 was controlled by preheating the surfaces in air at 1273 K. Surface oxidation of a freshly ground Si_3N_4 surface at 1273 K in air for 10, 15, and 20 minutes increased wetting respectively (see Fig. 4). Excessive oxidation of the Si_3N_4 surface caused spalling and non-uniform wetting of the $Ni/MoSi_2$ slurry coatings in subsequent treatments. Attempts to wet non-oxidized Si_3N_4 surfaces with the $Ni/MoSi_2$ slurry in vacuum $(7 \times 10^{-6}$ m bar) were unsuccessful even at 1773 K for 30 min, 1823 K for 60 min, or 1873 K for 60 min. Optimum wetting occurred with slurry coatings if the Si_3N_4 was oxidized at 1273 K in air for 20 min.

TABLE 3   Composition and Preparation of Slurry Coatings

Coating	wt %	Preparation
● Mo-Si Slurry:		
99.8% Mo powder (-75 μm)	63.0	Ball milled in WC mill
99.99% Si powder (-500 μm)	37.0	with WC balls for 21 hrs in toluene to reduce powder size distribution
● Mo-Si-Ni Slurry:		
Mo-Si Slurry Coating	50.0	Alcohol slurry mixed in
99+% Ni powder (-45 μm)	50.0	steel mills for 30 min. No reduction of powder size distribution
● Mo-Si$_2$-Ni Slurry:		
99+% MoSi$_2$ powder (-75 μm)	50.0	Alcohol slurry mixed in
99+% Ni powder (-500 μm)	50.0	steel mills for 30 min. No reduction of powder size distribution

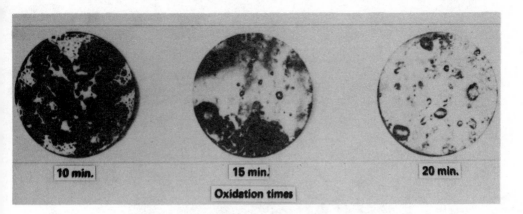

Fig. 4 Silicon nitride (yttrium oxide/alumina doped) pellets with
surface oxidation, done at 1273 K in air for three different times
before wetting studies (1X). Treatment with 50% molydisilicide/nickel
powder slurry done at 1773 K in vacuum for 20 min.

Bonding Conditions for Si_3N_4 to the Cr-Ni-W Steel

The premetallized surfaces (Ni-Mo-Si) of Si_3N_4 as described above were
then vacuum brazed to steel. The brazing temperature was selected for
each brazing alloy to get maximum diffusion with minimum alterations of the
steel's microstructure. The initial vacuum brazing alloy used, Braze A,
(58% Ag - 32% Cu - 10% Pd) was in the form of 0.0762-mm foil. This was
selected first because its solidus and liquidus temperatures (1097 K and
1125 K, respectively) are close to the desired brazing temperature
(1123 K.) Three other brazes were ordered (two in the powder form) to
investigate a range of braze compositions and temperatures (see Table 4).

A wafer (3 cm dia by 0.6 cm thick) of Si_3N_4 was vacuum-brazed with Braze A
to a disk (3.5 cm dia by ~1 cm thick) of steel at 1123 K in 1/2 h and
furnace-cooled. This configuration is a severe test of the bond strength
and ductility since there is a factor of four thermal expansion mismatch
between the two materials. The bond held and metallographic cross sections
disclosed microcracking into the Si_3N_4 surface as well as local pullout
regions at the steel-to-braze bonded interface (see Fig. 5). The failure
of parent material (Si_3N_4 and steel) at the interface indicates that an
extremely good bond had been developed. Such microcracking can be prevented in actual practice by designing the configuration so expansion
stresses put the bond into compression instead of shear. In this
application, the steel will fit over the Si_3N_4 hub and allow the steel to
cool down around the Si_3N_4, which puts the bond at their interface and the
Si_3N_4 itself into compression (see Fig. 6). Since the operating temperature (873 K) will always be below the bonding temperature (> 873 K), some
compressive stresses will always be on the bond and Si_3N_4 at operating
temperature.

Similar bonding tests with the second braze, Braze B, were not as successful even at higher brazing temperatures (< 1233 K). This was attributed
to the brittle intermetallics formed with this alloy system. The two

Braze

20 μm

Braze

Si₃ N₄

50 μm

Fig. 5 Microstructure of the vacuum brazed bond between Si₃N₄
and steel disks. Braze A (Ag-Cu-Pd) foil (70 μm thick) was used
to form this bond at 1123 K.

Fig. 6 Si$_3$N$_4$ turbine rotor and shaft before assembly and bonding.

remaining powder brazes listed in Table 4 were then used to bond Si_3N_4 wafers to steel wafers. The Si_3N_4 surfaces were preoxidized at 1173 K for 20 min in air and then coated with a 50 wt % $Ni/MoSi_2$ powder slurry and vacuum heat-treated at 1773 K for 20 min.

TABLE 4   Compositions and Solidus/Liquidus Temperatures of Brazing Alloys

| Braze Type | Composition wt % | | | | | | Temperature K | | Form |
	Ni	Cu	Ag	Mn	Pd	Ti	Solidus	Liquidus	
A	-	32	58	-	10	-	1097	1125	70 μm
B	15	15	-	-	-	70	1044	1233	-75 μm powder
C	2	42	56	-	-	-	1044	1166	-75 μm powder
D	9.5	52.5	-	38	-	-	1153	1198	-75 μm powder

These metallized surfaces were polished to remove any excessive lumps before slurry-coating with a thin layer (∿300 μm at full density) of braze powder mixed with acetone (3:1 volume ratio respectively). The polished faces of the steel wafers were placed over these slurry coatings of Braze C and Braze D and vacuum brazed for 15 min at 1173 and 1223 K, respectively. Both wafer sandwiches were intact after vacuum brazing. Metallographic examination of the Braze C bonded layer disclosed nearly continuous bonding with only isolated regions of fractured material (see Fig. 7). These latter regions of a different composition were very brittle and could not withstand the stresses generated by the thermal mismatch of the Si_3N_4 and steel during cooling. Closer examination of these two regions showed significant differences in their microstructure, bonding characteristics, composition, and microhardness. The brittle and undesirable region was high in Ni, Mo, Si and Cu, according to electron microprobe mapping (see Fig. 8). In comparison, the majority of the bonded region was high in only Ni, Cu, and Ag (see Fig. 9). A uniform dispersion of other elements (Al, Si, Fe, Cr, Ti, and Mo, respectively) was also detected by microprobe across the major braze regions in Fig. 9. Similar uniform distribution of elements (Al, Fe, Si, and Ti, respectively) was detected in the brittle second-phase regions in Fig. 8. Diffusion of Ni, Cu, and Mo into the Si_3N_4 was detected at all points along the interface. Diffusion of Al and Ti from the Si_3N_4 into the braze regions was also detected. Extensive diffusion of elements (Ni, Si, Cu, Al, and Ag, respectively) also occurred across the interface of Braze C to steel.

The Knoop microhardness along each interface was determined at 0.0196 or 0.05 N/min loading and parallel to the interfaces. These results are displayed in Figs. 7, 8 and 9. The brittle second-phase regions with microcracks had significantly high hardnesses in comparison with the rest of the braze regions (∿900 vs ∿100 HK 0.2).

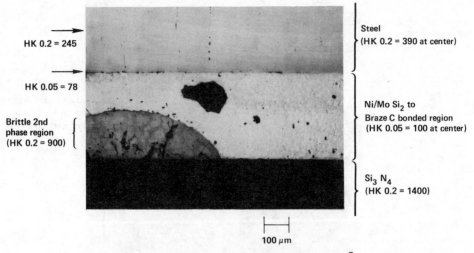

Steel
(HK 0.2 = 390 at center)

HK 0.2 = 245

HK 0.05 = 78

Brittle 2nd
phase region
(HK 0.2 = 900)

Ni/Mo Si$_2$ to
Braze C bonded region
(HK 0.05 = 100 at center)

Si$_3$ N$_4$
(HK 0.2 = 1400)

100 μm

Fig. 7 Cross-sectional view of bond region (3 x 10^{-3}m thick)
between Si$_3$N$_4$ and steel wafers. Bond region consists of
Ni/MoSi$_2$ and Braze C braze slurry coatings.

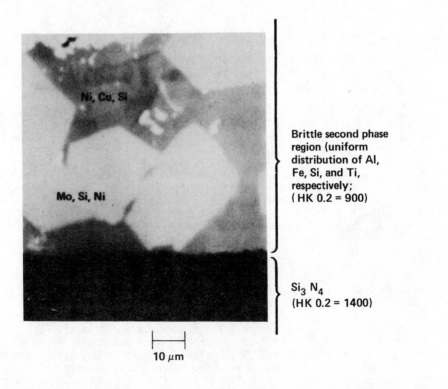

Ni, Cu, Si

Mo, Si, Ni

Brittle second phase
region (uniform
distribution of Al,
Fe, Si, and Ti,
respectively;
(HK 0.2 = 900)

Si$_3$N$_4$
(HK 0.2 = 1400)

10 μm

Fig. 8 Microprobe mapping results of minor brittle regions attached
to the Si$_3$N$_4$ surface of the Braze C brazed wafers.

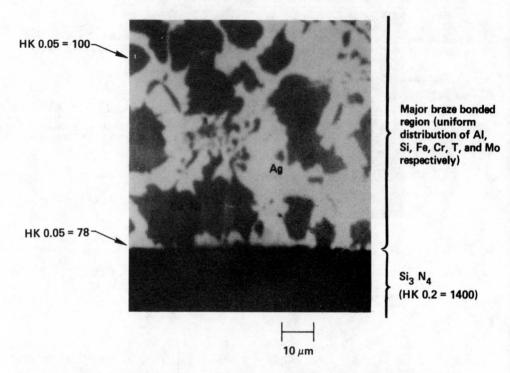

HK 0.05 = 100

Major braze bonded
region (uniform
distribution of Al,
Si, Fe, Cr, T, and Mo
respectively)

Ag

HK 0.05 = 78

$Si_3 N_4$
(HK 0.2 = 1400)

10 μm

Fig.9 Microprobe mapping results of major bonding region attached to both
the Si3N4 surface and the steel with the Braze C brazed wafers.

Microprobe evaluation of the Braze D wafers was performed across sections
shown in Fig. 10. Mn and Mo were uniformly dispersed in the braze region
while Ti was uniformly detected in the Si_3N_4 and the immediate braze
layer (see Figs. 11 and 12).

These microprobe results suggested the possibility of bonding this Si_3N_4
directly to the steel without using the premetallizing coating of $Ni/MoSi_2$
on the Si_3N_4 before brazing. A Si_3N_4 wafer with a preoxidized surface
was brazed directly to a steel wafer with the Braze C braze at 1173 K for
15 min in vacuum. The braze was selected since it had shown the greatest
promise in previous tests and was cheaper than Braze A. This bond was
very weak, which indicated the need for the premetallizing coating of
$Ni/MoSi_2$ on the Si_3N_4 before brazing.

The Si_3N_4 rotor and steel shaft as shown in Fig. 6 were bonded with Braze
C after preoxidizing and metallizing the Si_3N_4 bonding surface with
$Ni/MoSi_2$.   The rotor was heat-treated in air for 20 min at 1273 K to
oxidize the Si_3N_4 joint surface. The slurry coat of $Ni/MoSi_2$ (1:1 by
weight) was applied to the Si_3N_4 bonding surface and heat-treated in
vacuum for 20 min at 1773 K. This metallized surface was ground smooth
with a diamond tool and coated with the Braze C braze slurry. The steel
shaft was heated to 573 K and slipped over the Si_3N_4 rotor. This assembly
was placed in a vacuum furnace and heat-treated to 1173 K for 20 min to
complete the bond. Tests are in progress at an engine company on this
rotor/shaft bond.

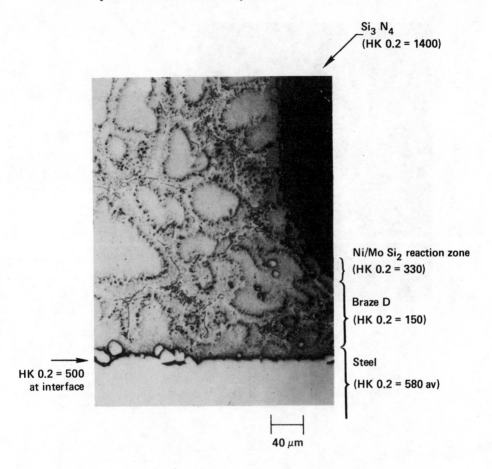

Fig. 10   Cross-sectional view of bond region between Si₃N₄ and
steel wafers. Bond region consists of Ni/MoSi₂ and Braze D
braze slurry coatings.

CONCLUSIONS

A method for diffusion bonding of high-density Si_3N_4 to steel has been
demonstrated.   This method employs multilayer bonding to reduce the
stresses anticipated from the large thermal expansion mismatch present
between these two materials during temperature cycling. Braze C was the
best braze alloy of the four tested in this study but others might be even
better. Metallizing of Si_3N_4 surfaces with Ni/MoSi₂ (1:1 ratio by weight)
was dependent on the preconditioning of these surfaces. A thin oxide film
was very desirable in wetting the Si_3N_4 with the melted Ni/MoSi₂ slurry
in vacuum. It was apparent from flat coupon test specimens that this bond
requires a bond design that puts the bond and the Si_3N_4 into compression
during cooling, or the stresses will shear the Si_3N_4 or bond. This was
accomplished by designing the steel joint to fit around the Si_3N_4 and
shrink-fit down onto the bond and Si_3N_4 during cooling.

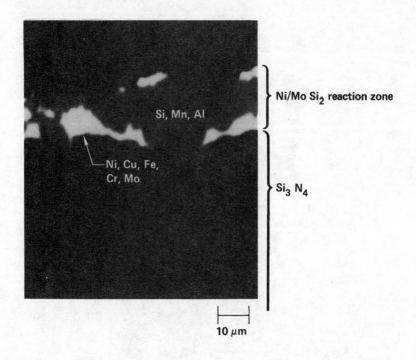

Fig. 11   Microprobe mapping results of braze region at Si_3N_4 and Braze D interface, showing two distinct phases of different compositions.   Note microcracks in phase with high Si, Mn, and Al content.

ACKNOWLEDGEMENTS

This work was performed under the auspices of the U.S. Department of Energy by the Lawrence Livermore National Laboratory under Contract W-7405-Eng-48. We acknowledge the support from Robert Graham and John Lewakowski at TACOM in making this study possible. The close cooperation from Isoroku Kubo from Cummins Engine Co. was most helpful in acquiring materials and submitting bonds for temperature cycle tests. We would also like to acknowledge the fine work provided by our Surface Science Section: microstructural preparation by Paul Curtis, microprobe analysis by Joe Balser and Don McCoy, and SEM (EDX) by Wayne Casey.

HK 0.2 = 500

Steel

Braze D layer
(HK 0.2 = 150)

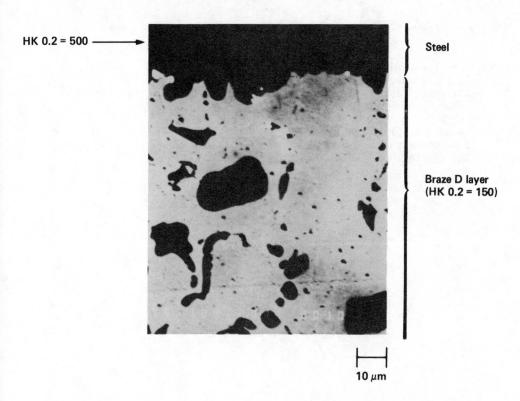

10 μm

Fig. 12 Microprobe mapping results of braze region at steel and Braze D
interface, showing at least two distinct phases of different
compositions in the braze region.

REFERENCES

Huffsmith, S.A. and Landingham, R.L. 1978, Metal Progress 122
Landingham, R.L. and Taylor, R.W. 1979, Materials Development and
    Evaluation of the Ceramic Helical Expander Program (CHEP), presented
    at the 4th Int. Meeting on Modern Ceramics Technologies (4th CIMTEC),
    Saint-Vincent, Italy, May 28 - June 1, 1979
Meyers, B., Landingham, R., Mohr, P. and Taylor, R. 1975, Proc. Symposium
    on the Environment and Energy Conservation (Environmental Protection
    Agency and Energy Research and Development Administration, Denver,
    Colorado, 1975). Also Lawrence Livermore National Laboratory,
    Livermore, California, UCRL-77905 (1975)
Mohr, P.B. 1977, Proc. Fifth Army Materials Technology Conference,
    Newport, Rhode Island, March 21-25, 1977
Myers, B., Deis, G. and Shell, T., Overall Efficiency of the Helical
    Expander for Brayton Cycle Hot Engines. Lawrence Livermore Laboratory,
    Livermore, California (to be published)

Taylor, R.W. 1977, Silica Vaporization and Condensation: a Potential
    Source of Fouling in Coal Combustion. Lawrence Livermore National
    Laboratory, Livermore, California, UCRL-79146
Taylor, R.W. and Shell, T.E. 1978, Ash Fouling and Erosion of Silicon-
    Based Ceramic Expanders in Coal-Fired Powder Plants. Lawrence
    Livermore National Laboratory, Livermore, California, UCRL-52390
U.S. Patent Office 1981, Patent No. 4,293,619 (October 1981)
Wells, W.M., Hanner, D.W., McElroy, J.L. and Robinson, E. 1967, High-
    Temperature Testing and Evaluation of Graphite Helical-Screw Expanders
    and Compressors. Lawrence Livermore National Laboratory, Livermore,
    California, UCRL-14660, pp 207-216

Inst. Phys. Conf. Ser. No. 75: Chapter 10
Paper presented at 2nd Int. Conf. Science Hard Mater., Rhodes

989

Alloy development of a Co-Mo-Cr-Si wear resistant alloy

A HALSTEAD(1) AND R D RAWLINGS(2)

(1) Brunel University, Uxbridge, Middlesex, England
(2) Imperial College, London, England

ABSTRACT

The microstructure and fracture of a Co based alloy has been investigated
and modifications made to enhance the alloy's mechanical behaviour.

The best combination of properties were obtained when the alloy had a fcc
solid solution component of the eutectic matrix. A marked deterioriation
occurred when a hcp solid solution was present with a Widmanstatten
precipitate. However, it was found that microstructural variations had
little effect on the erosion resistance.

Additions of iron were made to stabilise the fcc solid solution, these
gave acceptable values of $K_{1C} = 22.9$ MN$^{-3/2}$ increased the MOR by 40% and
decreased the critical flaw size by 70% when 15% Fe was added.

INTRODUCTION

The alloy studied belongs to a group of wear resistant alloys that gain
their strength and hardness from particles of an intermetallic Laves phase
not from the presence of carbides. The alloy designated T400 has a
composition 62% Co, 28% Mo, 8% Cr, 2% Si, 0.08% C, a very high bearing
strength (compressive strength 1800 MNm^{-2}) but little ductility (charpy
un-notched 4J).

T400 can be processed as a powder using plasma spraying or powder metallurgical techniques. Small components are castable by investment
and centrifugal methods and they can be overlayed on base metals by
oxyacetylene or arc welding processes.

The alloy was believed, Schmidt and Ferriss (1975) and Cameron and Ferriss
(1974), to be stable up to at least 1230°C and it was considered that the
microstructure could not be modified by subsequent heat treatment after
casting. However, later work by Halstead and Rawlings (1984) showed this
to be incorrect and the alloy can be both hardened and softened by heat
treatment. This lack of microstructural stability resulted because of the
allotropic nature of cobalt which can cause either the face centred cubic
form or the hexagonal close packed crystal structure or both to be

present. In addition heat treatment in the range 600°C - 1000°C was found to cause precipitation within the Co solid solution in either a spheroidal or Widmanstatten form.

This paper reports the role of microstructure in determining the erosion and fracture behaviour of T400 and the subsequent alloy modifications that were made to the T400 alloy to enhance the mechanical performance.

PROCEDURE AND RESULTS

Casting methods and heat treatments were chosen to obtain a range of micro-structural states, the key features of which are summarised in Table 1.

TABLE 1   Microstructural Conditions of T400 Specimens

Specimen No.	Cobalt solid solution	Primary Intermetallic Phase	
		Vol Fraction %	Mean Size μm
1	fcc	41 ± 7	25 ± 3
2	fcc/hcp	33 ± 4	24 ± 20
3	hcp + spheroidal ppt.	32 ± 6	16 ± 8
4	hcp + Widmanstatten ppt.	35 ± 4	29 ± 4

In the as-cast condition (specimen 3) T400 had large primary dendrites of the hard intermetallic $Mo(Co,Si)_2$ phase. The matrix was a lamellar eutectic of secondary Laves and Co solid solution (Fig. 1); both the fcc and hcp forms of the Co solid solution were present. Furthermore, the primary dendrites were generally separated from the eutectic by a layer of Laves free solid solution. All the microstructural modifications were variations within the Co solid solution, namely the totally fcc solid solution with a coarsened eutectic (specimen 1, Fig. 2), the hcp solid solution with a spheroidal precipitate (specimen 3, Fig. 3) and a hcp solid solution with a Widmanstatten precipitate (specimen 4, Fig. 4).

Fracture toughness (K_{1C}) and modulus of rupture (MOR) were chosen as the two fracture parameters to evaluate the alloys' performance. Tests were performed with specimens 37 x 5 x 2.5 mm^3 notched in the former case and un-notched and polished to a 1 μm finish in the latter. Testing took place under three point bend loading conditions with a span of 35 mm. The results of the test are summarised in Table 2.

TABLE 2   Mechanical Property Data for T400

Specimen No.	HV50 Kgmm^{-2}	K_{1C} MNm$^{-3/2}$	MOR MNm^{-2}	Critical Defect size mm^2
1	620	26.0	923	0.31
2	690	21.9	745	0.37
3	740	19.2	506	1.03
4	736	15.8	400	1.21

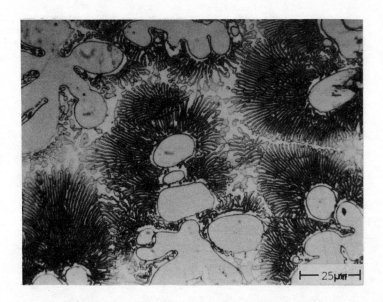

Fig. 1 T400 as-cast microstructure with lamellar eutectic, primary Laves phase dendrites and Laves phase free regions of Co solid solution.

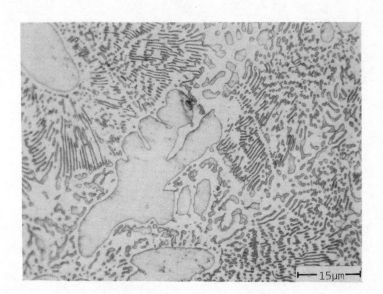

Fig. 2 T400 (specimen 1) primary Laves, coarse lamellar eutectic and fcc Co solid solution.

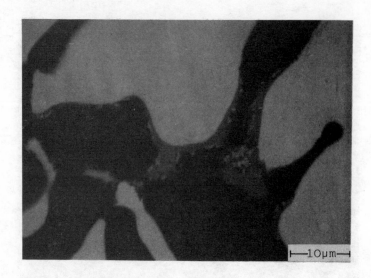

Fig. 3 Back scattered electron micrograph of T400.
Co solid solution composed of hcp crystal structure
spherical precipitate (specimen 3).

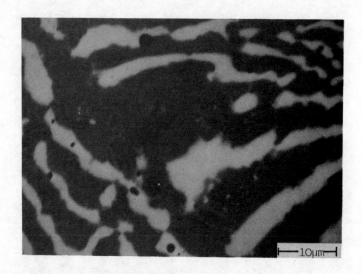

Fig. 4 Back scattered electron micrograph of T400.
Co solid solution composed of hcp crystal structure
and Widmanstatten precipitate (specimen 4).

The critical flaw size was calculated from the MOR and K_{1C} data using the following modified form of the fracture equation which takes into account the fact that inherent flaws are not through the thickness in nature.

$$MOR = \frac{1.68K_{1C}}{YA_c^{\frac{1}{4}}} \qquad\qquad 1$$

where Y is a constant dependent on the geometry of the specimen and the loading conditions and A_c is the area of the critical inherent flaw. Thus, flaw size can be related through the MOR and K_{1C} values giving the single parameter. The smallest critical inherent flaw size will always be exhibited by the alloy with the best overall fracture properties.

Erosion tests using dry silica sand of particle size less than 300 μm, and velocity of 15.7 m/s were carried out on selected microstructural conditions of T400. The weight loss after impact by 820 gm of erodant is presented as a function of the impact angle in Fig. 5.

DISCUSSION

The K_{1C} values lie in the range 15.8 - 26.0 $MNm^{3/2}$ and for most of the heat treated conditions the crack proceeded in a brittle manner through the primary Laves phase and matrix. However there was a preference for the crack path to favour the Laves particles through which it propagated in a cleavage mode (Fig. 6).

The only time the fracture toughness behaviour of the alloy was not dominated by the primary Laves phase was when the alloy had a hcp solid solution with a Widmanstatten precipitate (specimen 4). This matrix morphology led to the lowest K_{1C} values and was associated with a mixed failure mode, i.e. intergranular failure at the primary Laves phase/matrix interface and transgranular through the matrix. In this case the matrix played the major role in determining K_{1C}.

In contrast to the relatively small variation found in K_{1C} the MOR varied considerably from 400 - 723 MNm^{-2}. The primary Laves phase cleaved easily and it was considered to be the varying ability of the Co solid solution component of the matrix to blunt the microcracks by plastic deformation thereby hindering the linking that led to the wide range of values of MOR. The highest values were associated with the fcc solid solution when there was no precipitate present. The lowest values of the MOR corresponded to the hcp cobalt solid solution and Widmanstatten precipitation. In this situation the precipitate morphology was such that microcrack linkage through the matrix occurred at lower stresses than to cleave the primary Laves phase. The MOR results then clearly demonstrated the influence of the matrix in determining the fracture behaviour.

The erosion resistance of T400 was good and varied little with heat treat-ment. A maximum rate of erosion was observed between impact angles of 50° and 60°. The erosion behaviour of T400 has been compared to that of a glass-ceramic 'Silceram'. The erosion for this typical brittle material increased rapidly with impact angle to a maximum at normal incidence, this was consistent with other reported erosion data. In contrast ductile

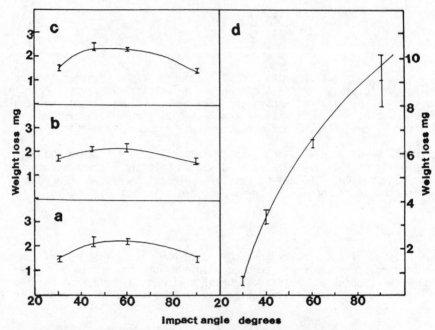

Fig. 5 Effect of microstructure on the erosion resistance of T400
(a) specimen 2 (b) specimen 1 (c) specimen 4 (d) erosion data
of 'Silceram'; a brittle glass ceramic.

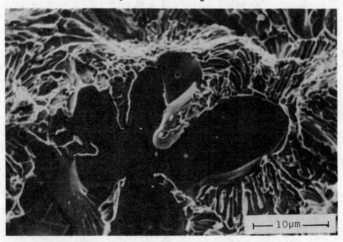

Fig. 6 Scanning electron micrograph of the fracture surface.
Transgranular cleavage and the brittleness of the failure
mode reflect the optical microstructure (Fig. 1).

materials such as aluminium and mild steel have been shown to reach a maximum at an oblique angle typically between 20° and 30°. Fig. 7 depicts the performance of aluminium and alumina (after Finnie, Wolak and Kabul, 1967) as an indication of the extremes in measured erosion.

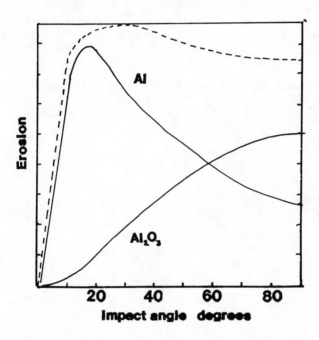

Fig. 7 Variation of erosion with impact angle for aluminium and alumina.

Summing the two extreme types of behaviour gave the dashed curve in Fig. 7, which is a much flatter curve showing no drastic variation with angle, although a slight peak was evident between 30° and 40°. This type of behaviour was very similar to that of T400 which indicated the advantage of the combination of the very brittle intermetallic embedded in the softer Co solid solution. The fact that the overall rate of erosion of T400 was by contrast very low indicated the benefit of using such a material where a high erosion resistance is required.

ALLOY DEVELOPMENT

The excellent wear resistance of T400, as reported by Cabot Corporation (1982), was mainly attributed to the large volume fraction of the Laves phase in the alloy. The present results indicate that the Laves phase also plays the major role in controlling the erosion resistance as microstructural variations involving the cobalt solid solution had little effect on the measured weight loss. Therefore the intermetallic compound has to be present in a reasonable amount for the alloy to function in a wear/erosion resistant capacity. It follows that the inherent resistance to crack propagation, as quantified by the critical stress intensity factor, will remain in the region of 20-25 MNm$^{-3/2}$ irrespective of any alloy modification.

In contrast the resistance to crack initiation and growth to a critical size can be altered significantly and the benefit of the fcc form of the solid solution has been demonstrated. Thus elements that stabilised the fcc crystal structure would be useful additions. Iron and nickel both increase the stacking fault energy, i.e. prevented the formation of the hcp crystal structure, but as iron may cause dilution when T400 is used as a hard facing alloy it represented a natural candidate for investigation.

Additions of 5, 10 and 15% Fe by weight were made to the as cast T400. The results of the microstructural and mechanical property data collected from the new iron bearing alloys have been listed in Table 3.

TABLE 3   Microstructural and Mechanical Property Data for the iron modified alloys

Specimen	Primary phase Volume Fraction %	Mean size µm	HV50 kgmm^{-2}	K_{1C} MN^{-2}	MOR MN^{-2}	Critical Defect size mm^2
T400 remelt	42	6	685	22.3	917	0.17
T400 5% Fe	35	8	646	20.8	965	0.11
T400 10% Fe	25	5	611	24.8	1280	0.07
T400 15% Fe	17	5	597	22.9	1279	0.05

As well as stabilising the fcc form of the Co solid solution the addition of iron significantly reduced the volume fraction of the primary Laves phase from 42% - 17% as the iron additions were increased from 5% - 15%. A comparison of the optical micrographs of the 5% and 15% Fe bearing alloys Figs. 8 and 9 illustrates the significant reduction in primary Laves phase that results. In addition, the areas of Co solid solution free from Laves phase are no longer evident and the new alloy structure consists solely of primary Laves and a very fine lamellar eutectic of secondary Laves and fcc Co solid solution.

The consistency of the K_{1C} values showed that there was no change in the alloys' resistance to crack propagation despite the fact that the volume fraction of the primary phase was substantially decreased and that the fcc form of the solid solution had been stabilised. The previous indication was that K_{1C} was controlled primarily by the volume fraction of primary phase with the fcc Co solid solution playing a smaller but significant part. If these factors were applicable to the iron bearing alloys the microstructural changes associated with the addition of iron would be expected to lead to an increase in K_{1C} and this was experimentally not observed. This apparent discrepancy was due to the absence of the Laves phase free regions of fcc solid solution. It would appear that these ductile regions, which separate the easily cleaved primary Laves particles from the lamellar eutectic are required to provide crack growth resistance. This was consistent not only with these results but also with the fall in K_{1C} for T400 when the solid solution was hcp and with the Widman-

Fig. 8 T400 5% Fe as-cast microstructure. The structure consisted of primary Laves in a fine lamellar eutectic.

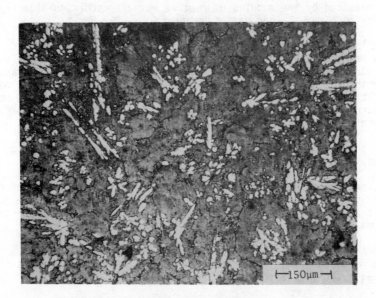

Fig. 9 T400 15% Fe as-cast microstructure. The increased additions of Fe reduced the volume fraction of primary Laves.

statten precipitate. Thus the primary Laves particles were important in determining K_{1C} only in the presence of the ductile solid solution regions. In the absence of these regions a crack may propagate with almost equal ease through either the primary phase or the fine lamellar eutectic.

It can be seen (Table 3) that retaining the fcc solid solution and reducing the volume fraction of the primary Laves phase has a significant effect on the MOR. The results showed that the strength for a constant matrix microstructure (i.e. fcc Co solid solution/Laves phase eutectic) was controlled by the volume fraction of the primary phase. At high volume fraction the particles can cleave and link easily to attain a critical flaw size. Whereas on low volume fractions although the large particles cleave the cracks found difficulty in extending into the lamellar eutectic and much higher stresses were needed to achieve this. So although a propagating main crack can traverse the eutectic without difficulty (as shown by the K_{1C} values), the modification in the micro-structure does increase the tensile strength and resistance to crack initiation as seen by the reduction in critical flaw size.

CONCLUSIONS

1)   The resistance to brittle crack propagation (K_{1C}) for T400 was between 15.8 - 26.0 MNm$^{-3/2}$ and because of the large volume fraction of the brittle Laves phase present, cannot be modified to any great extent.

2)   The alloys' resistance to crack initiation and growth to a critical size as measured by the modulus of rupture was affected significantly by the microstructure. Values varied between 400 - 923 MNm^{-2}.

3)   The lowest fracture parameters were obtained when the Co solid solution was hcp with a Widmanstatten precipitate. In this condition the failure occurred in a mixed intergranular/transgranular manner (trans-granular through the matrix and intergranular at the primary Laves par-ticles) at stresses below those needed to cleave the Laves phase. For all other microstructural conditions of T400 crack initiation and propagation took place predominantly in a transgranular mode.

4)   All the microstructural conditions of T400 which were investigated exhibited similar erosion resistance.

5)   The best room temperature properties were associated with an fcc Co solid solution. Iron additions of between 5% - 15% stabilised the fcc form of the Co solid solution and reduced the volume fraction of primary phase from 42% to 17%.

6)   Primary Laves particles and secondary Laves within the lamellar eutectic were equally detrimental in terms of the crack growth resistance and consequently iron additions do not affect the K_{1C} values.

7)   The modulus of rupture and the resistance to crack initiation was substantially increased by iron additions, i.e. 15 wt % Fe increased the MOR by 40% and decreased the critical flaw size by 70%.

ACKNOWLEDGEMENTS

This work was supported by SERC and the Cabot Corporation. The authors are indebted to Dr S. Grainger of Deloro Stellite, Swindon for his assistance and to Professor D.W. Pashley for laboratory facilities.

REFERENCES

Bansal, G.K. 1976, Am. Ceram. Soc. 59, 87
Cameron, C.B. and Ferriss, D.P. 1974, Cobalt 3, 49-53
Halstead, A. and Rawlings, R.D. 1984, Met. Sci. 18, 491-500
Schmidt, R.D. and Ferriss, D.P. 1975, Wear. 32, 279-289
Tribology wear resistant intermetallic materials. Stellite Division Cabot Corporation Kokomo Indiana 1982.

Inst. Phys. Conf. Ser. No. 75: Chapter 10
Paper presented at 2nd Int. Conf. Science Hard Mater., Rhodes

Wear mechanisms in low speed milling

A R CHAMBERS (1) AND E F SMART (2)

(1) Department of Engineering Materials, Southampton
    University of Southampton, England.
(2) Department of Metallurgy and Materials, Birmingham
    University, Birmingham, England.

ABSTRACT

When milling nimonic 90, fracture and flank wear dominate the wear process
of H.S.S. and cemented carbide tools. Fracture is shown to be dependent
upon the exit angle and a mechanism based on chip/tool adhesion and .work-
piece deformation in advance of the cutting tool is proposed. Flank wear
occurred by the attrition and diffusion wear mechanisms.

BM42 (H.S.S.) and WC - $12\frac{1}{2}$% Co (medium grain size) were more resistant to
fracture than WC - 10% Co (fine grain size) but less resistant to flank
wear.

INTRODUCTION

The wear mechanisms for high speed steel and cemented carbides have been
established for continuous turning operations (Trent, 1984). At high
cutting speeds, tool life is determined by the resistance of the tool
material to the temperature dependent mechanisms of diffusion and plastic
deformation whereas at low cutting speeds attrition, abrasion and adhesion
dominate the wear process. Fracture is a potential problem at all cutting
speeds if the selected tool material does not possess sufficient toughness
for the machining operation being performed.

Interrupted machining, milling being an example, is both more arduous and
difficult to analyse. In addition to the wear mechanisms occurring with
continuous turning the tool is subjected to mechanical and thermal shock
every time the tool enters and exits the cut. As a result, cutter teeth
used to mill steel frequently show cracks running perpendicular to the
cutting edge. These may eventually join together and cause a failure or
act as stress raisers through which fracture can be initiated by other
means (Zorev, 1963). An appreciation of these wear mechanisms has enabled
tool manufacturers to develop tool materials which are less sensitive to
fatigue.

The milling of high temperature materials, such as nimonic 90, is still a
major industrial problem because cutting edge fracture severely restricts
the use of cemented carbides. Consequently these materials are still
predominantly milled with high speed steel tools at low rates of metal
removal (Kirk, 1976).

Fig.1. Experimental Set-up

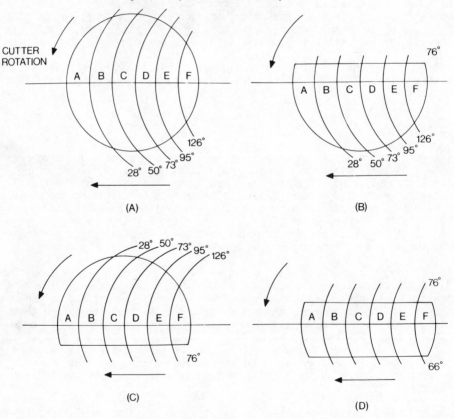

CUTTER
ROTATION

Fig.2. Procedure for obtaining combinations of
varying and constant exit and entry angles

The objective of this research was to investigate the mechanism of fracture and the other wear mechanisms which determine tool life when milling nimonic 90 at low speeds with a view to optimising the tool material composition.

EXPERIMENTAL

Single point milling was simulated by mounting a standard milling head containing one tool in the chuck of a VDF centre lathe with infinitely variable speed capability. The workpiece, which was generally in the form of a round bar, was secured in a block which was bolted to the lathe cross-slide (Fig.1).

With round bar, the entry and exit angles - which are defined as the angles between the tangents to the workpiece and the cutter at the points of entry and exit - and the length of cut arc vary continuously as the work is fed past the cutter. To differentiate between these variables, a bar of solution treated nimonic 90 was sectioned and positioned as shown in Figure 2. Thus it was possible to achieve milling with varying entry/ varying exit angles (Fig.2A), constant entry/varying exit angle (Fig.2B), varying entry/constant exit angle (Fig.2C) and constant entry/constant exit angle (Fig.2D). For these tests, the bar was divided into 5 sections (A - E) corresponding in length to 100 revolutions at a feed of 0.125mm/ rev. Section F was not tested. A different tool was used in each section and tool wear was measured after a fixed number of revolutions. The tools were examined optically and using a scanning electron microscope for evidence of the wear mechanism.

On the basis of the results of these tests, the experimental procedure was developed for a tool material comparison. The round bar, 62mm diameter was divided into 10 sections corresponding to 50 revolutions at the feed of 0.125mm/rev. (Fig.3). With a different tool being used in each section, tool lives were established by using each tool in its respective section until the end point of the test was reached. Depending upon the rates of wear, the end points were either the number of revolutions to give 0.4mm flank wear or a fixed number of revolutions after which wear was measured.

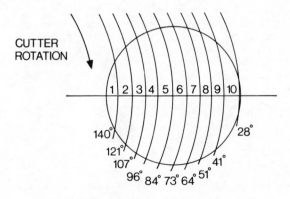

Fig.3. Experimental procedure for tool material comparison

Using this procedure, the performance of the following tool materials was evaluated. (Table 1).

TABLE 1 Tool Material Composition

Tool Designation	Composition	Grain Size (µm)
A	WC - 10% Co	0.7
B	WC - 12.5% Co	1.5
C	BM42 (High Speed Steel)	

The cutting speed (16 m/min), feed rate (0.125mm/rev), depth of cut (1mm), cutting fluid (dry), tool geometry (SPUN 120325) and workpiece material (nimonic 90) were test constants.

RESULTS AND DISCUSSION

Exit Angle Fracture

The results of the tests conducted with the bar sectioned and positioned as shown in Fig.2 are summarised in Table 2.

TABLE 2   Effect of Varying Entry and Exit Angles
on Tool Condition after 100 revolutions

Section	A	B	C	D	E
Varying entry/varying exit angle	unworn	unworn	chipped	chipped	chipped
Constant entry/varying exit angle	unworn	unworn	chipped	chipped	chipped
Varying entry/constant exit angle	chipped	chipped	chipped	chipped	chipped
Constant entry/constant exit angle	chipped	chipped	chipped	chipped	chipped

With continuously varying entry angle/continuously varying exit angle (Fig.2A) and constant entry angle/continuously varying exit angle (Fig.2B) significant tool fracture occurred on sections C, D and E within 100 revolutions cutting whereas the tools used in sections A and B were undamaged. Thus it can be deduced that the mechanism of fracture is not dependent upon the entry angle.

However, with the varying entry angle/constant exit angle 66° and constant entry/constant exit angle combinations significant fracture occurred in all the bar sections. It is apparent that this type of fracture is exit angle dependent and that the angle of 66° selected for these tests is especially severe.

Using the more detailed approach described in the experimental procedure, the dependence of tool life based on the 0.4mm flank wear upon exit angle is clearly demonstrated (Fig.4). For exit angles between 70 and 105° the life of material A was consistently less than 150 revolutions. As the exit angle was either decreased or increased from this critical range, the tool life improved as the life determining wear mechanism changed from fracture to chipping/flank wear and finally to flank wear.

A disadvantage of this test procedure is that the cut arc length is not a constant and hence differences in the maximum temperature and thermal cycling of the tool are inevitable. However, as sections of approximately equal arc length (sections 3 and 9 in Fig.3) cause significantly different tool wear, it can be assumed that with regard to fracture, the effect of cut arc length variation is insignificant in comparison with that of exit angle.

The Mechanism of Exit Angle Fracture

With an exit angle of 76°, flank wear was measured every 2 revolutions until the pattern of wear was firmly established and then every 5 revolutions until catastrophic failure occurred. For both A and B, the initial fractures occurred on either the first or second revolution and hence a fracture mechanism based on mechanical or thermal fatigue can be discounted. Subsequent revolutions progressively increased the damage at a rate which was dependent upon the composition of the tool material (Fig.5).

EFFECT OF EXIT ANGLE - TOOL `A` NIMONIC 90

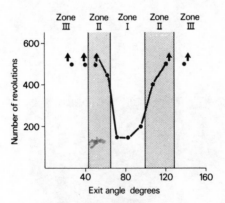

Zone I - severe fracture
Zone II - chipping/flank wear
Zone III - flank wear/micro-chipping

Fig.4.   The effect of exit angle
upon tool life

Observation of the cutting process and examination of tools, and chips provided evidence to suggest that exit angle fracture is associated with chip/tool adhesion. With favourable exit angles the frequency of chips adhering to the tool after exit was extremely low and the rake face contact area was completely covered by a thin layer of work material. The chips were being formed cleanly over a built-up layer in a manner similar to that which occurs in the continuous turning of nimonic 90.

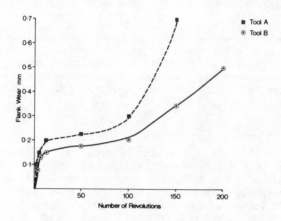

Fig.5.   Flank wear/damage curves
         for tool materials A and B

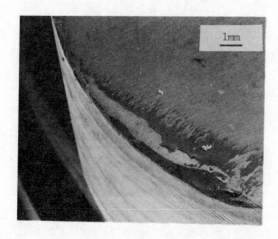

Fig.6.   Rake face of tool A showing layer
         removal in the vicinity of cutting
         edge.

However, with the critical angles the frequency of adhering chips was high (>50%) and the rake face layer was absent in the vicinity of the cutting edge; being found attached to the chip underside (Figs. 6 and 7). The appearance of the tools and chip undersides which replicate the tool surface (Fig.8) create the impression that with the critical exit angles the adhering chips are peeled from the tool in a direction opposite to that of chip flow.

Fig.7. Typical chip underside from
        critical exit angle region.

Fig.8. Detail of chip underside
        replicating rake face attrition.

Coinciding with these differences in chip and tool appearances, differences in the condition of the workpiece on the exit side were also observed. A collar or burr was invariably present with the critical angles but always absent with the favourable angles. Thus there is strong circumstantial evidence to suggest that exit angle fracture is linked to chip/tool adhesion, the detachment of the chip from the tool and collar formation which is indicative of workpiece deformation in advance of the cutting tool.

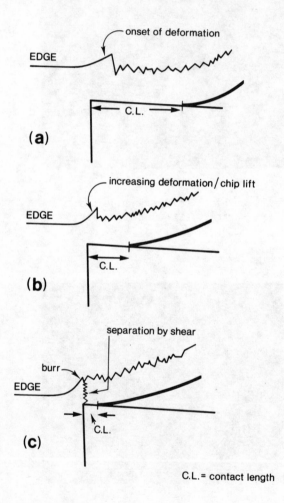

Fig.9. Change in chip formation as tool approaches exit

On the basis of this connection and in broad agreement with Pekelharing (1978) it is suggested that as the tool approaches exit, the unconstrained workpiece begins to deform with the result that the chip is gradually lifted from the tool (Fig.9). Thus the contact length begins to shorten and the remaining forces are concentrated over the cutting edge. As deformation and chip lift increase, the velocity of the chip in the direction of cutting will reduce until at the point of exit chip flow has ceased. Final deformation will either detach the stationary chip from the tool under the action of a tensile force or separate the chip from the workpiece by shear along the primary shear plane. In this case, the chip which is adhering to the rake face is detached by contact with the work material on entry to the next revolution.

The incidence of possible fracture will depend upon the tool material, the strength of the bond between chip and tool and the exit angle which controls the magnitude and line of action of the forces tending to cause workpiece deformation in advance of the cutting tool.

Exit Angle Fracture and the Tool Material

The results of the tool material comparison based on flank wear after 250 revolutions are shown in Fig.10. All three tool materials were prone to exit angle fracture but the critical angle range and severity of damage were dependent upon the tool material.

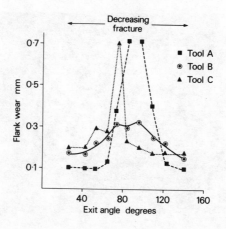

Fig.10. Tool Material Comparison

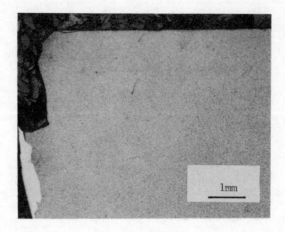

Fig.11.   Flank section of A showing
massive fractures

Fig.12.   Crumbling of Tool B in
critical exit angle zone

Within the critical angle range both A and B fractured immediately (Fig.5)
but, in terms of life, B was superior to A because it possessed a greater
resistance to fracture propagation.   Tool material A failed by a series
of massive fractures which propagated through the binder phase (Fig.11)
whereas B failed in a more controlled manner by a combination of minor
fractures and crumbling of the cutting edge (Fig.12).   The higher cobalt
content of B can be expected to contribute to its increased toughness but
as the fractures propagate through the binder phase with no evidence of
trans-granular fracture it is considered in this case that the improved
performance of B is more due to a coarser grain size than to a higher
cobalt content.

The poor performance of C (BM42 high speed steel) between exit angles of
65 and 80° can also be attributed to exit angle fracture. However, unlike
the cemented carbides, whose performance depended on resistance to further
fractures, C ultimately failed by plastic deformation. The increase in
flank contact as a result of edge fracture, increased the heat generation
on the flank face and hence accelerated the onset of plastic deformation.

Flank Wear
‾‾‾‾‾‾‾‾‾

As the exit angle was either decreased or increased from the critical
range, the tool failure criterion changed from fracture to chipping/flank
wear and finally, to flank wear. B was superior to A under conditions
of fracture but at exit angles were flank wear dominated, A was superior
to B, (Fig.13). In the intermediate region where stable edge chipping
assisted the rate of flank wear, it is not possible to differentiate
between A and B.

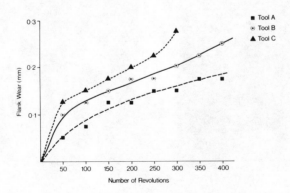

Fig.13.   Tool Material Comparison
          Exit angle 41°.

The change in the wear order can be explained by the wear mechanisms and the tool material properties.   In the absence of fracture, the flank wear surface of B was optically rough and in section the removal of individual and small groups of WC grains from the tool surface can be seen. (Fig.14).   This appearance is typical of attrition wear which occurs in the presence of a built-up edge, intermittent contact between chip and tool and when there is uneven metal flow (Trent, 1967).

The predominantly smooth wear surface of A indicates that it is more resistant to attrition wear.(Fig.15).   In view of the relationship between grain size and the rate of attrition, the greater resistance of A to attrition can be attributed to its finer grain size.   The smooth wear of A is characteristic of diffusion which is a mechanism whose rate is dependent upon the temperatures generated in cutting and hence the work material and the cutting parameters. (Trent, 1967).   At the comparatively modest cutting parameters employed in the tests, attrition wear is the potentially more damaging flank wear mechanism.

C, BM42 high speed steel, was more resistant to chipping than the cemented carbides and hence performed well in the intermediate region between fracture and flank wear.   However, as expected, C was clearly inferior to A and B in resistance to flank wear. With the employed test conditions, which in view of the cutting speed and absence of coolant, do not favour high speed steel, C was unable to maintain steady state flank wear beyond a maximum flank wear measurement of 0.25mm.

 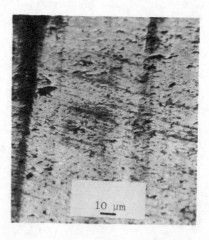

Fig.14.   Flank attrition of B   Fig.15.   Smooth flank wear of A
          with favourable exit   with favourable exit
          angles.   angles.

Rake Face Wear

With nimonic 90, tool life was determined by either exit angle fracture
or by flank wear and consequently rake face wear was of no practical
significance. However, the mechanisms of rake wear are of interest as
they illustrate differences between milling and continuous turning.

For the critical exit angle range, attrition resulting from detachment of
an adhering chip was the dominant wear mechanism of A, B and C. It is
interesting to note that in situations where strong tensile stresses are
exerted, high speed steel and fine grain cemented carbides are prone to
attrition.

Outside the critical exit angle range, the wear appearances were similar
to those encountered in continuous turning. Attrition was restricted to
the rear of the contact area where chip tool contact is intermittent and
wear by diffusion resulted in the majority of the contact area being
smoothly worn. The rates of diffusion wear were extremely slow, which
suggests that the cutting temperatures were comparatively low.

Practical Implications

The results show that cemented carbides can be used to successfully mill
nimonic 90 providing the critical exit angle is avoided. Depending upon
the geometry of the component this may be achieved by offsetting the work
either above or below centre.

In view of the fact that BM42, a comparatively tough tool material was
susceptible to exit angle fracture, it is unlikely that optimising the
composition and microstructure of a cemented carbide will produce a tool
material insensitive to the geometry of milling. In terms of resisting
exit angle fracture a medium grain size is preferable to a fine grain
size, but from the point of view of flank wear, the converse is true.
This highlights the difficulty experienced by those responsible for
selecting tool materials for specific applications and indeed the need for
caution in interpreting machinability test results.

The results strongly suggest that the mechanism of fracture is associated
with chip/tool adhesion, in which case the application of a cutting
lubricant or possibly a coating on the cutting tool may prove beneficial.

CONCLUSION

1. When milling nimonic 90, the incidence of fracture is dependent upon
    the tool exit angle. Tool life can be improved by positioning the
    work such that the critical angles are avoided.

2. The mechanism of exit angle fracture is dependent upon adhesion
    between chip and tool. Reductions in the strength of the bond by
    either a cutting fluid or a coating could be beneficial.

3. With regard to cemented carbides, grain size is an important
    structural variable. A medium size grain size affords resistance to
    fracture and a fine grain size resistance to low speed attrition wear.

ACKNOWLEDGEMENTS

The authors are extremely grateful for the financial support and encouragement given by Sandvik Hard Metals Research, Coventry, England.

REFERENCES

Kirk, D.C. 1976, Proc. Tools and Dies for Industry Conf. The Metals
    Society, London pp 77-92.
Pekelharing, A. 1978, CIRP ANNALS 27 5-10.
Trent, E.M. 1967, Iron and Steel Institute. Report 94 p 77.
Trent, E.M. 1984, Metal Cutting, Butterworths, London pp 96-112, 128-146.
Zorev, N. 1963, CIRP ANNALS 11, pp 201-210.

Discussion on Wear

Rapporteur: A BALL (1)
Editor: C A BROOKES (2)

(1) Department of Metallurgy and Materials Science,
University of Cape Town, South Africa
(2) Department of Engineering Science, University of Exeter, UK

Once again a thorough and enlightening review paper, on this occasion presented by Dr Buckley, preceded a vigorous discussion on the detailed and phenomenological behaviour of hard materials. His paper highlighted such fundamental effects as anisotropy in the mechanical properties and abrasive wear of similarly hard ceramic crystals sliding on one another and the nature of atomic bonding causing adhesion and wear between a soft metal (Cu) and a hard surface (diamond). Nevertheless, and despite the wealth of data emanating from this type of work, the ensuing discussions underlined the need for relevant wear models to apply to hard materials.

The main thrust of the work presented by Professor Ball was a comparison in the wear performance of specific materials when tested by different methods of abrasion and erosion. Dr Buckley asked whether it might not be better to bring the proper parameters from engineering systems to funda- mental experiments rather than to develop new tests for the study of wear to solve particular engineering problems. Professor Ball agreed that a laboratory test should incorporate the parameters relevant to the 'real' situation but that to ensure that the test is a true simulation of the engineering system necessitated a limited amount of work in the field. If the mode of erosion and the relative performance of a number of materials are the same for both tests then the laboratory test is validated and further work can proceed in the laboratory without recourse to the field test.

Dr Lee referred to Professor Ball's observation of droplet-like cobalt particles on the eroded surfaces and the possibility that melting had occurred through adiabatic shear. In his own work, on metal cutting at ballistic speeds, he had found that molten droplets were formed by the almost spontaneous oxidation of fine detritus and not by mechanical shearing. Was it possible that these cobalt droplets were also melted by rapid oxidation of extruded cobalt particles? The author replied that this possibility was currently under investigation both by calculation and experiment.

Dr Brookes observed that crack free indentations could be formed on these hard materials, provided the load was sufficiently small, and therefore we might anticipate changes in the rate of erosion as the particle size was reduced - small particles corresponding to low loads. Professor Ball replied that he considered the rate of strain to be more important than the magnitude of the stress and therefore the analogy with hardness indentations might be misleading. He pointed to the work (of I.M. Hutchings) which showed that the instantaneous strain rate produced by a small particle was greater than that by a larger particle impacting at the same velocity. This point was followed by reference to the possibility of work-hardening in hardmetals and the reply that this effect, in the binder phase, was of critical importance to the wear resistance of WC/Co alloys and particularly so in those alloys containing more than 8% of the binder. However, it was again cautioned that work-hardening behaviour should be considered under the high strain appropriate to erosion rather than, say, slow sliding or friction experiments.

Dr Horton asked whether erosion experiments had been carried out on metal bonded hardmetals other than WC/Co and, if so, how did their wear resistance compare? Professor Ball affirmed that he had tested WC/Ni hardmetals for resistance to cavitation erosion and particle erosion in both water and air streams. The nickel alloys were better than the cobalt alloys in both cavitation erosion and particle erosion in a water stream. But the cobalt alloys were better in dry conditions. Clearly the improved corrosion resistance of the nickel matrix was an important influence on the erosion rate and was the subject of continued research.

Dr Shaw commented on the interesting correlation between Shore scleroscope experiments and erosion resistance and presumed that the correlation should be strain rate dependent. He asked whether Professor Ball had any thoughts on suitable laboratory experiments that might explore this effect in a way that would lead to a more comprehensive understanding of impact deformation - strain rate relationships (other than erosion simulation tests). Professor Ball replied that he believed wear to be a high strain rate process and better correlations were obtained between wear rate and those indentation tests involving impact - such as in the Shore scleroscope. He considered that impact scratch and wire drawing tests should also give good indications of wear resistance.

A proper concern for the effects of composition and microstructure was readily apparent from the points raised in discussion of the papers dealing with hardened surfaces. Dr Tsipas asked Dr Habig whether he could comment on the percentage and distribution of Fe_2B and FeB layers in his alloys because he thought this could be important in their wear behaviour. Apparently, the FeB content was small but strong adhesive wear, under vacuum conditions, had been observed after tests on Fe_2B covered with a dense layer of FeB. Also, Dr Knotek suggested that the presence of Fe_2B/Fe or FeB/Fe eutectics - with some free Fe in the surface regions - could increase the adhesive wear due to the low melting point of the eutectic. Dr Habig replied that a thin iron rich layer was often found on boride coatings and that he thought adhesive wear could be increased as a result. There are many cases of bonded steels which show a high rate of 'running-in' wear which may be caused by the presence of iron. Furthermore, the low melting point of the eutectic can be dangerous if there is a marked increase in temperature due to frictional heating.

Dr Buckley asked if the generation of additional interfaces (chrome and nickel borides) increased the number of potential points for failure. Dr Tsipas replied that it depended on the nature and properties of those interfaces and that, for many applications, it was desirable to deposit successive layers of different compounds in order to achieve the optimum properties. Dr Habig added that he had also borided chromium rich steels and found that there was a nearly constant chromium content in the boride layer and the base material. He asked how the formation of a chromium boride layer under the iron boride layer be explained. Dr Tsipas replied that the formation of different layers depended on the free energy and diffusivity of elements in relation to the compounds to be formed but that the relevant experimental and theoretical work had not yet been carried out.

Dr Halstead was asked, by Professor Ball, of the applications of the Co/Mo/Cr/Si alloys and replied that they were used in devices where metal to metal contacts exist with comparatively little lubrication - e.g. journal/sleeve bearings, pistons and rings, seals and valves. In response to a question concerning the possibility of the face-centred cubic polymorph transforming to the close-packed hexagonal one during wear she said that she had not yet found any evidence to confirm this in her work - although it had been reported in the literature to occur during sliding. Dr Shaw asked how she determined the crystal structure of the various phases present in the microstructures shown since, presumably in such non-equilibrium structures, this would be difficult to infer from compositional information. Dr Halstead replied that X-ray diffraction methods were used to quantify the amounts of face-centred cubic and close-packed hexagonal cobalt whilst their presence was confirmed by transmission electron microscopy.

Inst. Phys. Conf. Ser. No. 75: Chapter 11
Paper presented at 2nd Int. Conf. Science Hard Mater., Rhodes   1019

Processing and microstructural development in liquid phase sintered and pure TiB$_2$ ceramics

P ANGELINI, P F BECHER, J BENTLEY, J BRYNESTAD,
M K FERBER, C B FINCH, AND P S SKLAD

Oak Ridge National Laboratory, Oak Ridge, Tennessee   37831

ABSTRACT

The fabrication of dense, high fracture strength and toughness TiB$_2$
ceramics requires that grain growth and secondary phase(s) be controlled.
This can be achieved by liquid phase assisted densification with nickel-
bearing additives or by use of submicron TiB$_2$ powders. This is modified
by other constituents, (i.e., densification is enhanced by Fe impurities,
retarded by oxygen and grain growth is reduced by carbon). A densification
model is proposed based on observations of microstructure evolution.

INTRODUCTION

Titanium diboride has several attractive properties including high elastic
moduli, electrical conductivity, hardness, and melting temperature which
make it attractive for technological applications. Potential applications
include cathodes in Hall-Heroult cells for aluminum smelting, valve trim
for erosive environments, armor, wear components and cutting tools.

However, the fabrication of TiB$_2$ components is complicated by its high
melting point (2980°C), and sintering temperatures >2100°C are usually
needed to achieve near theoretical density. When such high sintering
temperatures are employed, both normal and discontinuous grain growth
(i.e., development of large, idiomorphic grains) occur. Due to the
high Young's modulus and large anisotropy in both thermal expansion
and elastic constants, significant residual stresses are generated at
the grain boundaries of the dense TiB$_2$ during post-fabrication cooling.
The magnitude of the resultant residual stress intensity factor increases
with grain growth and results in microcrack generation (Becher and Ferber,
1985). In polycrystalline TiB$_2$ this occurs at grain sizes >15 microns
and results in decreased mechanical properties (Ferber and coworkers,
1983). Thus, grain growth during densification must be controlled to
optimize the properties of TiB$_2$.

At ORNL one of the areas of research on TiB$_2$ has been on the use of
metallic additives to achieve liquid phase assisted densification, lower
processing temperatures, and to control TiB$_2$ grain growth. For example,
with initial Ni additions of >5 wt %, relative densities greater than 97%
of theoretical are achieved at hot-pressing temperatures as low as 1400°C.
Most of the initial Ni is exuded during hot-pressing. The temperature
necessary to achieve full density can be regulated by controlling the

melting point of the metallic additive [and the boride phase(s) formed when the additive interacts with the TiB_2]. With this approach, grain growth is virtually eliminated and fully dense fine grained (≤5 μm) polycrystalline TiB_2 can be obtained [Tennery and coworkers, 1983(a)]. This process introduces an intergranular phase in the compacts through formation of TiB_2-metal eutectics. From the Ni-Ti-B phase diagram (Schöbel and Stadelmaier, 1965), one of the phases expected during the TiB_2-Ni liquid phase densification process was tau [$(Ni_xTi_y)_{23}B_6$]. The intergranular phase in hot-pressed TiB_2-Ni compacts was shown by Sklad and Bentley (1983) and Sklad and Yust (1983) to be Ni_3B. Intergranular Ni_3B was later shown by Angelini and coworkers (1984); Angelini and Bentley (1984); and Bentley, Angelini and Sklad (1984) to contain only ≈1% Ti. These data indicated that the bulk of the Ti component of the TiB_2 involved in the liquid phase densification process was incorporated in yet another phase or phases within the compact or in the exudant material. However, results on TiB_2-Ni specimens pressureless sintered at 1425°C (Angelini and coworkers, 1984) revealed that intergranular tau phase was indeed formed. The presence of tau phase implied that nearly all of the TiB_2 involved in the pressureless liquid phase densification process was accounted for. This paper summarizes mechanical properties, microstructural-compositional characteristics established in fabrication, phase evolution involved in the densification process, and proposes a model for the liquid phase densification process for various TiB_2 ceramics.

PROCEDURE

Dense TiB_2 compacts (Table 1) were prepared by vacuum hot-pressing TiB_2 powder, with and without Ni additions, in graphite dies at 1800°C and ~1425°C, respectively, as described by Tennery and coworkers [1983(a)], Tennery and coworkers [1983(b)], Ferber and coworkers (1983), and Angelini and coworkers (1984). For the most part, these studies utilized commercial TiB_2 powder which had a mean particle size of either 5 or 12 micrometers. Some initial studies on hot pressing of submicrometer (<0.1 micrometer) TiB_2 powders synthesized at ORNL (Brynestad and coworkers, 1985) without metallic additives were also conducted.

TABLE 1   Physical and Mechanical Properties of Hot-Pressed TiB_2-Ni and TiB_2 Specimens

TiB_2 grain size* μm	Hold time at temperature ks	Density† g/cm³	Fracture strength MPa Surface finish		Fracture toughness MPa·m$^{1/2}$	Nickel content wt %
			As ground	As polished		
5	7.2 (1425°C)	4.51	>710	(>870)‡	6.8	1.5
5	115.2 "	4.48	500		5.1	0.4
12	7.2 "	4.63	450		6.1	1.3
12	8.8 "	4.52	310		4.5	0.2
12	115.2 "	4.51	280		3.9	0.1
5	7.2 (1800°C)	4.42	430		5.8	≈0
40	7.2 "	4.44	110		3.7	≈0

*Mean intercept method used with grain size equal to 1.5 (intercept length).

†Theoretical density of 4.54 g/cm² used for TiB_2.

‡Strengths as high as 1250 MPa achieved with polished surfaces.

The as fabricated hot-pressed specimens were characterized as to grain size by image analysis, density by the immersion technique, and composition by electron microprobe and analytical electron microscopy (AEM). Oxygen contents of both the initial powders and the dense hot-pressed compacts were obtained by neutron activation analysis.

Electron microprobe analyses of TiB_2-Ni hot-pressed compacts, various exuded materials, and pure Ni_3B single crystal specimens were performed at an accelerating voltage of 12 keV. X-ray energy dispersive spectroscopy was used only to confirm the presence of elements with atomic numbers greater than 11. Data acquisition was performed with x-ray wavelength dispersive spectrometers and typical beam currents of approximately 72 na.

Elements analyzed included Ni, Ti, Si, O, C, and B. Measurements for N were not made. The instrument had a low sensitivity for N due to absorption of N x-rays by the hydrocarbon film protecting the detector. Measurements were standardized on elemental Ni, Ti, B, MgO for oxygen, and graphite for carbon. When analyzing exuded material, quantitative measurements of the Ni standard were also made to check the sensitivity and background levels of the elements of interest. Once the elements in each phase were identified, the cation contents were quantified by (1) correcting their integrated intensities due to matrix effects including mean atomic number (Z), absorption (A), and secondary x-ray fluorescense (F), and (2) determining the ratio of the modified intensity in the sample to its standard. The anion contents were then obtained by difference. This procedure is similar to that described by Quinto, Wolfe, and Haller (1983) in their work with WC and BN. The carbon film typically applied to prevent charging increased the carbon background and partially masked the boron signal and therefore was not applied to the specimens. Electrical charging of the specimen mount was eliminated by coating the entire upper surface of the epoxy mount with silver paint but leaving the polished specimen uncoated.

Analytical electron microscopy was performed with a microscope equipped with a field emission gun and capability for x-ray energy dispersive spectroscopy (EDS), electron energy loss spectroscopy (EELS), and convergent beam electron diffraction (CBED). The microscope was operated at an accelerating voltage of 100 kV. The electron probe diameter was about 10 nm when the microscope was operated in the transmission electron microscopy (TEM) mode. The specimen region is ion pumped and has a liquid nitrogen cooled anti-contamination device. Most of the EDS, EELS, and CBED analyses were performed with the specimen in a Be cup of a double tilt liquid nitrogen cooling holder. Quantification of x-ray EDS and EELS data was performed with standardless routines as described by Zaluzec (1979). Analytical electron microscopy of the highly reactive submicrometer TiB_2 powders prepared at ORNL was conducted by the method described by Angelini and Lehman (1983).

Fracture strength and toughness measurements of the TiB_2 compacts were obtained at 22°C and 70% relative humidity. The specimens for the fracture strength (four-point flexure) determinations were 3 × 3 × 13 mm in size and were cut with their longest dimension perpendicular to the hot-pressing direction (original compacts were 40 mm in diameter and 8 mm in thickness). The edges of the bars were beveled with 6 μm diamond to remove edge cracks. Fracture strength was determined with inner and outer load spans of 6 and 19 mm and crosshead speed of 0.017 mm/s. The fracture stress was determined from an equation described by Popov (1976):

$$\sigma_f = \frac{1.5 \ (L_0 - L_i) P}{bh^2} \qquad\qquad 1$$

where σ_f is the fracture stress, L_0 and L_i are the outer and inner load spans, P is the load at fracture, b and h are the specimen width and height. Fracture toughness was measured at 295 K by either the diamond indentation or the multiple controlled flaw fracture methods. In the former, tests were made at various load levels ranging from 25 to 50 N. The data were analyzed according to the equation given by Evans (1979).

$$\frac{K_c \ r^{1/2}}{H^{3/5} \ E} = F \ \frac{c}{r} \qquad\qquad 2$$

where K_c is the fracture toughness, H is the hardness, E is the Young's Modulus, r is one half the indent diagonal, c is the crack length, and $F(c/r)$ is a polynomial function. With the multiple controlled flaw fracture method, K_c values are obtained from the fracture strength and the final flaw size measurements (Cook and Lawn, 1983).

RESULTS AND DISCUSSION

Mechanical Properties

The fracture toughness (K_c) data are presented in Table 1 and Fig. 1. The data for the hot-pressed TiB_2 specimens (no Ni additive) show K_c that decreases as the TiB_2 grain size increases. This result demonstrates the detrimental effects which can occur with increasing grain size due to grain size effects on residual stresses and resulting microcracks (Ferber and coworkers, 1983) in the compacts. The results also clearly show that K_c increases as the residual Ni content in the compact increases to ≈ 1.5 wt %. The data for the hot-pressed TiB_2-Ni compacts converge to the value for the hot-pressed TiB_2 specimen without Ni additive as the residual Ni content decreases to nearly zero. These effects can occur if thermal expansion anisotropy related residual grain boundary stresses are reduced by the presence of intergranular solutes or phases. The enhanced diffusive processes which would be associated with such intergranular material can minimize residual strains resulting from the thermal expansion anisotropy.

Fracture strength results are presented in Table 1 and Fig. 2. The measurements were made with specimens prepared from the same compacts used in the fracture toughness measurements. The fracture strength data for the hot-pressed TiB_2 specimens (no Ni additive) increase with decreasing TiB_2 grain size. The data for the hot-pressed TiB_2-Ni compacts are also a function of grain size as well as residual Ni content. The fracture strength is significantly improved when the residual Ni content is increased to approximately 1.5 wt %. As discussed above and by Becher and coworkers (1985), these improvements in strength are related to reduction in thermal expansion anisotropy (TEA) residual stresses. However, when a large fraction of the nickel additive (>2 wt %) is retained, the mechanical properties are degraded due to the brittleness of the intergranular phase as shown by Becher and coworkers (1985). It becomes apparent then that a detailed analysis of the phases formed during densification can furnish an understanding of not only the densification mechanisms but also the

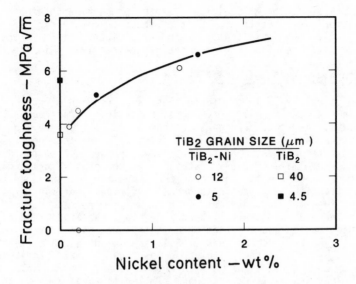

Fig. 1   Fracture toughness (K_c) versus residual Ni
content in hot-pressed TiB_2 ceramics (with and
without initial Ni addition).

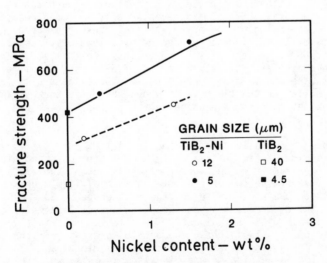

Fig. 2   Fracture strength versus residual Ni content
in hot-pressed TiB_2 ceramics (with and
without initial Ni addition).

microstructure evolution which influences the properties obtained in dense polycrystalline TiB_2. Factors which contribute to retained nickel content, phases evolved during densification, exudation, interaction of liquid phase with the TiB_2 and the effects of impurities are considered in the following.

Microstructural-Phase Development

In order to assess the role of liquid phase sintering on microstructure developed in the resultant TiB_2 compacts, extensive microstructural characterization was conducted on the ceramics as well as the liquid phase exuded from the ceramics during densification. In considering these observations reference was made to the Ti-B-Ni phase equilibrium diagram (Schöbel and Stadelmaier, 1965). Establishing the phases formed during densification allows one to understand the phase development occurring during densification, and the influence of additives and impurities. As discussed by Angelini and coworkers (1984) the primary secondary phase produced in the TiB_2-Ni ceramics shown in Table 1 was Ni_3B. Another secondary phase which appears dark in bright-field optical microscopy was observed by Tennery and co-workers (1983), and Sklad and Yust (1983) in hot-pressed TiB_2-Ni compacts. The phase was shown by Sklad and Yust (1983) to contain Ti by x-ray EDS and was tentatively identified by selected area electron diffraction as Ti_2O_3. A phase similar in appearance to that just mentioned was observed in this work in a TiB_2-Ni compact hot-pressed at 1425°C, with a hold time of 2 h at temperature. A backscattered electron image (BSE) and x-ray maps obtained by WDS for the elements O, B, and Ni in this specimen are presented in Fig. 3. The phase is difficult to differentiate in the BSE image since the effective BSE cross section is similar to that of the TiB_2 grains. The phase is rich in oxygen and contains little or no boron. An x-ray EDS analysis revealed only Ti for elements with $Z > 11$. The composition of the phase was evaluated by electron microprobe analysis by the difference procedure described previously. Both TiB_2 matrix grains and oxide secondary phase grains were analyzed. The results for the oxide grains are 32.5 at. % Ti and 67.5 at. % O by difference. The data indicated the matrix grains are TiB_2 and the oxide grains in this compact to be TiO_2 rather than Ti_2O_3. It should be noted that various types of titanium oxides may form depending on the oxygen partial pressure during hot-pressing.

As was mentioned previously, most of the initial Ni additive is exuded during hot-pressing of TiB_2-Ni compacts through the punch and die assembly clearances. The exudant which flows completely out of the die assembly may be representative of the composition of the initial liquid phase formed by reactions between the Ni additive and the TiB_2 powder. Some exudant also solidified in the clearance between the punch and die assembly, and some is found in contact with the compact. These latter materials would then be representative of the liquid phase composition present near the end of the hot-pressing process. A polarized light optical micrograph of a globule of exudant collected outside of the die assembly is shown in Fig. 4(a,b). Also shown [Fig. 4(c,d)] is a cross-sectional view of exudant material which solidified in the die-assembly clearance. Normal bright-field optical microscopy did not reveal much detail; however, polarized light micrographs show a complex microstructure. The exudants appear to have solidified by a eutectic reaction. Specimens shown in Fig. 4 were analyzed both quantitatively and qualitatively by electron microprobe analysis to determine the composition of the various phases.

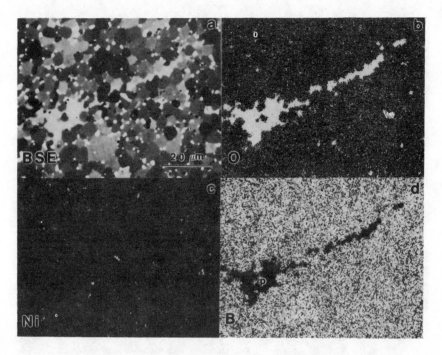

Fig. 3   Oxide phase present in a TiB_2-Ni hot-pressed compact
(specimen NT9OAB) (a) Backscattered Electron Image,
(b) O K_α x-ray map, (c) NiK_α x-ray map,
and (d) B K_α x-ray map.

First let us consider the exudant material found outside of the die
assembly [Fig. 4(a,b)]. A backscattered electron image and x-ray elemental
maps for the elements Ti, B, C, Si, and Ni are shown in Fig. 5(a) through
5(f). The data in Fig. 5 show that a significant amount of Ti is present
in the exudant. The quantitative electron microprobe results for the
compositions of the various phases are presented in Table 2. The exudant
contained phases with composition similar to Ni_3B and Ni_2B. These were the
major phases indicating the TiB_2-Ni system reacts rapidly to form a liquid
phase which is quickly exuded. Titanium is present in the exudant as large
TiC and TiB_2 crystals of up to 100 μm in length. The TiC may be formed
from the reaction of the Ti dissolved in the exudant with the graphite die.
There is also a Ti_xB_y intergranular phase in the exudant which tends to
delineate the Ni_3B phase. From AEM observation the intergranular Ti_xB_y
contains Ni rich grains in a relatively pure Ti_xB_y matrix. Another
important result is that an oxide phase was not detected in the exuded
material. These results imply that oxides which may form during liquid
phase densification reactions remain in the compact. These data are in
agreement with the observation that the oxygen content of hot-pressed
compacts (≈0.8 wt %) compares favorably with oxygen level of starting
TiB_2 powders (≈1 wt %).

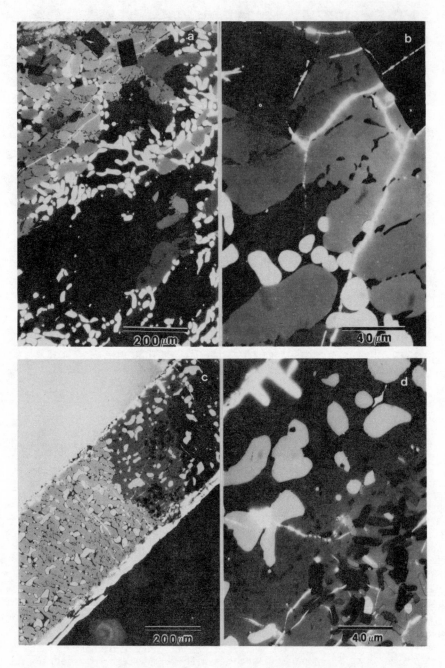

Fig. 4   Polarized light optical micrographs of (a,b) exudant found
outside of the die assembly, (c,d) cross-sectional view of exudant
solidified in punch-die clearance near compact.

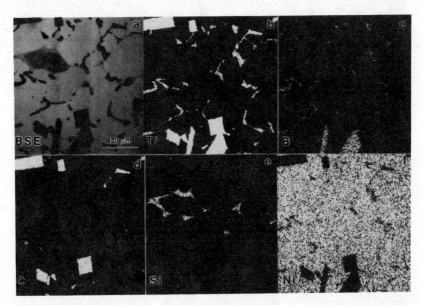

Fig. 5   Electron microprobe results for exudant globule located
outside of punch-die assembly (a) BSE image, (b) Ti Kα
x-ray, (c) B K$_\alpha$ x-ray map, (d) C K$_\alpha$ x-ray map,
(e) Si K$_\alpha$ map, and (f) Ni K$_\alpha$ x-ray map.

TABLE 2   Quantitative Electron Microprobe Analyses of Phases
Present in Exuded Globule (see Fig. 4(a,b) and Fig. 5)

	Composition at. %						Equivalent compound
	Ni	Ti	Si	O	C	B	
Exudant phases observed in polarized light							
Matrix (gray)	71	0.5				28.5	Ni$_3$B
Large grains (white)	65	.01				35	Ni$_2$B
Square (dark)		51			49		TiC
Intergranular Ti rich	13	27				60	— (mixture)
Large rectangular Ti$_x$B$_y$		34				66	TiB$_2$
Si rich phase (dark regions surrounding large white grains)	65	0.4	9.6			25	— (mixture)
TiB$_2$ standard		33				66	
Ni$_3$B single crystal standard	75					25	

The exudant which solidified within the graphite punch and die assembly clearances [Fig. 4(c,d)] and which was within a few millimeters of the compact did not contain large TiC or TiB_2 crystals. It did contain phases with compositions similar to Ni_3B (major component) and Ni_2B, intergranular Ti_xB_y, and a Si rich phase. Faceted TiB_2 crystals of ≈10 µm in length were observed. These may be TiB_2 powder particles which became entrained in the liquid exudant. However, grains of TiC were not observed in this type of exudant. Quantitative electron microprobe results are presented in Table 3. Of the various exudants, this particular type may be most indicative of the liquid phase solidified within the compact.

TABLE 3 Quantitative Electron Microprobe Analyses of Exudant Shown in Fig. 4(c,d)

Appearance of phase under polarized light	Composition at. %						Equivalent compound
	Ni	Ti	Si	O	C	B	
Matrix (gray)	72.6	0.4				27	Ni_3B
Large grains (white)	65	0.05				35	Ni_2B
Intergranular (Ti rich)	15	25				58	(Mixed phases)
Rectangular grains		34				66	TiB_2
Si rich phase (dark regions surrounding large white grains)	67.5	0.08				19.7	(Two mixed phases)
Bright (edge)				100			

Thus Ni_2B may also exist in the actual compacts although it has not yet been observed. The Ni_2B phase in the exudant is delineated by a Si rich phase. If a similar situation occurs within the compact, the presence of Ni_2B may be detected. The process which leads to the partitioning of the Si into the exudant is not known at this time; however, such partitioning plays a significant role in altering impurity distribution as very little Si (<500 ppm) impurity is present in the starting powders.

As seen in Table 1, the grain size of the hot pressed TiB_2-Ni mixtures are equivalent to the starting powder particle size. The virtual elimination of grain growth is a key factor in the mechanical performance of these materials as seen by the general decrease in mechanical properties with increase in grain size in both the TiB_2-Ni and TiB_2 compacts, Table 1. The ability to further refine grain size is thus important and initial results are briefly discussed below.

Hot pressing purer submicrometer TiB_2 powders in attempts to gain finer grain sizes resulted in a TiB_2 with extensive exaggerated grain growth, Fig. 6(a), which limited the final density achieved, (even at a hot pressing temperature of 1600°C vs 1400–1450°C for the TiB_2—Ni mixtures). In addition, increases in the oxygen content of the TiB_2 powder retard densification, Fig. 6(b), due to the promotion of particle coarsening and particle-particle neck formation by transport processes that do not result in shrinkage (e.g., surface or vapor transport). The addition of carbon to these powders does result in dense very fine grained TiB_2, Fig. 6(c).

Interestingly, even in these purer TiB$_2$ powders, segregation of iron and nickel impurities to grain boundaries occurs, and observations indicate that liquid phase assisted densification occurs (Finch and coworkers, 1985). These results then suggest a means to develop fine grained dense boride ceramics.

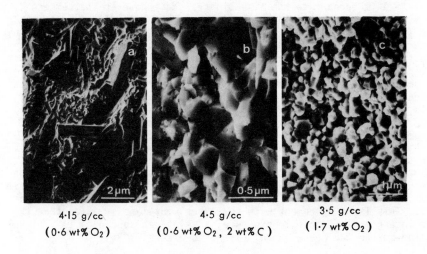

4·15 g/cc	4·5 g/cc	3·5 g/cc
(0·6 wt% O$_2$)	(0·6 wt% O$_2$, 2 wt% C)	(1·7 wt% O$_2$)

Fig. 6   Scanning electron micrographs of submicron TiB$_2$ powder hot-pressed with no initial Ni additive.

Liquid Phase(s) Sintering Process

The investigation of phases present in compacts and exuded materials can lead to an understanding of the reactions taking place during the liquid phase sintering process. The mass balance involved in the dissolution and solidification reactions is one aspect of an overall evaluation. From results presented thus far, oxygen is retained as a titanium oxide in the compact. No titanium oxide or other oxide was found in the exuded specimens. However, Ni$_3$B, Ni$_2$B, and Ti in the form of TiC, TiB$_2$, and intergranular Ti$_x$B$_y$ were observed in the exuded specimens. A plausible mechanism linking the various observations and a model for the liquid phase densification for the TiB$_2$-Ni system are now proposed.

In the case of TiB$_2$ hot pressed with 10 wt % initial additive of Ni the starting TiB$_2$ powder contained ≈1 wt % oxygen. If the TiB$_2$ were oxidized stoichometrically then TiO$_2$ and B$_2$O$_3$ could exist on the surface of the powder. This would result in 0.54 g of Ti in the form of TiO$_2$ and 0.24 g of B as B$_2$O$_3$ for each 100 g of TiB$_2$ powder. As the temperature is increased, and the Ni dissolves or reacts with the TiB$_2$ due to the presence of a low temperature tau phase ternary eutectic, both Ti and B enter the liquid phase. The liquid phase enhances the possibility for reactions to occur since diffusion is much more rapid in liquid than the solid state. Dissolved Ti (from TiB$_2$) in the liquid could reduce B$_2$O$_3$ to B and TiO$_2$ further increasing the B concentration while decreasing the Ti concentration of the liquid phase.

The reduction of B_2O_3 impurity by Ti results in 0.34 g of B entering the Ni rich liquid. If no other reactions occur, 0.6 g of B has entered the melt which would now contain 5.7 wt % B and practically no Ti. The corresponding wt % B values for the borides NiB, Ni_2B, and Ni_3B are 15.5, 8.4, and 6.1, respectively. The 5.7 wt % B from the mechanism just described is near that of stoichiometric Ni_3B. The effect of the Ti reacting with the B_2O_3 impurity essentially drives the system from near the tau phase field of the Ni-B-Ti ternary to the Ni-B binary. Note however that excess Ti was observed in the exudant (globule) in the form of TiB_2, TiC, and possibly other Ti borides. This implies that dissolution of TiB_2 powder continued beyond the amount of Ti needed to completely reduce the B_2O_3. At this stage the melt might well become richer in both Ti and B. However, most of the liquid phase has probably been exuded from the compact leaving an even more complicated system. Now the absolute amount of liquid phase and other secondary phase is much smaller, and other impurities, at even smaller levels, or mechanisms may begin to affect the system.

The same model can be applied to the results on pressureless sintered TiB_2-Ni specimens in associated studies (Angelini and coworkers, 1984). In this case the system can again be driven toward the Ni-B binary by the mechanism just described but since the liquid phase is retained, further dissolution of TiB_2 can occur. The Ti and B content in the liquid phase would then increase until its composition was in the tau—Ni_3B—TiB_2 phase field. This is in agreement with the experimental observation that the major intergranular phase in pressureless sintered specimens is a nickel-titanium-boride tau phase (Angelini and coworkers, 1984).

The model can also describe the results of the case where TiB_2 was hot-pressed with 10 wt % tau ($Ni_{20.3}Ti_{2.7}B_6$). Sklad and Yust (1983) reported that when the above composition was hot-pressed, the major intergranular phase in the compact was Ni_3B. The model can be applied in the following way. If the TiB_2 is oxidized to a level of ≈ 1 wt % this means that 0.8 g of Ti would be needed to reduce the B_2O_3 to B and TiO_2. The Ti content supplied by the tau melt is 0.94 g per 100 g TiB_2. If the Ti (from the tau phase) in the melt reacts with B_2O_3 to form TiO_2, then the liquid phase is enriched in boron resulting in a liquid phase composition within the α_{Ni}-tau-Ni_3B phase field with the largest phase fraction being Ni_3B. These reactions rapidly decrease the Ti concentration and increase the boron concentration in the liquid. In effect the system would be driven toward the Ni-B binary region. As the system dissolves more TiB_2, the liquid phase composition may enter the tau-TiB_2-Ni_3B phase field. Thus the model is supported by its consistency with observations for several TiB_2-metallic additive systems.

SUMMARY

The fracture toughness and fracture strength of hot-pressed TiB_2 liquid phase sintered with Ni increased with increasing residual Ni content (<1.5 wt %). These properties were also found to increase with decreasing TiB_2 grain size. On the other hand, submicrometer sized purer TiB_2 powder densified to >98% density by hot-pressing at a temperature of 1600°C with a 2 wt % carbon addition and resulted in 1—2 micron grain size. Segregation of Fe and Ni impurities to grain boundaries associated the densification of these purer powders. These observations reveal that the evolution of microstructure and secondary phases during densification play a key role in the properties of the dense TiB_2 ceramics.

Electron microprobe analysis of similarly formed TiB_2-Ni compacts revealed the presence of a titanium oxide. Analysis of exuded material revealed a multicomponent system with the major phases being Ni_3B and Ni_2B. Minor phases included TiC, TiB_2, an intergranular Ti_xB_y, and a Si rich phase. A model for the liquid phase sintering reactions was proposed. The model is based on reactions between the Ti component (which results during the dissolution of TiB_2 with Ni) and dissolved oxygen or B_2O_3 (present as either impurities or oxide surface layers on TiB_2 powder).

ACKNOWLEDGMENT

The authors acknowledge S B Waters for preparing hot-pressed specimens and mechanical property testing; T J Henson, K L Thompson, and L R Walker for electron microprobe analyses; N M Atchley for optical microscopy; C P Haltom for preparing thin sections of exudant and TiB_2 compacts for TEM analyses; A F Fisher for ion milling specimens; F C Stooksbury for typing the manuscript; and D N Braski and J I Federer for reviewing the manuscript. The research was sponsored by the Division of Materials Sciences, U.S. Department of Energy, Under contract DE-AC05-840R21400 with the Martin Marietta Energy Systems, Inc.

REFERENCES

Angelini, P., Becher, P.F., Bentley, J., Finch, C.B., and Sklad, P.S. 1984, "Processing and Microstructural Characterization of TiB_2 Liquid Phase Sintered with Ni and Ni_3Al", in Defect Properties and Processing of High Technology Nonmetallic Materials, Elsevier Science Publ Co.

Angelini, P. and Bentley, J. 1984, "Secondary Flourescence Effects on X-ray Microanalysis", Proceedings of the 42nd Annual Meeting of the Electron Microscopy Society of America, San Francisco Press, San Francisco, California, pp 582-84.

Angelini, P. and Lehman, G.L., 1983. Proceedings 41st Annual Meeting of the Electron Microscopy Society of America, Ed G.W. Bailey, San Francisco Press, San Francisco, California, pp 64-65.

Becher, P.F. and Ferber, M.K. 1985, "Grain-Size Dependence of Slow-Crack Growth Behaviour in Noncubic Ceramics", Acta Met. <u>33</u> (7) 1217.

Becher, P.F., Finch, C.B., and Ferber, M.K. 1985, "Effect of Residual Nickel Content on the Grain Size Dependent Mechanical Properties of TiB_2", J. Mater. Sci., in press.

Bentley, J, Angelini, P, and Sklad, P.S. 1984, "Secondary Fluorescence Effects on X-ray Microanalysis", in Analytical Electron Microscopy, Eds D B Williams and D C Joy, San Francisco Press, San Francisco, California, pp 315-17.

Brynestad, J., Banberger, C.E., Heatherly, D.E., and Land, J.F. 1985, "Syntheses of Submicron TiB_2 Powders", High Temperature Science, <u>19</u>, 41.

Cook, R.F. and Lawn, B.R. 1983, "A Modified Indentation Toughness Technique", J. Am. Ceram. Soc. <u>66</u> (11) C200.

Evans, A.G. 1979, Fracture Mechanics Applied to Brittle Materials, Ed. S.W. Freiman, ASTM STP No 678, Philadelphia.

Ferber, M.K., Becher, P.F., and Finch, C.B. 1983, "Effect of Microstructure on the Properties of TiB_2 Ceramics", J. Am. Ceram. Soc. <u>66</u> (1) C2.

Finch, C.B., Becher, P.F., Angelini, P., Baik, S., Bamberger, C.E., and Brynestad, J. 1985, "Effect of Impurities on the Densification of Submicron TiB_2 Powders", to be published.

Popov, H.P. 1976, Mechanics of Materials, Prentice-Hall, Englewood Cliffs, pp 119-130.

Quinto, D.T., Wolfe, G.J., and Haller, M.N. 1983, The Science of Hard Materials, Eds D J Rowcliffe, R K Viswanadham, and J Gurland, Plenum Publ. Co., New York, pp 947-69.

Schöbel, J.D. and Stadelmaier, H.H. 1965, "Die Nickelecke in Dreistoff-system Nickel-Titan-Bor", Z. Metallk. 19 715.

Sklad, P.S. and Bentley, J. 1981, "Analytical Electron Microscopy of TiB_2-Ni Ceramics", in Scanning Electron Microscopy, SEM, Inc, AMF O'Hare, Chicago, Illinois, pp 177-84.

Sklad, P.S. and Yust, C.S. 1983, "Characterization of TiB_2-Ni Ceramics by Transmission and Analytical Electron Microscopy", in The Science of Hard Materials, Eds. D.J. Rowcliffe, R.K. Viswanadham and J. Gurland, Plenum Publishing Co., New York.

Tennery, V.J., Finch, C.B., Yust, C.S. and Clark, G.W. 1983, "Structure-Property Correlations in TiB_2-Based Ceramics Densified Using Active Liquid Metals", in The Science of Hard Materials, Eds. D.J. Rowcliffe, R.K. Viswanadham and J. Gurland, Plenum Publishing Co., New York, pp 891-909.

Tennery, V.J., White, C.K., Padgett, R.A., and Finch, C.B. 1983, "Auger Electron Spectroscopy Analyses of Fracture Surfaces of TiB_2 Ceramics", in Advances in Ceramics, Vol 6, Eds. M. Yan and A.H. Heuer, Am. Ceram. Soc., Columbus, Ohio, pp 312-24.

Zaluzec, N.J. 1979, "Quantitative X-ray Microanalysis Instrumental Considerations and Applications to Materials Science", Ch. 4 in Introduction to Analytical Electron Microscopy, Eds. J.J. Hren, J.I. Goldstein and D.C. Joy, Plenum Press, New York, pp 121-67.

Inst. Phys. Conf. Ser. No. 75: Chapter 11
Paper presented at 2nd Int. Conf. Science Hard Mater., Rhodes

Aspects of trace elements in hardmetals

E KNY AND H M ORTNER

Metallwerk Plansee GmbH, A-6600 Reutte, Austria

ABSTRACT

Various aspects of trace elements in cemented carbides are discussed including specification, distribution and source of trace elements, behaviour of trace elements during processing; analysis, instrumentation and data evaluation; influence on mechanical properties.

Among the trace elements specifically considered are Ca, Mg, K, S and Si. It is shown, that the content of impurities in the raw materials used for production of hardmetal alloys has a significant influence on the transverse rupture strength of sintered compacts and on the number of B-size defects. A hypothesis is presented, that shows the effects and behaviour of the quite common silicate impurities containing Ca, Si and Al.

INTRODUCTION

The detrimental influence of trace elements on hardmetal properties is common knowledge among hardmetal producers, but surprisingly, very little information on the specific effects of trace elements can be found in the relevant literature. Among the few published investigations, Anderson (1976) shows the influence of trace elements on B-type porosity and on transverse rupture strength.

The transverse rupture strength of hardmetals is influenced by structural defects, surface flaws and the intrinsic strength of the alloy. For many of the defects, impurities and trace elements are believed to be responsible. This is shown in Table 1 where types of defects are listed together with their causes. Impurities and trace elements can be the cause of all types of defects but one.

TABLE 1   <u>Porosity and Defects in Hardmetals</u>
<u>and their Suspected Causes</u>

---

pores
- low carbon content
- insufficient comminution
- pressing defects
- uneven powder density
- entrapped gases
- impurities and trace elements

large WC-crystals
- trace elements
- insufficient comminution
- insufficient grain size distribution

mixed carbide clusters
- insufficient comminution
- insufficient grain size distribution

weak interfaces
- segregated trace elements
- alloying elements

unwanted phases and inclusions
- low or high carbon content
- heterogeneous impurities
- trace elements

EXPERIMENTAL

The behaviour and role of trace elements in hardmetals was investigated
by modern analytical methods such as Secondary Ion Mass Spectrometry
(SIMS) and Scanning Auger Electron Spectrometry (AES). While AES offers a
high lateral and depth resolution, the sensitivity to trace elements is
limited. Only elements segregating to grain boundaries such as phosphorus, chromium and titanium could be studied (Kny, 1984). SIMS has a
high sensitivity to trace elements in general, but a diminished lateral
resolution as compared to scanning AES. Trace elements in B-Type pores
were determined by this technique. Trace elements enriched in pores were
studied by microprobe analyses.

A step scan technique was used for SIMS to compare the composition of
pores with the bulk concentration of trace elements. The analysed area
was 1.4 µm in diameter, the step scan width was 3 µm; negative oxygen ions
of 14.5 KeV with 2.5 nA ion current were used. For bulk trace analyses
X-ray fluorescence spectrometry (XRS) and atomic absorption spectrometry
(AAS) were used.

The method of analysis for various elements together with their lower
limits of determination is given in Table 2. ISO standard procedures (ISO
1978, 1982) and house methods were applied (Ortner, 1984). Selective phase
analysis methods were employed to determine the distribution of elements
in binder- and WC-phase (Theiner, 1984). Metallography and transverse
rupture strength measurements (TRS) were carried out according to ISO pro-
edures (ISO 1978, 1982). In the case of TRS-measurements SPUN-indexable
inserts have been used instead of the regular bar shaped specimens.

RESULTS

Trace Element Content of Hardmetals

The trace element content of a series of commercially available hardmetals was investigated, Fig. 1.

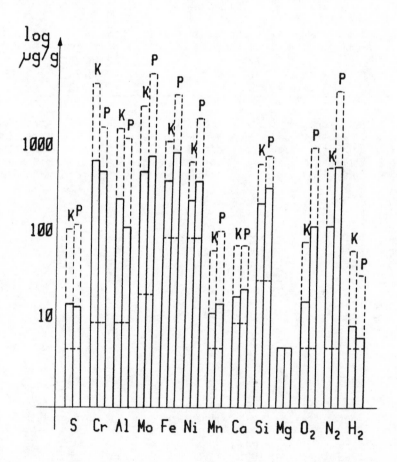

Fig. 1   Trace element content of 139 hardmetal grades from
    15 different producers. The mean, minimum and maximum
values are indicated. K designates WC-Co-grades and P grades
with mixed carbide additions according to the ISO nomenclature.

TABLE 2   Determination of Various Elements in Hardmetal Raw Materials
The lower limit of determination is indicated.

ELEMENT	MATRICES					
	TiC, (Ta,Nb) C mixed carbides		WC-Co Hardmetal with or without mixed carbides		Co-powder	
	Lower limit of determination in µg/g	Method	Lower limit of determination in µg/g	Method	Lower limit of determination in µg/g	Method
O	5	hot extraction	5	hot extraction	5	hot extraction
N	5	hot extraction	5	hot extraction	5	hot extraction
H	5	hot extraction	5	hot extraction	5	hot extraction
C	-	combustion anal.	-	combustion anal.	5	combustion anal.
S	5	combustion anal.	5	combustion anal.	5	combustion anal.
Ba	-		-		10	AAS
Al	50	FAAS	50	FAAS	10	AAS
As	-		-		10	AAS
Ca	10	FAAS	10	FAAS	10	FAAS
Co	10	FAAS	100	FAAS, XRFS	-	
Cr	100	FAAS	100	FAAS	10	FAAS
Cu	10	FAAS	10	FAAS	10	FAAS
Fe	10	FAAS	a) 10 b) 100	FAAS XRFS	10	FAAS
K	5	FAAS	5	FAAS	5	FAAS
Mg	5	FAAS	10	FAAS	5	FAAS
Mn	10	FAAS	10	FAAS	10	FAAS
Mo	20	FAAS	20	FAAS	-	
Na	5	FAAS	5	FAAS	5	FAAS
Nb	100	XRFS	100	XRFS	-	
Ni	10	FAAS	a) 10 b) 100	FAAS XRFS	10	FAAS
P	10	classical photometry	10	classical photometry	10	classical photometry
Si	-		-		10	AAS
Ta	-	XRFS	100	XRFS	-	
Ti	-	XRFS	a) 100 b) 100	XRFS FAAS	100	FAAS
V	100	FAAS	100	FAAS	100	FAAS
Zn	10	FAAS	10	FAAS	10	FAAS

Explanation of abbreviations used:

FAAS flame atomic absorption spectrophotometry
AAS   atomic absorption spectrophotometry in the graphite cell
XRFS X-ray fluorescence spectrometry
      (wavelength-dispersive and sequential)

Although the starting raw materials are generally quite pure, the resulting products exhibit up to 0.6% (m/m) impurities. These are introduced mainly during processing of hardmetal and powders by wear of stainless steel vessels, attritor mill surfaces etc. In some of the alloys, Cr appears to be added deliberately, probably as a grain growth inhibitor or to increase corrosion resistance. These alloys were excluded from evaluation. In all the other alloys, Cr, Fe and Ni were actual impurities. The impurities are not evenly distributed throughout the hardmetal after sintering. Most of them are concentrated in the binder phase. The distribution of various trace elements between the different phases is shown in Table 3 for a 6% Co steel cutting grade, based on actual determinations. This enrichment effect causes a ten-to-fifteen-fold increase of the impurity content of the binder phase as compared to the bulk composition.

TABLE 3   Enrichment of Trace Elements in the Binder Phase of
          a 6% Co-steel Cutting Grade Hardmetal

	Bulk content of impurities % (m/m)	Impurity content in binder phase % (m/m)	Impurity content in binder phase %-rel.
Fe	0.19	3.04	100
Ni	0.05	0.8	100
Al	0.017	0.3	100
Mo	0.152	0.5	approx. 20
Si	0.047	0.35	approx. 50
S	0.0005	0.01	100
Cr	0.136	1.1	approx. 50
	0.6	6.1	
	======	====	

Heterogenous Impurities in Hardmetals

Ca, Al, Si are usually introduced into the starting powders as ceramic inclusions. They are either residues from ore processing or contaminants of hardmetal powders. The increased use of Al_2O_3-coatings frequently increases the Al-content of recycled hardmetal powders. The nonmetals O, N, H, S, P originate from varying sources either during processing (O, N, H) or from the starting raw materials (S, P). The source of P is W or WC powder in many cases, while S is usually introduced by carbon black, which is used for carburising of WC. Most of these impurities are introduced in a heterogenous form. A few examples of heterogenous impurities in recycled hardmetal powders are shown in Figs. 2 and 3.

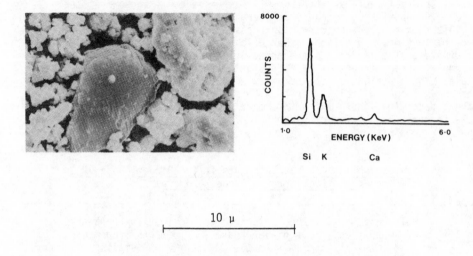

10 μ

Fig. 2   Si, Ca and K containing particle in recycled
hardmetal powder together with x-ray spectrum.

By using different sources of raw materials, the trace element content of
hardmetals can be varied. Recycled hardmetal powder is known to contribute substantially more trace elements than quality grade WC- and
Co-powders. The trace element content which is introduced in production
will not be altered using different raw materials. Different batches were
manufactured using either quality grade raw materials (Q) or a certain
amount of recycled hardmetal powder together with other raw materials (R).
The batches were sintered in such a way that the primary quality control
parameters (magnetic saturation, coercivity and density) all ranged within
internally specified tolerance limits. Composition and tolerance limits
of this grade which is used for steel cutting are given in Table 4.

The R-batches were manufactured in such a way, that 40% of the material
was added as recycled hardmetal powder. The impurity content of this
powder is shown in Table 5. The composition of the main elements was kept
constant. The trace elements present in the Q and R batches are shown in
Table 6. Most trace elements in the R-batches were increased in comparison to the average trace element content of the Q-batches.

B-type porosity in the Q-batches was significantly lower than for the
R-batches.

TABLE 4   Chemical Composition and Physical Properties of
          Experimental Grade used for TRS-testing

W + C	bal
Co	6.0 % (m/m)
Ti	2.0 % (m/m)
Ta	4.0 % (m/m)
Nb	1.0 % (m/m)
magn. saturation	110 ± 8 T 10^{-7} m^3/kg
coercivity	160 ± 15 Oersted
density	14.06 ± 0.06 g/cm^{-3}

Si ⟵   ⟶ Ti

Al,Ti,W

100 µm

Fig. 3   SE-micrograph of heterogeneous impurities in recycled hardmetal
  powder. Three different impurity particles can be seen in one frame.

This was observed repeatedly for samples from different sinterings.
Batches Q and R showed both < B02 porosity (< 140 B pores/cm^{-3}) but there
was still a significant difference between Q and R batches in the lower
range of B-type porosity. In the average 22 B-type pores were found in
the B-batches per cm^2. Only 7 B-type pores/cm^2 were found in the A-batches.

TABLE 5   Trace Element Content of Recylced Hardmetal Powder
WC, 8Co, 4Ti, 5Ta, 1 Nb

µg/g	Fe	Ni	Cr	Al	Ca	S	Si	Mo
	1900	500	60	114	0	5	470	1050

TABLE 6   Trace Element Contents of Q- and R-batches in µg/g

	Q 1	Q 2	mean values of impurities in R and relevant standard deviation	number of determinations
Fe	1300	900	1560 ± 400	15
Ni	400	400	550 ± 150	15
Cr	1800	115	1100 ± 480	11
Al	200	50	140 ± 66	14
Ca	100	20	70 ± 40	5
S	6	5	26 ± 18	14
Si	560	190	457 ± 342	10
Mo	55	160	716 ± 539	11

It was observed by SIMS that B-type pores were always lined with elements such as Si, Ca, Al, Mg, K. By using the step scan technique, a steep increase of the Ca, Mg and K signal was recorded when tracing across a pore as it is shown in Fig. 4.

For TRS-testing, 20 specimens were used in the as sintered condition and 20 specimens were HIP'ed after sintering. Both types of specimens were ground according to ISO-specifications. The TRS in the as sintered condition did not differ significantly between Q and R-batches. After elimination of residual porosity by HIP'ing, the difference in the TRS of Q and R-batches was more pronounced. In many of the fractured R-samples, inclusions and pores could be found as fracture origins. These pores were always lined heavily by Ca, S and K. As an example, a typical pore with an inclusion is shown in Fig. 5. The relevant TRS results are shown in Fig. 6.

B-type Pores in Hardmetals and their Origin

It can not be decided from the previous results whether the increased trace element content is the cause of the increased porosity and the decreased TRS, or whether this is just an unrelated phenomenon.

To clarify this, a 6% hardmetal grade identical to the investigated one was contaminated deliberately by Ca-Al-silicate powder similar in composition to the particles shown in Fig. 2.   A large number of pores developed after sintering.

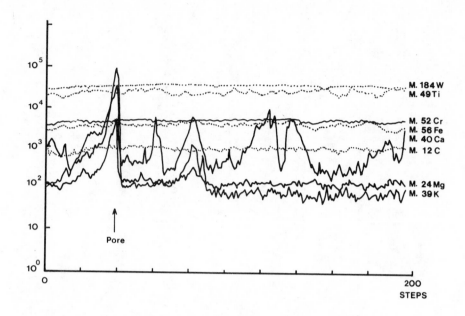

Fig. 4   Trace of SIMS-signals across a pore in the polished
surface of a hardmetal sample of type B-HIP'ed.

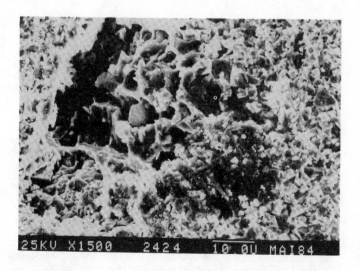

Fig. 5   Pore heavily lined with Ca and S on fracture surface of
hardmetal (type R). The pore acted as fracture origin.

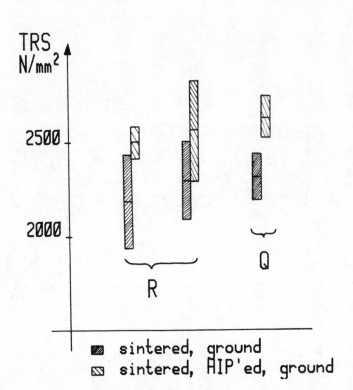

Fig. 6   TRS results of hardmetal batches Q and R.
Fracture sources in the Q-specimens were either not
detectable or were large WC-grains or mixed carbide clusters.

All the pores were filled with contaminating material. Scanning X-ray
analysis showed them to be filled with Ca, Al and S. The Si content of
the pores was low indicating that Si probably evaporated as SiS_2-
compound, Fig. 7. This is in agreement with J. Qvick's calculations that
CaS is a very stable compound under the conditions of W-carburisation.

Under the same conditions, free S is quickly lost as SiS_2 (Qvick, 1981).
It can be assumed that the conditions during hardmetal sintering are
similar to the conditions of tungsten carburisation, due to a high carbon
activity in both cases. The same mechanisms of Si loss can therefore be
assumed for the sintering of hardmetal. The scheme for the proposed
mechanism is presented in Fig. 8.

CONCLUSIONS

It is evident from the previous results that contaminated raw materials
give rise to an increase in B-type porosity and by this a decreased TR-
strength. HIP'ing can not increase the TR-strength to values obtainable
when using quality grade raw materials. Large pores survive the HIP'ing
process in the contaminated material. These pores are lined with trace
elements such as Ca, S and K.

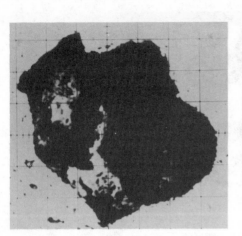

BSE

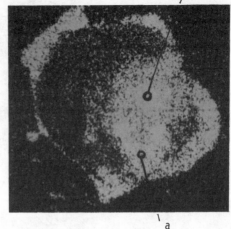

Ca-distribution

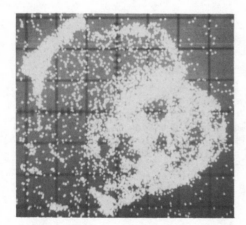

Al-distribution

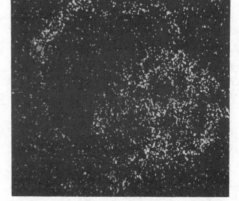

Si-distribution

Fig. 7   Deliberately produced B-type pore in hardmetal.
The pore is filled with a residue consisting of Ca, Al, S.
The Si-content of the pore is comparably low.
(a) and (b) give location of analysed spots.
(Both show high levels of Ca and S)

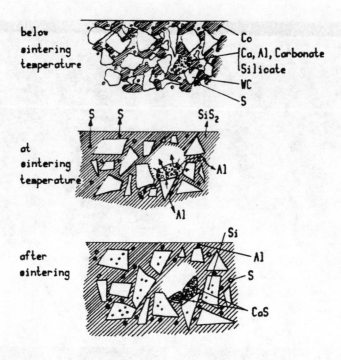

Fig. 8   The behaviour of Ca and S in the system WC-Co during sintering.

Whenever Ca-containing silicates are present in hardmetal powder, CaS will develop due to a thermochemical reaction with S. Si and S will be lost to a large extent during sintering. Some of the sulphur remains in the binder phase of the hardmetal. Al from heterogeneous impurities is dissolved in the binder phase. The silicate oxygen forms CO and is partly emitted. Pore formation is due to the evolution of Co and a change of the wetting behaviour of the binder phase in the vicinity of the CaS-inclusion.

ACKNOWLEDGEMENTS

The authors are grateful to M. Grasserbauer for SIMS-analyses and L. Schmid for continued experimental assistance. This research was supported in part by the "Forschungsförderungsfonds für die gewerbliche Wirtshcaft", Vienna.

REFERENCES

Anderson, P.B., 1976, Der Einfluss von Verunreinigungen auf die
    Eigenschaften von WC-Co-Legierungen am Beispiel von Gesteins-
    bohrlegierungen, Ph.D. Thesis, Technical University Graz
Kny, E., Grasserbauer, M., Wilhartitz, P., Stingeder, G.,
    Goretzki, H. 1984, Fresenius Z. anal. Chemie 317 782-787
ISO-International Standard, 1983, 7627/1-5
ISO-Draft International Standard, 1983, 7627/6
ISO-International Standard, 1978, 4883
ISO-International Standard, 1978, 4503
Theiner, J., 1984, Universität Wien, Elektrochemische Phasenisolierung
    an Hartmetallen, Diplomarbeit
ISO-International Standard, 1978, 4499
ISO-International Standard, 1982, 3327, 2nd edition
Ortner, H.M. 1984, Die chemischen Laboratorien von Plansee, Company
    Brochure, Metallwerk Plansee GmbH
Qvick, J., Snell, P., Noläng, B. and Richardson, M. 1981,
    10th Plansee Seminar, 1 717

Inst. Phys. Conf. Ser. No. 75: Chapter 11
Paper presented at 2nd Int. Conf. Science Hard Mater., Rhodes

1047

Trace element evaporation during the reduction of tungsten(VI)oxide — A comparative study between thermochemical calculations and experiments

JAN QVICK

Institute of Chemistry, University of Uppsala,
Box 531, S-751 21 Uppsala, Sweden

ABSTRACT

The evaporation of the trace elements sodium, silicon and phosphorus during the hydrogen reduction of tungsten(VI)oxide was studied experimentally and the results were compared with theoretical calculations. The evaporation of sodium is facilitated by a high reduction temperature and a low water content in the gas phase and is hindered by the presence of silicon and phosphorus. No measurable evaporation of silicon was obtained. The evaporation of phosphorus, as a single trace element, is somewhat facilitated by a high water content in the gas phase.

INTRODUCTION

In a recent investigation (Qvick, 1984) the trace element behaviour during the hydrogen reduction of WO_3 was studied by means of thermochemical calculations using the free energy minimization method. It was shown, among other things, that evaporation of Li, Na, K, Sn and P is possible during the reduction process while the amounts of Mg, Ca, Al and Si in the powder remain almost unaffected. Furthermore, the evaporation of the former elements was found to be strongly dependent on the water content in the gas phase as well as on the presence of other trace elements.

Experimental studies (Lassner, Petter and Tiles, 1975; Lassner, Schreiner and Lux, 1982) have shown a qualitatively similar behaviour to the theoretical predictions (Qvick, 1984) as regards alkali metals, Mg, Ca and Al. On the other hand, varying results as regards the evaporation of P have been reported. In one study (Lassner, Petter and Tiles, 1975) P was shown to be very stable during the reduction process, while a more recent study (Lassner, Schreiner and Lux, 1982) showed a higher degree of evaporation. However, the theoretical calculations (Qvick, 1984) indicated an even higher degree of evaporation of this element. Silicon, which according to the calculations (Qvick, 1984) should be stable in the condensed phase (as SiO_2) during the whole reduction process, was found to evaporate in both experimental investigations (Lassner, Petter and Tiles, 1975; Lassner, Schreiner and Lux, 1982). Quantitative comparisons between the theoretical and experimental studies were, however, not possible mainly because of a lack of reported data on experimental parameters such as reduction temperatures, hydrogen flow, reduction times *etc*.

In the present study the evaporation of Na, Si and P was studied experimentally. The aim of the investigation was to compare the experimentally obtained results with theoretical predictions (Qvick, 1984) as well as with earlier experimental results (Lassner, Petter and Tiles, 1975; Lassner, Schreiner and Lux, 1982). The evaporation of the above-mentioned trace elements was studied separately for each element as well as in mutual combinations. Since the earlier investigations showed that the evaporation of trace elements depends on the water content in the gas phase two different dewpoints of the hydrogen gas were used.

EXPERIMENTAL

The trace elements Na, Si and P were added to a water slurry of WO_3 (from Seco Tools AB) in a rotating evaporator. Thirteen different types of additions were made. Data for them as well as results from chemical analyses are presented in Table 1. The values for the trace element concentrations in Table 1 are given both as concentrations in the oxide powder and equivalent re-calculated concentrations based on the W powder. The latter concentrations should thus be compared with those obtained after reduction.

The reductions were carried out using a tube furnace with an alumina tube (Degussit AL 23) with an internal diameter of 20 mm. The powder was placed at the center of the tube on a boat (length 20 cm) made of heat-resistant stainless steel (Uddeholm NU Stainless 25, SS 2361-02). The reductions were performed for 12 grams of WO_3 at a time. The oxide powder was spread out on the boat to a length of 10 cm, which gave a layer thickness of $ca.$ 2 mm. The whole gas system including the reaction tube was helium leak tested and no leaks were detected (leak rate less than $5 \cdot 10^{-10}$ atm$\cdot$cm^3/s) either for the alumina tube or for the stainless steel pipework and fittings leading the gas into the furnace. All tube couplings had leak rates which were less than $1 \cdot 10^{-9}$ atm$\cdot$cm^3/s.

The reductions were carried out at atmospheric pressure at 800, 900 and 1000°C for each specimen presented in Table 1 at a hydrogen gas flow rate of 160 cm^3/min. The reduction times were established experimentally by re-oxidation of the W powder (minimum 99.6 % fractional reduction) and were found to be 6, 3 and 2 hours for the temperatures 800, 900 and 1000°C, respectively. The hydrogen gas used was of high purity (99.9997 %) and it was treated in two different ways in order to obtain the two different dewpoints already mentioned. For the reductions with a low water content in the gas phase (denoted L in the following), the hydrogen gas was passed through a catalytic purifier and subsequently dried at liquid nitrogen temperature (-196°C). For the reductions with a high water content (denoted H), the hydrogen gas was passed through a water bubbler placed in a water filled Dewar flask at room temperature. The water temperature varied between $ca.$ 18 and 23°C during the whole experimental sequence. The water content in the hydrogen gas was determined in experiments where the weight change of a molecular sieve was measured and a maximum difference of 0.5°C between the Dewar flask temperature and the hydrogen gas dewpoint was obtained. This procedure of saturating the gas with water was regarded as completely satisfactory for the present experiments.

Due to the high sensitivity to thermal shock of the alumina tube, leading to the formation of cracks, it was not possible either to push in or to pull out the boat at the reduction temperature. Therefore, the oxide powder was pushed into the furnace at 500°C and heating to 700°C was carried out under an argon atmosphere. At this temperature hydrogen of the desired

TABLE 1 <u>Additions of Trace Elements to the Tungsten Oxide and Chemical Analyses of the Oxide Powder</u>

Specimen No	Trace elements added	Addition made as	Concentration of trace element	
			obtained	re-calc.
			ppm	ppm
1	None		Na <10 Si 40 P 40	10 50 50
2-1	Na	NaCl	420	530
2-2	Na	Na_2WO_4	450	570
3	Si	$H_4[Si(W_3O_{10})_4]$	480	610
4	P	$H_3[P(W_3O_{10})_4]$	520	660
5-1	Na Si	NaCl + $H_4[Si(W_3O_{10})_4]$	460 450	580 570
5-2	Na Si	Na_2SiO_3	1020 640	1290 810
6-1	Na P	NaCl + $H_3[P(W_3O_{10})_4]$	370 550	470 690
6-2	Na P	Na_3PO_4	1120 540	1410 680
7	Si P	$H_4[Si(W_3O_{10})_4]$ + $H_3[P(W_3O_{10})_4]$	440 510	550 640
8-1	Na Si P	NaCl + $H_4[Si(W_3O_{10})_4]$ + $H_3[P(W_3O_{10})_4]$	380 410 490	480 520 620
8-2	Na Si P	Na_2SiO_3 + $H_3[P(W_3O_{10})_4]$	900 600 520	1130 760 660
8-3	Na Si P	Na_3PO_4 + $H_4[Si(W_3O_{10})_4]$	960 460 520	1210 580 660
8-4	Na Si P	Na_2SiO_3 + Na_3PO_4	2120 600 550	2670 760 690

humidity was led into the alumina tube and heating up to reduction temperature was carried out. The H_2 flow was started at 700°C since a free energy
minimization calculation using the computer program EKVICALC (Noläng, 1983)
showed that sublimation of NaCl in a dry argon atmosphere above 700°C might
result in a measurable loss of Na. The use of a wet hydrogen atmosphere instead leads to the formation of sodium tungsten bronzes or sodium tungstate
(Qvick, 1984). The cooling was carried out under H_2 down to 700°C and at
this temperature the gas flow was changed to argon. The powder was pulled
out into a cooling zone when the temperature had fallen below 600°C and the
furnace was opened when the W powder had reached room temperature. The rate
of temperature increase was approx. 450°C/h, and the cooling rate approx.
350°C/h. After each reduction the tungsten charge was mixed carefully.

In order to avoid contamination from one specimen to a later reduced specimen the following operation was carried out. When a series of three reductions for the same specimen had been performed in the order 1000, 900
and 800°C the alumina tube was taken out of the furnace and cleaned with
distilled water. After drying, the tube was helium leak tested and again
placed in the furnace (cracks were never detected when the above-mentioned
procedure with heating and cooling the specimens was followed). Prior to
the next reduction the boat was polished and placed in the furnace without
any powder. Then the furnace was heated to 1100°C with a flow of dry H_2
(flow rate *ca.* 250 cm³/min), at which temperature the furnace was held for
one hour. After this treatment the reductions of the next specimen were
carried out in the above-mentioned order. This cleaning procedure served
to reduce surface oxides and remove contamination from the boat.

Chemical analyses were carried out of all tungsten powders after reduction.

RESULTS AND DISCUSSION

The complete results of the chemical analyses of the W powders are given in
Table 2 together with the theoretically expected values determined in the
same manner as those presented earlier (Qvick, 1984). In the following,
typical experimental results as well as theoretical predictions of trace
element evaporation are presented and discussed.

Sodium

Sodium, as a single trace element, was added as NaCl or Na_2WO_4 to the oxide
powder (see Table 1). According to theoretical calculations (Qvick, 1984)
the behaviour and evaporation of Na during the reduction process are the
same no matter whether the addition is made as NaCl or Na_2WO_4. The present
experiments confirmed this result as regards evaporation, and consequently
probably also as regards behaviour, (see Table 2) and therefore only the
results from the Na_2WO_4 addition (specimen 2-2) are presented in more detail below.

Fig.1 shows the theoretically calculated concentrations of Na in the W powder (solid lines) as well as those obtained experimentally. It is obvious
that a low water content in the gas phase and a high reduction temperature
facilitate the evaporation of sodium in accordance with theoretical predictions (Qvick, 1984). The same observations have been made earlier both as
regards water dependence (in the case of K)(Lassner, Petter and Tiles,
1975) and temperature (Lassner, Schreiner and Lux, 1982).

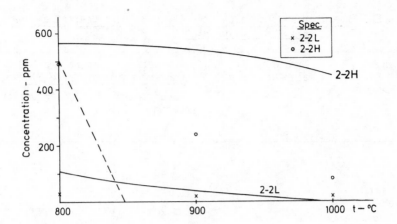

Fig. 1 Concentration of Na in specimen 2-2 *vs*. reduction temperature. Continuous curves indicate calculated sodium concentrations for reductions L and H and the dashed line the concentration of Na for the bronze $Na_{0.794}WO_3$ for reduction H. Experimentally obtained values are indicated as points.

However, it is also obvious that a higher degree of evaporation of Na was obtained experimentally than that predicted by the calculations. This observation might be explained in one or both of the following ways. According to the calculations (Qvick, 1984) it was found that sodium tungsten bronzes form in the early stages of the reduction and are stable during the process. Furthermore, the occurence of potassium tungsten bronzes during the reduction process has been experimentally proven (Neugebauer, 1975). Towards the end of the process when the water content in the gas phase has fallen below the equilibrium content of the last reduction step ($WO_2 \rightarrow W$) it was found that the bronzes disproportionate (Neugebauer, 1975; Qvick, 1984) into tungsten and tungstate according to eq.1.

$$\frac{2}{x}Na_xWO_3(s) + (\frac{6}{x}-4)H_2(g) \rightarrow Na_2WO_4(s \text{ or } 1) + (\frac{2}{x}-1)W(s) + (\frac{6}{x}-4)H_2O(g) \qquad 1$$

Theoretically it was found that the amount of Na added in the present specimen should form tungstate very rapidly. In practice, however, it might be possible for the Na-W bronzes to evaporate prior to tungstate formation according to eqs.2 and 3.

$$Na_xWO_3(s) + 3H_2(g) \rightarrow xNa(g) + W(s) + 3H_2O(g) \qquad 2$$

$$Na_xWO_3(s) + (3-\frac{x}{2})H_2(g) \rightarrow xNaOH(g) + W(s) + (3-x)H_2O(g) \qquad 3$$

A free energy minimization calculation carried out analogously to those presented elsewhere (Qvick, 1984) showed that evaporation of Na might take place to a much higher degree from the bronzes than is possible from the tungstate. The dashed line in Fig.1 shows the calculated result for reduction H for a bronze of the composition $Na_{0.794}WO_3$, which proved to be the

TABLE 2 Experimental and Calculated Trace Element Concentrations
after Reduction at 800, 900 and 1000°C

Specimen No	Trace element	Reduction temperature 800°C Concentration of trace element			
		Reduction L		Reduction H	
		experim.	calc.	experim.	calc.
		ppm	ppm	ppm	ppm
1	Na	20	10	10	10
	Si	40	50	50	50
	P	50	0	20	10
2-1	Na	20	110	430	530
2-2	Na	30	110	490	570
3	Si	590	610	600	610
4	P	200	0	140	10
5-1	Na	300	540	490	580
	Si	550	570	560	570
5-2	Na	600	730	1180	1290
	Si	790	810	780	810
6-1	Na	340	430	340	470
	P	260	190	270	210
6-2	Na	880	1370	1100	1410
	P	470	620	560	630
7	Si	530	550	570	550
	P	220	0	130	10
8-1	Na	400	440	420	480
	Si	550	520	540	520
	P	170	200	280	220
8-2	Na	870	1090	1060	1130
	Si	770	760	790	760
	P	380	490	440	510
8-3	Na	1060	1170	1160	1210
	Si	550	580	550	580
	P	420	530	460	540
8-4	Na	1700	2110	2230	2670
	Si	770	760	770	760
	P	540	670	530	690

Reduction temperature 900°C Concentration of trace element				Reduction temperature 1000°C Concentration of trace element			
Reduction L		Reduction H		Reduction L		Reduction H	
experim.	calc.	experim.	calc.	experim.	calc.	experim.	calc.
ppm	ppm	ppm	ppm	ppm	ppm	ppm	ppm
10	10	10	10	<10	0	10	<10
50	50	60	50	50	50	80	50
30	0	20	<10	10	0	20	0
20	40	210	500	10	0	50	410
20	40	240	540	20	0	80	450
610	610	590	610	610	610	600	610
70	0	40	<10	40	0	50	0
300	470	360	570	180	410	360	560
580	570	570	570	580	570	580	570
600	660	1000	1260	480	600	630	1170
800	810	800	810	800	810	770	810
100	380	230	470	10	240	60	460
140	170	170	210	30	110	80	210
540	1320	790	1410	30	1180	410	1400
280	590	430	630	30	530	260	630
540	550	530	550	530	550	520	550
100	0	40	<10	70	0	40	0
370	390	390	480	220	250	280	470
540	520	500	520	540	520	510	520
110	180	100	220	60	110	50	210
660	1040	860	1130	310	900	490	1120
760	760	760	760	770	760	750	760
160	470	250	510	40	400	100	500
770	1120	910	1210	290	980	490	1200
570	580	570	580	570	580	570	580
240	500	300	540	80	440	120	540
1300	2040	1700	2650	800	1870	1100	2540
770	760	780	760	780	760	770	760
360	650	420	690	110	590	340	680

most stable bronze for which calculations were carried out (Qvick, 1984). The calculation for reduction L showed complete evaporation of Na even below 800°C and is therefore not indicated in Fig.1. Thus, in practice it is probable that two competing reaction types occur for the Na-W bronzes, one involving tungstate formation (eq.1) and the other leading to evaporation of Na (eqs.2 and 3). Therefore, it can be anticipated that a larger amount of Na evaporates in practice than is predicted theoretically.

The second process that could lead to more complete sodium evaporation than predicted is as follows. The equilibrium calculations (Qvick, 1984) showed that a measurable evaporation of Na (from bronzes) should not be possible during the last reduction step ($WO_2 \rightarrow W$) since the oxygen activity in the gas phase is too high. In practice, however, it is probable that this reduction step does not occur at equilibrium and consequently a lower oxygen activity prevails in the gas phase than predicted. Thus, the evaporation of Na could begin earlier in the process than expected theoretically and the total evaporation of Na would be greater. However, according to theoretical calculations Na_2WO_4 will form (eq.1) at water vapour concentrations which are too high for a measurable evaporation of Na to take place. Therefore, the fairly large differences between calculations and experiments for the present specimens cannot be explained merely as a consequence of non-equilibrium conditions during the last reduction step.

It is worth pointing out that free energy minimization calculations using the program EKVICALC (Noläng, 1983) showed that Na-containing gaseous species (Na and NaOH) may react with the alumina tube to form solid $NaAlO_2$. Thus, a chemical transport of Na from the specimen to the tube might be possible in a closed system. However, in the present experiments since flowing H_2 was used and since the powder was never in contact with the alumina tube it can be anticipated that reactions of the above-mentioned type would be of negligible significance.

In certain cases it is possible that the discrepancies between calculated and experimental results are a consequence of low accuracy of thermochemical data. However, most data in the calculations for Na as a single trace element were taken from the JANAF tables (Stull and Prophet, 1971). The enthalpy of formation for Na_2WO_4 in these tables is for example in good agreement with later determinations (Gmelin, 1979). A further possible explanation of the high calculated concentrations is, of course, that some important gaseous Na-containing species was overlooked when the calculations were carried out. However, according to Gmelin (1979) the most important gaseous species in the vaporization of Na_2WO_4 is atomic Na, which was included in the calculations. The same result as that reported in Gmelin (1979) was also found in the theoretical calculations (Qvick, 1984).

Since the reductions were carried out using a tube furnace with the powder placed parallel to the flow of H_2 it is probable that a sodium concentration gradient appears in the tungsten powder, the tungsten situated on the gas inlet side containing less Na than that on the outlet side. Since the powder was mixed after reduction the concentrations obtained probably represent mean concentrations.

Silicon

Silicon did not show any measurable evaporation during the reduction process either when it was studied separately or when it was studied in combinations with Na and/or P (see Table 2). This is in agreement with the thermochemical calculations (Qvick, 1984) but it is in strong contrast to other experimental investigations (Lassner, Petter and Tiles, 1975; Lassner, Schreiner and Lux, 1982). The results in the present study as well as those in the theoretical calculations (Qvick, 1984) are, however, supported by industrial experience (Snell, 1983).

Phosphorus

The addition of P as a single trace element was made using the heteropoly-acid $H_3[P(W_3O_{10})_4]$. According to the thermochemical calculations (Qvick, 1984) phosphorus forms solid $W_2O_3(PO_4)_2$ during the reduction process and a large evaporation takes place towards the end of the process. Fig.2 (solid line) shows the theoretically expected concentration of P for specimen 4H. The evaporation of P for specimen 4L was predicted to be complete below 800°C.

From Fig.2 it is obvious that the experimentally-obtained results are higher than those predicted by the calculations. On the other hand, the evaporation in the present study was considerably larger than that in an earlier experimental study (Lassner, Petter and Tiles, 1975), where P, added as a heteropolyphosphorus compound, was found to have a constant concentration during the reduction process. The evaporation of P in a more recent experimental study (Lassner, Schreiner and Lux, 1982) was probably also lower than that obtained here, although quantitative comparisons are not possible.

The experimental results also showed that the evaporation of phosphorus (P as a single trace element) is probably favoured by a high water content in the gas phase (see Fig.2). This is in accordance with other experimental studies (Lassner, Petter and Tiles, 1975; Lassner, Schreiner and Lux, 1982) but it is inconsistent with the theoretical calculations (Qvick, 1984). However, the calculations indicated that a gaseous oxide (P_4O_6) is the most important gaseous species in the evaporation of P and therefore a high water content in the gas phase may under certain conditions favour the evaporation (see below).

It is possible that the differences between the theoretical and experimental results can be ascribed to a relatively slow evaporation reaction, due

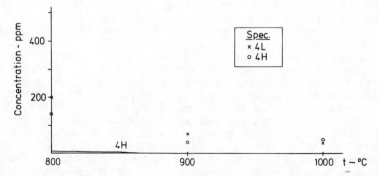

Fig. 2 Concentration of P in specimen 4 *vs*. reduction temperature. Continuous curve indicates calculated phosphorus concentrations for reduction H. Experimentally obtained values are indicated by points.

to kinetic impediments. However, the aim of the present study was to check the earlier thermochemical calculations. Consequently, the experiments were not adapted to draw conclusions on the kinetics of the process - a remark valid for all experiments in the present investigation. Furthermore, since all thermochemical data for $W_2O_3(PO_4)_2$ were estimated (Qvick, 1984), it is also possible that these were not sufficiently accurate. On the other hand, several ternary P-W-O compounds have been reported (Qvick, 1984) and it can not be excluded that other compounds, more stable than $W_2O_3(PO_4)_2$, form in the process. Since the evaporation was probably favoured by a high water content in the gas phase and the calculations showed that P_4O_6 is an important gaseous species it seems more probable that the evaporation of P takes place from a condensed phosphorus-containing compound with a P/O atomic ratio higher than 2/3 (such as in P_4O_6). Thus, it is possible that $W_2O_3(PO_4)_2$ does not form in the process.

Sodium with Silicon

The mutual influence of Na and Si was studied by using (a) NaCl + $H_4[Si(W_3O_{10})_4]$ (specimen 5-1) and (b) Na_2SiO_3 (5-2) as additions to WO_3. Table 1 presents the initial amounts of Na and Si. X-ray powder analysis of the dried water slurry showed that sodium tungstates ($Na_2WO_4 \cdot 2H_2O$ and Na_2WO_4) had already formed in the slurry in specimen 5-2. When the powder was heated to 750°C in air α-SiO_2 was also found in the specimen. The thermochemical calculations (Qvick, 1984) showed that Na-W bronzes and SiO_2 are stable during the reduction until the last step ($WO_2 \rightarrow W$) is complete. Then the Na bronzes disproportionate into tungsten and tungstate according to eq.1 and the tungstate reacts with SiO_2 to form $Na_2Si_2O_5$. According to the calculations the silicate was considerably more stable than Na_2WO_4.

Fig.3 shows the experimental and calculated concentrations of Na for the specimens 5-1 and 5-2. It should also be noted that the concentrations of Si did not change during the reduction process. It is obvious from Fig.3 that the evaporation of Na was facilitated by a low oxygen activity in the gas phase and by a high reduction temperature. This observation is in accordance with the results for specimen 2-2 (Fig.1) and also with theoretical calculations (Qvick, 1984). The sodium concentration values obtained experimentally were lower than those calculated; probable explanations of this were given in the section dealing with sodium.

Comparison between the experimental results for specimens 5-1 and 2-2 (compare Fig.3 to Fig.1), which both contained approximately the same amount of Na initially, makes it clear that the evaporation of Na is hindered by the presence of Si (in particular for reduction L). This result is in agreement with theoretical calculations (Qvick, 1984) as well as with earlier experimental studies (Lassner, Schreiner and Lux, 1982). However, if the specimens 2-2H and 5-2H are compared it is seen that more Na evaporated from the latter (at 800 and 1000°C), in spite of the fact that Si was present in it. This observation may seem puzzling but it can probably be explained in the following manner. When, towards the end of the reduction process, the water vapour content in the gas phase decreases, it becomes possible for the Na bronzes either to form tungstate according to eq.1 or to evaporate according to eqs.2 and/or 3. It is plausible to assume that the rate of these competing reactions is independent of the amount of Na bronze. Therefore, the rate depends only on the temperature and hydrogen amount, *i.e.* time at a fixed temperature and at a constant flow of H_2. Thus, the more Na that is added the more may evaporate (in absolute amount) from the bronzes prior to complete tungstate formation.

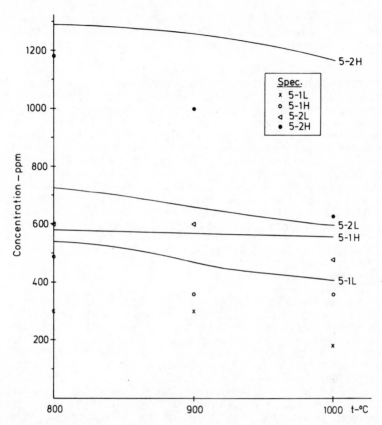

Fig. 3 Concentration of Na in specimens 5-1 and 5-2 *vs.* reduction temperature. Continuous curves indicate calculated concentrations of Na for reduction L and H. Experimental values are indicated by points.

It is, of course, also possible for Na to evaporate from the tungstate prior to the formation of a Na-Si containing compound. Sodium evaporation at this stage of the process would also lead to a lower concentration of Na in the final powder than predicted, since Na according to the earlier calculations evaporates easier from the tungstate than from the silicate. The formation of a Na-Si containing compound in the final stage of the reduction also requires an even distribution of Na_2WO_4 and SiO_2 in the powder. It is probable that this distribution is better obtained using Na_2SiO_3 as an addition rather than $NaCl + H_4[Si(W_3O_{10})_4]$.

In specimen 5-1 the initial Na/Si atomic ratio was *ca.* 1.2/1, while in specimen 5-2 it was *ca.* 1.9/1. (The deviation from the theoretical value 2/1 for specimen 5-2 is probably due to the fact that sodium tungstate was already formed in the slurry for this specimen and that sodium tungstate crystallized on the walls of the evaporator to a larger degree than did the Si-containing compound.) Since Na always seemed to evaporate prior to tungstate formation it is likely that the Na/Si atomic ratios in the tungsten powders obtained from specimen 5-1 do not correspond to the ratio in the stable Na-Si containing compound. On the other hand, in specimen 5-2 the initial Na amount was larger and a more even distribution of Na_2WO_4 and

SiO_2 was probably obtained. Therefore, the Na/Si atomic ratios in the tungsten powder obtained from this specimen probably reflect the ratio in the stable Na-Si containing compound more accurately than those from specimen 5-1.

The theoretical calculations showed that reduction 5-2H led to insufficient evaporation of Na from the tungstate (at least at the temperatures 800 and $900^{\circ}C$) to reveal the atomic ratio in the Na-Si containing compound formed in the process. For specimen 5-2L, however, the calculations showed a complete evaporation of Na from Na_2WO_4 at $800^{\circ}C$ but a negligible evaporation of Na from the silicate at 800 and $900^{\circ}C$. The Na/Si atomic ratios for these specimens were 0.93/1 and 0.92/1 for 800 and $900^{\circ}C$, respectively. This result supports the theoretical prediction (Qvick, 1984) that $Na_2Si_2O_5$ is formed during the reduction process. However, most of the specimens displayed a Na/Si atomic ratio lower than 1/1 which might indicate the occurrence of $Na_6Si_8O_{19}$ (Williamson and Glasser, 1965), but since this phase occurs only between approx. 700 and $800^{\circ}C$ in the *quasi*-binary system Na_2O-SiO_2 (Williamson and Glasser, 1965) a simultaneous occurrence of $Na_2Si_2O_5$ and SiO_2 seems more probable.

Sodium with Phosphorus

The simultaneous evaporation of Na and P was studied in specimens 6-1 ($NaCl + H_3[P(W_3O_{10})_4])$ and 6-2 (Na_3PO_4). It was shown by means of the X-ray powder technique that $Na_2WO_4 \cdot 2H_2O$ and Na_2HPO_4 had formed in the water slurry in specimen 6-2. The sodium hydrogen phosphate was transformed to $Na_4P_2O_7$ on heating the oxide to $450^{\circ}C$ in air. Theoretically (Qvick, 1984) it was found that Na-W bronzes and $W_2O_3(PO_4)_2$ are stable during the hydrogen reduction until the last step is complete. Then, the calculations showed, Na_3PO_4 would form *via* the formation of Na_2WO_4 (eq.1). According to the calculations Na_3PO_4 is considerably more stable than Na_2WO_4; this leads to a less complete evaporation of Na and P than when they are not combined.

The theoretical and experimental sodium concentrations for specimens 6-1 and 6-2 are shown in Fig.4. It is seen that the evaporation of Na is facilitated by a high reduction temperature as well as by a low water content in the gas phase, in agreement with thermochemical calculations (Qvick, 1984) and in agreement with the Na systems already presented. Furthermore, Fig.4 shows that the experimental concentrations of Na are, without exception, lower than the theoretical values. Probable explanations, which at least partly might explain this were given in the 'Sodium' section.

From Fig.4 it is also clear that the presence of P inhibits the evaporation of Na provided the water content in the gas phase is low (*cf.* 6-1L, Fig.4, and 2-2L, Fig.1). This inhibition of sodium evaporation is in agreement with the theoretical calculations (Qvick, 1984) and also with another experimental investigation (Lassner, Schreiner and Lux, 1982). On the other hand, no clear difference in evaporation of Na between 6-1H and 2-2H (*cf.* Fig.1) was observed. It seems probable that sodium evaporation in 6-1L was inhibited by the formation of a Na-P containing compound (to be discussed below) and that this did not form in 6-1H. On the other hand, the evaporation of P in 6-1H was lower than in specimen 4H (compare Fig.5 with Fig.2) which does suggest the formation of a Na-P containing compound (see below).

The experimental results in Fig.4 display a very large decrease in sodium concentration with temperature, much larger than that predicted by the calculations. This suggests that the Na-P containing compound formed evapor-

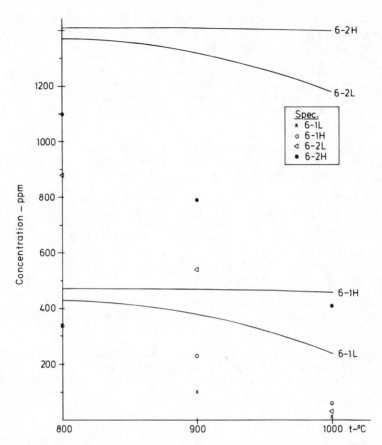

Fig. 4 Concentration of Na in specimens 6-1 and 6-2 *vs.* reduction temperature. Continuous curves indicate calculated concentrations of Na for reduction L and H. Experimentally obtained values are indicated by points.

ates more completely than is predicted for Na_3PO_4 (Qvick, 1984). The evaporation of Na in 6-1, however, did not seem to be so sensitive to water content in the gas phase as the evaporation (probably from Na_2WO_4) occurring in specimen 2-2. Thus, the sodium evaporation results for specimens 2-2H and 6-1H turned out to be very similar while the results for 2-2L and 6-1L were rather different.

It should furthermore be observed that sodium evaporation was very great for the specimens 6-2L and 6-2H (greater for 6-2H than for 2-2H). This observation can probably be explained in the same manner as the analogous observation for specimen 5-2H (see the section on sodium with silicon).

A comparison between the evaporation of Na for specimens with Na and Si (specimens 5-1 and 5-2, Fig.3) on the one hand and Na and P (specimens 6-1 and 6-2, Fig.4) on the other reveals that Si hinders the evaporation of Na to a greater extent than does P. This result is in accordance with theoretical calculations (Qvick, 1984) as well as with an earlier experimental investigation (Lassner, Schreiner and Lux, 1982).

The experimental and calculated phosphorus contents for specimens 6-1 and 6-2 are shown in Fig.5. It is obvious that the evaporation of P in these specimens was favoured by a low water content in the gas phase in contrast to the results for specimen 4 (see Fig.2), but in agreement with the calculations (Qvick, 1984). This observation supports the above assumption that a Na-P containing compound forms in the present case. Furthermore, a comparison of Figs.5 and 2 makes it clear that the evaporation of P is hindered by the presence of Na, in accordance with theoretical calculations (Qvick, 1984).

However, the concentrations of P are generally lower than predicted theoretically. This might be a consequence of evaporation of P (from the P-W containing compound) prior to the formation of the Na-P compound.

It is more difficult to assign a formula to this Na-P containing compound than was the case for the Na-Si containing compound in samples 5-1 and 5-2. For specimen 6-2L after reduction the Na/P atomic ratios are 2.52/1 and 2.60/1 for 800 and 900°C, respectively. According to Gmelin (1967) the reduction of Na_3PO_4 in hydrogen occurs *via* the formation of $Na_4P_2O_7$. Therefore, it seems possible that Na_3PO_4 and $Na_4P_2O_7$ are present simultaneously in the W powder obtained from specimen 6-2. Furthermore, since $Na_4P_2O_7$ is easier to reduce than is Na_3PO_4 (Gmelin, 1967) it is possible that the large differences between calculated and experimental results for specimens 6-1 and 6-2 are partly due to the formation of $Na_4P_2O_7$ instead of Na_3PO_4.

Specimen 6-1 showed Na/P atomic ratios considerably lower than those obtained for specimen 6-2, which might be due partly to evaporation of Na prior to tungstate and phosphate formation. However, if this were so the

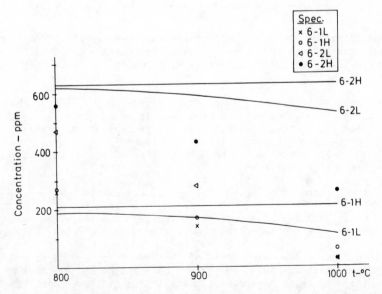

Fig. 5 Concentration of P in specimens 6-1 and 6-2 *vs.* reduction temperature. Continuous curves indicate calculated concentrations of P for reduction L and H. Experimentally obtained values are indicated as points.

phosphorus not bound to sodium should have formed the same compound as in specimen 4 and it should have been possible for larger quantities of P to evaporate, at least from specimen 6-1H. Thus, it is possible that other Na-P containing compounds were formed in the specimens 6-1 than in 6-2, for instance bronzes of the type $A_xP_8W_{8n}O_{24n+16}$ (A=K, Rb) (Giroult and co-workers, 1982), which possibly may occur also for Na.

Silicon with Phosphorus

The specimens containing Si and P together (specimen 7, see Table 1) did not show any marked differences after reduction from specimens in which Si and P were studied separately. Thus, Si remained completely in the W powder and the evaporation of P was favoured by a high water content in the gas phase, in accordance with specimen 4. Therefore, it does not seem likely that a compound containing Si and P formed during the reduction process.

Sodium with Silicon and Phosphorus

Sodium, silicon and phosphorus were simultaneously added to the oxide in four different ways which are specified in Table 1 (specimens 8-1 to 8-4). The behaviour of the specimens containing the three elements in combination was similar to that of specimens 5-1, 5-2, 6-1 and 6-2. Thus, the experimental evaporation of Na was larger than that predicted by the calculations and it increased with temperature, Na content and reduced water content in the gas phase. The evaporation of Si was not measurable and the evaporation of P was favoured by a low water content in the gas phase and a high reduction temperature. The evaporation of P was almost without exception greater than predicted theoretically.

The evaporation of Na was less in specimen 8-1 than in specimens 5-1 and 6-1 (all had almost the same initial concentration of Na in the oxide), due to the simultaneous presence of Si and P in specimen 8-1. The experimental and calculated concentrations of Na are shown in Fig.6. A comparison between this Fig. and Figs.1, 3 and 4 makes it clear that sodium evaporation

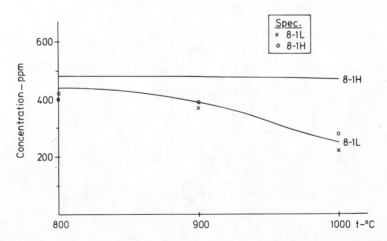

Fig. 6 Concentration of Na in specimen 8-1 *vs.* reduction temperature. Continuous curves indicate calculated concentrations of Na for reduction L and H. Experimentally obtained values are indicated as points.

is inhibited more effectively when Si and P are present simultaneously than when only one of them is present. It can be noted that the experimental sodium concentrations for specimen 8-1L are in good agreement with those calculated theoretically (see Fig.6). The better agreement for this specimen than for 5-1 and 6-1 is probably due to shorter diffusion distances for the formation of sodium silicates and phosphates in specimen 8-1. This implies that a smaller amount of Na evaporated from the tungstate prior to silicate or phosphate formation.

Specimens 8-2 and 8-3 which had approximately the same initial sodium concentrations as specimens 5-2 and 6-2 also showed that the evaporation of Na is hindered by the simultaneous presence of Si and P. Specimen 8-4, which had a very large initial concentration of Na, displayed a large evaporation of this element. However, due to the large initial concentration of Na it is not meaningful to compare the results for this specimen with the others.

ACKNOWLEDGEMENTS

The author is deeply indebted to Dr. M.W. Richardson for the construction of the furnace as well as for advice concerning the performance of the experiments. The chemical analyses were performed at Seco Tools AB, Fagersta and AB Sandvik, Hard Materials, Stockholm, which is gratefully acknowledged. Special thanks are due to Dr. P.O. Snell, Ms. A. Bruhn and Mr. C. Palm. Furthermore, the author is grateful to Dr. T. Lundström for valuable criticism of the manuscript. Financial support from the National Swedish Board for Technical Development (STU) and from the Swedish Natural Science Research Council is gratefully acknowledged.

REFERENCES

Giroult, J. P., Goreaud, M., Labbe, Ph. and Raveau, B. 1982, J. Solid
    State Chem. 44 407-414
Gmelins Handbuch d Anorg. Chemie 1967, Natrium, No 21, Ergänzungsband
    Lfg. 4. Verlag Chemie, Weinheim, pp 1570 and 1613
Gmelins Handbuch d Anorg. Chemie 1979, Wolfram, Ergänzungsband B3.
    Springer-Verlag, Berlin pp 157, 188-189 and 195
Lassner, E., Petter, H. and Tiles, B. 1975, Planseeber. Pulvermetall. 23
    86-100
Lassner, E., Schreiner, M. and Lux, B. 1982, Int. J. Refract. Hard Met.
    1 51-60
Neugebauer, J. 1975, Planseeber. Pulvermetall. 23 77-85
Noläng, B. 1983, Acta Universitatis Upsaliensis 691, Thesis,
    ISBN 91-554-1433-8. Available from Svensk Energidata, Agersta,
    S-740 22 Bälinge, Sweden
Qvick, J. 1984, Int. J. Refract. Hard Met. 3 121-131
Snell, P. O. 1983, Danit Hardmetal A/S, 21 Højvangen, DK-3060 Espergaerde,
    Denmark, private communication
Stull, D. R. and Prophet, H., eds. 1971, JANAF Thermochemical Tables,
    2nd ed., NSRDS-NBS 37, US Govt. Printing Office, Washington
Williamson, J. and Glasser, F. P. 1965, Science 148 1589-1591

Inst. Phys. Conf. Ser. No. 75: Chapter 11
Paper presented at 2nd Int. Conf. Science Hard Mater., Rhodes

1063

Current status and potential for development of W-free hard alloys

V A ZHILYAEV(1), E I PATRAKOV(2) AND G P SHVEIKIN(1)

(1) Institute of Chemistry, Urals Scientific Centre, U.S.S.R.
(2) Institute of Physics of Metals, Urals Scientific Centre, U.S.S.R.

ABSTRACT

The current status and tendencies in the development of materials for
metal cutting are analysed. Ways of substituting tungsten-containing hard
alloys with tungsten-free alloys are considered. It is shown that the
most promising substitutions for tungsten carbide are solid solutions of
carbides and nitrides of transition metals of the IV-VI groups. Some
physico-chemical foundations for obtaining tungsten-free hard alloys by
the method of liquid phase sintering are briefly described.

INTRODUCTION

In spite of the introduction of modern methods of fabricating metallic
details (stamping, precision casting, powder metallurgy) there remains a
need for additional machining. It is enough to say that about 20% of all
metal consumed in industry is reduced to chips. Metal cutting is performed mainly using materials of the following groups: high-speed steels,
hard alloys based on WC, W-free hard alloys based on TiC and metalloceramic materials. The productivity of materials of each group is
characterized by the specific volume of their usage, the range of speeds
of cutting and also the volume of removed chips. The analysis of these
data shows that even at moderate speeds of cutting the main portion of
metals (about 68%) is covered by hard alloys based on WC. First of all
this is explained by the universality of such materials, which is due in
turn to a unique combination of hardness, strength and toughness at
elevated temperatures. Meanwhile, the deficit of W and Co raw materials
encourages a search for, if not equivalent, at least acceptable sub-
stitutes for such alloys. This is, however, not a straightforward matter.

THE PROBLEMS OF INCREASING THE STRENGTH AND TOUGHNESS OF W-FREE HARD ALLOYS AND POSSIBLE SOLUTIONS

The most universal of existing cutting materials (after alloys based on
WC) are undoubtedly W-free hard alloys based on TiC. According to their
operational properties they have an intermediate position between high-
strength WC-containing alloys and brittle metalloceramic and superhard

materials (based on Al_2O_3, BN, B_4C etc.). The problem is to increase their relatively low strength properties while conserving their advantages (hardness, resistance to wear, oxidation and adhesion).

It is known, in the general case, that the strength of composite materials similar to hard alloys depends on uniform distribution of its components, on the strength of the constitutent phases and also on the cohesive strength between the phases. The scatter of strength properties of inter-phase boundaries must be minimum (Tresvyatsky, 1970).

Existing ways of strengthening composite materials may be divided into two groups. To the first group belong methods which give an increase in the strength properties of the composite as a whole (methods of structural strengthening), and to the second belong methods that involve increasing the strength of the components (methods of chemical strengthening).

Among the first group, the most widespread for hard metals are the following:

- to increase the dispersion of the refractory component of the alloy (generally the strength properties of a composite are inversely proportional to the square root of the mean grain size according to the Hall-Petch equation)

- to increase the volume fraction of the binder metal.

The second method is less effective because the strength increase of the alloy is achieved at the expense of hardness and wear resistance.

Methods of chemical strengthening are more versatile. Since the strength and toughness properties of the refractory component of the alloy are mainly determined by the presence of free electrons, existing methods of chemical strengthening delocalize some of the valence electrons forming the Me-Me- and Me-X- bonds. In practice this is achieved in the following ways:

- doping of a refractory component in the metal sublattice (e.g. $Ti_{1-n}Nb_nC$, $Ti_{1-n}Ta_nC$) or non-metal sublattice

    (e.g.) $TiC_{1-x}N_x$, $TiC_xO_yN_z$)

- doping of a refractory compound in both sublattices simultaneously (e.g. $Ti_{1-n}Mo_nC_{1-x}N_x$, $Ti_{1-n}V_nC_xO_yN_z$)

- using non-stoichiometric compounds (e.g. TiN_{1-x}, TaC_{1-x}).

The increase in the metallic character of the refractory component of the alloy also leads to improvement of the wetting by the binding metal (Ramqvist, 1965; Goretzki and Scheuermann, 1971). Since a thermodynamic condition of increasing strength of the interphase cohesion is that the contact angle between solid and liquid phases should decrease, the methods mentioned also increase the adhesive strength of the alloy.

Methods of strengthening the metal component of an alloy include:

- alloying the binder to the limits of substitutional solid solution or interstitial solution (solid solution strengthening),

  - the thermal treatment of the alloy in a region of strong
    temperature dependence of solubility of the alloying elements
    in the binder material (precipitation hardening),

  - the dispersion in the binder of very fine particles of refractory
    compounds with low solubility (dispersion strengthening).

The last method is not usually applicable in practice because of the rapid
coarsening of the strengthening particles in the presence of a liquid
phase (Voronkin and Gaidukova, 1973).

A combined application of the above methods for the strengthening of
alloys is effective only if negative changes do not occur in the microstructure during subsequent sintering. This condition is partially
satisfied for the alloys of WC-Co type (the composition of each phase of
the alloy is changed here only within the limits of existence of the solid
solution). The situation becomes more complicated in the case of W-free
hard alloys. During the process of liquid-phase sintering one observes
not only changes of chemical composition but also changes of phase
composition, rapid grain growth of the refractory components, changes of
the ratio between solid and liquid fractions and other undesirable
effects. The point is that systems which contain phases of variable
compositions are chemically much more active than the corresponding
systems based on the phases of stable (or practically stable) composition.
As a result, effects of strengthening which are expected according to the
optimum choice of compositions and to the ratio between the liquid and
solid components, appear to be lower or even completely suppressed during
the process of sintering.

That is why in recent years a new approach to the strengthening of W-free
hard alloys has been developed. The essence of the approach is that the
optimum composition of the alloy is not chosen in advance but is evolved
directly in the course of the liquid phase sintering. The composition of
the alloy should change in the process of sintering in such a way that the
most effective mechanisms of strengthening (the dispersion of solid phase
grains, the increase of the metallic character of the grain boundaries and
the strengthening of the binder) are not suppressed but initiated by the
presence of the liquid phase. It is evident from this that as the basis
of W-free hard alloys a multicomponent solid solution must be used that
should contain elements with a definite ultimate function. The most
promising elements are transition metals of groups IV-VI, N and O. The
composition changes of the solid solutions during the process of the alloy
formation are stimulated here by the interaction with the binder. Thus,
the problem of the strengthening of W-free hard alloys becomes much more a
chemical one.

The introduction of the described approach into practice requires the
solution of a number of mutually connected problems. The most important
are the following:

  - the invention of an industrial process for obtaining doped
    solutions of a predetermined composition of the
    $Me_{1-n}Me'_n C_{1-x}N_x$ type

  - improved understanding of the physico-chemical foundations
    of liquid phase sintering.

Here we briefly consider the main ways of solving such problems.

Complicated doped solid solutions of a predetermined composition may be obtained by a number of methods e.g. direct synthesis from the components, plasmochemistry, self-spreading high temperature synthesis, etc. One of the most economical and effective methods is the carbothermic reduction of a mixture of transition metal oxides in a controlled gaseous atmosphere (N_2 is preferred).

The physical foundation of liquid phase sintering (physics of sintering) has been analysed rather carefully by Humenik and Parikh (1956), Parikh and Humenik (1957), Kingery (1959), Eremenko, Naidich and Lavrinenko (1968), Huppmann and co-workers (1979), Kaysser, Huppmann and Petzov (1980), but these approaches do not take into account an important feature of W-free hard alloys - namely that they are chemically nonequilibrium systems. The present theories have limited validity for those systems where the changes of chemical composition during sintering are insignificant and the phase composition remains constant. Chemical effects dominate in the sintering of W-free hard alloys based on phases of variable composition. In this connection we will describe more carefully a number of general statements regarding the chemistry of sintering of such alloys and the role of separate components in the formation of their compositions and microstructure.

THE CHEMISTRY OF SINTERING OF W-FREE HARD ALLOYS

The elaboration of a physico-chemical foundation for liquid phase sintering should be based on the following general considerations. If the phase composition of the alloy is not changed (or should not be changed) during sintering, the basis of study should be the equilibrium phase diagram for the corresponding system; the kinetic aspects of the problem are subordinate here. When the phase composition changes during sintering, the kinetic approach becomes the main one, the aim of which is to reveal the mechanisms of dissolution and phase and structure formation. In general, both approaches - thermodynamic and kinetic - are complementary.

The nature of cubic (NaCl-type) carbides, nitrides and oxides of the transition metals of groups IV-VI as phases of variable composition, predetermines some specific features of their behaviour in contact with alloys based on the metals of the Fe-group. For example:

a)   the corresponding systems which contain stoichiometric solid phases are not quasibinary systems,

b)   the process of solid phase dissolution is of an incongruent nature.

The consequences of these facts are, firstly, the inevitability of chemical and phase composition changes in the system by interaction and, secondly, the increase of such changes with the complexity of the chemical composition of the solid phase. For instance, during the dissolution of carbide TiC_{1-x} ($x \rightarrow 0$) in a Ni-20 wt% Mo melt, carbon passes preferentially to this melt. As a result, a phase based on Mo_2C is crystallized primarily from the melt, and the composition of this congruently dissolving carbide approaches $TiC_{0.85}$ (Fedorenko, 1981). In the case of the carbonitride $TiC_{1-x}N_x$, and in particular the oxycarbide $TiC_{1-x}O_x$, the degree of incongruency for the process of dissolution under similar conditions increases. For instance, carbonitride $TiC_{0.5}N_{0.5}$ and oxycarbide $TiC_{0.5}O_{0.42}$ are converted into TiN (<0.5 wt% C) and $TiO_{\sim 1.3}$ (<1.0 wt% C) respectively, at 1720 K within an hour if they have grain sizes up to 10 μm. As the grain verge is enriched by N_2 or O_2 it stratifies (Fig. 1).

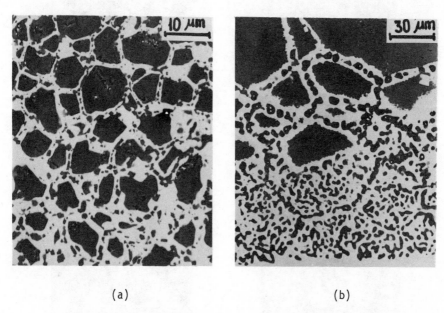

(a)   (b)

Fig. 1   Scanning electron micrographs of cross section of contact boundaries $TiC_{0.5}N_{0.5}$ / Ni - 20% Mo (a) and

$TiC_{0.5}O_{0.42}$ / Ni - 20% Mo (b)

(T = 1720 K, t = 1 hr, p = 10^{-2} Pa).

Analogous effects are also observed in the interaction between complicated carbides and carbonitrides (similar to $Me_{1-n}Me'_{n}C$ and $Me_{1-n}Me'_{n}C_{1-x}N_{x}$) with Ni-based alloys. In this case the incongruency of the dissolution process and phase composition changes are determined mainly by the transfer onto the melt not only of C but also of one of the metal components of the solid phase. This is shown in Fig. 2 by means of SEM and X-ray images of the contact boundary of carbonitride $Ti_{0.94}Zr_{0.06}C_{0.5}N_{0.5}$ with the melt Ni - 20% Mo (1720 K, 1 hr, 5 10^{-2} Pa).

Effective techniques for studying the kinetics of such processes are electron probe microanalysis (the study of composition), the scanning electron microscopy (the study of structure) and thermal analysis (the study of the metal phase state). An example follows of the study of the interaction between the components in the systems TiC-Ni-Mo and $TiC_{1-x}N_{x}$ -Ni-Mo which are bases for the production of W-free hard alloys.

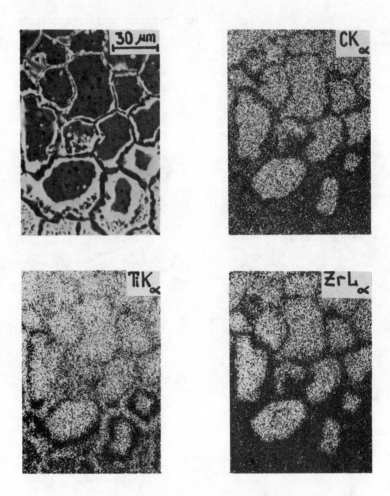

Fig. 2   SEM and X-ray images of cross section of
contact boundary $Ti_{0.94}Zr_{0.06}C_{0.5}N_{0.5}$ / Ni - 20% Mo

(T = 1720 K, t = 1 hr, p = 10^{-2} Pa).

It is known that the alloys based on TiC and $TiC_{1-x}N_x$ (x ⩽ 0.5) with
Ni-Mo- binders are characterized by a cored structure of the grains of
the refractory componenet (Moskowitz and Humenik, 1966; Suzuki, Hayashi
and Terada, 1971; Snell, 1974; Lindau and Stjernberg, 1976; Fukuhara and
Mitani, 1982). The core of carbide (carbonitride) is surrounded by a
shell of a cubic (NaCl-type) solid solution of composition $Ti_{1-p}Mo_pC_{1-x}N_x$
(K-phase), which is formed in the presence of liquid phase according to a
solution-precipitation mechanism (Suzuki, Hayashi and Terada, 1971;
Zhilyaev, Fedorenko and Shveikin, 1978). The direct diffusion of Mo from
the liquid binder alloy as was proposed in the literature (Moskowitz and
Humenik, 1966, 1978) does not occur (Zhilyaev, Fedorenko and Shveikin,
1978). It has been confirmed by local X-ray spectroscopy methods
(Zhilyaev and Patrakov, 1982) that the K-phase is practically free of N_2.

Experiments on infiltration of hot pressed $TiC_{0.96}$ by the melt, Ni - 30% Mo (Fig. 3) allowed an estimate of the following:

1. The composition of the K-phase crystallized from the melt can vary within very broad limits: 8-65 wt% Mo, 25-75 wt% Ti, 7-17 wt% C and up to 1.5 wt% Ni.

2. The metal phase can be enriched in Ti to a maximum (up to 13 wt%, when Ni_3Ti is crystallized from the melt) and depleted of Mo to a minimum (up to 0.2 wt%).

3. The initial step of the process of interaction is limited by the diffusion in the liquid phase.

The investigation of contact interaction in the system $TiC_{1-x}N_x$ - Ni(Mo) showed (Zhilyaev, Fedorenko and Shveikin, 1978, 1980) that composition variations of the K-phase in $TiC_{1-x}N_x$ - Ni(Mo) alloys are significantly narrower. In comparison with the carbide system it is more enriched by Mo (up to 70 wt%) and depleted of C (6-9 wt%). From experiments on the infiltration of hot pressed $TiC_{0.5}N_{0.5}$ by the melt, Ni - 25 wt% Mo it follows (Fig. 4) that the dissolution of carbonitride is controlled by the interphase reaction.

Such studies of the contact interaction and infiltration in these systems have made it possible to reveal the role of Mo, N and C in the processes of dissolution and phase and structure formation.

It was long ago established that Mo has a significant influence on the microstructure of such alloys (Fairhurst, 1968; Kieffer, Ettmayer and Lux, 1979). However, until recently the mechanism of such influence was not clear. Our investigation (Zhilyaev, Fedorenko and Shveikin, 1978) has shown that the role of Mo in improvement of wetting of TiC by Ni is not connected with its interphase activity. The cause of the contract angle decrease to zero in the presence of Mo is the formation of the K-phase shell on the grains of TiC (and also on $TiC_{1-x}N_x$ when $x \leqslant 0.5$). On the other hand, studies by electron microscopy show that the rate of growth of the K-phase shell on small grains of TiC is higher than on large ones (Pilyankevich and co-workers, 1978).

However, in alloys based on Ti carbonitrides this growth rate/grain size relationship is not observed (Mitrofanov, 1973). On the basis of these observations it can be proposed that the addition of Mo to the binder suppressed (at least in the initial stages of sintering) the mechanism of grain growth by Ostwald ripening (the growth of large grains as a result of dissolution and disappearance of small ones). Indeed, on the addition of Mo to the binder, the dependence of the oversaturation of the binder on the size of solid phase grains should be lost since it will be determined by the product of the activities of Ti, C and Mo in accordance with the equation:

$$G_{Ti_{1-n}Mo_nC_{1-x}} = RT \ln a_{Ti}^{1-n} a_C^{1-x} a_{Mo}^n \qquad 1$$

where G is the free energy of the K-phase formation and a_i is the activity of the corresponding element in the melt.

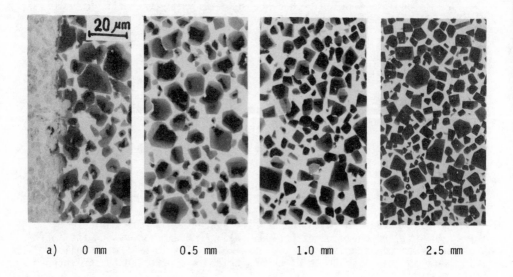

a)   0 mm   0.5 mm   1.0 mm   2.5 mm

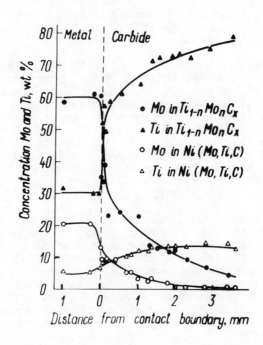

b)

Fig. 3   Infiltration of TiC skeleton by Ni –

30 wt% Mo melt (1720 K, 1 hr, 10^{-2} Pa)
accompanied by the microstructure (a) and
composition (b) changes.

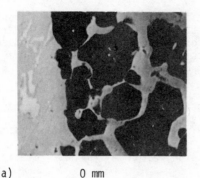 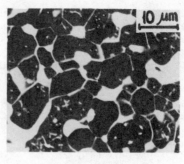

a)   0 mm   0.5 mm

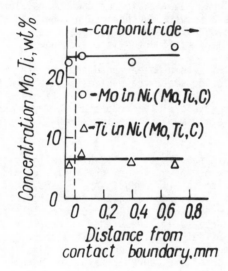

b)

Fig. 4   Infiltration of $TiC_{0.5}N_{0.5}$ skeleton

by Ni - 25 wt% Mo melt (1720 K, 1 hr, 10^{-2}Pa);
(a)   microstructure of the sample
(b)   melt composition.

Then, the higher the rate of solid phase dissolution (usually it is higher
for small grains), the faster (at the given activity of Mo in the binder)
the limiting product of the activites of Ti and C is achieved sufficient
to precipitate the K-phase.

The K-phase should precipitate, as observed experimentally, either
preferably on small grains (as in TiC-Ni-Mo, where the process is limited
by the diffusion in liquid phase), or on all the grains simultaneously (as
in $TiC_{1-x}N_x$-Ni-Mo, where the process is limited by the interphase
reaction). In both cases the number of particles in the alloy does not
change significantly and it remains fine grained.

According to the data of Fukuhara and Mitani (1982), further grain growth in these alloys is controlled by the interphase reaction on the boundary between the K-phase and the melt. Because of the absence of the K-phase in TiN-No-Mo alloys the interphase reaction on the nitride-melt boundary remains as the limiting step of grain growth during the whole sintering process (Fukuhara and Mitani, 1982).

No less important in the process of sintering is the role of N. It has been established in our studies that with increasing x in $TiC_{1-x}N_x$, up to x = 0.5, the rate of dissolution and solubility of $TiC_{1-x}N_x$ in the Ni-Mo melt diminishes and incongruency of the process increases (Zhilyaev, Fedorenko and Shveikin, 1980). This is the reason why the precipitating K-phase appears to be enriched by Mo.

On the other hand, since Mo has insignificant solubility in $TiC_{1-x}N_x$, the process of diffusional equalization of the compositions of core and shell is hindered. As a result, further grain growth is suppressed, the K-phase retains its favourable mechanical properties, the binder is not depleted of Mo and the separation of N is prevented.

Briefly, the presence of a thin Mo-enriched shell of the K-phase on $TiC_{1-x}N_x$ grains lead to the small grain size and the structural stability of the alloy (that is, the initial distribution of the grain shape and size remains stable). The complete wetting of the K-shell by the matrix melt gives a maximum of adhesive strength to the interphase boundaries.

In contrast to N-containing alloys, in TiC-Ni-Mo alloys, after the precipitation of the K-phase on the grains of the intial carbide a rapid homogenisation of core and shell compositions occurs, initially of C, and then of the metals (Zhilyaev, Fedorenko and Shveikin, 1978). Finally, the growth of grains increases, their shell loses mechanical strength (Nishigaki and co-workers, 1974), the binder is depleted of Mo and enriched by Ti. The initial equalization of the composition of core and shell by C explains the fact that the appearance of additional C in the system stabilizes the cored structure (Yamaya and Sadahiro, 1969).

The partial substitution of Ti in its C, N, O- containing compounds by transition metals (Zr, Hf, V, Nb, Ta, Mo, etc) leads to new effects. Certain additions promote the additional disintegration of the grains in the process of sintering (e.g. Zr, Nb), others promote the phase stratifications (e.g. V, Mo). Knowledge of the mechanisms and kinetics of formation of the microstructure of alloys based on solid solutions similar to $Me_{1-x}Me'_x C_{1-x}N_x$ permits control of their operational properties within broad limits. We consider that one of the most promising avenues in the development of W-free hard metals lies in an increasing sophistication of the composition and microstructure to be achieved by advances in the chemistry of sintering.

REFERENCES

Eremenko, V.N., Naidich, Yu.V., and Lavrinenko, I.A. 1968, Spekanie v prisutstvii zhidkoi fasi. Naukova Dumka, Kiev, 123 p
Fairhurst, W. 1968, Metallurgia 77 145-149
Fedorenko, V.V. 1981, Thesis, Institute of Chemistry, Sverdlovsk
Fukuhara, M., and Mitani, H. 1982, Powder Met. Intern. 14 196-200
Goretzki, H. and Scheuermann, W. 1971, Preprints of the 7th Plansee Seminar, Planseewerke, Reutte 4 paper 50
Humenik, M. and Parikh, N.M. 1956, J. Amer. Ceram. Soc. 39 60-63
Huppmann, W.J., Kaysser, W.A., Yoon, D.N. and Petzow, G. 1979, Powder Met. Int 11 50-61
Kaysser, W.A., Huppmann, W.J. and Petzow, G. 1980, Powder Met 23 86-91
Kieffer, R., Ettmayer, P. and Lux, B. 1979, Metal 33 466-471
Kingery, W.D. 1959, J. Appl. Phys. 30 301-306
Lindau, L. and Stjernberg, K.G. 1976, Powder Met. 19 210-213
Mitrophanov, B.V. 1973, Thesis, Institute of Chemistry, Sverdlovsk
Moskowitz, D. and Humenik, M. 1966, Modern Developments in Powder Metallurgy 3 83-94
Moskowitz, D. and Humenik, M. 1978, Intern. J. Powder. Met. and Powder Techn. 14 39-45
Nishigaki, K., Ohnishi, T., Shiokawa, T. and Doi, H. 1974, Modern Developments in Powder Metallurgy 8 627-643
Parikh, N.M. and Humenik, M. 1957, J. Amer. Ceram. Soc. 40 315-321
Pilyankevich, A.N., Shapoval, T.A., Vitryanyuk, V.K., Paderno, V.N. and Aronin, I.Ya. 1978, Poroshkovaya Metallurgia 8 49-53
Ramqvist, L. 1965, Int. J. Powder Met. 1 2-21
Snell, P.O. 1974, Planseeber. Pulvermet. 22 91-106
Suzuki, H., Hayashi, K. and Terada, O. 1971, J. Japan Inst. Metals 35 936-942
Tresvyatsky, S.V. 1970, Sovremennie problemi poroshkovoi metallurgii, Naukova Dumka, Kiev, 269-286
Voronkin, M.A. and Gaidukova, T.E. 1973, Poroshkovaya Metallurgia 12 38-40
Yamaya, S. and Sadahiro, T. 1969, J. Japan Soc. Powder and Powder Metallurgy 16 190-195
Zhilyaev, V.A., Fedorenko, V.V. and Shveikin, G.P. 1978, Proc. 5th Int. Conf. on Powder Metallurgy, Czechoslovakia, Gottwaldow, 189-200
Zhilyaev, V.A., Fedorenko, V.V. and Shveikin, G.P. 1980 Issledovaniya tekhnologii metallicheskikh poroshkov i spechennikh materialov. USC AS USSR, Sverdlovsk, 57-64
Zhilyaev, V.A. and Patrakov, E.I. 1982, Extended Abstracts of the 8th Conf. on local X-ray spectroscopy and applications. AS USSR, Chernogolovka, 185-187

Author Index

Subject Index